BOTANY

FOR DEGREE STUDENTS

PTERIDOPHYTA

(Vascular Cryptogams)

Dr. P. C. VASHISHTA
M.Sc. (Hons. Sch.) ; Ph.D.; P.E.S. (I)
Retd. Ex Principal, Government College for Boys Chandigarh and D.A.V. College, Hoshiarpur

Revised by

Dr. A.K. SINHA
M.Sc., Ph.D.; F.P.S.I.
Reader & Head (Retd.)
Department of Botany
Feroze Gandhi Postgraduate College
(Kanpur University), Raebareli, U.P.

Dr. ANIL KUMAR
M.Sc., Ph.D.
Reader
Department of Botany
Feroze Gandhi Postgraduate College
(Kanpur University), Raebareli, U.P.

S Chand And Company Limited
(ISO 9001 Certified Company)

S Chand And Company Limited

(ISO 9001 Certified Company)

Head Office: D-92, Sector–2, Noida – 201301, U.P. (India), Ph. 91-120-4682700

Registered Office: A-27, 2nd Floor, Mohan Co-operative Industrial Estate, New Delhi – 110 044, Phone: 011-49731800

www.**schandpublishing.com;** e-mail: **info@schandpublishing.com**

Marketing Offices:

Chennai	:	Ph: 23632120; chennai@schandpublishing.com
Guwahati	:	Ph: 2738811, 2735640; guwahati@schandpublishing.com
Hyderabad	:	Ph: 40186018; hyderabad@schandpublishing.com
Jalandhar	:	Ph: 4645630; jalandhar@schandpublishing.com
Kolkata	:	Ph: 23357458, 23353914; kolkata@schandpublishing.com
Lucknow	:	Ph: 4003633; lucknow@schandpublishing.com
Mumbai	:	Ph: 25000297; mumbai@schandpublishing.com
Patna	:	Ph: 2260011; patna@schandpublishing.com

First Edition 1971
Subsequent Editions and Reprints 1972, 73, 74, 75, 77, 80, 82, 83, 84 86, 87, 88, 90, 91, 92, 93, 95, 96, 97, 99, 2000, 2001, 2003, 2005
First Multicolour Illustrative Revised Edition 2006
Reprints 2008, 2010, 2012, 2014, 2015, 2016 (Twice), 2018, 2019, 2020, 2021 (Twice), 2022 (Twice), 2023

Reprint 2025

ISBN: 978-81-219-2785-7 **Product Code:** H6BPT68BOTN10ENZX06O

PRINTED IN INDIA

By Vikas Publishing House Private Limited, Plot 20/4, Site-IV, Industrial Area Sahibabad, Ghaziabad – 201 010 and Published by S Chand And Company Limited, A-27, 2nd Floor, Mohan Co-operative Industrial Estate, New Delhi – 110 044.

Preface to the Revised Edition

We take pleasure to present this thoroughly revised and improved multicoloured edition of the popular book ***Botany for Degree Students*–** **'PTERIDOPHYTA'** by **Dr. P.C. VASHISHTA** for the benefit of both undergraduate students as well as those pursuing post-graduate courses. While the original format of the book remains unchanged, efforts have been made to make the book examination oriented, student friendly and more useful by adding the topics generally asked in the university examinations.

The book has been thoroughly revised re-read and improved. In chapter 1, we have added a brief account of recent advances in Pteridology. A new chapter under the heading *'General Discussion'*, has been added. In this chapter, a few fossil pteridophytes, not described in earlier chapters, have been described. In addition, topics on Biotechnology and tissue culture studies in pteridophytes, Economic importance and Diversity and distribution of Pteridophytes have also been added. All the diagrams have been redrawn and several new and accurate diagrams have been added. As per the new format of question papers, all the three types of questions—essay type, short answer type and objective types—have been added.

During the course of the revision of this book we have been guided, assisted and helped by a large number of university and College professors. A few undergraduate and post-graduate students too offered their suggestions for the improvement and usefulness of the book. We have tried to incorporate all these suggestions. To each one of them, we offer our most sincere gratitude and thanks.

We are grateful to Shri Ravindra Kumar Gupta, Managing Director, Shri R.S. Saxena, Advisor, Shri Navin Joshi, General Manager (S & M), Dr. (Mrs) Swati Lohani, (Freelance Editor) and Mr. Shishir Bhatnagar, Senior Coordinator, S. Chand & Co. Ltd., New Delhi for their help and Cooperation in bringing out the book in the present form.

We hope that this revised and multi coloured improved edition of the book will prove much more useful to all its users. Suggestions and Criticisms for the improvement of this book will be gratefully acknowledged and included in the next edition.

A. K. SINHA
ANIL KUMAR

Preface to the First Edition

The study of vascular cryptogams forms an essential part of our syllabi for both undergraduate and post-graduate students. It is an interesting assemblage of plants that inhabited land for the first time and became established to a terrestrial mode of life. Their study reveals a striking array of characters that speak of their antiquity as early land plants. This book is designed so as to acquaint the students with the structure and reproduction of these remarkable vascular, but seedless plants.

The book, in its present form is meant for the degree students and covers the syllabi of most of the Indian Universities. It was decided to give a detailed account of the structure and reproduction of the representative types of vascular cryptogams so as to equip the undergraduates with a thorough information about them. Without such an information it is impossible to follow a comparative account and a generalised treatment of the subject at advanced levels. With this idea in mind the author deviated from the trend of giving brief life history descriptions that do not satisfy the inquisitive students. Tremendous amount of research has been carried, during the past twenty-five years or so, on the various aspects of these lowly organised vascular plants and many new and interesting facts have been brought to our knowledge. With the increase in the knowledge of the fossil history of the group some newer interpretations regarding the conquest of land and origin of vascular cryptogams have been advanced. Most of these hypotheses have been dealt with in Chapter One. Lot of experimental work has been done and is being done on the problems of apospory and apogamy. Interesting results have been obtained as a result of studies on tissue culture in India and abroad. A brief discussion on these topics has been included in the first chapter. Such an information is necessary for the students so as to equip them with modern trends in the subject. These valuable informations are given in the research papers that are not easily available to the students. Their brief description in the book will certainly go a long way to enhance the horizon of knowledge of the students in the colleges. Lot of additional information is incorporated in the chapters dealing with the life histories of representative types. This has added to the volume of the chapters concerned. The author has also deviated, at certain places, from old text book versions of some topics. This was necessitated on account of some recent work on these topics and newer interpretations. The authority and the year of publication have been quoted in all such cases.

This book is not an outcome of author's original research but is a compilation work incorporating the researches of the pteridologists in India and abroad. The matter has been compiled from standard texts, reviews, monographs and research journals in a manner suitable to the degree students. The language of the book is simple and easily understandable. It is profusely illustrated. Most of the figures have been redrawn or adapted from standard books, journals, monographs and research papers. Sources of all such figures have been duly acknowledged in their legends. Some figures have been drawn from actual specimens and prepared slides. These were drawn with the help of Camera Lucida or otherwise, by the author himself.

A short bibliography has been given at the end. It includes some of the recent publications; but is in no way a complete and a comprehensive compilation as is not needed for such an elementary type of book. To save space titles of papers have been omitted and only the names of the authors, years of publication and the names of journals have been given.

The author wishes to express his indebtedness to the authors and the publishers of standard texts, monographs, research journals and reviews from where the matter of this book has been compiled. The eminent among the authors who deserve special mention are D.H. Campbell; F.O. Bower; D.H. Scott; C.W. Wardlaw; Elizabeth, G. Cutter; I. Manton; P.N. Mehra; T.S. Mahabale; H.Y. Mohan Ram; D.W. Bierhorst; A. Arber; H.C. Bold; H.N. Andrews; G.M. Smith and K.R. Sporne.

My sincerest thanks go to Prof. B.R. Vasishta; Dr. S.S. Bir (Prof. of Botany, Punjabi University, Patiala). Dr. S.C. Verma (Reader in Botany, Punjab University, Chandigarh), Dr. D.S. Loyal (Reader in Botany, Punjab University, Chandigarh) and Dr. H.K. Palta (Lecturer in Botany, Punjab University, Chandigarh) who gave me all help in the form of suggestions and lending me some of their recent publications on the subject. To my class fellows Prof. R.P. Sood (Ramgarhia College, Phagwara), Prof. V.P. Chadha (Government College, Ludhiana), Prof. M.S. Sethi (Government College, Hoshiarpur) and Prof. B.N. Sood (G.M.N. College, Ambala Cantt.), I am indebted for many useful suggestions that they offered to make the book a success. I must thank my colleagues in the department, Shri P.S. Gill (Senior Lecturer in Botany) and Shri S.S. Sekhon for their thought-provoking discussions. I am grateful to my old student Shri T.N. Lakhanpal (Lecturer in Botany, Hans Raj College, Delhi) for going through a portion of the manuscript. Appreciation is also due to Shri Sarban Singh for undertaking the arduous task of typing the manuscript and to Shri Muni Lal Verma for taking great pains in drawing the figures. Last but not the least I offer my thanks to my publishers particularly Shri Shyam Lal Gupta and Shri Rajinder Kumar Gupta for bringing out this volume.

It is a sincere request to my colleagues all over India to point out the defects and give useful suggestions to improve the book.

Government College
Chandigarh,
January, 1971

P.C. VASHISHTA

Acknowledgements

The author wishes to express his sense of appreciation and indebtedness to the following—

Prof. B.R. Vashishta; Dr. S.C. Verma, Professor in Botany, Panjab University, Chandigarh; Dr. D.S. Loyal, Professor in Botany, Panjab University, Chandigarh; Prof. S.S. Bir, Head of Botany Deptt., Panjabi University, Patiala; Prof. C.G.K. Ramanujam from Osmania University, Hyderabad; Prof. S.B. Dalal of Vidharba Mahavidyalaya, Amravati; Dr. B.M.B. Sinha and Dr. A.K. Verma from Patna University; Prof. Abdul Rahim from Govt. College, Sahiwal (Pakistan); Miss Rangithakani from Vellore College for Women, Tamil Nadu; Prof. Mrs. S.S. Godbole of Bislop College. Nagpur, Prof. R.P. Diwedi of D.S. College, Aligarh; Prof. Shekhar Ranjan Saha of Comilla Victoria College, Comilla (Bangla Desh); Prof. Abu-Muhd Shafiuz-zaman of Govt. Rajindra College, Faridpur (Bangla Desh); Mustafa Kamal Pasha, University of Chitagong, Bangla Desh; Dr. M.L. Kasat and Prof. J.R. Soni of Dungar Post-graduate College, Bikaner (Rajasthan); Prof. Alokiasamy of St. Xavier's College, Palayam Kottai (Tamil Nadu); Prof. Subodh Kumar Ray of Balurghat College, West Dinajpur (Bangla Desh); Prof. M.S. Rawat of Govt. College Sehore; Prof, R.K.S. Rathore of R.B.S. College, Agra; Dr. V.C. Saxena and Dr. R.K. Sarabhoy of Agra College, Agra; Prof. R.S. Himanshu from J.C. College, Jamshedpur; Prof. K.S. Arya & Dr. K.L. Tokey from D.A.V. College, Chandigarh; Prof. Mrs. Surinder Inderjeet Sherne and Prof. Amar Single of Govt. College (Men), Chandigarh; Prof. S.S. Malik, D.A.V. College, Chandigarh; Prof. Charanjeet Chawla of Guru Gobind Singh College, Chandigarh; Prof. Mrs. Harjinder Kaur, Mrs. Saroj Ahuja, and Dr. Mrs. Kuldeep Rawla from Govt. College (Women), Chandigarh; Prof. P.S. Bhagwanani of Baroda; Dr. A.K. Choudhry of Hindu College, Moradabad; Dr. Y.P. Sharma of Hindu College, Moradabad; Prof. B.N. Naryana Rao of St. Philloonena's College, Mysore; Dr. A.K. Misra of C.M. College, Dharbhanga (Univ. of Bihar), Prof. M.S. Matharu of Govt. College, Kishangarh (Rajasthan); Prof. K. Lokendra Rao of Govt. Pingle College for Women, Waddepati—Warangal (A.P.); Prof. Benudhar Pradhan of B.J.B. College, Bhubneshwar; Prof. J.V. Suba Rao of Dharmvani College of Science, Hyderabad; Prof. M.K. Prasad of Maharaja College, Ernakulam; Prof. V.K. Sharma of Agriculture University, Ludhiana; Prof. S.P. Kanwal of D.A.V. College, Amritsar; Prof. Bhatnagar of D.A.V. College, Amritsar; Prof. Mrs. Jaya Kanwal of Govt. College for Women, Amritsar; Prof. V.P. Chadha, Govt. College, Ludhiana; Prof. K.P. Sood and Prof. Mangal Singh, Ramgarhia College, Phagwara; Prof. Inderjit Singh Klair, Govt. College, Sangrur; Prof. Balwant Singh, Lyalpur Khalsa College, Jullundur City; Prof. H.M. Sharma of D.A.V. College, Jullundur City; Prof. B.N. Sood of G.M.N. College, Ambala Cantt; Prof. H.C. Jain of S.D. College, Ambala Cantt; Dr. Ashok Kumar of D.A.V. College, Ambala City; Prof. S.K. Raina of Govt. College, Jammu; Prof. K.L. Narang and Prof. S.D. Sharma of D.A.V. College, Abohar, Prof. A.K. Garg, M.R. College, Fazilka; Prof. T. Sesaghiri Rao of S.K.B.R. College, Amalapuram, Distt. Godavari (A.P.); Prof. P.L. Vazi of Post-Graduate Govt. College, Raipur; Prof. P.K. Dharadiraj of H.D. Jain College, Arrah (Magadh University, Gaya); Prof. B.M. Joshi of Univ. of Jodhpur; Prof. K.C. Sheriar of Jai Hind College, Bombay; D. Mukherjee of Ramananda College, Bishnupur (W.B.); G.R. Nambiar of B.S. Mahavidyalaya, Dharmabad Distt. Nanded (M.S.); Prof. Darshan Kumar of S.D. College, Pathankot; Prof. R.S. Jain of Govt. College, Faridabad; Dr. A.K. Sharma of J.V. College, Barant.

I am extremely grateful to Prof. S. Bhambie, Professor of Botany, Kurukshetra University, Kurukshetra (Haryana) for sending me his publications and useful suggestions.

To Shri Muni Lal Verma of Amritsar, the author is grateful for taking keen interest in drawing sketches for this book. It is a pleasure to record my sincerest gratitude to late Shri Shyamlal Gupta, of S. Chand & Co. for bringing out the present edition in a new and attractive format.

Contents

SELAGINELLA FERTILE SPIKES

CHAPTER 1

The Vascular Cryptogams (Pteridophyta)

- Introduction
- Characteristic features
- Habitat of Pteridophytes
- Alternation of generations
- Abnormalties in the life cycle (Apogamy, Apospory)
- Origin of the vascular cryptogams
- Evolution of sporophyte
- Fossil record and geological time scale
- Classification
- Pteridology in India

Introduction

The word cryptogams is a synthesis of two Greek terms ***kruptos*** meaning '***hidden***' and ***gamos*** meaning '***wedded***'. This single term encompasses all plants that reproduce by means of spores and, do not produce seeds. The algae, fungi, bryophytes and pteridophytes are all cryptogams. The pteridophyta are treated as vascular cryptogams as they have a well developed conducting system. They form the subject-matter of this book. The lower cryptogams (algae, fungi and bryophyta) have no conducting system. The term pteridophyta also has a Greek origin. ***Pteron*** means a "feather" and the name was originally given to this group because of their pinnate or feather-like fronds. ***Vascular cryptogams are, therefore, an assemblage of seedless vascular plants that have successfully invaded the land and reproduce by means of spores.***

The Pteridophytes formed a dominant part of earth's vegetation in the historic past (Two hundred eighty to two hundred thirty million years ago). In the present day flora, excluding the non-vascular plants, they rank only next to the spermatophytes. Although they have been largely replaced by the spermatophytes in the modern day flora, yet they occupy an important and a crucial central position in the evolutionary history of the plant kingdom. No doubt lesser in number, the pteridophytes lend a distinct charm and physiognomy to the landscape. The elegant tree ferns of the warm humid forests of eastern Himalayas, Pachmahri and Nilgiri hills, the epiphytic ferns and the hanging club-mosses of the tropical forests attract ones attention. The lithophytic and terrestrial forms that grow in comparatively larger numbers are also a source of great attraction because of their beautiful foliage. Some of them (about a dozen species) grow in water and form a luxuriant hydrophytic component of the lakes (***Salvinia***), ponds and pools (***Azolla***, ***Marsilea***). In India about five hundred species of fern allies have been reported. Dixit (1984) has raised the number to one thousand.

Salvinia *Azolla*

The Pteridophyta have a long fossil history. Their first traces were recognised in the Silurian period of the late Paleozoic age (about 400 million years ago). They flourished well during Devonian, Mississippian and Pensylvanian periods of the late Paleozoic. ***Late Paleozoic can easily be regarded as the "Age of Pteridophyta."*** The tree ferns, giant horsetails and arborescent lycopods dominated the swampy landscapes of those ancient ages. The present-day lycopods (***Lycopodium***, ***Isoetes***, ***Selaginella***, ***Phylloglossum***) and Sphenopsids (***Equisetales***) are the mere relicts of the mighty lycopsid and sphenopsid group which once boasted of 120-foot ***Lepidodendròn***, ***Sigillaria***, and the colossal ***Calamites.*** Only the present-day ferns have somehow managed to conserve the former stature, diversity and glory of their ancestors. ***Psilotum*** and ***Tmesipteris***, our two surviving remnants of ***Psilopsids*** retain the primitive features of the first land plants.

Lycopodium

Isoetes

Surange (1966) has given a detailed and a comprehensive account of the Indian fossil pteridophytes. He has described one fossil member of Psilophyta, seven of Lycopodiophyta, twelve of Arthrophyta and sixty six of Pterophyta. He has observed that the Indian Paleozoic and Mesozoic pteridophytes resemble quite closely those described from far flung areas of the southern hemisphere. A number of fossil ferns have been described from the Rajmahal hills in Bihar. Bose and Sah (1967) and Sharma (1969, 1971, 1972, 1973) have explored these hills and described as many as eighteen species of ferns and one of ***Selaginellites*** from Brindaban and Dhokuti areas. Suthar and Sharma (1986) presented a complete restoration of ***Solenostelopteris jurassica*** of Schizaeaceae from the jurassic flora of Rajmahal hills. They described the anatomical details of different organs. Sharma, Suthar and Bohra (1986) have given a detailed account of the work done on Indian fossil pteridophytes (Lycopods, Sphenopsids and ferns). They have pointed that pteridophytes occur throughout the Gondwana system but more frequently during Mesozoic era and the ferns make the major bulk of the assemblage. The Sphenopsida played an important role in the constitution of Glossopteris flora during the Palaeozoic whereas the Lycopods never played any significant role in the constitution of Indian fossil flora.

Characteristic Features

(*a*) Sporophyte

The vascular cryptogams possess an independent sporophyte with a vascular system. It exhibits a great variation in form, size and structure. All the living pteridophytes are nearly herbaceous except a few woody tree ferns. They may be dorsiventral or radial in symmetry and have dichotomously or laterally branched stems that bear microphyllous or megaphyllous leaves. The roots are generally adventitious, the primary embryonic root being short lived.

The stem, root and the leaves grow by means of a single, two-sided or three-sided apical cell or by a plurality of such apical initials. A distinct apical meristem has been reported in a number of **Lycophyta** (***Lycopodium***, ***Isoetes***). Bhambie and Puri (1985) recognised three different configurations of shoot apical meristems as : (*i*) a single conspicuous pyramidal apical cell and its nearby derivatives, (***ii***) a group of conspicuous pyramidal or cubical apical initials surrounded by smaller cells and (***iii***) a lump of meristematic tissue having a central zone and capped by an outer layer. In roots, four kinds of apical organisations have been recognised: (*i*) a single conspicuous pyramidal cell, (***ii***) a gradation from one to several cubical cells, (***iii***) a group of inconspicuous initials and (***iv***) four tiers of initials. Bir, S. S. and Randhawa (1984, 1987) studied the apical organisation in rhizome, root and leaf of twelve different ferns belonging to different groups and pointed out that a single apical cell always occupies the tip of each growing organ. This single apical cell is replaced by a group of initial cells when the shoot apex becomes sufficiently grown up and well developed. They supported the classical concept of a single apical cell in ferns. The vasculature of the sporophyte may be a simple protostele, medullated protostele (siphonostele), dictyostele, or a polycyclic stelar organisation. Vessels are also present in some members (***Selaginella***, ***Equisetum***). Loyal and Singh (1978) recorded the presence

Equisetum

of vessels in about a dozen species of ***Marsilea***. They also recorded the presence of branched vessel members in the root-rhizome-petiole juncture in ***M. rajasthanensis.*** Presence of vessels has now been reported in a number of ferns such as ***Actiniopteris radiata***, ***Pteris vittata***, ***Nephrolepis cordifolia*** and ***Helminthostachys zeylanica.*** Agshe (1968) reported the presence of **sieve tubes *in Equisetum giganteum and E. arvense***. Loyal and Maheshwari (1979) noted the presence of well defined **sieve plates** with a number of pores in the sieve elements of rhizome and petiole of ***Marsilea minuta.*** The pteridophytes normally possess only **sieve cells** in the phloem and the presence of sieve tubes is considered to be an advanced feature. Secondary growth is absent in majority of the living representatives, ***Isoetes*** being an exception. Khandewal and Goswami (1977) reported the occurrence of periderm and cork cambium in the roots and rhizomes of ***Botrychium***, ***Helminthostachys*** and ***Ophioglossum.*** The sporophyte produces spores that may be **homosporous** (***Lycopodium***, ***Dryopteris***) or **heterosporous** (***Selaginella***, ***Stylites***, ***Isoetes***, ***Salvinia***, ***Azolla***, ***Pilularia***, ***Regnellidium***, ***Marsilea***). The spores are produced in special structures called the **sporangia** that are invariably subtended by leaf-like appendages called the **sporophylls**. The sporangia and the sporophylls may be scattered throughout the vegetative axis or may be restricted to a particular area. They are in many cases compacted to form distinct spore producing regions called the **cones** or the **strobili** (***Selaginella***, ***Equisetum***). The sporangia, in some cases, may be produced within specialised structures called the **sporocarps** (***Marsilea***, ***Salvinia***, ***Azolla***). Distinct segregation of vegetative and reproductive shoots and leaves has also been reported in some vascular cryptogams (***Equisetum arvense***, ***Pellaea atropurpurea, Matteuccia struthiopteris***). The development of sporangia may be **eusporangiate or lepto-sporangiate**. The spores are haploid and germinate to give rise to the **gametophyte generation,** which also leads an independent existence.

Marsilea

(*b*) Gametophyte

It is quite independent from the sporophyte and may be autophytic or saprophytic or partly autophytic and partly saprophytic. It is markedly dissimilar from the sporophyte. The homosporous pteridophytes generally have monoecious prothalli. They may be **protandrous** or **protogynous** (***Equisetum***). Dioecious prothalli have also been reported in ***Equisetum*** and some ferns (***Pteris***, ***Pteridium***, ***Schizea***). The prothalli in homosporous forms are either green, simple or branched and aerial structures; or they may be subterranean, colourless, tuberous and saprophytic (***Lycopodium***). Mehra and Asha (1986) have recognised seven

Pteris

Pteridium

types of mature prothalli in the pteridophytes as : (*i*) Cordate prothalli with a distinct cushion and two equal wings, (*ii*) Unequal cordate prothalli with two wings of different sizes and distinct central cushion, (*iii*) Lop sided prothalli with only one wing and with a cushion on lateral portion, (*iv*) Strap-shaped and unbranched prothalli, (*v*) Strap-shaped and branched prothalli, (*vi*) Filamentous and much branched prothalli, (*vii*) Subterranean, brown and mycorrhizic prothalli. The gametophytes are greatly reduced in the heterosporous vascular cryptogams (***Selaginella***, ***Isoetes***, ***Stylites***, ***Marsilea***, ***Salvinia***, ***Azolla***, ***Regenellidium***, ***Pilularia***).

The sex organs may be embedded or projecting. They resemble the bryophytes in general plan of structure. The archegonia have invariably four longitudinal rows of neck cells whose height varies from two to six cells. The neck canal cells vary from one (***Selaginella and Equisetum***) to 14 in number (***Lycopodium***). The antheridial development may be **Eusporangiate** or **Leptosporangiate.** The antheridia have a single layered jacket enclosing a variable number of spermatozoids. They may have short stalks or may be sessile. The number of cells in a jacket also varies from 3 to many. The gametophytes are **exosporic** in homosporous forms and **endosporic** in heterosporous forms. For a detailed account of the gametophytes and sex organs in ferns see chapter 8.

The zygote develops into an embryo within the archegonial venter and is surrounded and nourished by the prothallial tissue. The development of the embryo may be **endoscopic, exoscopic or lateral** (For a detailed account see chapter 8).

The alternation of generations is **heterologous** in all the vascular cryptogams.

Characteristic features of pteridophytes may be summarised as follows:

1. ***Pteridophytes are primitive land vascular plants*** both (***Lycopodium***, ***Selaginella***, ***Equisetum etc.***) and also fossil vascular plants (***Rhynia***, ***Asteroxylon***, ***Zosterophyllum*** etc.).
2. Usually pteridophytes grow in cool and shady habitat though few are xerophytic (***S. rupestris***) and many are aquatic (***Marsilea***, ***Salvinia*** and ***Azolla***).
3. Adult plant body is a sporophyte usually differentiated into leaves, stem and roots. Rhizoids (***Psilotum***) and Rhizophores (***Selaginella***) are also reported in a few pteridophytes.
4. Mature sporophytes are independent, though in early stage of development they are partially or completely dependent on gametophytes.
5. Vascular system is composed of both xylum and phloem and stele may be protostele (ectophloic, endophloic and amphiphloic) or Siphonostele (Solenostele and dictyostele).
6. Leaves may be simple, minute and sessile as in ***Lycopodium*** or petiolate and megaphillous as in members of ***Filicinae***.
7. Reproduction is through spores which are formed in sporangia. Sporangia develop either on the ventral surface or in axil of leaves (Leaves bearing sporangia are called sporophylls).
8. Plants may be homosporous (all spores are of same size and structure) or heterosporous (spores may be microspore or megaspore).
9. Sporangial development takes place either from :
   (*a*) Group of superficial cells (initials) as in ***Selaginella.*** Such type of development is known ***as Eusporangiate type***.
   (*b*) A single cell (initial) as in ***Marsilea***, this type of development is known as ***Leptosporangiate*** type.
10. Gametophytic body is known as ***Prothallus*** which is developed from haploid spore.
11. Prothallus is multicellular, autotrophic (independent) structure and may bear both sex organs : archegonia and antheridia.

12. Water is essential for fertilization. Diploid zygote is formed following fertilization.
13. Zygote develops into sporophytic plant body.
14. Alternation of Generation is an essential part of life cycle and is heterologous.

Habitat of Pteridophytes

Pteridophytes—the first land vascular plants, are represented by about 10,000 species distributed throughout the world, both in tropical and temperate climates. Most of the members of this group are terrestrial in nature, prefer to grow in cool and shady places, however aquatic and xerophytic habitat is not uncommon for pteridophytes. A short description of habitat of pteridophytes can be discussed as follows:

(*a*) Terrestrial habitat

Members right from fossil pteridophytes (***Rhynia sp.***) to most advanced pteridophytes (***Pteridium***) grow in terrestrial habitats. Among Psilophyta, terrestrial species are represented by ***P. nudum***, ***Osmunda sp.***, ***Ophioglossum sp.*** etc. Lycopods growing in such habits are ***L. clavatum***, ***L.cernuum***, ***L lucidum***, ***L. reflexum***, ***L.annotinum.*** Plants of ***Lycopodium*** are either creepers (***L. cernuum***), climbers (***L. volubile***) or mostly epiphytic in nature (***L. phlegmaria***, ***L.carinatum***). Most of the common species of ***Selaginella*** (***S. chrysocaulos***, ***S. kraussiana***, ***S. selagiuoides***), ***Isoetes*** (***I. coromandelina***) are terrestrial. Similar habitat is preferred by ***Equisetum*** sps, though its species ***E. debile*** can grow in hydrophytic, mesophytic as well as in xerophytic habitats. ***E. palustris prefers swampy habitats. E. arvense*** grow in open grasslands, railway embankments and exposed sandy and dry places.

Osmunda

Ophioglossum

Most of the ferns are terrestrial in nature, mostly grow in hilly regions (***Dryopteris***, ***Pteris***, ***Pteridium***, ***Adiantum*** etc.) prefering moist and shady environment.

(*b*) Aquatic Pteridophytes

Most of the members of family Marsileaceae and Salviniaceae grow in aquatic and semi-aquatic habitats. Plants of these groups are commonly known as aquatic ferns. Few examples of aquatic pteridophytes ***are M. minuta***, ***M. quadrifolia***, ***Salvinia natans***, ***S. auriculata***, ***Azolla filiculoides***, ***A. microphylla*** and ***A. mexicana***. ***Two species of Isoetes*** eg. ***Isoetes englemanni*** and ***I. Japonica*** are semi-aquatic pteridophytes. ***I. panchananii***, a very common plant of Pachmarhi is also considered as plant of amphibious habitat.

(*c*) Xerophytic Pteridophytes

Most of the xerophytic plants are reported from the genus ***Selaginella.*** Some common Indian xerophytic species of ***Selaginella*** are ***S. wightii*** and ***S. repanda***. ***S. lepidophylla*** is another xerophytic plant. ***Marsilea condensata*** and ***M. rajasthanesis*** are xerophytic pteridophytes as well.

Alternation of Generations

Hofmeister (1851) used this term in plants. He observed that in Mosses and Ferns there are two types of morphologically distinct individuals in the life cycle. Both alternate in a life cycle ***i.e.***, there are some events which lead one generation to produce the other and thus cause alternation. The actual phenomenon responsible for bringing about alternation was exposed by the significant discovery of "the periodic Reduction of Chromosomes" by Strasburger (1894). Strasburger discovered the process of **meiosis** in plants. This discovery revealed that the reduction in the number of chromosomes leads to the formation of a new individual in the life cycle. This individual has haploid number of Chromosomes in its nuclei. It bears sex organs and is concerned with sexual reproduction. It was given the name **'gametophyte'** and represented **gametophytic generation**. The haploid gametes unite **(fertilization or syngamy)** and establish a diploid nucleus or a ***Synkaryon***. Fertilisation establishes a diploid cell or the **zygote**, which is a pioneer structure of the diploid individual or the **sporophyte**. It germinates to form the embryo, which in turn develops into the sporophyte individual. This generation is termed the **sporophyte generation**. The sporophyte individual bears sporangia which produce spores as a result of **meiosis**. These spores are haploid and are also known as **meiospores**. The spores are the pioneer structures of the gametophyte generation. They germinate to give rise to the gametophyte individual known as the prothallus in vascular cryptogams.

Both the generations can reproduce vegetatively and effect an increase in the number of their individuals.

The above events, as displayed by the normal life cycle of a vascular plant, lead us to the following conclusions; *(i)* There are two distinct individuals in the life cycle of vascular cryptogams. *(ii)* These individuals are produced by the germination of the haploid spores and the diploid zygote. *(iii)* The one produced by the spores is the **gametophyte** plant. To it is attributed the function of sexual reproduction and is haploid. *(iv)* The zygote produces the embryo which gives rise to the diploid individual called the **sporophyte**. It bears haploid spores (or meiospores) after a process called the **meiosis**. *(v)* The two generations alternate with each other in the life cycle. *(vi)* Meiosis and syngamy (fertilization) are the two significant stages that switch on the life cycle from one generation to the other. *(vii)* The number of chromosomes is halved during meiosis and is doubled during syngamy or fertilisation.

Tmesipteris

In the vascular plants the sporophyte individual is a complicated, independent and a dominant generation, whereas the gametophyte is comparatively much reduced. The gametophyte in the homosporous forms, though inconspicuous and comparatively shortlived, is independent and may be surface living and green (autophytic) or subterranean and non-green (***Lycopodium***). They are always **exosporic** and thus are not enclosed by the spore wall. They are also infected by an endophytic or a mycorrhizic fungus in some genera (***Lycopodium***, ***Psilotum*** and ***Tmesipteris***). The heterosporous land plants, surprisingly enough, display a considerable reduction in their gametophytes. The gametophytes are reduced and **endosporic.** They have, as a rule, separate male and female prothalli. The male prothallus is extremely reduced and represented only by a single prothallial cell. The female prothallus, on the contrary, is well developed because it has to nourish the developing embryo. Such a dioecism is unknown in homosporous land plants, but is a regular feature in the heterosporous ferns. The life cycles of homosporous and heterosporous vascular plants are represented diagrammatically in Fig. 1.1, A, B.

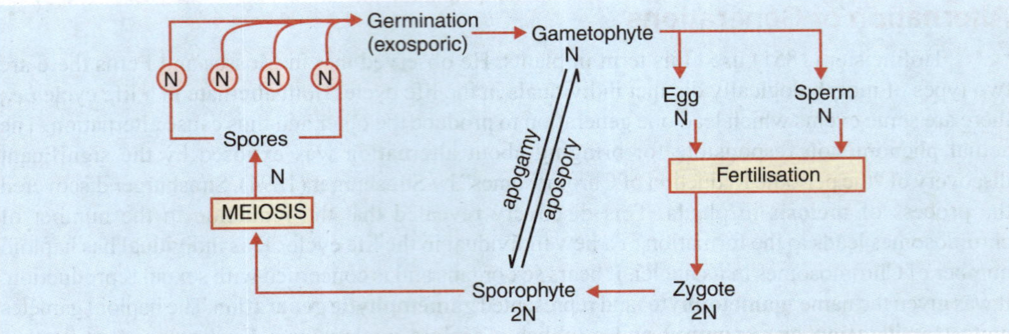

Fig. 1.1 [A]. Diagrammatic representation of the life cycle of a homosporous vascular cryptogam.

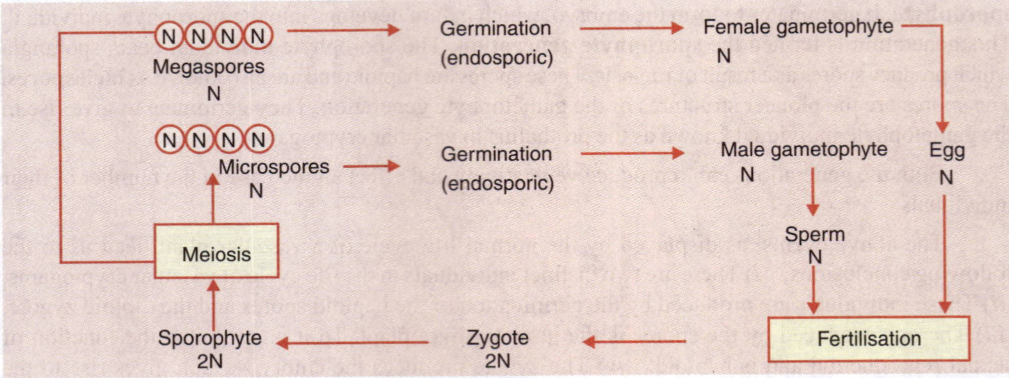

Fig. 1.1 [B]. Life cycle of a heterosporous vascular cryptogam.

The sporophyte in the vascular cryptogams is no doubt independent but it has to depend upon the gametophyte during its earlier stages of development. It achieves its complexity after establishing its independence. The sporophyte, however complex it may be, has never achieved complete independence. The huge and giant sporophytes of the gymnosperms, during their earlier stages of development, are completely at the mercy of the food stored in the gametophyte. This is true even in the case of apogamously formed sporophytes. This food in the homosporous plants is truly gametophytic in origin, but in the heteroporous plants the food is mainly drawn from the sporophyte plant and is only stored in the gametophytic tissue. The angiosperms, on the other extreme, tell a different tale. The food stored in the endosperm is derived from the sporophyte and is stored in a triploid tissue. The morphological complexities of the sporophyte of vascular plants coupled with their elaborate and well organised anatomical set-up equipped with efficient means of dispersal confer upon them the potentialities to live under varied environmental conditions. These capabilities are responsible for their being stamped as efficient land dwellers.

Both the homosporous and heterosporous vascular plants exhibit **heteromorphic** type of alternation of generations because the sporophyte and the gametophyte individuals present marked morphological and anatomical differences.

While studying the life cycle of a heterosporous individual a common feature that attracts our attention is the difference in the size of spores. It poses a serious question before us. Why the microspores (smaller in size) give rise to the male gametophyte and the megaspores to the female gametophyte? This question has not been satisfactorily answered. Two possible reasons have no doubt been suggested.

The first suggestion tends to explain the cause of difference in size. The megaspores are larger in size and produce female prothalli. The latter have to lodge the developing embryo sporophyte that requires lot of nourishment. The larger spores have greater nutritional store and can produce a well developed female prothallus that is capable of hoarding food sufficient for the development of embryo sporophyte. The microspores are smaller in size and thus have little food stored in their protoplasts. They cannot afford to produce massive prothallial tissues. They germinate to develop one or two-celled prothallus that bears a single antheridium. The function of the antheridium is to produce spermatozoids. After their liberation it has nothing to do and, therefore, perishes. This can be the cause of their smaller size. The microspores are produced in large numbers as they are likely to be wasted during dissemination.

The cause of their being unisexual appears to be a contrivance towards cross pollination. The monoecious or bisexual gametophytes of homosporous vascular plants have many chances of self-fertilisation, unless they are submerged under water. The dioecious prothalli have extremely rare chances of self-fertilisation and suffer from a great disadvantage if they grow under terrestrial conditions. Under such circumstances they have little chance of fertilisation because the spermatozoids cannot reach the archegonia. This difficulty is solved to a greater extent by the phenomenon of heterospory. The microspores that need not carry much food are light in weight and smaller in size. They can be easily carried by wind and other agents to the female prothalli. The chances of their wastage are eliminated to a great extent, on account of their large number per microsporangium. Those which fall on the female prothallus have absolutely no difficulty in effecting fertilisation as the spermatozoids have not to travel a long distance and require only a little moisture which is provided by dew drops. Chances of self-fertilisation under these circumstances are rare, and cross-fertilisation is usually effected. This leads to outbreeding and ultimately to variations and rapid evolution. Heterospory has, as a matter of fact, freed the sporophyte from aquatic environments. Such an emancipation from aquatic habitats has conferred on the sporophyte the freedom to grow under varied environmental conditions. Heterospory is, therefore, regarded as a necessary step towards the evolution of seed habit and must be significant development during the struggle to conquer land.

Origin of Alternation of Generations

Celakovsky (1874) listed three distinct types of alternation of generations in plants. These are : —

1. The Antithetic Alternation of Generations by which he meant the interpolation of a new phase between the successive gametophyte generations. Such an alternation, according to him, makes up the life cycles of the bryophytes and the pteridophytes.
2. The Homologous Alternation of Generations that was believed by him to be prevalent among algae and fungi.
3. Strophogenesis or the alternation of shoots.

Bower (1935) dropped the terms **antithetic** and **homologous** and replaced them with **Interpolation theory** for antithetic theory and **Transformation theory** for homologous theory. These two theories have been advanced to explain the origin of alternation of generations. Both of them are aimed at explaining the origin of sporophytic individual in the life cycle.

Interpolation or Antithetic Theory

It was put forth by Bower and later supported by Overton (1893), Scott (1896), Strasburger (1897), Williams (1904) and others. Bower discussed it in detail in 1935. The supporters of this theory believe that the gametophytes of the early land plants had an algal origin. They presumed that these early land plants migrated from amphibious situations; (where water was available in plenty for effecting sexual reproduction), to terrestrial habitats (where water was available only during rainy seasons and extensive dew). Under such circumstances these early land plants could not depend entirely upon the sexual process for propagation and had to adopt another method. This was accomplished by the development of zygote into embryo and ultimately into a diploid individual called the **sporophyte**. Development of the sporophyte involved certain changes that can be listed below :—

(*i*) Development of jacketed sex organs especially the archegonium.

(*ii*) The retention of the fertilised egg within the archegonium that could provide it with sufficient food.

(*iii*) The zygote or the fertilised egg divided mitotically and not meiotically to form a diploid embryo, which in primitive land plants distinguished into a sterile jacket enclosing a number of diploid cells or spore mother cells.

(*iv*) The diploid spore mother cells underwent meiosis to form tetrads of haploid spores. Such a primitive sporophyte is exemplified by ***Riccia***.

Marchantia

(*v*) Gradually the primitive sporophytes underwent progressive sterilisation and gave rise to complex sporophytes of liverworts like ***Marchantia***, ***Anthoceros*** and ***Mosses***. These sporophytes were dependent wholly or partially upon the gametophytes for food.

(*vi*) Enhanced sterilisation of the sporophyte led to the development of an independent sporophyte that developed vascular tissue and possessed root, stem and leaves. These sporophytes also bore sporangia that produced spores and disseminated them efficiently.

The events of apogamy and apospory were dismissed by Bower as 'ex post facto' events. He stated, "They illustrate the potentialites of plants of present day rather than evolutionary features of the remote past, and will have no part in producing that normal alternation which is characteristic of the Archegoniatae."

Homologous or Transformation Theory

This theory was put forth by Pringsheim (1876-77) and has a large number of supporters. Stebbins (1960) has also supported it. The proponents of this theory claim that the sporophyte is a modification of the gametophyte. They do not believe progressive sterilisation to be the cause of origin of photosynthetic tissues in the sporophyte. They do not regard the sporophyte and gametophyte individuals to be as different as postulated by the adherents of the antithetic theory. According to this theory the primitive land plants exhibited isomorphic alternations ***i.e.***, they had free living sporophytes and gametophytes that resembled each other in all respects except their genetic constitution and functions. They derived support from the following events :

1. Existence of isomorphic alternation of generations in the algae ***e.g.***, ***Cladophora***, ***Ulva***, ***Ectocarpus*** and ***Dictyota***. Both the generations are photosynthetic.
2. Presence of Chloroplasts in the sporophytes of some bryophytes.
3. Presence of tracheids in the gametophytes of some lower cryptogams ***e.g.***, ***Psilotum***.
4. Occurrence of phenomenon of apospory and apogamy. These deviations from the normal life cycle suggest that either generation can arise as an outgrowth from the other. These phenomena lend support to the view that sporophyte can be a modification of the gametophyte.

The supporters of this theory further believe that the free living sporophyte later became attached to the gametophyte and partially dependent upon it. Such a step leads to the reduction in complexity of the sporophyte as in the bryophytes. Further reduction leads to the origin of simple and completely dependent sporophytes of ***Riccia***. This permanent retention of the sporophyte on the gametophyte is considered to be responsible for the development of morphological differences in the two generations and led to the development of heteromorphic alternation of generations in the bryophytes. In the vascular cryptogams also the embryo is retained within the gametophyte.

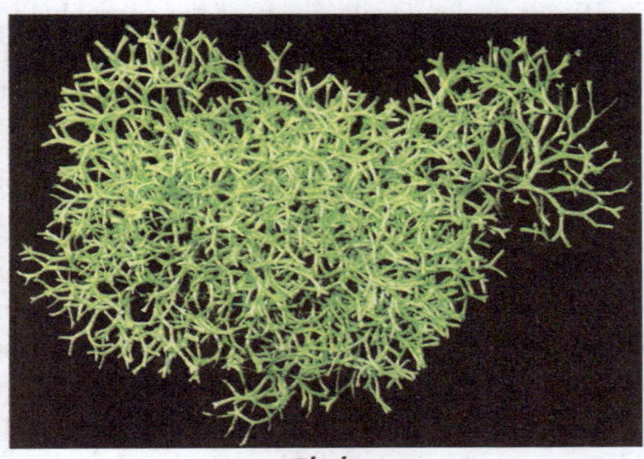

Riccia

Modern evidence provides more support for the homologous origin of alternation of generations (Stebbins, 1960) but even if we accept this theory we have still to account for the morphological differences between the sporophyte and the gametophyte. There must be certain factors which initiate the development patterns responsible for the distinct morphological characteristics of the sporophyte and the gametophyte. Two hypotheses have been advanced to explain these differences.

A. Lang's Hypothesis

Lang ***introduced his hypothesis in the year*** 1909. ***He emphasized that the spore and the zygote, which are the first cells of the gametophyte and the sporophyte respectively develop under different environmental conditions.*** The spores after dissemination are free from the influence of the sporophyte and are freely exposed to external environments. The zygote, on the other hand, is retained within the venter of the archegonium and is not free from the influence of the surrounding gametophytic tissue. It is, therefore, subject to various restraints and other influences. Lang considered that the different conditions during early development of the two pioneer structures of the alternating generations may be sufficient to account for the differences between them. This hypothesis has recently been put to experimental tests and some favourable results have been achieved. To test the validity of Lang's hypothesis, Bell (1959) argued that if this were true, then by interchanging the positon of spore and zygote, it should be possible to change their sequence of development. It was not possible to produce a sporophyte from the spore, but surgical experiments on zygote under reduced or complete absence of archegonial restraint, yielded results contrary to the hypothesis. Ward and Wetmore (1954) and Jayasekera and Bell (1959) were successful in reducing the archegonial restraint on the zygote by cutting away archegonial tissue. De Maggio and Wetmore (1961) dissected embryos of various ages or even the zygotes from the archegonial venter and were successful in germainating them under experimental conditions free from the influence of the gametophyte. In most of these experiments the freed tissue underwent abnormal development, but ultimately was able to produce a sporophyte. It, however, shows that gametophytic influences are

Todea barbara

significant in the development of the sporophyte, but not necessary. In ***Todea barbara*** the freed zygotes produced small thalloid structures of gametophytic nature, but it was not possible to achieve the formation of sex organs on these thallus-like structures. The older embryos, however, produced normal sporophytes.

B. Blackman's Hypothesis

Blackman (1909) ***attached no importance to the difference environmental conditions to which the zygote and spore are prone. He expressed the opinion that differences in the chromosome numbers and other intrinsic differences between spore and the zygote are responsible for morphological differences in the structures produced by their germination.*** Bell (1963) made cytochemical and ultrastructural studies on the fern egg cell and concluded that there are many cytochemical and cytological differences between the spore and egg cell.

It appears that the views expressed in both the hypotheses are partially correct. Environmental, cytoplasmic, nutritional and hormonal and genetic factors coupled together appear to be responsible for the distinct developmental differences between the two generations. None of these hypotheses have taken into consideration the effect of organic and inorganic nutrition and of hormones on the developmental pattern of the spore and the zygote. Recent experimental data collected on a number of ferns indicate the importance of hormonal and nutritional factors in controlling the development of the zygote. Bistow (1962) experimented with the callus tissue derived from the sporophyte of ***Pteris cretica***. He found that the callus remained undifferentiated on a medium containing sucrose and high concentration of auxin. In a medium containing sucrose alone the callus developed into a sporophyte. If the callus is grown in a medium containing only mineral nutrients it differentiated into a gametophyte. Whittier (1962) and Whittier and Steeves (1960, 1962) could induce the apogamous production of sporophytes from a number of otherwise normal fern gametophytes by the incorporation of suitable sugar concentrations in the culture medium. Wetmore ***et al*** (1963) demonstrated that when prothalli of ***Onoclea***, ***Osmunda*** and ***Todea*** are planted erect on a medium containing one per cent sucrose, cylindrical and sturdy growths, containing a vascular strand of xylem elements are produced. De Maggio (1964) was able to induce the formation of sporophytic buds and roots on the prothalli of ***Lycopodium obscurum*** grown in a culture medium containing coconut milk and sucrose. These experiments reveal the importance of nutritional factors in determining the morphological differences between the two alternating generations of the vascular cryptogams.

Onoclea

Modern Concept

The studies on tissue culture and the effect of sugar or lack of sugar on the development of sporophytic or gametophytic structures from calli lead us to postulate two hypotheses :—

(A) ***The* Gene Block Hypothesis *which postulates the presence of 4 gene blocks in fern systems and any one of them can be stimulated into action.*** These gene blocks control the formation of various organs in fern sporophytes and gametophytes. A gene block controlling the formation of root, leaf or gametophyte can be stimulated into action in any cell whether it is of a sporophyte or that of a gametophyte. Every gene block has a master gene for that block and it is necessary to activate this master gene in order to trigger into action any sub-division of a gene block; **(B)** The second hypothesis concerns the phenomenon of alternation of generations. These studies lead us to disbelieve the **Antithetic** and **Homologous theories** of alternation of generations; but instead confirm our belief in a third theory called the **Genetic theory of alternation of generations.** ***This theory postulates that a change in the genetic system in the early progenitors of land plants brought into being two distinct and independent individuals in the life cycle of the land plants***. These individuals may be identical or dissimilar in appearance. Bell (1970) opined that the totality of evidence leads us to perceive the life cycle as an integrated sequence of events, each key event coded by a specific set of the genome. No one questions the action of genes in the organisation of a living system but what is not clear is the location of genes switches. Smith (1979) is of the opinion that certain metabolic changes may be responsible for determining the developmental pathways during the germination of spore, rather than specific gene action. One of the major mysteries in cell biology is the missing link between the macromolecules and visible structural organisation.

Abnormalities in the Life Cycle

The normal life cycle of a vascular plant has two alternating generations. These are the diploid sporophyte and the haploid gametophyte. Both alternate regularly in the life cycle and this alternation is brought by two significant steps known as the fertilisation and the meiosis. The regular alternation of chromosome number is sometimes impaired by the occurrence of two common phenomena known as apogamy and apospory. They will be dealt with separately.

Apogamy

It was first reported by Farlow (1874) in ***Pteris cretica***. ***Apogamy can be defined as the development of a sporophyte directly from the gametophyte without the intervention of sex organs and gametes. The sporophytes thus formed usually have the same chromosome number as the gametophyte i.e. haploid number for the species***. Apogamy occurs in nature and has also been induced under experimental conditions. It is a common and a widespread phenomenon in the ferns. Natural apogamy has been reported in more than 50 species of ferns belonging to 20 genera. In some species of ferns apogamy appears to be a necessity and is a regular process. It is perhaps due to the inherited constitution of the plant. Natural apogamy is commonly known in ***Dryopteris***, ***Pteris***, ***Pellaea***, ***Adiantum***, ***Osmunda***, ***Todea***, ***Athyrium***, ***Cheilanthes***, ***Polystichum***, ***Asplenium***, etc. To this list can be added a number of genera and species in which apogamy has been experimentally induced. Cases of parthenogenetic development of the egg into the sporophyte are not included under apogamy, because apogamy is the development of vegetative tissue of the prothallus into the sporophyte. Apart from the ferns apogamy has also been induced in some species of ***Lycopodium*** and ***Equisetum***. It has been estimated that 12 percent of the cytologically investigated homosporous ferns are apogamous.

Athyrium *Polystichum*

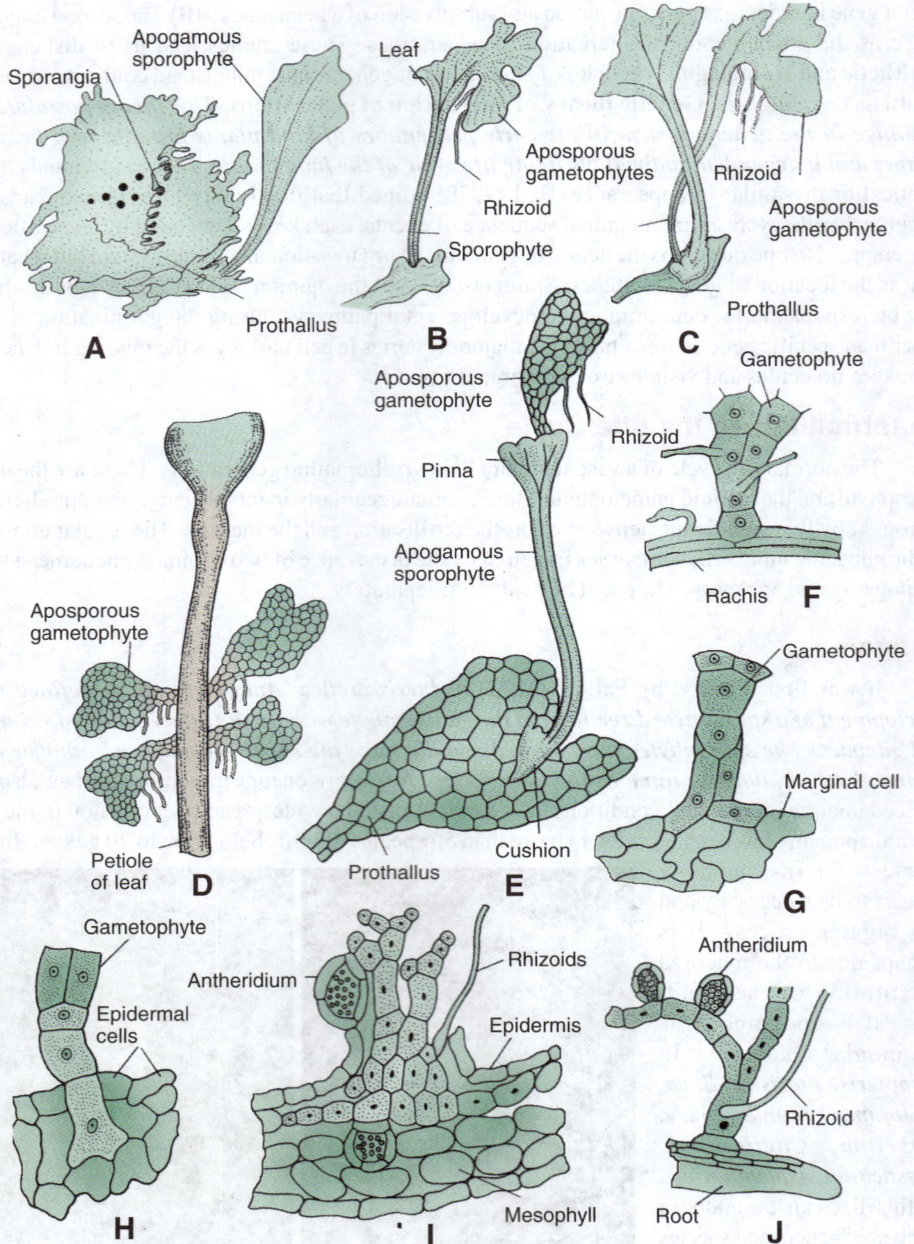

Fig. 1.2 (A–J). Apogamy and Apospory in ferns. **A.** Apogamous sporophyte arising from the gametophyte of ***Phyllitis scolopendrium*** (After Lang); **B–C.** Aposporous gametophytes arising from the primary leaves of sporophytes of ***Osmunda regalis*** (After Lang); **D.** Aposporous gametophytes arising from the petiole of ***Pteris vittata*** (After Palta); **E.** Apogamously produced sporophyte of ***P. vittata*** bearing aposporous gametophyte (After Palta); F–J Aposporous gametophytes of ***Pteridium aquilinum*** var. ***latiusculum***; **F.** Prothallus arising from epidermis of petiole; **G.** Prothallus arising from marginal cells of a pinna; **H.** Prothallus arising from leaf epidermal cells; **I.** Prothallus arising from mesophyll cell of pinna; J. Prothallus arising from root epidermis. (F–J. After Takahashi, 1962).

Causes of Apogamy

Regarding the cause of apogamy, several explanations have been put forth. Lang (1898) induced the formation of sporophytic buds (Fig. 1.2 A), roots, sporangia (Fig. 1.3) and tracheids in various fern prothalli by avoiding watering of the prothalli from above. Brown (1923) summarised literature regarding the induction of apogamy by avoiding fertilisation of the egg. Many workers regard failure of normal fertilisation as a cause of apogamous production of sporophytes. Mottier (1931), however, demonstrated that in ***Matteuccia struthiopteris*** failure of fertilisation does not induce apogamy. Brown (1923) induced apogamy in ***Phegopteris polypodioides*** by avoiding normal fertilisation. Other conditions favouring apogamy have also been suggested. These are : culture in bright light and at higher temperatures (Nathansohn, 1900); by lowering the vitality of the prothallus by fungal and algal attack : and failure of formation of functional sex organs under various unfavourable nutritional conditions. Willams (1938) suggested that in addition to the environmental factors there must also be some internal factors such as the nature of inherent susceptibility due to abnormal nuclear composition and behaviour, that bring about apogamy.

Fig. 1.3. Apogamy ***in Phyllitis scolopendrium.*** Note the sporangia have developed from the prothallial tissue that has also developed tracheids (After Lang).

Ageing of the prothallus has also been regared as one of the factors influencing apogamous developments on the prothalli of some ferns. Recent work (Whittier and Steeves, 1960) on ***Osmunda, Adiantum and Pteridium*** has shown that apogamy can be induced by growing the prothalli on an agar culture medium rich in glucose. Wetmore and his associates (1963) demonstrated that when prothalli of ***Onoclea***, ***Osmunda*** and ***Todea*** are planted erect on a medium containing one per cent sucrose, cylindrical and radially symmetrical growths with vasucular strands are produced. De Maggio (1964) induced the formation of sporophytic buds on the prothalli of ***Lycopodium obscurum*** grown in culture media containing coconut milk and sucrose. These experiments reveal the effect of nutritional factors in inducing apogamy. Loyal and Chopra (1973) induced apogamy in ***Regnellidium diphyllum***.

Regnellidium diphyllum

Cytology of Apogamy. Work during the last 15 or 20 years has revealed that one out of every 15 species of ferns has what is called the "apogamous life cycle" in which both generations have the same chromosome number. The question now arises as to how this apogamous type of life cycle is maintained cytologically. Two methods are known. According to one method, which is rare, both the processes of spore formation and fertilisation are eliminated from the life cycle. The sporophyte or the fern plant produces a prothallus or the gametophyte which arises as a bud from the leaf and develops into a full-fledged prothallus. It is produced from the diploid tissue by mitosis and is, therefore, dipoid in constitution. This diploid prothallus does not bear sex organs but instead gives rise to sporophytic buds that develop into diploid sporophytes. Such a method, therefore, involves the occurrence of both apospory and apogamy in the same plant. It is not a common method and has since been reported in ***Athyrium filix foemina*** var. ***clarissiana***, ***Dryopteris filix-mass*** var. ***cristata apospora*** (Farmer and Digby, 1907); and in ***Trichomanes kraussiana*** (Georgevitch, 1910). Sarbadhikari (1939) discovered it in ***Osmunda javanica.*** It has been induced in ***Pteris vittata*** by Palta.

Drypteris flix-mas

The second method which is very common among the ferns involves the formation of spore mother cells that actually double their chromosome number by the fusion of daughter nuclei. During this method the number of spore mother cells is reduced. Normally the sporangia in the leptosporangiate ferns possess 16 diploid spore mother cells, but in apogamous ferns there are only 8 spore mother cells with their nuclei having double the chromosome number (tetraploid). They will undergo meiosis and produce 32 diploid spores. These spores will produce diploid gametophytes. So the number of chromosomes in both the generations is same. These gametophytes will bear sporophytic buds that will develop into full-fledged diploid sporophytes. Sex organs are usually not produced by such gametophytes and there is thus no fertilisation. Such a method is known to occur in a large number of ferns and is by far the commonest.

In India it has been reported by Mehra (1944) in ***Adiantum lunulatum*** ; and by Mehra and his students (see Mehra, 1961) in thirty species of Himalayan ferns (***Pteris cretica***, ***P. aspercaulis***, ***P. blumeana***, ***P. biaurita*** ; ***Dryopteris paleacea***, ***D. fibrillosa***, ***D. odontoloma***, ***D. attrata*** ; ***Adiantum caudatum*** and others). Mehra (1961) has given a vivid description of the apogamous and aposporous Himalayan ferns and has discussed their bearing on speciation and phyletic relationship among the ferns.

Asplenium

Evans (1965) has reported an interesting case of apogamy in a species of ***Polypodium.*** In this case he reported the formation of **32 mitospores** in the sporangia. The sporangia have 16 spore mother cells which do not undergo meiosis, but instead divide mitotically into 32 spore. These spores are reniform in shape and occur in diads rather than tetrads. He reported this sequence to occur regularly in all sporangia. The spores germinate readily to form prothalli which bear numerous apogamous sporophytic buds. The prothalli bear stomata but produce no sex organs. Virginia, M. Marzenti (1967) reported apogamy in ***Asplenium curtissii*** and ***A. plenum***. In both the species the spore mother cells do not undergo reduction division and act directly as spores. These spores germinate to produce gametophytes that bear antheridia containing viable sperms, and also bear apogamous sporophytes.

Vittaria

A very interesting fern gametophyte (***Gametophyta appalachina***) was discovered by a bryologist A. J.Sharp from the Appalachian region of North America. It is a ribbon like and branching prothallus that reproduces mainly by means of marginal clusters of gemmae. It has not been seen to produce sporophytes under natural conditions. Wagner regards it as a species of ***Vittaria*** (shoe string fern) that has somehow lost its sporophyte during evolution. Stokey (1951) cultured these prothalli for three years and was able to induce the formation of small young sporophyte directly from the gametophyte. These apogamously produced sporophytes had certain diagnostic features of the ***Vittariaceae***.

Investigations on the cytological basis of apogamous life cycles in ferns enable us to learn a lot about speciation among ferns, and the evidence collected so far seems to indicate that majority of apogamous ferns arose by hybridisation or reticulate evolution.

Cytological studies. Cytological studies on pteridophytes were initiated after the First World War. The first pteridophyte whose chromosome number was studied was ***Psilotum triquetrum.*** Okabe (1929) found that the haploid chromosome number ***n*** was 52 while the polyploid of the same species had 104 chromosomes for the haploid state. Dopp (1938) published a review of the work done before the War. Manton (1950) is credited with studying the chromosome numbers of almost all the pteridophytes in Britain and published his observation in a magnum opus "Problems of Cytology and Evolution of in Pteridophyte". Manton's researches stimulated world wide as a result of which cytological information is available for more than 2000 species and more is added everyday.

Chromosomes number of a few genera of pteridophytes is given below :

(i) ***Low chromosome number pteridophytes***

Selaginella : ***n*** = 9, 10 ; ***Hymenophyllum*** : ***n*** = 13, 19

Osmunda : ***n*** = 22 ; ***Isoetes*** : ***n*** = 10, 11.

(ii) ***High chromosome number pteridophytes.***

Psilotum : ***n*** = 52, 104 ; ***Tmesipteris*** : ***n*** = 52

Lycopodium : ***n***= 23 to 264 ; ***Equisetum*** : ***n*** = 108

Ophioglossum eliminatum : ***n*** = 30

Ophioglossum reticulatum : ***n*** = 720.

Apospory

The phenomenon of apospory was recorded by Druery (1884) in a fern called ***Athyrium filix-femina*** var. ***clarissima***. He observed the development of prothalli from the stalk of the sporangium as well as from the spore case of this fern. Later apospory was noticed in many ferns and some workers induced it under cultural conditions. During this phenomenon gametophytes or prothalli develop from the vegetative tissue of the sporophyte and not from the spores. Such prothalli are diploid ***i.e.***, they have the same chromosome number as the sporophyte. Normally the haploid spores germinate and give rise to the haploid gametophyte, but during apospory spores are not required. ***Apospory thus is the development of gametophytes from the vegetative parts of the sporophyte without the***

intervention of spores. Bower reported it in two species of ***Trichomanes*** in 1885. In this case aposporous gametophytes were produced from soral regions of the leaf and from leaf tips. Later apospory was reported by several authors in many genera and species of ferns. Druery and Neurnberg (1938) and Steil (1939, 1951) reviewed the literature on apospory and apogamy in ferns. Bristow (1962) developed gametophytes from a callus tissue derived from the sporophyte of the fern ***Pteris cretica***. This callus developed into gametophytes when grown in media containing only mineral nutrients. De Maggio and Wetmore (1961) isolated zygotes and undeveloped embryos from the archegonia of ***Todea barbara*** and cultured them. They were able to induce these isolated zygotes to develop into thalloid structures that resembled the gametophytes of this fern. Bell (1959) reviewed many cases of apospory in ferns and was tempted to state that the phenomenon of apospory must be general among ferns. Wetmore and De Maggio (1963) have also reviewed cases of apospory and apogamy among ferns.

Cytology and Artificial Induction of Apospory

In ***Osmunda javanica*** Sarbadhikari (1936) reported the formation of gametophytes from any vegetative part of the young or old sporophyte. He observed filamentous as well as heartshaped prothalli arising from the sporophyte. These prothalli bore antheridia containing spermatozoids but did not produce archegonia. Later apogamous sporophytes also developed from these aposporously produced gametophytes. Lawton (1936) recorded the presence of both antheridia and archegonia on the aposporously produced gametophytes of ***Osmunda regalis*** and ***Cystopteris fragilis.*** The gametophytes were diploid and so were the sex organs and the gametes. As a result of normal fertilisation tetraploid embryos and sporophytes were produced. Such tetrapoloid sporophytes have been induced to produce tetraploid aposporous gametophytes. This may lead to the formation of octaploid sporophytes. Lawton also observed cases where the diploid gametes of diploid aposporus gametophytes cross fertilise with haploid gametes from haploid gametophyte and produce triploid sporophytes. Lawton (1932) also induced apospory in ***Aspidium marginale*** and ***Woodwardia virginica*** and produced tetraploid sporophytes. Manton (1932) induced apospory in ***Osmunda regalis*** and produced diploid, triploid and tetraploid gametophytes and sporophytes.

Cystopteris

W.N. Steil (1944) induced the formation of aposporous gametophytes on the young leaves of apogamously produced sporophytes of ***Tectaria trifoliata.*** He was able to induce the formation of gametophytes from the margins of the young leaves and from the hair borne on the petiole of the first formed leaf of the sporophyte. In both the cases the gametophytes were produced in large numbers.

Tectaria trifoliata

Goebel (1907) and Palta (1973) induced the formation of prothalli from the petiole and laminar surface of the pinnae in ***Pteris vittata*** (Fig. 1.2, D). Goebel also observed the formation of thalloid structures bearing antheridia, rhizoids, stomata, and even vascular strands on the detached juvenile leaves of ***Alsophila van-geertii*** and ***Ceratopteris thalictroides***. These thalloid structures resembled gametophytes in possessing antheridia and rhizoids and sporophytes in possessing stomata and vascular strands. He designated them as "Mittelbildungen".

Development of tracheids and stomata are no longer regarded as indicative of aposporous or apogamous developments, but are considered to have developed under the influence of chemical environments under which the organs are cultured. So formation of stomata and tracheids do not give these aposporously developed gametophytes the characteristics of sporophytes. They can be regarded as purely gametophytic structures because they bear antheridia and rhizoids.

Aneimia densa

Beyerle (1932) observed the formation of undifferentiated outgrowths on the isolated leaves of some ferns. He observed them to grow into prothalli in ***Aneimia densa***, ***Pteris trenula***, ***Anogramma leptophylla***, ***Polypodium heracleum***, ***Cibotium schiedei*** (leaf margin), ***Dicksonia fibrosa*** (leaf margin), and ***Drynaria heraclea*** (small and dying leaves).

Woronin (1908) reported the development of aposporous gametophytes on the attached primary leaves of ***Pellaea nivea.*** Koehler (1920) induced the formation of prothalli on the leaves of ***Platycerium bifurcatum*** grown in dim light. Under strong light the leaves produced shoot buds.

Brown (1918) reported an interesting case of regeneration from the petioles of the leaves of ***Phegopteris polydioides.*** Under experimental conditons a cellular mass of green cells developed on the cultured petioles. It grew into a prothallus that bore rhizoids, true leaves, and structures intermediate between leaves and prothalli. It is a queer case of development of sporophytic and gametophytic structures together. Such instances clearly indicate that there is no inherent difference between the two alternating generations (sporophyte and gametophyte), and consequently no clear relation to the chromosome number.

Phegopteris polydioides

Charles Morlang (1967) induced apospory in three species of ***Asplenium*** (***A. platyneuyron***, ***A. rhizophyllum and A. montanum***). He cultured the leaves cut from the sexually produced sporophytes of these ferns, under controlled conditions. The leaves were observed to produce two types of neoplastic growths. These were two dimensional growths and three dimensional growths. The former developed into a normal heart-shaped prothallus that produced both types of sex organs that bore gametes. They had a diploid chromosome complement. The three dimensional growths developed into sporophytes.

Takahashi (1962) induced aposopory (Fig. 1.2, F—J) in ***Pteridium aquilinum*** var. ***latiusculum.*** He was able to induce the growth of gametophytes from petiolar epidermis (Fig. 1.2 F) ; from margins of pinnae (Fig. 1.2 G), from epidermal cells (Fig. 1.2, H), from mesophyll cell (Fig. 1.2, I) and from root epidermis (Fig. 1.2, J).

Causes of Apospory

Several factors seem to influence the aposporous development of gametophytes from vegetative tissues of the sporophyte :—

1. Briston (1962) demonstrated that mineral nutrition is responsible for the formation of prothalli from callus tissue obtained from the sporophytes of ***Pteris cretica.*** When he supplied sucrose to such a callus tissue, it developed into a sporophyte.
2. Goebel (1902) and Beyerle (1932) demonstrated that there is a pronounced relation between the stage of development of sporophytic cells (under culture) and the kind of organs regenerated. They observed that in ***Ceratopteris thalictroides*** aposporous gametophyte developed on decapitated young sporophytes with one or two leaves whereas in older sporophytes only shoot buds developed. Beyerle (1932) observed that in ***Davallia canariensis*** and in ***Nephrolepis biserata*** the prothalli develop on leaf tips and shoot buds at the basal and older parts of the leaf.

Drynaria rigidula

3. In some ferns ***e.g., Drynaria rigidula, Polypodium aureum*** and ***P. heracleum*** the leaves develop prothalli under dim light and sporophytic buds under strong light. Koehler (1920) also demonstrated that in ***Platycerium bifurcatium*** prothalli develop on leaves grown under dim light whereas the same leaves produce sporophytic buds on exposure to strong light.
4. The work done in Punjab University Botany Laboratories (Mehra and Palta, 1969 to 1973) on tissue culture has revealed interesting results. A few examples are being cited here. They obtained root callus tissue from the roots of a tetraploid ***Cyclosorus dentatus*** on Knudson's medium +2% sucrose +2, 4—D. The callus was fragile and its cell suspensions were obtained in sterile distilled water. The isolated cells of the callus were then placed on three different media :

Cyclosorus dentatus

A. Knudson's medium basal.

B. Knudson's medium +1% sucrose.

C. Knudson's medium + 2% sucrose.

In A the root cells behaved as spores and germinated to give rise to a prothallus. This shows that under mineral nutrition the callus tissue is induced to form gametophytes because basal Knudson's medium contains only minerals. In B the callus cells gave rise to structures that were partly gametophytes and partly sporophytes. They called such structures as Gameto-sporophytes. In C the callus cells gave rise to the complete sporophytes (regeneration).

Apospory has also been induced in ***Pteris vittata*** by Palta (1973). Sexually produced normal sporophytes produced gametophytes from the juvenile leaves. These gametophytes bore normal sex organs in Knudson's medium (Km) without sucrose. In this very species he was able to produce apogamous sporophytes in KM with 2% sucrose. Later this apogamous sporophyte was placed in KM without sucrose. After 3 weeks' time the apical pinnae of the same sporophyte produced aposporous gametophytes (Fig. 1.2, E). So with a change in the nutritional set-up of the environments apogamy and apospory can be induced at any time.

A careful perusal of the discussion regarding the life cycles of the ferns and fern allies (in the previous pages) it becomes obvious that :

1. It is a mystery as to why a particular cell or a group of cells develop into a gametophyte while others develop into sporophytes.
2. The occurrence of normal morphological and cytological alternation of generations, besides the phenomenon of apospory and apogamy in polyploids, indicate that haploidy and diploidy alone do not determine the characteristics of a generation (as advanced by Blackmans's hypothesis).
3. Lang's hypothesis completely fails in certain taxa of ferns and fern allies. The gemmae produced by the rhizome and gametophytes of ***Psilotum*** and ***Stromatopteris*** develop into same structure on which they are produced, even if they are detached. The rhizomes and the prothalli of both these plants are subterranean and saprophytic and contain the same fungus and their nutritional status is apparently identical. Experimental evidence provided by some surgical experiments also go against this hypothesis.
4. It is the same genome which is expressed in both gametophyte and sporophyte. This genome is compartmentalised into 4 concrete sets of gene blocks, each determining root, rhizome, leaf and gametophyte. In each case the master gene initiated the activity first with the subordinate genes expressing themselves in a sequential order resulting in the expression of an organ or a different generation.
5. The fern system is highly plastic and can be manipulated at will by using different experimental techniques. It is possible to invoke the gametophyte or the sporophyte from either generation simply by using different concentrations of sucrose in the culture medium (Mehra, 1975).
6. Some metabolic and nutritional factors may be responsible for manipulating the activity of the genes.

In some ferns. ***e.g. Actinostachys*** and ***Schizaea*** the sporophytes are very similar but the gametophytes are very different, whereas in majority of the higher ferns the gametophytes are very similar but sporophytes are quite different and dissimilar. It means that a certain set of expressions has evolved in the sporophyte and another in the gametophyte, with a considerable overlap (Bierhorst, 1971). Bierhorst (1971) has further opined that the "degree of overlap or the degree of differential expression of certain genomic segments, has occasionally changed and is of some morphological significance." It has also been noticed that there are close similarities in the developmental, patterns of some sporophytic and gametophytic structures such as antheridia, archegonia, sporangia and the hair or the trichomes. The later stages of development between the anteridia and sporangia of ***Equisetum*** are quite similar. In some ferns the establishment of initials of archegonia, antheridia and sporangia is similar. The divisions which establish these initials are on a similar pattern. Similarly the structure and development of trichomes on the sporophytes and prothalli of certain ferns is very similar. In the words of Bierhorst "these are all subtle expressions of apogamy and differ only in degree from those which drastically upset the life cycle..."

Schizaea

Parthenogenesis

Selaginella

This may be defined as ***the phenomenon where the egg develops into embryo without the act of fertilisation.*** This act is very common in algae, fungi and ferns. According to Farmer and Digby (1907) in homosporous leptosporangiate ferns, apospory was followed by neither apogamy nor fertilisation but by parthenogenesis. There are various species of ***Selaginella*** where parthenogenesis have been reported. In ***S. atroviridis*** with unisporangiata strobili the egg develops into embryo without fertilization. This phenomenon has also been observed in ***S. intermedia*** and ***S. langere*** (Hieronymus, 1911). Bruchman (1912, 1919) and Goebel (1915) observed parthenogenesis in a few other species of ***Selaginella*** with bisporangiate strobili (***S. anacardia***, ***S. rubicaelis*** and ***S. helvetica***). In these cases archegonial neck remains closed and does not permit spermatozoids to enter. The egg, in these species develops directly into embryo. In ***Marsilea drummondii*** too the megaspores on germination produced diploid megagametophyte (1970 Strasburger), which without fertilization developed into sporophytes.

ORIGIN OF THE VASCULAR CRYPTOGAMS

A number of theories have been put forward to explain the origin of the vascular cryptogams. All of them tend to explain the origin of the independent sporophyte and are based on two assumptions :

1. The vascular cryptogams originated from the bryophytes. Bower (1935), Lignier (1903), and Zimmermann (1930, 1938) regard Bryophytes and Pteridopytes as two divergent lines originating from a hypothetical group of archegoniate land plants. Campbell (1895) and Smith (1958) believed them to have originated from an anthocerotean type of a bryophyte.
2. The adherents of the second view believe algae to be the ancestors of the vascular cryptogams. The proponents of this idea trace the origin of the pteridophytes from various algal groups. Scott (1900), Bohlin (1901), Lotsy (1909), Church (1919), Fritsch (1916, 1945), Eames (1936), Arnold (1947), Gregus (1955), Andrews (1956), Axelrod (1959) and Mehra (1968) are the chief proponents of the idea of algal origin of the pteridophytes.

Anthocerotean Theory

It was put forth by Campbell (1859) and later supported by Smith (1938). Presence of intercalary meristem and photosynthetic tissue in the sporophyte of ***Anthoceros*** lead Campbell (1895, 1899) to put forth the view that such sporophytes are very near to the independent sporophytes of the vascular cryptogams. He conjectured, "Were the foot of the sporophytes in ***Anthoceros*** prolonged into a root penetrating the earth, it would become quite independent of the gametophyte, and were a special assimilative organ or leaf developed, a condition directly comparable to the sporophyte of the lower Pteridophytes or ferns would result". The discovery of the rootless, leafless and dichotomously branched psilophytales led Smith (1938) to advance his views favouring Campbell's views. Smith followed the origin of vascular cryptogams from ***Anthoceros***-like ancestors along the following assumptions :

(*i*) The intercalary meristem of ***Anthoceros***-like sporophytes shifted to the apex and initiated dichotomous branching (Fig. 1.4., A—D).

(*ii*) Dichotomous branching (Fig. 1.4) necessitated the formation of terminal sporangia like those of the Psilophytales (***Rhynia*** and ***Psilophyton***).

(*iii*) Metamorphosis of the anthocerotean columella into a conductive strand.

(*iv*) Occurrence of conducting tissue in the lower portions of the sporophyte of ***Anthoceros fusiformis*** (Campbell, 1924) and restriction of spore formation to apical regions alone.

Rhynia-Terminal sporangia

Smith extended his observations to the gametophytes also and pointed out the resemblances between sex organs of ***Anthoceros*** and the Eusporangiate vascular crypotogams (***Lycopodium*** and ***Equisetum***). In both, the sex organs are embedded and have no stalk. The only difference between the archegonia of ***Anthoceros*** and the pteridophytes is the presence of a jacket layer around venter of the former and its absence in the latter. This is due to the divisions of the archegonial initial of ***Anthoceros*** into three jacket initials and an axial cell. In pteridophytes this division is absent and the initials function directly as an axial cell. According to Smith obliteration of these divitions of the archegonial initials would lead to the pteridophytean archegonium, which is homologous to the cover cells and axial row of cells of the anthocerotean archegonium. Regarding the antheridia, Smith pointed out that the entire anthocerotean antheridium is equivalent to the fertile region of the antheridium of eusporangiate pteridophytes.

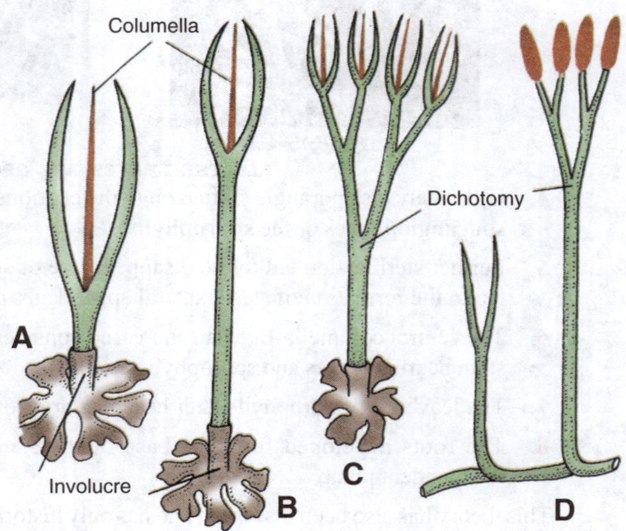

Fig. 1.4. (A-D). A scheme illustrating origin of early land plants from *Anthoceros-like* ancestors (After Smith)

Campbell (1939) campared the simple gametophytes of ***Anthoceros*** with the green algae. Smith also pointed towards some superficial resemblances between the embryos of ***Psilotum*** and ***Tmesipteris*** with those of ***Anthoceros.***

This theory has not received wide acclaim and has almost been discarded by the modern botanists.

The Strobilus Theory

It was put forth by Bower in 1894 and supported the origin of sporophytes of vascular cryptogams from the liverworts. According to this theory the simple sporophyte of the liverworts evolved into an elaborate and a complex structure along the following lines :—

1. There was progresive sterilisation of the sporogenous tissue in the liverwort sporophyte. It led to the formation of alternating sterile and fertile regions.
2. The fertile regions or sporangia became superficial in position.

3. The sterile parts gave rise to lateral projections called the enations. The enations were either below the sporangia or besides them. At this stage the entire sporophyte looked like a strobilus. The sporangia lie in the axils or at the bases of the enations. Such a condition was visualised in ***Phylloglossum***, ***Isoetes*** and those species of ***Lycopodium*** where all the leaves are sporophylls.

Phylloglossum

Isoetes

ALL THE LEAVES ARE SPOROPHYLLS

4. Ultimately the sporangia shifted on to the enations. The enations could then be called the **sporangiophores** or the **sporophylls**.
5. Further sterilisation led to the disappearance of sporangia on the sporophylls. This step led to the formation of sterile lateral appendages called the **leaves.**
6. The central columella-like part increased considerably in size and became an axis or the stem bearing leaves and sporophylls.
7. The leaves and sporophylls later became large and elaborate.
8. The roots developed from the base of the stem and stamped the sporophyte as an independent plant.

This theory has also been discarded and has only historical importance. Discovery of a number of extinct plants belonging to the psilophytales does not support the existence of such a simple sporophyte as pictured by Bower. All these primitive land plants possess branched stems that were not small. The enation theory is even now believed because it explains the origin of micophyllous leaves, but Bower has postulated that the microphylls which originated as enations later developed into large megaphylls in his strobilus theory. This latter part has not appealed the morphologists. Later, Bower, however, realised this and regarded his enation theory to explain only the origin of microphyllous leaves. He proposed another sequence leading to the origin of megaphyllous leaves, which later became the basis of Zimmermann's ***telome concept***.

Protocorm Theory

This theory was put forth by **Treub *in 1884.*** He regarded the protocorm of the ***Lycopodium*** embryos as a primitive organ and considered it to be an ancestral feature that has been retained in some contemporary species (***Phylloglossum*** and ***Lycopodium***). He postulated that it might have been of common occurrence in the ancestors of the vascular cryptogams. This idea did not receive the general approval of the morphologists and is now of historical importance only. Bower (1907, 1935) regarded the protocorm as "opportunist growth" and not a primitive feature. He attached only ecological importance to the protocorm which helped in establishing the young sporophytes of certain lycopods. Browne (1913) regarded the protocorm as a modified and a reduced stem. The opponents of this theory state that (*i*) protocorm is not present in many primitive species of ***Lycopodium*** ; (*ii*) it is a resting stage that has only an ecological importance ; (*iii*) ***Phylloglossum*** is not a primitive but a reduced plant ; and (*iv*) that protocorm is also present in ***Ophioglossum*** and some monocotyledons.

Phyton Theory

This theory was put forward by Celakovsky **in 1901**. He denied the existence of the stem as an independent member His theory was closely related to the ***strobilus theory*** of Bower because he postulated that the early land plant, when in a strobilus stage was nothing but a cluster of leaves or sporophylls. Axis developed later from the sporophyll bases. Schoute (1931) in his paper entitled "On Phytonism" severely criticised Celakovsky's memoir and stated that his theory has only constructed a scheme of thought without giving any decisive proof in support of his ideas, which can be explained otherwise also.

Wolff (1759) and Gaudichaud (1841) were the earlier authors who gave the idea of 'phytonism' to the plant morphologists. Wolff made a very frank statement that he saw nothing in the plant except the leaves and stem. He considered the roots to be portions of the stem. The stem was regarded to be a mere prolongation of the leaf stalks. His views were later elaborated by Gaudichaud (1841) and Celakovsky (1901).

Algal Origin of the Vascular Cryptogams

The consensus of opinion now favours the view that early land plants originated from an algal ancestor. The algal origin of the vascular cryptogams and the seed plants is supported by the following facts :

1. The chloroplasts of both the green algae and the vascular cryptogams contain chlorophyll ***a*** and ***b.***
2. Most of the carotenoid pigments in these groups are also similar.
3. Their carbohydrate reserves are similar. Their starch grains contain a mixture of two types of glucose macromolecules, amylose and amylopectin.
4. Their cell wall structure and composition are also similar ***e.g.*** the cells in both the groups are characteristically surrounded by a pectic wall layer that contains galacturonic acid and a wall layer containing cellulose.
5. The flagella in the male gametes of vascular cryptogams like those of green algae are of whip-lash type.

Scott (1900) and Eames (1936) do not specify any particular algal group which gave rise to the first land plants. Church (1919) and Arnold (1947) believed that primitive land plants originated from a marine algal group growing near the upper limits of the tide levels. They perhaps meant the brown algae which are the only modern group of algae that grow profusely in the intertidal zone and possess motile reproductive cells. Bohlin (1901), Fritsch (1916) and Lotsy (1909) traced the origin of vascular plants from a chaetophoraceous green alga. Fritisch (1945) suggested that such an alga must have been an erect and parenchymatous form with isomorphic alternation of generations. All these authors believed that bryophyta and the vascular cryptogams are two parallel lines of evolution and that there is no phylogenetic connection between them. A brief account of the various theories regarding algal origin of pteridophyta is given below :—

1. Church's Hypothesis

Dr. Church in his publication entitled "Thalassiophyta and the sub-aerial Transmigration" conjectured a polyphyletic origin for the vascular land plants. He wrote his essay without any reference to the fossil evidence. He imagined the transmigration to have taken place at a time much earlier than the Devonian. Church postulated the origin of land plants from a hypothetical or imaginary group of marine algae which he called as **Thalassiophyta.** The Thalassiophyta were assumed to have originated along several independent lines from the flagellated green algae. While giving a description of the process of the transmigration, Church (1919) imagined a universal ocean covering the entire face of the earth. He supposed this ocean to the inhabited with a countless number of minute and free swimming organisms. This stage was called the **Plankton stage.** Later the sea bottom started rising at certain places and the plankton stage was succeeded by the **'Benthic Stage,'** which included a number of attached and rooted sea weeds. This process went on for ages and the attached sea weeds gradually attained a very high degree of organisation. This uplift of sea bottom was accompanied by a large number of environmental changes that included increase in the amount of atmospheric oxygen and light intensity. The uprising sea bottom later emerged out of the sea and that was the first sight of land. This brought in the conditions of desiccation and insolation. The sea weeds which emerged out of water along with the land were now under terrestrial environments and were exposed to free oxygen, light and desiccation. This led to struggle for existence among the exposed algae and brought in adaptations among them. Natural selection played its role and more adapted changes accompanied by variations were brought about in these sea weeds. Church believed that land plants originated under tropical conditions. Under the changed set up of environments the exposed and terrestrial sea weeds attained a high degree of complexity and had exhibited differentiation into stem and leaves. Roots appeared in the form of holdfasts. Conducting strands in the form of phloem elements had already appeared. These sea weeds possessed heteromorphic alternation of generations. When the transmigration had occurred these highly organised plants merely required to become adapted to subaerial environments. These included the appearance of xylem elements and stomata to permit a ready stream of transpiration. The roots which hitherto served the only function of attachement were required to take on a new function of absorption. The spores needed a resistant cell wall to equip them for the changes and chances of dissemination by air. According to Church the sporangial tetrads of the early vascular plants originated from the unilocular tetrasporangia of the algal ancestors. The female sex organs or the archegonia were postulated to have evolved independently along various lines. Both the gametophytes and the sporophytes were believed to be already leafy and autophytic. It appears from the above description that brown algae were considered to be the probable ancestors of the land plants.

The geologists do not regard the land to have emerged from the sea. They believe that land was already in existence when the oceans appeared. Church's hypothesis, therefore, is not considered to be tenable. Moreover, it has been urged that the highly organised transmigrant brown algae that were used to marine environment were too specialised to adapt entirely to new mode of existence on dry land.

2. Greguss's Hypothesis

Greguss put forth his views in 1955. He traced the origin of vascular plants from the three major divisions of algae, ***i.e.***, Chlorophyta, Phaeophyta and Rhodophyta. He did not take into consideration the pigment constitution, the products of photosynthesis, and the presence or absence of cilia in the reproductive units. He laid stress on the branching of the organisms ***i.e.***, whether it is dichotomous, monopodial or verticillate. He derived mosses from Chlorophyceae and liverworts from the Phaeophyceae. In the same way he traced the origin of the ***Rhynia*** and ***Horneophyton*** from the Chlorophyceae and ***Psilotum*** and ***Tmesipteris*** from the Phaeophyceae. Lam (1957) criticised Greguss's hypothesis and regarded it as based on unsound footings. He almost disregarded the phylogenetic relationship of the plant groups while tracing their origin ***e.g.***, Psilophytales and Psilotales that are quite closely related to each other have been regarded as descendants of two different algal groups.

3. Andrews' Hypothesis

Andrews (1956, 1959) also believed in the polyphyletic origin of vascular plants. His conclusions are based on the discovery of some peculiar fossils that are not vascular plants but can be regarded as descendants of marine algae that made several attempts to conquer land. These are ***Nematothallus***, ***Crocalophyton*** and ***Protosalvinia.*** The former was discovered by Lang (1937) from the upper Silurian of England and possessed firm cuticle covering its body. Their spores were also cutinized and had a resistant wall. ***Crocalophyton*** possessed tracheid-like structures in its body. ***Protosalvinia*** was discovered from Upper Devonian of Ohio. Its body was made up of distinct short stalk bearing a spherical and bilobed head. It was covered with cuticle and possessed distinct conceptacles bearing spore tetrads. The spore walls were thick and resistant. Occurrence of such plants led Andrews to believe that a number of algal groups attempted to conquer land. All these algal groups gave rise to the different types of vascular plants. His belief was strengthened due to diversity in the morphology and structure of the various groups of land plants such as Lycophyta, Psilophyta, Articulatae (Sphenophyta) and Pterophyta. He considered that morphological and structural variations presented by these main division of vascular cryptogams are due to their origin along various independent lines of evolution from different algae rather than from 3 or 4 groups. He gave a vivid description of the important fossils from the early Palaeozoic in 1959 and then commented on Zimmermann's **telome concept.** He supported his concept regarding the evolution of the pteropsids and sphenopsids, but disregarded his concept for the evolution of the lycopsids and dismissed it as a purely hypothetic approach.

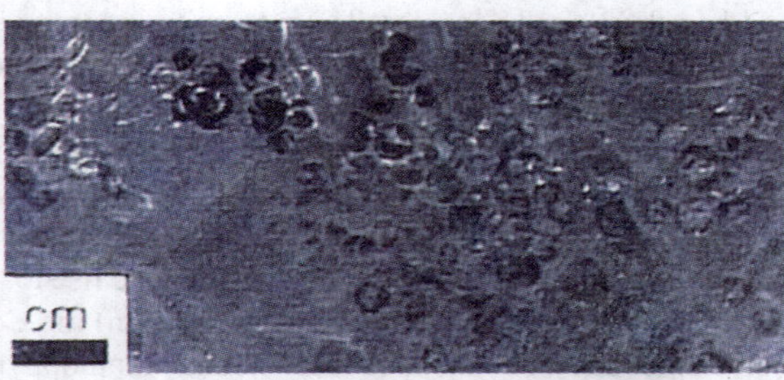

***Protosalvinia*– a fossil descendent of marine algae.**

4. Leclercq's Hypothesis

Leclercq (1954, 1956) also believed in the polyphyletic origin of the vascular-cryptogams. She postulated that various divisions (Psilophyta, Lycophyta, Sphenophyta, Pterophyta) originated along independent lines running down separately into an ancestral algal complex. As a result of her palaeopalynological studies of the early land plants that inhabited the earth from the upper Devonian down to the Cambrian, she concluded that the land plants originated in the pre-Cambrian period. Her studies on the form and structure of the spores in the upper and middle Devonian revealed a striking correlation between these spores and the megaflora. She also studied the spores from lower horizons like Ordovician and Cambrian period and concluded that land plants exists in those ages. They must, therefore, have originated in the Pre-Cambrian periods. She also reviewed the extinct land plants belonging to the middle Devonian, upper Silurian and Cambrian periods. She found that the Rhyne fossils (***Rhynia***, ***Horneophyton***) that were discovered by Kidston and Lang were of quite simple organisation as compared to those which occurred in earlier periods like the upper Silurian. The fossils belonging to this period were quite complex and elaborate and, therefore, led her to conclude that simple psilophytes like ***Rhynia*** and ***Horneophyton*** are not primitive and ancestral, but represent the last remnants of a dying race that might have flourished in the earlier ages (Pre-Ordovician). She concluded that the psilophytes like ***Rhynia*** and ***Horneophyton*** are the last and recent in the series that had an independent origin from the algal stock.

5. Axelrod's Hypothesis

Axerlrod (1959) followed the lines of Leclercq and Andrews and regarded the vascular cryptogams to have originated along different phyletic lines of evolution from diverse algal stock

(Fig. 1.5). He regarded ***Rhynia*** and its allies as 'bradyletic types and not ancestral to the other Pteridophytes. Discovery of sixty different types of spores with cutinised and resistant walls from the Cambrian period led him to believe that land plants originated somewhere far down in the pre-Cambrian age from diverse algal stocks. The idea is based on the assumption that these spore types belonged to the early land plants that might have colonised the land in the Cambrian period. The transmigrants evolved from these algal stocks were believed to have passed through a 'primitive psilophytic stage' and had developed vascular elements quite independently. In his phyletic scheme illustrating the origin of vascular land plants he visualised five independent lines originating from the pre-Cambrian algal stock. These five lines represent the five hypothetical transmigrants from where the five major divisions of vascular land plants originated along five independent lines. These are the Psilophyta, (including ***Rhyniaceae***, ***Psilophytaccae, Zosterophyllaceae***, leading to the Psilotales), Lepidophyta, Arthrophyta (Sphenophyta), Pterophyta (Filicophyta) and Pteridospermophyta. From those transmigrants or the ancestral stock which gave rise to the Pteridospermophyta, there also arose at a very early stage (somewhere in the Ordovician) the Coniferophyta. The Pteridospermophyta gave rise to the Gnetophyta, Cycadophyta and Anthophyta. Axelrod's phyletic scheme regarded the Pteridophyta (Psilophyta, Lepidophyta, Filicophyta, Arthrophyta) and the Pteridospermophyta to have originated independently and directly from the independent algal stocks.

6. Merker's Hypothesis

Merker (1961) postulated that the land plants along with the Bryophytes originated independently from a group of pre-Cambrian archegoniates which in turn were derived from different algal sources. He recognised the following five evolutionary off-shoots from the ancestral archegoniates :—

1. Extinct ***Rhyniaceae***, Psilotales and ***Ophioglossaceae.***
2. Bryopsida which include the liverworts and mosses.
3. The Sphenopsida.
4. The Lycopsida.
5. The Pteropsida and Pteridosperms. Gymnosperms and Angiosperms were believed to have evolved from the Pteridosperms.

Merker considered that the gametophytes and the sporophytes in the ancestral archegoniates were almost alike and independent. Such a condition persisted in the first evolutionary line of origin, which included ***Rhyniaceae***, Psilotales and ***Ophioglossaceae***. During the course of evolution the sporophytes either progressed or retrogressed. The former lines are represented by Lycopsids, Sphenopsids and Pteropsids. The latter line is represented by the Bryopsida. He regarded ***Rhyniaceae*** as reduced and simplified forms and not a primitive and ancestral group of land plants.

7. Lam's Hypothesis

Lam (1955) suggested a diphyletic origin for the vascular land plants in his essay on the "Phylogeny of cormophyta." Lam believed them to have arisen from the Thallophyta along two lines during the Mid Cambrian. These lines were Psilopsida and Lycopsida, respectively. Three divergent lines emerged from the Psilopsida. These are the Sphenopsida, the Pteropsida and the Cycadopsida which arose through ***Archaeopteris.*** From the primitive members of the Cycadopsida arose the Protoangioperms (***Welwitshia***, ***Ephedra***, ***Gnetum*** and ***Casuarina***) and the Angiosperms. The Lycopsid plexus gave rise to the Coniferopsida that include the Cordaitales, Ginkgoales, Coniferales and Taxales.

There are two well founded objections to this hypothesis (Mehra, 1968). To derive Coniferopsida from the Lycopsid stock does not appear to be justified. The second objection is that Lam grouped

Casuarina with other members the Gnetophyta. ***Casuarina*** is very much an Angiosperm and should not have been grouped with the Gnetophyta. It is, however, known from biochemical analysis (Alston and Turner, 1966) that ***Casuarina*** is a unique angiosperm that produces compounds called biflavonyls. These compounds are not produced by any other angiosperm, but are produced by many Cycads, conifers and other gymnosperms. In its lignin chemistry also ***Casuarina*** resembles gymnosperms. These chemical features which are unique for this genus cannot be ignored in determining its phyletic relations.

Gnetum **Casuarina**

PROTOANGIOSPERMS (Gnetum, Casuarina) ORIGINATED FROM PRIMITIVE MEMBERS OF CYCADOPSIDA

8. Mehra's Hypothesis

Professor P. N. Mehra in his fifteenth Albert Charles Sward Memorial Lecture entitled "Conquest of Land and Evolutionary patterns in Early land plants" gave a vivid description of the evolutionary patterns in the land plants and summarised the views regarding the origin of land plants. While concluding his lecture he proposed his own hypothesis explaining the origin of land plants (1968). He is against polyphyletic origin of vascular plants and believed in the diphyletic origin.

He postulated that the green algae (Chlorophyta) must be regarded as probable source of origin of land plants because of their similar pigmentation, photosynthetic activities and flagellated cells. Mehra agrees with Fritsch (1945) in regarding ***Frischiella*** like Chaetophoraceous green algae as the probable ancestors of land plants. There was an organisation of an apical cell with three cutting faces in these ancestral algae. Such a differentiation leads to the development of: (*i*) radial symmetry; (*ii*) cortication of the thallus; and (*iii*) development of sterile jacket around the sex organs. Development of these structures enabled these transmigrating plants to bear desiccation. These plants were erect, non-vascular, had similar sporophytes and gametophytes (Isomorphic alternation) that were independent and photosynthetic, both had apical cell organisation and were radially constructed. Such a group of plants could live under subaerial conditions. Mehra called them Protoarchegoniatae. From these protoarchegoniatae there diverged two lines of evolution: (A) the Psilophytaceous line of evolution; and (B) the Lycopodiaceous line of evolution.

(A) Psilophytaceous Line of Evolution

The Protoarchegoniate plants developed cuticle. The gametophytes possessed jacketed sex organs that developed from branch systems, endowed with apical cell segmentation (as is found in some mosses). These sex organs were projecting type, superficial and jacketed. The sporophytes were independent and their terminal ends were sporiferous, ***i.e.***, the sporangia were terminal and had a distinct sporangial wall. These developments gave rise to a **terminal sporangiate archegoniate plexus** (plants). In these plants the archegonia after fertilization retained the oospore. The independent sporophytes of this plexus followed two lines of development:

(a) Along one line the sporophytes, developing from the oospores enclosed within the archegonia, did not become free and remained attached to the gametophyte, ***i.e.*** they became dependent upon the gametophyte. The gametophytes of this line of evolution followed two courses of development. Along one line the sex organs did not become embedded, whereas along the other line the sex organs became embedded. The sporophytes with the former type of gametophytes gave rise to the liverworts and the mosses, whereas those of the other type gave rise to the Anthocerotae. Both the lines of evolution ended blindly. The spermatozoids were biciliate.

(b) This line of evolution led to the appearance of vascular tissues in the sporophytes which remained independent and became well established to life on land. The gametophytes had embedded sex organs with multiflagellate spermatozoids. The sporophytes, however, bore terminal sporangia and were mostly dichotomously branched. This led to the appearance of the **Rhyniaceous Stock.** This stock fanned out into the following four lines of evolution :

(i) **Prosphenopsid line :** It was linked with the Rhyniaceous stock by plants like ***Protohyenia***, which gave rise to ***Hyenia*** and ***Calamophyton*** of the early and middle Devonian periods. From them arose the Sphenophyta (Equisetales).

Hyenia and ***Calamophyton***

(ii) **Pro-pteropsid line :** It was linked with the Rhyniaceous stock through ***Protopteridum*** like plants. They had terminal sporangia but also bore short, sterile branches with **Circinate ptyxis.** Such plants gave rise to the Filicales or Filicophyta.

(iii) **Progymnospermopsid line :** It is represented by ***Archaeopteris***, ***Callixylon*** type of plants that showed secondary growth. Some of them had also established heterospory. They gave rise to the pteridosperms on the one hand and to Gnetophyta, Cordaitales, Coniferales and Ginkgoates along other independent lines of evolution. The Pteridosperms gave rise to the Bennettitales and the Cycadales.

(iv) **Psilophytaceous line.** It bifurcated into two lines of evolution. One culminated in ***Rhyniaceae*** whereas the other gave rise to the Psilotales represented in the modern flora by ***Psilotum*** and ***Tmesipteris.***

(B) Lycopodiaceous Line of Evolution

It started directly from the proto-archegoniatae. The sporophytes remained free and developed vascular system and lateral sporangia as specified by ***Baragwanathia*** and ***Drepanophycus.*** The gametophytes developed embedded sex organs with biciliate spermatozoids. This line was termed by Prof. Mehra as **Lateral sporangiate plexus.** From this plexus arose the pro-lycopsida represented by ***Baragwanathia*** and ***Drepanophycus.*** From this pro-lycopsid stock there fanned out the various orders to the lycophyta, ***e.g.***, the Lycopodiales, Selaginellales, Lepidodendrales and Isoetales through Pleuromiales.

Professor Mehra has not indicated the origin of angiosperms in his scheme. The problem of origin of angiosperms is still an abominable mystery.

9. Nayar's Hypothesis

Prof. B. K. Nayar (1976) suggested a **monophyletic** origin for the pteridophytes. He also believed in the algal origin of the pteridophytes and suggested that some littoral plants (algae) with isomorphic or homologous alternation of generations were the probable ancestors of the land plants. The main points of his hypothesis can be listed as under :—

1. The immediate ancestors of pteridophytes were algae with isomorphic alternation of generations.
2. These were erect growing filamentous algae with a tendency to become multiseriate filament in the older regions but retaining the uneseriate condition in the younger apical regions.
3. They had a hemispherical apical cell with a single flat posterior cutting face.
4. Such algae came to live under terrestrial conditions. Due to adverse growing conditions on land both the gametophyte and the sporophyte plants became short and stumpy.
5. To avoid excessive transpiration both the generations became subterranean and developed rhizoids all over the surface to absorb water.
6. Both were independent of each other and had to "per force develop efficient photosynthetic mechanisms for survival". To meet this requirement both the generations followed different lines of evolution.
7. The short and stumpy sporophyte is believed to have developed special aerial branches that were green, unbranched and cylindrical. These were considered as progenitors of fronds.
8. These unbranched aerial branches bore simple sporangia at their tips.
9. The central elongated cells of these branches became specialised to form a protostelic vascular cylinder, which performed the dual function of conduction and mechanical support. This protostelic vasculature extended upto the tuberous basal or lower part.
10. The aerial branches became flattened to form an expended lamina and thus appeared like a simple frond with an undivided lamina with a single median vein. Later modifications resulted in the appearance of a reticulate venation. Such a frond bore a terminal sporangium. It is also supposed that in some cases the frond branched dichotomously forming a primitive fan-shaped leaf. Each lobe of such a leaf bore a terminal sporangium.
11. With the development of aerial photosynthetic organs there arose the necessity for the establishment of absorbing organs or the roots. The roots developed their own vasculature which became continuous with that of the aerial organs. Nayar compared such a primitive land plant with young sporelings of the present day ferns. He has opined that the vascular strands developed independently in the leaves and the roots and then became connected.
12. Ultimately the sporophyte of the ancestral land plant looked like having a tuberous rhizome with simple unbranched aerial organ expanding into a simple leaf with a terminal sporangium. Such a simple structure later on evolved into various groups of pteridophytes.

The gametophyte of the primitive algal ancestor, as pointed out above, became short and stumpy and erect growing structure covered with rhizoids all around. It became subterranean and developed mycorrhizic associations with a fungus. The sex organs later developed all around the prothallus. All the different forms of prothalli met with among the pteridophytes are supposed to have been derived from this primitive prothallus. Such a tuberous, subterranean and saprophytic thallus underwent a series of reductions, "so much so that the two generations though morphologically similar in the primitive ancestor became markedly different as evolution progressed." Later in the year 1982, Nayar again elaborated his hypothesis. In this communication he related the heterotrichous ***Fritschiella***—a Chaetophoraceous green alga, to the 'serpoerectus' ***Equisetum*** prothallus which he considers as the "base" or the "missing link" for the evolution of both the gametophytic as well as sporophytic generations of land plants.

Majority of the botanists are sure of a genetic relationship between green algae and the vascular cryptogams or the land plants on the whole, but there is less agreement on the way or method of origin of land plants. Stages intermediate between algae and primitive land plants have not been recognised. As discussed above the question of monophyletic and polyphyletic origin of land plants is yet a subject of lively discussion. Such a discussion becomes more complicated when we realise that there is no tacit agreement among the botanists regarding the relationships of land plants among themselves.

EVOLUTION OF THE SPOROPHYTE

According to a general belief the earliest and the primitive sporophyte of the land plants was axial in nature. The leaves and branches were evolved later by a modification of the central stem. These primitive land plants were supposed to have rootless, leafless and dichotomously branched green stems. The vascular system was a simple protostele. Some of the ultimate branches terminated in sporangia. This assumption is supported by the discovery of some simple plants in the upper Silurian and Devonian periods of earth's history (***Rhynia***, ***Horneophyton***, ***Psilophyton***, ***Asteroxylon*** and others). Discovery of such plants led may botanists to propose theories explaining the appearance of leaves, branches, root and sporangia on the sporophyte. These theories tend to explain the process of elaboration of the primitive sporophyte into the modern sporophyte possessing roots, leaves and variously disposed sporangia. Of all these theories **Telome theory** of Zimmermann (1930) has attracted great attention and has a large number of supporters.

Asteroxylon

Telome Theory

This theory was first proposed by Zimmermann in 1930 and later elaborated by him in 1938 (Manual of Pteridology, 558–618). Zimmermann postulated that all the vascular plants evolved from a very simple, leafless and a dichotomously branched ancestral type like ***Rhynia***. Such a plant is made up of sterile and fertile branches. The fertile branches terminated in sporangia. **The simple and terminal parts of a dichotomously branched stem were designated as telomes** (Fig. 1.5). The telomes meet each other at the point of forking. Portions of the stem below the forking were called **mesomes**. The telomes were classified as sterile or vegetative telomes also called **phylloids** and **fertile telomes.** The fertile telome consists of a terminal and a stalked sporangium usually traversed by a single vascular strand up to its base. As evolution proceeded there was a change from dichotomous to sympodial branching as a result of the fusion of two or more **telomes** called the **telome trusses or syntelomes**. The fertile telome trusses were called the **sporangial trusses** and sterile telome trusses were called the **phylloid trusses.** Both the types of **syntelomes** also occurred together to form **mixed telome trusses** or **mixed syntelomes.**

Zimmermann postulated that the primitive vascular cryptogams originated from the green algae. The unicellular green algae divided in all planes to form a parenchymatous thallus. Later **meristematic** tissues developed and erect radially constructed branches came into existence. It was followed by the appearance of distinct alternating generations. The sporophyte branched dichotomously and possessed a central conducting strand. Such algal ancestors, according to Zimmermann, led to the evolution of early vascular plants of the upper Silurian and Devonian periods. They possessed heteromorphic alternation of generations and their sporophytes were simple and dichotomously branched. Some of the branches terminated in sporangia. They possessed no leaves and had no roots. The vascular system was a simple protostele. The aerial and simple stems possessed stomata. The function of roots was performed by hair or rhizoids. These sporophytes had subterranean portions called the rhizomes. All these hypothetical steps are diagrammatically represented in Fig. 1.5.

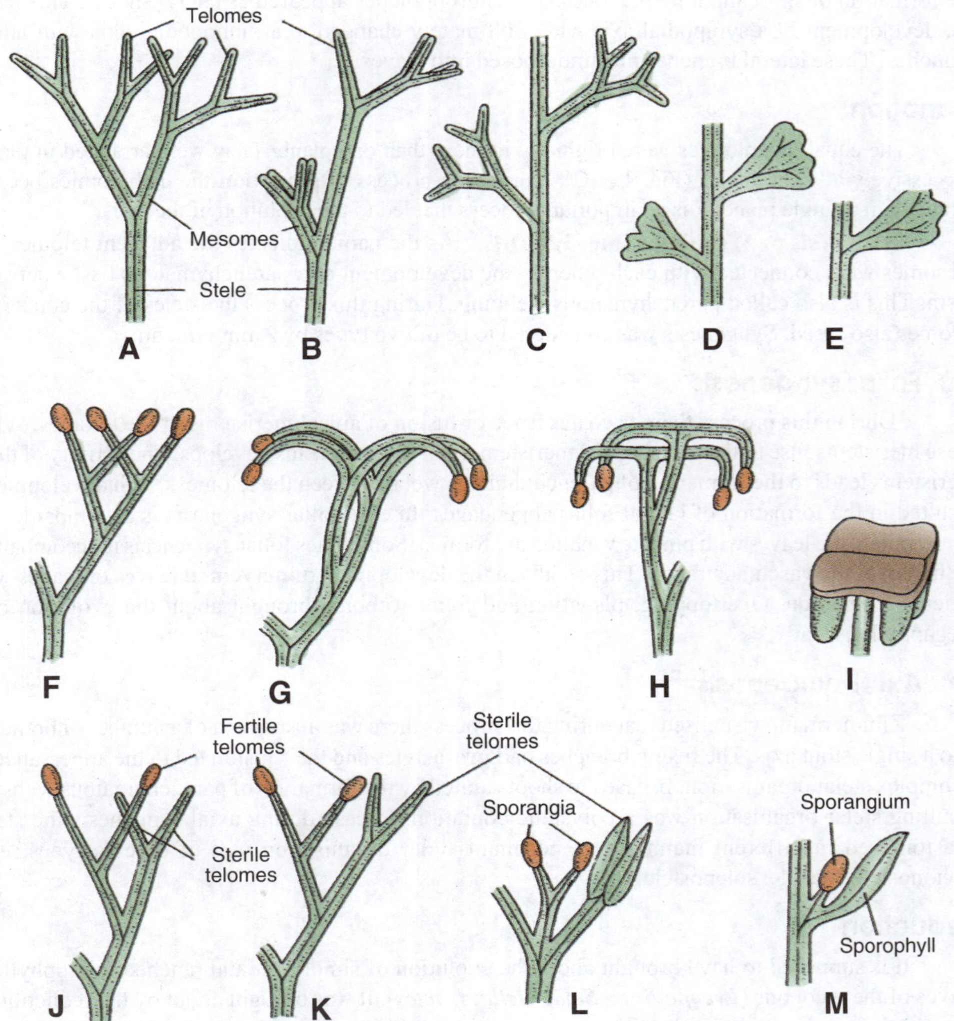

Fig. 1.5 (A–L). Telome concept of Zimmermann. A. Illustrates hypothesis of telomes and mesomes; B. Overtopping; C. Planation; D–E. Webbing and syngenesis; F–I. Illustrate the origin of the sporangiophores of Sphenophyta by recurvation and fusion and flattening of the telcomes and mesomes; J–M. Illustrating the origin of Lycophyta.

These early land plants evolved into the higher vascular cryptogams along three independent lines of evolution. These were the Lycopsid, Sphenopsid and Pteropsid trends of evolution. During the course of this evolution certain **elementary processes** of progressive differentiation took place in the sporophytes. These can be listed as : (*i*) Overtopping (*ii*) Plantion, (*iii*) Webbing or syngenesis, (*iv*) reduction, and (*v*) recurvation. All these elementary processes of organogenesis were believed to have occurred in varying degrees in the different taxonomic groups. These will be discussed separately.

Overtopping

The equally dichotomising axes developed unequal dichotomy (Fig.1.5 A-B). This resulted in the formation of short and long branches. The short branches appeared as lateral shoots. This led to the development of a sympodial axis which ultimately changed to a monopodial axis with lateral branches. These lateral branches metamorphosed into leaves.

Planation

The equal dichotomies were originally in more than one plane. They were arranged in planes successively at right angles (Fig. 1.5, C). During the process of **planation** the dichotomies became arranged in a single plane. It is an important process that led to the evolution of the leaf.

Syngenesis or Webbing : (Fig. 1.5, D-E). As the name indicates the adjacent telomes and mesomes were connected with each other by the development of a parenchymatous tissue between them. This is also called parenchymatous webbing. During this process the steles of the concerned telomes also fused. Syngenesis was considered to be of two types by Zimmermann:-

(*a*) Foliar syngenesis

During this process there is coalescence or fusion of apical meristems of the telomes. When these meristems fuse to form a marginal meristem, a lamina with veins develops. The activity of these meristems leads to the formation of parenchymatous webs between the telomes. Such developments resulted in the formation of lateral foliar appendages. In case foliar syngenesis is accompanied by overtopping, the leaves with pinnate venation are formed. Sometimes foliar syngenesis is accompanied by fusion of the vascular strands. This results in the development of **netveined** leaves or leaves with reticulate venation. Overtopping, planation and foliar webbing brought about the evolution of a megaphyllous leaf.

(*b*) Axial syngenesis

Zimmermann visualised that during this process there was absorption of a number of branches into a single stout axis. The fusing branches had **protosteles** and their fusion led to the appearance of a complex stelar organisation. In case the shoots united by the formation of parenchymatous webs the resulting stelar organisation was a polystelic condition. In case during axial syngenesis the steles anastomosed in different manners, the resultant stelar organisation was of diverse types, ***e.g.***, siphonostele, eustele, solenostele.

Reduction

It is supposed to have brought about the evolution of simple and unbranched microphyllous leaves of the lycopods (***Lycopodium***, ***Selaginella***, ***Isoetes***). It was brought about by the reduction of the syntelome to a single needle-like lateral appendage (Fig. 1.6, **A-C**).

Recurving

During this process the fertile telomes were supposed to become reflexed (Fig. 1.6, **H, I**). As a result the sporangium assumes an inverted position. Zimmermann called this process as **incurvation.** Wilson (1953) recognised two processes —

(*a*) Recurvation : During this process the sporangia bent downwards as in Sphenophyta.

(*b*) Incurvation : This led to the shifting of the sporangia to the ventral surfaces of the foliar appendages thus bringing about a condition found in the ferns.

Wardlaw (1952), however, pointed out that all these changes can be embraced within one term, which he calls as **recurvation.**

While discussing the above listed five elementary processes of organogenesis as postulated by Zimmermann, it becomes evident that microphyllous leaves originated as a result of the process of **reduction.** The megaphyllous leaves of ferns (Filicophyta) originated as a result of combined processes of overtopping, planation and foliar syngenesis. We shall now discuss the origin of the fertile leaves or the sporophylls as visualised by Zimmermann.

Fig. 1.6 (A–J) A–C. Illustrating reduction of syntelomes to single needle-like lateral appendages (C); D–G. Illustrating origin of microphylls; H–I. Recurvation (All after Zimmermann, 1938).

Origin of Sporophylls

The origin of sporophylls in the three main divisions of the vascular cryptogams has been visualised differently by Zimmermann.

Sphenophyta or Arthrophyta

Recurvation and syngenesis were supposed to be the elementary processes that led to the origin of the sporangiophores in the members of this division (Fig. 1.6 **F-1**). Recurvation led to the downward bending of the sporangia. It was followed by fusion and flattening of the telomes and mesomes to form a peltate disc. Intermediate stages leading to this condition existed in the fossil members of this division, ***e.g.***, ***Calamophyton***, ***Hyenia***, ***Eviostachya*** and ***Protocalamostachys***. Planation followed by reduction, however, led to the development of the sterile leaves of this division. The intermediate stages are met with in the fossile genera ***Calamophyton*** and ***Asterocalamites.*** The Telome

Calamophyton-fossil member of Div. Sphenophyta

theory gives a satisfactory explanation of the evolution of sporangiophores and sterile leaves in the Sporophyte.

Lycophyta

The sporangia are usually borne in the axils of microphylls in the lycophyta. Such a position was derived by Zimmermann along following steps :

(*i*) There was aggregation of fertile and sterile telomes (Fig. 1.6, J).

(*ii*) There was reduction in the number of mesomes and the sporangia (Fig. 1.6 L, M). This led to the development of a single leaf with a single sporangium in its axil.

The bifid tips of the sporophylls and leaves of the extinct genus ***Protolepidodendron*** may be cited as an example of an intermediate type.

Filicophyta or Pterophyta

The elementary processes of overtopping, reduction and foliar syngenesis lead to the development of the megaphyllous sporophylls of the ferns. As a result of overtopping, planation and syngenesis, a pinnate sporophyll was evolved. The sporangia were marginal in position (Fig. 1.7, **C, D**). Later recurvation led to the shifting of the sporangia from the margins to the ventral side (Fig. 1.8, E-H). Intermediate types were provided by the extinct genera ***Pseudosporochnus***, ***Stauropteris*** and ***Botryopteris***.

Zimmermann postulated that the roots developed from the creeping rhizomes of the primitive or ancestral types before the evolution of the leaves.

Botryopteris

Merits of the Telome Theory

It is an outcome of a master mind that has portrayed, in a skilful manner the origin and evolution of the sporophytes of land plants. Zimmermann has based his theory mostly on account of his comparative study of the fossil as well as living genera of the vascular plants. Many of his assumptions are correct and based on exact phyletic relationships between the various groups of plants both living and extinct. His five 'elementary processes' of planation, overtopping, syngenesis, reduction and recurvation give us a unified concept of the manner in which evolution might have proceeded in the land plants. These processes explain in a simple and lucidway as to how the primitive land plants led to the evolution of both the simple and the complex land plants of today. It explains in a nice manner that the entire sporophyte is an axis that has an underground portion called the root and an aerial part called the shoot. The appendages of shoot that is the sporophylls, sporangia and sterile leaves are nothing but modified parts of the stem. While building up his theory Zimmermann has taken into consideration the morphological aspects of the various groups of extinct and living groups of plants.

Demerits of the Telome Theory

This theory is open to criticism in its application to the origin of Lycophyta (Lycopsida). The origin of microphyllous leaves of the Lycophyta by reduction of telome trusses is not exemplified by any living or extinct vascular plant. Andrews (1960) has expressed his views on the Telome theory in the following words. "Zimmermann's scheme for the pteropsids or at least some pteropsids, has much supporting evidence; his concept for the articulates may be valid, but we are only on the verge of understanding the origin of this group; his content for the lycopsids is, so far as I am aware, purely hypothetical."

Enation Theory of Bower

Bower (1935) postulated a theory explaining the origin of microphyllous leaves (Fig. 1.7 D-G). According to enation the microphyllous leaves originated as superficial outgrowth from the stem. These outgrowths were originally not supplied by any vascular strand. Later these outgrowths or enations increased in size, the vascular strand reached up to their bases only. Later the leaf trace entered the enations of the lateral outgrowths and traversed its whole length. This vascular supply in the form of a single unbranched strand enervating the individual outgrowth gave them the characteristics of the microphyllous leaves. According to this theory microphylls originated by a process of progressive elaboration and not by reduction as visualised by Zimmermann. ***Psilophyton*** is an extinct vascular cryptogam belonging to the order Psilophytales, represents the first stage in the process. Another extinct member of the same order called ***Asteroxylon*** represents the next stage where the leaf trace reaches the base of the lateral appendage. In ***Drepanophycus***, another extinct member, the leaf trace enters the lateral appendage.

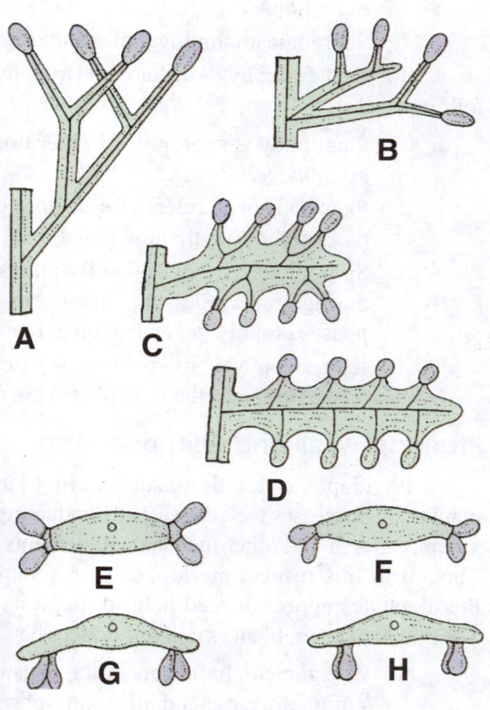

Fig. 1.7 (A–H). Illustrates the various steps by which the meagaphyllous sporophylls of Fillicophyta originated. A, B. Planation and overtopping; C, D. Syngenesis; E–H. Stages leading to the shifting of sporangia to margin and then to the ventral side of the sporophylls. (After Zimmermann).

Drepanophycus-an extinct vascular cryptogam

Pteridophyta in relation to Bryophyta

Pteridophytes have been placed between Bryophytes and Spermatophytes as they posses characters few of which are similar to both resemble the Peridophytes Bryophytes in the following characters :

1. Both are terrestrial in habit.
2. The sporophytes of Psilolates are root less and leafless like sporophytes of Bryophytes.
3. The development of embryo stage after fertilization is similar in both.

4. Structure of archegonia–the female reproductive organ is fundamentally similar in both the groups.
5. In both a single ovum or egg cell is protected by a layer of sterile investing cells.
6. In pteridophytes too like the bryophytes, the antheridium,– the male reproductive organ usually consists of a single layered jacket enclosing a mass of androcytes or antherozoid mother cells.
7. Antherozoids are ciliated in both.
8. Presence of water in liquid state is necessary for fertilization.
9. Sexual Reproduction is of oogamous type.
10. Both the groups show clearly defined heteromorphic alternation of generation; with some exceptions.
11. Sporangia are multicellular and spore development is same in both the groups.

But the pteridophytes differ with Bryophytes in some respects. The points of differences are as follows :

1. Plant body is sporophyte in Pteridophytes whereas the adult plants is a gemetophyte in bryophytes.
2. Plant body of pteridophytes are differentiated into leaf, stem and root where as in Bryophytes the plantbody is thalloid in nature.
3. Sporophytes are nutritionally independent of gametophyte except for very short period during embryonic stage in case of Pteridophytes, however in Bryophytes sporophytic phase is totally dependent on gametophytes.
4. In Pteridophytes sporophytes are strengthened with well developed vascular organizations/ tissues whereas the Bryophytes are devoid of vascular tissues/organizations.

Pteridophytes and Gymnosperms

Pteridophytes are designated as first land vascular plants because its dominant sporophyte plant body possesses well developed conducting tissue system. The vascular tissues are composed of xylem, phloem and other mechanical elements required for conduction of water and food material. Though in this respect pteridophytes resemble the higher plants but for seed bearing capacity. Pteridophytes approach seed habit through its heterosporous species (***Selaginella***, ***Marselia*** etc). The points of resemblances and differences between Pteridophytes and Gymnosperms are as follows:

1. The gametophytic generation is camparatively reduced in case of Gymnosperms as it is completely dependent upon sporophyte, whereas in pteridophytes the gametophytic generation is partially independent.
2. Root system of Pteridophytes are camparatively less developed and primitive where as in Gymnosperms roots are much more developed and complex and provide better anchorage to plant for efficient use of soil nutrients.
3. Heterosporous pteridophytes resemble with Gymnosperms but spore germinate ***in situ***, whereas the megaspore is retained within megasporangium in case of Gymnosperms.
4. Secondary growth in Pteridophytes are normally absent, though unique type of secondary growth is reported in ***Isoetes***, ***Stylites*** and ***Botrychium***. In Gymnosperms Secondary Growth is common and prominent.

Botrychium

5. Seed habit is dominant in Gymnosperms whereas in living Pteridophytes it is absent.
6. Fern and cycadales resembles in respect of multiflagellate and motile antherozoids but higher gymnospernes (Coniferales and Gnetales) have non-flagellated antherozoids.
7. Motile Sperm reaches the egg through neck canal of archegonia in pteridophytes (zooidogamy) whereas in Gymnospherms fertilizations is effected by pollen tube (Siphonogamy).
8. Archegonia of Gymnosperms lack neck canal cells and ventral canal cell. However they exist in pteridophytic archegonia.
9. Ferns resemble with cycadales as both possess compound leaf and circinate vernation.
10. Phloem lack companion cells in Pteridophytes as well as in Gymnosperm.
11. Both the pteridophytes and gymnosperms generally lack vessels in xylem (except in the order Gnetales).

FOSSIL RECORD AND GEOLOGICAL TIME SCALE

The remains of plants and animals, which lived in the remote past, in many cases, are preserved in the rocks in one form or the other. These are called the **fossils**. A brief knowledge of the different types of fossils is essential to know the type of plants that lived in ages bygone and to form an idea of evolutionary process that has gone on for times immemorial and is still going on. A fossil in reality is a clear manifestation of the existence of prehistoric living beings. The discovery of new fossil types, in future, will help us to piece, together a continuous story of the development of plant kingdom from an age more than one billion years ago to the modern times. Profound changes have been effected in the plant world during this long interval of time. Diverse types of plants and plant groups have arisen, flourished and became extinct during these millions of years. Without the fossil record modern botainsts can never know about such plant groups that dominated the earth's crust in the past. The fossils can throw light on the origin of most of the present day groups of aquatic plants to the terrestrial situations. The fossil records also help us to know about the changes in the earth's climate and their effect on the flora and fauna. There are variety of forms of fossils depending upon the conditions under which fossilization occurred. There are some factors involved during the fossilization process. These can be listed as under :—

1. The nature of the plant or animal material ***e.g.***, whether it was a thin and delicate plant part of a tough and woody portion.
2. The kind of entombing material.
3. The rapidity with which the plants were covered and thus prevented from exposure to organisms of decay.
4. The condition of the plant material at the time of fossilization.
5. The alterations that the rocks have been subjected to during the time that has elapsed since their formation.

Common types of fossils are :

1. Dead and Preserved Bodies

The actual dead and preserved bodies, or parts of bodies, of plants and animals, with the original tissue intact, and enclosed either in ice or in amber or else mummified in various ways form this class of fossils.

2. Compressions

Such fossils are formed when the whole plant or part of a plant gets buried under sediments of different types. As a result of the weight of these sediments the plant part becomes flattened resulting in the formation of a carbonaceous film revealing the outlines and external features of the plant or plant part. A very little or no cellular detail is preserved in such fossils. They usually reveal the external characters of the plant or its part. In many cases, in addition to the carbonaceous film, cuticle is also preserved. It is possible to remove this cuticle and mount it on a slide after carefully clearing it. Such preparations enable us to study microscopical details of the epidermal layer such as structure and distribution of stomata, shape of epidermal cells and kinds of hair, trichomes and scales that cover the epidermis. The study of fossil cuticles have revealed many useful data to the palaeobotanists. Volcanic ash, sand stone, shale and diatomaceous earth are some of the materials that cause compressions. Fine grained shale and diatomaceous earth yield good compressions. Coal is also a kind of compression formed by an extensive accumulation of plants and plant parts, in body of water, that is gradually compressed. Spores, pollen grains, cuticle, bark fragments, pieces of wood and resins are the common plant parts preserved in coal. In rare cases the coal consists of several layers of cuticle preserved in a remarkable manner. These cuticles can be easily separated and mounted on slides for microscopical examinations. Such a coal is also called the **paper coal.**

3. Impression or Imprints

They are formed when the sediments around the compressed plant or plant part become hardened and when split open at the region of the fossil, they very clearly reveal the outline of the plant or its part. They are aptly regarded as negatives of the compression. It is just as we put our foot on a wet clay. The foot leaves its impression on the clay and as the clay hardens, the print of the foot becomes permanently inscribed on it. Such fossils only reveal the outline of a plant or its part. The cellular details are not preserved. In some cases the sediments all around the plant or plant part harden and form a three dimensional **mold.** The plant material within decays and disintegrates leaving a space which is filled with other sediments. These filling sediments harden to form a cast completely enclosed by the mold. The cast has the same external appearance as that of the plant or its part and reveals only the external appearance or the outline. The casts and molds help us to reveal or reconstruct the external features of the plants. They do not preserve the organic material and cellular details. The plant impressions are found in larger number in the entire Gondwana series in India.

4. Petrifactions

They are considered to be the best type of fossils available to the palaeobotanists because they reveal the internal structure of the part of a plant that is petrified. The process of petrifaction consists in the molecule-by-molecule replacement of the plant part by minerals in solution that ultimately harden. It so happened that the entire forest or a group of plants growing in a locality got buried under the ash from nearby volcanoes. This ash and other exudates of volcanic eruption covered the entire forest which got completely entombed. The water charged with minerals filtered down through the sediments covering the forests and the plant parts became infiltrated and replaced to a greater degree by silica and other minerals. The organic material making up the plant parts, in some cases, becomes completely replaced by the silica and their compounds like calcium, magnesium, iron carbonate, iron pyrite, and phosphates of calcium and magnesium, and Sulphur or its compounds. The plants and plants parts that have become infiltrated by the matrix become harder than the ash enclosing them. They are more resistant to erosion and as the surrounding or entombing sediments are washed away by water currents, the petrified stumps and logs stand out distinct as spectral and magnificient reminders of the distant part.

These petrified specimens reveal the detailed internal structure of the various organs of the plant body, that is in some cases, excellently preserved. A number of techniques have been evolved to study these petrified specimens. The most common one and a standard technique is to cut thin sections of them. In this method a small piece of the petrifaction is cut with diamond edged saw. The exposed surface from which the section is to be prepared is polished. It is mounted on a glass slide with a transparent adhesive. The fixed specimen is again cut so as to leave a very thin section adhering to the slide. It is further ground and polished so as to make the rock section almost transparent. It can now be studied under the microscope. There are a number of other methods, out of which **peel technique** is a recent one. In this method the surface of the petrifiaction is wetted with acetone. A prepared acetate sheet is then gently laid on the surface of the fossil. After some time the acetate hardens. It can now be gently pulled apart. The sheet carries along with it a thin section of the petrifiaction, which can be studied under the microscope.

A remarkable example of petrifaction from India is afforded by ***Azolla intertrappea*** discovered by Birbal Sahni and H.S. Rao (1941) from Tertiary Chert beds from Chindanwara. In this specimen the megaspores with attached massulae and glochidia are very clearly preserved.

5. Encrustation Fossils

Such fossils are neither compressions nor petrifactions. These fossils are usually found in hot springs whose waters contain abundant mineral matter. The fallen twigs, leaves, spores and the algae and becteria that live in such hot springs become coated with the carbonates and are encapsuled. These carbonates becomes precipitated on cooling. Such precipitates on examination reveal the encapsuled organisms in an unaltered state. During encrustation decaying organisms and air are excluded so that the organisms remain as such even for millions of years. The organisms preserved in this manner can be embedded in celloidin and sectioned and stained for detailed examination.

6. Calcification

This process is used by some algae. They secrete a heavy coat of calcium carbonate around them. This coat preserves them for millions of years and these calcified algae can be studied for their details that are well preserved. It is now believed that the "coral reefs" found in the oceans are built more by the secretion of calcium carbonate by marine algae rather than by the corals.

7. Diatomaceous Earth

Deposits of diatomaceous earth are found in many parts of the world. It is believed that such regions were earlier inundated by the oceans that were inhabited by countless members of tiny unicellular algal organisms called the diatoms. These algae were encapsuled by beautifully sculptured frustules or shells made of silica (secreted by these algae). After the death of these diatoms, their siliceous shells sank to the bottom of the sea. The mass of these shells went on accummulating at the bottom for millions of years and formed the diatomaceous earth. Microscopic examination of this earth reveals the presence of variously ornamented shells of diatoms.

8. Fossil Faeces or Coprolites

Fossil faeces belonging to the upper Eocene from the Bridger formation of Wyoming have yielded a large number of bacteria (belonging to the genus ***Escherichia***) and some algae (***Staurastrum***). Large numbers of zygospores and conjugating cells of the desmids have been found in the preserved animal faeces. It is believed that some animals drank water from the pools containing these organisms. These organisms passed undigested through intestines along with the excreta. The excreta or the faeces, when exposed for long hours to the sunlight and the heat, got hardeneal and the organisons (algae and bacteria) entrapped in it got fixed. Later the faeces were covered by volcanic ash or other sediments, which released silica and resulted in the silicification of the unicellular organisms (Bacteria and desmids), enclosed within the faeces.

9. Siliceous Colloids

These are formed by the dissolution of siliceous shells of diatoms, radiolarians and sponges and are found at the bottom of the sea. Some organisms get embedded in this siliceous colloid and are preserved. In some cases even the flagella of some unicellular flagellates have been nicely preserved. Wetzel, Otto (1953) described a unicellular flagellate ***Ophiobolus utinensis*** from the Upper Cretaceous flints and chalks of Europe. In this case Wetzel was able to photograph even the flagella.

10. Actual Remains

Such plant fossils are of comparatively young age and do not decompose completely due to low temperature or vacuum. All materials that are resistant to natural decomposition are preserved like this, ***e.g.***, cuticle, cutin, sporopollinin in the walls of pollen grains and spores.

11. Chemical Remains

Chemical materials like amino acids and hydrocarbons may be obtained in a natural state or in a modified form from the rocks in which organic matter of the petrified plant part or organ has been incorporated.

12. Products of Activity of Organism

Sometimes products excreted by living organism are obtained from the excavations, ***e.g.*** secreted lime deposits of algae. Such products have been found on pottery and other manufactured goods discovered during excavations.

Pseudofossils

These are unreal fossils which appear like real ones and are not the actual remains of plants and animals. They resemble the leaves and stems of plants or the filaments of algae but in actual practice they are outright fakes or artifacts formed by the peculiar arrangement or deposition of the mineral particles in the crevices of rocks. One such fossil named as ***Dendrites*** resembles the fern leaves but on closer examination it was found to be nothing but mineral formations simulating a fern leaf.

Coal Formation

Coal is one of the major fuels in daily use. It is, in fact, the changed form of vegetation which was existing on this earth. Due to various earth movements thick forests at the past ages got buried inside the earth and the remarkable change in temperature and pressure conditions, thus produced, changed them into thick carbonaceous seams which we mine or sometimes quarry in the form of coal. The thickness of the coal seams varies from exposure to exposure.

Geologically speaking, the coal deposits of the world are divisible into two groups as follows:

1. Carboniferous coal, ***i.e.***, coal of carboniferous period of Palaeozoic era, produced about 220 to 275 million years ago.
2. Tertiary coal : It is comparatively quite younger, produced in nature 70 million years ago.

Based upon the percentage of carbon contents, coal has been classified into four parts as below:

1. Peat coal.
2. Lignite coal.
3. Bituminous coal.
4. Anthracite coal.

Anthracite coal is the most pure variety whereas Peat coal is poorest so far as the percentage of carbon contents is concerned.

The Tertiary coal of India is generally of Anthracite quality. The mass with 100% carbon contents is known as diamond which is the hardest known mineral for the present.

India is sufficiently rich in coal deposits and it is distributed in almost all the states of the country. The presently known coal deposits of India point to the fact that the country can depend upon coal energy for at least one hundred years more.

Geologic Time Scale

With the help of new tools and techniques the geologists have been able to make fairly an accurate estimate of the age of the earth and the ages of various rock strata, by comparing the rates of accumulation of salt in the sea and of decomposition of radio-active substances such as uranium and thorium.

On the basis of their study geologists have divided the geological history of earth into 4 major eras (according to recent changes and modifications, see Kulp, J.L., 1961 ; and Eicher, D.L., 1968) ; these in turn are sub-divided into periods ; each period has been split into still smaller units of time intervals called the epochs (see Table 1).

TABLE 1

Geological Time Scale (*Modified after Kulp and Eicher*)

Era	Period	Epoch	Important Plant Events	Beginning in Millions of Years
(1)	(2)	(3)	(4)	(5)
COENOZOIC				
	Quaternary	*Pleistocene*	Speciation of herbaceous plants	One
		Pliocene	Spread of herbaceous dicots	Thirteen
		Miocene	Rise of herbaceous angiosperms	Twenty-five
	Tertiary	*Oligocene*	Dispersal of woody angiosperms	Thirty-six
		Eocene	Rise of woody angiosperms now extinct	Fifty-eight
		Paleocene	Modernization of angiosperms	Sixty-three
MESOZOIC				
	Cretaceous		Earliest known Pines	Hundred thrity-five
	Jurassic		Origin of Angiosperms	Hundred and eighty
	Triassic		Spread of Conifers. Rise of Cycads.	Two hundred and thirty
PALAEOZOIC				
	Permian		Rise of conifers ; coal swamp extinct	Two hundred eighty
	Pennsylvanian	*Carboniferous*	Widespread coal swamps, formation of coal beds	Three hundred ten
	Mississippian		Development of coal swamps with gymnosperms, lycopsids, sphen-opsids and ferns	Three hundred and forty-five

(1)	(2)	(3)	(4)	(5)
	Devonian		Flourishing psilopsids, lycopsids, sphenopsids, ferns, seed plants and bryophytes. Evidence of brown algae and dinoflagellates	Four hundred five
	Silurian		*Cooksonia*, the oldest known vascular plant. Record of some non-vascular land plants.	Four hundred twenty-five
	Ordovician		Marine red and green algae	Five hundred
	Cambrian		Evidence of lime secreting algae.	Six hundred
PRECAMBRIAN				
	Late Precambrian		Eukaryotic organisms	Twelve hundred
	Middle Precambrian		Stromatolites of algal origin. Blue green algae, Bacteria, origin of eukaryotic cell	Two thousand Five hundred
	Early Precambrian		Bacteria like cells and unicellular alga like organisms of age 3,000 million years	Four thousand five hundred
			Graphites of possible organic origin	Five thousand
			Origin of procaryotic cell origin of earth cell, origin of earth	

CLASSIFICATION

The classification of the vascular cryptogams has undergone vast changes in the recent past. On the basis of the presence or absence of seeds the older taxonomists place the vascular plants in two divisions : **Pteridophyta** and **Spermatophyta**. The former includes the ***primitive vascular plants which bear no seeds.*** According to this traditional system of classification the Pteridophytes are divided into the following classes and important orders :—

DIVISION Pteridophyta

1. Class Psilophytineae

(*i*) Order **Psilotales**. Examples ***Psilotum*** and ***Tmesipteris.***

(*ii*) Order **Psilophytales**. Example ***Rhynia.***

2. Class Lycopodineae

(*i*) Order **Lycopodiales**. Examples ***Lycopodium***, ***Selaginella***.

(*ii*) Order **Isoetales**. Example ***Isoetes***.

(*iii*) Order **Lepidodendrales**. Example ***Lepidodendron.***

3. Class Equisetineae

(*i*) Order **Equisetales** represented by ***Equisetum***.

(*ii*) Order **Sphenophyllales** represented by ***Sphenophyllum***.

4. Class Fillicineae

It includes the following important orders:

(*i*) Order **Filicales**. Examples ***Dryopteris***, ***Marsilea***, etc.

(*ii*) Order **Marattiales** represented by ***Marattia***.

(*iii*) Order **Ophioglossales** represented by ***Ophioglossum*** and ***Botrychium***.

(*iv*) Order **Osmundales** represented by ***Osmunda.***

(*v*) **Fossil orders**.

Later some fern-like seed-bearing fossil plants (Cycadofilicales) were discovered in 1903. This discovery eliminated the distinction between the two divisions—**Pteridophyta** and **Spermatophyta.** On the basis of origin and development of vascular plants Sinnott (1935) introduced a term ``Tracheophyta" for all those plants having vascular arrangements including Thallophyta and Bryophyta. So in the revised systems of classification proposed by Tippo (1942) all the vascular plants are placed in a single division **Tracheophyta**. It is segregated into four taxas which are Psilopsida, Lycopsida, Sphenopsida and Pteropsida. As to the ranks of these groups, the botanists differ. Haupt (1953) consider them as classes of the division Tracheophyta. Tippo assigns them the ranks of sub-phylla of the phylum Tracheophyta. The terms phylum and sub-phyllum being not in accord with the International Code of Nomenclature, Wardlaw (1955) suggested the rank of sub-divisions for the four groups. The outline of this revised system of classification is given below :

DIVISION Tracheophyta

1. Sub-division Psilopsida

Class Psilophytineae

(*i*) Order **Psilotales** (*ii*) Order **Psilophytales**

2. Sub-division Lycoposida

Class Lycopodineae

(*i*) Order **Lepidodendrales** (*ii*) Order **Lycopodiales.**

(*iii*) Order **Pleuromeiales.** (*iv*) Order **Isoetales.**

3. Sub-division Sphenopsida

Class Equisetineae.

(*i*) Order **Equisetales.** (*ii*) Order **Sphenophyllales.**

(*iii*) **Fossil orders.**

4. Sub-division Pteropsida

The term is used here in a restrictive sense to cover the megaphyllous pteridophytes only.

Class Filicineae.

(*i*) Order **Ophioglossales.**

(*ii*) Order **Filicales.**

(*iii*) Order **Osmundales.**

(*iv*) Order **Marattiales.**

(*v*) **Fossil orders.**

The increasing knowledge about the vascular anatomy and the position of sporangia in the above mentioned sub-divisions has revealed marked divergence among them from one another. The modern writers thus consider that the differences between the Psilopsida, Lycopsida, Sphenopsida and Pteropsida series are of the magnitude of a division. Smith (1955), Bold (1957), Benson (1957) and Zimmermann (1959) have given these groups the rank of a division. According to the recommendations of **I.C.B.N.** (1952) the name of the division should end in the suffix **phyta**, of a sub-division in **phytina** and of a class in **opsida.** On this basis Smith (1955) divides the vascular cryptogams into 4 divisions, **Psilophyta, Lepidophyta, Calamophyta** and **Pterophyta**. Some writers suggest other names. Given below with slight modifications is one of the widely accepted modern classifications of the vascular cryptogams :—

I. Division Psilophyta

In it are included the most primitive and oldest known land inhabiting vascular plants which are rootless. The division comprises the following classes :—

1. Class Psilophytopsida.

(*i*) Order **Psilophytales.** It includes the most primitive fossil vascular cryptogams e.g. ***Rhynia***, ***Horneophyton***, ***etc.***

2. Class Psilotopsida.

(*i*) Order **Psilotales**. It includes two living genera which are allied to the Psilophytales. These are ***Psilotum*** and ***Tmesipteris***.

II. Division Lycophyta or Lepidophyta

Bold used the term **Microphyllophyta** for this division. On the basis of the presence or absence of ligules, this division may be divided into the following two classes :—

1. Class Eligulopsida.

(*i*) Order **Lycopodiales**. The order includes **homosporous, non-ligulate,** forms such as ***Lycopodium*** and ***Phylloglossum.***

2. Class Ligulopsida. The class includes **heterosporous** froms with **ligulate leaves**. The four orders recognisable in this class are :—

(*i*) Order **Selaginellales** represented by ***Selaginella.***

(*ii*) Order **Isoetales** represented by ***Isoetes.***

(*iii*) Order **Pleuromeiales**. e.g. ***Pleuromeia***.

(*iv*) Order **Lepidodendrales** represented by ***Lepidodendron***, ***Stigmaria***, ***Lepidocarpon***.

III. Division Arthrophyta or Calamophyta or Sphenophyta

The division includes two classes :—

1. Class Sphenophyllopsida. This class includes the well-known extinct genus **Sphenophyllum** which is placed in the following order:—

(*i*) Order **Sphenophyllales.**

2. **Class Calamopsida.** This class includes both fossil and living forms. The living genus ***Equisetum*** is referred to the following order :—

   (*i*) Order **Equisetales.**

IV. Division Fillicophyta or Pterophyta

The classes included in this division are given below with the important orders :—

1. **Class Eusporangiopsida.** The class includes the following two orders :—

   (*i*) Order **Ophioglossales**. The order is represented by two important genera ***Ophioglossum*** and ***Botrychium.***

   (*ii*) Order **Marattiales.** It includes six genera of which ***Marattia*** is well-known.

2. **Class Protoleptosporangiopsida.** The class includes the following order :

   (*i*) Order **Osmundales.** It is represented by three living genera of which ***Osmunda*** is widely distributed.

3. **Class Leptosporangiopsida.** It is divided into three orders : —

   (*i*) Order **Filicales.** Almost all the members of this order are **homosporous.** The green, thallus-like prothallus bears sex organs on the lower surface. The sporangia are borne in groups called the **sori** on the margin or abaxial face of the sporophyll.

   (*ii*) Order **Marsileales.** It includes **heterosporous** forms ***e.g. Marsilea, Pilularia, Regnellidium.***

   (*iii*) Order **Salviniales.** This order includes the heterosporous, free-floating aquatic forms ***e.g. Salvinia*** and ***Azolla.***

4. **Class Primopteropsida.** The class includes the many extinct genera of ferns. It includes the orders:

   (*i*) **Cladoxylales,** and

   (*ii*) **Coenopteridales**

   These were probably the ancestors of modern ferns and appeared in the Middle Devonian and survived up to the end of the Palaeozoic.

This system has been followed in this text.

Reimers (1954) in Engler's ***Syllabus der Pflanzen familien*** classified Pteridophytes or the seedless vascular plants as follows :—

PTERIDOPHYTA — Seedless Vascular plants

Class :—A. **PSILOPHYTOPSIDA**

Order **Psilophytales** (fossil)

Class :—B. **PSILOTOPSIDA**

Order **Psilotales**

Class :—C. **LYCOPSIDA**

Orders 1. **Protolepedodendrales** (fossil)

2. **Lepidodendrales** (fossil)

3. **Lycopodiales**

4. **Selaginellales**
5. **Isoetales**

Class :—D. **SPHENOPSIDA**

Orders 1. **Hyeniales** (fossil)
2. **Sphenophyllales** (fossil)
3. **Calamitales** (fossil)
4. **Equisetales**

Class :—E. **PTEROPSIDA :—** It includes 4 sub-classes.

Sub-class 1. **Primofilices** (fossil)
Orders 1. **Cladoxylates** (fossil)
2. **Coenopteridales** (fossil)

Sub-class 2. **Eusporangiatae**
Orders 1. **Marattiales**
2. **Ophioglossales.**

Sub-class 3. **Osmundidae**
Order **Osmundales**

Sub-class 4. **Leptosporangiatae**
Orders 1. **Filicales**
2. **Marsileales**
3. **Salviniales**

Cronquist, Takhtajan and Zimmermann (1966) classified the Pteridophytes into five divisions :—

1. Division :— ***Rhyniophyta***
   Class :— ***Rhyniatae***
   Order :— ***Rhyniales***
2. Division :— ***Psilotophyta***
   Class :— ***Psilotatae***
   Order :— ***Psilotales***
3. Division :— ***Lycopodiophyta***
   Class :— ***Lycopodiatae*** (Orders :—***Asteroxylales***, ***Drepanophycales***, ***Protolepidodendrales*** and ***Lycopodiales***
   Class :— ***Isoetatae*** (Orders :—***Lepidodendrales***, ***Pleuromiales***, ***Isoetales***, ***Selaginellales***)
4. Division :— ***Equisetophyta***
   Class :— ***Sphenophyllatae*** (Orders :—***Sphenophyllales***, ***Pseudoborniales***)
   Class :— ***Equisetatae*** (Orders :—***Calamitales***, ***Equisetales***)
5. Division :— ***Polypodiophyta***

Class :— ***Polypodiatae*** (Sub-classes :—***Protopteridiidae***, ***Archaepteridiidae***, ***Ophioglossiidae***, ***Noeggerothiidae***, ***Marrattiidae***, ***Polypodiidae***, ***Marsilleidae***, ***Salviniidae***)

These sub-classes are further divided into orders.

Banks (1968) divides the Tracheophyta into seven sub-divisions like Rhyniophytina, Zosterophyllophytina, Psilophytina, Lycophytina, Sphenophytina, Trimerophytina and Pterophytina. According to Banks, subdivision Pterophytina includes ferns, progymnosperms, gymnosperms and angiosperms.

Doyle (1971,) divided Pteridophyta into four classes :—

1. Class **Psilopsida** (Psilopsids)
2. Class **Lycopsida** (Lycopsids)
3. Class **Sphenopsida** (Horsetails)
4. Class **Pteropsida** (ferns)

He included these classes under sub-division ***Embryophytina*** and division **Chlorophyta** (green plants). The seed plants were included under class **Spermopsida**.

Bierhorst (1971) followed the following system of classification :—

Division :— ***Tracheophyta***
Class :— ***Rhyniopsida***
Order :— ***Rhyniales***
Class :— ***Zosterophyllopsida***
Order :— ***Zosterophyllales***
Class :— ***Lycopodiopsida***
Orders :— ***Asteroxylales***, ***Lycopodiales***, ***Protolepidodendrales***, ***Selaginellales***, ***Lepidodendrales***, ***Isoetales.***
Class :— ***Cladoxylopsida***
Order :— ***Cladoxylales***
Class :— ***Equisetopsida***
Order :— ***Hyeniales***, ***Pseudoborniales***, ***Sphenophyllales***, ***Equisetales.***
Class :— ***Coenopteridopsida***
Order :— ***Coenopteridales***
Class :— ***Filicopsida***
Orders :— ***Noegerathiales***, ***Filicales***, ***Marsileales***, ***Salviniales***
Class :— ***Ophioglossopsida***
Order :— ***Ophioglossales***
Class :— ***Marattiopsida***
Order :— ***Marattiales.***

Sporne (1975) suggested a system of classification in which he has divided pteridophytes into six classes instead of five. These classes are Psilopsida, Psilotopsida, Lycopsida, Sphenopsida, Pteropsida and Progymnospermopsida. Sporne's system of classification is actually a modification of the Reimer's (1954) system.

According to Parihar (1996) the classification system of Pteridophytes has undergone vast changes during the last five decades and modern system of classification divides Pteridophytes into seven divisions as : Rhyniophyta, Trimerophytophyta, Zosterophyllophyta,Psilotophyta, Lycopodiophyta, Equisetophyta and Polypodiophyta. It was Bold (1973) who has differentiated Psilophytes into living and fossil groups. The living Psilophytes were placed under division Psilotophyta and extinct psilophytes were divided into three divisions ***i.e*** Rhyniophyta, Trimerophytophyta and Zosterophyllophyta. Taylor (1981) followed almost the similar pattern and proposed three divisions as Rhyniophyta, Zosterophyllophyta and Trimerophyta for early vascular plants.

The *characteristic features* of seven new divisions of Pteridophyta are described below :

(*i*) Rhyniophyta : Rootless plant body which was differentiated into two parts : an under ground rhizome which bear rhizoids and an errect aerial leafless portion arising from underground rhizome; aerial branches were dichotomously branched, some of which bear terminal sporangia. Sporangia are homosporous, thick walled and their development is eusporangiate. Spores were in tetrads. Stems cylindrical, vascular anatomy show lobed protostete and xylem was centrarch.

(*ii*) Trimerophytophyta : Plant body has monopodial branching in its main axis where as lateral axis has dichotomous or trifurcate branching. Surface of both axis were smooth, punctate or spiny. Branching differentiated into fertile and sterile axes. Fertile branches bear terminal sporangia which were aggregated at tips of dichotomous branches. Sporangia homosporous and development in eusporangiate. Stem solid, with protostele and centrarch xylem.

(*iii*) Zosterophyllophyta : Plant body rootless, leafless and branched dichotomously or pseudomonopodially, Protostelic stem with exarch xylem, xylem strands elliptic or terete with scalariform tracheids; Globose or reniform sporangia borne an lateral stem tips which were often aggregated into terminal clusters.

(*iv*) Psilotophyta : Plant body is differentiated into an underground rhizome and aerial shoot, underground rhizome covered with rhizoids, true root absent;aerial shoots bear minute scale like lateral appendages. Two to three lobed synangia associated with fertile appendges are found on aerial shoots; Plants homosporous. Multiciliate antherozoids present. Development of embryo is exoscopic and are without suspensor.

(*v*) Lycopodiophyta : Plant body well differentiated into root, stem and leaf, leaves microphyllous with single vein, Vascular tissue system is devoid of leaf gaps; Sporangia borne on adaxial surface of fertile leaf known as sporophyll, Plants homosporous as well as heterosporous, Antherozoids bi-to multiciliate, Development of embryo is endoscopic.

(*vi*) Sphenophyta : Plant body differentiated into leaf, stem and root, whorls of simple leaves are found on nodes of stems; Internodes of stem and branches longitudinally ridged; Leaf gaps absent in vascular arrangement of stems, Sporangia borne on sporangiophore; Number of sporangia on sporangiophore may vary from one to many, Antherozoids are multiciliate, Embryo development is exoscopic and embryo lacks a suspensor.

(*vii*) Pterophyta : Plant body differentiated into leaf stem and root (with few exceptions), Leaves are prominent and megaphyllous, Leaf gaps found in protostelic stem. Sporangia usually borne on margin or abaxial surface of fertile leaves; Antherozoids multiciliate and embryogeny may be either exoscopic or endoscopic.

A summary of the characters of living pteridophytes viz. Psilophytes, horsetails, club-mosses and ferns is given below :

	Psilotophyta	Sphenophyta "horsetails"	Lycopodiophyta "clubmosses"	Pterophyta "ferns"
Sporophyte	Rootless.	Roots present.	Roots present.	Roots presents.
Stem	Dichotomously branched rhizomes, with erect, aerial branches.	Rhizomatous, with erect, aerial branches.	Creeping, and/or erect or pendulous (or, in ***Isoetes***, a corm).	Creeping, rhizomatous, or erect (becoming tree-like in size in the Cyatheaceae).
Leaves	Spirally arranged, scale-like in ***Psilotum***	Whorled.	Spirally arranged.	Spirally arranged, often compound.
Sporangia	Thick-walled, homosporous	Thick-walled, homosporous, aggregated in a terminal strobilus	Thick-walled, homosporous (***Lycopodium*** and segregate genera) or heterosporous; aggregated in a terminal strobilus (that in ***Isoetes*** represents the entire plant).	Thick-or thin-walled (leptosporangiate), homosporous or, in the water ferns, heterosporous. Terminal, on axes, or on lower (abaxial) surfaces of leaves, variously aggregated.
Gametophyte	Remarkably similar to the rhizomes of sporophytes; subterranean, non-photosynthetic, cylindrical, covered with rhizoids and bearing both antheridia and archegonia. Mycorrhizal.	Either male, or initially female, becoming hermaphroditic,; sex ratio of germinating spores apparently controlled by light intensity and crowding.	Free-living (***Lycopodium***) or endosporic (***Isoetes, Selaginella***)	Free-living or endosporic (water ferns; Marsileales, Salviniales)
Ontario representation	None; one of two genera, ***Psilotum***, subtropical USA	***Equisetum*** sole genus; 9 out of 14-15 spp. worldwide	***Isoetes***, ***Lycopodium***, ***Selaginella***	Abundant and diverse.
Ecology	Grow on ground, or as epidphytes.	Herbaceous perennials, variously evergreen or dying back to the ground each year.	Widely distributed, epiphytic, terrestrial, or (***Isoetes***) aquatic plants.	Widely distributed, epiphytic, terrestrial, or (Marsileales, Salviniales) aquatic plants.

Pteridology in India

It was Linnaeus (1753) who described a number of Indian pteridophytes in his book "Species Plantarum" Swartz (1801, 1806) presented a more comprehensive account of Indian ferns in the book "Genera et Species Filicum". Later a number of Himalayan ferns were described by D'Don (1825), Hooker and Greville (1827-1831), Hooker (1846-1864) and Hooker and Baker (1865-1868). In 1863 R.H. Beddome published two of his well known books "Ferns of Southern India" and "Handbook to the ferns of British India". It appears surprising because the country possesses a rich pteridophytic flora. More than a decade later "Ferns of Northern India" by C.B. Clarke (1880) and "Ferns of North Western India" by C. W. Hope (1899–1902) were published. Generic accounts of ***Salvinia*** and ***Azolla*** were the first to be described and figured by W. Griffith (1851). The first Indian botainst to publish a paper on ***Equisetum*** was Prof. S.R. Kashyap (1914) who was followed by Ghosh (1916) at Punjab University, Lahore. At the same time at Lucknow University, Birbal Sahni initiated studies on ***Nephrolepis*** Researches on pteridophytes have been periodically reviewed by P. Maheshwari and Kapil (1963), T. S. Mahabale (1973) and Chowdhary (1971). B.D. Sharma in 1994 published an excellent account of the development of studies on pteridophytes in India. His article entitled "Pteridophyta-Morphology, Anatomy and Reproductive Biology" presented and excellent review of papers on Indian pteridophytes. In the same year H.K Cheema (1994) presented a comprehensive review of Morphogenetic studies on Pteriodophytes in India. Comprehensive studies on ***Tmesipteris*** were carried out by Prof Birbal Sahni in 1925 who was probably the first to study Psilotales. Later in 1950 Ninan summarised the contributions on ***Psilotum*** especially with regard to their chromosome number.

N. P. Chowdhary (1937), Mahabale (1937, 1948) Ninan (1958) Manton (1950) and D D. Pant (1976-1980) made significant contributions on ***Lycopodium***. The genus ***Selaginella*** was extensively studied by Alston (1945), Panigrahi and Chowdhary (1962), G. P. Mazumdar (1942), and Panigrahi (1958)

Indian scientists who studied ***Isoetes*** include Pant and Srivastava (1962), Usha Sharma (1961), Bhambie (1957), and Ninan (1958). Mahabale (1945, 1946), Ninan (1955) and Bir (1980). Cytomorphological studies on ***Isoetes*** have been made by Srivastava, Rai and Srivastava (2002).

The group Filicales of pteridophytes has attracted a large number of workers from different regions of the country. The genera of ophioglossaceae was studied by Mahabale (1933 ***Ophioglossum***), and L.N. Rao (1962, ***Botrychium***). Cytological studies on the two generat were carried out by Verma (1957) and Ninan (1958). The phenomenon of Apogamy and apospory was studied by Sarbadhikary (1937).

T. S. Mahabale (1933-1972) and his students at Poona made extensive collections of pteridophytes and described prothalli of a large number of ferns and fern allies.

Mehra and Loyal (1956) initiated studies on the effect of colchicine on the prothalli of ferns. Similarly the effect of NAA, kinetin and Gibberellic acid on ferns was investigated by Wardlaw and Metra (1958). ***Marsilea*** was studied in detail by K. M. Gupta (1962) and a monographic account of the genus was published by him. The nature of sporocarp of the genus was also discussed by Puri and Garg (1953), Mahabale and Giriji (1948), Mahabale (1956) and Bhardwaja and his associates (1967-1980). Cytological Studies on the genus were carried out by Mehra and Loyal (1959).

The impact of Cytological studies on the phylogeny of ferns was ably discussed by P.N. Mehra, A. Abraham and G. Panigrahi (1962). Most of the research, thereafter centered around ferns which occupy more than 60% of the pteridophytic flora. These studies include morphological, taxonomical, cytotaxonomic and palynological studies.

Revision of ferns was carried out by Mehra and co-workers (1963), Mehra and Bir, (1964, 1965); Mehra and Palta, (1971), Mehra and Verma, (1970), Panigrahi and Dixit, (1976). During the decade between 1960-1970 critical and experimental approach to taxonomy was initiated. Surange (1966) gave an excellent account of fossil pteridophytes of India. Chowdhary (1970) published a monographic account of ***Lycopodium*** which was subsequently revised by Pant and Pandey (1985). B.K. Nayar (1954-1995), who joined National Botanical Research Institute (NBRI) (the then National Botanic Garden) in 1955 after a brief stint in Gauhati University, developed a strong centre of Research in Pteridology first at NBRI, Lucknow and later at Calicut University. His major research areas at NBRI were morphology of gametophytes, palynology and morphoanatomical studies of pteridophytes, particularly ferns. He also conducted studies on taxonomy and floristics of Indian ferns, He setup a culture laboratory at NBRI where he and his students studied gametophytes of over two thousand taxa grown under uniform conditions At Calicut, he concenterated on study of organisaton and activity of shoot apex of ferns including organogenesis, stelar morphology and ontogeny and the evolution of pteriodophytes from alga like ancestors. Prof. Nayar made extensive studies on the vasculature of juvenile plants of ferns as well. One of the of the major contributions of Prof. Nayar and his students was the elucidation of evolutionary trends of gametophytic generation of pteridophytes. According to them, the gametophytes of ***Equisetum*** are the most primitive. At Calicut, he also undertook bioecological studies of tropical forests of Silent valley and Sahargiri regions of Kerala. Another subject of his interest was the spread, ecology and eradication of freshwater weed ***Salvinia molesta***. His students include Ms. Surjit Kaur, P.B. Khare, P. Chandra, S. Devi, Subhash Chandra, P. Madhusudan etc.

Subhash Chandra (1996) at NBRI in addition to extensive studies on stelar morphology Filicopsida, undertook indepth studies on taxonomy and nomenclature of Indian ferns. Chandra has also undertaken critical examination of nomenclatural aspects of new taxa reported since the classical work of Beddome. In collaboration with Mritunjaya Srivasta, Subash Chandra (2002) edited and published articles of reputed Pteridologists in the book entitled "Pteridology in the New Millennium". G. Panigrahi (1962-2002) conducted studies on taxonomy and phylogeny of pteridophytes at Botanical Survey of India. He was ably succeeded by R.D. Dixit (personal communication).

Pteridophytes of Northern Bihar were studied and reported by Alka Kumari and Ranjit Bahadur Srivastava (2002).

A brief account of annual interaction with pteridophytes in India was described by Bera, Patra and Ghorai (2002).

Singh (2002) gave a detailed account of economic importance of Indian Pteridophytes.

An excellent review of reproductive biology of the gametophytic generation of homosporous ferns was published by S.C. Verma (2002) in "Pteridology in the New Millennium."

Distribution, endemism and species radiation of ***Isoetes*** of India were studied in detail by Shukla, Srivastava and Shukla (2002). Simuitaneously P.B. Khare (2002-2006) completed several projects on Morphology and taxonomy of ferns and pteridophytic flora of Kumaon and Sikkim Himalayas with special reference of threatened taxa. His group also conducted studies on reproductive biology of triploid, tetraploid and pentaploid cytotypes of ***Pteris vittata.***

Rajesh, Nair and Madhusudan (2003) reported ***Bolbitis crispatula*** as a new record of Peninsular India while Hameed *et al* described Filmy ferns of South India in detail. Pharmacognostic studies on medicinal properties are being carried out by Gayathri, Asha and Subramaniam (2005). The immunomodulatory and antioxidant properties of ***Selaginella*** species have been carried out by them. A lot of studies on diverse aspects of pteriphytes are being carried out at almost all the Universities of the country where Botany is taught at post-graduate level. Important centres of research on Pteridophytes include Botanical Survey of India, National Botanical Research Institute and Maharashtra Association for cultivation of Science besides Universities of Punjab, Patiala, Delhi, Allahabad, Lucknow, Calcutta, Aligarh, Meerut, Kalyani, Jodhpur, Jaipur, Calicut etc.

QUESTIONS

ESSAY TYPE

1. Describe the characteristic features of pteridophytes. Give a brief account of the classification of pteridophytes also.
2. Write an essay on Alternation of generations in pteridophytes Or What do you understand by alternation of generations? Discuss with particular reference to pteridophytes.
3. What do you know of Antithetic and Transformation theories? Explain these with reference to alternation of generation.
4. Give an illustrated account of origin of vascular cryptogams.
5. Give a detailed account of various systems of classification of pteridophytes. Which system of classification is regarded as the best and why?
6. Discuss the classification of pteridophyta.
7. Give an account of classification of pteridophytes given by Sporne.
8. "Pteridophytes are said to form a bridge between Bryophytes and Gymnosperms". Explain giving reasons.
9. Give a diagrammatic representation of the geological Time scale, indicating therein the chief events and the appearance and disappearance of major plant groups.
10. With the help of suitable sketches write a short essay on Telome theory.
11. Describe the occurrence of Apogamy in Pteridophyta and discuss its significance.
12. Give an outline classification of Pteridophytes. Mention the main features of various classes and orders with examples. **(Kumaon 2004)**
13. Describe important features of Pteridophytes and Gymnosperms. How these groups are differentiated from each other?
14. Give the outline of classification of Takhtajan and Zimmerman (1966) and discuss the general characters of Pteridophytes. **(Kumaon, 2000)**
15. Give the classification of Pteridophytes based on Eames and Sporne's Classification.
16. Give an account of classification of Pteridophytes. Name the orders of the genera included in your course.
17. Give a detailed account of any system of classification of pteridophytes.
18. Write an essay on phylogeny and evolutionary tendencies in pteridophytes.
19. What is Telome theory? Discuss its importance in tracing the evolution in pteridophytes. **(Rohilkhand, M.Sc. 2000)**
20. "Pteridophytes were the simplest vascular plants." Discuss.
21. (*a*) What are fossils? How are they classified? Add a note on the importance of fossils. **(Kerala, 2000)**

    (*b*) What are fossils? Write various types of fossils studied by you. **(Nagpur, 2002)**
22. Describe the modern system of classification of pteridophytes upto the rank of orders.
23. Define Heterospory? Trace its origin in pteridophytes and point out its segnificance.
24. Pteridophyta is an intermediate between between Bryophyta and Spermatophyta. Elaborate the statement. Mention the origin and evolution of vascular structure in Pteridophyta. **(Guwahati, 2000)**
25. Write a brief essay on the origin of Alternation of generations.
26. Write a brief account of apospory, under the heads

    (*i*) Cytology and artificial induction,

    (*ii*) Causes of apospory and

    (*iii*) Significance of apospory.
27. Write an essay on the origin of vascular cryptogams.
28. Give a detailed account of the origin of alternation of generation.
29. What do you know of the origin of vascular cryptogams? Discuss in detail.
30. (*a*) Discuss in detail the evolution of the sporophyte in pteridophytes.

    (*b*) With the help of suitable diagrams, describe the evolution of stele in pteridophytes. **(Kumaon, 2004)**
31. What is the function of elaters?
32. What is columella?
33. What is heterospory? **(Kuvempu Univ., 2005)**
34. What do you understand by heterospory? **(Garhwal Univ., 2005)**
35. Explain the importance of study of fossils. **(Garhwal Univ., 2005)**
36. (*a*) Define obligate apogamy.

    (*b*) Define synangia, give example.

    (*c*) What is sori? **(M.S. Univ., Tirun., 2005)**
37. Elucidate the sporangial development in pteridophytes. **(M.S. Univ., Tirun., 2005)**

38. Write the primitive and advanced characteristic features of pteridophytes. **(M.S. Univ., Tirun., 2005)**
39. Define a living fossil. **(M.S. Univ. Tirun, 2005)**
40. Write short notes on alternation of generations. Give example. **(M.S. Univ. Tirun., 2005)**

SHORT ANSWER TYPE

41. Write short notes on :
    (*a*) Apogamy.
    (*b*) Apospory. **(Kerala, 2000; Nagpur, 2002)**
    (*c*) Apogamy and Apospory **(Lucknow, 2000)**
    (*d*) Geological Time Scale. **(Poorvanchal, 2000; Nagpur, 2002)**
    (*e*) Alternation of generations in Pteridophytes.
    (*f*) Telome theory.
    (*g*) Causes of Apogamy and Apospory.
    (*h*) Classification of Pteridophytes.
    (*i*) Salient features of Pteridophytes.
    (*j*) Habitat of Pteridophytes.
    (*k*) Petrifaction. **(Kerala, 2000; Nagpur, 1996; Bangalore, 2001)**
    (*l*) Importance of heterospory. **(Garhwal, 2000)**
    (*m*) Affinities of pteridophyta. **(Nagpur, 2002)**
    (*n*) Coal formation.
42. Describe briefly
    (*a*) Alternation of generation in pteridophyta.
    (*b*) Range of habitat in pteridophyta.
    (*c*) Process of fossilisation.
    (*d*) Heterospory. **(Kerala, 2000, 2001)**
    (*e*) Geological Time Scale.
    (*d*) Give a brief account of apogamy and apospory in pteridophytes.
    (*e*) Demerits of Telome theory. **(Garhwal, 2000)**
    (*f*) Antithetic theory of origin of alternation of generations.
    (*g*) Homologous theory of origin of alternation of generations.
    (*h*) Parthenogenesis.
    (*i*) Cytological studies in pteridophytes.
    (*j*) Six characters of Pteridophytes. **(Lucknow, 2002)**
43. Define apogamy.
44. What is the essence of the Telome Concept?
45. Define apospory.
46. The plant body of a pteridophyte is gametophyte or sporophyte.
47. Name any pteridologist of our country.
48. Differentiate between pteridophytes and gymnosperms on the basis of a single character.
49. Write one important character of pteridophytes not found in bryophytes.
50. What is apospory? What is its significance. **(Himachal Pradesh, 2000)**
51. What is apogamy?
52. Differentiate between.
    (*a*) Homospory and Heterospory.
    (*b*) Apogamy and Apospory.
    (*c*) Pteridophytes and Gymnosperms. **(Himachal Pradesh, 2000)**
    (*d*) Eusporangium and Leptosporangium.
53. Write two characters of resemblance between pteridophytes and Gymnosperms. **(Lucknow, 2001)**
54. Write explanatory notes on the resemblances and differences between pteridophytes and gymnosperms.
55. Water drop is not needed for fertilisation in Pteridophytes. Is it correct or wrong?
56. What is eusporangiate development?
57. What is meant by impression (fossil).
58. Name the dominant pteridophytes of the Devonian period.
59. Name the person who put forth the Telome concept.
60. (*a*) List various types of fossils.
    (*b*) Name the important geological eras.
61. How are petrifactions formed during fossilisation process.
62. What are the affinities of pteridophytes with gymnosperms? **(Garhwal, 2000)**
63. Write two characters of resemblances between pteridophytes and gymnosperms. **(Lucknow, 2001)**
64. Explain the application of Telome Concept in the interpretation of origin of microphyll and megaphyll. **(Kerala, 2001)**
65. What is the significance of heterospory?
66. What do you think were the essential characters of the first vascular plants?
67. Suggest the possible steps along which the evolution of sporophytic generation might have taken place among the land plants.
68. Write notes on :
    (*a*) Rhizophore
    (*b*) Apospory
    (*c*) Protonema
    (*d*) Elaters
    (*e*) Sporangiophore
    (*f*) Prothallus

OBJECTIVE TYPE

69. **Select the correct answer :**
    (*i*) Spores of pteridophytes are
        (*a*) Haploid

(*b*) Diploid
(*c*) Triploid
(*d*) Tetraploid.

(*ii*) If all the spores are of same size and shape, the plant is said to be as
(*a*) Aposporous
(*b*) Homosporous
(*c*) Heterosporous
(*d*) None of the above.

(*iii*) Seed habit originated in
(*a*) Algae
(*b*) Fungi
(*c*) Bryophyte
(*d*) Pteridophyte.

(*iv*) Which one of the following is a leptosporangiate character?
(*a*) Sessile Sporangia
(*b*) High spore production
(*c*) Presence of stomium
(*d*) Single layered sporangial jacket.

(*v*) Heterologous alternation of generation is present is
(*a*) Algae
(*b*) Fungi
(*c*) Bryophytes
(*d*) Pteridophytes.

(*vi*) Living example of near telome is
(*a*) ***Osmunda***
(*b*) ***Marsilea***
(*c*) ***Adiantum***
(*d*) ***Psilotum.***

(*vii*) Which of the following is regarded as the age of pteridophytes?
(*a*) Early Palaeozoic
(*b*) Middle palaeozoic
(*c*) Late palaeozoic
(*d*) All of the above.

(*viii*) Sporophytic generation is dominant in
(*a*) Algae
(*b*) Fungi
(*c*) Bryophyte
(*d*) Pteridophyte.

(*ix*) Antithetic theory was discussed in detail by
(*a*) Alexopoulos (*b*) Andrews
(*c*) Bower (*d*) Mehra.

(*x*) Which of the following bryophytes is regarded as the ancestor of pteridophytes?
(*a*) ***Riccia***
(*b*) ***Marchantia***
(*c*) ***Funaria***
(*d*) ***Anthoceros.***

(*xi*) Telome theory was proposed by
(*a*) Eanaes
(*b*) Zimmerman
(*c*) Mehra
(*d*) Sahni.

(*xii*) Pollentube is not formed in
(*a*) Pteridophytes
(*b*) Gymnospems
(*c*) Angiosperms
(*d*) None of the above

(*xii*) Prothallus occurs in
(*a*) Pryophyta
(*b*) Pteridophyta
(*c*) Gymnosperms
(*d*) Angiosperms

60. State whether true or false:
(*i*) Fossils are those species which are now extinct.
(*ii*) ***Impressions*** are those fossils in which anatomical details are preserved.
(*iii*) Pteridophytes are first land vascular system.
(*iv*) Pteridophytes do not reproduce by spores.
(*v*) Gametophyte is dominant in pteridophytes.
(*vi*) Xylem makes its first appearance in pteridophytes.
(*vi*) Apogamy is defined as the phenomenon of formation of gametophytes without gametes.
(*viii*) Germination of spores is exosporic in homosporous pteridophytes.
(*ix*) Antithetic theory was discussed in detail by Bower (1935).
(*x*) Water is needed for fertilisation in pteridophytes.

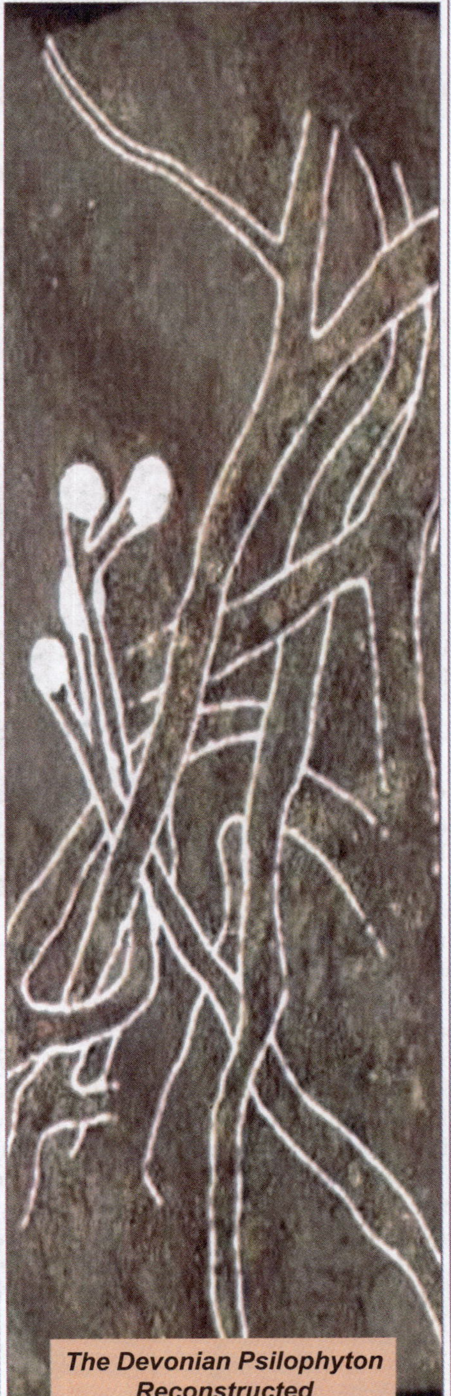
The Devonian Psilophyton Reconstructed

CHAPTER 2

Division–Psilophyta Class–Psilophytopsida

- Introduction
- Classification
- Family Rhyniaceae (***Rhynia, Horneophyton and other members***)
- Family Psilophytaceae (***Psilophyton, Paulophyton, Trimerophyton, Taeniocrada***)
- Family Pseudosporochnaceae (***Pseudosporochnus***)
- Family Asteroxylaceae (***Asteroxylon***)
- Evolutionary trends in Psilophytales

Introduction

The division Psilophyta includes living as well as extinct representatives. The division shares most of its characteristics with the **Psilophytales** (discussed below). They have a simple vasculature with xylem lacking vessels and fibres. The tracheids may be spiral, annular, scalariform or even pitted. There is no secondary growth. Mycorrhizal infections are common in the rhizomes (***Psilotum***).

It is divided into two classes ;

***(i)* Class Psilophytopsida :** It includes all the extinct members of this division. It is further divided into a single order **Psilophytales** which is further spilt up into families.

(*ii*) Class Psilotopsida : It includes the order **Psilotales** which is represented by two living genera ***Psilotum and Tmesipteris.***

Class : PSILOPHYTOPSIDA

Characteristic features :

1. *All the plants are extinct.*
2. *Only sporophyte are known so far.*
3. *Plant body is distinguished into rhizomes and aerial branches which are more or less dichotomously branched.*
4. *Roots are completely lacking, but at places unicellular rhizoids are present in tufts.*
5. *Aerial branches are naked or has spirally arranged small appendages.*
6. *Thick walled homosporous sporangia are present and their position is terminal or lateral.*
7. *Stele is simple protostele.*

Order Psilophytales

They are the simplest extinct vascular plants that were discovered among the rocks of the Early Devonian period of the Palaeozoic age. Among these plants, the first to be recorded in 1858 was Sir William Dawson's ***Psilophyton princeps*** that was discovered from the Gaspe Sandstone. Later it was also recorded in beds of various Devonian horizons in many countries, such as the United States, Scotland, Norway and Belgium. It is the first plant with which our knowledge of the group **Psilophytales** started.

Psilophyton princeps, **first extinct vascular plant to be recorded in 1858**

The group Psilophytales was actually recognised and named when the Rhynie discoveries came into light. Our present accurate 'knowledge of these simple vascular plants is mostly due to investigations of the Rhynie chert-bed. This locality is in Aberdeenshire, and the bed was discovered by the geologist Dr. Mackie in 1913. The plants have been thoroughly worked out by Dr. Kidston and Prof. Lang, whose results are recorded in their remarkable series of memoirs that were published in 1917 to 1921.

The Psilophytales are most prevalent in the lower and the middle Devonian period. They have been recorded by some in the upper Devonian rocks but the discoveries are not entirely free from doubt. They are characterised by : — 1. A simple plant body distinguished into a rhizome and an aerial region. 2. Absence of roots. 3. Rhizoids arise in tufts from rhizome. 4. The stems branch dichotomously and are naked or covered with small spinous outgrowth or reduced leaves. 5. Vasculature is a protostele. 6. Sporangia are terminal and may occur solitary or in groups or in simple recemes. They are not associated with any kind of sporophylls. They have a single spore cavity. 7. The wall of the sporangium is several cell layers thick. 8. They are homosporous, spores arranged in tetrads. These characteristics are shared by majority of Psilophytales, but variations are there. In ***Asteroxylon*** the sporangia are now reported to occur singly a little above the leaf (cauline). For this reason it has been included by some authors under the Lycophyta (Bierhorst, 1971).

CLASSIFICATION

Krausel and Hirmer have put forth a scheme of classification of the group. They divide it into 9 families including more than 20 genera. Out of these 9 families five are most firmly established.

1. Rhyniaceae

Cooksonia

The plants are rootless and leafless with dichotomously branched aerial shoots. Sporangia are terminal. It includes ***Rhynia***, ***Horneophyton***, ***Sporogonities***, ***Cooksonia***, ***Yarravia*** and ***Hicklingia.***

2. Zosterophyllaceae

The rhizome is profusely branched. Aerial shoots are devoid of leaves. Sporangia are borne on the apices of lateral branches. It includes ***Zosterophyllum***, ***Pectinophyton bipectinatum*** and ***Bucheria.***

3. Psilophytaceae

Plant body as in Rhyniaceae. The aerial shoots are provided with spinous outgrowths. Sporangia are terminal in position and borne on small branch ramifications. It includes ***Psilophyton***, ***Dawsonites***, ***Paulophyton***, ***Trimerophyton***, ***Taeniocrada*** and ***Loganella.*** Andrews and Andrew (1972) reported the presence of 4 distinct species of ***Psilophyton*** from the trout valley formation in Northern Maine. The two common and often reported species are ***P. princeps*** (Dawson) Hueber, and ***P. forbesi*** Andrews. ***Paulophyton Jongmansi*** discovered by Krausel in 1957 from Naumurian of Rurh region also belongs to this family.

Taeniocrada

4. Asteroxylaceae

It includes ***Asteroxylon*** and the plant body consists of a branched rhizome. Some of the branches function as roots. The aerial stems are covered with leaves. They are simple, oval and slightly dorsiventral. The smaller aerial twigs are naked and spinous. The wood is star-shaped.

5. Pseudosporochnaceae

It is typified by a single genus ***Pseudosporochnus.*** The stem is thick and bulbous below, dividing up above into numerous fine branches. Its appearance is remarkably like that of an alga, but the stem is known to be vascular. The sporangia are oval and borne on the tips of the branches. The branching is dichotomous. Branches that are without sporangia are filiform and most probably serve as leaves.

Banks (1968) recognised two sub-divisions **Rhyniophytina** and **Trimerophytina.** In the former he included one order (Rhyniales) and two families (Rhyniaceae and Cooksoniaceae). He included the genera ***Rhynia***, ***Horneophyton***, ***Hicklingia***, ***Eogaspesia***, ***Taeniocrada***, ***Hedeia***, ***Yarravia*** and ***Hostimela in the family*** Rhyniaceae. The genus ***Cooksonia*** was included in Cooksoniaceae. The sub-division **Trimerophytina** included order **Trimerophytales** and family ***Trimerophytaceae*** with two genera ***Trimerophyton*** and ***Dawsonites.***

Bierhorst (1971) recognised the class **Rhyniopsida** in which he included one order **Rhyniales** and two families Rhyniaceae and Trimerophytaceae. The Rhyniaceae includes ten genera ***e.g. Rhynia***, ***Horneophyton***, ***Hicklingia***, ***Eogapesia***, ***Taeniocrada***, ***Hedeia***, ***Yarravia***, ***Hostimella***, ***Nothia*** **and** ***Cooksonia. Trimerophyton and Dawsonites*** have been included in family Trimerophytaceae.

Heuber and Banks (1967) and Banks (1968a) have emended the diagnosis of ***Psilophyton princeps*** Dawson (1859) and gave it the name ***Dawsonites.*** Bierhorst (1971) also agrees, with Heuber and Banks.

Bierhorst has included the family **Zosterophyllaceae** under a separate class **Zosterophyllopsida** and order **Zosterophyllales**. He included the genera ***Zosterophyllum***, ***Bucheria***, ***Gosslingia***, ***Crenaticaulis*** and ***Psilophyton ornatium.*** The family Asteroxylaceae has been included under the class

Lycopodiopsida and order **Asteroxylales**. The families Asteroxylaceae and Kaulangiophytaceae represented by ***Asteroxylon*** and ***Kaulangiophyton*** have been included in the order Asteroxylales.

Rhynia

Laclercq and Banks (1962) included the family Pseudosporochnaceae under the order **Cladoxylales** and Bierhorst (1971) has followed these two authors.

The families have been discussed briefly in this chapter.

The genus ***Rhynia*** is named after the locality Rhynie in Aberdeenshire district of Scotland and possesses two species, ***R. gwynne-vaughani*** and ***R. major. R. major*** is larger than the former . The genus was discovered in 1917 by Kidston and Lang from Rhynie chert bed of Middle Devonian. According to Parihar (1996) the chert deposites at Rhynie, originally thought to be of middle Devonian are now beleived to be lower Devonian (Late Siegenian or Emsian epoch – about 390 to 374 million years ago.). Accounts appeared in various papers that were published in the Transactions of the Royal Society of Edinburgh. These plants were discovered in the form of petrifactions and grew in dense aggregations along with ***Horneophyton*** and ***Asteroxylon.*** The plants were preserved with exquisite perfection and revealed all possible details that a fossil can.

Rhynia gwynne-vaughani was a small (Fig.2.1, A) herbaceous plant, possibly about 18 cm. high, consisting of slender cylindrical aerial stems about 2 mm in diameter and branches arising from a basal rhizome-like portion (Fig. 2.1 A) that was buried in the peat formed by the dead remains of other plants of the same kind. There is not much difference in the structure of the rhizome and the aerial stem except that the former bears at

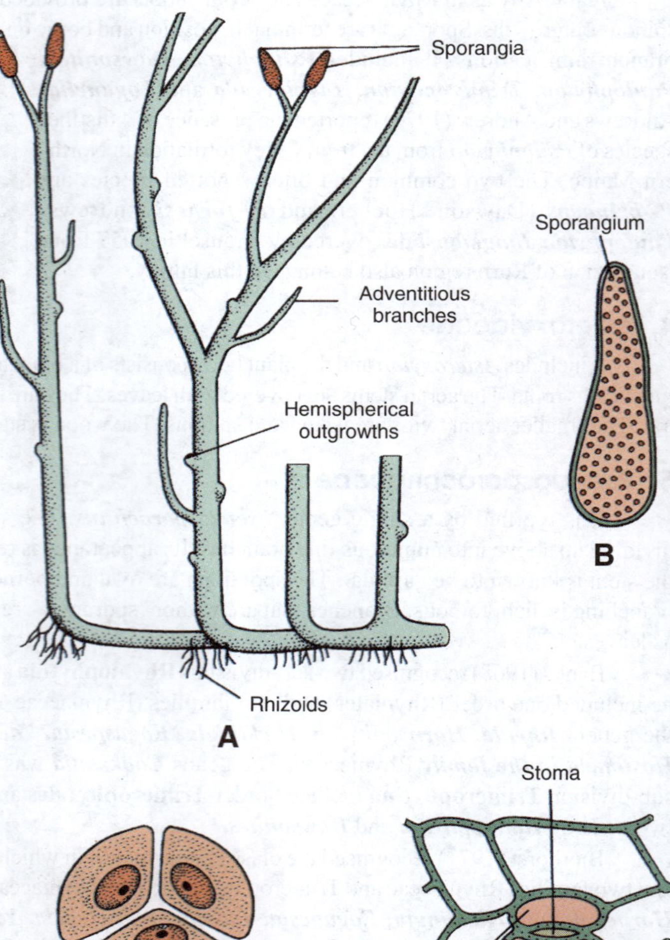

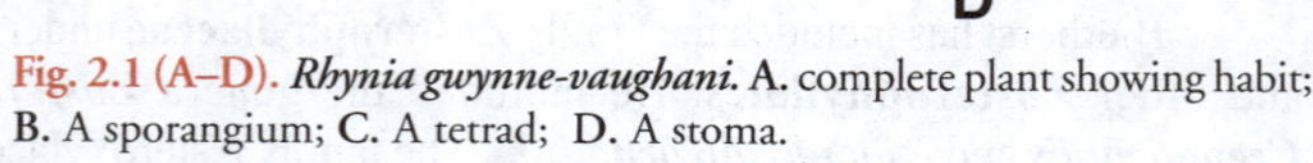

Fig. 2.1 (A–D). *Rhynia gwynne-vaughani.* A. complete plant showing habit; B. A sporangium; C. A tetrad; D. A stoma.

places tufts of rhizoids on its underside . There are no roots and the rhizoids performed the dual functions—absorption and anchorage. The aerial stem was dichotomously branched and tapers gradually towards its apex and is not perfectly smooth but bears a number of hemispherical outgrowths (Fig. 2.1, A). These outgrowths have often been regarded as rudimentary leaves but this interpretation is doubtful. They appeared very late in the life of the plant, usually as new formations beneath the stomata . The aerial shoots may end in pointed tips or they may bear oval sporangia (Fig. 2.1, A, B). Stomata are present all over the surface of the aerial shoots. Adventitious branches also arose from the aerial shoots.

According to Mercker (1955, 1959) the rhizomatous parts of the plant represent the gametophyte and the upper erect part is the sporophyte. The two evidences that support Mercker's view are : (*i*) the presence of some flask-shaped cavities in the subterranean creeping axis that have been regarded as disintegrated sex organs, and (*ii*) presence of groups of four cells with openings in the centre, in a section of the subterranean part from Munich, resemble the surface view of embedded archegonia. Mercker states that the presence of these groups of 4 cells "reminds us very much of the general type of sunken, more or less neckless archegonia." Similar groups of 4 cells are also present in the pictorial material presented by Kidston and Lang referring to **R. *major.*** Mercker regards the prostrate regions of ***Psilophyton***, ***Asteroxylon*** and ***Rhynia*** as richly branched gametophytes. He assumes that the primitive land plants like ***Rhynia*** must be possessing massive and well developed gametophytes as are found in the present-day eusporangiates (***Lycopodium***).

Another feature of importance in this species is that the hemispherical outgrowths mentioned above were sometimes the seats of the formation of additional branches, quite apart from the normal forking of the stem. Both these structures are absent in **R. *major.*** These extra or adventitious branches thus formed were usually without any vascular connection with the main stem. They often become detached and afford a ready means of vegetative propagation. Sometimes these swellings also produced rhizoids. These hemispherical outgrowths go on becoming smaller and sometimes appear as mere eruptions made of only few cells. These smaller ones seem to have a small pore leading into a central channel or a canal leading to an inter cellular space below. Lemoigne (1968) regarded these structures as archegonia and assumed that the plants of this species represent gametophytic generation of ***R. major.*** Beirhorst (1971) regarded these structures as ***hydathodes*** or "secretory structures of some kind".

Internal Structure or Anatomy of the Stem (Fig. 2.2)

A transverse section of the aerial stem reveals a distinct epidermal layer interrupted at places by the presence of stomata. The cells of the epidermis also secrete a thick cuticle. Next is the cortical tissue differentiated into an outer and an inner cortex. The cells of the outer cortex are larger than the cells of the epidermis. The inner cortex possesses smaller cells with intercellular spaces between them. These intercellular spaces are connected with stomata through spaces between the outer cortical cells. The inner cortex was most probably the seat of photosynthesis.

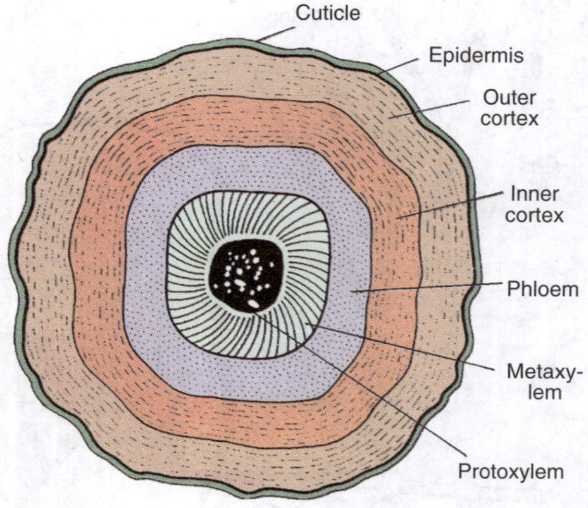

Fig. 2.2. Diagrammatic T.S. of stem showing a protostele.

The centre of the stem is occupied by a vascular strand of a simpler organisation. The centre is occupied by xylem tracheids (Fig. 2.2) that showed traces of annular and sometimes of spiral thickenings. It has also been observed in certain cases that the tracheids in the centre are smaller than the tracheids surrounding them. It reflected on the endarch differentiation of the xylem tissue. The cells surrounding the xylem tissues were thin-walled and without any intercellular spaces. In longitudinal sections these cells appear to be elongated with oblique end walls. This tissue appears to be stelar in nature and is most probably phloem. There are no traces of sieve plates or pores in the cells of phloem and it is in this respect that this tissue differs from the phloem of the living Pteridophytes. Satterthwait and Schopf (1972) reported the occurrence of numerous pores (3-8 μ in diam.) distributed along the lateral cell walls in this zone of tissue. In face view these pores appear to be composed of circular sub-units. They reported the presence of sieve tubes and parenchyma cells in the phloem zone and also observed the presence of pit fields in the lateral cell walls and inclined end walls of sieve tubes. The pericycle and endodermal layers were lacking.

In the rhizome there is no distinction between the outer and inner cortex and the xylem strands consist of but fewer tracheids.

The growing point consists of numerous dividing cells. There is no apical cell.

Sporangia

They are club-shaped or cylindrical structures (Fig 2.1) borne terminally on the fine aerial branches (4 m.m. long and 1.4 m.m. broad). Their wall is three layered : an outer epidermal layer, a middle three cell rows thick layer and an inner layer. The outer layer is heavily cutinised. The sporangium contained large number of spores arranged in tetrads like the modern ferns. They are homosporous and measured 40 μ in diameter. The spore wall must have been tough so as to preserve them so nicely. The spores were trilete. The development of the sporangium was of **eusporangiate type** as it developed from a group or a row of superficial cells that divided periclinally to form an outer primary wall layer and an inner layer of primary sporogenous cells.

RHYNIA MAJOR

It is wholly a leafless plant (Fig. 2.3) distinctly differentiated into a rhizomatous portion bearing tufts of rhizoids and an aerial portion quite naked and devoid of any hemispherical outgrowths as are found in the other species. Branching is by forking dichotomy. No adventitious branches are present. The estimated height is about 20 cm. or even more and diameter 6 mm. The sporangia borne at the ends of the branches (Fig. 2.3, B) are about 12 m.m. long and 4 m.m. in breadth. The spores were about 65 μ in diameter.

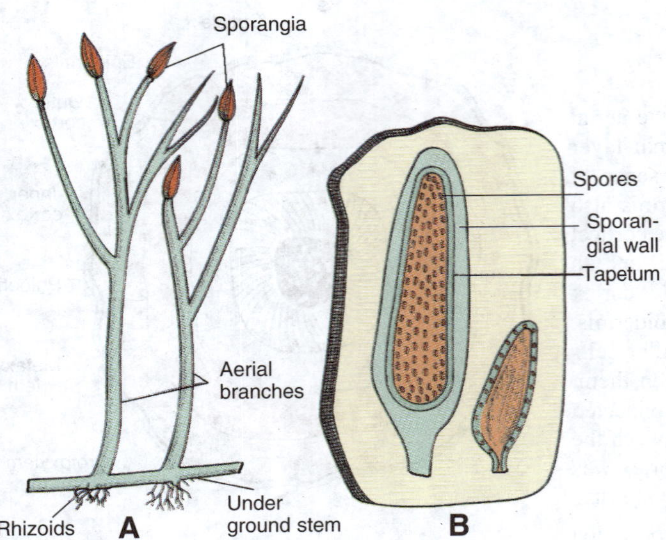

Fig. 2.3. ***Rhynia major.*** A. Reconstruction of a complete plant; B. sporangia. (After Kidston and Lang).

Internal structure too is very simple and agrees in all respects to that of ***Rhynia gwynne-vaughani.*** Stele is a typical protostele. Its larger size, absence of hemispherical outgrowths and the absence of adventitious branches are the points that differentiate this species from the smaller ***Rhynia gwynne-vaughani***.

Pant (1960) regarded the small-sized ***R.. gwynne-vaughani*** to be the gametophytic part of ***R. major*** because :—

(*i*) The sporangia assigned to ***R. gwynne-vaughani*** were not found in organic connection with this plant and might have been the immature sporangia of ***R. major.***

(*ii*) The hemispherical bulges and the adventitious shoots of ***R. gwynne-vaughani*** were the young embryos that were clearly demarcated from the main axis and possessed haustorial cells, like those present in the foot of the young embryos of ***Psilotum.***

(*iii*) Resemblance between the axis of R. ***gwynne-vaughani*** and the vascular gametophytes of ***Psilotum.***

(*iv*) Presence of a canal like channel in some of the smaller hemispherical protuberances which could be archegonia (see also Lemoigene, 1968).

(*v*) The hemispherical projections resemble the young sporetlings of the Psilotales.

(*vi*) The development of the hemispherical buldges is similar to the young sporophytes of Psilotales.

Forms of *Rhynia*

The chief arguments against such an interpretation are :—

1. Absence of definite sex organs in connection with the axis. In case embryos could be preserved there should have been no difficulty for the sex organs to be preserved (Mehra, 1968).
2. Presence of stomata with guard cells on the axis.
3. Presence of a vascular strand.

Thus the taxonomic position of ***Rhynia*** may be given as

Family : **Rhyniaceae**

Genus : ***Rhynia*** Kidston and Lang.

Horneophyton (= ***Hornea lignieri***)

This is as simple a plant as ***Rhynia*** and was discovered from the same chert bed. It is much smaller in size and shows certain characteristics of great importance and singular interest. The aerial shoots are only two millimetres in diameter and arise from a more or less tuberous or swollen rhizome, from the underside of which arise numerous rhizoids. The aerial shoots are leafless, naked and dichotomously branched. The rhizoids are nonseptate.

Anatomy of the aerial stem is the same as in ***Rhynia*** but differs in the rhizome. The xylem of the vascular cylinder of the aerial stem terminates in the upper part of the tuberous rhizome in a group of approximately isodiametric cells, the walls of which have the same colour as the tracheids, in contrast to the other cells of the rhizome which appear to have been parenchymatous. So there is no vascular tissue in the rhizome itself. More than one shoots arise from the rhizome but without any vascular connection.

Horneophyton-Sporangial tips

The sporangium or the fructification of ***Horneophyton*** is of chief botanical interest and remarkable in more than one ways. The sporangia are borne terminally upon certain branches of the stem. There is very little differentiation in the wall, although it is thick. A well defined tapetum could no doubt be recognised over-arching a central columella

(Fig. 2.4 B, C). In fact the sporangia are nothing but enlarged tips of the branches. In certain cases the sporangia too are dichotomously forked (Fig. 2.4, B, C). This is a point of interest that shows that the sporangium is not a distinct organ but just the end of a branch, modified for spore bearing purposes.

Another important sporangial feature is the possession of a central sterile column of cells called the **columella**. So it may be added that the spore bearing layer overarches the columella, recalling the situation met with in ***Sphagnum***-like mosses. This feature has strongly suggested that ***Horneophyton*** and its alies, though vascular in nature may have had affinities with the moss stock.

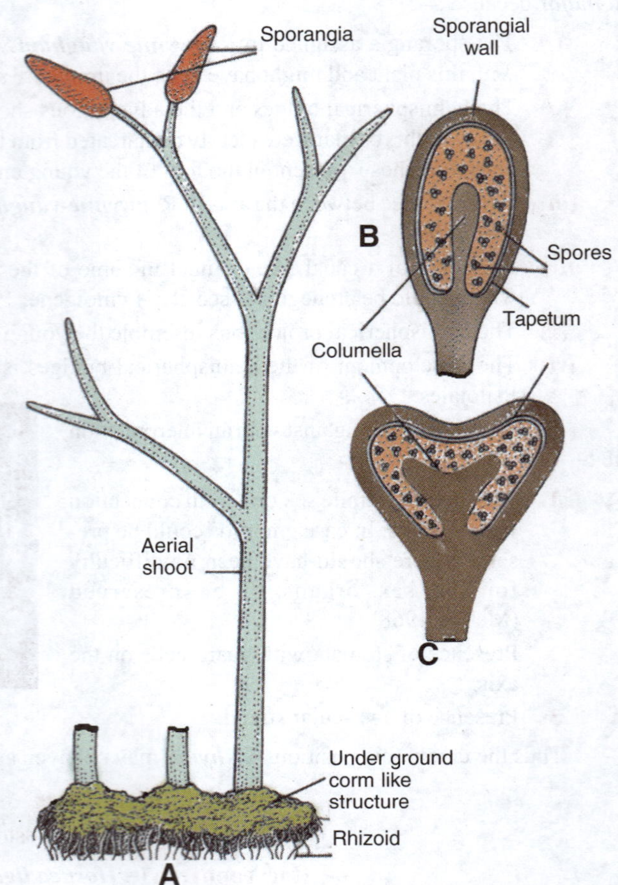

Fig. 2.4. *Horneophyton lignieri*. A. A complete plant; B. A normal sporangium with columella; C. An abnormal dichotomously branched sporangium.

In the forked sporangia the columella too is forked (Fig. 2.4 C). The spores are formed in tetrads and are 50μ in diameter. The cells of the columella are elongated and continuous with the phloem of the stem below. It served for the conduction of food as in the recent plants. Tapetum is a nourishing tissue for the developing spores.

Brief account of the other members of the family Rhynicaeae

1. Sporogonites exuberance

This is another fossil, from Norway, of lower Devonian age, and was described by Dr. Halle. The plants are smaller in size, the stem is naked and the sporangia terminal. The sporangium is much more like the moss sporangium. It is not more than 9 mm in lengths and the diameter varies from 2 to 4 mm. There is a well developed columella made up of slender cells, occupying the base of the sporangium and extending right in the centre as a column of stele cells. It is distinctly overarched by a sporogenous layer. The spores are formed in tetrads. Their diameter varies from 20 to 25μ. Regarding the mechanism of dehiscence, nothing is known as yet.

In the stem no trace of vascular strand is seen. The absence is most probably due to faulty preservation. It makes the position of ***Sporogonites*** doubtful. The resemblance of the sporangium which a moss capsule and absence of vascular strand place ***Sporogonites*** among the Bryophytes. But no member of Bryophytes has been reported from the Devonian age with certainty so ***Sporogonites*** is placed in Psilophytales, closely resembling ***Horneophyton.*** If presence of vascular system is reported the assignment to Psilophytales becomes sure. The discovery of ***Horneophyton*** and Halle's ***Sporogonites*** have thrown much light on the analogy of these early vascular plants with the mosses. This is a point of considerable interest, for previously the fossil record has failed to throw any light on the history of the Bryophytes.

Another species of the ***Sporogonites*** discovered in the upper Silurian of Australia is ***Sporogonites chapmani.***

2. Cooksonia

(Fig. 2.8, C) It was found in the lower Devonian of Wales (Lang, 1935) but has not been fully worked out. There are two species, ***viz.***, ***C. hemispherica*** and ***C. pertoni.*** The stems are naked, straight and dichotomously branched. These features relate it to ***Rhynia.*** The spore sac is much broader than long. The upper surface is more or less rounded. Much less has been known regarding the anatomy of the stem and the internal structure of the sporangium. The sporangium is terminal no doubt. It has also been discovered from Bohemia by Obrhel (1962).

3. Yarravia

(Fig. 2.8, C) Lang and Cookson discovered the fructification called ***Yarravia*** in 1935 from Monograptus beds of Australia. These beds have been estimated to belong to the middle Silurian period. The chief point of interest lies in the fructifications that are closely joined together at the base but free at the ends. Three to five oval sporangia have been discovered in lateral union as above. These sporangia are borne terminally on a smooth, straight and leafless stalk. There are two species differentiated with respect to the shape and size of sporangia. These species are ***Y. subspherica***, ***Y. oblonga.***

4. Hicklingia

(Fig. 2.8, H) It was discovered in Caithness in Scotland in the middle old Red Sand Stone. The plants had indistinct rhizomatous portion giving off numerous aerial stems that branch dichotomously. The branching may become monopodial in further branches. Sporangia are terminal and are oval and like a coil in shape. The aerial stems are leafless and about 15 cm. long and about 2 mm. in diameter.

5. Hedeia corymbosa

H. corymbosa Cookson was collected from the Upper Silurian of Australia, and also from Lower Devonian. The plant is known in the form of terminal aggregation of more or less dichotomously branched branches arranged in a corymbose manner. The tips of branches were swollen to form sporangia.

Hedeia

6. Nothia aphylla

Nothia aphylla is the name given by Lyon (1964) to the unattached, leafless and dichotomously branched fructifications formerly attached to ***Asteroxylon maekiei*** by Kidston and Lang. The sporangia in this genus are terminal with distinct thickened dehiscence cells at the apex. The dehiscence of sporangia in ***Nothia*** is considered in to be transverse as compared to longitudinal in other rhynialean genera.

DISCUSSION

In the words of Dr. D.H. Scott, "the ***Rhyniaceae***, simple as they are, show such a remarkable combination of characters and they have been placed by different botanists in three of the great sub-kingdoms of plants. Kidston and Lang, whom I have followed, assign them to the Pteridophytes, because the spore bearing generation was a vascular plant; my late friend Dr. Newell Arber, in whose time only ***Rhynia*** had been described referred the genus to Thallophytes, on the ground of its undifferentiated thalloid stem : the family has also been included under Bryophytes, on account of the characters presented by ***Horneophyton*** and ***Sporogonites.*** This is a very extraordinary position and shows that Rhyniaceae are a unique race of plants."

Rhyniaceae can very well be treated as a synthetic group of plants combining the characters of vascular cryptogams and the bryophytes on one hand and retaining some of the characters of the thallophytes on the other side. Such a conclusion is justified by the structural facts actually known to us. The question arises whether the Rhyniaceae were really primitive plants. Their environments were quite unfavourable and it is just possible that the plants became reduced due to environmental stresses.

According to Leclereq (1954) Devonian Rhynie flora can not be considered to be the representative of earliest land plants. This is particularly true for **Rhynia** as **Asteroxylon** (more complex form) co-existed with it. Fossil records of Silurian vascular plants (**Yarravia, Hedia corymbosa, Cooksonia** and **Zosterophyllum** *sp.*) also confirms this view. These findings of fossil records make it doubtful wheather to consider **Rhynia** as most primitive land vascular plant or not. Whatever the case may be the facts remain that Rhyniaceae are a group of simplest and the ancient of the plants known to us. The character which make **Rhynia** a simple and primitive plant are :

1. Plant body is not differentiated into stem and roots.
2. Presence of uniform dichotomous branching in whole plant.
3. Leaves are absent and appendages (hemispherical outgrowth) can not be considered leaf.
4. Stele very simple with a slender strand of tracheids.
5. Sporangia were the modified branch tips.

Out of the living Pteridophytes the ***Rhyniaceae*** bear resemblances with the Psilotales. The order includes two living representatives, ***i.e.***, ***Psilotum*** and ***Tmesipteris***. The aerial stems of ***Psilotum*** are devoid of true leaves like ***Rhynia*** and are in the same fashion, dichotomously branched. There aare no roots. The underground rhizomes are provided with tufts of unseptate rhizoids. ***Tmesipteris*** does bear leaf-like appendages but they are flattened laterally and merge with the stem by strong decurrent bases. The sporangia are borne on short branches that arise laterally from the stem.

The bullbous rhizome of ***Horneophyton*** recalls the protocorm of some members of the Lycopodiales, e.g., ***Phylloglossum.*** Protocorms always develop before the roots and are connected to the soil by means of rhizoids.

The lowly organised body of the ***Rhyniaceae*** attaches it with algae. It is only the appearance of vascular system that separates them from the Thallophytes. Algae having as complex a plant body as that of ***Rhynia*** do exist. They have a well developed rhizome and upright dichotomously branched aerial parts bearing tetrasporangia. The more commonly accepted opinion is the early pteridophytes(Psilophytales) had an algal ancestor rather than a bryophyte.

The fructifications of ***Rhyniaceae*** suggest a close relationship with the Bryophytes. The overarching sporogenous layer in ***Horneophyton*** and ***Sporogonites*** and the presence of columella are suggestive of affinities with the Bryophytes. The sporangia are terminal. We cannot extend our comparisons to the gametophytes as the prothalli of Psilophytales have not been discovered. ***Rhyniaceae*** lack annulus and no definite mechanism of dehiscence has been found out.

Anthoceros

Smith has drawn a line of similarity between the tuberous rhizome of ***Hornephyton*** and the foot of some species of ***Anthoceros.*** Anthocerotean sporophyte has been very closely homologised by the supporters of this view. They believe that the ***Rhynia*** like sporophytes have originated from sporophytes of ***Anthoceros.*** (For details see Chapter 1).

Recent views believe in the origin of both bryophytes and pteridophytes from a common stock, most likely from algae with alternation of generations, since this is shown by all Archegoniatae. No higher plants can be derived from the bryophytes; they form a blind alley of evolution. Bryophytes and Pteridophytes form parallel branches of the evolutionary tree, stemming from a common green algal ancestor. Such an ancestor belonged most probably to the heterotrichous Chaetophorales. (For details see Chapter 1).

Bierhorst (1971) while supporting the monophyletic origin of the entire embryophytes has suggested, "a group of avascular archegoniates, probably aquatic, which so far has completely escaped detection by those searching the rocks, must be assumed to have given rise to either three or four lines of archegoniates : (1) Anthocerotae, (2) Hepaticae and Musci, and (3) all vascular plants as a single lineage or as two separate ones, whose basal stocks were the rhyniophytes, on one hand, and zosterophyllophytes, on the other." He has further stated that "Rhyniales probably did in reality give rise to all other groups of vascular plants except the zosterophyllales and the lycopods and did so in the upper Silurian or Lower Devonian times." He further speculates that, "the common ancestral form of the Rhyniales and the zosterophyllales may well have had no sporangium in the ordinary vascular plant sene. This ancestral plant may have had scattered about its axes sporogenous tissue in the form of hypodermal patches, large or small or even single cells. These may have become aggregated or defitinised terminally to produce the rhyniophyte sporangium and laterally to produce the zosterophyllophyte sporangium."

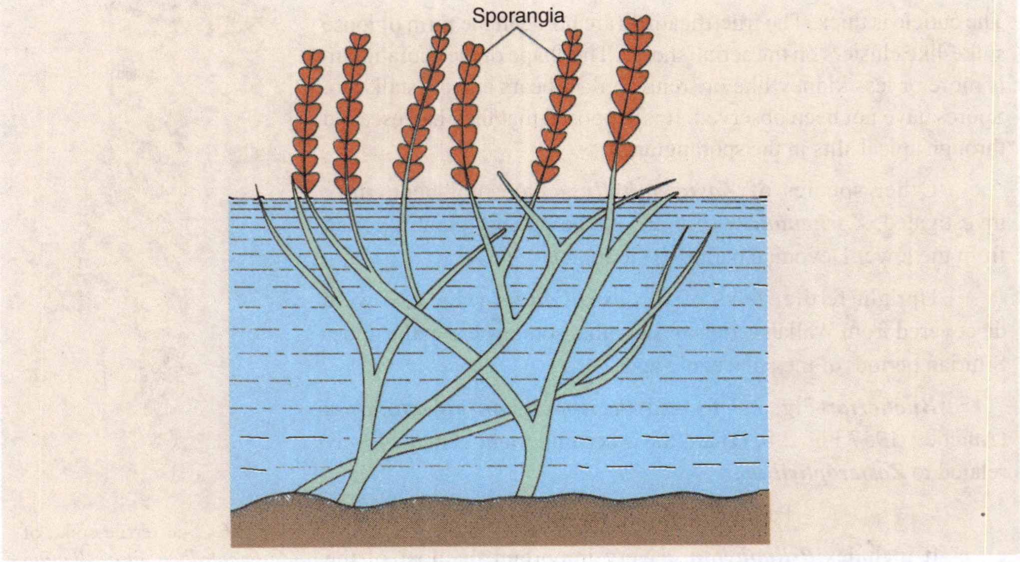

Fig. 2.5. Restoration of ***Zosterophyllum rhenanum*** (After Krausel and Weyland).

Family : **Zosterophyllaceae**

Zosterophyllum is the genus that has been properly studied and includes three species :

1. ***Zosterophyllum myretonianum.***
2. ***Z. australianum***
3. ***Z. rhenanum***

Zosterophyllum myretonianum.

Z. australianum

Z. rhenanum

The first species is the best known of the three and has been explored from the lower old Red Standstone of Scotland. The plant of ***Z. myretonianum*** is small and consists of a branched rhizomatous portion giving out naked aerial branches. The branching of the rhizome-like portion is peculiar in that some of the branches grow away from the main rhizome whereas others run backwards in the opposite direction and there are others that run backward forming a complicated system. The diameter of the rhizome is 2 mm. It is traversed by a central strand. The xylem is composed mainly of tracheids with annular thickenings.

The aerial stems are straight, upright and naked. Branching is not profuse. The stem tapers towards the apex. There are no leaves. The epidermal layer has not shown any traces of stomata. Absence of leaves shows that stomata should be present in the aerial stems. The cuticle is thick. The fructifications are borne in the form of loose spike-like clusters on the aerial shoots. The shape of the sporangium is more or less kidney like or 'reniform'. It bears a short stalk too. Spores have not been observed. It is supposed that they had escaped through apical slits in the sporangium.

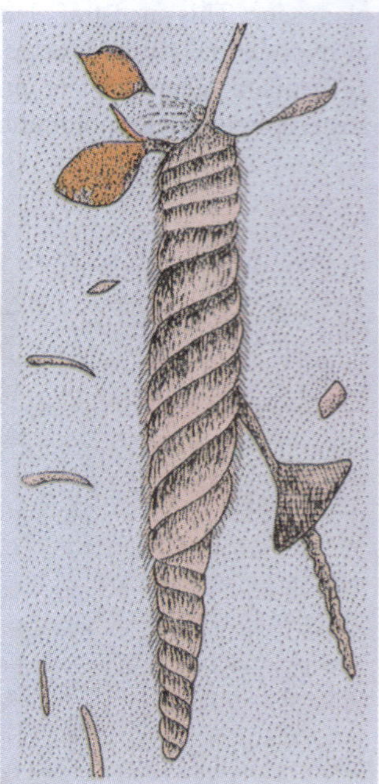

Fig. 2.6. The fertile spike of *Zosterophyllum austra-lianum* (After Cookson, 1949)

Other species of ***Zosterophyllum*** have not been fully investigated. ***Z. rhenanum*** (Fig. 2.5) is an aquatic plant reported from the lower Devonian period in Germany.

Upright fertile stalks of ***Z. australianum*** (Fig. 2.6) were discovered from Wallah series of Australia, somewhere in the upper Silurian period of the palaeozoic age.

Bucheria (Fig. 2.8 E) and ***Pectinophyton bipectinatum*** (Ananiev, 1957 Fig. 2.8, D) are the other important genera closely related to ***Zosterophyllum.***

Family : **Psilophytaceae**

It includes ***Psilophyton,*** a very important member of the Psilophytales. This plant was reconstructed from fossil bits of plants that were discovered by Sir J. W. Dawson from the lower Devonian Rocks of the Gaspe Canada. The name ***Psilophyton princeps*** was given to this reconstruction (Fig. 2.7). Now it has been discovered in certain other parts too. Dawson pictured his ***Psilophyton princeps*** as having upright forked stems of somewhat slender proportions, springing from a creeping rhizome. The aerial stouter stems were ornamented with spines, but for the upper fertile branches, that were devoid of them. The tips of the young branches were curled in a circinate manner, similar to that of the Ferns. The stem was less than 3 feet in length and varied in diameter from 1 to 7 mm.

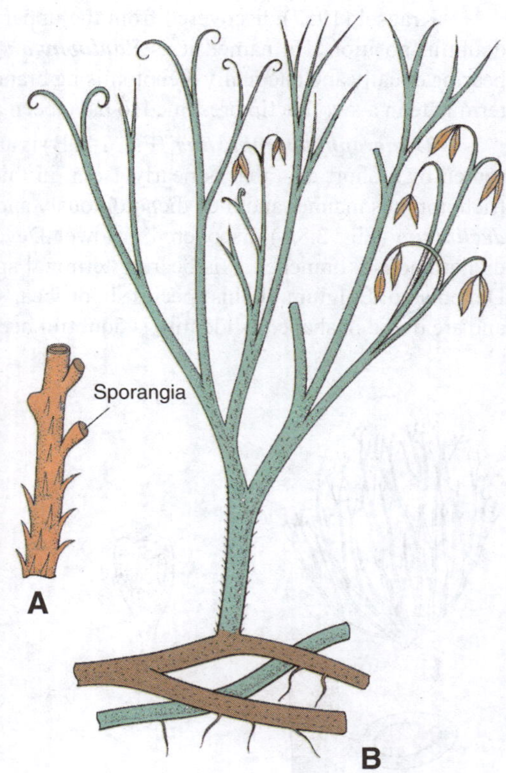

Fig. 2.7. A. Reconstruction of *Psilophyton princeps*; B. Part of an unknown taxon.

The spines varied from 0.5 to 2.5 mm in length. These spines have been regarded as foliar in nature by several authors. Arber compared them with the hemispherical outgrowths of the stems of ***R. gwynne-vaughani***, but this view is not plausible.

The surface of the spines seems to be covered with the cuticle and epidermis, same as in the stem.

Aerial branches were dichotomously forked. Branching was profuse as compared with ***Rhynia***. The sporangia were borne terminally in pairs on the fertile aerial shoots (Fig. 2.7). Sporangia were long, oval, sac-like structures, much similar to those of ***Rhynia.*** The length of the sporangium varies from 4 to 5 mm. In breadth they vary from 1 to 1.5 mm. Exact structure of the sporangial wall is not visible but it seems to be composed of two layers. The spores are arranged in tetrads and have more or less smooth walls. They vary in diameter from 60 to 100µ. They are all of the same size.

Some evidence as to the anatomy of the rhizome is available. Dawson figured sections that revealed similarities with ***Rhynia.*** Edwards in his unpublished work has described the epidermis with stomata that are more numerous than in ***Rhynia***. There are also evidences that two guard cells surround the elongated stomata. This shows that the main seat of photosynthesis is the outer cortex as in ***Rhynia.***

There are other species of ***Psilophyton*** that have been discoverd from other parts of the world. They number twenty. One of them ***P. wyomingense,*** has been found at Bear Tooth Butle, Wyoming, U.S.A., in the lower Devonian rocks. But out of all the other species ***Psilophyton princeps*** has only been studied in detail; others lack such perfect details regarding their structure.

Psilophyton

Krausel (1957) discovered from the upper carboniferous of the Ruhr region a psilophyte of doubtful position. He named it as ***Paulophyton jongmansi*** (Fig. 2.9). It has a 3 cm. broad axis bearing equally and unequally dichotomising branches. The ultimate branches are equal in length and terminate in a swollen tip presumed to have been a sporangium.

Trimerophyton robustius, (Fig. 2.8, B) is another member related to ***Psilophyton*** and is represented by a short axis that is nearly 1 cm. in thickness. It bears lateral branches that branch in a trichotomous manner and then dichotomously and ultimately bear terminal sporangia. ***Taeniocrada dechniana*** (Fig. 2.8A) discovered in lower Devonian of Rhineland possessed ribbon-shaped and dichotomously branched axis bearing terminal sporangia. ***T. langi*** was discovered from the lower Devonian of Belgium. In this species shoots bear stalked sporangia (Fig. 2.8, F,G) that arise laterally and are ovoid in shape. Its identity is doubtful because spores have not been found in the sporangia.

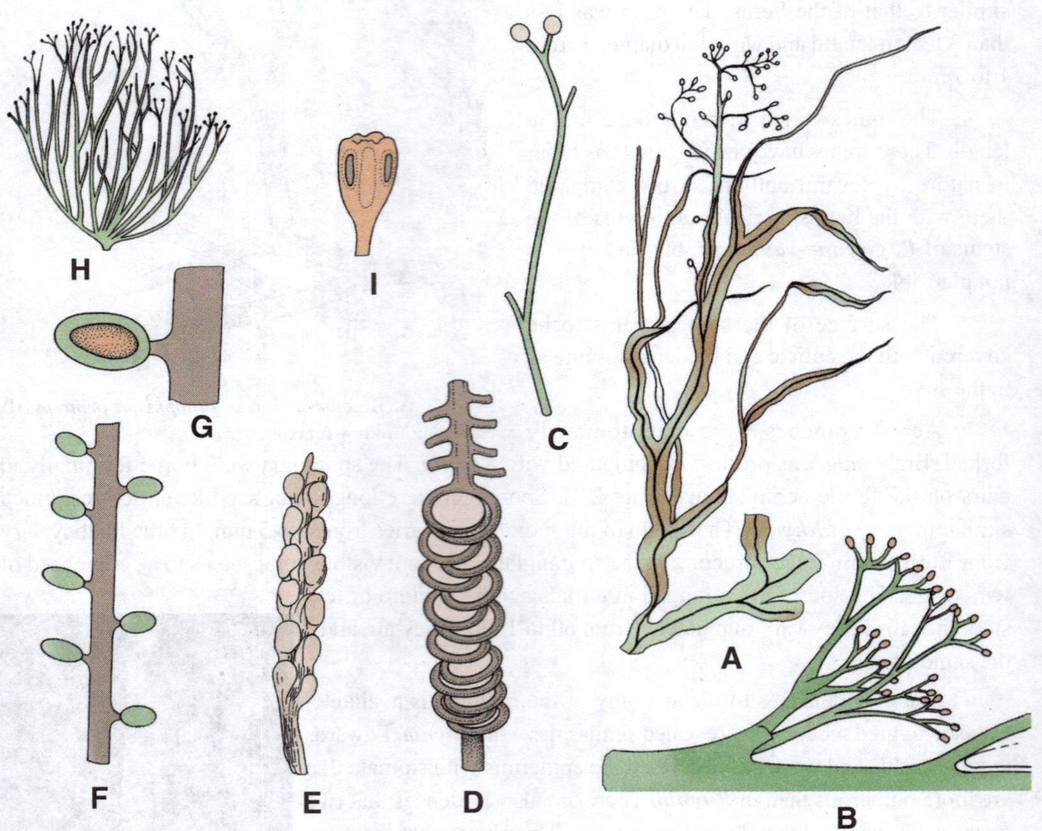

Fig. 2.8 (A–I). A. *Taeniocrada decheniana*; B. *Trimerophyton robustius*; C. *Cooksonia* sp; D. *Pectinophyton bipectinatum*; E. *Bucheria ovata*; H. Hicklingia (after Banks, 1968 a); I. *Yarravia*– a Sporangium (after Andrews, 1959); F, G. *Taenocrada langi* (from Andrews)

The family **Pseudosporochnaceae**

It is characterised by the genus ***Pseudosporochnus*** (Fig. 2. 10). It was reported from the Bohemian Middle Devonian period. It has also been reported from Norway and Scotland but the fossils from Bohemia are nicely preserved and have been properly investigated.

The plant is tree-like in habit and has attained the size of 2 metres or even more. The rhizome is represented by a swollen base. The aerial part is in the form of an upright, straight and stout trunk. It branches profusely and there is a crown of branches at the top of the trunk. These branches may again fork dichotomously. The tips of branches are linear or filiform and most probably represented leaves. The sporangia are terminal and oval. Anatomy is not known.

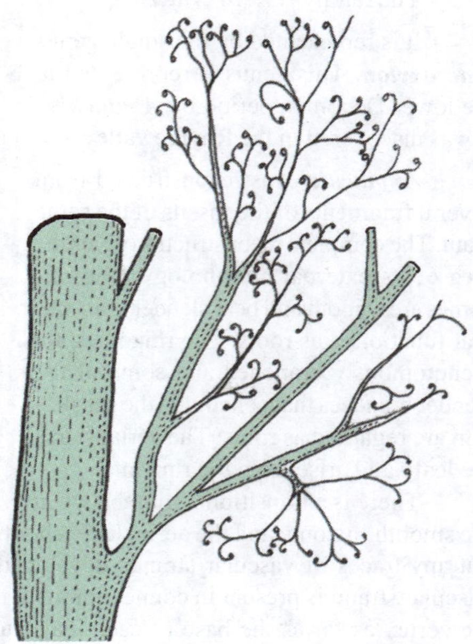

Fig. 2.9. *Paulophyton jongmansi* (After Krausel).

Pseudosporochnus nodosus

Laclercq and Banks (1962) discovered from Middle Devonian of a quarry in eastern Belgium a species of ***Pseudosporochnus.*** They named it as ***P. nodosus***. It is a small tree (Fig. 2.10 A) with main trunk upto 8 cm. in diameter and several feet tall. The main stem bears a crown of primary branches that divide twice or thrice to produce secondary and tertiary branches. The ultimate divisions, referred to as fronds, may be wholly sterile or a mixture of sterile and fertile "pinnae". The fertile "pinnae" are terminated by a pair of sporangia. The stele has been partly preserved and consists of a lobed xylem. Small ovoid vascular strands (Fig. 2.10 B,C) are also seen to arise from the peripheral portions of the stele. They pass into the bases of the fronds. The cortical portion possesses bands of thick walled and sclerotic cells.

Laclereq and Banks place it in the order Cladoxylales, because they considered it to be far advanced than Psilophytes.

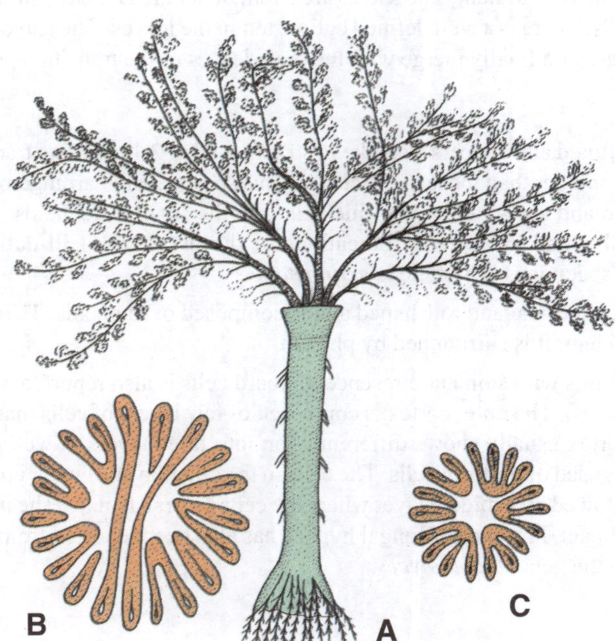

Fig. 2.10. **Pseudosporochnus nodosus** (A–C). **A.** Restoration of the entire plant; **B.** Cross section of the xylem from branches of the first order; **C.** Cross section of the xylem from branches of the second order. (From Laclercq and Banks, 1962).

The family **Asteroxylacease**

It is represented by the single genus *Asteroxylon*. This genus is represented in the lower Devonian period by ***A. mackiei.*** It was discovered in the Rhynie valley.

A. mackiei was reconstructed from several fragments of the fossils of the same plant. The complete reconstruction gives the idea of its external morphology. The rhizomes are smooth and bear slender branches that functioned as roots. The rhizomes are dichotomously branched and some of the slender branches that penetrated the substratum are regarded as roots. The aerial stems are leafy and arise from the rhizome.

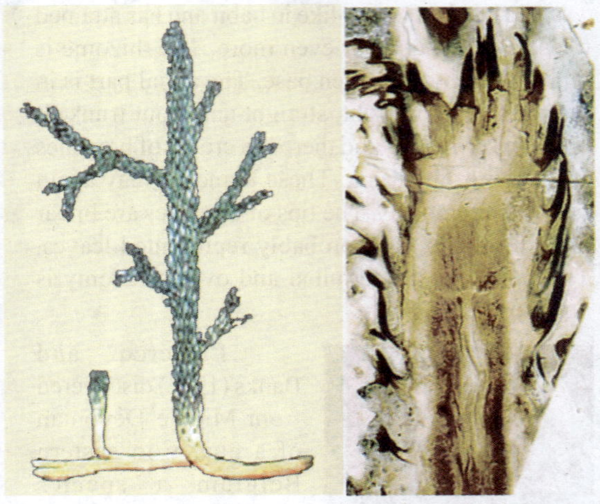

Asteroxylon *A. mackie*-aerial shoots

There is a transition region between the smooth rhizome and the aerial leafy shoot. On the transition region there were small scales without any traces of vascular strands developed in relation to them. On the aerial parts, however, a vascular strand is present in connection with the xylem of the shoot which extends upwards through the cortex as far as the base of each leaf into which, however, it does not extend. The leaves are dorsiventral and oval in section.

The aerial stem branched laterally and these branches forked dichotomously (Fig. 2.11, A). Dichotomy has also been recorded in the main stem. The diameter of the stem is about 1 cm. near the base. It is 1 mm. in the outer extremities of the stem. The leaves are 5 mm. in length and have stomata distributed among the epidermal cells. There is a well defined cuticle too in the leaves. The leaves go on becoming smaller towards the base and finally merge with the scale leaves of the transition zone.

Anatomy

Xylem in the stem is deeply fluted and appears star-like in structure (Fig. 2.11 B). The tracheids at the ends of the xylem rays are smaller than those at the centre. This shows exarch arrangement. The tracheids are spirally thickened and show an annular thickening in the smaller tracheids. The phloem extends as a thin layer all round the xylem. Indications of the presence of ill-defined endodermis are also found in some specimens. The stele is ***actinostelic.***

In the rhizome the xylem is cylindrical and rodshaped and is composed of tracheids. There is no distinction into proto and metaxylem. It is surrounded by phloem.

There is on the outside epidermis with stomata. Presence of guard cells is also reported. Cortex has two distinct layers (Fig. 2.11, B). The outer cortex is composed of few layers of cells that are tangentially extended. The inner cortex usually shows differentiation into three layers in well preserved specimens. Usually it is composed of uniform cells. The cells in the outer layer of inner cortex are more or less compact and are followed by a middle layer where the cells are trabaculate. The inner cells of the inner cortex are also compact. Presence of fungal hyphae has also been reported from the cortical cells. They are attributed to the genus ***Paleomyces.***

Spore Bearing Members

Previously some naked sporangia bearing branches (Fig. 2.11, A) were attributed to ***Asteroxylon.*** Lyon (1964) has named those fructifications as ***Nothia aphylla*** (**Rhyniaceae**). Distinct kidney-shaped sporangia have now been reported to arise from the aerial axis a little above the leaf axil. They are cauline The spores were all of one type (homosporous).

Asteroxylon is slightly advanced than other members of the Psilophytales in possessing leafy stems and star-shaped xylem with distinction into proto and metaxylem.

Anatomically and to some extent in its habit and in position and shape of sporangia ***Asteroxylon*** is homologous with ***Lycopodium.*** The rootless rhizome is the only character suggestive of Psilophytean character. Banks (1968) and Bierhorst (1971) have included ***Asteroxylon*** in the lycopods. Bierhorst has created the order ***Asteroxylales*** with two families ***Asteroxylaceae*** and ***Kaulangiophytaceae.*** The former includes ***Asteroxylon*** and the latter ***Kaulangiophyton.***

Asteroxylon elberfeldense is not regarded as belonging to the genus ***Asteroxylon*** because :—

1. It has been shown to possess terminal leafless fertile shoots, whereas the genus ***Asteroxylon*** has cauline sporangia.
2. It has a medullated stele which has pitted and scalariform tracheids in the xylem.

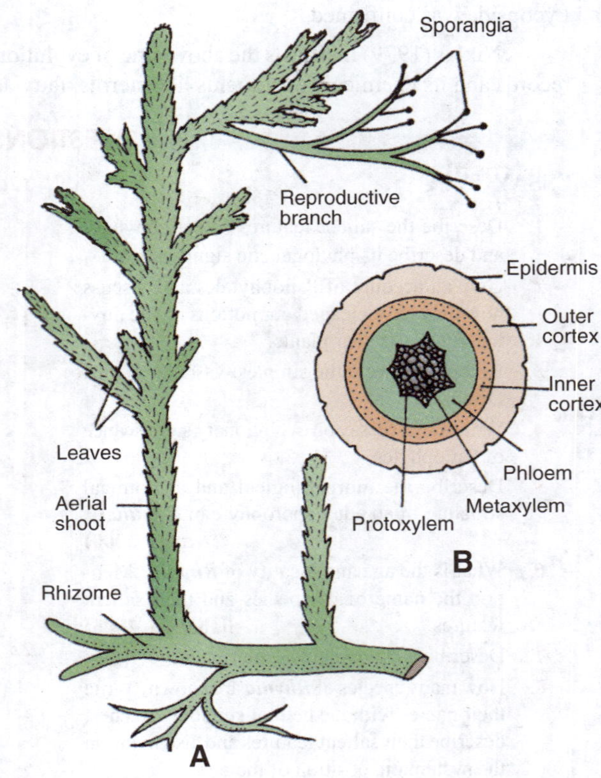

Fig. 2.11. A. Reconstruction of *Asteroxylon mackiei*; B. T.S. Stem showing actinostelic protostele.

Evolutionary trend in Psilophytales

On the basis of morphological and anatomical features, Bank (1968) described three lines of evolution within Psilophytales and these are: Rhyniophytina, Zosterophyllophytina and Trimerophytina.

(a) Rhyniophytina : Which include ***Rhynia, Horneophyton*** and ***Cooksonia***, represent one line of evolution as these plants bear terminal sporangia that dehisced longitudinally.

(b) Zosterophyllophytina : Group of plants represent another line of evolution because sporangia were lateral not terminal, globose or some what reniform and dehisced along the distal edge.

(c) Trimerophytina : ***Trimerophyton*** and ***Psilophyton*** also represent a separate line of evolution where sporangia terminate dichotomously or trichotomously branched lateral axes.

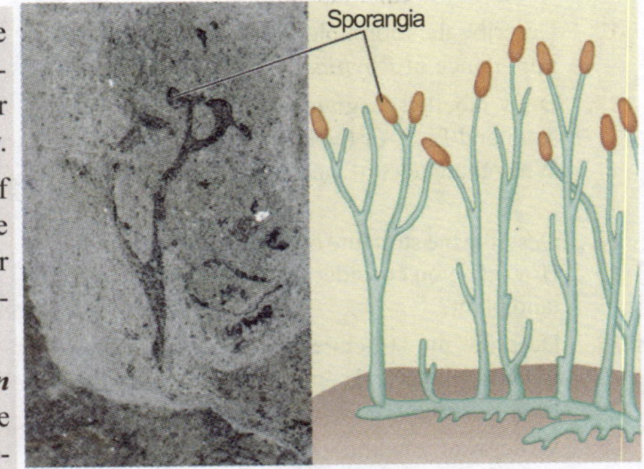

Cooksonia-Terminal sporangia

Banks' perception about evolutionary specialization came after the transfer of ***Asteroxylon*** to Lycopsida was confirmed.

Niklas (1979) reaffirms the above line of evolution in Psilophytales following studies on fossil records and its chemical constituents like sterols, fatty acids, lignin and aromatic acids.

QUESTIONS

ESSAY TYPE

1. Describe the salient features of Psilophytales and describe its phylogenetic significance.
2. Give an account of Psilophytales and discuss how they have cleaned our notions on the phylogeny of vascular plants.
3. Psilophytes were the simplest vascular plants. Discuss.
4. Write a short essay on evolutionary significance of Psilophytes.
5. Describe the morphological and anatomical structure of an adult sporophyte of ***Rhynia.*** **(Awadh 2000)**
6. What is the age and locality of ***Rhynia***? Mention the name of its species and their salient features. **(Allahabad, 2003)**
7. Describe a Devonian vascular plant.
8. How many species of ***Rhynia*** are known? Write their names with the help of suitable sketches, describe their salient features and also mention the systematic position of the genus. **(Lucknow, 2001; Allahabad, 2002)**
9. Mention the age and locality of ***Rhynia*** and point out the similarities and differences between ***R. major*** and ***R. gwynne vaughanii.***
10. Give an illustrated account of the morphological structure of ***Rhynia.***
11. Describe the morphology and anatomy of ***Rhynia.*** How can you consider it as the most primitive vascular land plant.
12. Describe the morphological and phylogenetic importance of ***Rhynia.***
13. Draw labelled diagrams of the T. S. of aerial stem and L. S. of the sporangium of ***Rhynia.*** What characters of land plants does it show? **(Poorvanchal, 2000)**
14. Describe the structure of sporophyte of ***Rhynia.*** How can you consider it as the most primitive land plant?
15. Describe the structure of the earliest known vascular plants and comment briefly upon their fructifications.
16. Write what you know about ***Rhynia.***
17. Write the following about ***Rhynia***
    (*a*) Classification
    (*b*) T. S. aerial stem
    (*c*) L. S. Sporangium
18. Mention the characteristic features of Psilophytales. Write a short account of early vascular plants studied by you.
19. Are Psilophytales starting or resulting points? Discuss. **(Awadh, M.Sc, 2000)**
20. Describe in detail the external morphology of ***Rhynia*** plant and anatomy of its reproductive part with the views on its gametophyte.
21. Describe briefly the morphology and anatomy of the vegetative and reproductive parts of ***Rhynia.***
22. Comment on the common features of Psilophytales and Psilotales. **(Calicut, 2001)**
23. With suitable illustrations, give an account of ***Rhynia***. How does it differ from ***Psilotum*** ?
24. Discuss the evolutionary importance of ***Rhynia.*** Place it in its systematic position.
25. Describe the distribution and structure (both external and internal) of aerial branch of ***Rhynia***.
26. Describe the life history of ***Rhynia.***
27. Give a detailed account of the morphology and anatomy of ***Asteroxylon.***
28. Describe ***Psilophyton*** and ***Pseudosporochnus***, with labelled diagrams.
    or
    Describe a Devonian Vascular plant. Illustrate your answer with suitable diagrams. **(Lucknow, 2005)**
29. Discuss the various genera of ***Psilophytales.*** Give a diagrammatic account of ***Rhynia*** and ***Asteroxylon.***
30. "***Zosterophyllum*** occupies a special position". Discuss the statement.
31. How many genera have been placed in the class Psilophytopsida? With illustrated diagrams, describe two genera in detail.
32. Write an account of ***Rhynia.*** **(Mysore Univ. 2004)**

33. Give an account of reproduction in ***Rhynia.*** **(Nagpur Univ. 2005)**

34. Are ***Rhyniales*** the oldest vascular plants in Psilophytales. Explain. **(M.S. Univ., Tirun, 2005)**

SHORT ANSWER TYPE

35. Write short notes on
    (*i*) ***Rhynia gwynne vaughanii.*** **(Rohilkhand M.Sc. 2000)**
    (*ii*) Sporangium of ***Rhynia.*** **(Allahabad, 2000)**
    (*iii*) L. S. Sporangium of ***Rynia.***
    (*iv*) ***Rhynia*** **(Kumaon 2000; Kanpur, 2000, 2001; Nagpur, 2002)**
    (*v*) ***Rhynia*** Rhizome.
    (*vi*) Morphology and Anatomy of ***Rhynia.***
    (*vii*) ***Gametophyte*** of ***Rhynia.***
    (*viii*) Anatomy of ***Rhynia*** stem.
    (*ix*) ***Rhynia*** as the connecting link between Bryophyta and Pteridophyta.
    (*x*) Phylogenetic importance of ***Rhynia.*** **(Poorvanchal, 2000)**
    (*xi*) Conclusions derived from the morphology of ***Rhynia*** Sp. **(U.P. College, 2000)**
    (*xii*) Morphology of ***Rhynia.***
    (*xiii*) Structure of ***Rhynia*** plant an d its sporangia.
    (*xiv*) Structure of sporophyte in ***Rhynia.***
    (*xv*) Plant body of ***Rhynia.***
    (*xvi*) Sporophyte of ***Rhynia.*** **(Gauhati, 2000)**
    (*xvii*) Morphology of aerial shoot of ***Rhynia.***
    (*xviii*) ***Horneophyton.***
    (*xix*) ***Zosterophyllum.***
    (*xx*) ***Psilophyton.***
    (*xxi*) ***Yarravia.***
    (*xxii*) ***Pseudosporochnus.***
    (*xxiii*) ***Asteroxylon.***

36. Mention two features of Psilophyta which are not found in other pteridophytes.

37. Mention the age and locality of ***Rhynia.***

38. When, where and by whom was ***Rhynia*** discovered ? **(Allahabad, 2000; U.P. College, 2000)**

39. Name the extinct plant studied by you which is devoid of true roots and leaves.

40. Describe the rhizoid of ***Rhynia*** in not more thirty words.

41. ***Rhynia*** is a pteridophyte. Write two reasons for it. **(Lucknow, 2001)**

42. In which period ***Rhynia*** plant was found.

43. Is ***Rhynia major*** a vascular plant ?

44. ***Rhynia*** has both sporophytic and gametophytic plant. Answer briefly.

45. In which type of fossil ***Rhynia*** is found. **(Allahabad, 2003)**

46. (*a*) What are the reasons for regarding ***Rhynia gwynne Vaughanii*** as gametophyte and sporophyte both? Discuss.
    (*b*) What recent changes have been suggested in plant of ***Rhynia major***? Discuss.

47. What is the new name of ***Rhynia major?*** **(Allahabad, 2001)**

48. Write systematic position of ***Rhynia.***

49. Explain as to why ***Rhynia gwynne Vaughanii*** is a gametophyte.

50. Give labelled diagrams of the following:
    (*i*) Fertile branch of ***Rhynia.***
    (*ii*) Sporangium of ***Rhynia.***
    (*iii*) T. S. ***Rhynia*** stem.
    (*iv*) Morphology (external) of ***Rhynia.***
    (*v*) L.S. Aerial shoot of ***Rhynia.*** **(Allahabad, 2004)**
    (*vi*) ***Rhynia.***
    (*vii*) Internal structure of the stem in ***Asteroxylon.***
    (*viii*) Reconstruction of ***Zosterophyllum.***
    (*ix*) T. S. Stem of ***Asteroxylon.***
    (*x*) Structure of Sporangium in ***Horneophyton.***
    (*xi*) Reconstruction of ***Psilophyton.***
    (*xii*) ***Pseudosporochnus***–restoration of entire plant.
    (*xiii*) Reconstruction of ***Asteroxylon.***

51. (*a*) Name two species of ***Rhynia.***
    (*b*) Comment that ***Rhynia*** is a most primitive vascular plant.

52. Why ***Rhynia*** is considered as a connecting link between bryophytes and pteridophytes.

53. Is ***Rhynia*** primitive?

54. Describe the importance of ***Rhynia*** in evolution.

55. Describe the important characters of ***Zosterophyllum.***

56. Give a brief account of ***Horneophyton.***

57. How many genera have been placed in the family Rhyniaceae? Name the important genera.

55. Describe the important characters of ***Psilophyton.***

58. Give a brief account of the family Pseudosporochnaceae.

59. Give a brief account of the internal structure of ***Asteroxylon.***

60. Classify the following genera to a recognised system of classification giving suitable reasons.

(*a*) ***Rhynia.***

(*b*) ***Asteroxylon.*** **(Awadh, M.Sc. 2001)**

(*c*) ***Horneophyton.*** **(Awadh, M.Sc. 2001)**

(*d*) ***Zosterophyllum.*** **(Awadh, M.Sc. 2001)**

(*e*) ***Psilophyton.***

(*f*) ***Pseudosporochnus.***

61. Which plant have you studied belonging to middle Devonian? **(Kanpur, 2000)**

62. Name the geological period of ***Rhynia.***

63. Differentiate between.

(*a*) ***Rhynia*** and ***Psilotum.***

(*b*) ***Rhynia major*** and ***R. gwynne vaughanii.***

(*c*) ***Rhynia*** and ***Asteroxylon.***

(*d*) ***Psilophyton*** and ***Zosterophyllum.***

64. Name the fossil genera discovered from Rhynie chest bed.

65. Comment whether

(*a*) ***Rhynia gwynne vaughani*** is a sporophyte.

(*b*) ***Rhynia gwynne vaughani*** is a gametophyte.

(*b*) Mention two characters of ***Rhynia*** which are not found in pteridophytes studied by you. **(Kanpur, 2004 ; Allahabad, 2004)**

66. Mention the characters of ***Rhynia*** which are not found in pteridophytes studied by you. **(Kanpur, 2004 ; Allahabad, 2004)**

OBJECTIVE TYPE

67. Fill in the blanks :

(*i*) ***Rhynia*** was discovered by ------------------------- for the first time.

(*ii*) A Devonian vascular plant is named as -------------------------.

(*iii*) Fossil of ***Rhynia major*** was discovered by ------------------------.

(*iv*) ***Rhynia major*** is found in ----------------.

(*v*) ***Rhynia*** was discovered from ------------------------- **(Kanpur, 2001)**

(*vi*) ***Rhynia*** is a -------------------------- pteridophyte.

(*vii*) ***Zosterophyllum rheananum*** is an ------------------------- fossil.

(*viii*) Leafy stems and star shaped xylem is found in the genus --------------------.

(*ix*) A well defined tapetum and definite columella is present in the sporangium of -------------------------.

(*x*) ***Psilophyton*** was discovered from the ------------------------- Rocks.

68. Select the correct answer :

(*i*) ***Rhynia*** was discovered by

(*a*) Arnold

(*b*) Kidston and Lang

(*c*) Campbell

(*d*) Birbal Sahni.

(*ii*) ***Rhynia*** was discovered from

(*a*) India (*b*) England

(*c*) Scotland (*d*) China.

(*iii*) ***Rhynia*** was first discovered from.

(*a*) India (*b*) America

(*c*) China (*d*) Holland.

(*iv*) Which one of the following is the fossil pteridophyte of the middle Devonian period?

(*a*) ***Lycopodium*** (*b*) ***Rhynia***

(*c*) ***Selaginella*** (*d*) ***Lygodium.***

(*v*) Protostelic stem is present in

(*a*) ***Selaginella*** (*b*) ***Equisetum***

(*c*) ***Pteris*** (*d*) ***Rhynia.***

(*vi*) In ***Horneophyton*** the sporangia are borne terminally on specified branch and have

(*a*) Spores without columella.

(*b*) Spores, tapetum and columella

(*c*) tapetum and columella but no spores.

(*d*) Spores and columella but no-tapetum.

(*vii*) ***Zosterophyllum myretonianum*** was discovered from.

(*a*) Red sandstone of scotland

(*b*) Rhynie chest bed

(*c*) Rajmahal hills of India

(*d*) Russia.

(*viii*) Which of the following is an aquatic plant reported from the lower Devonian of Germany ?

(*a*) ***Zosterophyllum rhenanum***

(*b*) ***Z. australtianum***
(*c*) ***Z. myretonianum***
(*d*) ***Pectinophyton bipectinatum.***

(*ix*) ***Psilophyton*** was discovered by
(*a*) A. C. Seward
(*b*) J. W. Dawson.
(*c*) C. A. Arnold
(*d*) Birbal Sahni.

(*x*) In the stem of ***Asteroxylon***, the xylem is
(*a*) Simple solid cylinder
(*b*) Trabeculate
(*c*) Deeplyfluted and appears starlike in structure
(*d*) Protostelic.

(*xi*) * ***Rhynia*** belongs to a geological period known as
(*a*) Cretaceous
(*b*) Palaecozoic
(*c*) Devonian
(*d*) Coenozoic.

(*xii*) The first fossil of ***Rhynia*** was found in
(*a*) England
(*b*) Scotland
(*c*) Netherland
(*d*) New Zealand

(*xiii*) ***Rhynia*** lacks
(*a*) Root
(*b*) Rhizoids
(*c*) Rhizome
(*d*) Sporangia

(*xiv*) ***Aglaeophyton major*** was previously known as
(*a*) ***Psilophyton sp***
(*b*) ***Rhynia major***
(*c*) ***Rhynia gwynne-vaughani***
(*d*) None

* It has been discussed in the previous pages that four plants were originally described from Rhynie Chertbet by Kidston and Lang (1917). These were : ***Rhynia major*, *R. gwynnevaughani*, *Horneophyton*** and ***Asteroxylon.*** However, a reinvestigation by two German palaeobotanists, Winnifred and Renate Renny in 1980's was conducted on Rhynie Chert bed. Further studies on *R. major* revealed that these do not have similar tracheids as in modern plants, as claimed earlier. Hence, it was thought proper to rename ***Rhynia major*** as ***Aglaophyton major.***

Psilotum nudum

CHAPTER 3

Division–Psilophyta Class–Psilotopsida

- Introduction
- Distinguishing features
- Description of ***Psilotum***/Swartz
- Description of ***Tmesipteris***/Berhardi
- Difference between ***Psilotum*** and ***Tmesipteris***
- Affinities of Psilotales
- Significance and Phylogeny of Psilophyta

Introduction

The order Psilotales is closely related to the extinct order Psilophytales and comprises two genera. These are ***Psilotum*** (2 species) and ***Tmesipteris*** (10 species). Both these genera are placed in family ***Psilotaceae***. It has now been proposed to refer ***Tmesipteris*** to a separate family ***Tmesipteridaceae.*** Bierhorst (1971) has raised the status of ***Psilotaceae*** and has placed it in the class ***Filicopsida*** somewhere near the family ***Stromatopteridaceae.*** Lovis (1977) accepted Bierhorst's contention in considering the

Psilotales as ferns and treated them as first order of 'ancient ferns' followed by ***Ophioglossales, Marattiales, Osmundales***, etc., but far apart from ***Stromatopteridaceae.*** He opposes its relationship with the ***Rhyniales.*** Pichi Sermolli (1977), Wagner (1977), Ching (1978a) and Love ***et al*** (1977) treated Psilotales as a separate group outside the Filicopsida. In this text both these genera are included in the family ***Psilotaceae.***

Distinguishing Features

1. Plant body (sporophyte) distinguished in two parts: the dichotomously branched rhizomes and aerial branches.
2. Rhizome bears mycorrhizal association with fungal hyphae.
3. Leaf-like or scale like lateral appendages are spirally arranged on aerial branches.
4. Stele is either solid protostele or medullated protostele.
5. Plant is homosporous, sporangia thick walled, terminating short lateral branches.
6. Antherozoids are multiflagellate.

PSILOTUM/SWARTZ

Distribution

It grows as an epiphyte in tropical and subtropical regions of the world and is abundant in Florida, Bermuda, Jamaica, Mexico, a few Pacific Islands and Hawaii islands. In India ***P. nudum*** (L.) Beauv. has been collected from Bengal, Assam, Pachmarhi hills, Kulu (Himachal Pradesh) and Madhya Pradesh. At Kulu it has been found growing from the crevices of rocks and on soil. It is quite common in South India in Andhra Pradesh, Tamil Nadu (Madras), Maharashtra (Bombay) and Kerala. ***Psilotum nudum*** (= ***P. triquetrum***) and ***P. flaccidum*** (= ***P. complanatum***) are the only species discovered so far. The terrestrial forms of ***P. nudum*** grow erect and are dwarf; whereas the epiphytic forms grow pendent (75-100 cm. high). ***P. flaccidum*** differs from ***P. nudum*** in possessing flattened stems.

P. triquetrum

Plant Body

The plant body is distinguished into two parts (Fig. 3.1) :—**(a) Rhizome** which is mostly hidden below the humus (epiphytic species) or the soil and is dichotomously branched and bears scales. The branching of the rhizome may, in some cases, be lateral or even irregular. This variation in the branching pattern is considered to be due to injury or variations in the meristematic activity. The rhizome lacks roots, harbours an endophytic

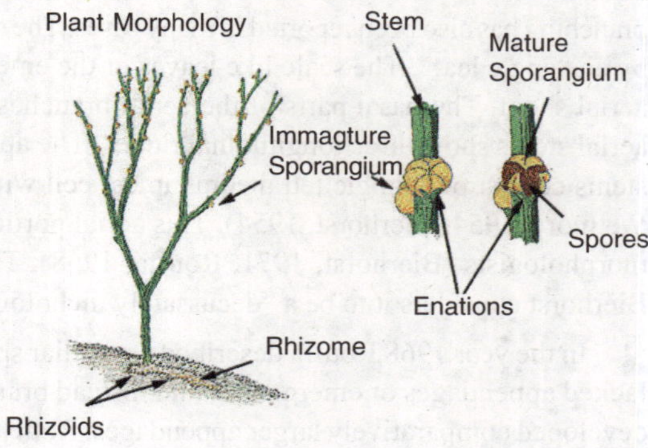

Fig. 3.1 (a) Plant body of *Psilotum nudum.*

fungus and bears numerous filamentous rhizoids that perform the twin function of anchorage and absorption. The endophytic fungus within the cortical cells sends out hyphae that come out through the rhizoids and probably help in the function of absorption. The rhizoids are one to three cells in height and are covered with cuticle. They arise from the epidermal cells of the rhizome. The epidermal cell gives out a lateral protuberance which grows out into a 1-3 celled rhizoid that is separated by a cross wall from the epidermal cell. Bierhost (1954) is of the opinion that branching of the rhizome is dependent upon the type of obstacles encountered by its growing tips. **(b) An erect or pendent green aerial portion**, which arises from the rhizome. The tip of any branch of rhizome turns upwards and by apical growth grows into an aerial shoot. Light is considered to be the causative factor in inducing the rhizome tip to grow into aerial branches, but sometimes tips that are deep seated in the soil also grow up and develop into an aerial branch. If we bury the growing tip of the aerial shoot again in soil it will develop into rhizome rather than giving rise to the aerial shoot. Experimental studies have revealed that after burying the aerial apex swells up and numerous rhizoids arise from the aerial portions already developed, the formation of new scale leaves stops, and the buried part grows into a rhizome. The young aerial shoots can be made to grow under aseptic cultural conditions into a fungus free rhizome. The aerial branch is green in colour and dichotomously branched in three dimensions. It is the main photosynthetic organ because the leaves are mere emergences without any vasculature and are spirally arranged on the aerial stem. Trichotomous branching has also been reported in many cases. The dichotomies and trichotomies are always subtended by an "angle leaf". The scale like leaves or the emergences are inserted along the three ridges of the aerial shoot. The basal parts of the aerial branches are distinctly ribbed. Near the upper region the aerial stems show three longitudinal ridges. The apical meristems of both the rhizome and the aerial stems consist of a single tetrahedral apical cell with three cutting faces. (Ford, 1904; Marsden and Wetmore, 1954; Bierhorst, 1954). This aerial portion is regarded as a frond or leaf by some modern morphologists (Bierhorst, 1971, Rouffa, 1968). The leaf-like appendages are regarded as pinnae. Bierhorst considers it to be a "decussately dichotomous frond with spirally arranged distal pinnae."

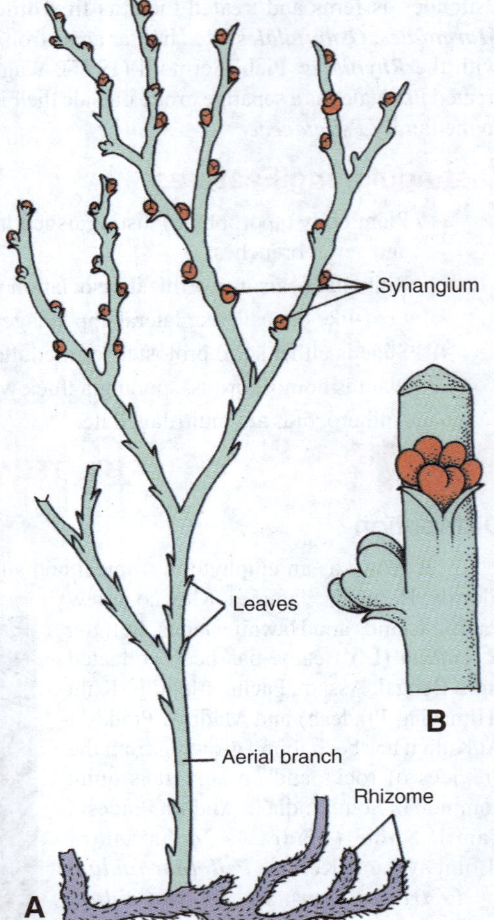

Fig. 3.1 (A–B). *Psilotum nudum*. A. A complete plant showing Synangia; B. Part of stem bearing synangia. (B. After Pritzel)

In the year 1968 Rouffa described a peculiar specimen of ***P. nudum*** in which the aerial branches lacked appendages or emergences and instead branched in a lateral fashion. These lateral branches developed comparatively larger appendages. Roth (1963 a) described pseudodichotomy in ***P. nudum***.

Anatomy of Stem

The outermost boundary of stem is formed by a single layered epidermis with thick outer walls. The epidermis is lined by a layer called cuticle which is made up of cutin and waxes. It is secreted by the epidermal cells and extends into the stomatal pore. The stomata have no subsidiary cells and have two prominent outer ledges (Fig. 3.3). They are restricted to the grooves between longitudinal ribs.

Cortex

It is differentiated into three well marked regions (Fig. 3.2, A).

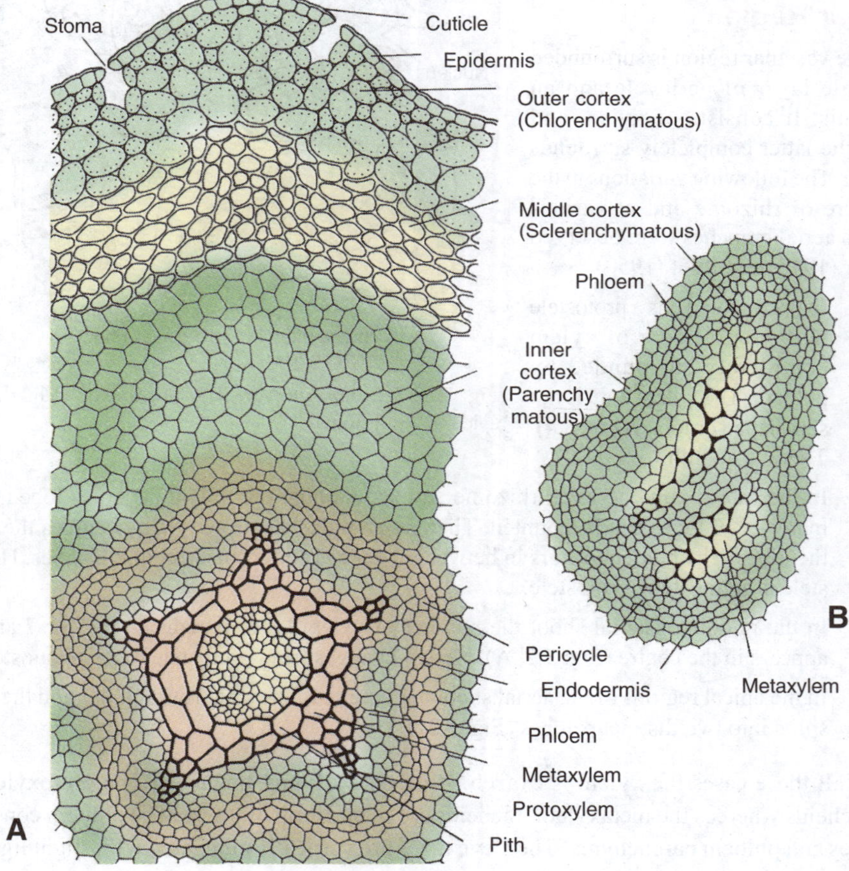

Fig. 3.2. *Psilotum nudum.* A. T.S. part of aerial stem near the upper region showing actinostele; B. Portion of T.S. of stele of aerial stem cut near the apex.

1. The outer cortex which is composed of thin-walled slightly lobed, loosely arranged and chlorenchymatous, elongated cells. The cells may contain starch grains.
2. The middle cortex is composed of vertically elongated chlorophylless cells that have thickened cell walls and enclose comparatively smaller intercellular spaces. They contain few or no starch grains. In the basal portions of the aerial stems this region is composed of lignified cells.

3. The inner cortex is composed of several layers of cells which show gradual reduction of the thickness of cell walls as we proceed towards the stelar region. The cells contain starch grains. This region is also called the storage region.

The cortex becomes entirely chlorenchymatous in the younger portions of the aerial axes.

The endodermal cells have distinct casparian strips. The pericycle is single layered and consists of thin-walled cells.

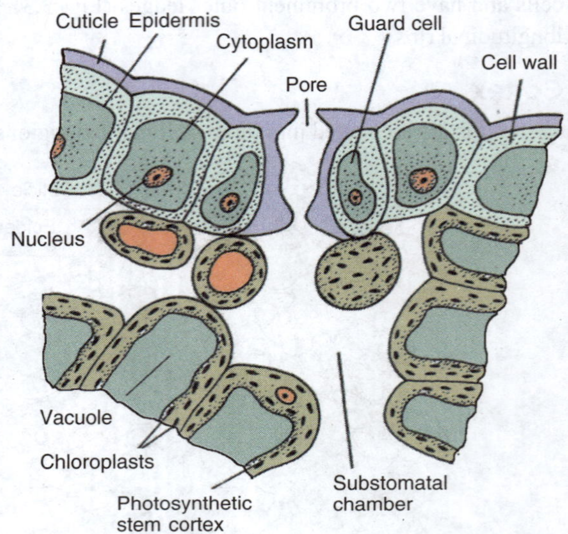

Fig. 3.3. *Psilotum nudum*. T.S. portion of stem showing epidermis and stoma.

Vascular Region

The vascular region is surrounded by a single layer of pericycle and an endodermis. It consists of xylem and phloem; the latter completely surrounds the former. The following variations in the vasculature of rhizome and at various levels of a aerial branches were observed by Bower (1935) and Pilot (1950) :

(*i*) In the rhizome it is a protostele with central core of xylem completely surrounded by phloem. The xylem core is not starshaped or lobed (Fig. 3.4). There is no pith.

(*ii*) In the transit zone between rhizome and aerial shoot the xylem becomes lobed and as many as 10 lobes can be counted. There is no pith. The protoxylem occupies the tips of the lobes. The phloem occurs in between the lobes and forms irregular patches. Here the stele is actinostelic protostele.

(*iii*) In the middle of aerial shoot the number of xylem lobes is reduced to 5 or 7 and pith appears in the centre (Fig. 3.2, A). Here it changes from protostele to a siphonostele.

(*iv*) In the apical regions of the aerial shoot the outline of the stele is triangular and the xylem splits into two distinct patches (Fig. 3.2, B).

In all these cases the xylem is **exarch**. It consists of only tracheids. The protoxylem has spiral tracheids whereas the metaxylem tracheids are scalariform or pitted. The phloem consists of sieve tubes and phloem parenchyma. The sieve tubes are vertically elongated with slight lignification at the angles and have sieve plates in both lateral and terminal walls. They lack nuclei at maturity, but some spherical bodies have been reported in the cells (Ford, 1904; Stiles, 1910, Essau, 1953).

In the region of the rhizome a transverse section reveals a well defined epidermis (Fig. 3.4, A), followed by a broad zone of cortex. The cortex has an outer mycorrhizic zone (Fig. 3.4, A), middle zone of starch storing cells and an inner cortex whose cells contain phlobaphene or tannin anhydrides. There is a distinct endodermis surrounding a single layered pericycle which in turn encircles the protostele (Fig. 3.4, A). The protostele has a central core of exarch xylem surrounded completely by phloem.

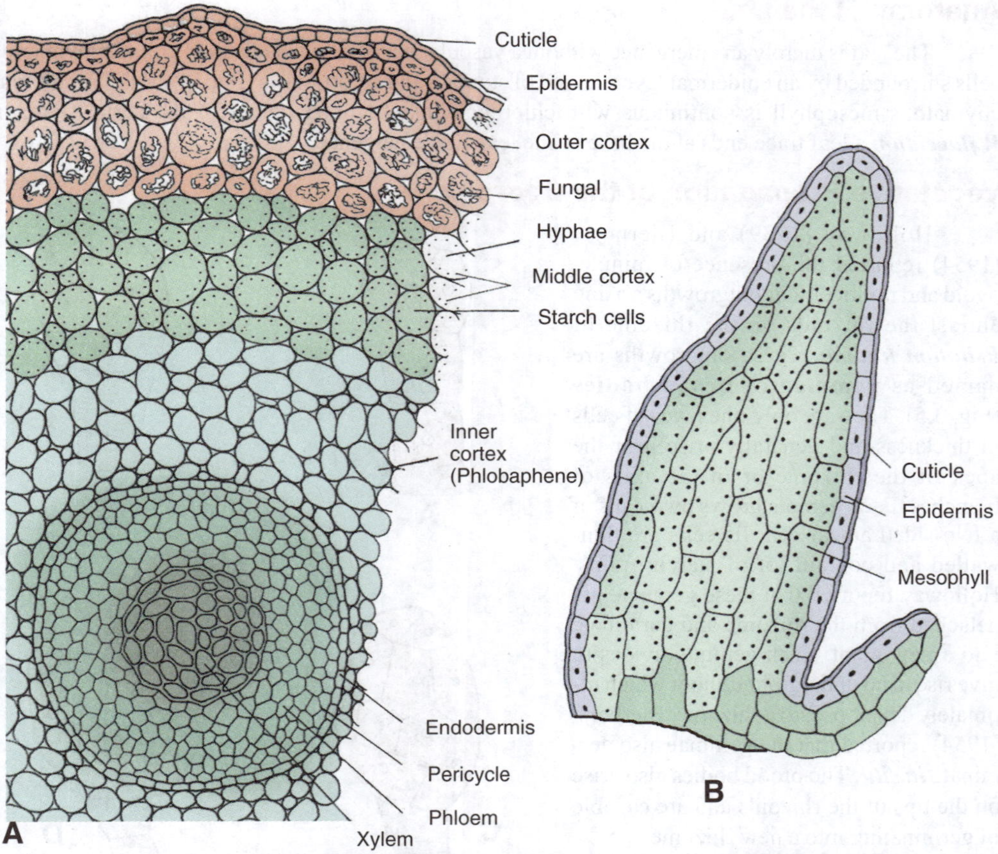

Fig. 3.4. ***Psilotum nudum.*** **A.** T.S. portion of rhizome; **B.** L.S. leaf.

Styles (1910) reported secondary growth in ***Psilotum flaccidum.*** Boodle, L.A. (1904) also reported the presence of secondary xylem in the rhizome of ***Psilotum*** (Fig. 3.4_a C). He observed that in some parts of the parenchymatous zone surrounding the scalariform tracheids may become meristematic and produce secondary xylem tracheids (x^2). He did not report the presence of any definite cambium, but reported that the radially disposed tracheids and the adjacent parenchyma elements clearly indicate (See Fig. 3.4_a C) the secondary nature of the tissue external to the group of primary xylem (x^1).

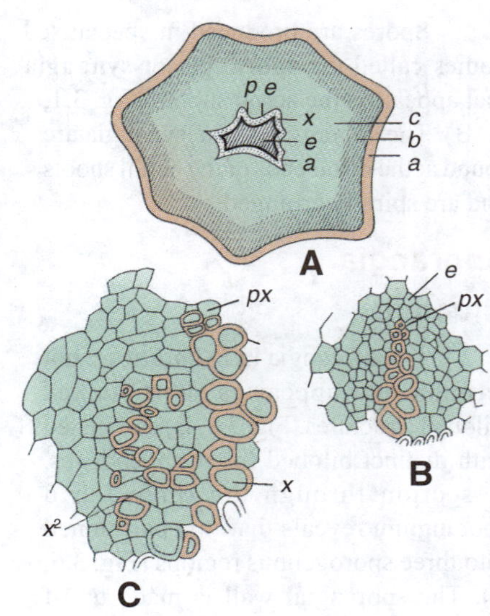

Fig. 3.4a. (A-C). A. Diagram of transverse section of aerial shoot of ***Psilotum triquetrum.*** *a–c*, cortex; *p*, phloem; *e*, endodermis; *s*, stereome; *x*, xylem; *a*, gap in xylem; **B.** Enlarged view of one of the angles of the xylem shown in A. *px*, protoxylem; **C.** Part of transverse section of an approximately triangular rhizome stele showing a portion of the metaxylem x^1; *px*, protoxylem elements; x^2, secondary xylem. (Alter Boodle 1904)

Anatomy of the Leaf

The leaf is merely an emergence without a vascular supply. It is made up of chlorenchymatous cells surrounded by an epidermal layer devoid of stomata and chloroplasts (Fig. 3.4, B). The chlorenchymatous mesophyll is continuous with chlorenchymatous outer cortex of the aerial stem. In *P. flaccidum* a leaf trace ends at the base of the leaf-like emergence.

Vegetative Propagation of the Sporophyte

Holloway (1939) and Bierhorst (1954) reported the presence of minute, ovoid and multicellular outgrowths arising amidst the rhizoids on the rhizome of ***Psilotum nudum.*** These outgrowths are named as **gemmae** or **brood bodies** (Fig. 3.5). They are only one layer of cells in thickness and generally arise near the apex of the rhizome or in the axils of branches. Each gemma grows by means of a two-sided apical cell. Its cells are thin-walled and contain lot of starch grains. Holloway reported that these gemmae get detached from the rhizome and germinate into a new plant. Each gemma divides to give rise to an underground shoot which ultimately develops into a rhizome. Bierhorst (1954) reported that the gemmae also germinate ***in situ.*** The brood bodies also arise on the tips of the rhizoids and are capable of germinating into a new rhizome.

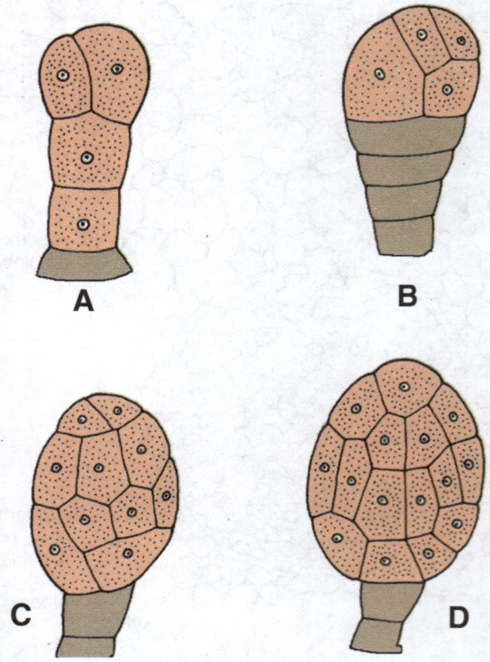

Fig. 3.5. Gemmae or Brood bodies from the rhizomes of *Psilotum nudum.*

Reproduction by Spores

Spores are produced in specialised bodies called the **sporangia** or **synangia** that appear on the aerial shoots (Fig. 3.1, A, B). The sporangia or the synangia are found at the distal ends of the aerial shoots and are spirally arranged.

Sporangia

Structure

The sporangia in ***Psilotum*** are not terminal but appear as short-stalked trilobed structures (Fig. 3.1, B) associated with distinct bilobed foliar appendages. A section through the three lobed sporangium reveals that it is partitioned into three sporogenous regions (Fig. 3.6, E). The sporangial wall is made up of three or four layers of cells (Fig. 3.6, E).

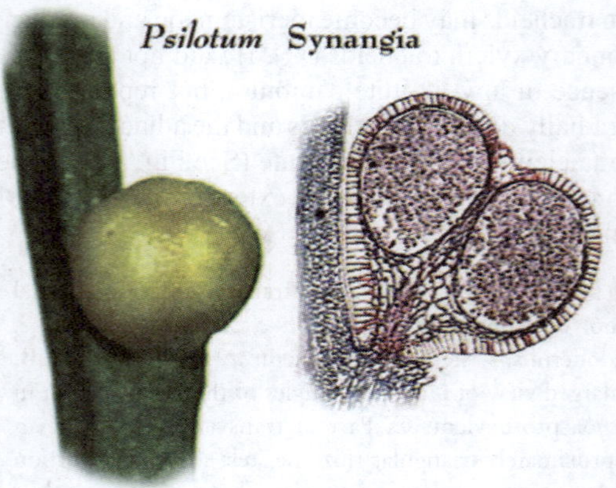

Fig. 3.5 (a) Synangia of *Psilotum.*

In some cases the wall is made up of even six layers (Bierhorst, 1971). At maturity the outer wall layer develops thickening material. Its cells became longer than broad except at the point of line of dehiscence (Fig. 3.6_a G). A **nonfunctional annulus** is also discernible in the form of a patch of thick walled cells at one end of the line of dehiscence. It is positioned on the apex of the sporangia. The inner layers of the sporangial wall extend inwards to form septa of the trilocular sporangium (Fig. 3.6, E). Each of the three chambers contains a sporogenous tissue. Some of the cells of this tissue become **spore mother cells** (Fig. 3.6, E) whereas the rest disintegrate and serve as a nourishing fluid. The spore mother cells undergo meiosis and form tetrads of haploid spores.

The fertile sporangial apparatus of ***Psilotum*** is supplied by a single trace from the stele of the aerial shoot. It enters the base of the synangium without giving off any branches to the bilobed foliar appendages. Within the sporangial base it divides into three parts and later fades out.

Dehiscence

The sporangium dehisces along three apical lines (Fig. 3.7, A) passing through the median line (Fig. 3.6_a G) of each sporangial lobe (loculicidal).

Development

The development of the sporangium is of the **eusporangiate** type. According to Bower (1935), the sporangium arises as an outgrowth on the adaxial side of the bilobed appendage (Fig. 3.6, A). Bierhorst

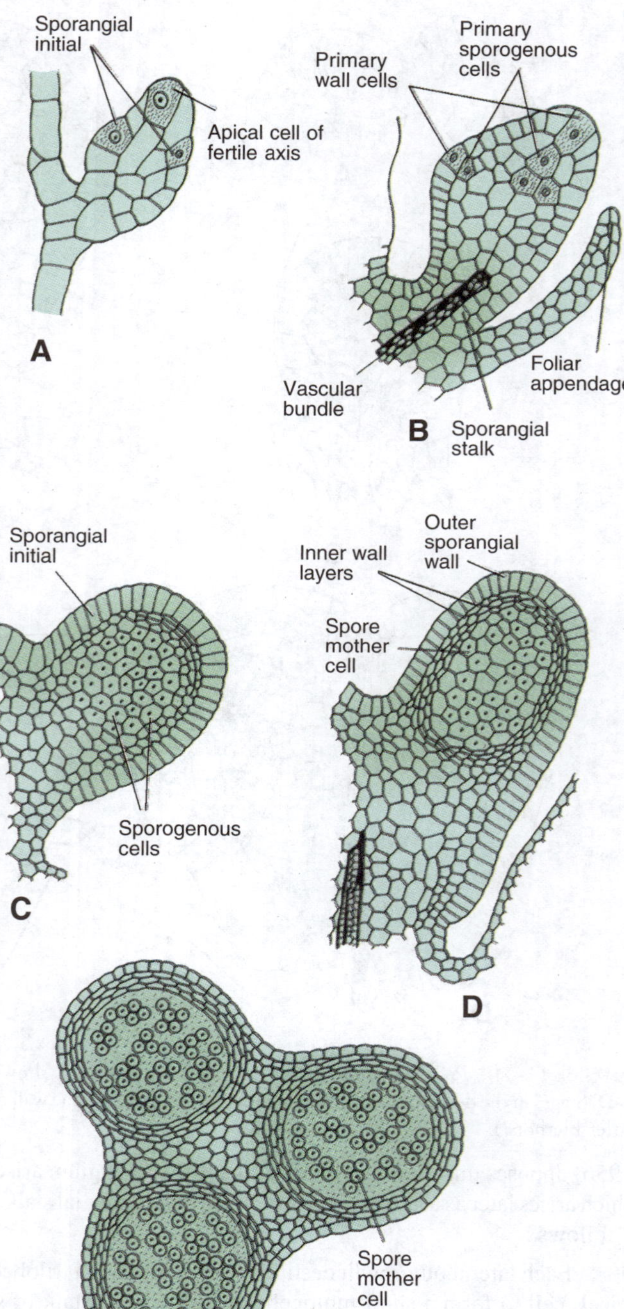

Fig. 3.6 (A–E). ***Psilotum nudum.*** A-D. Stages in development of sporangium; E. Mature synangium showing three loculi. (After Bower).

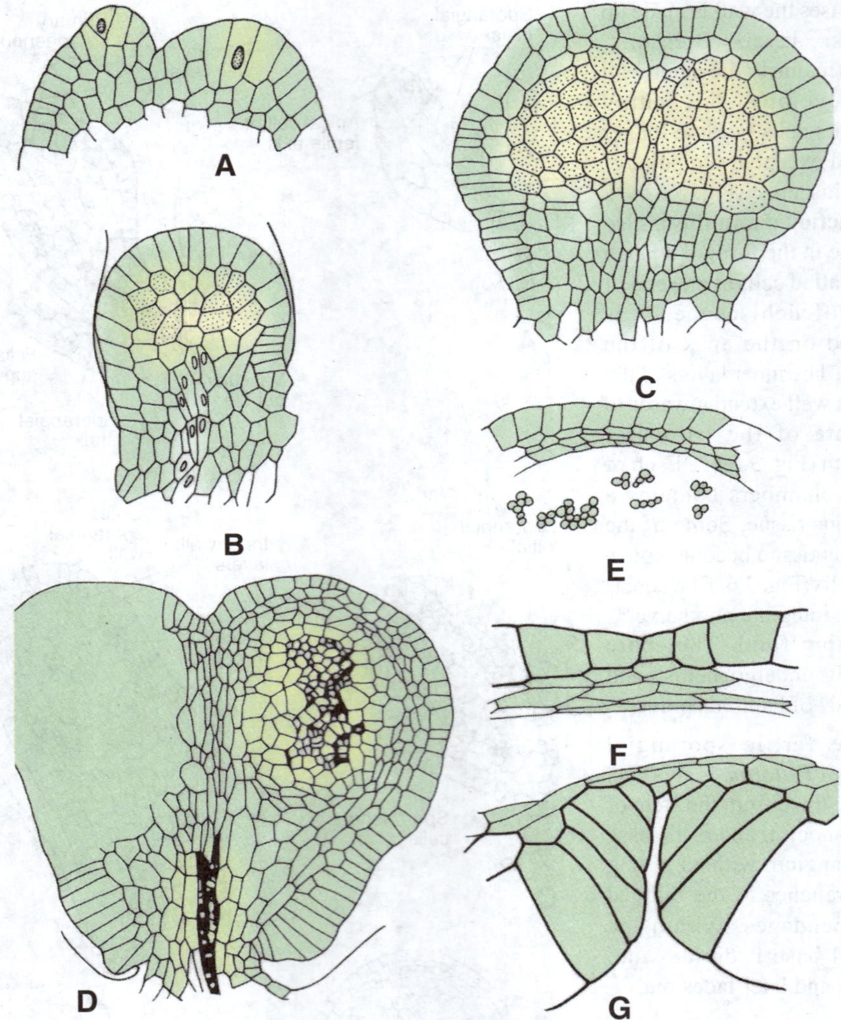

Fig. 3.6a (A–D). ***Psilotum nudum.*** A. L.S. Aerial shoot apex showing a fertile appendage towards its left; B–D. Stages in the development of Synangium ; E. T.S. Through the wall of sporangium; G. Shows line of dehiscence. (After Bierhorst).

(1956) opposes this view and believes that the sporangium arises earlier than the bilobed appendage which arises later as a lateral outgrowth from the sporangial stalk. Bierhorst describes the development as follows :

Each lateral outgrowth destined to develop into a trilobed sporangium, grows by means of an apical cell to form a short multicellular fertile axis (stalk of sporangium). Three separate surface initials make their appearance at the summit of the fertile axis and indicate the sites of three sporangial locules (chambers). Each initial divides periclinally to form an outer primary wall initial and an inner primary sporogenous cell (Fig. 3.6, B). The former undergoes periclinal and anticlinal divisions to form a 3 to 4 layered thick wall (Fig. 3.6, D). The primary sporogenous cell divides to form a mass of sporogenous cells, some of which later on develop into spore mother cells. The development of sporangium is, therefore, eusporangiate.

The sporangial stalk and the partition walls between the sporangial locules are traversed by a vascular strand, which does not extend into the foliar appendages.

Morphology of the Sporangium

Three divergent views have been advanced to interpret the spore producing structures in ***Psilotum***. According to one view the trilobed sporangium is regarded as a synangium, ***i.e.***, it is considered to be a fusion product of three sporangia (Bierhorst, 1956). The second view regards the forked appendage as the sporophyll bearing a trilocular sporangium. This view was put forth by Solms-Laubach (1884) and was supported by Bower (1908), Schoute (1938), Seward (1910) and Velonovsky (1910). The third view regards the sporangium as occupying a terminal position on a short lateral branch. According to this view the sporangium is trilocular and cauline in origin with a terminal position. This view was advanced by Juranyi (1871) and was supported by Sachs, Strasburger and Goebel. Eichler also supported this view.

Presence of a vascular strand in the sporangium supports its being axial in position.

Some regard the sporangial stalk as comparable to the sporangiophore in the Equisetales.

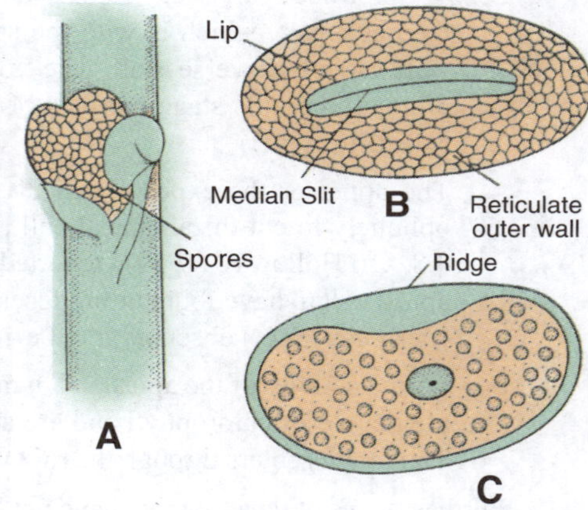

Fig. 3.7 (A–C). *Psilotum nudum*. **A.** Showing a dehiscing sporangium; **B.** A spore; **C.** A spore seen in optical section. (A and B after Wettstein, C after Darnell Smith).

Bierhorst (1956) regards that the foliar appendage arising in connection with the sporangium is a lateral outgrowth of the sporangial stalk. He regards the sporangial stalk as the fertile axis bearing the sporangium and the leaf-like appendage. Bierhorst regards this fertile axis as representing a reduced structure derived from a more elaborate branch system. He observed that the spore producing tissue or the archesporium originates as three separate archesporia derived separately from the young sporangial outgrowth. He also observed that a separate vascular strand supplies each one of the three locules. These two observations led him to conclude that the trilobed sporangium is actually a synangium formed by the fusion of three sporangia.

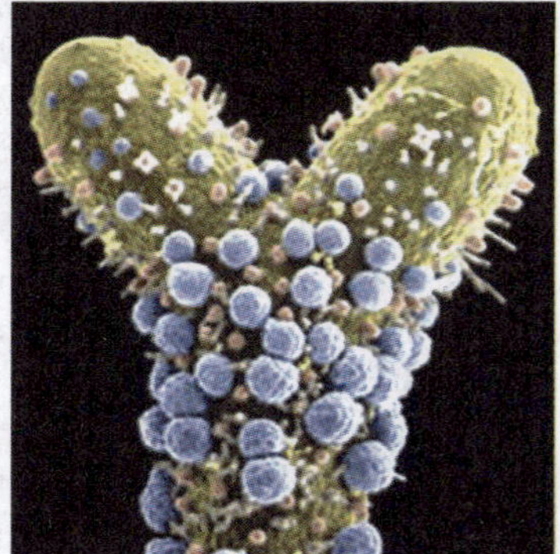

Fig. 3.7 (a) Whisk fern (*Psilotum nudum*), reproductive stage. SEM by Karen Renzaglia. Color enhancement by Steve Mueller, IMAGE.

Gametophyte Generation

It starts with the spore and has been described by Darnell Smith (1917) and Bierhorst (1953) in ***P. nudum.***

The Spore

Psilotum is homosporous. The spore tetrads may be tetrahedral or even isobilateral. The two curved ends of the mature kidney-shaped spores are joined by a narrow **ridge** traversed by a median **slit** that runs along 3/4 of its length (Fig. 3.7, B). The slit is bordered on either side by a thick and smooth border called the lip. Oblong, sickle-shaped and irregular spores have also been recorded in ***P. nudum*** (Sinha and Verma 1977). The spore has an outer thin and reticulate **exine** and an inner **intine.** It is uninucleate, about 0.65—0.32 mm. in size and with granular and foodladen cytoplasm. The exine is **spinulose.** The spore is monolete.

Germination of spore

The spores germinate very slowly and it takes 3 to 4 months for a spore to germinate in artificial cultures. The earlier stages of germination (Fig. 3.8) as studied by Darnell Smith (1917) can be listed as below :

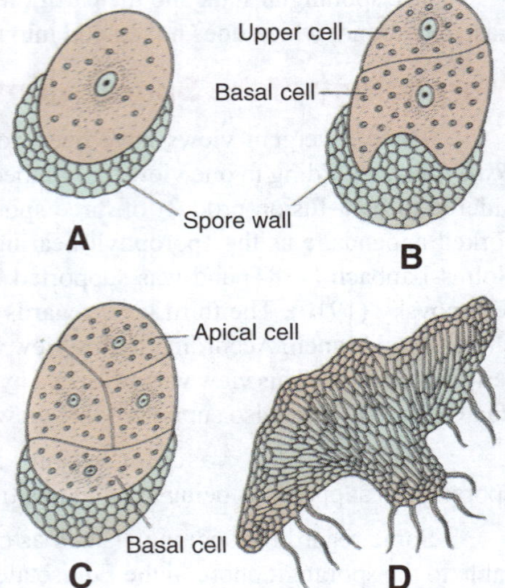

Fig. 3.8. ***Psilotum nudum.*** Earlier stages of spore germination. (After Lawson)

1. The exine ruptures along the median slit.
2. The initine protrudes out as a small globular out-growth which gradually increases in size.
3. The protruded part is later separated from the basal part, which is with the spore wall, by a transverse wall. The young prothallus at this stage consists of two cells (Fig. 3.8 B).
4. The upper or the exposed cell divides obliquely to cut off an apical cell (Fig. 3.8, C). Holloway (1939) reported the apical cell to have four cutting faces but Bierhorst (1953) observed it to be a pyramidal cell with three cutting faces.
5. By the activity of the apical cell a mass of thinwalled cells is formed (Fig. 3.8, D). They are devoid of chlorophyll and are attacked by an endophytic fungus at this early stage. The fungus enters through the rhizoids.

Further stages of development have not been studied.

Mature Prothallus (Fig. 3.9, A, B)

It grows as a saprophyte with an associated endophytic fungus and is a cylindrical (rarely bilaterally symmetrical in some portions) and imperfectly dichotomously branched structure. Usually the size of the gametophyte varies from 1—15 mm. in length and 0.5—1.8 mm in diameter. Under natural conditions they are found growing in the crevices of the rocks or on the tree trunks. They are colourless or yellowish brown in colour. Numerous rhizoids arise from the surface of the gametophyte.

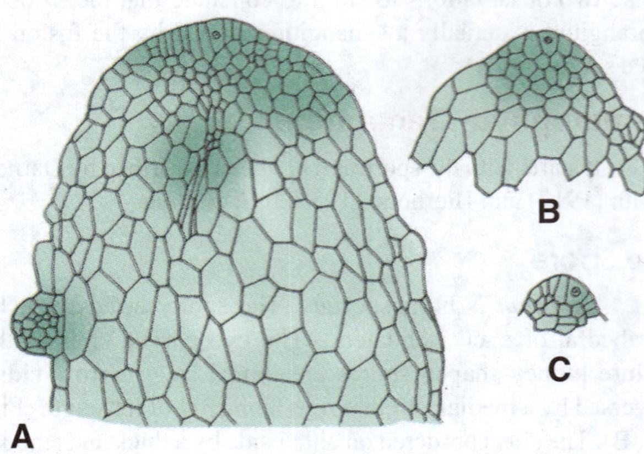

Fig. 3.8 [a]. ***Psilotum nudum.*** A. L.S. of a large, well developed, procambium producing apex of the prothallus; **B, C.** L.S. of smaller apices of the prothallus. (After Bierhorst, 1953)

The prothallus is mostly composed of colourless and hexagonal cells. The cells are strongly cutinized along their outer, radial and also inner corners of their radial walls. The cells on the outer surface of the prothalli are resistant to decay. The outermost layer of cells gives off rhizoids at frequent intervals. The rhizoids are short, filamentous outgrowths that are two or three cells in length. The cells in the rhizoids are also cutinised and are resistant to decay. The growing apices of the prothalli are cospicuous by their nearly white colour. There is no indication that a green colour develops on the gametophytes on exposure to light (Bierhorst, 1953). The branching patterns observed in the gametophytes may be equal or unequal dichotomies. In some cases trichotomous and tetrachotomous branching may also occur (Bierhorst, 1953). The prothalli are monoecious and the sex organs are uniformly distributed over the surface and length of the fully developed specimens. Holloway (1938, 1939) and Bierhorst (1953) have recorded the presence of well defined, scalariform or scalariform-reticulate tracheids, surrounded by phloem and endodermis in the centre of the prothallus. Cells of the vascular strand and cells in its vicinity lack endophytic fungus which is present in all other cells including the sex organs and the rhizoids. Such a vascular strand is a characteristic feature of the sporophyte and their presence in the gametophyte of ***Psilotum*** raised interest in the minds of botanists. Cytological investigations have revealed that the prothalli with such a vascular strand are diploid. Manton (1950) has recorded a wild growing diploid gametophyte from Ceylon. The haploid chromosome number for ***Psilotum*** has been reported to be 50–52 whereas prothalli with 100 chromosomes have also been reported (Bierhorst, 1953) in a race of ***Psilotum***.

Barber (1957) reported prothalli with ***n*** = 93, 95, 104, 102-104 chromosomes. Ninan (1956) and Abraham and Ninan (1965) reported prothalli with n = 104 from South India. Brownlie (1965) reported n = 210 in a specimen from New Caledonia. This indicates that polyploidy is common in ***Psilotum***. Such polyploid gametophytes have also been reported in other pteridophytes (ferns), but the presence of a distinct and well differentiated

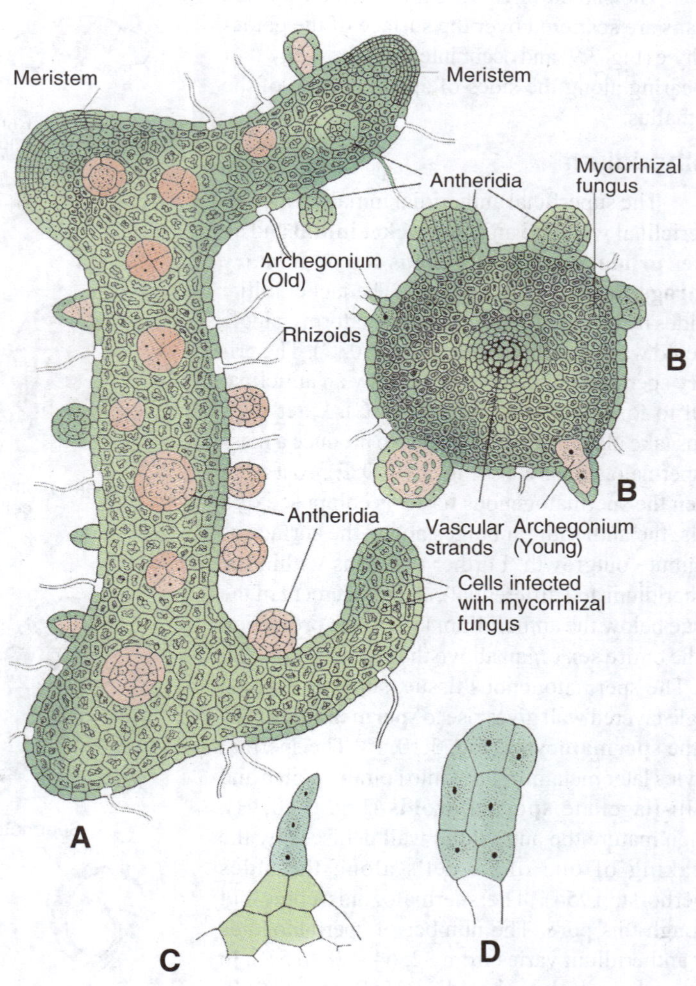

Fig. 3.9. ***Psilotum nudum***. A. Mature gametophyte; B. T.S. mature gametophyte showing sex organs and a central vascular strand; C–D. Gametophyte gemmae.

vascular strand in the prothallus is an unusual character. Holloway (1939) studied the growth of prothallus in ***P. nudum*** and described the occurrence of interrupted vascular system that was essentially a protostele. He found that the protostele was periodically interrupted by intervening regions of parenchyma. Holloway found that growth in these prothalli was intermittent. An active period of apical growth accompanied by the differentiation of vascular tissue was followed by an inactive period of growth when there was no vascularisation. Again a period of active growth followed by differentiation of vascular tissue was there. This resulted in an interrupted vasculature. Bierhorst (1953) atributes this discontinuity to be due to the injury of the apical cell or region by the soil particles through which it grows. The vascular strand originates within the apical region (Fig. 3.8_aA).

Sexual Reproduction

The gametophytes are monoecious and sex organs are scattered over the surface of the gametophyte (Fig. 3.9) and occur intermingled. They start appearing along the sides of apical portion of the prothallus.

Antheridium

The superficial antheridial initial divides by a periclinal wall into an outer **jacket initial** and an **inner primary spermatogenous** cell or **primary androgonial cell** (Fig. 3.10, B). The jacket initial divides by repeated anticlinal walls to form a single layered wall of antheridium (Fig. 3.10, C). The primary spermatogenous cell divides by an anticlinal wall to form two cells (Fig. 3.10, C). Later divisions take place in various planes to produce a mass of spermatogenous cells (Fig. 3.10, D, E). At a stage when the spermatogenous tissue is composed of 8 cells, the antheridium bulges above the surface as a minute outgrowth. Further divisions within the antheridium and differential cell enlargement of the tissue below the antheridium lead to the projection of the entire sex organ above the surface (Fig. 3.9, B). The spermatogenous tissue, surrounded by a single layered wall gives rise to **sperm mother** cells or the **spermatocytes** (Fig. 3.10, F). The spermatocytes later metamorphose into spirally coiled and multi-flagellate **spermatozoids** (Fig. 3.10, H). When mature the antheridial wall dehisces by the breaking of one of its cells along the sides (Bierhorst, 1954). The spermatozoids come out through this pore. The number of spermatozoids per antheridium varies from 32, 64, 128 to 512. In some abnormal antheridia (Holloway 1939, Bierhorst 1954) the cells of antheridial wall divide repeatedly and grow inwards and crush some of the spermatocytes. This results in lessening of the number of spermatozoids. The ratio of presence of such abnormal antheridia is one in ten.

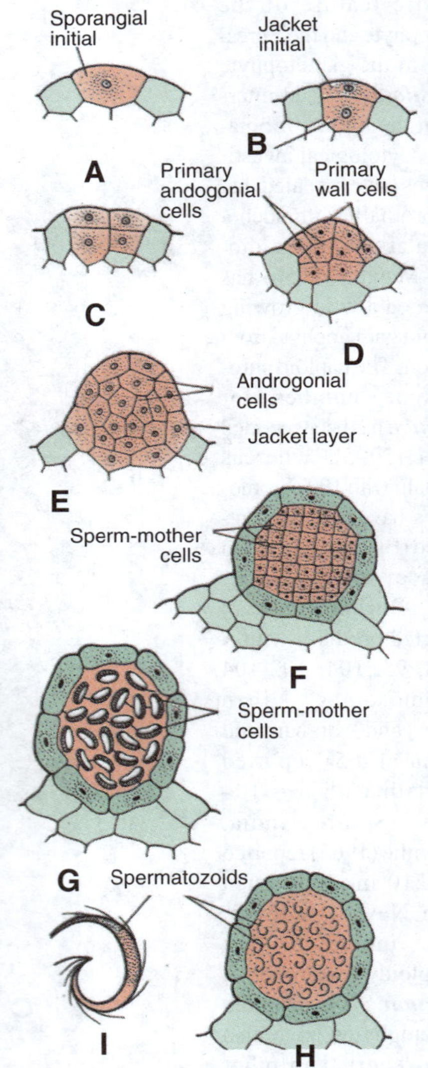

Fig. 3.10. Various stages in the development of antheridium in *Psilotum nudum.* (After Lawson)

The antheridia which develop on the older portions of the prothallus are smaller in size and produce only thirty two spermatozoids.

A mature antheridium (Fig. 3, 10, G, H) is more or less a globular structure without any stalk. It consists of a single layered wall with enclosed spermatocytes.

Archegonium

It develops from a single superficial cell (Fig. 3.11, A) which divides periclinally into an outer **cover cell** and an inner **central cell.** The cover cell **divides** by two intersecting anticlinal walls to form four cells, which act as initials of the four **rows** of the archegonial neck (Fig. 3.11, C). These four cells divide by repeated transverse walls to form four longitudinal rows of the neck, each row from 4—6 cells high. The **central cell** divides (when the neck is only three cells long) by a transverse wall into an upper **primary canal cell** and a lower **primary ventral cell** (Fig. 3.11, D). The former divides by a transverse wall to form two neck canal cells. In most cases the wall separating the two neck canal nuclei has not been observed (Fig. 3.11, E). The **primary ventral cell** divides transversely (Fig. 3.11, F) into an upper **ventral canal cell** and a ***lower*** **egg cell.** Bierhorst (1954) has reported a case where the central cell functions directly as the egg cell.

The necks of the archegonia are straight, and the venter is embedded in the tissue of the gametophyte.

A mature archegonium consists of 4 longitudinal rows of neck cells, each row 4 to 6 cells high. The neck encloses a neck canal (Fig. 3.11, F) with two **neck canal nuclei** or two **neck canal cells.** There is a single **ventral canal cell** and a single **egg cell.** There is no sterile wall surrounding the venter as this portion is embedded in the tissue of the gametophyte.

As the archegonium matures the cell walls between upper tiers of neck cells become cutinised and the upper part of the neck breaks away (Fig. 3.11, G).

Bierhorst (1954) states that this breaking off of the neck takes place after the process of fertilisation. Earlier workers believed it to have taken place before the act of fertilisation.

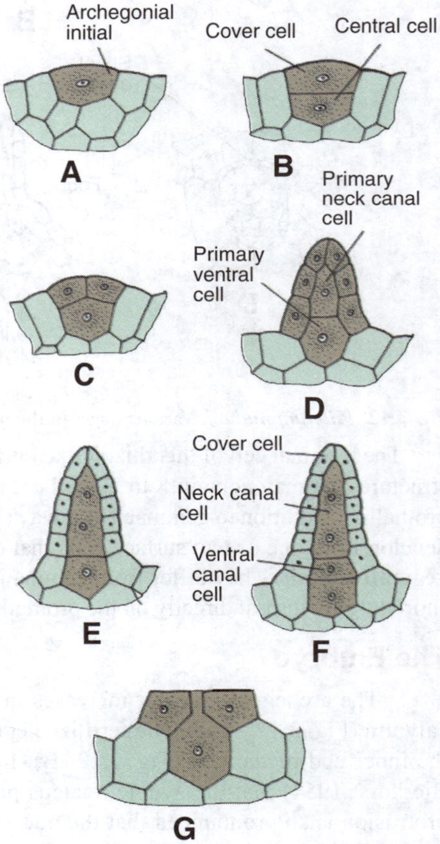

Fig. 3.11. Stages in the development of Archegonium in ***Psilotum nudum.*** (After Bierhorst)

Prior to fertilisation, it appears, the usual disintegration of axial row of cells (neck canal cell and ventral canal cell) takes place. They absorb moisture, become mucilaginous and force open the uppermost tier of neck cells thus permitting the spermatozoids to enter the archegonial neck and reach the egg.

Fertilisation consists in the union of spermatozoid and the egg, followed by the fusion of the male and female nuclei (syngamy). There is every possibility that fertilisation many not even take place in the diploid or the polyploid gametophytes.

Vegetative Propagation of the Gametophyte

Holloway (1939) and Bierhorst (1953) reported the development of gemmae on the surface of the prothallus (Fig. 3.9, C, D). They appear at the tips of short, erect and multicellular outgrowths.

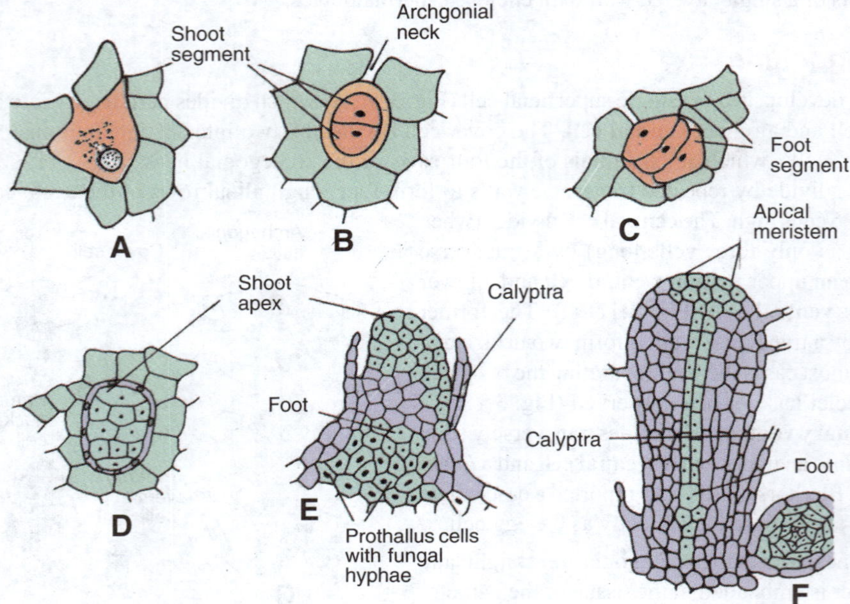

Fig. 3.12. *Psilotum nudum.* Various stages in the development of embryo (A–E after Bierhorst ; F after Holloway).

The terminal cell of this rhizoidlike outgrowth divides repeatedly to form an ovoid gemma-like structure that may contain 8 to 12 cells at maturity. These gemmae germinate to give rise to new prothalli. In addition to gemmae Holloway (1939) reported the formation of small outgrowths which develop from one or two surface cells that divide to form bud-like outgrowths. He named them as **vegetative buds.** They differ from gemma in their mode of origin. They do not arise on the tips of short outgrowths but directly on the prothallus.

The Embryo

The archegonial venter increases in size and the open neck is closed by the formation of calyptra (Fig. 3.12_a A,B). The fertilized egg or the zygote nucleus increases in size and moves to the inner end of the cell (Fig. 3.12_a B). In some cases it has been observed (Holloway, 1917 ; Bierhorst 1954) that the zygote nucleus protrudes into the archegonial neck and sometimes the protrusion is so pronounced that the nucleus appears to be an "attenuated beak." The nucleus is surrounded by a delicate membrane. The zygote divides first by a transverse wall into an **epibasal cell** or **shoot cell** (Fig. 3.12, B and Fig. 3.12_a C) towards the neck of the archegonium and a **hypobasal cell** or **foot segment** below. These cells are distinct by their larger nuclei. The nuclei contain distinct nucleoli.

The epibasal cell gives rise to the shoot system (both rhizome and the aerial branches) and the hypobasal cell divides in various planes to give rise to a bulbous foot. The epibasal and the hypobasal cells, each divide by a longitudinal or an oblique wall to give rise to a four celled stage (Fig. 3.12_a D). In Fig. 3.12_a D only the lower cell has divided. After this four celled stage the further divisions are somewhat irregular. The hypobasal cells divide by ten longitudinal divisions which divide the foot into a number of cells that are longer (Fig. 3.12_a E) than broad. Later some of the foot cells divide in

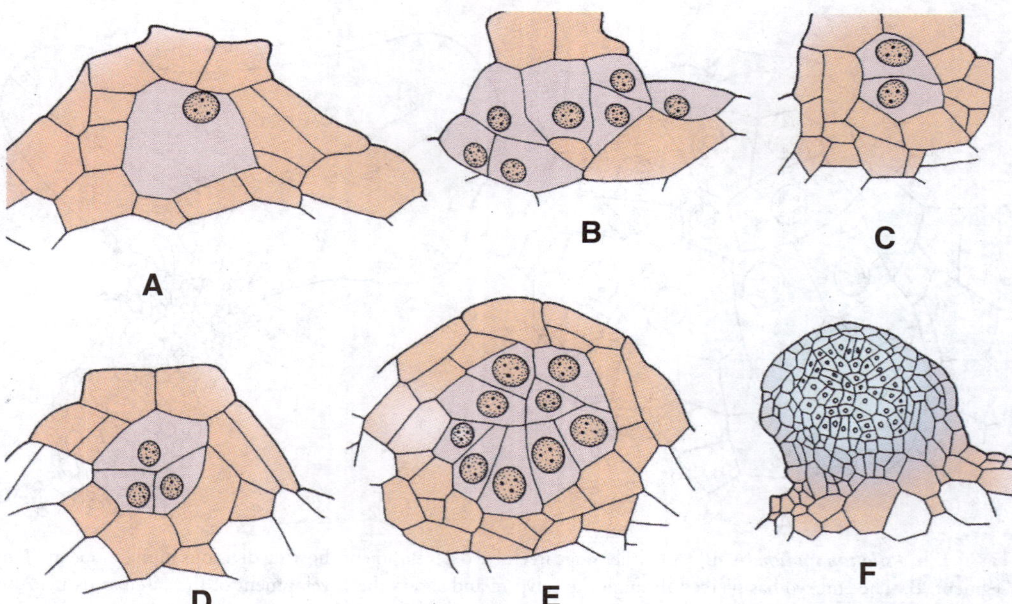

Fig. 3.12a *Psilotum nudum* (A-F). Stages in the development of embryo. A–B. Showing zygote enclosed within the calyptra; C. First transverse division of zygote into upper or epibasal (towards archegonial neck) shoot segment and lower or hypobasal foot segment; D-E Later stages of development; F. Showing well developed foot and shoot segment. (After Bierhorst)

a transverse plane and form finger like rows of cells which grow into the tissue of the gametophyte below and crush its cells. Rest of the foot cells undergo longitudinal divisions which increase the number of potential haustorium like processes (Fig.3.12$_a$, F and Fig. 3.12$_b$, B). The foot becomes bulbous and secures the sporophyte to the gametophyte and absorbs nutrition from it till the sporophyte becomes independent. While the foot is developing the epibasal cell divides in three different planes. (Fig. 3.12, C, D and 3. 12$_a$ E). The first division of the epibasal cell or the rhizome initial, as mentioned earlier, is oblique to the first division of the zygote. The second division is nearly perpendicular to the first. Two subsequent divisions lead to the formation of a three sided apical cell of the young sporophyte (Fig. 3.12 D). At this stage the embryo bulges out as a prominent outgrowth on the surface of the gametophyte (Fig. 3.12, E). As a result of the activity of this apical cell the shoot soon grows vertically and becomes prominent (Fig. 3.12$_b$ B). The surrounding cells of the gametophyte divide to form a calyptra-like outgrowth enveloping the young embryo. By further growth the embryo pierces through the calyptra (Fig, 3. 12 E,F). The young shoot soon establishes itself as a young axis with vascular tissue. It is the future rhizome which is at first unbranched, but undergoes branching at a later stage. The young rhizome detaches itself from the foot by the foot by the formation of a separation layer at the original boundary between the foot and shoot.

The rhizome continues to grow in length and branches repeatedly in a dichotomous manner. The tips of the ultimate branches turn upwards and develop into aerial branches that come out of the humus and grow erect.

The embryo of ***Psilotum*** is remarkably different from most of the other pteridophytes in lacking initials of suspensor, leaf and root. Apogamy has been reported in ***P. nudum*** (Bierhorst, 1971).

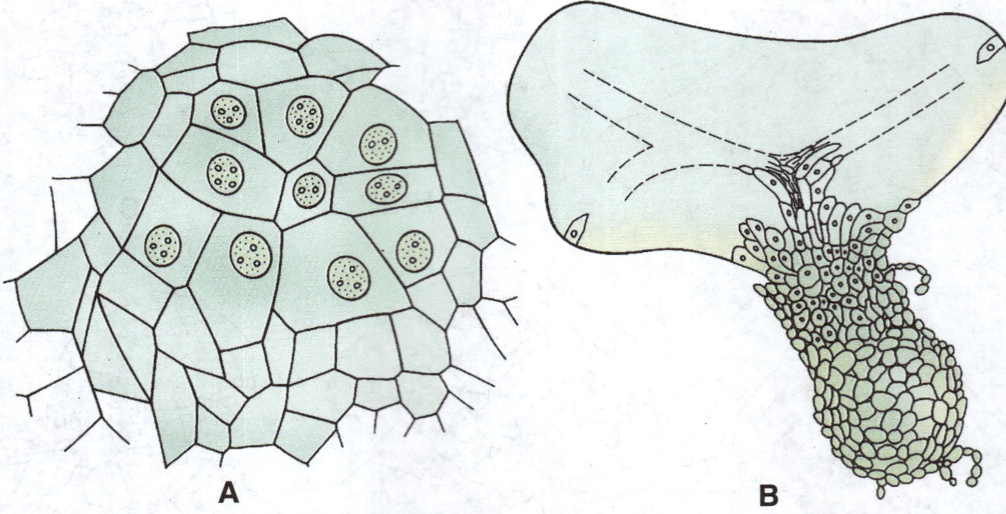

Fig. 3.12b. ***Psilotum nudum*** (A–B). A. Earlier stage in embryo development showing divisions in the shoot and foot segment; B. The embryo has pierced through the calyptra and shows the development of the vascular tissue. Also note the massive foot. (After Bierhorst, 1954)

TMESIPTERIS / BERNHARDI

Tmesipteris with about 10 species (Bierhorst, 1971) is a genus quite restricted in distribution. It is confined, to some islands in the South Pacific ocean, Australia, New Zealand, Tasmania, Polenesian islands and New Caledonia, Polenesian islands and New Caledonia. Majority of the species are epiphytes growing on tree trunks, aerial trunked stems of some tree ferns or in the humus formed by organic debris on the trees. It has also been collected from the root mantle of palms. ***T. tannensis*** (Fig 3.13 A) usually grows as pendulous epiphyte hanging freely from tree trunks. It has also been recorded growing on heaps of humus. ***T. vieillardi*** is a terrestrial species that has been found growing erect on ground in New Caledonia (Sahni, 1925 and Dangeard, 1926). The two other species that are commonly found include ***T. lanceolata*** and ***T. forsteri.***

T. lanceolata

Plant Body

The plant body consists of a creeping rhizome that (Fig. 3.13 A) branches repeatedly and gives rise to sparingly branched aerial shoots Fig. 3.13A. The branching of the rhizome is usually dichotomous. The dichotomies are equal or sometimes unequal, and irregular due to occasional injuries. The aerial shoots usually show only one dichotomy. The aerial axis in ***T. forsteri*** is highly branched. They bear foliar appendages which are reduced and scale like at the base (Fig. 3.13 A) but become distinct leafy structures towards tip. These leaves or the foliar appendages are arranged in two rows of sub-opposite pairs in species with flattened aerial shoots e.g. ***T. lanceolata*** and ***T. tannensis.*** In ***T. vieillardi*** the aerial shoots

T. tannensis

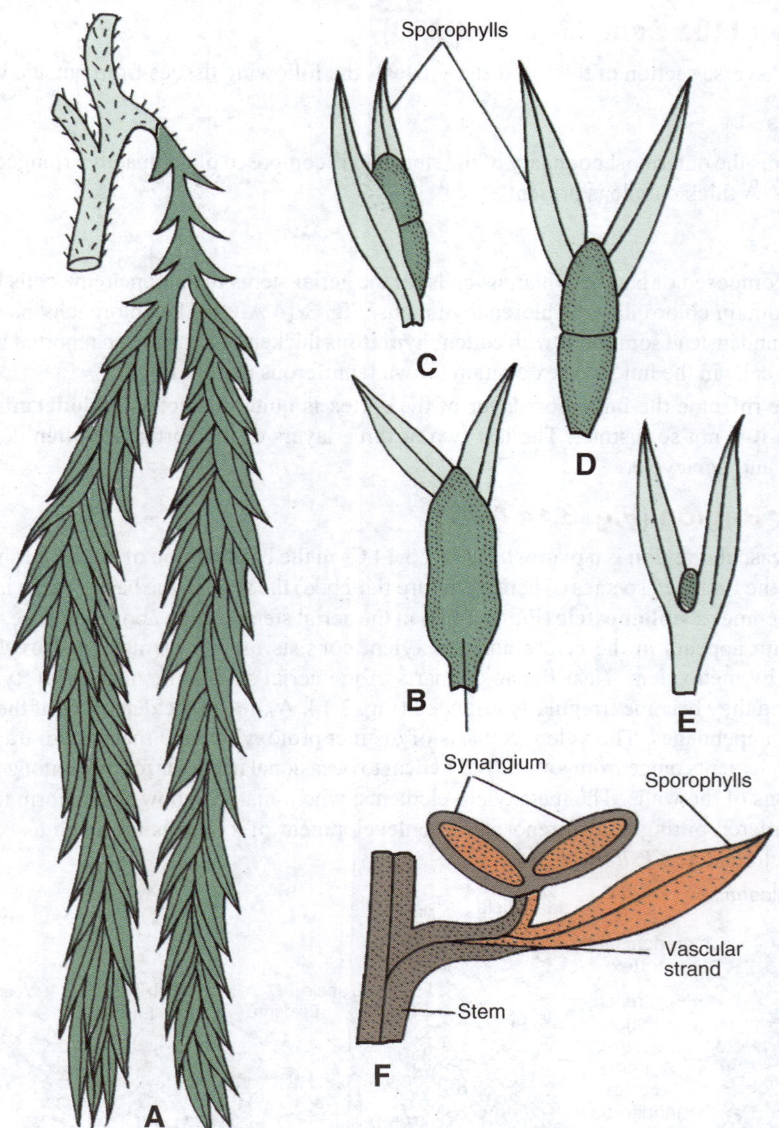

Fig. 3.13 (A–F). A complete sporophyte of *T.tannensis* showing habit; B-F. Fertile appendages as seen from different directions; B. from back; C. from side; D. from above; E. Adaxial view; F. Radial section showing vascular supply. (After Pritzel, F. After Eames)

have a radial symmetry and bear leaves in pseudospirals. The leaves are flat, lanceorlate and with a mucronate tip. They are about half an inch long. The smaller leaves at the base are **avascular** whereas the larger upper ones are traversed by a single distinct vascular strand. The leaf bases are continuous with the stem decurrent). The aerial shoot is terminated by a leaf.

The roots are absent and instead the rhizomes are covered with rhizoids that are evenly distributed all over the surface. The rhizoids may be unicellular or two to three cells in length. The rhizoids are covered by a cuticle.

The aerial shoots arise from the tips of rhizome. The tip turns up and grows into an aerial axis.

Anatomy of the Stem (Fig. 3.14 A-C)

A transverse section of the aerial stem reveals the following tissues from outside within:

Epidermis

It forms the outermost boundary of the stem and is composed of compactly arranged parenchymatous cells. A thick cuticle is present.

Cortex

It is composed of parenchymatous cells. In the aerial stems the parenchyma cells beneath the epidermis contain chloroplasts (Chlorenchymatous. Fig. 3.14 A-C). The chlorenchyma is absent in the rhizome and instead some cells with **collenchymatous** thickenings have been reported by Bierhorst (1971). The cells in the inner cortex contain brown tanniferous materials.

In the rhizome the innermost layer of the cortex is quite distinct as **endodermis** but in the aerial stems it is not so distinct. The last two or three layers of the cortex are often designated as endodermis and **pericycle.**

Vascular Region (Fig. 3.14 A–C)

The vascular region is a **protostele** (Fig. 3.14 C) in the basal region of the aerial shoot and the rhizome. In the flattened species (where stems are flattened) the stele in the basal region is U-shaped. The stele becomes a **siphonostele** (Fig. 3.14 B) in the aerial stems a little above the base. In this case a distinct pith appears in the centre and the xylem consists of five or more **protoxylem** groups surrounded by metaxylem. Near the apical parts of the aerial shoots the number of xylem groups decreases and they become irregularly arranged (Fig. 3.14. A) due to the departure of the leaf traces to the foliar appendages. The xylem consists of distinct protoxylem and metaxylem tracheids. The protoxylem elements range from **annular** to **helical** to occasional irregular forking among the systems of thickenings of the walls. The metaxylem elements, whem mature, show scalariform thickenings. Circular bordered pitting is also reported. The development of xylem is **mesarch** as compared to **exarch** development in ***Psilotum***.

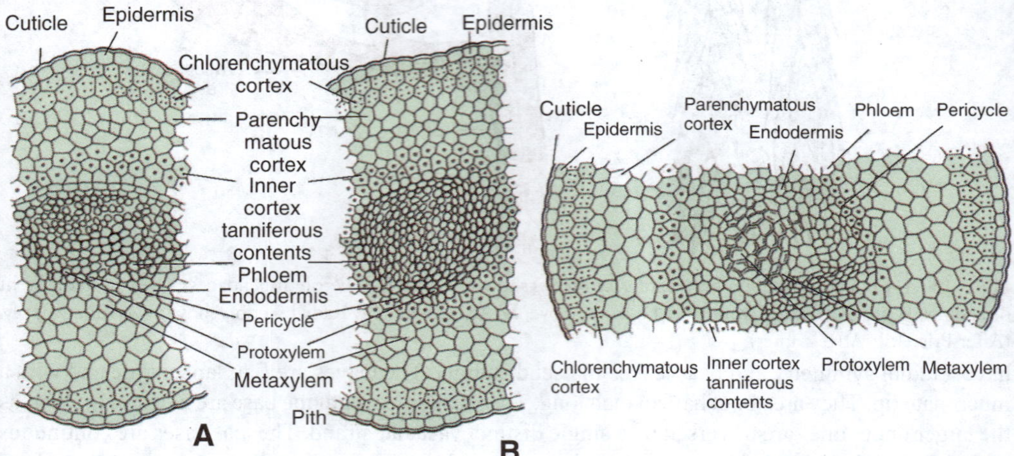

Fig. 3.14 (A–C). Transverse sections of the aerial shoot cut at various levels. **A.** T.S. near the apex showing irregularly arranged xylem groups; **B.** T.S. above the base showing a siphonostele with a distinct pith surrounded by xylem. Note the five protoxylem groups; **C.** T.S. at the base showing protostele with protoxylem in the centre (After Bower).

Phloem completely surrounds the xylem and consists of **sieve elements** with tapering end walls. The **sieve tubes** have numerous sieve areas on their lateral walls (Sykes, 1908).

Anatomy of the Leaf (Fig. 3.15)

The leaf in ***Tmesipteris*** shows a basal meristematic zone (intercalary meristem) which adds to its length. It grows in width by marginal meristematic activity. The terminal pinna in ***Tmesipteris*** develops directly from the apex of the aerial shoot, which is consumed in its formation.

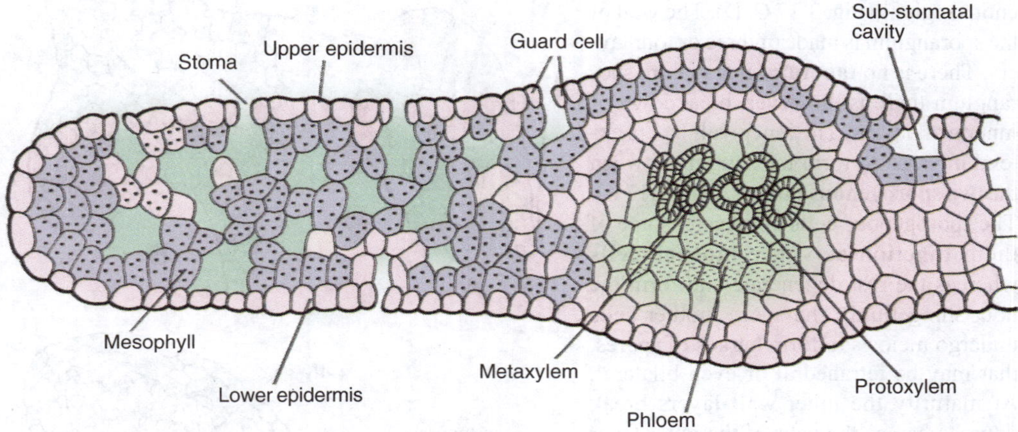

Fig. 3.15. *Tmesipteris tannensis.* V.S. through leaf (After Sykes)

The leaves in ***Tmesipteris*** are larger then those of ***Psilotum*** and show a comparatively well marked internal organisation. The leaves have distinct upper and lower epidermal layers. The epidermis is made up of a single layer of cells whose outer walls are cutinised. Stomata are present on both the epidermal surfaces. The region between the two epidermal layers is called **mesophyll** and is composed of lobed parenchyma cell that contain chloroplasts.

The single vascular strand is located in the centre and consists of several protoxylem tracheids surrounded incompletely by metaxylem elements. Phloem surrounds the xylem. No distinct endodermis is present.

The Fertile Appendages (Fig. 3.13. B-F)

The fertile appendages or the sporangia bearing structures which have been named as **fertile pinnae** by Bierhorst (because he designates the aerial shoots as fronds) and usually known as **sporophylls** are arranged in the same manners as the leaves. They can arise anywhere on the aerial axis. In ***Tmesipteris*** their position has been observed to vary with species. In ***T. lanceolata*** they arise near the base of the aerial shoot. In ***T. tannensis*** they arise at different places along the length of the aerial stem. In ***T.vieillardi*** they occur in the form of a strobiloid cluster at the tip of the stem.

Each sporophyll consists of a short lateral branch or the **fertile axis** (Fig. 3.13 F) terminating in a bilobed sporangium with two foliar appendages situated below the sporangium. These foliar appendages may grow more and extend beyond the sporangium. The vascular strand of the fertile axis divides into three strands just at the point where the foliar appendages arise. Out of the three strands two lateral ones enter the foliar appendages whereas the median one continues its course upwards and again divides into three strands. Two out of these supply the margins of the sporangium and the third enters the septum. The fertile axis arises laterally from the aerial stem and has also been called the sporangiophore. It takes the place of a sterile leaf.

Structure of the Sporangium (Fig. 3.16 C)

It is a bilobed structure that is more than two millimeters long. A section through the sporangium (Fig. 3.16 C) reveals that it is partitioned into two sporogenous regions (Fig. 3.17 C, D). The wall of the sporangium is made of three or four layers. There is no **tapetum.** In a mature sporangium there is, however, breakdown of inner wall layers. The inner wall layers extend inwards to form the septum between the two **sporogenous masses.** (Fig. 3.17 C). The sporogenous cells divide and some of them function as **spore mother cells** whereas the rest degenerate and form the nourishing fluid. The spore mother cells undergo meiosis to form tetrads of spores, that may be tetrahedral or even bilateral. At maturity the inner wall layers break down, whereas the cells of the outer layer become thickened. The dehiscence of the sporangium is **loculicidal.**

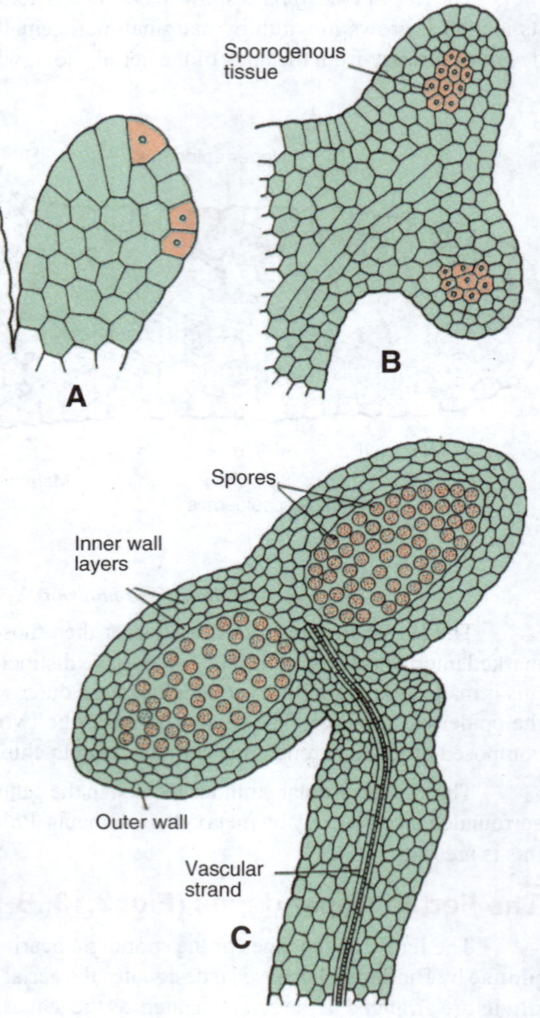

Fig. 3.16 (A–C) *T. tannensis.* A. A very young sporangium; B. a later stage in development; C. A mature synangium (A-B. after Bower, C. after Foster and Gifford).

Development of Sporangium (Fig. 3.16 A-C & Fig 3.17 A-D)

The development of sporangium is of **eusporangiate** type. The fertile axis grows by means of an apical cell (Fig. 3.17 A, B.) into a fertile axis. Two separate surface initials make their appearance at the apex of the fertile axis and indicate the sites of two sporangial locules. Each initial divides periclinally to form an outer **primary wall initial** and an inner **primary sporogenous initial** or cell. The former undergoes periclinal and anticlinal divisions to form a three or four layered thick sporangial wall. The primary sporogenous cell divides to form a mass of sporogenous cells (Fig. 3.16 B and 3.17 C). The sporogenous cells develop into **spore mother cells** which undergo meiosis to form tetrads of spores. Some of the sporogenous cells may disintegrate to form nourishing fluid. The leaf like appendages at the base of the sporangium develop later and usually overgrow the sporangium (Fig. 3.13 B-F).

Morphology of the Sporangial Stalk

Regarding the nature of the sporangial stalk various views have been put forth. These can be summarised as below :

1. Miss Sykes (1908), regards the sporangium as axial in position. This view is supported by the presence of a vascular bundle.

2. Sahni (1925) regards the sporangiophore as a shoot structure.
3. In 1921 Miss Margret Bensen explained that the sporangiophore of ***Tmesipteris*** is an axial structure and the two lobes are cladodified branches of the axis.
4. One view holds that the leaf is also a modified branch of the aerial system. It is possible that sometimes the leaf may have preceded the specialised sporangiophore in its differentiation.

The discovery of Rhyniaceae, in which the sporangium is not an organ by itself, but the modified end of a branch, seems to support the view that the sporangiophore is a fertile branch.

The structure of the spore is similar to that of ***Psilotum***. It germinates into a prothallus or gametophyte. The stages of germination are not known.

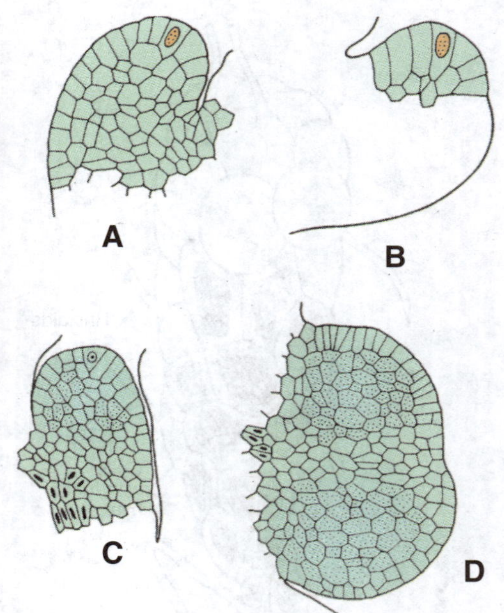

Fig. 3.17 (A–D). *T. tannensis*. A-B. Young fertile appendages with apical cells; C-D Synangia in later stages of development (After Bierhorst)

Gametophyte (Fig. 3.18 A-D)

The prothallus of ***Tmesipteris*** is subterranean. It is found usually in moist, sandy soil and is devoid of chlorophyll. Its mode of nutrition is saprophytic. It varies in form and may be cylindrical, bent, curved and sometimes twisted in curious forms. It is branched and each branch is provided with an apical meristem. Numerous rhizoids arise from its surface. The rhizoids are long and unicellular and afford immense absorptive surface.

The presence of endophytic fungus (***Chladochitrum tmesipterides***) is necessary for the development of the prothallus. There is a distinct epidermal layer composed of slightly thickened cells that are sharply defined. The rest of the tissue is uniform throughout. There is no differentiation into vegetative and reproductive regions. A more distinct pattern of distribution of gamentangia (sex organs) is seen in a species of ***Tmesipteris*** from New Caledonia where they occur in distinct zones along the axis. The surface cells within the fertile zones are relatively shorter than those in the infertile areas.

Sex Organs (Fig. 3.19 and 3.20)

They are distributed all over the surface of the prothallus.

Antheridium (Fig. 3.19). The antheridia are distributed all over the surface of the prothallus and are invariably accompanied by archegonia. The antheridium is larger in size than the archegonium. It is almost spherical in shape and is almost completely above the epidermal surface. A mature antheridium is composed of an outer wall one cell in thickness. The antheridial wall encloses a mass of spermatogenous cells which number 128-512. The contents of these cells metamorphose into multiciliated spermatozoids. The jacket has a single opercular cell that breaks down to allow exit of spermatozoids.

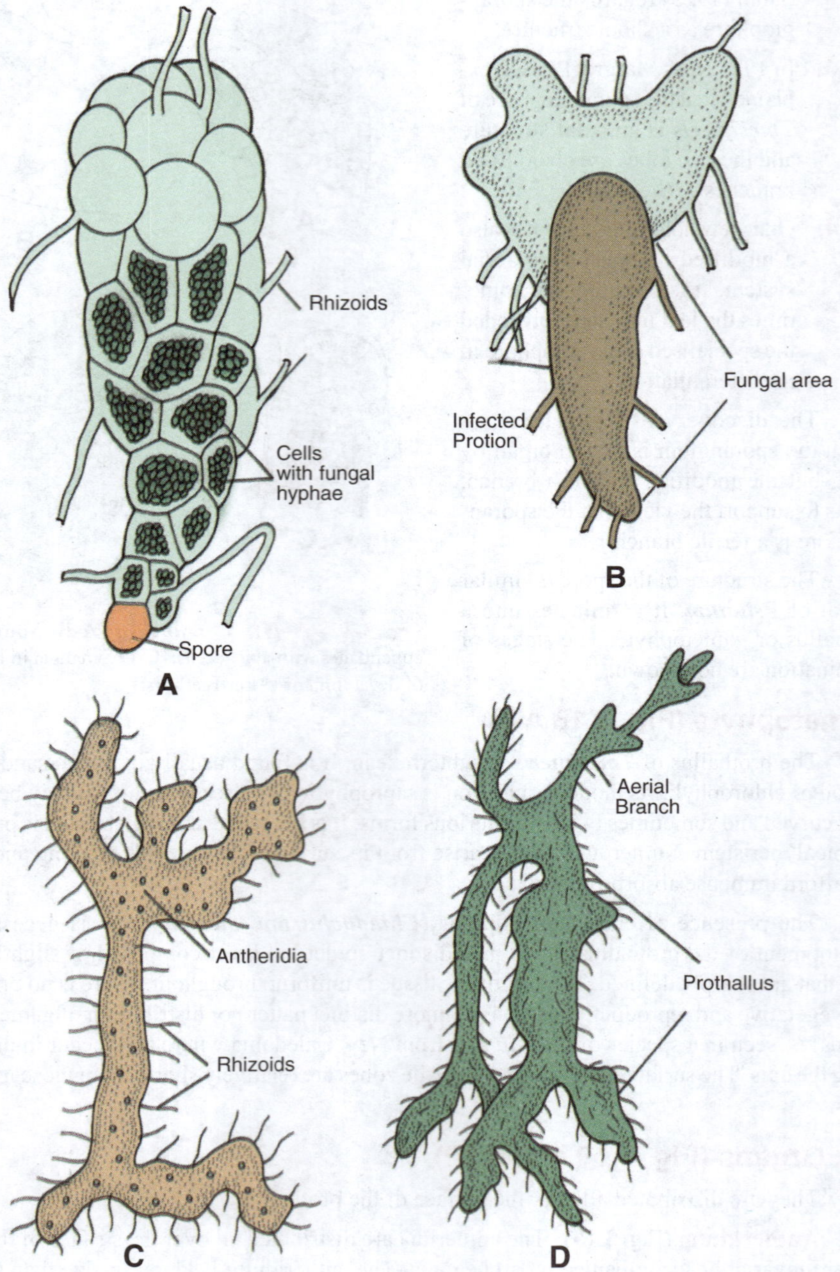

Fig. 3.18 (A-D) Gametophytes of *T. tannensis.* A. Young prothallus; B. Later stage showing forking at the distal end; C. Mature prothallus showing irregular shape and branching; D. Prothallus with a young sporophyte (After Holloway).

The antheridium develops (Fig. 3.19. A) from a single superficial cell that divides periclinally into an outer **jacket initial** and inner **spermatogenous *cell.*** The nucleus of the spermatogenous cell undergoes division and contains several free nuclei. Meanwhile the jacket initial divides by anticlinal wall formation to give rise to a single layered jacket. Delicate walls are then laid around the free nuclei in the spermatogenous cell to form a spermatogenous tissue. They divide further to form **sperm mother cells** whose granular contents and larger nuclei distinguish them from other cells. At this stage the antheridium assumes a spherical shape and enlarges in sieze. The nuclei undergo change and appear dense, coiled and crescent shaped bodies. They metamorphose into multi-ciliated and coiled **spermatozoids** (Fig. 3.19 E).

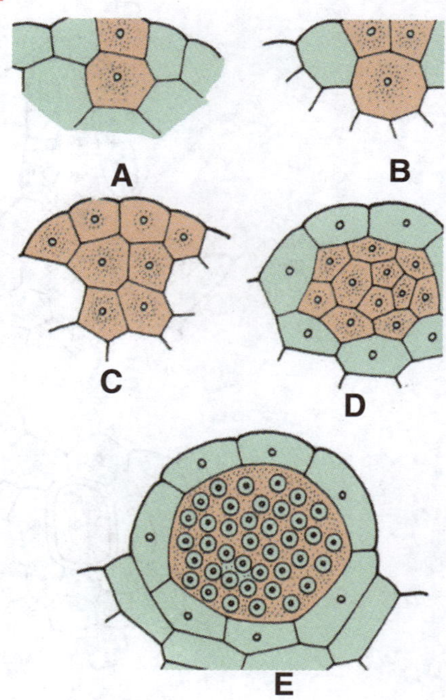

Fig. 3.19 (A-E). Various stages in the development of an antheridium in *T. tannensis* (After Holloway)

Archegonia (Fig. 3.20)

Their size is comparatively smaller than the antheridia and are produced in larger numbers. The archegonium neck is made up of 4 longitudinal rows of cells that project above the surface of the prothallus. The venter remains embedded in the tissue of the prothallus. Each row of the neck is 4 celled high ***i.e.*** the neck consists of four tiers of four cells each. The venter contains the egg cell. The neck encloses a neck canal which contains two nuclei.

The archegonium develops from a superficial cell of the prothallus. It enlarges in size and divides by a periclinal wall into an outer **cover cell** and an inner **central cell.** The cover cell divides twice anticlinally and then by a series of periclinal walls to form an archegonial neck made up of four tiers of four cells each. The neck protrudes above the surface. The central cell divides periclinally into a lower **primary ventral cell** and an upper **primary neck canal cell.** The former divides into an **egg cell** and a **ventral canal cell.** The primary neck canal cell nucleus divides into two neck canal nuclei. No wall is laid down between them.

At maturity the upper three tiers of the neck disorganise (Fig. 3.20. C), the fourth tier forming a flat disc, slightly concave in the middle, but otherwise quite parallel with the surface of the prothallus. This results in the great shortening of the neck at the time of fertilisation. This is an adaptation to meet the unusual condition of subterranean fertilisation.

As soon as the antheridium dehisces, the spermatozoids come out in the surrounding thin film of moisture. They swim towards the archegonia presumably attracted by some chemical substance. They enter the open neck and one of them reaches the egg to effect fertilisation. After fertilisation the zygote is completely enclosed in the venter by the formation of calyptra (Fig. 3.21A).

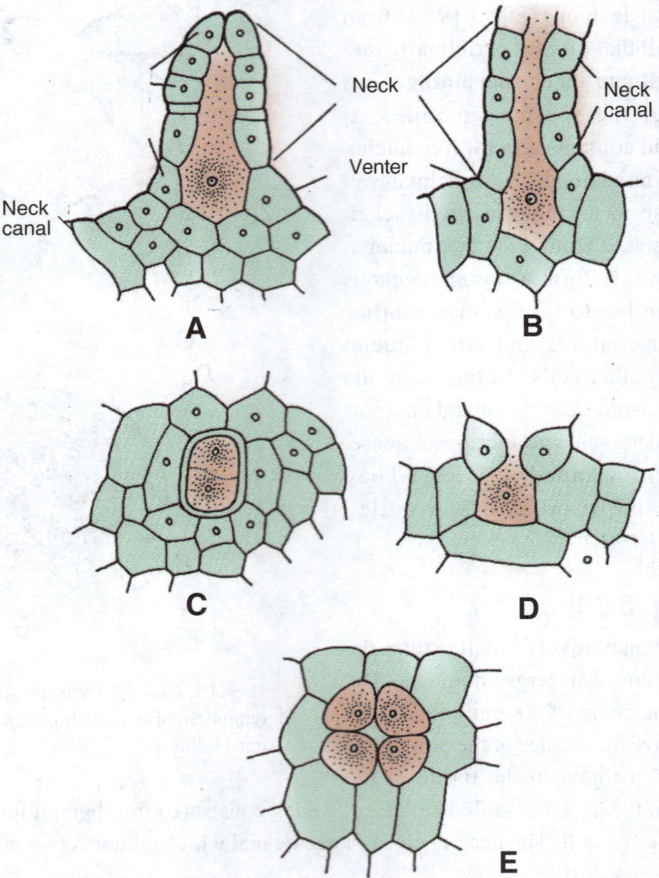

Fig. 3.20 (A-E) *T. tannensis* Archegonium. A. Mature archegonium (After Holloway); B. Mature archegonium (After Lawson); C. Archegonium after fertilisation when the projecting part of the neck has been shed (After Holloway); D. Same as C (After Lawson); E. Archegonium as viewed from surface during fertilization (After Lawson)

Embryogeny (Fig. 3.21 A-D and 3.22 A-E)

The development of embryo is known only for ***T. tannensis.*** The zygote elongates (Fig.3.21 A) and then divides by a transverse wall (Fig. 3.22, A) into an **epibasal cell** (towards the archegonial neck) and a **hypobasal cell** (Fig. 3.22 A). The hypobasal cell divides repeatedly to from the **foot** whereas the epibasal cell divides repeatedly to form the sporophyte axis. After a few divisions of the epibasal cell two growing points or apical cells (Fig 3.22, D) are established at the flanks. The establishment of two apical cells result in the formation of a branched rhizome (Fig. 3.23 A, B). In the foot region further divisions lead to the formation of lobed structures that extend into the gametophyte tissue (Fig. 3.21, D and Fig. 3.22 D, E). In the epibasal region the two apical cells divide while still contained in the gametophyte tissue. Later they emerge out of the calyptra and branch into two. The branched rhizome is subterranean. In ***Tmesipteris*** it develops into green aerial shoots quite early as compared to ***Psilotum*** where green aerial parts arise quite late, may be after several years.

The embryology in ***Tmesipteris*** is **exoscopic.** The young sporophyte of ***Tmesipteris*** (Fig.3.21 D, 3.22 E) appears to be quite similar to that of ***Anthoceros*** in possessing a haustorial foot.

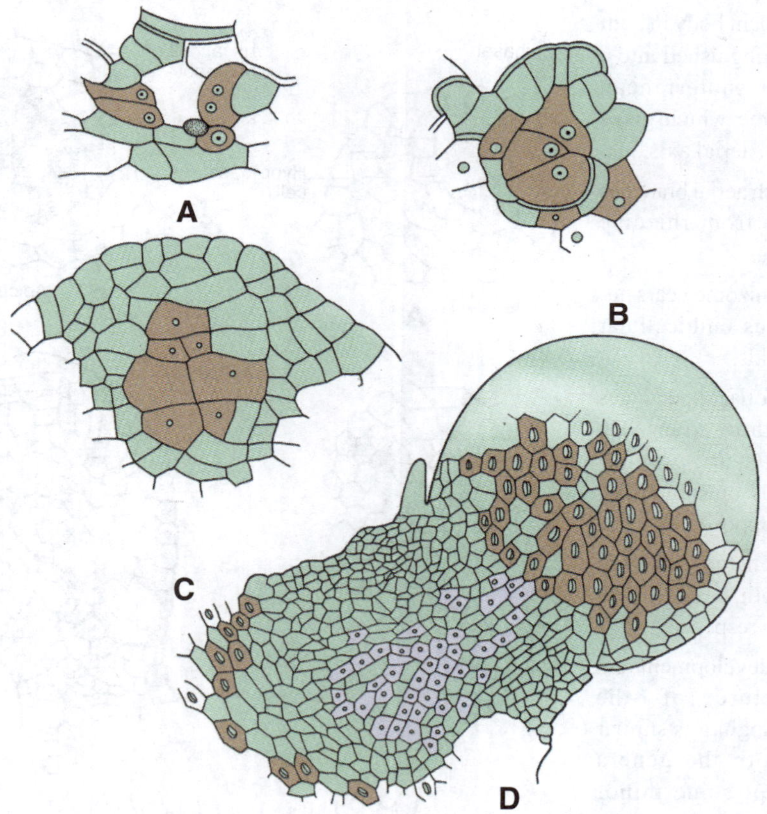

Fig. 3.21 (A-D) *T. tannensis.* A-C. Earlier stages in the development of embryo; D. Mature embryo showing lobed foot, the central strand and surrounding tissue (After Holloway).

Chromosome Number

Barber (1957) in his article Polyploidy in Psilotales' has mentioned the Chromosome numbers in ***Tmesipteris tannensis*** as $n = 200+$. He recognised six new species of ***Tmesipteris***, five out of which have $\boldsymbol{n} = 204 - 210$ and one has $n = 102–105$. The basic chromosome number for the family is considered to be x = 13 (Chiarugi, 1960, Ninan, 1956).

Taxonomic Position

Division	:	*PSILOPHYTA*
Class	:	*PSILOTOPSIDA*
Order	:	*PSILOTALES*
Family	:	*PSILOTACEAE*
Genera	:	*Psilotum* and *Tmesipteris*

Psilotum and Tmesipteris Compared

Resemblances

The two genera resemble each other in the following respects :—

1. Both may grow in soil or as epiphytes.

2. The plant body in both is distinguished into a lower subterranean rhizome which gives rise to aerial axis.
3. In both aerial branches grow from rhizome apices.
4. The rhizome bears numerous multicellular rhizoids.
5. The foliar appendages in both are arranged in spirals or pseudo-spirals or sometimes just-opposite.
6. The anatomy of stem in both is similar in many respects.
7. The development and structure of the sporangium is similar in both the genera except some minor differences.
8. The prothallus in both is subterranean and saprophytic. It is covered with numerous rhizoids.
9. The structure and development of sex organs in both is similar.
10. The development of embryo in both is similar to a greater extent.
11. The basic chromosome number in both is believed to be x = 13.
12. Polypoidy is prevalent in both the genera.

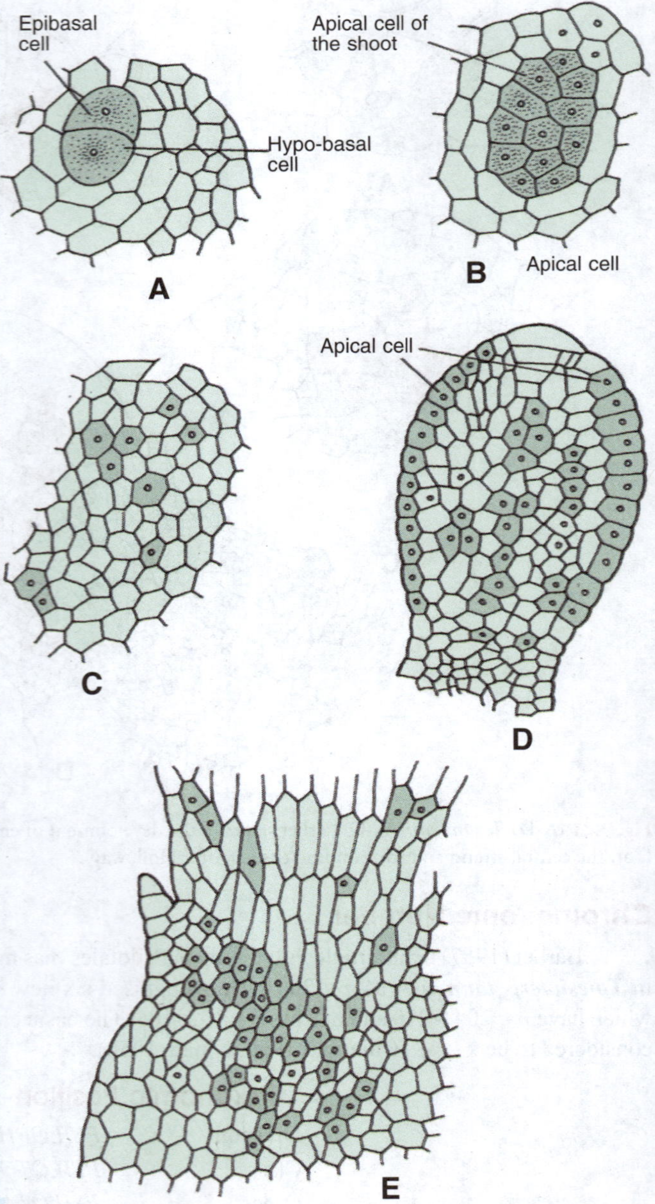

Fig. 3.22 (A–E) *T. tannensis* Embryology **A.** Two celled embryo with an epibasal and a hypobasal cell; **B.** Embryo with a distinct apical cell of the shoot; **C.** The epibasal cell has produced an irregular lower part that will develop into a foot; **D.** Young embryo showing lower foot and the epibasal hemisphere with apical initials that will grow into a shoot; **E.** Basal portion of the mature embryo showing a lobed foot. (After Holloway)

Differences

Tmesipteris	*Psilotum*
1. The aerial branches are sparingly branched and show only one dichotomy.	The aerial shoots are profusely branched and show repeated dichotomy.
2. The foliar appendages are larger in size, green, and possess a distinct single vein in the centre.	The foliar appendges are minute and **avascular.**
3. Stomata are present in the leaves.	The stomata are absent.
4. The xylem development is **mesarch.**	The xylem development is **exarch.**
5. The xylem is not star-shaped but consists of varying number of xylem groups near the apex, it is a siphonostele in the middle and the xylem is U-shaped or a protostele at the base.	The xylem is star-shaped and is usually a protestele and not siphonostelic.
6. The sporangium is bilobed.	The sporangium is trilobed.
7. The tapetal layer is absent.	The tapetum is present.
8. Two foliar appendages are associated with the sporangium.	Only one bilobed foliar appendage is associated with the sporangium.
9. The lateral fertile axis that bears the Sporangium and foliar appendages is long and distinct.	The fertile axis is abbreviated.
10. The prothallus is smaller in size, thin and dark brown.	The prothallus is comparatively larger in size and is thicker.
11. The prothallus is wholly subterranean.	The prothallus may be wholly or even partially subterranean. It projects out slightly from the crevices in the rocks.
12. The antheridia are much more numerous than the archegonia.	The archegonia are numerous than the antheridia.
13. The archegonia occur in dense crowded groups.	The archegonia are widely separated from each other.
14. The antheridia and archegonia are larger in size.	The antheridia and archegonia are comparatively smaller in size.
15. The zygote enlarges sufficiently before dividing.	The zygote does not show much enlargement.
16. The embryonic shoot possesses two apical cells.	The embryonic shoot possesses only one apical cell.
17. The aerial shoots arise early from the embryonic shoot.	The aerial shoot sometimes takes several years to develop from the young subterranean rhizome.
18. The aerial shoot is terminated by a leaf.	The aerial shoot is not terminated by a leaf.

Distinct Features of the Prothallus

1. The gametophytes of the Psilotaceae are subterranean. 2. The colour is light brown or dark brown. 3. They lack chlorophyll. 4. The body is usually cylindrical. It branches and each branch has a terminal meristem. 5. The prothallus tissue is uniform except the epidermal layer that is quite distinct. 6. Presence of tracheidal cells or even distinct vascular strands is a unique feature of the Psilotaceous gametophyte. 7. The prothallus is saprohytic and is attacked by an endophytic fungus. 8. Surface of the prothallus is beset with numerous unicellular rhizoids. 9. The prothallus is monoecious. 10. The sex organs are irregularly distributed and intermingled. 11. The spermatozoids are multiflagellated. 12. The antheridia bulge above the surface of the prothallus. 13. The archegonial neck is composed of four longitudinal rows of cells and projects above the surface of thallus. The venter is embedded. 14. Disorganisation of the tiers of neck cells is a common feature.

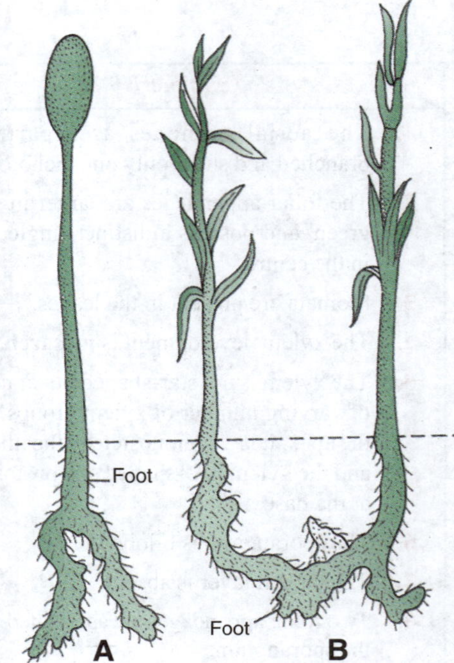

Fig. 3.23 (A-B). Young sporophytes of *T. tannensis* **A.** Only one of the embryo apices has grown into a young shoot; **B.** Both the apices of the embryo have grown into two young plants. (After Holloway)

Affinities of Psilotales

Regarding the affinities of this group there are still doubts in the minds of botanists. Four theories have been put forth to explain the position of the group :

1. Kidston and Lang (1917–1921) propounded a theory suggesting close relationship between Psilotales and Psilophytales. This theory is also supported by Holloway and Bower.
2. Arber suggested their late origin along independent lines from the algae.
3. The hypothesis maintained by A.P.W. Thomas and formerly suggested by Bower and Scott, of an affinity with the Sphenophyllales.
4. The old fashioned view that Psilotales are essentially Lycopods (Lawson, 1917).

Considering the first theory, the Psilotales resemble the Psilophytales in : (*i*) having rootless rhizomess, (*ii*) protostelic stem, (*iii*) small size of the leaves, and (*iv*) dichotomously branched shoots.

They differ only in the position and number of sporangia. Kidston and Lang pointed out that not only do the Psilotales resemble the fossil Psilophytales in vegetative characters, but that even in the fructifications a reliable basis for comparison is supplied by the repeatedly forked aerial shoots of Rhyniaceae with their large terminal sporangia capable of showing the dichotomy of the axis (***Horneophyton***). Considering the similarity of the various genera in the two orders, one might speculate that the present members of the Psilotales are living remnants of Devonian flora that has come down to the present day without much chang from the very ancient times.

Lawson (1917) expreses himself favourably towards the affinity with the Sphenophyllales while pointing out that the gametophyte generation can offer no positive evidence.

Affinity with Sphenophyllales is ruled out because there are many important differences between them and the Psilotales. The leaves in Sphenophyllales are whorled as compared to spiral arrangement in the Psilotales. The branching in the Sphenophyllales is monopodial as compared to the dichotomous branching in the Psilotales.

Regarding their affinities with the Lycopods as suggested by Lawson in 1917, after studying the gametophytic generations of the Psilotales, there seems no justification because of some major differences between their prothalli. Apart from some superficial resemblances like their subterranean existence, presence of mycorrhiza and absence of chlorophyll, there are striking differences also. The Psilotaceous prothallus does not show any tissue differentiation so characteristic of the Lycopods. The antheridia in the Psilotales are so prominently bulged out whereas in Lycopods they are completely embedded in the prothallus. The spermatozoids are multiflagellate in the Psillatales.

Tmesipteris

Arber agrees with Thoday and others in that the Psilotales are an entirely distinct group from either the Lycopods or the Sphenophyllales. Arber regards it as quite an independent race that had an algal ancestor.

Bierhorst (1968, 1969, 1970) has related ***Psilotales*** (Psilotaceae) to some filicalean families like ***Schizaeaceae*** and newly created ***Stromatopteridaceae***, which is a segregate from ***Gleicheniaceae***. He has included the family ***Psilotaceae*** (***Psilotum*** and ***Tmesipteris***) under the order ***Filicales*** and has placed it next to ***Stromatopteridaceae***. Such a conclusion has been drawn as a result of following similarities between the two families (Bierhorst, 1968) :—

1. The axial gametophytes are underground. **2.** The under ground sporophytic stem and the gametophytes are indistinguishable in both. **3.** The gametophytes bear septate rhizoids in both. **4.** Alternate periods of growth in both the sporophyte and the gametophyte produce short and long cell. **5.** The archegonia in both the families dehisce by the decapitation of the neck. **6.** The embryos are rootless. **7.** Spores are monolete. **8.** Similarities in the development and wall structure of the sporangia. **9.** The terminal leaf of ***Tmesipteris*** and terminal pinna of ***Stromatapteris*** develop from the apical meristem and consume it. **10.** Large superficial antheridia with lateral opercular cell which breaks down at maturity. Discontinuous vascular tissue in smaller axes of the sporophyte. **11.** Synchronous xylem maturation in smaller subterranean stems. **12.** Mesarch xylem without protoxylem elements in the large subterranean stems. **13.** Successive dichotomies with determinate apices in the vertical subterranean stem system. **14.** A laxity of chromosome coiling at meiosis so that chromosomes in anaphase resemble early prophase ones. **15.** Nucleoli at mitosis persist up to anaphase stage.

Significance and Phylogeny of Psilophyta

The affinities of the two classes of Psilophyta have been discussed in detail in the relevant chapters. The division Psilophyta is of great evolutionary significance as it includes two closely related classes of extinct and living plants (Psilophytopsida and Psilotopsida). They are the oldest known and simple vascular plants that possess no roots and are either leafless or possess rudimentary leaves. They have been regarded as connecting links between the aquatic algae and the terrestrial and more complex vascular plants like the lycopods, horsetails, and the ferns. Some Psilophytales resemble some algae in (*a*) their dichotomously branched axes ; (*b*) absence of leaves ; and (*c*) absence of roots. They differ from algae in possessing : (*a*) definite cuticle ; (*b*) stomata ; (*c*) cutinised spores ; and (*d*) a vascular system with well defined xylem and phloem.

This division of plants offers a number of evidences that have a great evolutionary importance. These can be listed as below :

1. They throw light on the origin of root, as suggested by Prof. Lignier from that portion of the axis which became negatively geotropic and entered the soil where it became adapted to perform the functions of anchorage and absorption. The rootless and subterranean rhizomes of the Psilophytales and Psilotales offer a great support to this because they grow beneath the surface of the soil and serve to fix the plant and absorb nutrition.
2. They also suggest the origin of microphyllous and megaphyllous leaves of lycopods and ferns. The leaf-like emergences of ***Asteroxylon*** and ***Psilotum*** might have developed into small single veined and microphyllous leaves of the Lycopods. In ***Pseudosporochnus*** and ***Protopteridium***, the leaves are formed by flattening of the branch system and are supposed to have led to the development of larger and megaphyllous leaves of ferns.
3. The anatomy of the shoot in the Psilophytales also suggests that protostele is the primitive type of vasculature and that siphonostele has originated from it by the appearance of pith. It also suggests that annular tracheids are the simplest and the spiral and scalariform tracheids are advanced.
4. A comparative study of the Psilophytales and the higher vascular plants also suggests the establishment of four distinct subdivisions, namely, the Psilopsida, Lycopsida, Sphenopsida and the Pteropsida. The Psilophytales indicate clearly the three lines of evolution which lead to the development of Lycopsida, Sphenopsida and the Pteropsida.

Psilophytales= *Protopteridium*

Certain members of the Psilophytales, ***e.g.***, ***Asteroxylon*** possess small leaves that arise as emergences and might have given rise to the Lycopsid line of development.

A few Psilophytales, ***e.g.***, ***Pseudosporochnus*** and ***Protopteridium*** exhibit a tendency towards the formation of leaves by the flattening and webbing of branch system and might have led to the development of large leaved vascular plants or the Pteropsida.

This line of evidence, therefore, suggests that the plants with microphyllous and megaphyllous leaves had separate lines of evolution. It suggests that the old division pteridophyta should be split up at least into two subdivisions, the Lycopsida and Pteropsida. The seed plants are included in the Pteropsida, along with the ferns, because they have megaphyllous leaves. The ferns are closer to the seed plants (spermatophyta) than to the Lycopsida and the Sphenopsida.

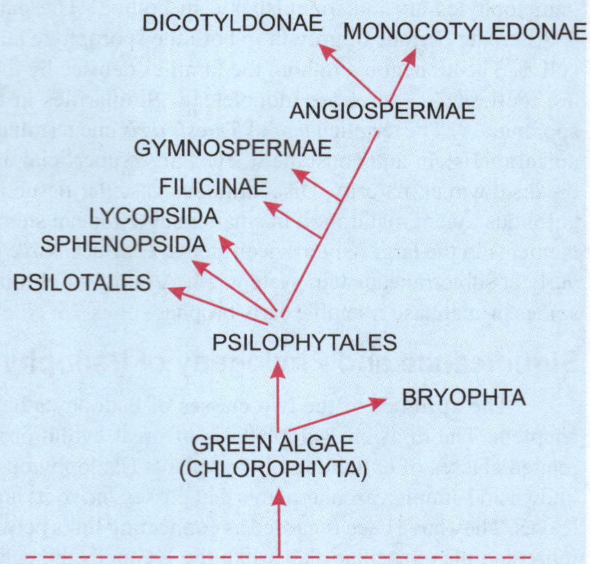

Fig. 3.24. Phylogenetic scheme illustrating relationship among various groups of vascular plants (After Fuller and Tippo)

The above evidences lead us to conclude that the Psilophytales have originated from the green algae and have in turn given rise to the three subdivisions of the vascular plants, namely, the Lycopsida, Sphenopsida and the Pteropsida. It is also believed that the bryophyta have also originated from the green algae and have followed a different course of evolution that has ended blindly. It is also believed by some botanists that bryophytes may have originated as a result of retrogresive evolution accompanied by reduction, from the Psilophytales.

The Psilotales are regarded as the last living descendants of the Psilophytales. They have persisted through ages unchanged and have retained in a remarkable manner their simple structure. They have been aptly regarded as the 'living fossils'.

Fuller and Tippo have drawn out a scheme (Fig. 3.24) of phylogenetic relationship among the various groups of vascular plants.

CHROMOSOME NUMBERS :— The haploid chromosome number in ***P. nudum*** has been reported to be $n = 50$ (Rosen, 1896) or $n = 52$ (Okabe 1929). Diploid prothalli with $n = 100$ (Manton, 1950), 104 (Barber 1957, Sorse 1965, Walker, 1965, Ninan 1956, 1965) have also been reported from various localities in India and abroad. Brownlie (1965) reported $n = 210$ in ***P. nudum*** from New Caledonia. Sinha and Verma (1971) reported $2n = 312$ in a hexaploid ***P.nudum*** from Patna. The same authors reported (1978) $2n = 468$ from Patna.

QUESTIONS

ESSAY TYPE

1. Describe the chief characteristics of ***Psilotum*** and discuss the phylogeny of Psilotales in the light of recent researches.
2. Describe the distinguishing features of Psilotales. What is the structure of Synangium? Discuss the distribution of ***Psilotum***.
3. Justify the creation of two distinct monotypic families of Psilotopsida.
4. Give a comparative account of the gametophytes of ***Psilotum***, ***Lycopodium*** and ***Ophioglossum***.
5. Where ***Psilotum*** is found in India? Mention the salient features of ***Psilotales*** and discuss its systematic position.
6. Discuss the primitive characters of ***Psilotum.***
7. Describe the life history of ***Psilotum*** and pointout the features of its primitiveness.
8. "***Psilotum*** combines the characters of simplicity and primitiveness". Discuss.
9. Describe the structure of synangium of ***Psilotum*** with the help of a suitable diagram and comment on its morphology.
10. Describe the synangium of ***Psilotum.***
11. Give an account of stelar structure of ***Psilotum*** and comment on the affinities of the genus.
12. Give a comparative account of the spore producing organs of ***Psilotum***, ***Selaginella*** and ***Equisetum.*** **(Kerala, 2001)**
13. Describe the external and internal structure of ***Psilotum*** stem. **(Bangalore, 2001)**
14. Give an illustrated account of asexual and sexual reproduction in ***Psilotum*** or ***Tmesipteris***.
15. With the help of labelled diagrams give an account of the Sporophyte of ***Tmesipteres.***
16. Describe the life-history of ***Tmesipteris***.
17. Describe the structure of prothallus in ***Tmesipteris.***
18. Give an illustrated account of the structure and morphology of sporangium of ***Tmesipteris***.

SHORT ANSWER TYPE

19. Classify the following and write their salient features :
    (*a*) ***Psilotum.***
    (*b*) ***Tmesipteris.***
    (Awadh, M.Sc. 2000)
20. Write short notes on.
    (*a*) ***Psilotum.***
    (*b*) Prothallus of ***Tmesipteris.***
    (*c*) Development of synangium in ***Psilotum.*** **(Kanpur, M.Sc. 2000)**
    (*d*) Aerial shoot of ***Tmesipteris.***

(*e*) Synangium in ***Psilotum.***

(Gauhati, 2000, 2002).

(*f*) Rhizome in ***Tmesipteris.***

(*g*) External morphology of ***Psilotum***.

(*h*) Stele of the aerial shoot of ***Psilotum.***

(*i*) Vascular bundle in ***Psilotum.***

(*j*) ***Prothallus*** of ***Psilotum.***

(Nagpur, 2002; Bangalore, 2001)

(*k*) Vegetative reproduction in ***Psilotum.***

(*l*) External morphology of ***Tmesipteris.***

21. Draw a labelled diagram of the following ;

(*a*) T. S. stem of ***Psilotum***.

(*b*) Synangium.

(*c*) T. S. aerial shoot of ***Psilotum nudum.***

(*d*) T. S. leaf in ***Psilotum.***

(*e*) T. S. aerial shoot of ***Tmesipteris.***

(*f*) Prothallus of ***Psilotum*** and ***Tmesipteris.***

(*g*) T. S. leaf of ***Tmesipteris.***

22. Is the so called rhizome of ***Psilotum*** a gametophyte or sporophyte? Explain.

23. Describe the structure of sporangium in ***Psilotum.***

24. Compare the stem anatomy of ***Psilotum*** with that of ***Equisetum***.

25. Why is the sporangium of ***Psilotum*** called synangium?

26. Mention two features of gametophyte (prothallus) of ***Psilotum***,

27. Draw T. S. of synangium of ***Psilotum***, label the parts and describe its structure.

28. Give in brief the structure of gametophyte in ***Psilotum.***

29. Describe the distribution and structure of aerial branch of ***Psilotum.***

30. Give an account of the stele of the aerial shoot of ***Psilotum.***

31. Write about morphological nature of spore producing organs of ***Psilotum***.

(Himachal Pradesh, 2000)

32. Give an account of the state of the aerial shoot of ***Tmesipteris***.

33. Compare the synangia of ***Psilotum*** and ***Tmesipteris.***

34. Differentiate between ***Tmesipteris*** and ***Psilotum.***

35. How many species of ***Tmesipteris*** are known? Name at least two species.

36. What is the morphological nature of sporangial stalk in ***Tmesipteris***?

37. Describe, briefly, the external morphology of ***Tmesipteris*** sporophyte.

38. Name the spore bearing organ of ***Psilotum.***

(Kuvempu Univ., 2005)

39. Describe synangium of ***Psilotum***.

(Mysore Univ., 2004)

40. Discuss the development of antheridium and archegonium of ***Psilotum.***

(M.S. Univ., Tirun, 2005)

OBJECTIVE TYPE

38. Select the correct answer :

(*i*) ***Psilotum*** consists of

(*a*) Root and stem

(*b*) Root and leaves

(*c*) Stem and leaves

(*d*) Root, Stem and leaves.

(*ii*) Apical growth in ***Psilotum*** takes place by

(*a*) Apical cell

(*b*) Middle cell

(*c*) Basal cell

(*d*) Lateral cell.

(*iii*) The common characters shared by ***Rhynia*** and ***Psilotum*** are:

(*a*) Scale leaves

(*b*) Photosynthetic stem

(*c*) Rhizoids

(*d*) Sporangia.

(*iv*) Spore producing organ in ***Psilotum*** is known as

(*a*) Strobilus

(*b*) Sporangium

(*c*) Synangium

(*d*) Sorus.

(*v*) ***Psilotum*** belongs to the order

(*a*) Psilophytales

(*b*) Psilotales

(*c*) Lycopodiales

(*d*) Equisetales.

(*vi*) Three chambered sporangium is present in

(*a*) ***Lycopodium***

(*b*) ***Selaginella***

(*c*) ***Psilotum***

(*d*) ***Equisetum.***

(*vii*) An Angle leaf is present in the genus

(*a*) ***Psilotum***

(*b*) ***Lycopodium***

(*c*) ***Selaginella***

(*d*) ***Equisetum.***

(*viii*) In the rhizome of ***Psilotum***, the stele is
- (*a*) Solid protostele with central core of xylem surrounded byphloem
- (*b*) Mixed protostele
- (*c*) Actinostele
- (*d*) Dictyostele.

(*ix*) In ***Psilotum*** actinostele is present in the
- (*a*) Rhizome
- (*b*) Basal portion of the stem
- (*c*) Middle portion of the stele
- (*d*) Apical region of the aerial shoot.

(*x*) In which of the following genus, the prothallus has well defined scalariforim reticulate tracheids surrounded by phloem and endodermis?
- (*a*) ***Psilotum***
- (*b*) ***Lycopodium***
- (*c*) ***Selaginella***
- (*d*) ***Equisetum.***

(*xi*) In which of the following roots are absent?
- (*a*) ***Lycopodium***
- (*b*) ***Selaginella***
- (*c*) ***Tmesipteris***
- (*d*) ***Equisetum.***

(*xii*) ***Psilotum*** differs from ***Tmesipteris*** in the
- (*a*) The structure of sex organs
- (*b*) Prothallus
- (*c*) The presence of rhizoids in both
- (*d*) The structure of synangium.

(*xiii*) The synangium in ***Tmesipteris*** is
- (*a*) One lobed
- (*b*) Two lobed
- (*c*) Three lobed
- (*d*) Four lobed

(*xiv*) The presence of a fungus is essential for the development of prothallus in
- (*a*) ***Psilotum***
- (*b*) ***Tmesipteris***
- (*c*) ***Lycopodium***
- (*d*) ***Selaginella.***

CHAPTER 4

Lycophyta–Eligulopsida

Phylloglossum –Erect penduncles bearing terminal strobilii

- Introduction
- Characteristic features
- Classification of Lycophyta
- Description of ***Lycopodium*** Linn
- ***Lycopodium selago*** and ***L. phlegmaria***
- ***L. clavatum*** and ***L. annotinum***
- ***L. cernuum*** and ***L. inundatum***

Introduction

The division Lycophyta is represented both by living and extinct genera and has a very long evolutionary history as it extends from the palaeozoic era to the present day. The living members of this sub-division are represented by five genera and 900 species that are widely distributed over the surface of the globe and occur under varied climatic conditions. The five living genera are :—

(1) *Lycopodium,*

(2) *Phylloglossum,*

(3) *Selaginella,*
(4) *Stylites* and
(5) *Isoetes*.

Selaginella

Characteristic features

1. Sporophytic plant body consists of roots, stems and spirally arranged microphyllous leaves.
2. Stems protostelic rarely polystelic.
3. May be homosporous or heterosporous.
4. Sporangia thick walled, usually develop on sporophyte.
5. Antherozoids biflagellate or multiflagellate.

Classification

Lycopodium

The division Lycophyta has been divided into two classes:

1. Class **Eligulopsida** which includes only one order :

(*a*) **Order Lycopodiales :** It comprises two families and has both living and extinct plants.

Family : **Lycopodiaceae.** It includes living as well as extinct plants ; herbaceous, cambium absent, sporophyll with sporangia arranged in loose strobili ; homosporous ; exosporic gametophytes; sperms biflagellate. The family includes ***Lycopodium***, ***Phylloglossum*** and ***Lycopodites*** (extinct).

2. Class **Ligulopsida**. It includes the following four orders :—

(*b*) Order **Selaginellales :** It includes both living and extinct forms. It is characterised by microphyllous and ligulate leaves which may be isophyllous or anisophyllous; strobili loose; heterosporous; gametophytes endosporic; secondary growth absent; spermatozoids biflagellate. It includes family **Selaginellaceae**.

(*c*) Order **Lepidodendrales :** Includes only extinct plants that had tree-like habit and possessed vascular cambium, microphyllous and ligulate leaves, distinct strobili and two types of spores (heterosporous). Some of the genera included in the family Lepidodendraceae are : ***Lepidodendron***, ***Stigmaria***, ***Sigillaria***, ***Lepidocarpon.***

(*d*) Order **Isoetales :** Includes both living and extinct forms. They are characterised by sporophytes with a cormlike stem; perennial root producing meristem; secondary growth; small and ligulate leave; two types of spores; endosporic gametophytes; and multiflagellate spermatozoids. It includes the family **Isoetaceae**.

(*e*) Order **Pleuromeiales :** It is an extinct order with tree-like sporophytes that bore ligulate microphylls near their apices. They were heterosporus. It includes family **Pleuromeiaceae** with **Peluromeia** as one of the genera.

The order **Lycopodiales** is characterised by herbaceous and dichotomously branched stems, microphyllous leaves, absence of secondary growth, sporangia axillary or on adaxial side of the leaf, homosporous, exosporic gametophytes, biflagellate sperms, protostelic or siphonostelic (***Phylloglossum***) stele; exarch or mesarch (***Phylloglossum***) xylem. The order includes the family ***Lycopodiaceae*** that includes two living genera, ***Lycopodium*** and ***Phylloglossum***. The life histories of ***Lycopodium*** and ***Phylloglossum*** are discussed in this chapter.

LYCOPODIUM Linn.

Distribution and Occurrence

Lycopodium is a cosmopolitan genus with about four hundred species that are found under varied habitats. Some of them grow in the colder climates of the arctic region. There are others that inhabit temperate, tropical and sub-tropical regions of the world. They are very common in the tropical zones. In India the genus is represented by thirty-three species (Chowdhry, 1930, Panigrahi, 1962). Most of the species grow in Eastern Himalayas and in South Indian mountains. Mehra and Bir (1964) recorded twelve species from Darjeeling and Sikkim Himalayas. Bir and Vasudeva (1973) collected ***L. cernuum*** from Pachmarhi. Clarke (1880) reported two species from the western and eleven species from the eastern parts of the Himalayas. As many as nine species are found in Kodaikanal (South India), namely ***L. wightiana***, ***L. setaceum***, ***L. hamiltoni***, ***L. cernuum***, ***L. clavatum***, ***L. phlegmaria***, ***L. serratum***, ***L. complanatum and L. phyllanthum***. The stem is copiously covered with sharp and pointed leaves. Their moss-like appearance and club-shaped strobili, earn them the name 'club mosses'.

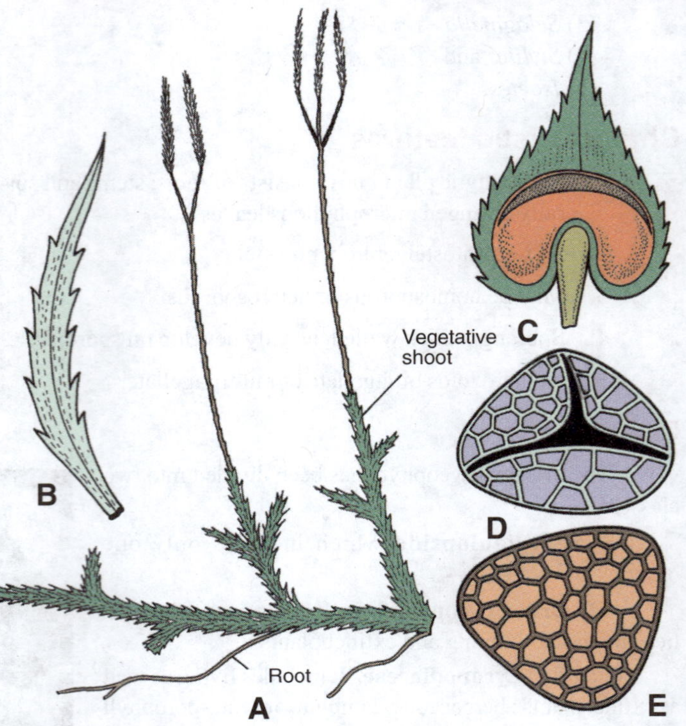

Fig. 4.1 (A–E). *Lycopodium clavatum*. A. Portion of a plant showing habit; B. A sterile leaf; C. A sporophyll bearing sporangium; D. A spore seen from above showing triradiate ridge; E. Spore showing reticulations of exine.

Habit

Most of the tropical species are epiphytic and grow hanging from the tree trunks e.g., ***L. phlegmaria*** (Fig. 4.3), ***L. squarrosum***, ***L. linifolium*** and ***L. funiform*** (all Indian species). ***L. phlegmarioides*** and ***L. carinatum*** are also epiphytic. The temperate species grow in open wood-lands and on moist and acidic soils. ***L. lucidulum*** and ***L. reflexum*** are erect and shrubby forms. ***L. annotinum*** is also an erect and terrestrial form. It is sparsely branched and not bushy. ***L. clavatum***, ***L. inundatum*** (Fig. 4.1, 4.4) and ***L. cernuum*** are common creeping forms. ***L. drumondii*** and ***L. carolianum*** have short creeping stems with erect strobilus bearing branches. ***L. volubile*** (Fig. 4.2) is a climber. The creeping forms give out erect branches at intervals. The erect and shrubby forms also have a creeping stem or the so called rhizomes which may be above ground in ***L. densum*** and subterranean in ***L. obscurum***. The erect branches in these species are much branched and appear bushy.

L. annotinum *L. clavatum*

L. inundatum *L. lucidulum*

L. volubile *L. squarrosum*

Stem

The stem is dichotomously branched. The two branches formed as a result of dichotomy may be equal and continue to grow and undergo further dichotomy, or the two branches of a dichotomy may be unequal, ***i.e.***, one of them remains small and the other grows to a greater length. The smaller or the weaker branches may ultimately stop growth and bear strobili (Fig. 4.1). This type of branching (unequal dichotomy) is called **pseudomonopodial** (***L. clavatum***, ***L. densum***, ***L. obscurum***). In prostrate species the main axis or the rhizome is completely or partially subterranean.

Fig. 4.2 (A-D). *Lycopodium volubile*. A. Portion of a plant; B. Fertile branch with two strobili; C. Sterile branch with larger leaves in two lateral rows; D. Portion of branch with two lateral rows of large leaves and two rows of small leaves.

Leaf

The leaves are simple, sessile and small in size and possess a single median vein that fails to reach the leaf apex. Such leaves are called **microphyllous** or **microphylls**. The leaves are scale like in ***L. densum***. The leaf margin may be entire or serrate (***L. serratum***). The leaf base is decurrent in ***L. complanatum***. The leaves lack a ligule. Usually the leaves are 2—10 mm. long but in some cases their length reaches 2 to 3 cm. long. The leaves are spirally arranged and densely cover the branches. In some cases the leaf arrangement may be decussate (***L. alpinum***) or even whorled (***L. verticillatum***, ***L. nikoense*** and ***L. cernuum***). The phyllotaxis appears to be opposite in the mature plant of ***L. complanatum*** (Bhambie, 1965).

L. alpinum *L. serratum* *L. cernuum*

The leaves may be **isophyllous** or **anisophyllous** (***L. volubile***). The branches in ***L. volubile*** bear four rows of leaves : two lateral rows of large and falcate leaves, ventral row of small hair-like leaves, and a dorsal row of medium sized acicular leaves. The creeping axis bears long and needle-like leaves. In addition to the normal leaves the bulbils in ***L. selago*** develop thick and fleshy leaves. In ***L. clavatum*** the strobilus bearing or fertile branches bear scale-like vegetative leaves that are yellowish green in colour. The sterile branches bear large linear and green leaves that are dentate.

Root

The roots are dichotomously branched, adventitious and arise endogenously from the lower side of the prostrate stems in an acropetalous manner. The ephemeral primary root of the young sporophyte soon dies.

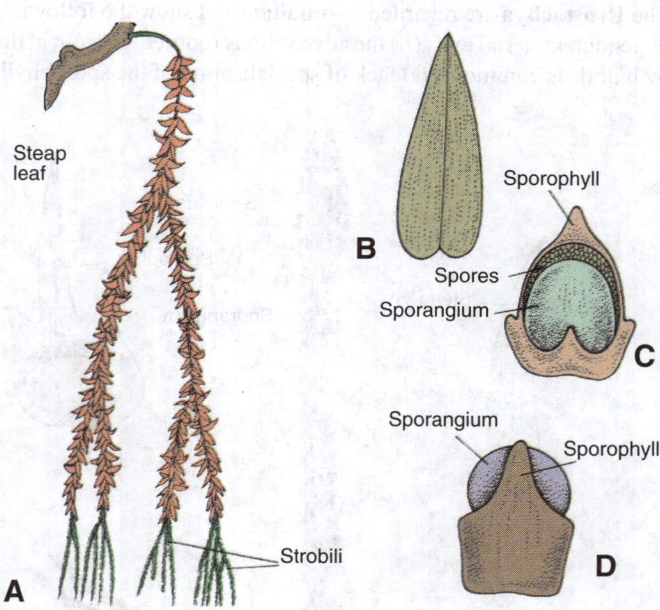

Fig. 4.3 (A–D). *Lycopodium phlegmaria.* A. The plant hanging from the host tree and showing pendent branches and strobili; B. A leaf ; C-D. Sporangia with sporophylls.

In the erect forms the roots come out from the basal part of the stem. There are no roots elsewhere along the entire legnth of the stem. In such cases the adventitious roots initiate near the shoot apex but do not come out of the stem, instead they grow downwards through the cortex and emerge out of the stem only near its base. A series of transverse sections of such stems shows numerous roots in the cortex (Fig. 4.12). Such roots are called **cortical roots** and are found in ***L. selago***, ***L. pithyoides***, ***etc***. Stockey (1907) reported as many as 52 cortical roots at one level of stem in ***L. pithyoides***. This is due to their acropetalous origin. The roots on coming out of the stem (aerial roots) branch freely in a dichotomous manner. In ***L. clavatum*** and ***L. cernuum*** (prostrate forms) the creeping axis bears adventitious roots all along its length. In these cases the root primordia take more direct course through the cortex.

Rothmaler (1944) suggested, on account of the variations in the genus, to split it up into four genera. He named these genera as ***Lycopodium***, ***Lepidotis***, ***Diphasium***, and ***Huperzia.*** Rothmaler's treatment is also supported by the recent cytological investigations of the various species. His four genera show a complete cytological disparity. Pritzel (1900) divided the genus into two sub-genera which he named as ***Rhopalostachya*** and ***Urostachya***. The former includes all those species that have prostrate main stems giving out erect secondary branches (***L. clavatum*** and ***L. cernuum***). The latter includes the erect and epiphytic species (***L. pithyoides***, ***L. selago***, ***L. selago***, ***L. phlegmaria***, etc.).

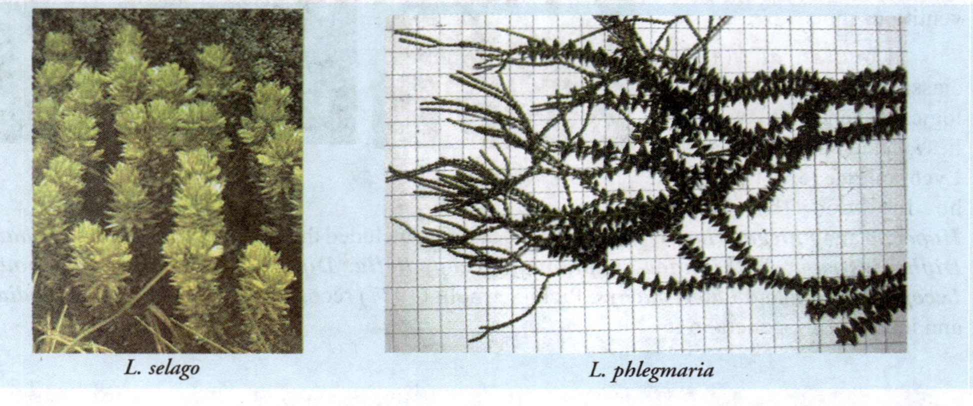

L. selago *L. phlegmaria*

The **Urostachya** are regarded as primitive and show the following characters :— (*i*) they have erect or pendulous aerial axis; (*ii*) the adventitious roots emerge out at the base; (*iii*) vegetative propagation by bulbils is common; (*iv*) lack of specialisation of the sporophylls.

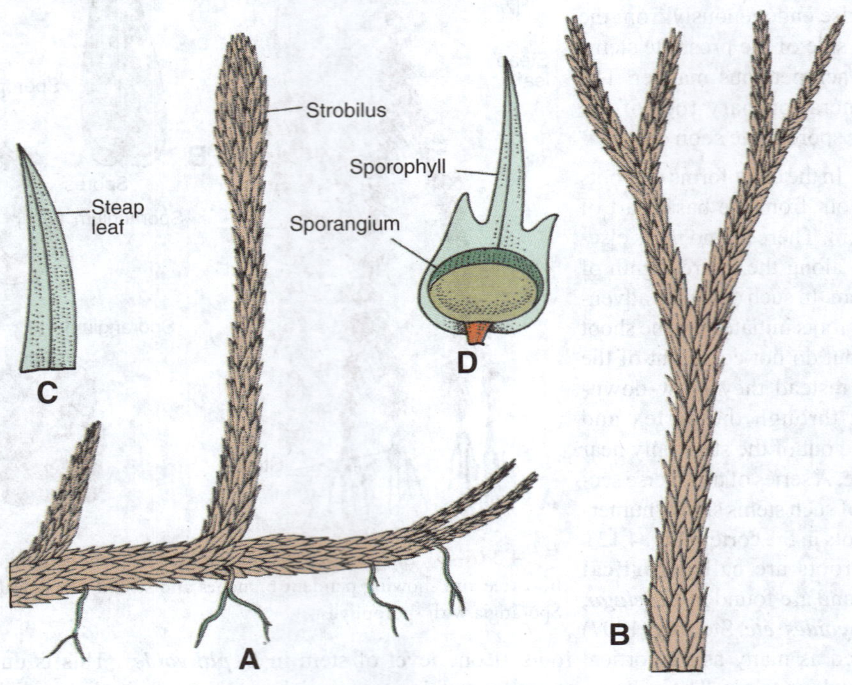

Fig. 4.4. *Lycopodium*. External morphology. A. *L. inundatum*; B. *L. selago* ; note that the sporangia occur scattered and do not form a distinct strobilus; C. A sterile leaf from A; D. A sporophyll with sporangium from A.

The Rhopalostachya show the following features:— (*i*) the main stem prostrate; (*ii*) bulbils rare; (*iii*) sporophylls specialised; (*iv*) the roots may emerge at any point so that the whole axis is covered by adventitious roots.

Huperzia *Diphazium*

Love ***et. al.***, (1977) classified the genus on cytological grounds into two families ***i.e.***, the **Huperziaceae** and **Lycopodiaceae**. In the former he included the genera ***Huperzia*** and ***Phlegmariurus***, whereas in the latter he included the genera ***Lycopodium***, ***Diphazium***, ***Diphasiastrum***, ***Palhinhaea***, ***Lepidotis***, ***Lycopodiella***, ***Diphasiopsis***, ***Pseudolycopodium***, ***Lycopodiastrum*** and ***Lateristachys***. Pichi-sermolli (1977) recognised only family ***Lycopodiaceae*** and included six genera in it.

Growth

The growth of the shoot takes place by means of an apical meristem which consists of a group of apical cells (Fig. 4.5). The derivatives of these cells differentiate into the primary meristematic tissue : the **protoderm**, the **ground meristem** and the **procambium**. The protoderm gives rise to the outer protective layer of the epidermis. The ground meristem differentiates the cortex, endodermis, pericycle and the pith. The procambium gives rise to the vascular tissue (xylem and the phloem). Wetmore (1943) has shown that the procambium is centrally located and extends very close to the shoot apex. Its cells are enlongated and usually divide in a longitudinal plane and only frequently in transverse plane.

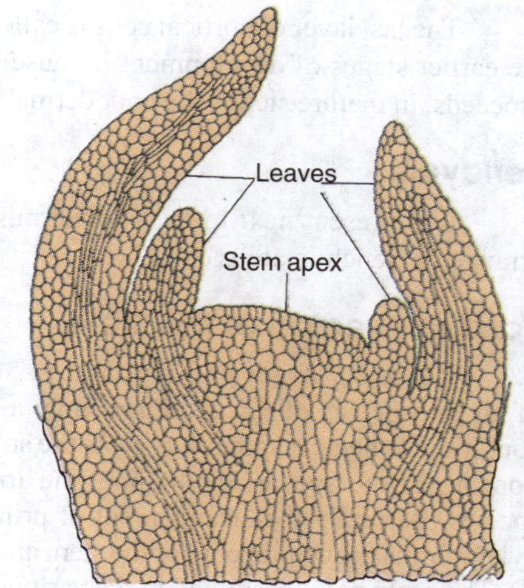

Fig. 4.5. Longitudinal section of stem-apex of *Lycopodium reflexum.*

ANATOMY

Stem

A transverse section of the stem reveals :

Epidermis

It is a single layer of parenchymatous cells with cutinised outer walls and stomata.

Cortex

It is present next to the epidermis and shows a great variation in its thickness and structure. Usually three types are recognisable :—

1. In some cases the whole of the cortex is made up of thin-walled cells. They enclose small or larger intercellular spaces. Such a cortex is called homogenous (***L. selago***).
2. In some species the cortex in the mature stems is made up wholly of sclerenchymatous cells. There are no intercellular spaces between the cells.
3. In others (***L. clavatum***) the cortex is differentiated into three zones (Fig. 4.6, A).
   (*i*) The outer zone of thick-walled cells (**Hypodermis**)
   (*ii*) The middle zone of large and thin-walled cells. These cells enclose lot of intercellular spaces.
   (*iii*) The inner zone of thick-walled cells.

In ***L. cernuum*** the outer and inner cortex are thin-walled and the middle cortex is sclerenchymatous (Fig. 4.6, B).

The stem diameter and its position also effect the composition of the cortex. In the underground stems the cortex is usually entirely sclerenchymatous. In stems with smaller diameter the cortex may be entirely parenchymatous or chlorenchymatous and the cells are usually loosely arranged. The sclerenchyma is either absent or greatly reduced. The above ground stems with larger diameter usually show differentiation into chlorenchymatous, parenchymatous and sclerenchymatous cortex.

The last layer of cortical cells is called the **endodermis**. Endodermis is clearly recognisable in the earlier stages of development because of the presence of casparian strips. As the development proceeds, in mature stems, the endodermal cells become thickened.

Pericycle

It is present next to the endodermis and is composed of one or many layers of compactly arranged parenchymatous cells.

Vascular Region or the Stele

It consists of only primary tissue ***i.e.***, the primary xylem and the primary phloem. The sporelings of all the species of ***Lycopodium*** show a central protostele with radiating arms, ***i.e.***, **actinostelic** condition. When the sporelings grow the stelar structure goes on becoming complex and shows variations in various species. Sometimes the arrangement differs at different levels in the shoot of same species. Basically the arrangement of primary xylem and phloem is a **protostele**. In this case the xylem forms the central core of the stem and pith is absent. The phloem surrounds the xylem (Fig. 4.6, C). The protostele, as studied in the various species of ***Lycopodium*** shows variations with respect to its shape and arrangement of vascular tissue (xylem and phloem). It may be :—

(*i*) **actinostele** (Fig. 4.6, C) in which the xylem appears star-shaped with protoxylem situated at the (Fig. 4.6, C) tips of the star-shaped projections and is exarch (***L. selago***, ***L. serratum. L. phlegmaria***) ; (***ii***) **plectostele** in which the xylem, in a transverse section, appears to be in form of separate plates of (***L. volubile***, ***L. clavatum***) variable sizes with phloem in between them (Fig. 4.6, A); and (***iii***) **mixed protostele** in which the xylem and phloem are uniformly distributed and in a transverse section, it appears as if strands of xylem are embedded in the phloem (Fig. 4.6, B) which forms the ground mass (***L. cernuum***). The xylem is exarch.

L. lateralis

In a plectostele the separate bands of xylem and phloem appear interconnected in a longitudinal section. They remain free only up to a certain distance up and below. In a mixed protostele the xylem occurs in the form of irregular tracheids scattered in the ground mass of phloem (Ogura, 1938). It is also found in ***L. lateralis*** and ***L. drummondii.***

There is no doubt that the arrangement of vascular elements varies with species but on the other hand it has also been observed (Wardlaw, 1924, Bierhorst, 1971) that the diameter of the stem also plays a significant role in effecting this variation. In some species the stems with smaller diameter have actinostelic protostele whereas in the same species the stems with larger diameter have plectostelic protostele ***e.g.***, ***L. densum***. In ***L. phlegmarioides*** the stele may be plectostelic in some parts and a mixed protostele in parts with larger diameter.

The protoxylem elements are composed of annular or spiral trachieds. Sometimes the rings in the annular tracheids are inter-connected with each other by one or two oblique or vertical bands of thickening. Bierhorst (1960) has reported the tracheids of the metaxylem to have circular bordered pits (***L. selago***, ***L. serratum*** and ***L. lucidulum***) or sometimes sclariform pits. The end walls of the tracheids are oblique and possess elongated pits. The phloem is composed of sieve cells and phloem parenchyma. The sieve cells are long and tubular with sieve areas on the lateral as well as end walls. The sieve elements develop centripetally from outer angles towards the centre. The vessels are absent.

Strands of vascular tissue called the **leaf traces** extend outwards from the stele and run obliquely upwards through the cortex. Only one strand enters the leaf to form a single median vein. Similarly stele is connected with each branch by a **branch trace**. In a transverse section the leaf traces appear as distinct vascular strands.

There are no leaf gaps.

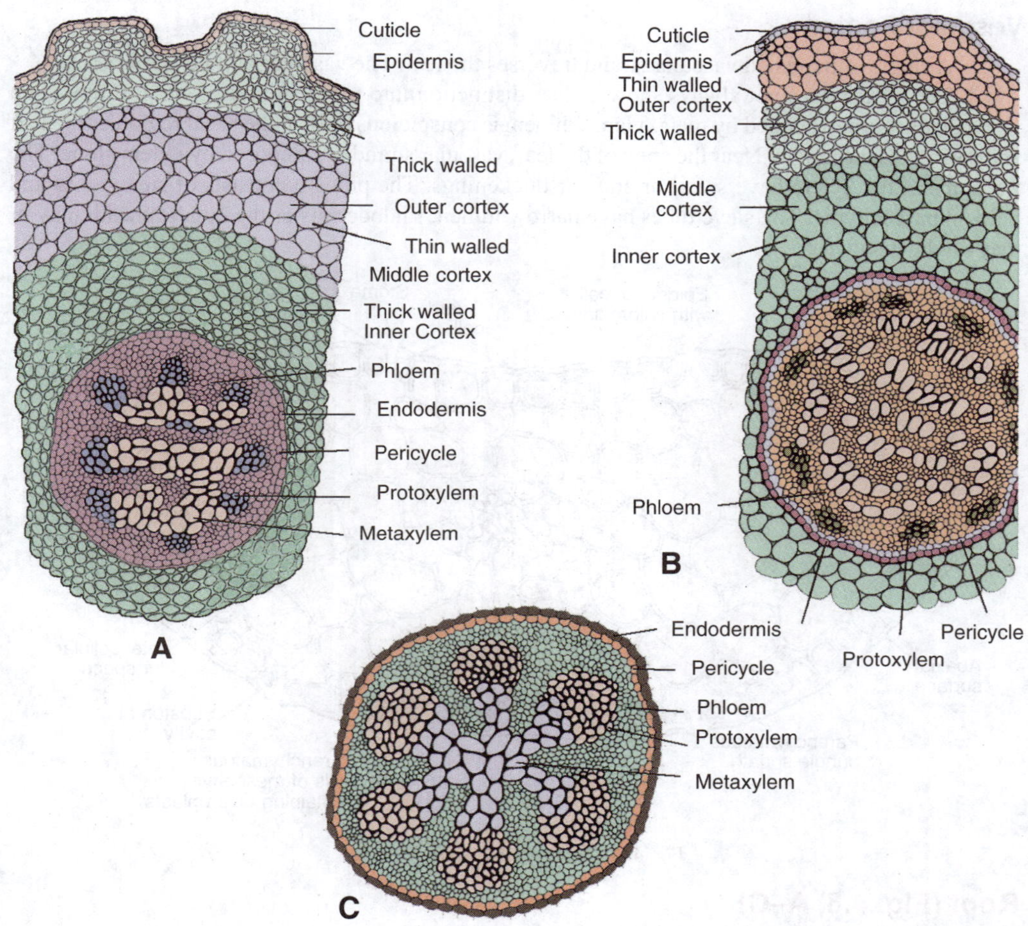

Fig. 4.6. *Lycopodium*–Stem anatomy. *A. L. clavatum.* T.S. portion of stem. Note inner sclarenchymatous cortex and plectostelic-protostele; *B. L. cernuum.* The middle cortex is sclerenchymatous. The stele is a mixed protostele; *C. L. serratum.* Actinostelic protostele

Leaf

In ***Lycopodium clavatum*** the leaf has s triangular outline in a transverse section. In ***L. selago*** (Fig. 4.7) the leaf is almost flattened. A transverse section reveals the following structure :—

Epidermis

The epidermis is a single layer of compactly arranged cells with sinuate and thick outer periclinal walls which are also covered by a layer of cuticle. It is perforated by stomata. The cuticle extends over the guard cells. The stomata may be on both the sides of the leaf in isophyllous species and only on lower epidermis in some anisophyllous species (***L. volubile***, ***L. complanatum***).

Mesophyll

It is undifferentiated and consists of loosely arranged chlorophyllose cells with small or large intercellular spaces. The cells may be round or oval or slightly lobed.

Vascular Region

A single unbranched vascular strand traverses the leaf. The vascular bundle is **mesarch.**

The xylem does not always show a clear distinction into proto and metaxylem. Protoxylem when distinct is surrounded by metaxylem. Phloem is conspicuous only in the basal region of the leaf and surrounds the xylem. Near the apex of the leaf, vascular strand is composed by xylem alone. The tracheids in the xylem have spiral or annular thickenings. The phloem consists of sieve tubes and phloem parenchyma. The sieve tubes have narrow lumen. Endodermis or the bundle sheath may or may not be distinct.

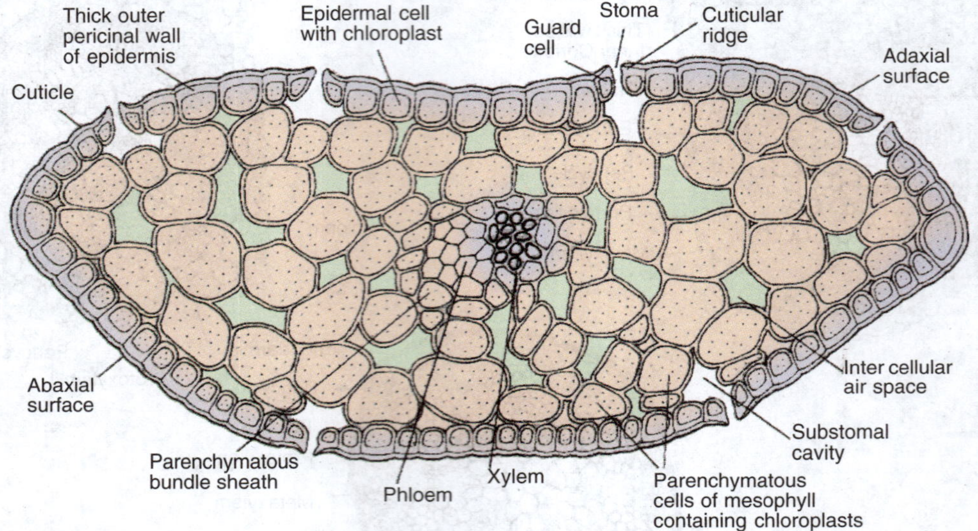

Fig. 4.7. *Lycopodium selago*, V.S. leaf.

Root (Fig. 4.8, A–C)

The adventitious roots arise from the pericycle. A transverse section of the aerial root reveals the following structure:

Epidermis (Fig. 4.8, A)

It consists of a single layer of thin-walled cells and is covered by numerous root hairs. The root hairs are arranged in pairs and their origin is characteristic for the genus ***Lycopodium***. An epidermal cell divides either by an anticlinal or by an oblique wall into two daughter cells. These are called the **root hair initials**. Both of them give out root hairs which therefore, lie in pairs. Epidermis is persistent in the aerial roots.

Cortex

In mature roots the cortex forms a broad zone and can be differentiated into two distinct regions. The outer cortex which is composed of several layers of thick-walled cells (Fig. 4.10) and the inner parenchymatous cortex. Endodermis and pericycle are not distinguishable.

Stele

The stele may be diarch, tetrarch, or polyarch. In most of the prostrate or creeping species ***e.g.***, ***L. clavatum***, the stele is polyarch. There is no pith and the stele consists of six to ten plates of xylem arranged in a radial fashion. The xylem is exarch. The metaxylem of each radial plate of xylem unites in the centre to give a star-shaped appearance. The phloem is present in the form of patches in between the radiating arms of xylem. This structure is just like the actinostelic prostostele.

In most of the erect and epiphytic species the stele is diarch or tetrarch. In ***L. pithyoides*** it is diarch and the xylem is crescent-shaped with protoxylem at the ends ***i.e.***, exarch. The phloem occupies the region between the two ends of the crescent-shaped xylem ***i.e.***, on the concave side. There is only one group of phloem.

In ***L. selago*** the stele is diarch in one part of the root and tetrarch in the other. The diarch condition (Fig. 4.8, A, B) is similar to the one described for ***L. pithyoides***. In tetrarch condition the stele reveals two separate bands of xylem with protoxylem at either end (exarch). The phloem lies between the xylem bands. This condition is somewhat similar to the plectostelic protostele. In some cases more than two groups of protoxylem occur on the convex sides. Sometimes the entire convex side is covered with protoxylem tracheids.

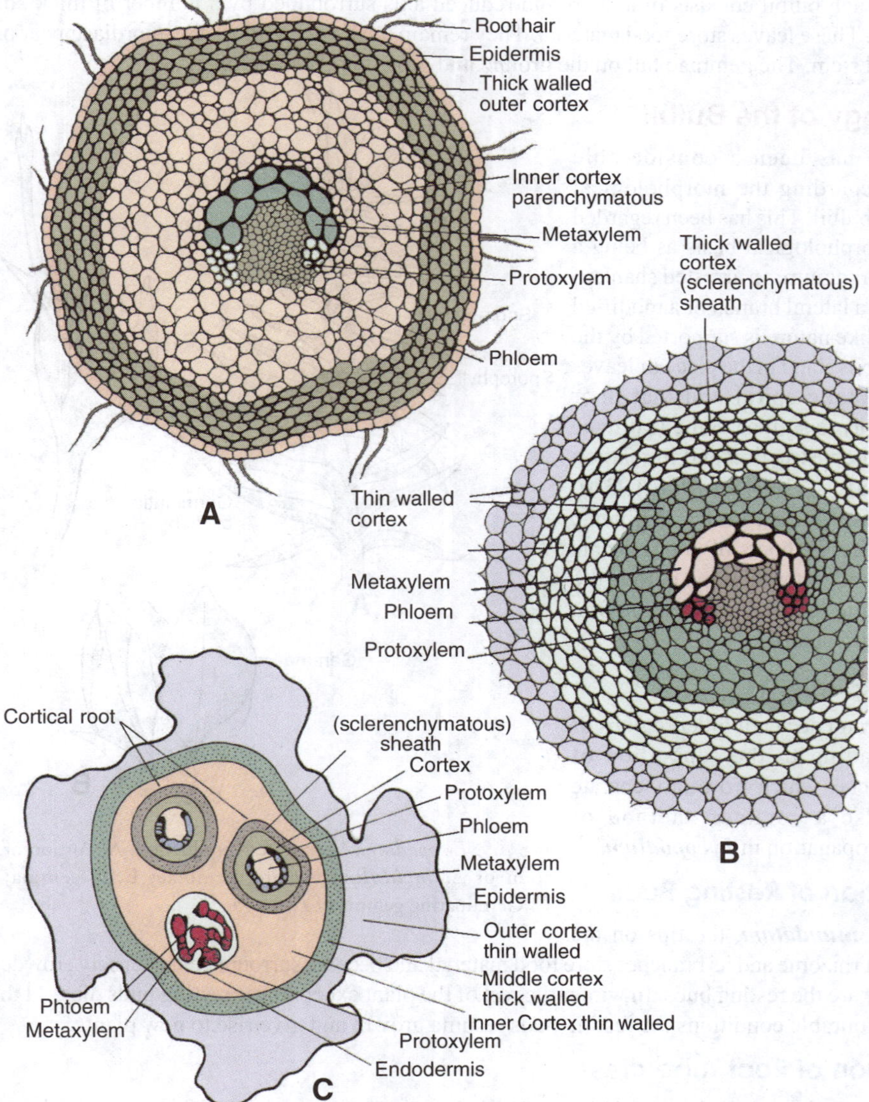

Fig. 4.8. ***Lycopodium selago.*** T.S. Root. **A.** T.S. Aerial Root; **B.** T.S. Cortical Root; **C.** Diagrammatic T.S. through stem showing one stele and two cortical roots.

The cortical roots (Fig. 4.8, B and C) are almost similar in their internal organisation to the aerial roots. In the cortical roots the epidermis is not persistent and the root hairs are absent or appear as small protuberances (trichoblasts). The cortex is usually wholly sclerenchymatous or may be partly so.

VEGETATIVE PROPAGATION

The sporophyte of ***Lycopodium*** reproduces vegetatively by the following methods :—

1. By the Formation of Gemmae or Bulbils (Fig. 4.9, A–C)

These are modified vegetative structures that arise as lateral out-growths from near the stem apices and take the place of leaves. They have been reported in ***L. selago***, ***L. phlegmaria*** and ***L. lucidulum***. Each bulbil consists of a short and reduced axis surrounded by a number of thick and fleshy leaves. These leaves store food material. They remain on the plant till root primordia appear on the shortened stem. The gemmae fall on the ground and grow into a young plant.

Morphology of the Bulbil

There has been a considerable discussion regarding the morphological nature of the bulbil. This has been regarded on purely morphological basis as being a modified sporangium, an arrested shank of a dichotomy, a lateral branch or a modified leaf. Its leaf-like nature is supported by the facts that bulbils stand in the place of leaves in the phyllotatic spiral and that their vascular supply is similar to the leaf trace in anatomical construction and in relation to the axial stele (Smith, 1930). Experiments on ***L. selago*** (S. Williams, 1933) show that if shoots of young plants are decapitated immediately below the apex, the young leaf rudiments transform into regenerative buds.

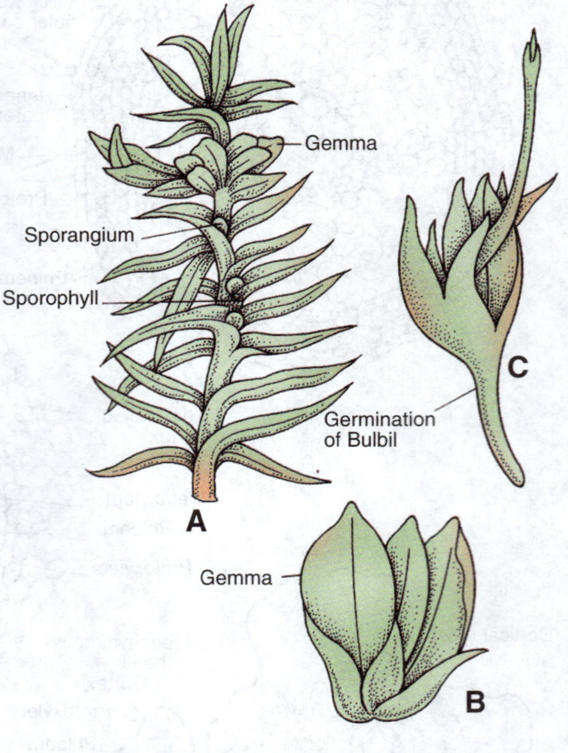

Fig. 4.9. *Lycopodium*. Vegetative reproduction. A. Portion of stem of *L. lucidulum* bearing gemmae; B. A gemma; C. Germinating gemma of *L. selago*.

2. Fragmentation

Death and decay of older regions of the stem leading to the separation of younger branches which grow into separate plants is also a common method of vegetative propagation in ***Lycopodium.***

3. Formation of Resting Buds

In ***L. inundatum***, the tips on the apical buds of rhizome and its branches store food material and become surrounded by densely crowded leaves. These are the resting buds. In winter the rest of the plant except these resting buds dies. At the advent of favourable conditions the resting buds resume growth and give rise to new plants.

4. Formation of Root Tubercles

In ***L. ramulosum*** (Holloway, 1917), tubercles originate from the parenchymatous region of the root cortex. They consist of a group of cells with stored food material and protected by thick walls and have the capacity to germinate into new plants.

5. Formation of Adventitious Buds

Such buds have been induced on isolated bulbil leaves. In the leaves epidermal cells proliferate near the base and grow into buds. Decapitation of stem near its apex also induces the formation of such buds. In this case the leaf rudiments change into such buds. Such experiments of adventitious bud induction were carried out on ***L. selago*** by Williams (1933). Goebel reported the formation of such buds from isolated leaves of sporelings in ***L. inundatum.*** These buds are capable of germinating into new plants.

ORGANISATION OF THE STROBILUS

In the simple and primitive species of ***Lycopodium*** all of which belong to the ***Urostachya*** (***L. selago***, ***L. lucidulum***, ***L. squarrosum***, etc.) every leaf on the plant is a sporophyll or at least potentially so. In the species belonging to ***Rhopalostachya*** and in some species belonging to ***Urostachya*** (***L. phlegmaria***) the leaves near the apices of the branches bear sporangia and are called the **sporophylls**. The lower leaves in these species are sterile and function merely as foliage leaves. Aggregation of sporophylls is called the **strobilus**. The arrangement of the sporophylls may be loose when they form a lax or a loose **strobilus** or they may be arranged compactly to form a compact or a definite **strobilus**. The compact type of strobilus is a stem with short internodes bearing sporophylls. It develops from an apical meristem and its vasculature is similar to that of vegetative axis.

In ***L. selago*** and ***L. reflexum*** (Fig. 4.4, B) the sporophylls are similar to the foliage leaves and are arranged in a loose manner at intervals all along the erect branches. Such an arrangement gives rise to alternating fertile and sterile regions. Such a condition is regarded by many morphologists as a strobilus comprising the entire plant. The supporters of this view regard the foliage leaves as secondarily sterilised sporophylls. Such a conclusion is based on the observation that such leaves also bear abortive sporangia. The same is the condition in ***L. lucidulum***. In ***L. squarrosum*** the strobilus is terminal in position but can be hardly distinguished from the sterile region because the sporophylls are similar to the foliage leaves and are loosely arranged.

In ***L. phlegmaria*** (Fig. 4.3) the sporophylls are smaller than the foliage leaves and are arranged in a compact manner to form distinct strobili (Fig. 4.3) at the distal ends of the stem and its branches. The strobili in this species are dichotomously branched and pendulous.

In the section ***Rhopalostachya*** there is a great variatioin among the sterile leaves and the sporophylls. In ***Lycopodium inundatum*** (Fig. 4.4, A) the sporophylls differ only slightly from the sterile leaves and the strobilus is not very distinctly marked off from the vegetative shoot. The sporophylls are slightly modified (Fig. 4.4, D) so as to protect the sporangium.

In ***Lycopodium annotinum*** the strobilus is quite distinct from the vegetative shoot. The sporophylls are quite distinct from the sterile or vegetative leaves. Each sporophyll has an abaxial outgrowth or a flange that protects the sporangia belonging to the lower sporophylls. The strobili of this species terminate normal leafy branches and the sterile leaves are all alike.

Lycopodium annotinum

In *L.* ***clavatum*** (Fig. 4.1) the strobili are quite distinct and compact and are borne at the tips of special erect shoots. There is segregation of fertile and sterile branches. The sporophylls possess a small flange on ventral side. It protects the sporangium below (Fig. 4.10, D).

In ***L. volubile*** (Fig. 4.2, A–D), the strobili terminate younger branches and are compact and branched (Fig. 4.2, B). The sporophylls possess a small bulge or a flange on the ventral (abaxial) side. This outgrowth affords some protection to the sporangia below.

In some species with distinct strobili the strobilar apex, under special growth conditions, continues its growth to produce a vegetative extension of the strobilus ***e.g.***, ***L. obscurum*** and ***L. flabelliforme***. The same is the case in ***L. drummondii*** where the strobiler apex resumes growth at a later stage to produce a vegetative region which again bears a strobilus and this sequence is repeated many times.

There is a considerable variation regarding the position of the sporangium with respect to the sporophyll. In ***L. selago***, ***L. phlegmaria*** (Fig. 4.10, E) ***L. lucidulum*** and ***L. inundatum*** the sporangia are **axillary** in position. In ***L. cernuum*** and ***L. clavatum*** (Fig. 4.10, A and D) the sporangium arises on the dorsal (adaxial) surface of the sporophylls, that is, it is **'foliar'** or **'epiphyllous'**. In ***L. squarrosum*** the position is **sub-foliar**, that is, in this case the sporangium does not arise exactly on the sporophyll, but a little towards the sporophyll. Every sporophyll is supplied, from the stele of the stem by a single trace.

A careful perusal of the facts cited above reveals that in the simpler species where there is no or very little differentiation between vegetative and reproductive regions, the stems are dichotomously branched and possess a radial type of stele. The cone bearing species on the other hand have stem with pseudo-monopodial branching and most of them show the parallel-banded type of stele.

SPORANGIUM

The sporangium is mostly reniform or kidney-shaped (Fig. 4.1 C 4.2 C D; 4.10) and sometimes sub-spherical. It is usually raised on a short and massive stalk (Fig. 4.10). The stalk of the sporangium is indistinct in some species. The sporangia arise singly and may be exposed or protected. They are exposed in ***L. squarrosum***, ***L. lucidulum*** (Fig. 4.9, A), ***L. selago*** etc., and protected in ***L. inundatum***, ***L. cernuum***, ***L. clavatum etc***. In the latler case they are protected by flange-like outgrowth from the sporphylls above. They vary in colour from orange to yellow. The usual length of sporangium is two milimeters or a little more.

Structure of the Sporangium

The sporangium is clearly distinguishable into two (fig. 4.11, F, G) prominent parts : (*i*) the **stalk** and (*ii*) the **capsule**.

The stalk may be short and distinct as in ***L. phlegmaria*** (Fig. 4.10, E) or may be just represented by a cushion of cells (***L. cernuum***). It may be narrow as in ***L. selago*** or broad as in ***L. alpinum***.

The stalk bears a curved and reniform capsule which is not lobed and is unilocular (Fig. 4.10, A). It has a three or more layered wall surrounding a mass of spore mother cells (Fig. 4.11, G). In ***L. selago*** the sporangial wall is three layered at the apex and may be more than three layers near its base. The innermost layer of wall is called the **tapetum**. The tapetum forms a complete investment of the sporogenous tissue. Its cells are prominent due to the more granular nature of their cytoplasm and bigger nuclei. These cells contain nutritive substances. The tapetal layer maintains its identity throughout the life of the sporangium. The sporogenous cells separate and round off to form **spore mother cells** (Fig. 4.11, G) which undergo meiosis. The tetrads are tetrahedral and after secreting their own walls they separate into spores. The number of spores per sporangium varies considerably.

Development of the Sporangium (Fig. 4.11, A–G)

The sporangium in ***L. selago*** originates from a single transverse row of superficial cells (Fig. 4.11, A, B) on the dorsal or upper surface of the sporophyll near its base or close to its point of attachment with the stem. In a radial section only one of the cells in a row is visible and it appears as if the sporangium develops from a single cell. As a matter of fact, all the cells in this transverse row take part in the development of the sporangium. Each of these cells divides periclinally to form a middle row of **archesporial cells** (Fig. 4.11, B, C, D) covered by a single row of **primary wall cells** above and a row of cells below (Fig. 4.11, C). In a median radial section (Fig. 4.11, B) this stage shows a row of three cells; (***i***) an upper cell which contributes to the sporangial wall (primary wall cell), portion or whole of the sporangial stalk and part of the tapetum, (***ii***) a middle cell which is archesporial in nature, and (***iii***) a lower cell that contributes to the **tapetum** in the lower or the basal region of the sporangium. This lower cell as it appears in median radial section or a row of lower cells as they appear in a tangential section constitute the **sub-archesporial cells**. Since the sporangium originates from a group of cells the development is of the **eusporangiate** type.

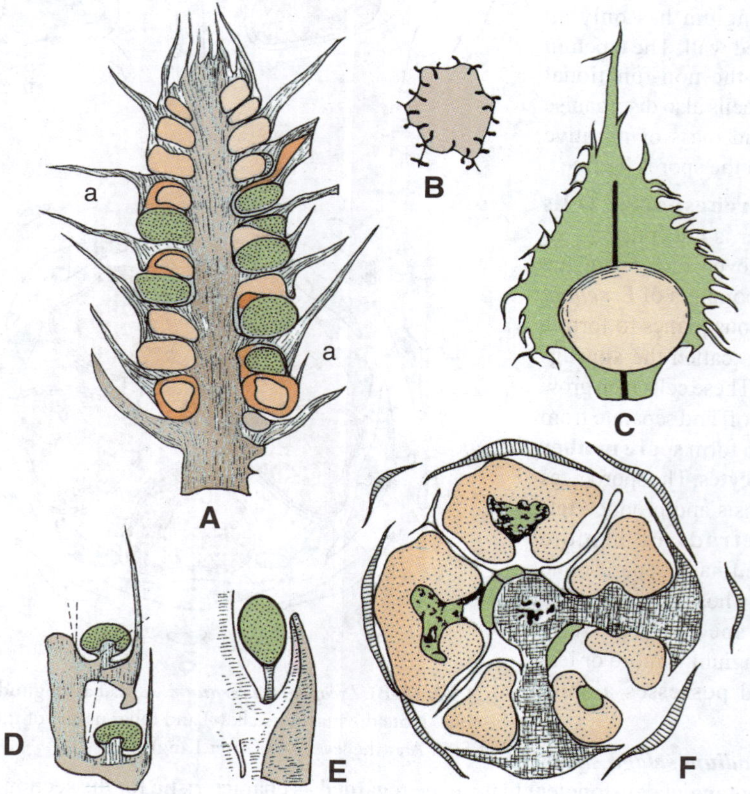

Fig. 4.10 (A–F). A. *Lycopodium cernuum* longitudinal section of strobilus: *a,* band of lignified cells; B. *L. cernuum.* Cell from sporangium wall; C. *L. cernuum.* Sporophyll and sporangium ; *lt,* vascular bundle; D. *L. clavatum.* Part of radial longitudinal section of strobilus ; *P,* sterile tissue; E. *L. phlegmaria.* Sporophyll and stalked sporangium; F. *L. clavatum.* Transverse section of strobilus ; *p,* sterile pad, a,b,c,d, sporangia (After Seward, 1910).

The primary wall cells undergo periclinal and anticlinal divisions to form a three or more layered thick sporangial wall. The inner-most layer of wall is differentiated as the **tapetum.** Its cells are prominent and contain stored food material. This layer provides nutrition to the developing sporogenous tissue. The tapetum in the basal region of the sporangium differentiates from the subarchesporial cells. The tapetal layer in ***Lycopodium*** remains intact and does not undergo disorganisation to form the **periplasmodial** fluid. It has been observed that in ***L. selago*** and other species the outer wall cells of the maturing sporangium elongate anticlinally and develop thickenings on their inner and radial walls. The inner wall layers break down completely or nearly so, so that the mature sporangium has only an outer thickened wall. The tapetum and some of the non-functional spore mother cells also disorganise to form a fluid mass or nutritive fluid in which the spores lie free.

The **archesporial cells** whose number is not constant, although it is seven in most of the developing sporangia of ***L. selago*** divide in various planes to form a group of cells called the **sporogenous tissue**. These cells then grow in size, round off and separate from one another to form **spore mother cells** or **sporocytes**. The sporocytes undergo meiosis and produce **tetrahedral tetrads** of haploid spores. These finally separate and secrete a wall whose ornamentation varies with species. The fully grown sporangium is more or less reniform and possesses a short stalk.

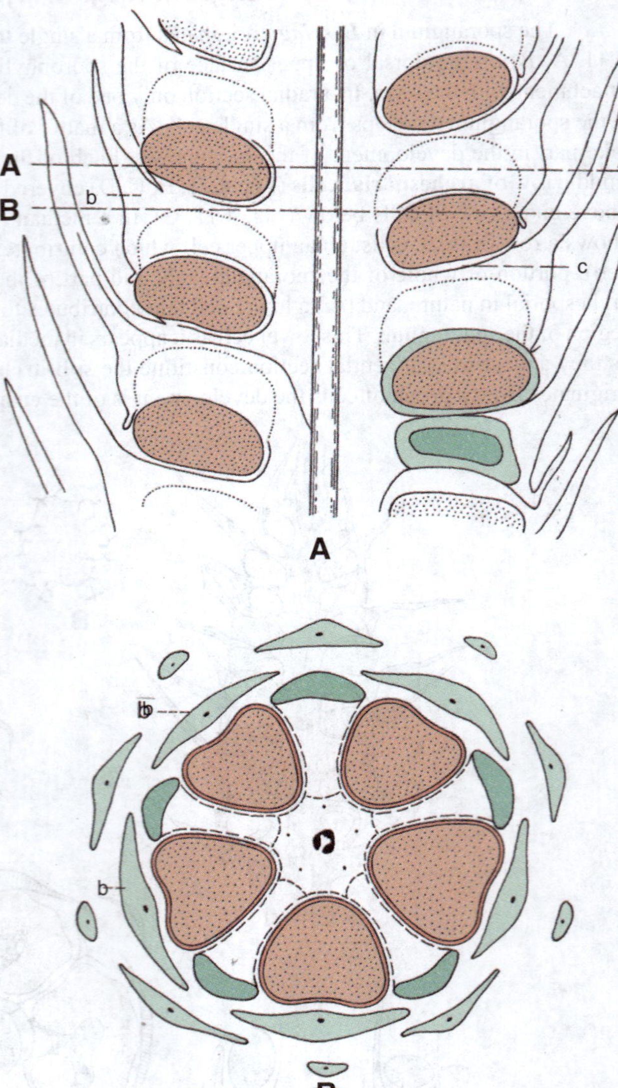

Fig. 4.10a (A-B). *Lycopodium cernuum*. A. Radial longitudinal section of the cone. Note the massive stalk (*c*) and foliar nature of the sporangium; B. T.S. of A at the level A-A (After Lang)

Lycopodium selago shows the type of sporangial development that can be regarded as characteristic for the section ***Urostachya***. There are, however, differences in the number of cells in the archesporial row and in the number of layers in the sporangial wall. In ***L. dichotomum*** the sporangial wall is 4-7 layers thick.

In ***L. inundatum*** which belongs to ***Rhopalostachya*** there are two tangential rows of archesporial cells as compared to one in ***L. selago***.

However, in ***L. clavatum***, and ***L. alpinum***, both of which belong to ***Rhopalostachya*** group there are three tangential rows of archesporial cells. The sporangium thus is more massive. In these species the **subarchesporial** pad is comparatively well developed and extends in between the sporogenous tissue as small processes. The sporangial stalk is thick and short as compared to the long and narrow stalk in *L. selago*. Bower regarded ***L. inundatum*** as an intermediate form linking the two types of sporangial development in ***L. selago*** and ***L. clavatum***.

Dehiscence of Sporangium

In species with compact strobili the dehiscence of the sporangium is preceded by elongation of the internodes of the central axis. As a result the sporophylls spread out and expose the sporangia. The sporangia dehisce along a line of cells running across the upper surface of the kidney-shaped sporangium. This line of cells is differentiated in the outermost layer of sporangial wall and may be designated as the **stomium**. The inner walls of the cells constituting the stomium are thick and lingified. The cells making up the bulk of sporangial wall, other than the stomium have their side walls thickened only. As the exposed sporangia lose water and dry, a condition of stress and strain develops in the cells of the sporangial wall. These stresses lead to the appearance of a slit in the stomium and cause the sporangial wall to split open from top to the base of the sporangium in the form of two valves. The loose spore mass projects out of the open slit. The air currents scatter the spores and disseminate them.

GAMETOPHYTE

It starts with the spore that germinates to give rise to a new individual called the **gametophyte** or the **prothallus**.

Structure of the Spore

Lycopodium is homosporous. The spores are unicellular and range in diameter between 0.03—0.05 mm. The spores occur in tetrahedral tetrads and hence have a tetrahedral shape with a rounded or semicircular base (Fig. 4.1, D, E). Each spore has a triradiate ridge and is enclosed within a thin or thick spore wall. The spore wall has two layers (***Afzelius et el***, 1954). The inner layer as revealed by ultramicroscope studies is granular and is surrounded by an outer layer which is made up of concentric lamellae. Each spore has a single haploid nucleus surrounded by cytoplasm which is filled with reserve food material (fats and oils). The chloroplasts may or may not be present.

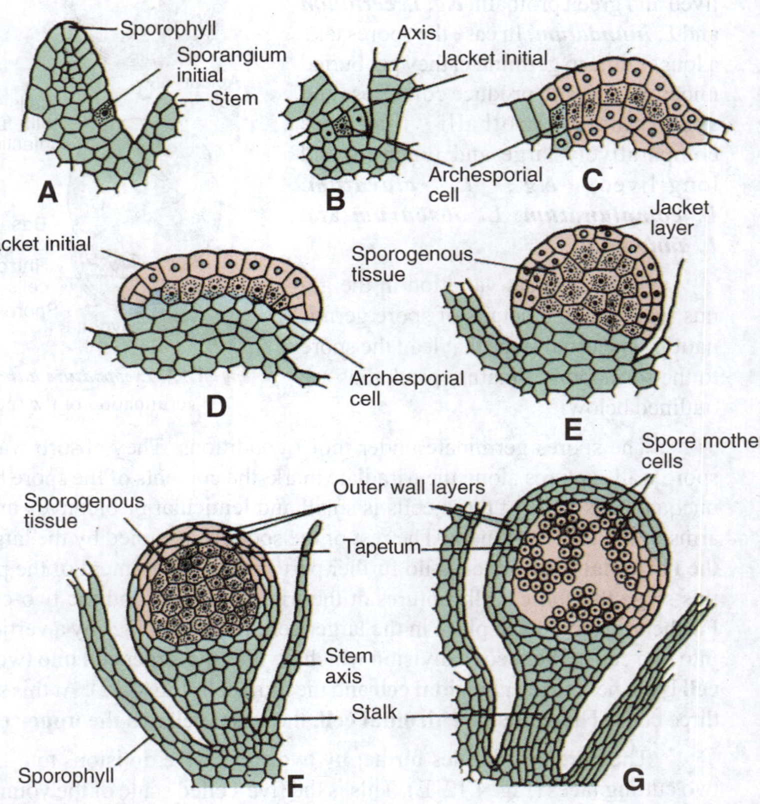

Fig. 4.11 (A-G). *Lycopodium selago*. Various stages in the development of sporangium. For explanation see text (After Bower)

The spore wall may be smooth or it may be variously sculptured (Fig. 4.1, D, E). Sculpturing may be in the form of pits or reticulate ridges as in ***L. complanatum***, ***L. clavatum*** (Fig. 4.1, D, E). In ***L. selago*** and ***L. phlegmaria*** the outer spore wall is beset with rounded or peg-like outgrowths. In ***L. inundatum***, ***L. cernuum*** and some other species the outer layer of the spore wall is thrown into slightly raised ridges that alternate with shallow grooves. In some species of ***Lycopodium*** the spore wall is thick and cutinised.

The spores of ***Lycopodium*** are used as reducing agents in gunpowder. They are also used as coating for pills.

Germination of the Spore (Fig. 4.12, A–F)

The time taken by the spores to germinate varies from a few days after their liberation from the sporangium to several years (3 to 8 years). This unusual delay in germination of spores has been attributed, by some workers (Bruchmann ; Barrows, 1935) to their thick and cutinised walls. In case the spores germinate within a few days of their liberation from the sporangium, they produce aerial short-lived and green prothalli, ***e.g. L. cernuum*** and ***L. inundatum***. In case the spores take a longer time to germinate, they get buried under the soil and produce colourless and subterranean prothalli that are comparatively large and tuberous and long-lived, ***e.g.***, ***L. clavatum***, ***L. complanatum***, ***L. obscurum*** and ***L. annotinum.***

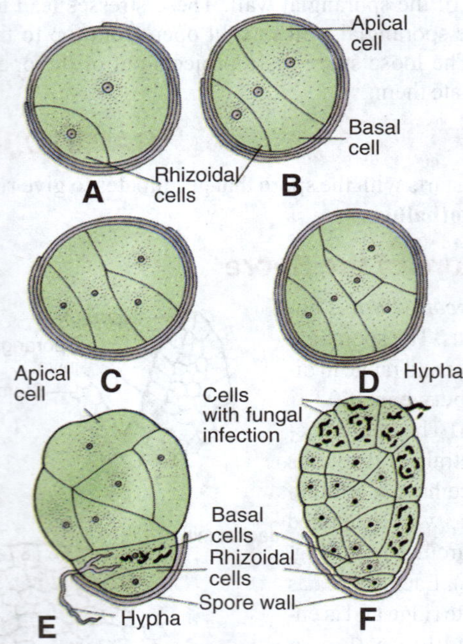

Fig. 4.12. ***Lycopodium annotinum.*** Various stages in the germination of the spore. (After Bruchmann)

There is great variation in the genus regarding the details of spore germination. The usual steps that lead the spore to the formation of mature prothallus are outlined below:

The spores germinate under moist conditions. They absorb water and swell up. Before the spore wall ruptures along the triradiate mark, the contents of the spore have already divided into two unequal cells. One of these cells is small and lenticular or biconvex and lies under one of the three arms of the triradiate mark. The rest of the spore is occupied by the larger cell. The smaller is called the **rhizoidal cell** and takes no further part in the development of the prothallus. It is colourless. At this stage the spore wall ruptures at the triradiate mark and the two-celled prothallus projects out. Further division taken place in the larger cell. It soon divides by a vertical or an oblique second wall into two cells. The second division which divides the larger cell into two cells differentiates the **basal cell** lying next to the **rhizoidal cell** and the third cell (Fig. 4.12). At this stage the young prothallus has three cells. These are the **rhizoidal cell,** the **basal cell** and the **upper cell.**

The third cell divides further by two successive divisions to differentiate an **apical cell** with two cutting faces (Fig. 4.12, E). This is the five-celled stage of the young prothallus. At this stage the cells are devoid of chloroplasts and endophytic fungus. In the species destined to produce aerial, green prothalli, the rest of the stages leading to the formation of a mature prothallus are completed within the season depending upon the attack of the mycorrhizal fungus. In case the mycorrhizal fungus fails to enter the five-called prothallium, further growth is arrested and the young prothallus may die. In the species with subterranean prothalli, further growth stops at the five-celled stage. A resting period of about a year intervenes between this stage and further development. During this interval the mycorrhizal infection takes place (Fig. 4.12, E, F). This infection takes place through the **basal cell** (Fig. 4.12, E, F) and it is believed that physiologically, the fungus supplies certain substances vital for the proper development of the prothallus. Further growth is resumed and is marked by the activity of the apical cells which cut off six-segments. Later the apical cell is replaced by a group of meristematic cells. These cells divide repeatedly and contribute to the further development of the prothallus.

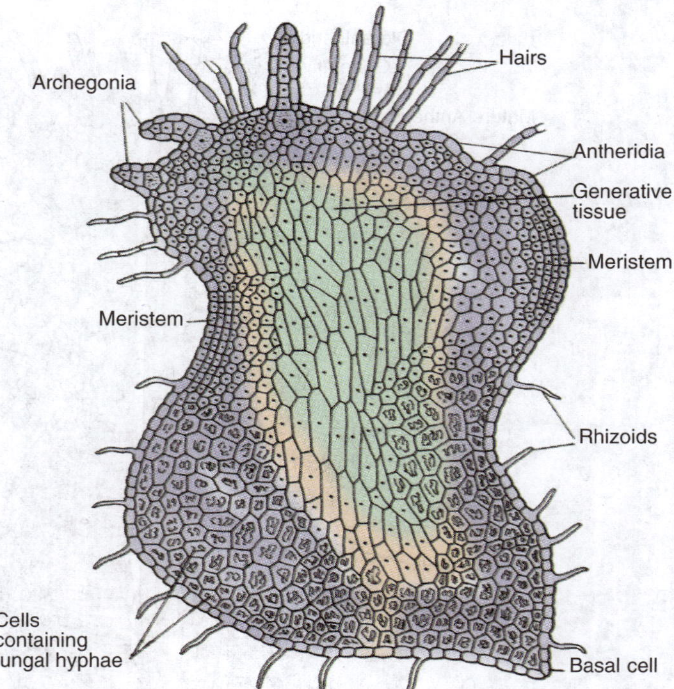

Fig. 4.13. *Lycopodium selago.* A mature prothallus showing details of internal structure. The generative region bears antheridia and archegonia (After Bruchmann)

Wetmore and Morel (1951) cultivated the spores of ***L. cernuum*** in culture solution containing minerals and glucose. The spores first develop chloroplasts and later divide into two unequal cells. The smaller remains inactive and the larger divides by oblique walls to cut off a two-sided apical cell, which cuts off segments to form a mass of undifferentiated cells called the **primary tubercle**. Further development is arrested here and continues further only if an endophytic fungus attacks the cells. Resumption of further growth results in the characteristic shape of the prothallus of ***L. cernuum***. It takes six months under controlled conditions.

Structure of Mature Prothallus

The structure of the mature prothallus, in various species of ***Lycopodium***, was described by Treub (1884, 1889) and Bruchmann (1898–1916). As a result of their studies three intergrading types of prothalli have been discovered. These three types will be discussed in detail.

First Type

It is exemplified by ***L. cernuum*** and ***L. inundatum***. The prothallus in ***L. cernnum*** (Fig. 4.17, A–E) has an erect cylindrical body only two to three millimeters long. It grows at the surface of the ground and consists of a colourless basal portion buried in the soil and a conspicuously lobed aerial crown that is green and bears sex organs at the bases of the green lobes. The rhizoids are restricted to

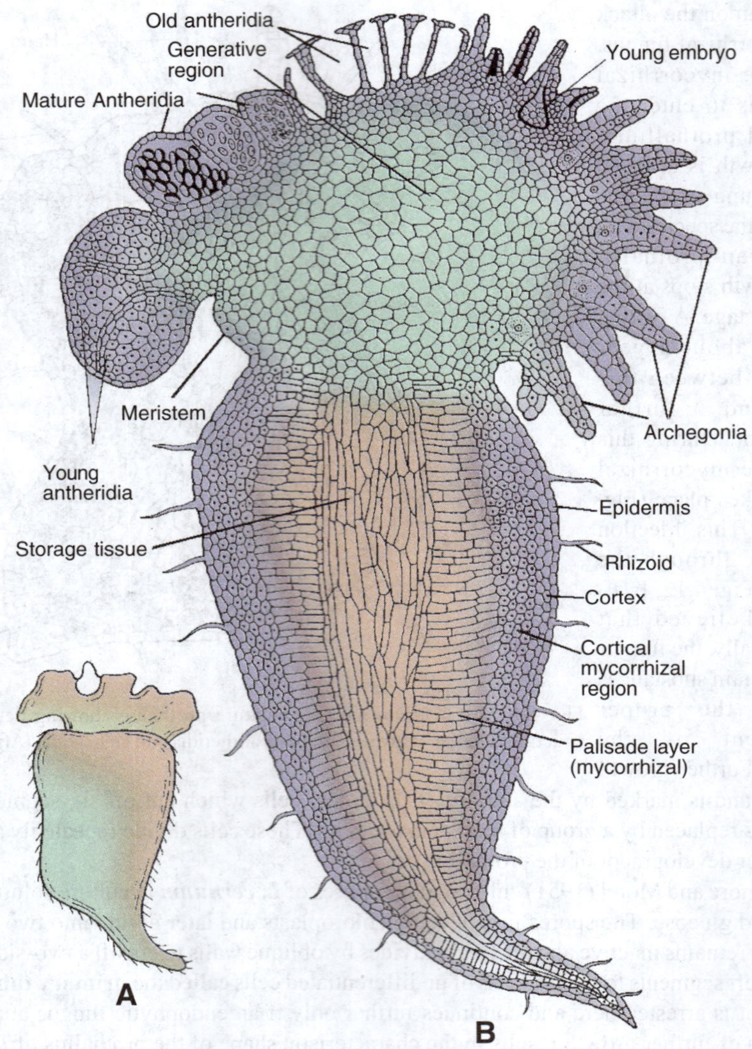

Fig. 4.14 (A-B). ***Lycopodium complanatum.*** A. A mature prothallus; B. Section of a mature prothallus showing details of internal structure. (After Bruchmann).

the lower buried portion which also contains an endophytic fungus. This prothallus is independent because it manufactures its own food. The prothallus of this species reaches maturity within six months and grows by means of a meristematic zone that is present below the apex and is marginal in position. It forms a rim around the cylindrical upper part of the prothallus. The sex organs develop from cells cut off by the meristem and the youngest is present near the meristem. The portion of prothallus bearing the sex organs is called the **generative zone**. This zone is green and is made up of undifferentiated parenchymatous cells, which are devoid of fungal hyphae.

Second Type

This type includes subterranean prothalli that have acquired a saprophytic mode of nutrition. It is represented by ***L. complanatum*** (Fig. 4.14), ***L. clavatum*** (Fig. 4.15), ***L. annotinum*** and ***L. obscurum***. The prothalli in all these species are tuberous, yellowish brown or even colourless. They store sufficient food material and vary in dimensions from 12 to 18 mm. long. Their shape also varies and may be top-shaped (***L. clavatum***), carrot-shaped or conical (***L. complanatum***) and discoid. They have two distinct regions : (***i***) the lower conical region, and (***ii***) the upper broad **generative region**. The conical region shows distinct zones. The outermost layer of cells resembles epidermal tissue and bear numerous rhizoids. Next to it is the several layered **cortical zone** which is composed of tightly packed parenchymatous cells that are filled with the endophytic fungal hyphae. Next to this zone is the **palisade zone** which is made up of closely packed columnar cells that also contain hyphae (Fig. 4.14). Next to this is the **central storage zone** whose cells are filled with stored food (Fig. 4.14). The upper broad region of the prothallus is called the **generative region** and is a broad expanse of parenchyma cells unaffected by endophytic fungus. It has a marginal meristem (Figs. 4.14 and 4.15) and bears sex organs on its flattened surface. The archegonia are found near the margins and the antheridia are located centrally. In ***L. complanatum*** (Fig. 4.14) the upper broad region is irregularly lobed and bears sex organs. In ***L. clavatum*** the upper generative region is irregularly cup-shaped (Fig. 4.15) with a depressed centre surrounded by a broad rim. In ***L. obscurum*** De Maggio (1967) induced the formation of tracheids in the prothallus.

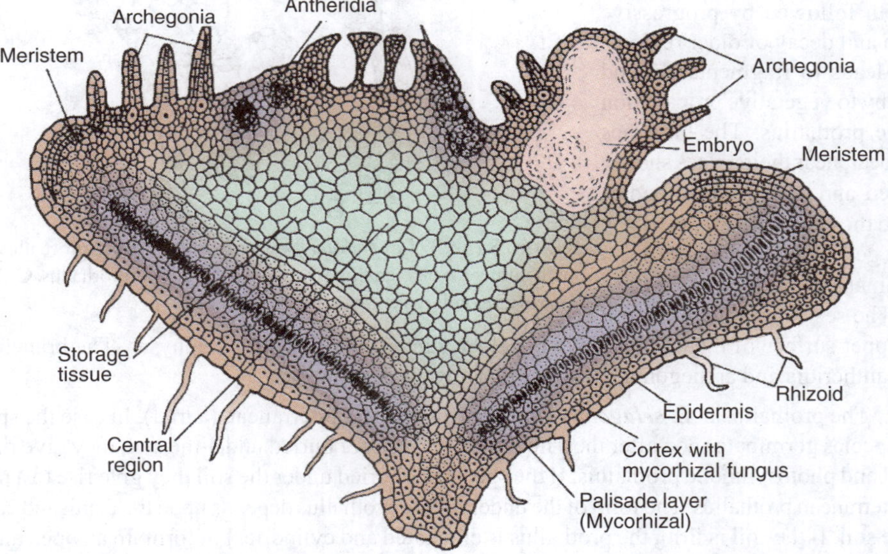

Fig. 4.15. ***Lycopodium clavatum.*** Section through the prothallus showing detailed internal structure. (After Bruchmann)

Misra (1978) described the geophilous gametophytes of ***L. serratum*** that are cylindircal to ovoid in shape and brownish in colour. They are monoecious and saprophytic. The endophytic fungus in this species is not restricted to any special region but is found throughout the cells of the prothallus.

Third Type

This type includes colourless, subterranean, saprophytic and repeatedly branched monopodial prothalli that are found only in the epiphytic species of ***Lycopodium***. The typical type is represented by ***L. phlegmaria*** (Fig. 4.16, A—C). The prothallus grows in the humus collected on the branches of the trees. The branches arise irregularly and ramify into the dead bark. The prothallus is attached to the substratum by means of absorbent rhizoids that grow in all directions. The growth of the prothallus and its branches is apical and takes place by the activity of two prismatic initial cells that lie side by side. In this case the endophytic fungus is found in the inner core of cells. These cells also store oil as reserve food. This centrally located mycorrhizal region is surrounded by one or more layers of cortical cells that are devoid of endophytic fungus and are colourless. The rhizoids arise from the outermost layer. The prothallus is marked by unlimited apical growth followed by progressive death and decay of older regions. This leads to fragmentation and thereby to vegetative propagation of the prothallus. The branches also bear, near their apices shortly stalked and multicellular bodies called the **gemmae** (Fig. 4.16, C). They get easily detached and can germinate to produce new prothalli. The sex organs are borne on the upper surface of the branches and are always intermingled with paraphyses. The branches bear both antheridia and archegonia but antheridia always develop first.

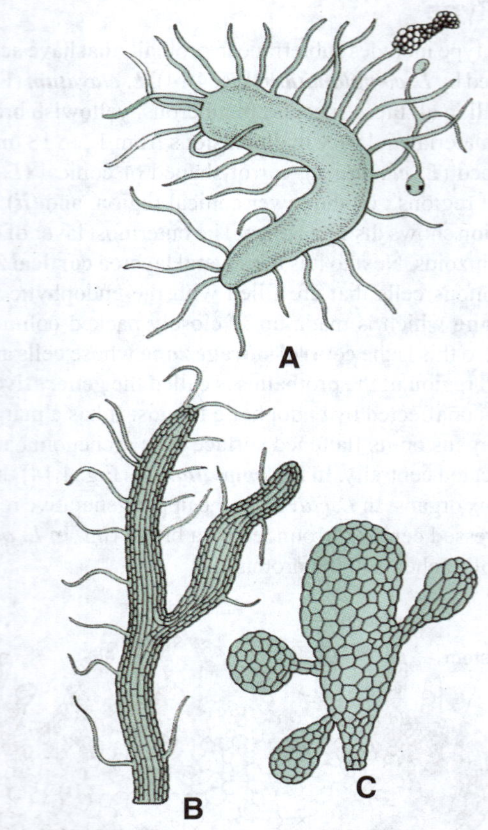

Fig. 4.16 (A-C). *Lycopodium phlegmaria*. A. Mature prothallus covered with numerous filamentous outgrowths; B. A sterile prothallus; C. A gemma bearing secondary gemmae. (After Gobel)

The prothallus in ***L. selago*** may be subterranean or epiterranean (aerial). In case the spores of this species germinate soon after their liberation and are not buried under the soil, they give rise to an aerial and photosynthetic prothallus. If the spores are buried under the soil they give rise to a pale and a subterranean prothallus. The form of the underground prothallus depends upon the depth and character of the soil. If the soil is firm, the prothallus is elongated and cylindrical in form. In an open and loose soil the prothalli are compressed and flattened and taper below into a conical point (Fig. 4.13). The upper region is broad and cup-shaped and bears sex organs. It is also called the **generative zone** and lacks endophytic fungus. It is lobed and the lobes bear sex organs on their side facing inwards. The outer side bears rhizoids alone. This indicates a feeble transition to dorsiventrality. The meristematic zone occupies the margin of the cup. **Mycorrhizal zone** is confined only to the lower conical region. The rhizoids cover the entire surface and arise from an outer layer called the epidermis (Fig. 4.13). The cells containing endophytic fungus also store food material. The antheridia are borne in a regular sequence from the centre to the marginal meristem and are followed by archegonia. The lobes bearing the sex organs also bear paraphyses. The underground prothalli of ***L. selago*** when growing near the surface of the soil may emerge out partly and become green in colour. Such prothalli are partly colourless and partly green. So we derive the following conclusions from the study of the prothallus of ***L. selago***:—(***i***) they are green and photosynthetic when aerial. (***ii***) they may be partly green and partly colourless or they can be partly subterranean and partly epiterranean. (***iii***) their form may be obconical below and flat and lobed above as in prothalli growing in loose soil. (***iv***) they assume a cylindrical form in firm soil and when growing in rotten wood they give out branches that show a dorsiventral symmetry. (***v***) they exhibit both holophytic and saprophytic nutrition. (***vi***) the aerial region of the thallus is lobed.

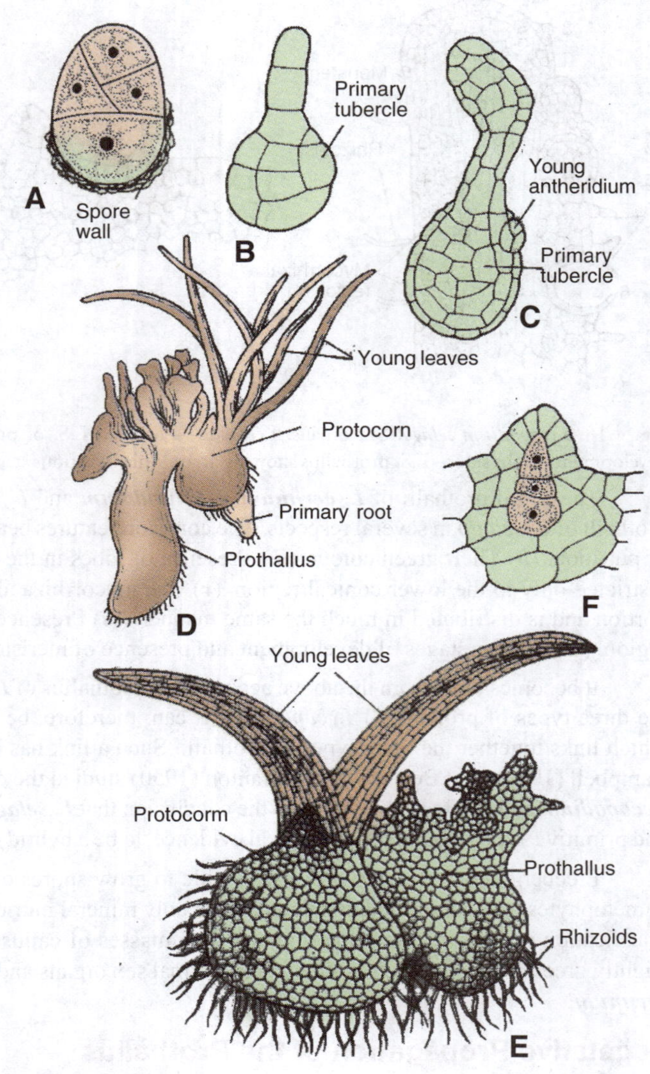

Fig. 4. 17 (A–E). *Lycopodium cernuum.* Gametophytic generation. A, B. Younger stages of development; C. An older stage with the first antheridium; D. A mature prothallus with attached young sporophyte; E. The same as D; F. A mature archegonium. (After Treub).

The subterranean gametophytes of ***L. selago*** resemble the gametophytes of epiphytic species like ***L. phlegmaria*** and ***L. billardieri*** in possessing slender and cylindrical branches and in the presence of paraphyses among the sex organs. The underground prothallus of ***L. volubile*** which is very much like that of ***L. clavatum*** resembles ***L. selago*** in that it sometimes emerges out of the soil and develops chlorophyll. ***L. clavatum***, ***L. complanatum***, ***L. annotinum***, ***L. volubile*** and other subterranean species have at least of a few common stages in their ontogeny with that of ***L. selago***. These are: (***i***) the presence of a two-sided apical cell in the earlier stages of development; (***ii***) further development depending upon the attack of endophytic fungus; (***iii***) formation of a meristematic zone whose position is marginal; (***iv***) the obconical or conical shape of the lower region; and (***v***) presence of lobes in the upper region. The prothalli of all these subterranean species resemble with that of ***L. selago*** in their saprophytic mode of nutrition.

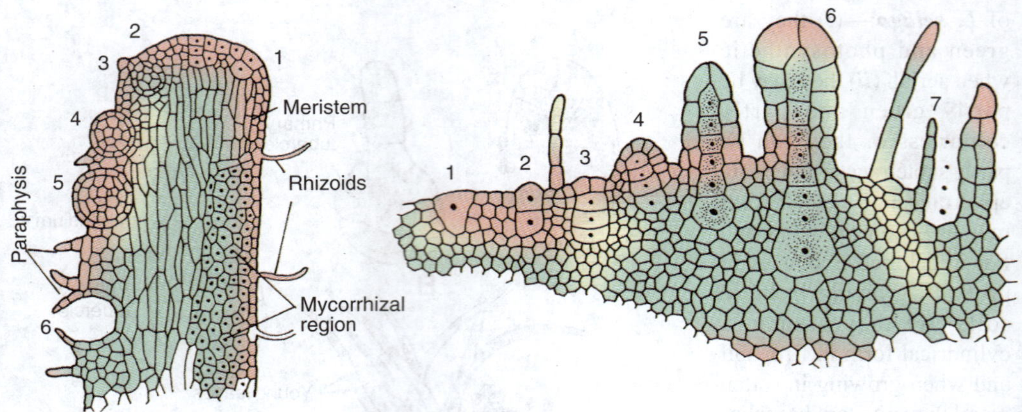

Fig. 4.18. ***Lycopodium selago.*** Sex organs. Left side :– Median L.S. of prothallus showing stages in antheridial development. Right side :– L.S. prothallus showing archegonia in various stages of development. (After Bruchmann)

The aerial prothalli of ***L. cernuum***, ***L. inundatum*** and ***L. salakense*** resemble that of aerial prothalli of ***L. selago*** in several respects. The common features bearing :—(*i*) Their autophytic mode of nutrition. (*ii*) Their green colour. (*iii*) Presence of lobes in the upper distal region. (*iv*) Rhizoids restricted only to the lower conical region. (*v*) The mycorrhiza also restricted to the lower conical portion and is distributed in much the same manner. (*vi*) Presence of sex organs in the upper lobed region. (*vii*) Earlier stages of development and presence of meristem.

It becomes clear from the above account that prothallus of ***L. selago*** bears resemblances to all the three types of prothalli of ***Lycopodium***. It can, therefore, be regarded as an intermediate type which links together the three types of prothalli. Such a link has been pointed out by Lang (1899), Campbell (1939), and Bower (1935). Manton (1950) studied the cytology of four British species of ***Lycopodium*** and as a result she came to the conclusion that ***L. selago***, ***often*** considered to be a central and primitive type is found on cytological evidence, to be a hybrid of very high chromosome number.

Freeberg and Wetmore (1957) were able to grow spores of some species with subterranean gametophytes on a cultural medium containing only mineral nutrients and sucrose. The spores (after scarification or acid treatment) germinated into masses of callus tissue with erect lobes that were slightly green in colour. Such prothalli bear normal sex organs and resembled very much those of ***L. cernuum***.

Vegetative Propagation of the Prothallus

1. **By the formation of gemmae** (Fig. 4.16, C). They are borne on the branches of the prothallus in ***L. phlegmaria*** and arise singly or in clusters near the tips of the branches. A gemma develops from a single superficial cell of the branch and is usually a large club-shaped and shortly stalked structure. In starved prothalli the gemmae are few celled and thick-walled and their cells store food material. They are capable of perennating and germinate into new prothalli on the advent of favourable weather.
2. **By the progressive decay of the older parts** of the prothallus and thereby separating the younger branches, which behave as individual prothalli (***L. phlegmaria*** and ***L. billardieri***).
3. In ***L. inundatum*** adventitious buds arise from the injured parts of the lobes.

Sex Organs

The exosporic gametophytes of ***Lycopodium*** are essentially monoecious but protandrous. The antheridia and archegonia, in most of the species of ***Lycopodium***, are formed in distinct patches or clusters on the upper surface of the prothallus. In ***L. lucidulum*** Spessard (1922) reported intermingling of the antheridia and archegonia.

Antheridia

The antheridia in ***Lycopodium*** develop earlier than the archegonia and the youngest lie near the marginal meristem. In ***L. lucidulum*** there is intermingling of the two kinds of sex organs. The antheridia are mostly embedded (Fig. 4.18, 4.19) in the tissue of the prothallus.

Development (Fig. 4.18 and 4.19)

The antheridium arises from a single superficial cell which divides by a transverse wall into an outer **primary wall cell** and an inner **primary spermatogenous cell**. The **primary wall cell** divides by anticlinal wall to form a one-layered wall or the jacket of the antheridium. A triangular **opercular cell** is ultimately differentiated in the middle of the jacket layer and marks the point of exit of spermatozoids. The **primary spermatogenous cell** divides repeatedly to give rise to a mass of androgonial cells. The **androgonial cells** divide further and give rise to a mass of **spermatocytes** or **androcytes** or **sperm mother cells** (Fig. 4.19, E) whose number varies considerably. In ***L. phlegmaria*** a single section of the antheridium contains up to two hundred **spermatocytes**. The development of spermatozoid within the spermatocyte has not been studied in detail.

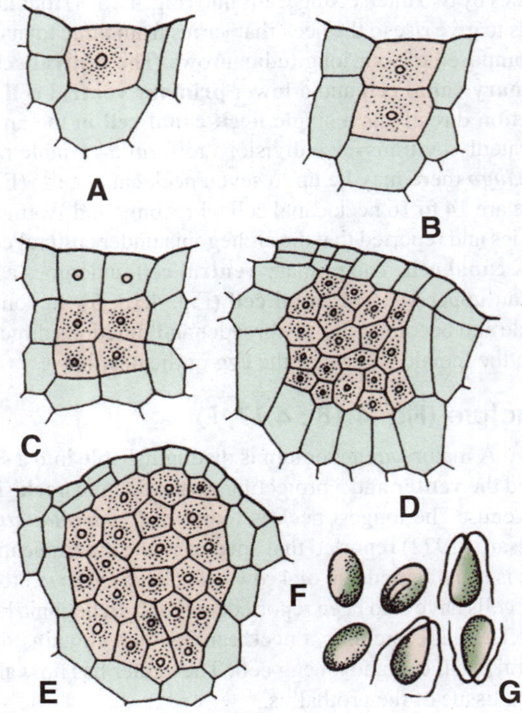

Fig. 4.19 (A–G). *Lycopodium clavatum*. A–E. Various stages in the development of antheridium; F. Spermatocytes; G. Spermatozoids. (After Bruchmann)

Bruchmann has studied the structure of the spermatozoids in ***L. clavatum***. It is oval in shape (Fig. 4.19, G) and about twice as long as broad. It is pointed at the anterior end which bears two flagella. The cytoplasm forms the major portion of the spermatozoid and the nucleus lies on one side. It is interesting that unlike all other pteridophytes the nucleus in the spermatozoids of ***Lycopodium*** does not form a major part of its body. In this respect it resembles the male gametes of some algae. Spessard (1922) studied the spermatozoids of ***L. lucidulum*** and described the nucleus as elongated and curved. Treub described the spermatozoids of ***L. phlegmaria*** as elongated and twisted.

Structure of Mature Antheridium (Fig. 4.18 and 4.19, E)

A mature antheridium is either completely embedded in the tissue of the prothallus (***L. selago***) or slightly projecting. It has a single-layered antheridial wall with a distinct triangular **opercular cell**. The wall encloses a large number of spermatocytes whose protoplasts metamorphose into biflagellate spermatozoids.

It dehisces by the mucilaginisation of the single opercular cell or a group of jacket cells. The spermatocytes containing spermatozoids absorb water, swell up and cause the rupture of the antheridial wall.

Archegonia (Fig. 4.18 and Fig. 4.19 a)

Development. The archegonial initial is superficial. It divides by a periclinal wall into an outer **primary cover cell** (Fig. 4.18, 2, Fig. 4.19_a B) and inner **central cell**. The **primary cover cell** divides by two intersecting walls into (Fig. 4.18, 3) four **neck initials** which divide further by transverse walls to give rise to the neck that varies from three to many cells in height. (Fig. 4.19_a D–F). The neck is composed of four longitudinal rows. The **central cell** divides by a transverse wall into an upper **primary canal cell** and a lower **primary ventral cell** (Fig. 4.18, 3 and 4.19_a D). The former may function directly as a single **neck canal cell** in the archegonia of aerial prothalli or it may divide repeatedly by transverse divisions to form a variable number of neck canal cells (Fig. 4.19_a E). In ***L. selago*** there may be up to seven neck canal cells (Fig. 4.18, 6 and 4.19_a F). In ***L. complanatum*** there are 14 to 16 neck canal cells. Freeburg and Wetmore (1957) cultured the prothalli of these two species and reported that the archegonia under cultural conditions have very short necks and only one **neck canal cell**. The primary **ventral cell** undergoes a transverse division to form a lower egg cell and an upper **ventral canal cell** (Fig. 4.18, 5). In some cases the **primary ventral cell** does not divide and becomes slightly broader and functions directly as the egg cell. Its cytoplasm and nucleus form the female gamete or the **egg** or the **oosphere.**

Structure (Fig. 4.18 ; 4.17, F)

A mature archegonium is distinguishable into a slightly broader and embedded lower portion called the **venter** and a projecting part called the **neck.** The length of the neck varies considerably in the genus. The longest neck is found in ***L. complanatum*** (Fig. 4.14) and ***L. selago*** (Fig. 4.18, 6). Spessard (1922) reported that species with subterranean prothalli bear long necked archegonia. The neck is usually made up of 4 rows of cells, each row 3 to several cells high; five longitudinal rows of neck cells have also been reported and such archegonia have very long necks (***cf.***, bryophyte archegonia). The neck encloses a neck canal which contains one to 16 neck canal cells. The venter has a ventral canal cell and an egg cell. The venter has no wall of its own but is protected and surrounded by the tissue of the prothallus.

Dehiscence

The archegonium when mature dehisces by the disorganisation of the **neck canal cells** and the **ventral canal cell** and opening the uppermost tier of the neck cells. This creates an open passage for the spermatozoids to enter the neck and reach the **oosphere.**

Fertilisation

It is affected in the presence of water through which the biflagellate spermatozoids swim and reach the open archegonial necks. Bruchmann (1909) reported that the spermatozoids are attracted by a chemical stimulus. He stated that the archegonial necks secrete free citric acid or its salts whose smell attracts the spermatozoids. Only one spermatozoid is able to reach the egg and effect fertilisation. The male and female nuclei fuse together (syngamy) and form a diploid fusion nucleus. So the act of fertilisation switches on the diploid generation. The diploid nucleus with its surrounding cytoplasm and a wall is called **the ospore.**

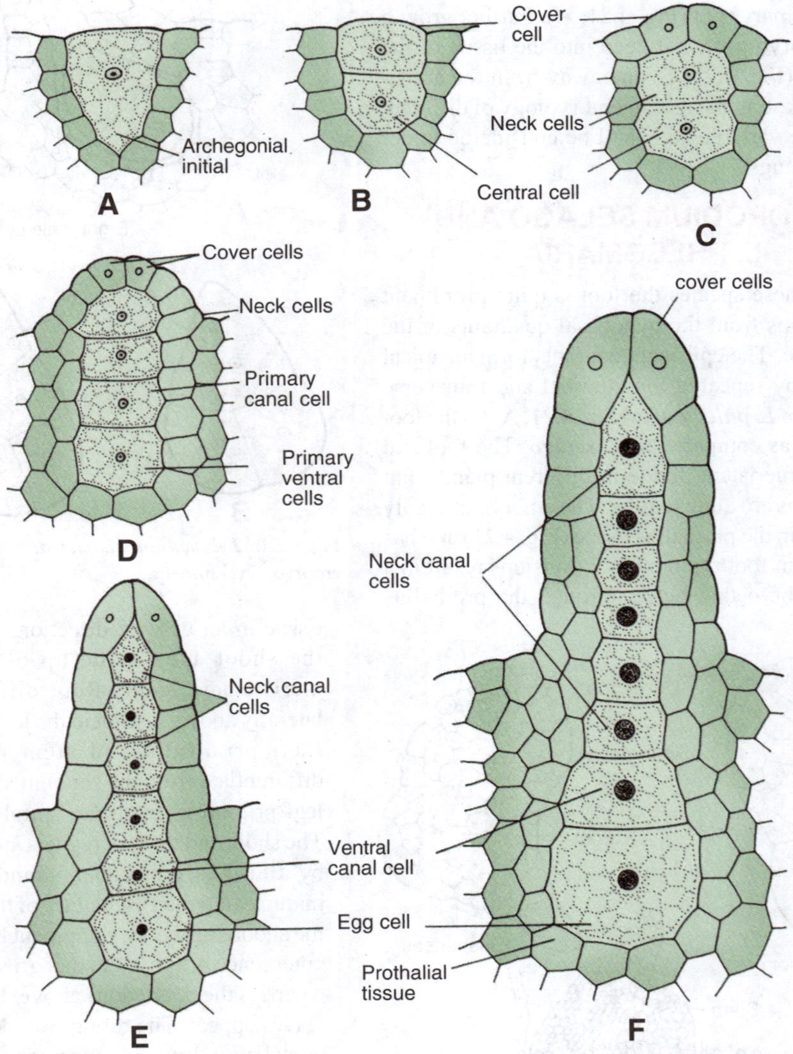

Fig. 4.19 (a). ***Lycopodium selago (A-F).*** Various stages in the development of archegonium.

THE EMBRYO

The embryology in all the species of ***Lycopodium*** is **endoscopic** (***i.e.***, the shoot apex is directed away from the archegonial neck). The earlier stages of development, in all the species, are similar. The nucleus of the oospore divides into two daughter nuclei that are separated by a transverse wall (Fig. 4.20, A). This first division of the oospore determines the polarity of the embryo. It results in the formation of an **epibasal** and a **hypobasal cell**. The cell next to the archegonial neck is **epibasal** and away from it is the **hypobasal cell**. The epibasal cell enlarges to function as the suspensor. Rarely it may divide once or twice. The **hypobasal cell** or the **embryonal cell** gives rise to the embryo and divides by transverse and longitudinal walls to form two quadrants (Fig. 4.20, C). The

hypobasal quadrant which is next to the suspensor cell divides further and gives rise to the foot (Fig. 4.21, A). The **epibasal quadrant** forms the shoot and the primary root (Fig. 4.21, A). Further growth of the embryo pushes it deep into the tissue of the prothallus (Fig. 4.21, C) and away from the archegonial neck. Later stages in embryology of different species are variable and shall be considered under three headings:—

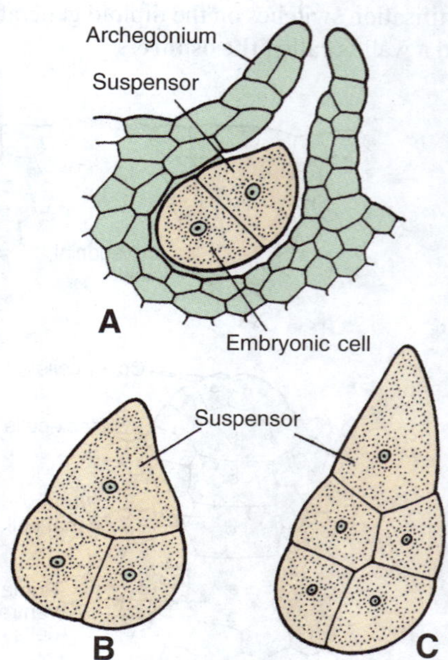

Fig. 4.20. ***Lycopodium clavatum.*** Early stages of embryo development.

LYCOPODIUM SELAGO AND L. PHLEGMARIA

In these species the foot is quite prominent and develops from the hypobasal quadrants of the octant stage. The epibasal quadrants form an apical meristem by repeated longitudinal and transverse divisions. In ***L. phlegmaria*** (Fig. 4.21, A–C) the foot is massive as compared to ***L. selago***. The foot and the apical meristem divide in different planes, but the divisions are quite regular. The embryo is deeply embedded in the prothallus tissue (Fig. 4.21) and has large cells in foot region. Later divisions result in a curved embryo that bursts through the prothallus tissue in an upward direction. Its apex is the shoot apex made up of a row of meristematic cells. Root differentaties laterally above the foot in the leaf quadrant. Leaf primordia and stem primordia differentiate from the shoot meristem. The leaf primordia arise in a spiral sequence. The shoot and the root regions are traversed by a distinct procambial strand that later matures into the vasculature of the root and the shoot. Further development leads to the emergence of first stem that grows erect an d carries the cotyledons above. Later other leaves appear. The root grows down but is sort-lived and is soon replaced by adventitious roots.

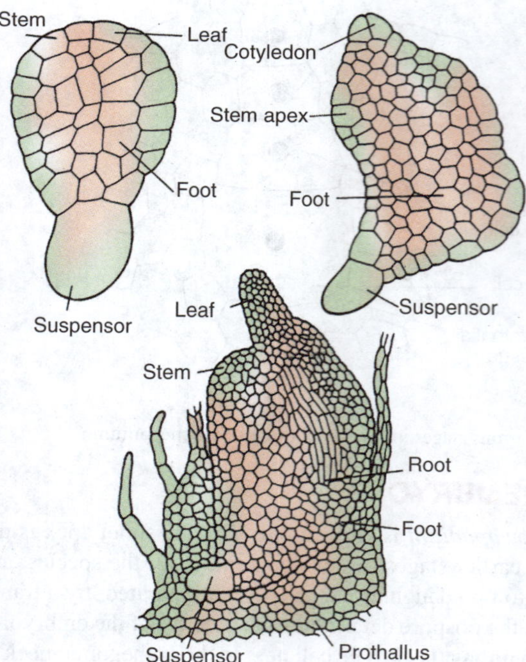

Fig. 4.21 (A-C). ***Lycopodium phelgmaria.*** Later stages of embryo development. (After Treub)

L. CLAVATUM AND L. ANNOTINUM

In these species the foot is massive and haustorial and the growth of the apical region of the embryo is comparatively very slow and the first two leaves are opposite and not spiral as in ***L. selago***. They develop late and are scale-like. The procambial strands also develop late. In these creeping species first stem is soon replaced by the horizontal stem which arises from the base of the former and bears colourless scale leaves that are devoid of midrib and vascular supply. The development is very slow and this young sporophyte depends for years together on the subterranean, gametophyte for nutrition. It takes many years for it to come above ground and become independent.

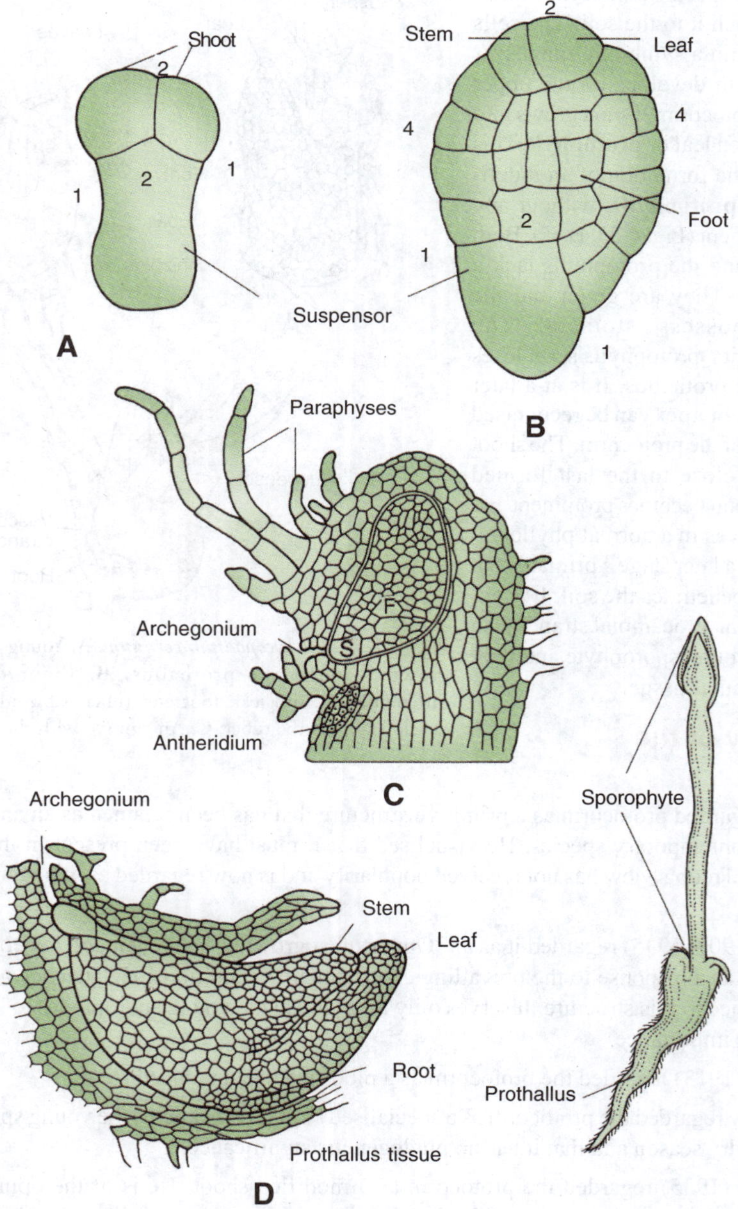

Fig. 4.22 (A-E). ***Lycopodium selago.*** A. Young embryo as seen in L.S. The wall 1-1 has divided it into suspensor segment and shoot segment; B. L.S. through older embryo showing suspensor, foot, stem and leaf segments. Positions of walls 1-1, 2-2 and 4-4 are clear; C. Later stage of embryo still enclosed the prothallus tissue; D. Mature embryo breaking through the prothallus tissue, S = suspensor; F = foot; E. A young sporophyte attached to prothallus. (After Bruchmann).

L. CERNUUM AND L. INUNDATUM

These species are characterised by the development of a tuberous mass of cells called the **protocorm** from the epibasal quadrant (Fig. 4.23). It is made up of undifferentiated mass of parenchymatous cells. It bears numerous rhizoidal hair that attach it to the soil. The cells at the base contain a symbiotic fungus. A conical outgrowth develops on the upper surface of the protocorm. It soon grows into a cylindrical green leaf or **protophyll**. This is followed by the formation of an indefinite number of **protophylls**, without any regular arrangement (Fig. 4.23, B, C). Both the protocorm and the protophylls lack a vascular strand. They are green and the protophylls possess stomata. The protocorm with its protophylls soon loses contact with the prothallus. It is at a later stage that the **shoot apex** can be recognised at the distal end of the protocorm. The shoot apex develops close to the last formed protophylls. It soon becomes prominent and bears young leaves in a normal phyllotactic sequence. At a later stage a primary root also arises and penetrates the soil. It is exogenous in origin. Procambial strands also develop in the young sporophyte and later mature into vascular tissue.

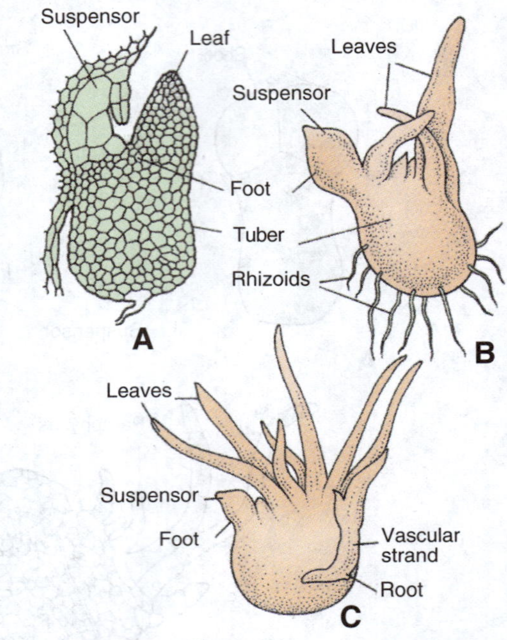

Fig. 4.23 (A-C). *Lycopodium cernuum*. **A.** Young embryo has grown through the prothallus; **B.** Embryo showing differentiation into leaf, foot and tuber, Rhizoids have also developed from the tuber; **C.** An embryo with three leaves.

Morphology of the Protocorm

Treub regarded protocorm as a primitive structure that has been retained as an ancestral feature by some contemporary species. He visualised that it must have been present in the ancestral vascular plants. Treub's view has not received popularity and is now regarded as of historical importance only.

Bower (1907, 1935) regarded it as an 'Opportunist growth' that developed in certain species of ***Lycopodium*** as a response to the prevailing environmental conditions. He attaches only an ecological importance to this structure. It serves only to establish the young sporophytes to soil and has no phylogenetic importance.

Browne (1913) regarded the protocorm as a modified and a reduced stem.

Holloway regarded the protocorm as a specialised structure that helps the young sporophyte to perennate over dry season and that it has no phylogenetic significance.

Wardlaw (1935) regarded the protocorm as a modified shoot. He is of the opinion that its regular appearance in certain species of ***Lycopodium*** is not due to any environmental effect but to certain genetic factors that must be involved in its persistence in these species. He attributed the development of protocorm to certain metabolic conditions of the prothallus and young sporophyte. According to him, carbohydrate and nitrogen ratio (C/N) is responsible for the development of localised swellings in the embryo. In case the embryo received a balanced nutrition and C/N ratio is normal the embryo produces no swellings and produces normal enlongating leafy shoot. In case the embryo receives unbalanced nutrition and C/N ratio is high the organisation of the young shoot apex is delayed and protocormous outgrowths develop. Wardlaw states that in ***L. cernuum*** and ***L. laterale*** (protocormous species) the genes which determine protein synthesis at the shoot apex are ineffective in the beginning. This leads to the delay in the development of shoot apex and aids in the formation of a new structure called the **protocorm**. Later the green **protophylls** and **protocorm** as a result of their photosynthetic activity produce certain essential metabolic products that lead to the growth of the shoot. The ineffective gene is thus switched on to activity.

Deviations from Normal Life Cycle

Occurrence of the phenomenon of apospory and apogamy in ***Lycopodium*** has not so far been reported in nature. Recently Freeberg (1957) was able to induce apogamy in ***L. complanatum***, ***L. cernuum*** and ***L. selago***. He was able to grow these species under cultural conditions by germinating their spore on culture media. He got full-fledged gametophytes that bore sex organs and produced after normal fertilisation, embryo and young sporophytes. During his experiments he found that under conditions unfavourable for fertilization (absence of water) the gametophytic tissue produces complete sporophytic plants apogamously. The young sporophytes arise apogamously from the tips of the branches of prothalli. Gametophytic tissue was also induced to produce a single leaf. Such apogamously

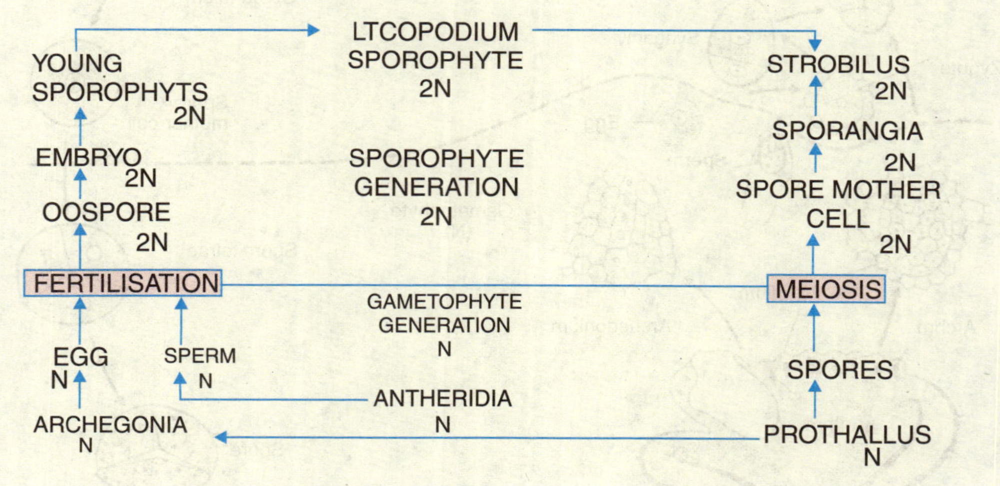

Fig. 4.24 A. Graphic representation of life cycle of *Lycopodium.*

produced sporophytes were also observed, in rare cases, to produce from the stem a gametophyte like outgrowth. Only one such case of formation of aposporous gametophyte was observed by Freeberg (1954). De Maggio (1964) induced apogamy in the gametophytes of ***L. obscurum*** by supplying coconut milk or sucrose or both to the nutrient medium. He was able to induce the formation of root buds in the gametophytic callus of this species.

Economic Importance

The spores of ***Lycopodium inundatum*** and some other species yield 50% fixed oils. They are much used as cover for pills, as a diluent for insufflations, and as a dusting powder for abraded surfaces. In industry they are used for making pattern molds ; because of their inflammability they are used in flares, fireworks and tracer bullets. Some species have a medicinal value also. ***Lycopodium*** is widely used in the homoeopathic system of medicine.

Chromosome Numbers

Cytology of a number of species has been worked out (Dunlop 1949; Manton, 1950; Mehra and Verma 1957; Ninan, 1958) and the haploid number of chromosomes in the various species has been confirmed to vary between ***n*** =14 to ***n*** = 136. Dunlop (1949) and Manton (1950) reported 2***n*** = 260 in ***L. selago***. They found that in this species during meiosis there are a number of unpaired chromosomes that show this species to be a hybrid. Such observations go against the supposition that ***L. selago*** is an intermediate type.

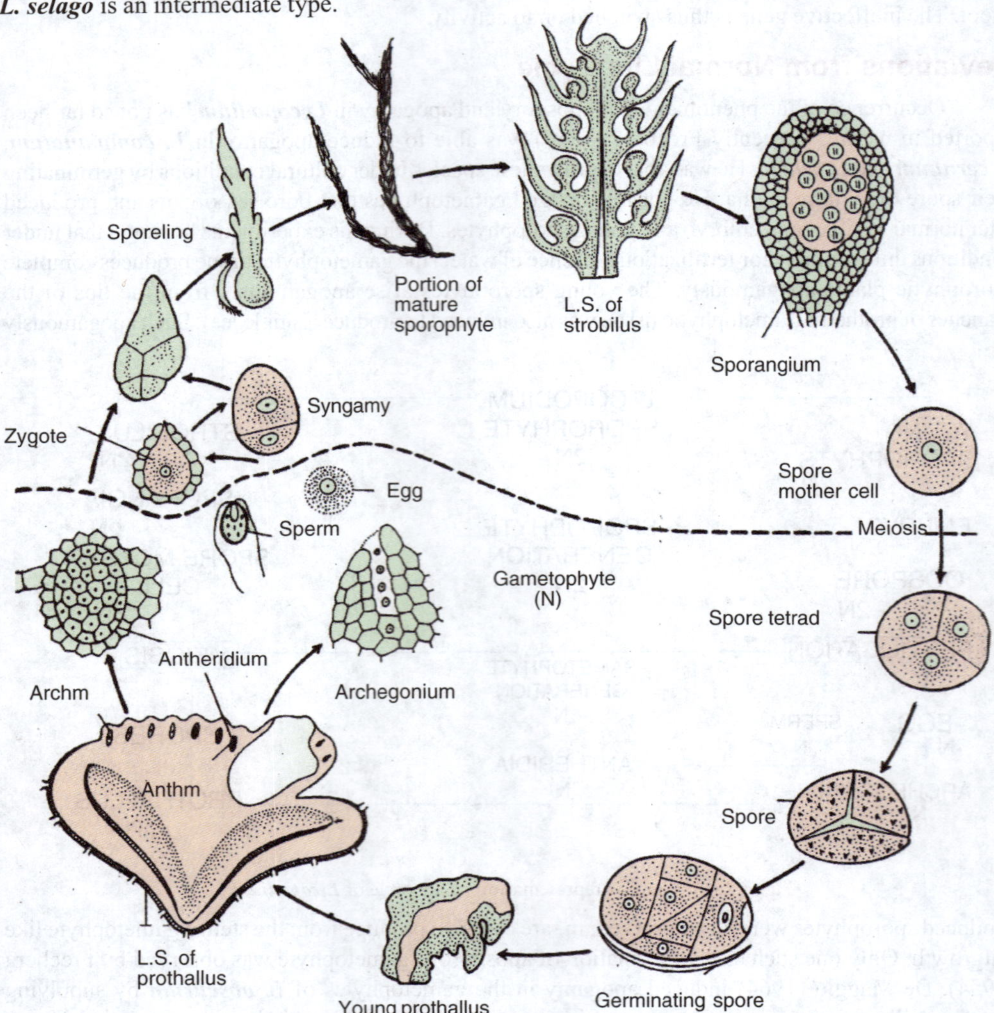

Fig. 4.24 B. Diagrammatic life cycle of *Lycopodium*

Taxonomic Position

Division	:	***Lycophyta***
Class	:	***Eligulopsida***
Order	:	***Lycopodiales***
Family	:	***Lycopodiaceae***
Genus	:	***Lycopodium***

PHYLLOGLOSSUM Kunze

It is the second monotypic genus of the lycopodiaceae and is represented by ***Phylloglossum drummondii***. In its habit it resembles the young protocormous sporophytes of ***Lycopodium cernuum***

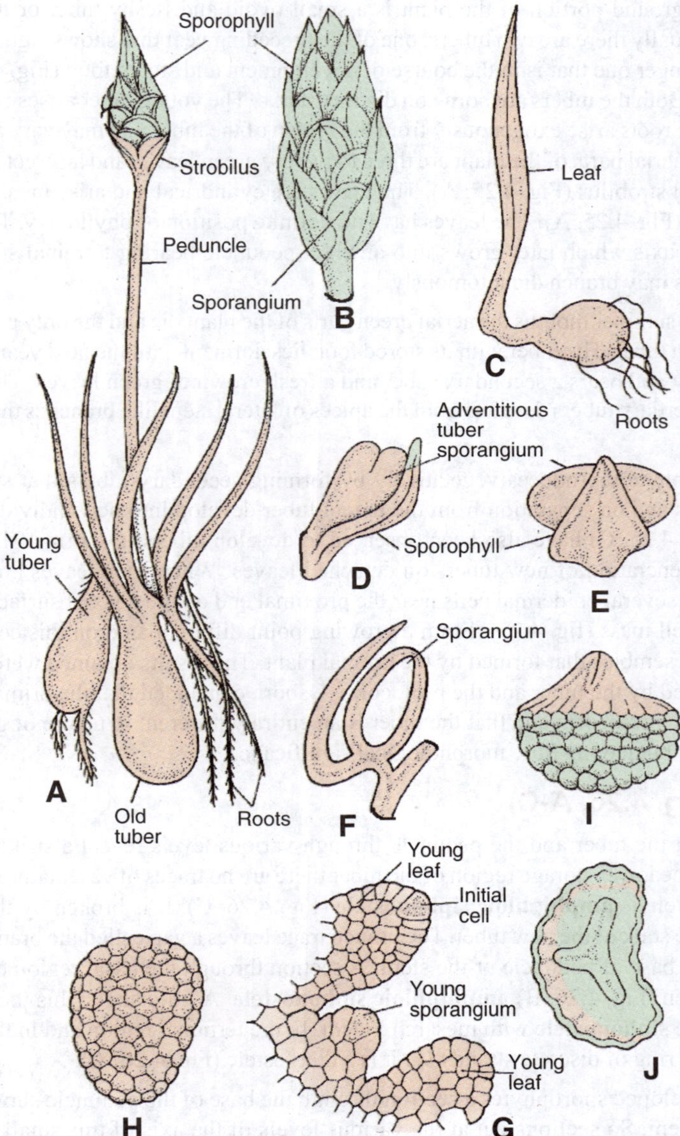

Fig. 4.25 (A–J) *Phylloglossum drummondii.* A. a complete plant showing habit. Note the young and old tuber and position of leaves and roots; B. Portion of spike enlarged; C. Isolated leaf with adventitious tuber; D. Lateral view of the sporophyll with an axillary sporangium; E. Sporophyll and sporangium as seen from distal end. The sporangium is foliar; F. Radial section of the sporophyll bearing a foliar sporangium. The sporangium has dehisced; G. L.S. Young strobilus; H. Spore as viewed from below; I. Spore as seen from side; J. Spore as seen from above. (All after Bertrand)

and *L. laterale*. It has a restricted distribution and occurs only in New Zealand, Tasmania and Southern and Western Australia. The sporophytes of this species (Fig. 4.25, A) make their appearance above ground only during winter and reach a maximum height of four centimetres. This interesting genus has been studied in detail by Bower (1885) and Bertrand (1885, 1886).

The underground portion of the plant is a small ovoid and fleshy tuber or the 'protocorm' (Fig. 4.25, A). Usually there are two tubers, one of the preceding year that shows signs of exhaustion, and the other younger one that is in the course of development and stores food (Fig. 4.25, A) for the succeeding year. Both the tubers are borne on distinct stalks. The young tuber arises from the apex of the older one. The roots arise exogenously from the crown of the tuber and may vary in number from one to three. The aerial parts of the plant are three to twenty green leaves and an erect green peduncle bearing a terminal strobilus (Fig. 4.25, A). The leaves are cylindrical and arise in a whorl from the apex of the tuber (Fig. 4.25, A). The leaves have no definite position or phyllotaxy. They surround a small and simple axis which later grows into an erect peduncle bearing terminal strobilus. In rare cases the strobilus may branch dichotomously.

During hot summer months the aerial green parts of the plant die and the only part that survives is the subterranean tuber. The tuber with its stored food lies dormant until the next year. During winter it resumes activity and bears a secondary tuber and a fresh crown of green leaves. The secondary or sometimes even tertiary tubers develop from the apices of lateral stem like branches that grow out and downwards.

The sporophyte reproduces vegetatively by forming secondary tubers that strike their own adventitious roots and on separation from the parent tuber develop into new individual plants. Isolated leaves (Fig. 4.25, C) have also been observed to develop tubers (Osborn, 1919). Osborn has described the regeneration of new tubers on detached leaves. When such leaves are laid down on moist soil, one or several epidermal cells near the proximal end on the abaxial surface proliferate to form a primary cell mass (fig. 4.27). From a growing point differentiated on this cell mass a tuber develops which resembles that formed by the normal plant. The results obtained were applied to the problems presented by the tuber and the protocorm. Osborn concluded that the primary cell mass is homologous to the protocorm and that the tuber is an entirely different structure of great biological value to the plant, but having little morphological significance.

Anatomy (Fig. 4.26, A-C)

Sections of the tuber and the peduncle through various levels reveal a striking variation in stelar system. In the lower storage region of the tuber there are no traces of vasculature. Near the apex of the tuber the stele is **amphiphloic siphonostele** (Fig. 4.26, C). It is broken by the departure of branch trace to the stalk of the new tuber. The branch trace leaves a gap called the branch gap. Above this region is the base of peduncle or the stem. A section through the basal region of the peduncle shows an unbroken (Fig. 4.26, B) **amphiphloic siphonostele**. A little above this the stele becomes simple **ectophloic siphonostele** with mesarch xylem. In the terminal region and in the strobilus the peduncle shows a ring of discrete strands ***i.e.*** it is a dictyostele (Fig. 4.26, A).

In less developed sporophytes a section through the base of the peduncle shows a protostele with mesarch xylem. So sections cut at the various levels of the axis of this small plant show all important types of steles ***i.e.*** **protostele**, the **siphonostele** (both ectophloic and amphiphloic) and the **dictyostele**. Presence of all these types of steles within a short height of four centimetres is suggestive of the fact that all these types are modifications of one fundamental type of stele, the **protostele.**

The anatomy of the root is very simple. A transverse section reveals a monarch or a diarch stele surrounded by the cortex.

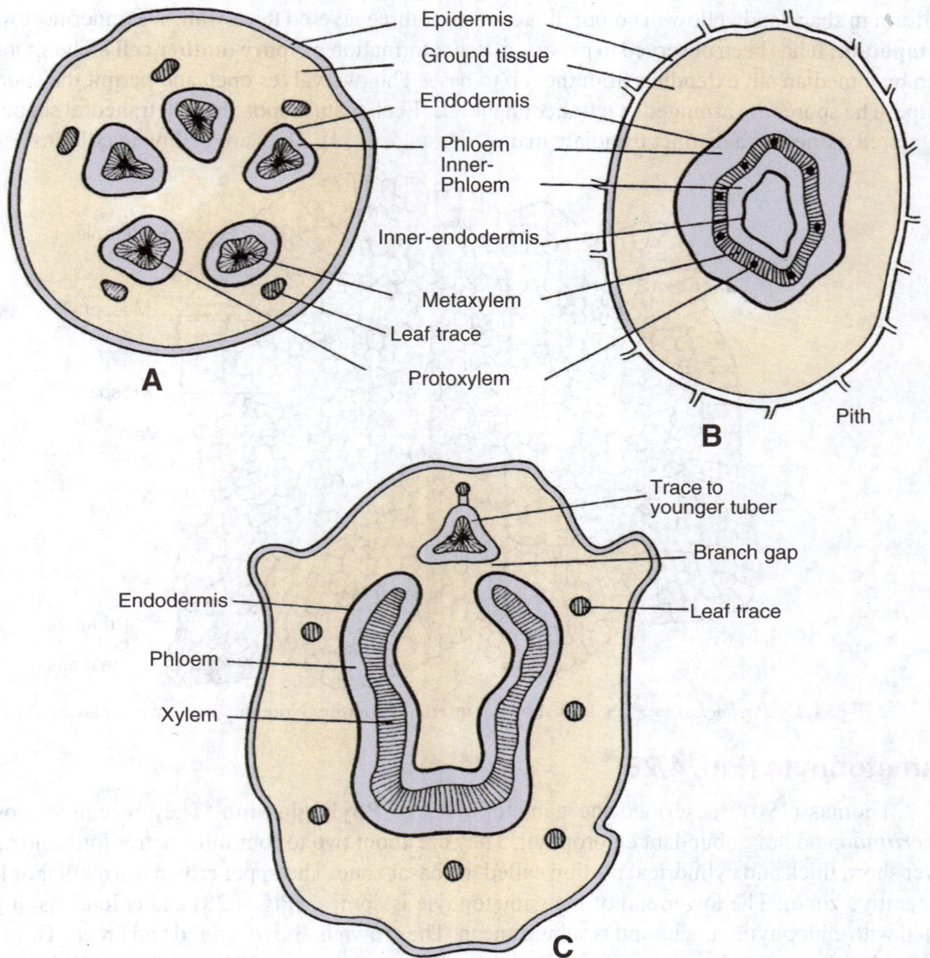

Fig. 4.26 (A–C) Anatomy of the tuber and penduncle of ***Phylloglosum drummondii.*** A.T.S. Through upper part of the peduncle showing dictyostelic condition; B. T.S. Through middle portion showing amphiploic siphonostele; C.T.S. Through top of the tuber showing siphonostele with a branch gap.

A transverse section of the cylindrical leaf (Fig. 4.27) also reveals a single central vascular strand surrounded by an undifferentiated mesophyll which in turn is enveloped by a distinct layer of epidermis.

Organization of the Strobilus

The strobilus is terminal in position (Fig. 4.25, A, B) and is quite distinct. It has a central axis with sporophylls arranged without any definite pattern. They appear to arise in whorls in the lower part of the strobilus. The sporophylls are small and scale like and yellowish in colour. They have a broad base with narrow and pointed (Fig. 4.25, D–F) apex and may be sessile or subsessile. The sporophylls in the middle of the cone appear to be subsessile and those at the base sessile. The sporangia arising in the axils of lower sessile sporophylls are shortly stalked and cauline in origin.

The middle subsessile sporophylls bear sessile sporangia that are foliar in origin. Each sporangium is reniform in shape and yellow in colour. It has a two or three layered thick wall. The innermost wall is the **tapetum**. It has been observed to persist up to the formation of **spore mother cells**. The sporangia open by a median slit extending from the top to base. The two valves open and permit the spores to escape. The spores are arranged in tetrahedral tetrads. Each mature spore has a tetrahedral shape with sculptured exine and a distinct triradiate mark (Fig. 4.25, H., J). They are yellow in colour.

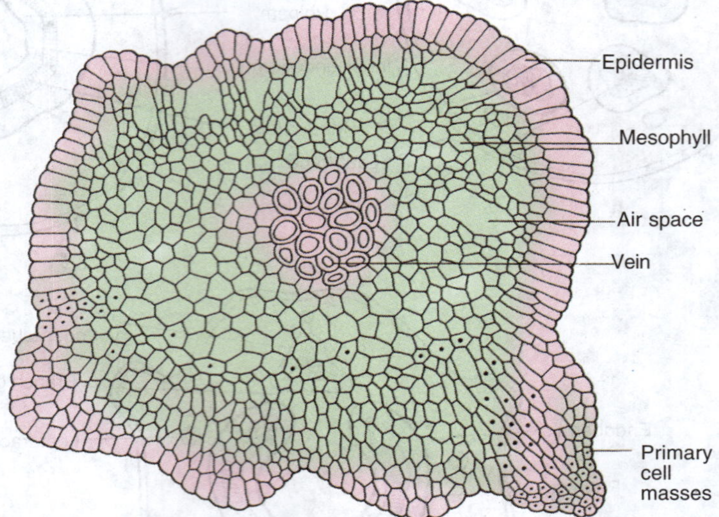

Fig. 4.27. ***Phylloglossum***. T.S. Leaf showing internal structure. Note the primary cell masses.

Gametophyte (Fig. 4.28)

Thomas (1901) described the gametophytes of **Phylloglossum**. They resemble those of ***L. cernuum*** and have abundant chlorophyll. They are about two to four millimetres long and have a lower short, thick and cylindrical portion called the basal zone. The upper crown is irregular or lobed (generative zone). The lower end of the gametophyte is conical (Fig. 4.28) and colourless. It is infected with endophytic fungus and is subterranean. The crown is above ground and green. The antheridia and archegonia arise in crown and have same structure as ***Lycopodium***. Thomas did not study the spermatozoids.

Bertrand stated that M. L. Crie has described the germination of spores. According to Crie, the spores germinate readily and produce a colourless gametophyte. He has also described the spermatozoids to be bicilliate. Crie's account is different from Thomas.

Embryo

The embryo development has not been studied in detail. Thomas (1901) and Sampson (1916) have given brief descriptions of the embryo. They describe it to be similar to ***L. cernuum***. The embryo is attached laterly to the gametophyte. It has a prominent foot which attaches it to the gametophyte. The leaf and the stem apex break through the prothallial tissue and come out laterally. At this stage the embryo appears cylindrical. The lower part swells up, becomes fleshy and forms the tuber, the upper end grows into the first green and cylindrical leaf. The leaf has a vascular strand. The tuber has often been regarded as a ***protocorm***. Eames (1986) is not in favour of this view. He regards it a true tuber of the first season because it bears a true leaf that is traversed by a distinct vascular strand. The tuber bears no root for the first season and has only one leaf. Roots appear later and so do other leaves.

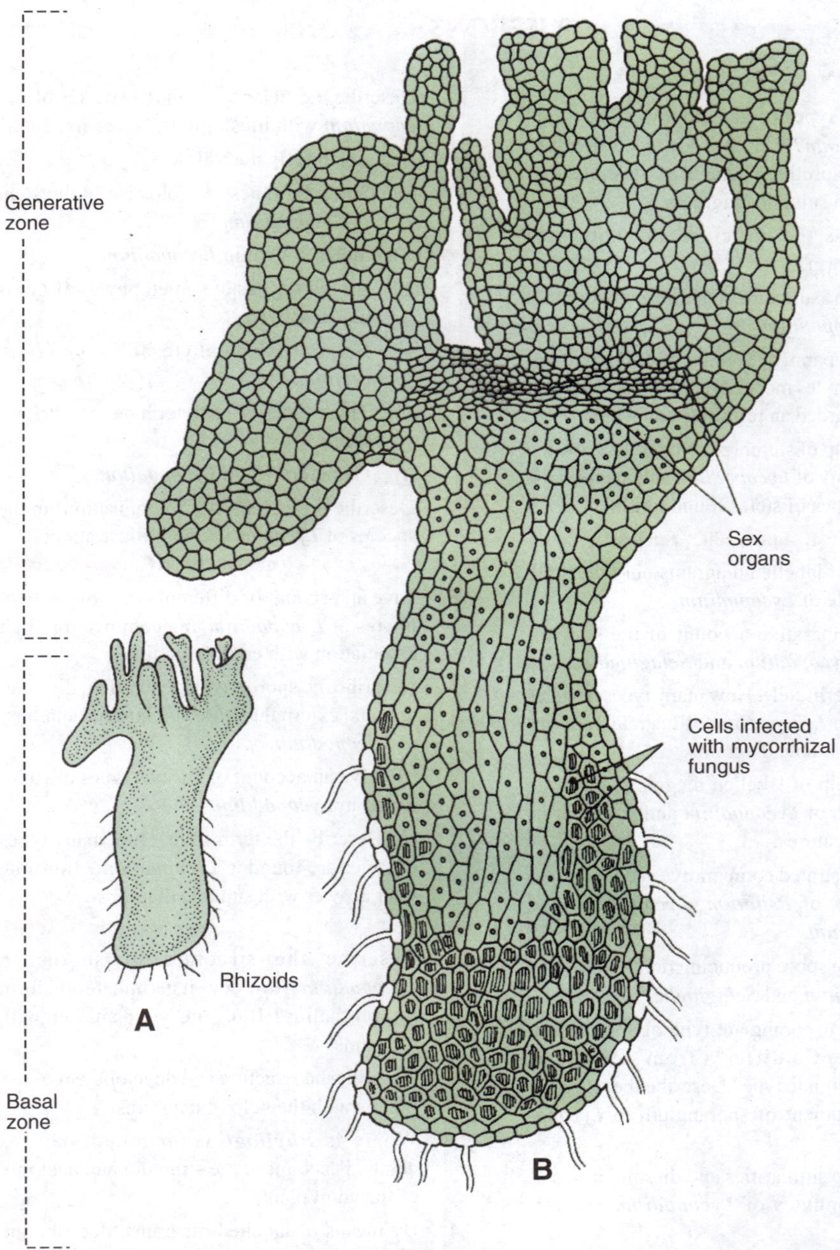

Fig. 4.28 (A-B). A diagrammatic (A) and hypohetical (B) drawing of the gametophyte of *P. drummondii*

Taxonomic Position

Division	:	***Lycophyta***
Class	:	***Eligulopsida***
Order	:	***Lycopodiales***
Family	:	***Lycopodiaceae***
Genus	:	***Phylloglossum***

QUESTIONS

ESSAY TYPE

1. How many types of gametophytes are found in *Lycopodium*? Why it is always difficult to collect such prothalli in nature? Illustrate your answer with suitable diagrams.
2. (*a*) Discuss the structure and nature of Protocorm.
   (*b*) Describe structure of nature gametophytes of ***Lycopodium.***
3. Write a comparative account of different types of gametophytes met in ***Lycopodium***. Which of them is regarded an primitive and why?
4. With the help of suitable sketches describe the stem anatomy of ***Lycopodium*** and mention the different types of steles found in it. **(Kanpur, 2002; Lucknow, 2001)**
5. By means of labelled diagrams only illustrate the life cycle of ***Lycopodium.***
6. Give a comparative account of the gametophytes of ***Lycopodium*** and ***Selaginella.***
7. Define the term stele. How many types of steles are found in ***Lycopodium***? Illustrate your answer with suitable diagrams.
8. With the help of labelled diagrams, illustrate the lifecycle of ***Lycopodium*** and point out its important features.
9. Give an illustrated comparative account of gametophytes, of ***Psilotum***, ***Lycopodium*** and ***Ophioglossum.***
10. Describe the spore producing organs (strobili) of ***Lycopodium*** and ***Selaginella.***
11. How does eusporangiate type of sporangium development differ from that of Leptosporangiate type? Describe the structure and development of sporangium in ***Lycopodium.***
12. Describe the similarities and dissimilarities of the gametophytes of ***Lycopodium*** and ***Selaginella.***
13. Give a comparative account of spore bearing organs of ***Lycopodium***, ***Selaginella*** and ***Equisetum***. Support your answer with labelled sketches. **(Lucknow, 2001)**
14. Give a brief account of the internal structure of the stem of ***Lycopodium.*** **(Lucknow, 2001)**
15. Describe the stelar types in the species of ***Lycopodium*** with illustrations. **(Kerala, 2000)**
16. Describe briefly the following
   (*a*) Different types of protostele in the stem of ***Lycopodium.***
   (*b*) Stelar system in ***Lycopodium.***
   (*c*) Any two types of gametophytes of ***Lycopodium.***
   (*d*) Mature gametophyte of ***Lycopodium cernuum.***
   (*e*) Three types of gametophytes of ***Lycopodium.***
   (*f*) Gametophyte of ***Lycopodium.***
17. Describe the stelar types (organisation) in the species of ***Lycopodium*** with illustrations. **(Kerala, 2001; Lucknow, 2002)**
18. Give an account of different types of gametophytes of ***Lycopodium*** and comment on their association with endophitic fungi.
19. Describe the sporophytic generation, distribution and external feature and internal anatomy of ***Lycopodium***.
20. (*a*) Give an account of various types of steles found in ***Lycopodium.***
   (*b*) Describe the term stele. How many types of steles are found in ***Lycopodium*** ? Illustrate your answer with suitable diagrams. **(Allahabad, 2004)**
21. Describe the structure of spores in ***Lycopodium.*** How does it germinate and form the prothallus? Illustrate your answer with diagrams.
22. Describe the structure and development of sex organs with the help of diagrams.
23. Where is ***Phylloglossum*** found? Give a detailed account of the structure and anatomy of the adult plant.
24. By means of labelled diagrams, describe the anatomy of the tuber, peduncle and leaf of ***Phylloglossum***.
25. Describe the gametophyte and embryo of ***Phylloglossum.***

SHORT ANSWER TYPE

26. **Answer the following:**
   (*i*) Name a pteridophyte which has mycorrhiza in its gametophyte.

(*ii*) Name a pteridophyte in which vegetative reproduction takes place by gemmae.

(*iii*) Which pteridophyte is called "club moss". **(Kanpur, M.Sc. 2000)**

(*iv*) Name a homosporous pteridophyte.

(*v*) Name a pteridophyte which has plectostele. **(Kanpur, 2000)**

(*vi*) (*a*) Where is protocorm formed?

(*b*) What is protocorm?

(*vii*) Name the pteridophyte having different types of steles.

(*viii*) From which place will you collect ***Lycopodium.*** **(Kanpur, 2000)**

(*ix*) Where does ***Lycopodium phlegmaria*** grow in nature?

(*x*) Name a pteridophyte whose spores were used as explosive during the Ist World War.

(*xi*) Mention the names of two subgenera of ***Lycopodium.***

(*xii*) Why is it always difficult to discover and collect the prothalli of ***Lycopodium*** in nature?

(*xiii*) How can you explain the occurrence of ***Selago*** condition as ***Lycopodium***?

(*xiv*) Name a pteridophyte and its species which shows the presence of protocorm in it. **(Allahabad, 2001)**

(*xv*) Write botanical name of club moss.

(*xvi*) Give the function of protocorm.

(*xvii*) What is selego condition? **(Poorvanchal, 2000)**

(*xviii*) Is protocorm a sporophyte or gametophyte?

(*xix*) Name an epiphytic species of ***Lycopodium.*** **(Lucknow, 2001)**

(*xx*) What are the types of steles in ***Lycopodium.***

(*xxi*) Mention the stelar types in ***Lycopodium.***

(*xxii*) What is the name of common species of ***Phylloglossum.***

(*xxiii*) How does ***Phylloglossum*** reproduce vegetatively?

(*xxiv*) What is the shape of spore of ***Phylloglossum***?

(*xxv*) Name a pteridophyte which has primary cell masses in the leaf.

(*xxvi*) Which chemical substance is evolved during the disintegration of neck canal cells and ventral canal cells in ***Lycopodium*** ? **(Allahabad, 2003)**

27. Write short notes on

(*i*) Prothallus of ***Lycopodium*** **(Kanpur, 2000; Awadh, 2000)**

(*ii*) Protostele. **(Meerut, 2000)**

(*iii*) Economic importance of ***Lycopodium.***

(*iv*) Gametophytes of ***Lycopodium clavatum.***

(*v*) Protocorm **(Kumaon, 2000; Bangalore, 2001)**

(*vi*) Types of steles in stems of ***Lycopodium*** species.

(*vii*) Steles in ***Lycopodium.*** **(Devi Ahilya, 2001)**

(*viii*) Morphology of Protocorm.

(*ix*) Spore producing organs of ***Lycopodium.***

(*x*) Gametophyte of ***L. cernuum.*** **(Poorvanchal, 2000)**

(*xi*) Stelar organisation in the stem of ***Lycopodium.***

(*xii*) ***Lycopodium*** strobilus. **(Bangalore, 2001; Meerut, 2000)**

(*xiii*) Different types of gametophytes of ***Lycopodium.*** **(Devi Ahilya, 2001)**

(*xiv*) Stem anatomy of ***Lycopodium.***

(*xv*) Vegetative reproduction in ***Lycopodium.***

28. Differentiate between.

(*a*) Cones of ***Lycopodium*** and ***Selaginella.***

(*b*) Sperms of ***Lycopodium*** and ***Equisetum.***

(*c*) Root and Stem of ***Lycopodium.***

(*d*) T. S. Stems of ***L. cernuum*** and ***L. clavatum.***

(*e*) Reproductive bodies of ***Lycopodium*** and ***Equisetum.***

(*f*) Gametophyte of ***Lycopodium cernuum*** and ***Equisetum debile.***

(*g*) Leaves of ***Lycopodium*** and ***Selaginella.***

(*h*) Prothallus of ***Lycopodium*** and ***Selaginella.*** **(Lucknow, 2000)**

(*i*) ***Phylloglossum*** and ***Lycopodium.***

(*j*) Gametophytes of ***Lycopodium*** and ***Phylloglossum.***

29. Draw labelled diagram of the following.

(*a*) L. S. Strobilus of ***Lycopodium.***

(*b*) T. S. Stem of ***Lycopodium.***

(Allahabad, 2000)

(*c*) Prothallus (gametophyte) of ***Lycopodium clavatum.***

(Kanpur, 2001; Lucknow, 2000)

(*d*) V. S. Gametophyte of ***Lycopodium cernuum.***

(*e*) T. S. ***Lycopodium clavatum*** stem.

(*f*) Strobilus of ***Lycopodium***.

(*g*) V. L. S. gametophyte of ***Lycopodium cernuum.***

(*h*) T. S. Root of ***Lycopodium.***

(*i*) T. S. Rhizome of ***Lycopodium***

(*j*) Sporophyte of ***Phylloglossum.***

(*k*) Gametophyte of ***Phylloglossum.***

(*l*) ***Lycopodium*** stem showing plectostele

(Kanpur, 2004)

30. Answer the following going two important points:

(*a*) Antheridium of ***Lycopodium.***

(*b*) ***Lycopodium*** leaves.

(*c*) Structure of stele in ***Lycopodium*** roots.

(*d*) Sporangium of ***Lycopodium.***

31. (*a*) Trace the steler variations in ***Lycopodium.***

(*b*) Describe any type of gametophyte of ***Lycopodium.***

32. Compare the gametophyte of ***Lycopodium*** with that of ***Equisetum.***

33. What is the chromosome number of ***Lycopodium***?

34. Why are those species of ***Lycopodium*** which show the selago condition regarded as more primitive than the others?

35. Describe briefly the gametophyte of ***Phylloglossum.***

36. How does prothallus of ***Lycopodium*** reproduce vegetatively?

37. What is protocorm? Discuss its morphology.

(Kanpur, 2004)

38. Explain the sporophyll of ***Lycopodium***.

(M.S. Univ., Tirun, 2005)

39. Discuss the development of sex organs in ***Lycopodium***.

40. Explain the sporophyll of ***Lycopodium***.

(M.S. Univ., Tirun, 2005)

41. Discuss the development of sex organs in ***Lycopodium***. **(M.S. Univ., Tirun., 2005)**

OBJECTIVE TYPE

42. Select the correct answer.

(*i*) In ***Lycopodium*** stem the stele is

(*a*) Solenostele

(*b*) Dictyostele

(*c*) Siphonostele

(*d*) Plectostele.

(*ii*) Protocorm formation occurs in

(*a*) ***Lycopodium***

(*b*) ***Selaginella***

(*c*) ***Equisetum***

(*d*) ***Marselia.***

(*iii*) Actinostele is found in

(*a*) ***Lycopodium cernuum***

(*b*) ***L. clavatum***

(*c*) ***L. serratum***

(*d*) ***L. phlegmaria.***

(*iv*) Plectostele is a type of stele present in

(*a*) ***Lycopodium***

(*b*) ***Selaginella***

(*c*) ***Equisetum***

(*d*) Marsilea.

(*v*) Mycorrhiza is found in the prothallus of

(*a*) ***Selaginella***

(*b*) ***Equisetum***

(*c*) ***Pteris***

(*d*) ***Lycopodium.***

(*vi*) ***Lycopodium*** produces the spermatozoids with

(*a*) Two flagella

(*b*) Four flagella

(*c*) Many flagella

(*d*) No flagella.

(*vii*) Biflagellate spermatozoids are found in

(*a*) ***Lycopodium***

(*b*) ***Selaginella***

(*c*) ***Equisetum***

(*d*) Pteris.

(*viii*) Stele in ***Lycopodium*** is

(*a*) Dictyostelic

(*d*) Protostelic

(*c*) Solenostelic

(*d*) Siphostelic.

(ix) On which side is the sporangium seen on the sporophyll in ***Lycopodium***?

(*a*) Abaxial
(*b*) Basal,
(*c*) Axillary
(*d*) Adaxial.

(x) Wall of sporangium in ***Lycopodium*** usually consists of

(*a*) Single layer
(*b*) Two layers
(*c*) Three layers
(*d*) Many layers.

(xi) ***Lycopodium*** is commonly called as

(*a*) Horsetail
(*b*) Club moss
(*c*) Quillwort
(*d*) Stonewort.

(xii) Fleshy tuber and underground portion of the sporophyte is characteristic of

(*a*) ***Lycopodium***
(*b*) ***Selaginella***
(*c*) ***Phylloglossum***
(*d*) ***Equiseatum.***

(xiii) Where is ***Phylloglossum*** found naturally?

(*a*) India
(*b*) England
(*c*) Russia
(*d*) Tasmania.

(xiv) In ***Phylloglossum*** the sporophylls are arranged on the axis

(*a*) Alternately
(*b*) Opposite to each other
(*c*) Without any definite pattern
(*d*) Sometimes alternately and sometimes opposite.

(xv) Spores are Tetrahedral in

(*a*) ***Lycopodium***
(*b*) ***Phylloglossum***
(*c*) ***Selaginella***
(*d*) ***Equisetum.***

(xvi) Homosporous condition is found in

(*a*) ***Lycopodium***
(*b*) ***Marsilea***
(*c*) ***Selaginella***
(*d*) ***Cycas***

(xvii) How many types of prothalli are found in different types of ***Lycopodium***?

(*a*) 1 (*b*) 2
(*c*) 3 (*d*) 4

(xviii) Many types of steles occur in

(*a*) ***Marsilea***
(*b*) ***Lycopodium***
(*c*) ***Rhynia***
(*d*) ***Equisetum***

Isoetes hystrix

CHAPTER 5

Lycophyta–Liguloposida

- Isoetales–General
- Isoetales
- Description of ***Isoetes*** Linn
- Description of ***Stylites***
- Differences between ***Isoetes*** and ***Stylites.***

ISOETALES–General

It includes a single family **Isoetaceae**. ***Isoetes*** and ***Stylites*** are the living, and ***Isoetites*** is the extinct member of the family. The fossil history of the genus is meagre and on that account the living member ***Isoetes*** is regarded as an isolated type possibly of considerable antiquity and as pointed out by Seward and others closely akin to such extinct genera as ***Pleuromia*** and ***Sigillaria. Some*** species of the extinct genus ***Isoetites*** have been described but their identity is doubtful. Only ***Isoetites choffali*** described by Saporta from Lower Cretaceous of Portugal is slightly better known. ***Stylites*** has two species found in Peruvian Andes. It differs from ***Isoetes*** in having elongated stem and unbranched roots. Rauh and Falk reported it in 1959. ***Stylites andicola*** and ***S. gemmifera*** are the two species reported from central Peru.

ISOETES Linn.

It includes about 75 species (Airy Shaw in Willis 1966) that are familiarly known as **'quillworts'** due to their narrow and elongated grass-like leaves with broad and spathulate bases. Majority of the species are found in Europe and North America. In India, seven species have been recorded. The most common out of them is ***Isoetes coromandelina*** (Fig. 5.1). The others are ***I. samapathkumarani***, ***I. indica***, ***I. mirzapurense***, ***I. panchananii***, ***I. dixetii*** and ***I. sahyadrii.*** Bierhost (1971) mentions 40 species.

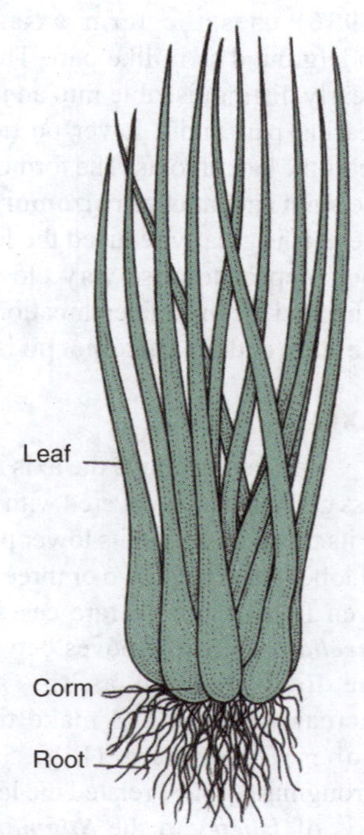

Fig. 5.1. *Isoetes coromandelina.* A complete plant showing habit.

The Sporophyte

Habit and Habitat

The common Indian species ***I. coromandelina*** grows wild along the Coromandel Coast, Madras, Kerala, Bombay, Varanasi, Baroda, Serampur in Bengal, along the banks of river Jamuna at Delhi, Allahabad, Rewa. Padampur, Eastern India, Lucknow, Faizabad, Indore and in an uncultivated land at Meerut. ***I. panchananii*** is very common at Pachmarhi (Bhambie and Sharma, 1956; Bir and Vasudeva, 1972, 1973). Pant and Srivastva (1962) have given a complete range of distribution of various species in India along with their distinguishing characters and key for identification. The Indian species are terrestrial as well as amphibious. ***Isoetes coromandelina*** is perennial with a distinct growing season that starts about mid-October after the monsoon and continues till March. It is mostly amphibious in existence and is common along the edges of ponds, pools and rivers. It also grows in shallow water with portion of leaves projecting above water. ***I. panchananii*** is also amphibious. The plants vary in size from 20 to 60 cms.

The America species ***I. englemanni*** is a partially submerged, aquatic ***Isoetes japonica*** grows submerged in water. ***I. butleri*** is terrestrial.

Isoetes coromandelina

The sporophyte of ***Isoetes***, when growing under natural conditions, gives the appearance of a liliaceous or a graminaceous monocotyledon due to its characteristic linear leaves that arise from a subterranean and bulbous axis (Fig. 5.1). The underground portion of sporophyte has been given the names of "Corm" and 'rhizome', because it is thick, reduced and looks like a condensed rhizome. Eames

(1936) uses the term **axis** for this underground stem like part. The axis is clearly distinguishable into an upper leaf bearing part and a lower portion which bears the lateral roots. The former is called the **stem** and latter as **rhizomorph**. Since the axis is greatly reduced the leaves and roots appear to arise very close to each other and the line of demarcation between the **stem** and the **rhizomorph** is obscure.

I. butleri

Axis

In young plants the axis is more or less globular and is covered with the leaves at its top and roots at its lower parts (Figs. 5.1 and 5.2). In older plants it becomes lobed. The number of lobes vary from two or three and even four or five in rare cases (***I. saccharata*)**. The grooves between the lobes deepen as the axis increases in age and make them (lobes) prominent (Fig. 5.2). Brongniart (1828) related the lobed axis of ***Isoetes*** to the ***Stigmarian*** axis. This view is not valid as the lobes in ***Isoetes*** do not develop due to the action of supplementary vegetative cones at the basal side (as in ***Stigmaria***), but only due to the local formation of the secondary tissues by the cambium. Bhambie (1965) reported plants of ***I. coromandelina*** with four lobed axis. According to him this condition is quite prevalent in this species and appears to be a normal feature. The shape of the axis in such plants is roughly cylindrical in the upper and measures about 1.5 cm. in diameter. Such four lobed axes of ***Isoetes*** has been compared with those of fossil lycopods like ***Pleuromia*** and the stigmarian base. In the words of Bhambie (1965), "though the stelar lobes in ***Isoetes*** do not coincide with those of the cortical lobes of the axis, the four lobed axes of ***I. coromandelina*** show more resemblance with the basal region of ***Pleuromia*** and stigmarian base". Four or even Five lobed axes have also been reported in some other species ***e.g. I. lacustris***.

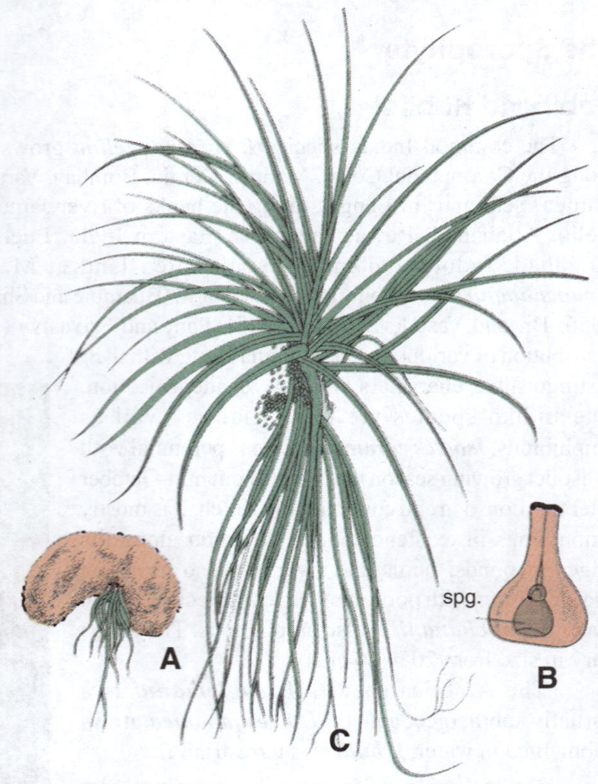

Fig. 5.1a (A—C). A. Axis of *I. lacustris.* Note the lobes and dichotomously branched roots; **B. Base of sporophyll of I.** ***lacustris.*** I, ligule ; spg., sporangium partially covered by velum; C. A complete plant of *I. echinospora* (After Motelay and Vendryes)

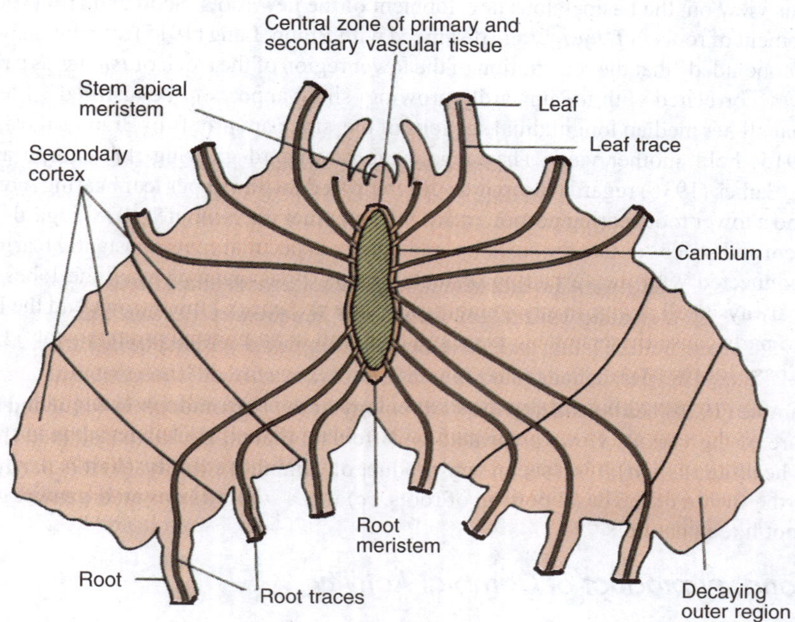

Fig. 5.2. ***Isoetes coromandelina.*** V. S. through the axis (diagrammatic) showing internal organisation.

The place where the lobes arise is determined by the phyllotaxis of the young plants. In distichous plants the grooves alternate with the leaf orthostichies under which the lobes are formed. Such a position was first observed by A.L. Braun in 1847. Once the lobes are formed they maintain their position even on change of phyllotaxis. The lobes grow more in breadth by the dilatation of the tissue and vary from 0.5—4 centimetres in width. Grooves develop by the appearance of longitudinal slits between the lobes. The tissues are sloughed off in the form of characteristic 'caps or shoulders' (Eames, 1936) that appear on the upper margins of the axis (Fig. 5.2). This sloughed off tissue includes the bases of the leaves and the roots of the previous year and undergoes decay at the end of the growing season. The growth in length of the axis is very slow and is effected by a group of the meristematic cells that lie sunken in a groove formed by the greater growth of the surrounding tissue.

Karrfalt and Eggert (1972) studied the development of roots and lobes in some species of ***Isoetes*** (***I. saccharata*** and ***I. tickermanni***). They ascribe their formation to two developmental processes. They found that there is a row of tangentially elongated cells at the base of the furrows. They perform the mechanical function of separating the lobes. " Progressive rupture of these cells from surface to deeper layers maintains the depth of the furrow. This layer of elongated cells becomes double and by meristematic activity produces a new lobe in the region between the two layers. New lobes also appear by the development of a furrow between the two bands."

Dichotomous branching has also been observed in some plants, but the dichotomy never becomes prominent due to the very slow growth of the axis. It is evident only by the presence of two crowns of leaves.

Morphological Nature of the Axis

Regarding the morphological nature of the axis, several views have been advanced. Braun (1847) believed that the lower part of the ***Isoetes*** stock base was nothing but a telescoped main root.

He based his view on the basipetalous development of the new roots. Scott and Hill (1900) studied the development of roots in ***I. hystrix*** and disagreed with Braun. Lang (1915) after intensive anatomical studies concluded "that the recognition of the lower region of the stock of ***Isoetes*** as a rhizomorph in some way correlated with the upwardly growing shoot appears to be justified". Liebig (1931) declared that all her median longitudinal sections of the stock are proofs of Braun's thesis. West and Takeda (1915) hold another view. They regard the downward growing rhizomorph as an organ ***suigeneris***. Eames (1936) regards the axis as differentiated into an upper leaf bearing portion called the **stem** and a lower root bearing portion called the **rhizomorph**. Schoute (1938) regards the ***Isoetes*** axis as a stem and states "....that the stock base of ***Isoetes*** is peculiar in its constantly tearing furrows, a feature connected with the formation of local humps of secondary tissue, the lobes. Roots are formed ,as always in all plants, in any suitable place; the presence of the furrows and the lack of new higher stem parts cause these suitable places to be found in an unusual arrangement." He used the term stock base for the axis.

Bhambie (1970) has considered the morphology of the **rhizomorph** in detail and has given a vivid picture of the various views presented by different authors. He has considered this problem under five headings as : (***a***) it is a secondary product of cambium activity, (***b***) it is partly or wholly formed by the fusion of the basal portion of roots, (***c***) it is a root, (***d***) it is an organ ***sui generis*** and (***e***) it is shoot like in nature.

(a) Secondary product of Cambial Activity

The proponents of this view believe that rhizomorph is a product of secondary activity and is formed by the cambial activity of the plant. Hofmeister (1862) and Schoute (1938) support this hypothesis. Hofmeister did not consider rhizomorph as a fundamental part of the plant body and regarded it as later development due to cambial activity. Schoute also regarded the rhizomorph lobes as humps of tissue formed by cambial activity. Anatomical studies (Bhambie, 1957, 1962, 1970) have revealed that the stele of the axis is anchor-shaped and has two distinct regions : (*i*) The upper region which is cylindrical and forms the stem stele and gives off leaf traces and (*ii*) the rhizomorphic stele which sends off roots traces and is bi-,tri, or tetra- radiate depending upon the number of lobes of the rhizomorph. Both the steles are connected in the central region of the axis. The lobes of the rhizomorphic stele have upturned free ends that alternate with the cortical ridges of the axis and centrally fused. Three distinct types of meristems have been recognised. These are : (*i*) the apical meristem which forms the stems stele;

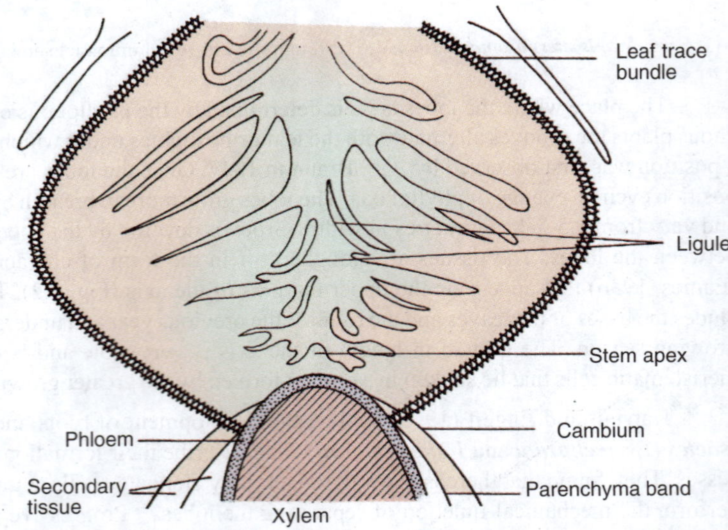

Fig. 5.3. ***Isoetes coromandelina.*** L. S. Through the axis (upper part) showing the stem apex, vascular region and arrangement of young leaves. Also note the ligules. (After Bhambie)

(*ii*) the lateral meristem which originates in periphery of the primary phloem and completely surrounds the stem stele and is located in the form of strips between the rhizomorphic lobes and (*iii*) the basal meristem present all along the ridges of the rhizomorphic stele. It adds new tissue to the rhizomorphic stele (Bhambie 1962) and gives rise to the roots also. It extends upto the upturned edges of the rhizomorhpic stele. The stem and the rhizomorph have their own meristem. In the light of these observations it is difficult to believe that the rhizomorph is a new structure and is not a fundamental part of the plant body.

(b) Fusion product of Roots

This view was put forward by Sachs (1882) and De Bary (1884) and later modified by Scott and Hill (1900). Sachs and De Bary were of the opinion that the rhizomorph is formed by the fusion of the basal parts of the roots whereas Scott and Hill believed that it is partly formed by the fusion of the bases of roots and partly by the activity of the cambium. The view has been opposed by Bhambie who is of the opinion that the anatomical studies strongly support that rhizomorph is a "solid structure having its own entity." He further remarks, " Moreover, the root primordia arise just below the rhizomorph and later their steles become connected with the stele of the rhizomorph (Bhambie, 1963 b)."

(c) Organ Sui-genesis

This view was put forward first by West and Takeda (1915) and later supported by Bowers (1935), Sporn (1963) and Wardlaw (1965). They believed that rhizomorph is neither a stem nor a root, but an organ of indeterminate character.

(d) Root like nature

It was proposed by Stokey (1909) who stated that the tuberous body (axis) is not wholly stem but a "contracted stem and main root". She has not given any anatomical reasons for such a conclusion. The present day studies on the anatomy of the rhizomorph do not support such a conclusion.

(e) Shoot like nature

This view was first put forward by Von Mohl (1840) who regarded the rhizomorph as a descending organ of the axis which gives rise to the roots. Lang (1915) regarded it as a downwardly growing rhizophore whose apex behaves like the stem apex. Eames (1936) also regarded it as resembling the rhizophore which is shoot like in nature. Seward (1910), Smith (1955), Stewart (1947) and Andrews (1961) also regard it as shoot like in nature. Laclercq (1930), Weiss (1931-32), Eames (1936), Andrews (1961) and Philips and Gilbert (1966) compared the rhizomorph of **Isoetes** with root bearing regions of some fossil lycopods and confirmed the shoot like nature of this structure. Bhambie (1970) also agrees with the proponents of axial nature of the rhizomorph. His conclusions are based on the detailed anatomical studies of the genus.

GROWTH

There are two schools of thought regarding the organisation of the shoot apex in ***Isoetes***. These are :

Apical Cell Hypothesis

It was put forward by De Bary (1884), Hofmeister (1862) and Nageli (1878). These botanists asserted that there is a single apical cell at the apex of the axis. This was later supported by Van Tiegham (1891), Scot and Hill (1900) and by Karrfalt (1977) in the shoot apex of ***I. hystrix.***

Many-celled Meristem Hypothesis

Hegelmaier (1874) described the presence of a superficial meristematic layer. Farmer (1890) observed a single layer of columnar cells covering the entire stem apex in ***I. lacustris***. Smith (1900), West and Takeda (1915) and Lang (1915 a, 1916 b) also observed a group of meristematic cells at the shoot apex. Paolillo (1963) and Bierhorst (1971) described that there is a dome of meristematic cells having a few centrally located apical initials at the crest as in ***Lycopodium*** that are responsible for the growth of the plant. Bhambie (1957, 1972, 1988) studied the shoot apex in ***I. coromandelina*** and ***I. englemanni*** and discovered a hump of meristematic cells. He distinguished two more or less well demarcated regions in this apical hump. A superficial layer of columnar meristematic cells that divide mainly by anticlinal divisions and an inner tissue that divides in an irregular fashion. Histochemical studies in ***I. setcea*** by Michaux Ferriere (1976 - 1978) have confirmed Bhambie's observations.

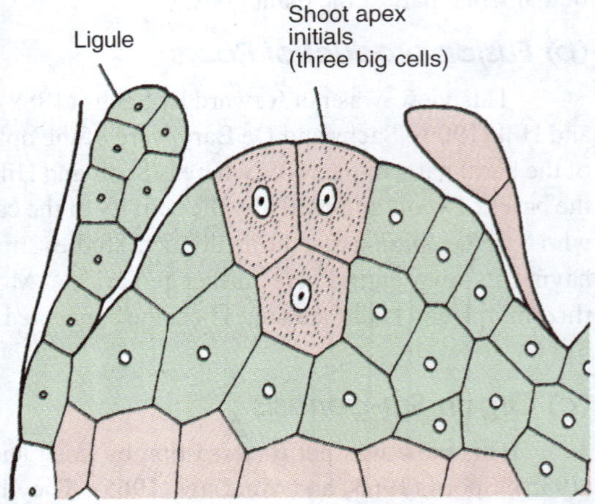

Fig. 5.4. *Isoetes sampathkumarani.* A section through stem apex showing three big shoot apex initials. (After Usha Sharma)

Usha Sharma (1960) studied the shoot apex in ***Isoetes sampathkumarani*** and observed a superficial layer of meristematic cells that divide anticlinally and give rise to the outer epidermal layer (Fig. 5.4). She also observed an inner mass of irregularly dividing cells. For a detailed information on the apical meristems in pteridophytes reference may be made to Bhambie, S. and V. Puri (1985, Trends Pl. Res. pp. 55-81).

Root

In the embryo the first root arises laterally and exogenously (Farmer, 1890). It is provided with root hair and with a root cap. The position of the later roots that arise endogenously is always related to the place of the grooves. These roots, according to most of the authors, are formed in series along the sides of the grooves and are, therefore, in the same position as the orthostichies. This is because the grooves appear in the same position as the orthostichies. In each groove or a furrow the roots arise in an acropetalous succession, ***i.e.***, the youngest is near the stem apex or the end of the groove. West and Takeda (1915) reported that all the roots in furrow are of the same age—a statement that is very difficult to believe. They have not given any observations in support of their statement. Scott and Hill (1900) have clearly shown the acropetalous development of the roots in ***I. hystix.*** West and Takeda state that in ***I. japonica*** new roots develop on the sides of the grooves near the base . When new roots develop in the next season the older ones are pushed outwards towards the ends of the groove (stem apex). Accordingly the younger roots are near the centre of the groove and older ones at the ends of the grooves.

I. hystix. *I. japonica*

The roots branch dichotomously after emerging from the **rhizomorph** (lower part of the axis) ground tissue. The root apex has been reported to consist of a group of initials (Farmer, 1890, Campbell, 1891) that give rise to the root cap, epidermis and cortex. According to these authors, the vascular tissues arise from a single initial. According to Bhambie (1963, 1985 p. 68) the root apex "has a group of centrally located inconspicuous initials which produce a cap distally and a body of the root proximally. The root cap has a columella." For details refer to Bhambie and Puri (1985).

Leaf

The leaves are microphyllous as they are traversed by a single median and unbranched vein. The leaf traces leave no gap in the vascular cylinder of the stem. They vary in length from two centimetres to two feet (***I. englemanni***). They are arranged in close spirals near the upper end of the axis, and range in number from a few to 20. Each leaf has a broad spoon-like base and a narrow and linear upper part. The leaf tips are pointed and straight or recurved and sometimes four-angled. The basal, broad portions of the leaves are underground and colourless, whereas the aerial and linear parts are green and photosynthetic. The leaf bases of outer leaves overlap those of the inner leaves and thus give a bulbous appearance (Fig. 5.1). Usually the first formed or the older leaves and those produced at the end of the growing season are sterile and the rest are fertile, ***i.e.***, they bear sporangia and are called **sporophylls**. In some species abortive sporangia have also been noted at the bases of the sterile leaves. So every leaf is a potential sporophyll and the entire plant can be regarded as a **strobi-**

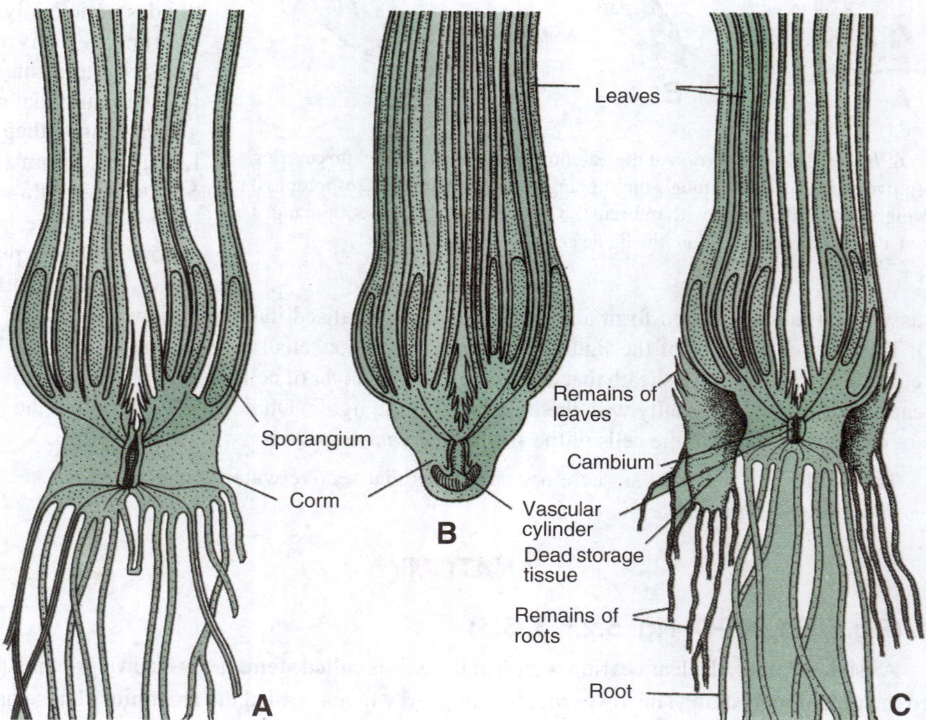

Fig. 5.5 (A-C). V. S. diagrams through the entire plant of *I. coromandelina* illustrating form and structure. A. V. S. cut at right angles to the basal groove of the "corm" of a young plant, showing sporangia, the apex and the stele of the corm; B. V. S. of the plant showing part of the groove where the stele appears like an inverted 'T'; C. Same as A but the plant is older and shows storage region of the corm and dead remains of older tissue.

lus. Each leaf has, therefore, a lower colourless sporangium bearing region (Fig. 5.5) or fertile region and an upper sterile or photosynthetic region. At the junction of the basal, spoon-shaped part of the leaf and the upper green part or the blade there is a small depression on the adaxial side. From this pit or depression arises a small ligule (Fig. 5.6).

Development of Leaf and Ligule

Leaves originate as small outgrowths that arise by the periclinal divisions of the surface cells of the apical meristem. The outgrowth assumes a crescentic outline in the beginning and soon a group of apical initials appears and initiates growth in length. During earlier stages of growth when the leaf mound is only few cells high, a prominent cell distinguishes itself on the adaxial side near the base of the leaf mound. This cell is the initial of the **ligule** (Smith, 1900). This cell divides periclinally and by repeated division gives rise to a filamentous structure that soon grows longer than the leaf primordium (Fig. 5.7, A). Later this filamentous structure, by divisions in various planes and by cell differentiation assumes a tongue-shaped form and shows highly specialised internal structure (Figs. 5.7 and 5.11). The base or the foot of the ligule is to some extent socketed in the tissue of the leaf. This socketed part is covered by a sheath that consists of a single layer of cells continuous externally with the leaf epidermis and internally with the epidermis of the ligule. Dunlop (1949) reported the presence of **casparian strips** in the cells of the sheath in ***I. macrospora.***

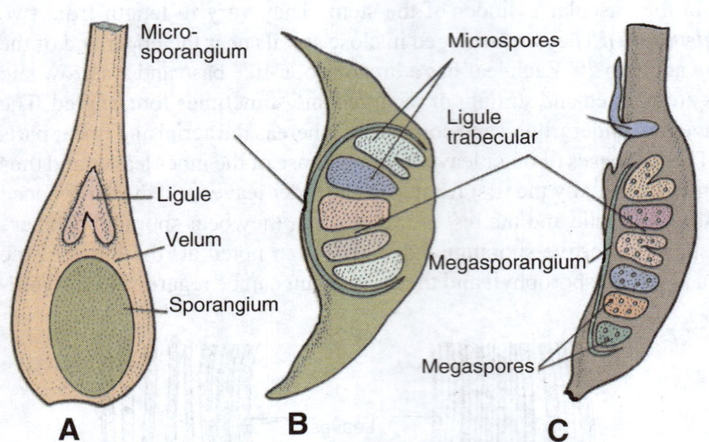

Fig. 5.6. *Isoetes.* A. Lower portion of the leaf showing ligule and the velum covering the sporangium; B. T. S. through the sporangial region of the microsporophyll showing a microsporangium with trabeculae; C. L. S. through the megasporangial part of a megasporophyall. Note the ligule and the velum.

The ligules are regarded as secretory structures that secrete water or mucilage and keep the leaves and the sporangia moist.

ANATOMY

Axis (Fig. 5.6a, A—D, Fig. 5.2,5.3, 5.5)

As stated before, the leaf bearing region of the axis is called **stem.** It has its own growing point called the **apical meristem**. The lower root bearing part which is called the **rhizomorph** also has its own growing point called the **basal meristem** or the **root meristem**. Growth in length of the axis is very slow, but increase in diameter is considerable. This circumferential increase is due to the activity of cambium also called the lateral meristem ***i.e.***, this axis exhibits secondary growth. The anatomy of the axis is complicated and is variously interpreted. The stelar arrangement in the axis is typically

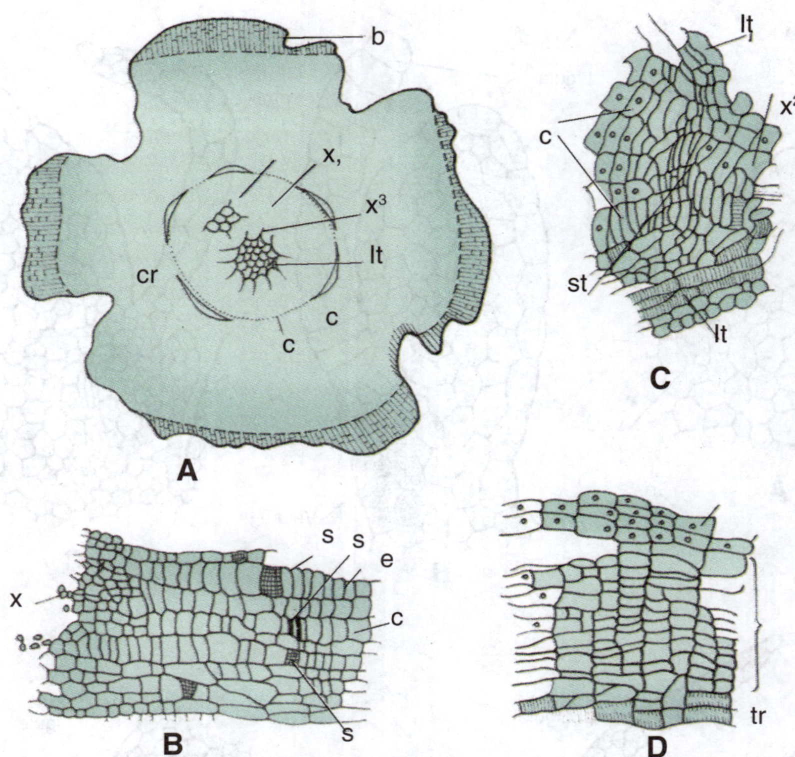

Fig. 5.6a (A—D) ***Isoetes lacustris.*** **A.** T.S. stem (upper leaf producing part of axis) showing epidermis (b), cortex (cr) extra stelar cambium (c), central stele (x), leaf traces (lt) xylem (x, x^2) and parenchyma zone (a) internal to cambium; **B.** L.S. portion of A showing cambium (c), phloem (s) and xylem (x); **C.** T.S. portion of A showing cambium (c), xylem (x, x^1), leaf traces (lt); **D.** T.S. portion of stem showing details of cambium and its derivative cells (d) and tracheids (tr). The derivative cells will distinguish into secondary phloem and xylem elements. (From Seward)

protostelic with a central core made up of xylem (tracheids and parenchyma) surrounded by phloem, which in turn is surrounded by an extra stelar layer of cambium. External to the cambium are the cortex and epidermis. The stele of the axis in a longitudinal section appears like an inverted 'T' or is anchor-shaped (Fig. 5.5 B). The vertical arm of the 'T' represents the stele of the upper region of the axis (stem) and the horizontal arm represents the stele of the rhizomorph. There is a prominent constriction between the two steles (Fig. 5.2 and 5.5 B). The xylem is composed of tracheids and xylem parenchyma. The phloem is made up of a few layers of prismatic cells. The cambium is absent opposite the apical meristems of stem and the rhizomorph and is, therefore, not a complete cylinder (Fig. 5.2)

The extra-stelar cambium which originates very early, cuts off cells both towards its inner side and outer side. (Fig. 5.2). Those cut off towards the inner side become changed into phloem (West & Takeda, 1915) or sometimes into phloem and xylem elements both (Scott and Hill, 1900). In some species the inner cells cut off by the cambium develop into tracheids and parenchyma (Stokey, 1909). Esau (1953) examined the cells cut off towards the inner side in ***I. howellii*** and found them to be secondary phloem. The **sieve cells have sieve areas** on their lateral and end walls.

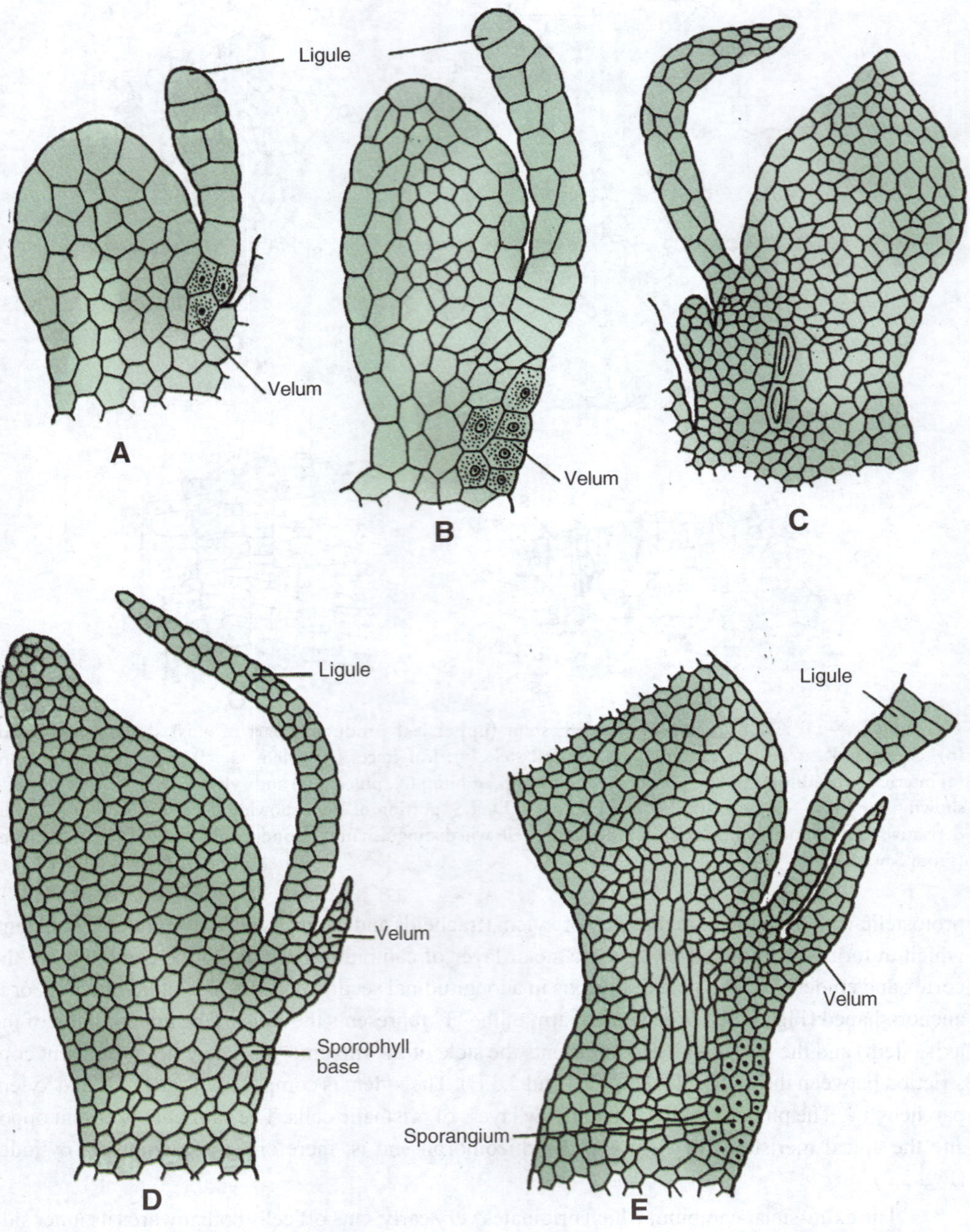

Fig. 5.7. (A—E). *Isoetes coromandelina.* Various stages in the development of the ligule and the velum. A younger sporangium is also seen developing below the velum in E. (After Ekambaram and Venkatanathan.)

The cells cut off towards the outer side develop into secondary cortex. The cells of the secondary cortex store food material and form the storage parenchyma. Every year new storage parenchyma is formed by the cambium and the older one is pushed outward and is ultimately sloughed off as characteristic 'caps' or 'shoulders'. The sloughed off tissue also includes portions of secondary vascular tissue. The surface layers that persist become suberised. Numerous leaf traces arise from the stele of the stem, one going to each leaf (Fig. 5.2). In the rhizomorph root traces pass from the stele to the roots (Fig. 5.2). The root traces are formed by the activity of the basal meristem or the root producing meristem. According to Bhambie (1962), this meristem also produces the elements of the stele of the rhizomorph.

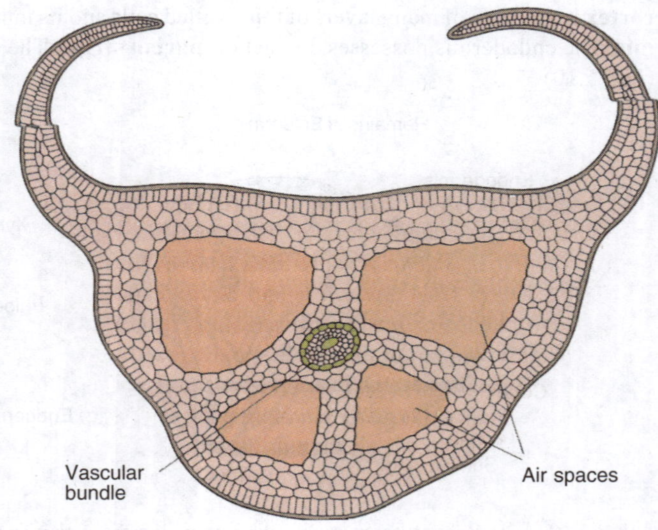

Fig. 5.8. ***Isoetes coromandelina.*** T.S. through leaf.

Leaf

A transverse section of the leaf (Fig. 5.8) shows an outer layer of compactly arranged cells called the **epidermis**. It is covered by a thin cuticle. Next to it is the mesophyll that is not differentiated into palisade and spongy parenchyma. The cells contain chloroplasts (Fig. 5.9) which range in number from one to four per cell. The chloroplasts are very close to the nucleus. The mesophyll is traversed by four longitudinal air spaces that are separated from each other by several layered thick partitions. The air spaces or the lacunae are divided by a series of partitions or diaphragms (Fig. 5.8). There is a single central collateral vascular bundle.

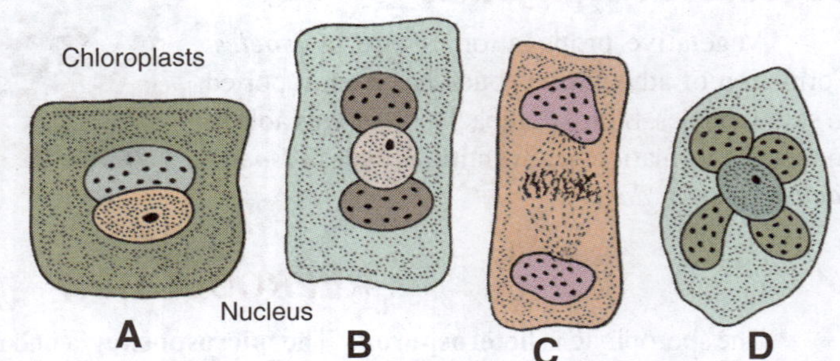

Fig. 5.9 (A-D). ***Isoetes melanopoda.*** A. Cell from a leaf with the chloroplast; B. Cell with two chloroplasts; C. same as B; D. Cell with four chloroplasts. (After Ma)

In the young sporophytes the leaf has only two lacunae. In the submerged species the epidermis lacks stomata whereas in the amphibious and terrestrial species stomata are present.

Root

A transverse section of a root (Fig. 5.10) reveals the usual tissue systems, *viz.*, the epidermis, the cortex, the endodermis and the stele. The **epidermis** is single layered and temporary. The

cortex is made up of many layers of thin-walled cells and its innermost layer is called the **endodermis**. The endodermis possesses distinct **casparian strips**. The cortex surrounds a large air cavity. (Fig. 5.10).

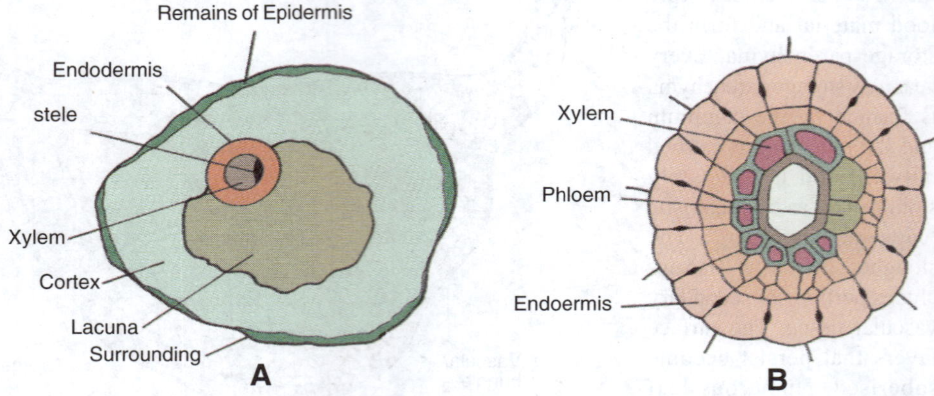

Fig. 5.10 (A-B). Root of *Isoetes sp*. A. A diagrammatic cross-section showing internal structure; B. stele shown in detail.

The vascular cylinder is present on one side of the cavity and is attached to the cortex on that side (Fig. 5.10). The stele consists of primary xylem and phloem that are collateral in arrangement. The phloem is oriented towards the cavity.

Isoetes duriei

Vegetative Propagation

Vegetative propagation is rare in ***Isoetes***. Formation of adventitious buds has been reported in some species. Solms-Laubach (1902) was able to induce the formation of adventitious buds in ***Isoetes lacustris*** and ***I. duriei.***

REPRODUCTION

The sporophyte is **heterosporous**. The **microsporangia** and **megasporangia** arise at the bases of **microsporophylls** and **megasporophylls** (Fig. 5.5). The megasporophylls form the outermost whorl (Fig. 5.5) and are followed by an inner whorl of microsporophylls which in turn surround the innermost sterile leaves that may in some species bear abortive sporangia. At the end of the growing season and during the resting period the megasporophylls and the microsporophylls degenerate or die whereas the innermost sterile leaves may continue to live. They act as outermost whorl during the next season and generally disappear before the new leaves appear.

Arrangement of Sporophylls

The arrangement of sporophylls has been studied in detail in some species by various authors. A brief resume is given here. The Indian species ***Isoetes coromandelina*** is very interesting. The first account of the arrangement of sporophylls in the species was given by Pfeiffer (1922). She found that majority of the plants bore only megasporophylls. In some specimens she found groups of microspores

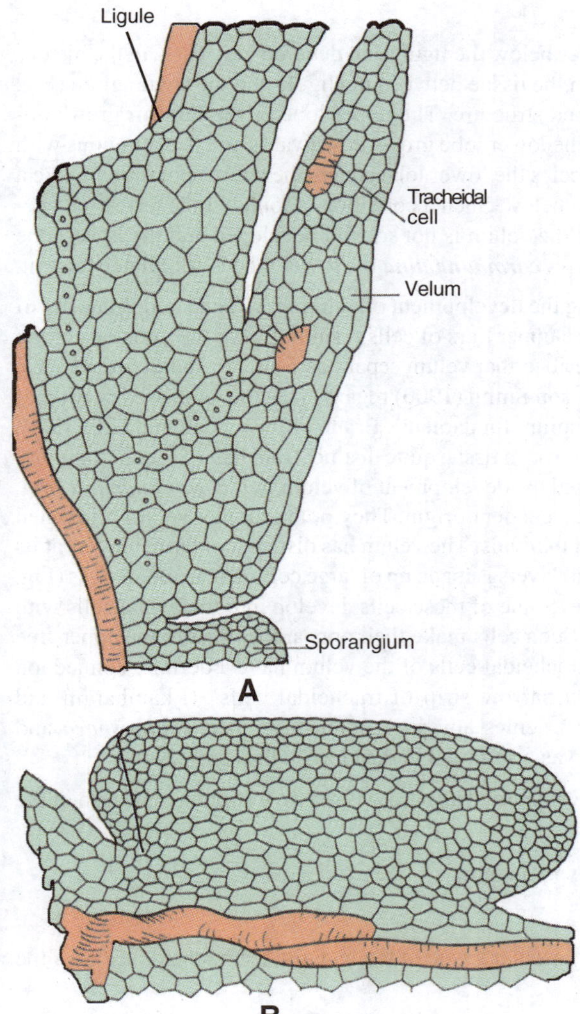

Fig. 5.11 (A—B). *Isoetes coromandelina*. A. portion of a sporophyll (in Longitudinal section) showing a portion of ligule, velum and upper part of sporangium; B. A young sporangium in section. There is no distinction of internal tissues. (After Ekambaram and Venkatanathan)

sticking to the bases of sporophylls. She was unable to observe the complete and intact microsporangia. Ekambaram and Venkatanathan (1933) investigated the species in detail and found that the first two or three leaves that appear during the resumption of growth by the axis are sterile or possess only abortive sporangia. The leaves that arise later are all fertile. They observed two kinds of plant: (*i*) bisporangiate plants and (*ii*) megasporangiate plants. The megasporangiate plants are found in abundance and bear only megasporophylls with megasporangia at their bases. The megasporophylls are arranged in close spirals. In the bisporangiate plants there are only one or two microsporophylls and the rest are all megasporophylls. These plants are less common and were encountered during early rainy season. The ratio of bisporangiate plants to megasporangiate plants is 2 or 3 to 50. There is no external difference between the two kinds of plants. The first microsporophylls, develops after the formation of two or three megasporophylls. In case there are two microsporophylls, the second one does not develop immediately after the first but it comes up later, *i.e.*, after the formation of a few megasporophylls. The megasporophylls that develop late have megasporangia with immature megaspores or do not form megaspores at all. This is probably due to onset of unfavourable conditions.

In ***Isoetes lacustris*** innermost leaves are sterile, the middle ones are microsporophylls and the outermost are megasporophylls. Gluck (1924) studied various species of ***Isoetes*** and observed a number of variations in the arrangement of the microsporophylls and the megasporophylls. His result reveal that there is no regularity in the arrangement of the sporophylls with respect to each other. Wilson Smith (1900) also observed irregularity in their arrangement. West and Takeda (1915) studied in detail the morphology and ontogeny in ***Isoetes japonica***, and found no regular sequence in the development of sporophylls. They are irregularly distributed.

The Velum

It is a characteristic structure that arises below the ligule and develop very early in the history of development of the leaf. It originates from the tissue between the ligule and the sporangium (Fig. 5.7, A—E and 5.11, A). The velum is a lobed structure. The upper lobe or the **labium** grows obliquely upwards and remains short whereas the lower lobe grows downwards and is continuous with tissues of the leaves at the sides. In some species the lower lobe covers the entire sporangium, like a pouch, leaving only a small slit-like aperture below. Such a condition is found in the terrestrial species, ***e.g., Isoetes hystrix***. In aquatic species the velum is not so well developed and the lower lobe only partially covers the sporangium. In ***Isoetes coromandelina*** the lower lobe is completely absent.

Several conflicting accounts regarding the development of velum have appeared. Majority of them describe that velum originates from the upper tiers of cells resulting from the divisions of the sporangium initials. Hofmeister (1862) described that velum separates from the sporangium by the earlier divisions of the sporangial initial. Wilson Smith (1900) regards it as the 'sterilised portion of a sporogenous tissue, or a part of the 'sporangium fundament'. Braun (1863), Scott and Hill (1900) describe that in ***I. hystrix*** velum has an origin from a tissue quite distinct from that of the sporangium. Ekambaram and Venkatanathan (1933) studied the development of velum in ***I. coromandelina*** (Fig. 5.7 and 5.11) and observed that it has an independent origin. They point out that velum originated much earlier than the appearance of sporangial initials. The velum has distinct epidermal layers at its free portion. The tissue between the epidermal layers is made up of large cells that are colourless (Fig. 5.11, A) and lose their contents when mature. Some of these cells develop into tracheidal cells with spiral or annular thickenings (Fig. 5.11, A). Such cells make their appearance first in the upper free portion and later in the lower portion. The tracheidal cells of the velum have a definite connection with the vascular bundle of the leaf by "a narrow strip of tracheidal cells" (Ekambaram and Venkatanathan, 1933). Wilson Smith (1900) denies any such connection in ***I. echinospora*** and ***I. englemanni***. In ***I. coromandelina*** velum was also found to develop on the sterile leaves.

Ekambaram and Venkatanathan (1933) regard it as purely vegetative structure because of the following characteristic features :

1. It originates quite independent of the sporangium.
2. Its formation in sterile leaves.
3. Presence of connections between the vascular tissue of the leaf and the tracheidal cells of the velum.

Sporangia

(*a*) Position and Structure (Fig. 5.7, 5.11 to 5.14). In ***Isoetes coromandelina*** the sporangia originate from the group of cells below the velum. They have no ontogenetic connection, whatsoever with the velum. In some species the lower lobe of the velum covers the entire sporangium except near its base. In others the sporangium is only partially covered. In ***I. coromandelina*** the lower lobe of the velum is absent and the sporangium is naked.

The sporangia are sessile and are almost oblong-rounded in shape and vary in length from 4 to 10 mm. They arise at the base of the sporophylls and are, therefore, foliar in nature. They occur singly and are adaxial. The mature sporangium is surrounded by a sterile jacket of three or four layers of cells. The innermost layer is made up of large and transparent cells with distinct nuclei. This is called the **tapetum**. In some species (***I. nuttallii***) the tapetum is two-layered. The cavity of the sporangium is incompletely divided by transverse partitions of sterile tissue called the **trabeculae** (Fig. 5.14). These trabeculae are numerous in megasporangia and comparatively fewer in microsporangia. The incompletely divided sporangial cavity is full of sporogenous tissue which later distinguishes into

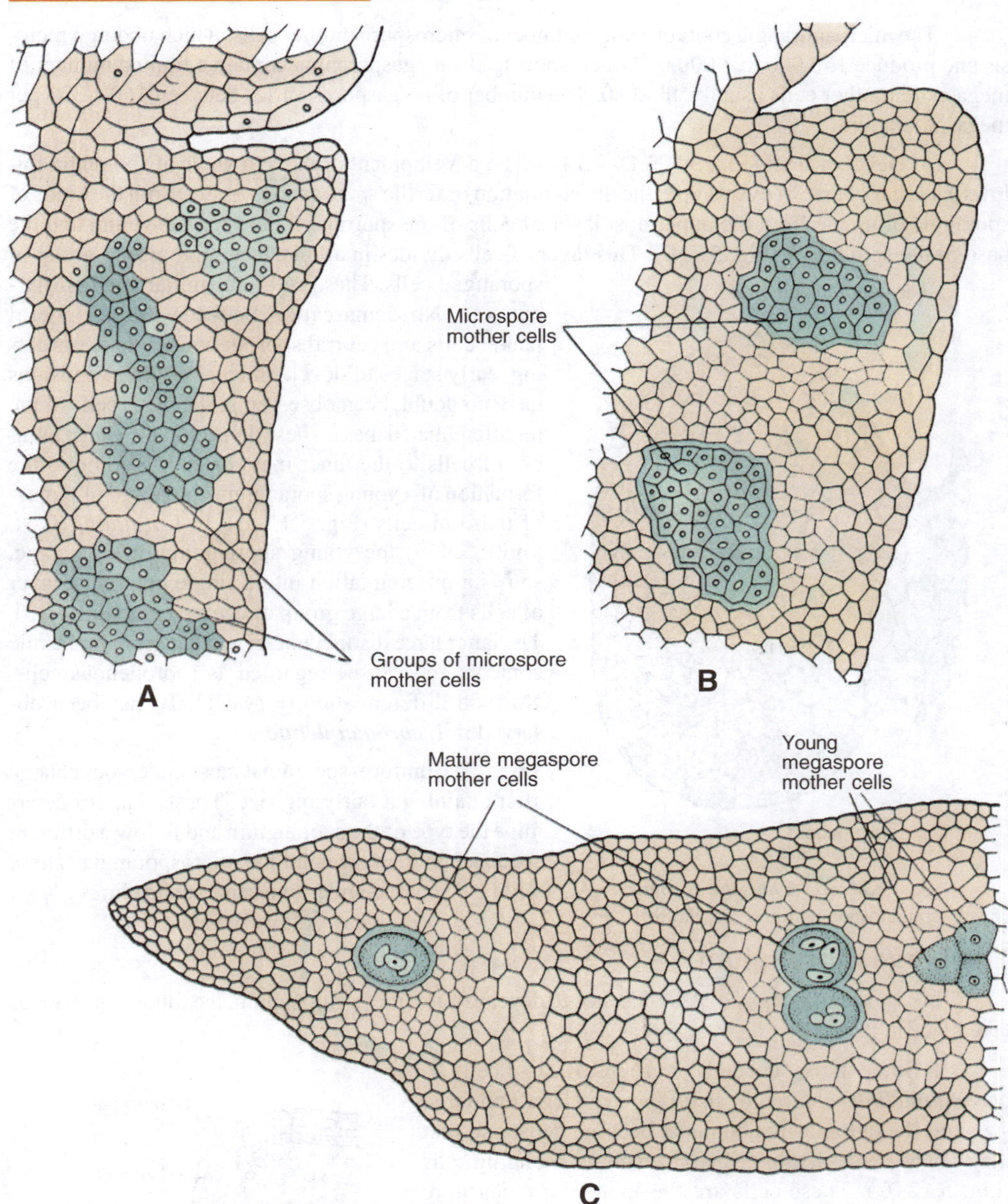

Fig. 5.12 (A—C). *Isoetes coromandelina*. Stages in the development of sporangia. A, B. Earlier stages in the development of a microsporangium. Groups of microspore mother cells are quite distinct; C. Early stage in the development of a megasporangium. The wall layers and the trabeculae have not differentiated as yet. (After Ekambaram and Venkatanathan).

spore mother cells or the **sporocytes**. The sporogenous cells adjacent to the trabeculae and sides of the sporangium become modified into tapetal cells. The rest of the sporogenous cells function as potential sporocytes. At this stage the micro and megasporangia follow a different course of development.

The microsporangia contain a large number of microspore mother cells, which undergo meiosis and produce 150,000 to 1,000,000 microspores. The megasporangia contain a limited number of megaspore mother cells usually 40 to 80. The number of megaspores varies between 160—320 per megasporangium.

(*b*) Development (Fig. 5.7, 5.11—5.14). The development of the two kinds of sporangia follows a similar course of events upto the differentiation of fertile sporogenous tissue. Both the kinds of sporangia originate from the uppermost layer of cells of the sporophyll between the velum and the base of the sporophyll (Fig. 5.7, E). This layer of cells divides in all planes to give rise to group of sporangial cells. These cells are similar and undifferentiated. No demarcation between the peripheral jacket cells and central sporogenous cells exists during early stages of development. Periclinal divisions have no doubt, been observed in the outer cells of the undifferentiated mass. These divisions add a large number of cells to the inner mass of cells leading to the formation of a young sporangium composed of a mass of uniform cells (Fig. 5.11, B). In I. ***nuttallii*** (G.M. Smith, 1955) the young sporangium, at this stage, shows a differentiation into a single outermost layer of cells from a large group of central cells (Fig. 5.13). The latter have distinct nuclei and denser cytoplasmic contents and can be regarded as sporogenous cells. No such differentiation (Fig. 5.11, B) has been observed in ***I. coromandelina***.

Microsporangium wall

Microspore mother cells

Fig. 5.13. *Isoetes nuttalli.* A young microsporangium (After Smith)

The uniform sporangial mass undergoes change after attaining a fairly big size. These changes determine the type of the sporangium and follow a different sequence in the mega and the microsporangia. These will be discussed under two separate headings :

The Megasporangium

In case the sporangial mass is destined to develop into a megasporangium, the following changes take place in it :—

1. Some individual cells or groups of 2 to 3 cells (***I. coromandelina***) enlarge in size. Their nuclei attain large size and the cytoplasmic contents become denser so that they can be easily distinguished from the surrounding cells (Fig. 5.12, C). These cells are the 'potential megaspore mother cells'.

2. All the potential megaspore mother cells are not capable of reaching maturity. Their differentiation is followed by a considerable enlargement of the sporangium. The megaspore mother cells also enlarge considerably.

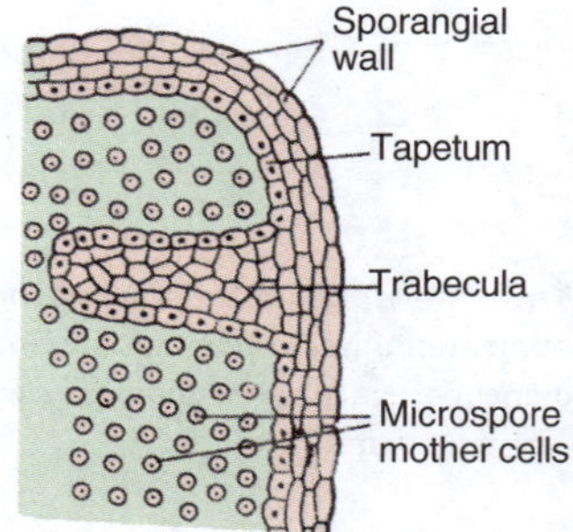

Fig. 5.14. *Isoetes.* T.S. microsporangium showing a trabecula, tapetum, microspore mother cells and sporangial wall.

3. Some of the megaspore mother cells stop growing. Their contents becomes less dense and vacuolated. Besides this, a number of megaspore mother cells become abortive. These abortive mother cells divide and redivide to form

groups of sterile cells. The functional mother cells keep on increasing in size and become rounded or elliptical in shape.

4. After the functional mother cells have reached maturity and are ready to divide and produce spores., there is differentiation of the sterile tissue into tapetum, trabeculae and jacket cells. The outermost three or four layers of sterile cells differentiate into the jacket of the sporangium and the tapetum. The abortive mother cells form strands of sterile cells that become differentiated as trabeculae. The sterile cells around the trabeculae also differentiate into one to two or many layers of tapetum (Fig. 5.14). The tapetal layers around the trabeculae are continuous with the tapetum of the sporangial wall. In ***Isoetes coromandelina*** some sterile cells surrounding the megaspore mother cells degenerate and are completely dissolved to form cavities of the sporangium. After the formation of these cavities the trabecular cells elongate and form distinct trabecular strands.

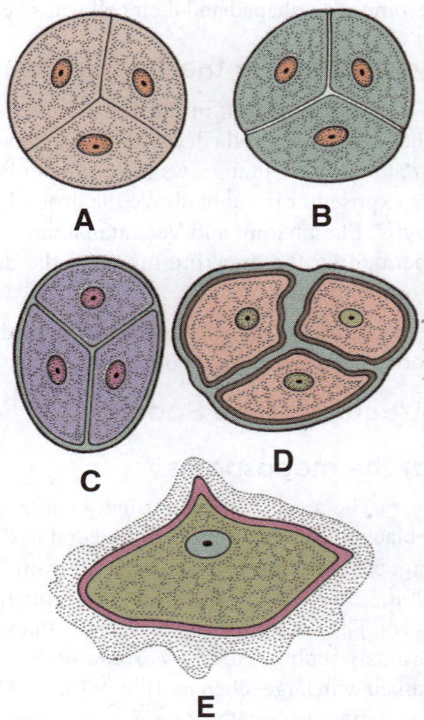

Fig. 5.15 (A-E) ***Isoetes coromandelina.*** Stages in Megasporogenesis. (After Ekabaram and Venkatanathan)

5. The megaspore mother cells undergo meiosis and form tetrads of haploid megaspores. The four haploid nuclei formed as a result of meiosis separate into haploid cells by cytokinesis. Cytokinesis is accomplished by the formation of a fine cell plate between the nuclei (Fig. 5.15, A). Later the cell plate splits (Fig. 5.15, B) thus separating the four cells still enclosed within the parent wall of megaspore mother cells (Fig. 5.15 D). Each of the four daughter cells secretes a wall around itself and contains a plastid body. These cells now grow in size and their cell walls also undergo changes. These cells become thick-walled megaspores. They are arranged in tetrahedral tetrads. As the spores differentiate the plastids disappear. The megaspores wall increase in thickness and is finally differentiated into three layers. The megaspores are now mature. The wall of the megaspore mother cell disappears so that the megaspores lie free in the cavity of the megasporangium.

The Microsporangium

In the microsporangium the sterile sporangial mass differentiates into groups of microspore mother cells and sterile cells (Fig. 5. 12, A, B). All the microspore mother cells are functional and undergo reduction division to give rise to an exceedingly large number of microspores. There is absolutely no degeneration of the microspore mother cells. The sterile cells around the microspore mother cells differentiate into trabeculae. Wall layers and tapetum are formed in the same fashion as in the megasporangium. After the formation of microspores the sterile cells around them lose their contents and ultimately disintegrate so that the microspores lie free in the cavities between the trabeculae.

The microspore mother cells undergo meiosis. After the completion of meiosis, the haploid daughter protoplasts separate, by the formation of cell plates and their resultant splitting (Fig, 5.16). Later they mature into microspores that separate and are set free from the surrounding microspore mother cell wall. At this stage they are more or less semilunar in shape (Fig. 5.16, D, E). Later they become bean-shaped and their cell walls become differentiated into three layers.

Dehiscence of the Sporangia

The sporangia in ***Isoetes*** are indehiscent and the two types of spores are liberated only when the sporophylls decay after the end of the growing season. Osborn (1922) reported that in ***I. drummondii***, that grows in vernal pools, with its axis buried completely in the soil, the spores are exposed or brought above the ground by expansions of mucilage cells at the bases of sporophylls. Ekambaram and Venkatanathan (1933) reported that in ***I. coromandelina*** the spores are liberated by the breaking up or by the decay of the portions of the sporangia attached to the sporophyll base.

Spores are mostly wind disseminated. They are also carried by water in aquatic species. Earthworms have also been reported to carry the spores from place to place.

Structure of the Spores

(a) The megaspores

The megaspores are usually white, grey or black in colour. They are tetrahedral in shape (Fig. 5.15, E) and vary in diameter from 200-900μ. The wall is differentiated into three layers. The outermost layer is usually thick and variously sculptured. In ***I. coromandelina*** it is marked with large tubercles (Fig. 5.15, E). There is a distinct triradiate ridge. The outer wall is called **exosporium**. Next to it is the **mesosporium** which is followed by **intine** or **endosporium**. According to some authors the original cell wall that separates the spore tetrads develops, in each spore into an outermost sculptured layer called the **epispore**. Later exine or exosporium, mesine or mesosporium and entine or endosporium develop. According to them the spore wall is four layered. The spore protoplasm shows a spongy structure and contains food reserves in the form of starch. The nucleus, in the mature spore, lies at its apex (Fig. 5.15, E) or at its base (***I. braunii***). It has no central vacuole and has dense cytoplasmic contents.

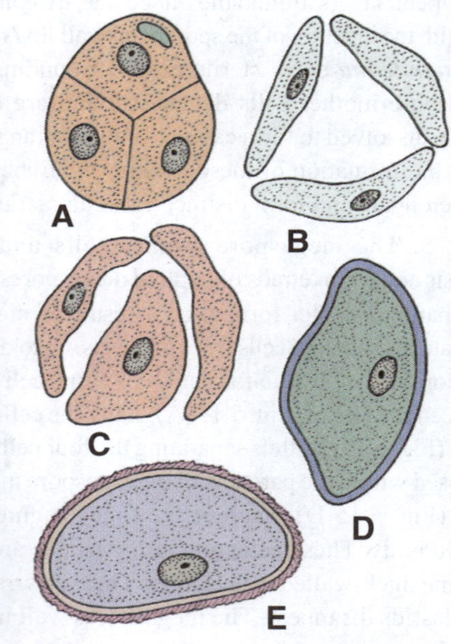

Fig. 5.16 (A-E). Microsporogenesis in *Isoetes coromandelina* (After Ekambaram and Venkatanathan)

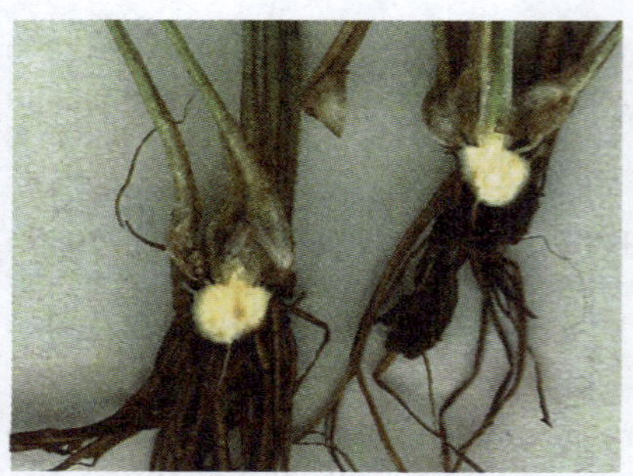
I. braunii

(b) The microspores

They are smaller in size and vary from 20—45μ in diameter. The microspores are semilunar in shape when young but become kidney-shaped at maturity. The cytoplasm is thin and finely vacuolated. The nucleus lies on one side. The microspore wall is three layered. The outermost layer or exine is marked with small densely packed spines (Fig. 5.16, E). They are usually ash coloured or brown.

GAMETOPHYTES

Heterospory leads to the formation of two types of gametophytes in ***Isoetes.*** The microspores germinate to give rise to the male or **microgametophytes** and the megaspores to female or **megagametophytes.** Both types of gametophytes are endosporic. The spores may start germinating immediately after their release from the sporangia or the germination may be delayed depending upon the climate conditions. Quite contrary to ***Selaginella*** the spores in ***Isoetes*** start germinating only after their release from the sporangia.

Microgametophyte

The first division of the nucleus results in the formation of a small **prothallial cell** and a large cell that has been interpreted as **antheridial initial** (Fig. 5.17, B). The antheridial cell undergoes a series of successive divisions to cut off **4 wall cells** surrounding a single **central cell** (Fig. 5.17, C—G). The central cell divides twice to form four **spermatocytes** or **sperm mother cells** (Fig. 5.17, H—J). The four wall cells enclosing 4 spermatocytes represent the antheridium. The wall cells remain intact until the microspore wall cracks (Fig. 5.17, N). This generally happens after two weeks of developmental period. The contents of the sperm mother cells metamorphose into multiflagellate spermatozoids (Fig. 5.17, H, L). The microgametophyte of ***Isoetes*** is the most reduced type known among the vascular cryptogams. One gametophyte produces only four spermatozoids.

After the cracking of microspore wall the exposed wall cells of the antheridium degenerate thus liberating the ***spermatozoids.*** The prothallial cell also degenerates. The multiflagellate sperms of ***Isoetes*** are in striking contrast to biflagellate sperms of other lycopods (***Selaginella***, ***Lycopodium***).

The spermatozoid of ***Isoetes*** (Fig. 5.18) has been described by Dracinshi (1932). The free swimming spermatozoid consists of two large and spherial vesicles (Fig. 5.18). The vesicles are soon lost. Each vesicle contains fat and vacuole elements. Starch is considered to be absent. The nucleus pointed at either end is enclosed within a spongy plasmatic structure. The nucleus does not extend to the pointed anterior end of spermatozoid. The

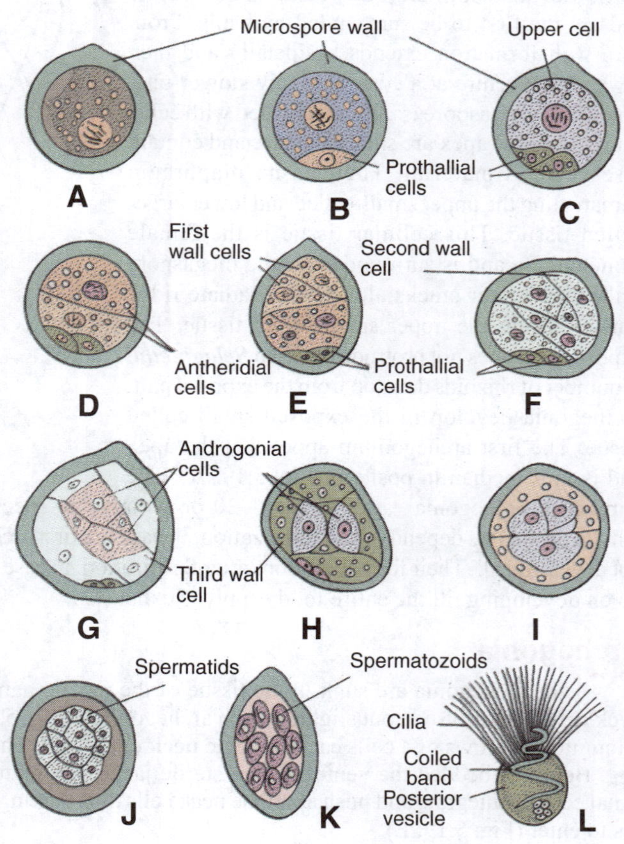

Fig. 5.17 (A-L). *Isoetes lacustris.* Various stages in microspore germination and formation of male gametophyte. For explanation see text (After Liebig)

flagella bearing band or the stalk is homogeneous and bears about 25 flagella. The flagella arise from the **basal granules** situated on the stalk. Four basal granules are grouped together at the anterior end whereas the rest are scattered in a single row along the stalk.

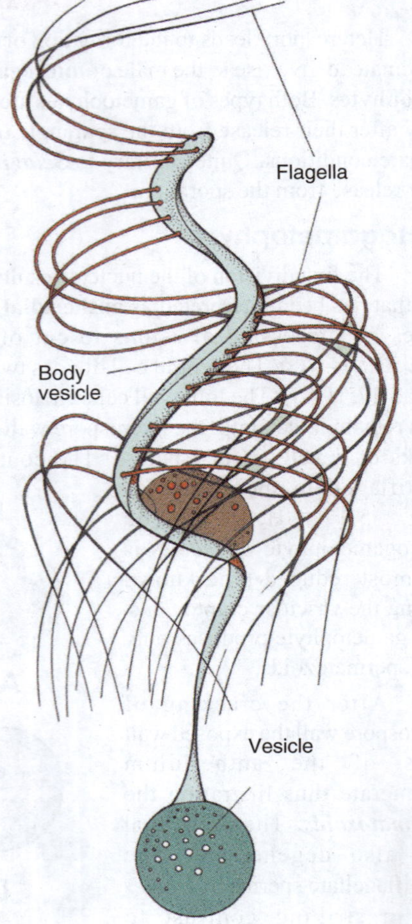

Fig. 5.18. Spermatozoid of *Isoetes.*

Megagametophyte (Fig. 5.19, A—D)

The nucleus of the tetrahedral megaspore divides by free nuclear divisions until 30—50 free nuclei are formed. The nuclei migrate through the cytoplasm and get distributed along the periphery of the spore protoplast. The larger number is near the apex (triradiate ridge). Wall formation starts after 30—50 free nucleate stage. The nuclei at the apical portion of the spore are quite close to each other and are the first to be surrounded by walls. From here wall formation extends basipetally and then towards the centre at a comparatively slower rate. The whole megaspore is ultimately filled with cells. The cells at the apex are smaller in size and contain reserve food material. There is no diaphragm demarcating the upper small-celled and lower large-celled tissue. This cellular tissue is the female gametophyte and is surrounded by the megaspore wall, which later cracks along the triradiate ridge thus exposing the upper small-celled tissue. The exposed part does not protrude out as in ***Selaginella***. A number of rhizoids develop from the exposed part. Archegonia develop in the exposed small-celled tissue. The first archegonium appears at the apex and is thus median in position (Fig. 5.19, A). The number of archegonia varies from 3—20 or 30 in some species. The number of archegonia in a female prothallus depends upon fertilization. In case the first archegonium is fertilized the others may not develop at all. Their formation stops after fertilization. In case fertilization is delayed the archegonia go on developing till the entire food supply is exhausted.

Archegonia

The archegonia are sunk in the tissue of the female gametophyte and only the upper tier of neck cells may protrude out slightly. Each archegonium (Fig. 5.19, D) consists of a neck made up of 4 longitudinal rows of 4 cells each, a single **neck canal cell.** The venter has a **ventral canal cell** and **egg**. Being embedded the venter has no sterile jacket. At maturity the neck canal cell and ventral canal cell distintegrate and push apart the neck cells thus making an open channel for the spermatozoids to enter (Fig. 5.19, E).

The archegonium develops (Fig. 5.19, A—E) from a single superficial cell which divides periclinally into an upper and a lower cells. The outer cell divides anticlinally/periclinally to form four tiers of four cells constituting the neck. The lower divides periclinally to form a **neck canal cell** and a **primary ventral cell.** The latter divides into a **ventral canal cell** and **an egg cell.**

Fertilisation has not been observed in ***Isoetes*** but it takes place.

The Embryo

The fertilized egg in ***Isoetes*** almost fills the cavity of the venter. The fertilised egg divides by a transverse or oblique transverse wall (Fig. 5.20, B). Both the cells take part in the formation of the embryo and no suspensor is formed. The cell towards the neck of the archegonium can be regarded as the **hypobasal cell** and the one away from the neck as the **epibasal cell**. The next two divisions are at right angles to the first wall and to each other thus forming an **octant stage.**. The hypobasal tier gives rise to the foot alone. The epibasal tier gives rise to the first leaf and the root and also the shoot apex after some delay. Due to the absence of suspensor the development of embryo is **exoscopic**, but further developmental changes suggest it to be **endoscopic.** The octant stage by further development forms an oviod mass of undifferentiated cells. Later division lead to the widening of the embryo in the plane of the basal wall (Fig. 5.20, D). Later the first leaf originates from one half of the epibasal region (Fig. 5.20, D). From the base of the first leaf a large cell becomes prominent and divides first by periclinal walls to distinguish a ligule. A semicircular ridge appears at the same time at the base of the root. This is the primordium of the **velum**. Between the base of the velum and the first leaf there is a small cleft. In this cleft there distinguishes a small group of cells. This group of meristematic cells is destined to give rise to the shoot apex at a later stage. The second leaf develops facing the first with the shoot apex lying between them. Campbell (1918, 1939) observed that the vascular strands of the first leaf and the first root coincide with each other. The foot is small in young embryo but it becomes massive and well developed in mature embryos. During its growth, the embryo elongates downwards and erodes the prothallial tissue. The uneven growth of the embryo makes it curved and as this growth continues, accompanied by rapid absorption of prothallial tissue, the polarity of the embryo becomes reverse. The foot which in the young embryo is towards the archegonial neck, comes to lie on the lower side (Fig. 5.20, D) and the long axis of the embryo lies parallel to the

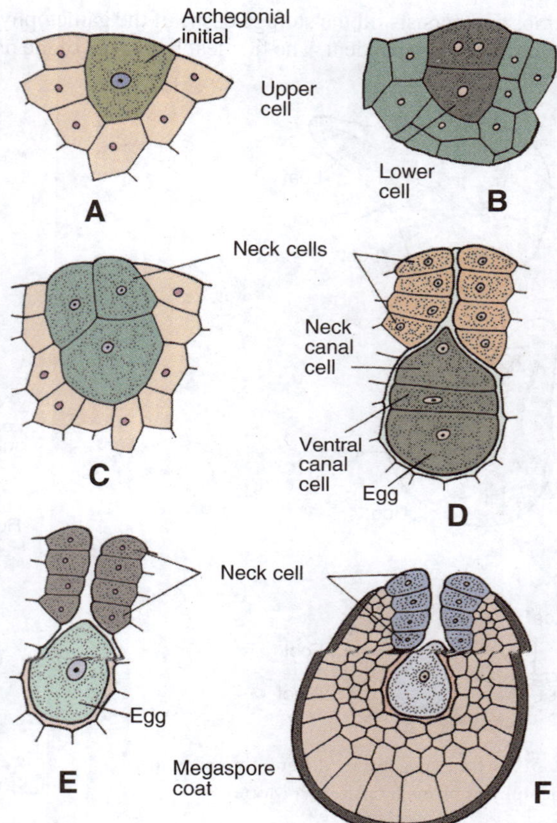

Fig. 5.19 (A-F). *Isoetes echinospora.* A-D Stages in the development of an archegonium; E. A mature archegonium; F. Mature archegonium enclosed in female gametophyte. Megaspore wall is also shown.

Fig. 5.20 (A-D). *Isoetes echinospora.* Stages in the development of embryo. (After Campbell)

surface of the prothallus (Fig. 5.20, D). The young leaf elongates rapidly and in a horizontal direction. It comes out of the prothallus in horizontal position and later turns up. The root curves downwards and as the sporophyte grows it straightens out and all its traces of horizontal position are lost. The foot develops and persists till the storage tissue of the gametophyte lasts. It perishes after the young sporophyte becomes independent. The first leaf bursts out of the prothallus after seven days of the growth of the embryo.

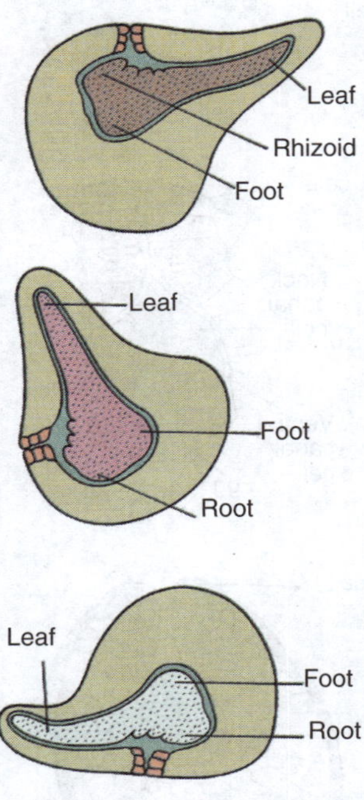

Fig. 5.21. Embryogeny in differently oriented gametophytes of *Isoetes* (After La Motte).

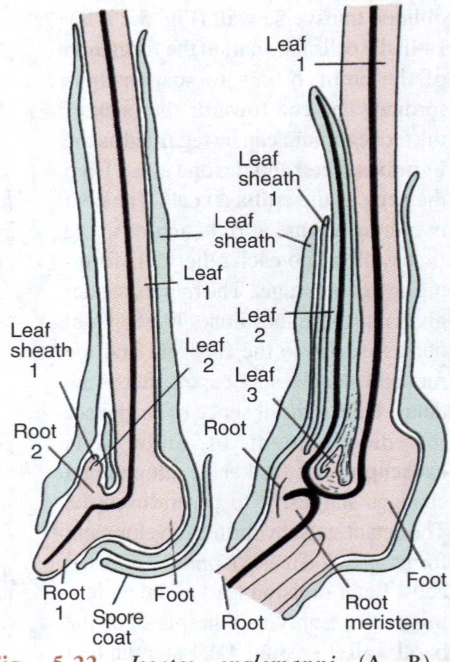

Fig. 5.22. *Isoetes englemanni* (A—B). L.S. Young sporophyte (After Baldwin)

Taxonomic Position

Division	:	***Lycophyta***
Class	:	***Ligulopsida***
Order	:	***Isoetales***
Family	:	***Isoetaceae***
Genus	:	***Isoetes***

Systematic Position

Bower and many other investigators regard ***Isoetes*** as a member of the ligulate Lycopsida because:

1. It has ligulate and microphyllous leaves. The ligule arises in a definite pit (ligule pit).
2. It is heterosporous.
3. The gametophytes are reduced.
4. Germination of spores is similar to ***Selaginella.***
5. Structure of male and female prothalli is also similar to ***Selaginella.***
6. The sporangia are massive and elongate.
7. The arrangement of roots in definite rows is similar to some extinct lycopods like ***Nathorstiana.***
8. There is a very close correspondence in root structure with other lycopods.

Cambell (1939) included it under the eusporangiate ferns because of the following features :

1. Early leaf and root formation.
2. Delayed shoot formation.
3. Relationship of vascular strands of root and leaf.

These features bring ***Isoetes*** closer to ***Botrychium virginianum*** and to some Marattiaceae.

The present consensus of opinion regards ***Isoetes*** as a greatly reduced and highly specialised ligulate lycopod. Some researchers regard the whole plant as single strobilus with roots arising from its base and other reduced vegetative parts telescoped with it. A definite evolutionary sequence leading to the reduced plant body of ***Isoetes*** and ***Stylites*** can be visualised if we insert ***Pleuromia*** and ***Nathorstiana*** as intermediate forms between tree-like extinct ***Lepidodendrons*** and the ***Isoetales*** (Magdefrau, 1932). The reduction series may thus be summarised as :

Lepidodendrales → Pleuromeiopsis → Nathorstianella → Nathorstiana → Pleuromeia → Stylites → Isoetes

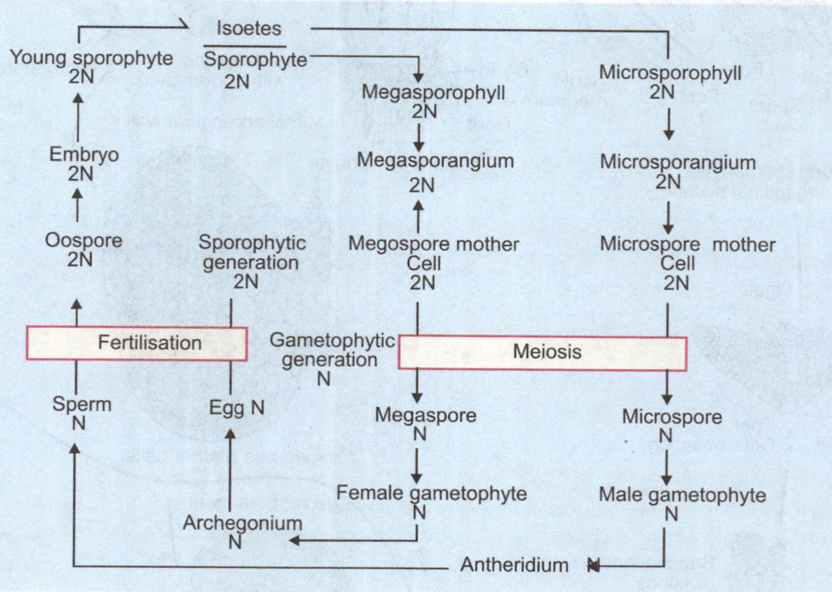

Fig. 5.23. *Isoetes.* Graphic representation of life cycle

Stylites

The other genus of Isoetales is ***Stylites***. The genus was discovered by Amstutz in 1957 from high Andes of Peru. Two species ***Stylites andicola*** and ***S. gemmifera*** have been described. The species generally grow around the boggy margins of a lake at an altitude of 4750 meters. The genus closely resembles ***Isoetes*** but differs in certain respects. The differences have been described below.

Stylites andicola *S. gemmifera*

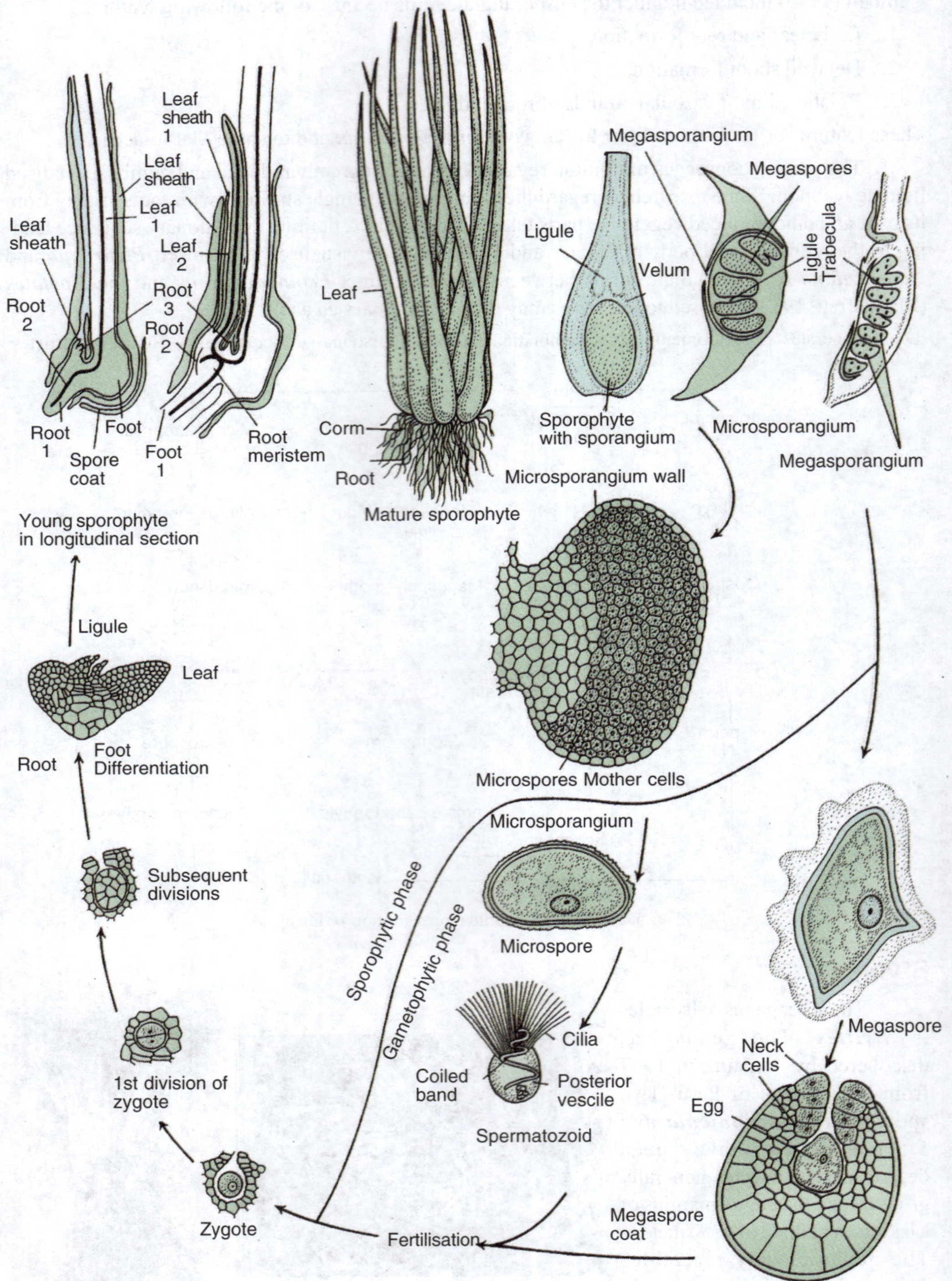

Fig. 5.24 : *Isoetes.* Pictorial Life cycle

Isoetes	*Stylites*
1. Small, herbaceous and perennial, plant body consists of a short fleshy upright corm like axis unbranched.	Larger, perennial, the axis is several inches long cylindrical with limited power of vertical growth; axis bifurcates at the crown; dichotomously branched, at least three times to form characteristic tussocks.
2. Roots are mostly unbranched but if branched, it is dichotomous.	Roots are borne in a single deep vertical groove on one side of the lower half of axis. The number of grooves may vary from 1– 3.
3. Leaf : Leaf has a thickened and expanded colourless base and a tapering distal portion which is oval-shaped and pointed.	Similar to *Isoetes*.
4. Vegetative Reproduction : Rare	Common by means of small adventitious buds.
5. Stele : Cylindrial upper part with anchor shaped arm.	Cylindrical upper part with 1– 3 ridges along the xylem cylinder from which root traces aries.
6. Secondary growth – Common and active.	Common but less active.
7. Sporangium – Situated on the adaxial surface mean the expanded base of sporophyll.	Sporangium situated on the adaxial surface far above the sporophyte.

However, the creation of ***Stylites*** has been questioned. According to Rubitzki and Borchert (1964) genus ***Stylites*** has many similarities with ***I. triquetra*** described from Andes of Columbia. It is assumed that ***Stylites*** can be derived from ***Isoetes*** (Bierhorst, 1971) by only contracting the lobes of anchor shaped stele and extending the root producing meristem vertically. Osmund and Raven (1983) reported that ***Stylites*** lacks stomata and absorbs carbon dioxide through roots. In conclusion it may be said that ***Stylites*** is a divergent form of ***Isoetes.***

QUESTIONS

ESSAY TYPE

1. Describe briefly the formation and nature of secondary tissues in ***Isoetes***.
2. Describe in detail the salient features, and classification of Isoetales with special reference to morphological nature of axis and anatomy of sporophyte of ***Isoetes.*** **(Awadh, M. Sc., 2001)**
3. Give an account of the important features of order isoetales and discuss its affinities. **(Kanpur M. Sc., 2000)**
4. Describe briefly the distinguishing features of Isoetales. What do you know about the distribution of ***Isoetes*** in India.
5. Describe the life-history of ***Isoetes*** pointing out the features of special interest.
6. Describe the structure of Rhizome of ***Isoetes.*** How does it differ from that of ***Lycopodium.***
7. Give a brief account of the axis of ***Isoetes*** and discuss its morphological nature.
8. Describe the structure of sporophyte ***of Isoetes***. How does it differ from that of ***Selaginella***?
9. Give an account of reproduction in ***Isoetes.***

SHORT ANSWER TYPE

10. How would you differentiate between
    (*i*) ***Isoetes*** and ***Stylites***,
    (*ii*) Spore producing organs in ***Isoetes*** and ***Phylloglossum***.
    (*iii*) Spore producing organs of ***Isoetes*** and ***Selaginella.***
11. Give a brief account of sporophyte of ***Isoetes.***
12. Give the characteristic features of ***Isoetes.*** **(Awadh, M.Sc. 2000)**
13. ***Isoetes*** shows secondary growth. Discuss.
14. Write short notes on.
    (*a*) ***Isoetes***, **(Rohilkhand M.Sc. 2000)**

(*b*) Axis in ***Isoetes***,

(*c*) Velum,

(*d*) Leaf and ligule,

(*e*) Microspore in ***Isoetes***,

(*f*) Megaspore in ***Isoetes***.

15. Give a brief account of the structure and position of sporangia in ***Isoetes***.

16. Draw labelled sketches of the following:

(*a*) Anatomy of leaf in ***Isoetes***,

(*b*) Microsporangium of ***Isoetes***,

(*c*) Megasporangium of ***Isoetes***,

(*d*) Megaspore and microspore of ***Isoetes***.

17. Briefly describe the development of embryo in ***Isoetes***.

OBJECTIVE TYPE

18. Select the correct answer:

(*i*) Which of the following is commonly called as ***Quillwort***?

(*a*) ***Psilotum***

(*b*) ***Lycopodium***

(*c*) ***Selaginella***

(*d*) ***Isoetes***.

(*ii*) The most common species of ***Isoetes*** is

(*a*) ***I. indica***

(*b*) ***I. coromandelina***

(*c*) ***I. mirzapurense***

(*d*) ***I. dixetei***.

(*iii*) Which of the following is the other member of the ***Isoetales*** besides ***Isoetes***?

(*a*) ***Lycopodium***

(*b*) ***Stylites***

(*c*) ***Selaginella***

(*d*) ***Psilotum***.

(*iv*) ***Isoetes*** is

(*a*) ***unisporous***

(*b*) ***Bisporous***

(*c*) ***Homosporous***

(*d*) ***Heterosporous***.

(*v*) Velum in ***Isoetes*** arises

(*a*) above the ligule

(*b*) below the ligule

(*c*) between the ligule and axis

(*d*) as an extension of ligule.

(*vi*) At maturity the microspore of ***Isoetes*** is

(*a*) Oval

(*b*) Circular

(*c*) Kidney shaped

(*d*) Globular.

(*vii*) Sperms in ***Isoetes*** are uninucleate and

(*a*) Uniflagellate

(*b*) biflagellate

(*c*) multiflagellate

(*d*) aflagellate.

Selaginella rupestris
–Erect Habit

CHAPTER 6

Lycophyta — Ligulopsida (Contd.)

- Order Selaginallales
- Description of ***Selaginella*** Palisto-de Beauvios
- Heterospory
- Heterospory and Seed Habit

SELAGINELLALES

The order is represented by single living genus ***Selaginella*** included in the family **Selaginellaceae**. An extinct genus ***Selaginellites*** was discovered from the Carboniferous. It is represented by ***Selaginellites crassicinctus***, ***S. canonbiensis, S. sussei, S. polaris, S. elongatus*** and ***S. primaevae.***

SELAGINELLA Palisto-de Beauvois

Distribution

It is widely distributed genus with 700 species found under diverse climatic conditions. Majority of the species grow under shady and damp places especially in the tropical regions of the world. They are rare in the temperate regions and almost absent in the alpine zone. Some are xerophytic (S. ***lepidophylla***). ***S. oregana*** is epiphytic. Some are cultivated as ornamentals due to their delicate and feathery fronds (***S. picta, S. pygmea, S. kraussiana***).

Seventy-one species of ***Selaginella*** grow in the tropical and subtropical forests along the Himalayas and in the plains of India. Some xerophytic Indian species include ***S. wightii***, ***S. repanda***, etc. Majority of the species grow in the Eastern Himalayas (***S. biformis***, ***S. adunca***, ***S. monospora***, ***S. caulescens*** ; ***S. ciliaris***, ***S. aitchtsoni***, ***S. sanguinolenta***, ***S. jacquemonthi***, ***S. atroviridis***, ***S. nepalensis*** etc.) and comparatively few in the North-Western Himalayas (***S. chrysocaulos***, ***S. chrysorrhizos***, ***S. subdiaphana***, ***S. adunca***, ***S. sanguinolenta***, ***S. pellidissima***, etc.). ***S. exigua***, ***S. ciliaris***, ***S. bryopteris***, ***S. radicata*** and ***S. rupestris*** grow in Pachmarhi hills (central India). ***S. rodrigaesiana*** has been reported from Malabar (Panigrhi, 1978). ***Selaginella involvens***, ***S. radicata*** and ***S. catarctarum*** have been reported from Kodaikanal. Some species also grow in the upper Gangetic plains ***e.g***. ***S. bryopteris***, ***S. adunca***, ***S. subdiaphana*** and ***S. chrisocaulos***. Some other species that have been reported from eastern India include ***S. willdenovii***, ***S. minutifolia***, ***S. pubescens*** and ***S. wallichii*** (Panigrahi and Dixit, 1966-1968). Haines (1924) reported eleven species from Bihar and Orissa. A number of them grow in U. P. S. ***hallei*** and ***S. amesiana*** are the extinct species reported from Sweden and Illinois.

S. ciliaris

S. pubescens

S. sanguinolenta

Selaginella subdiaphana

MORPHOLOGY OF THE SPOROPHYTE

Habit

Many species of ***Selaginella*** are herbaceous perennials, a few are annuals, ***e.g. S. pygmaea.*** Majority are dorsiventral and grow prostrate (***S. chrysocaulos, S. chrysorrhizos, S. pellidissima, S. kraussiana,*** etc.), a few are radial and grow erect (***S. selaginoides***—Fig. 6.2, A, ***S. rupestris***, etc.), rarely scandent (***S. wildenovii***, ***S. adunca***) and suberect. Some dorsiventral species are **caulescent** with erect stems arising from creeping rhizomes (***S. umbrosa***). They range in size from a few centimetres (***S. pygmaea***) to a few metres (***S. pentagona***) in length and are capable of

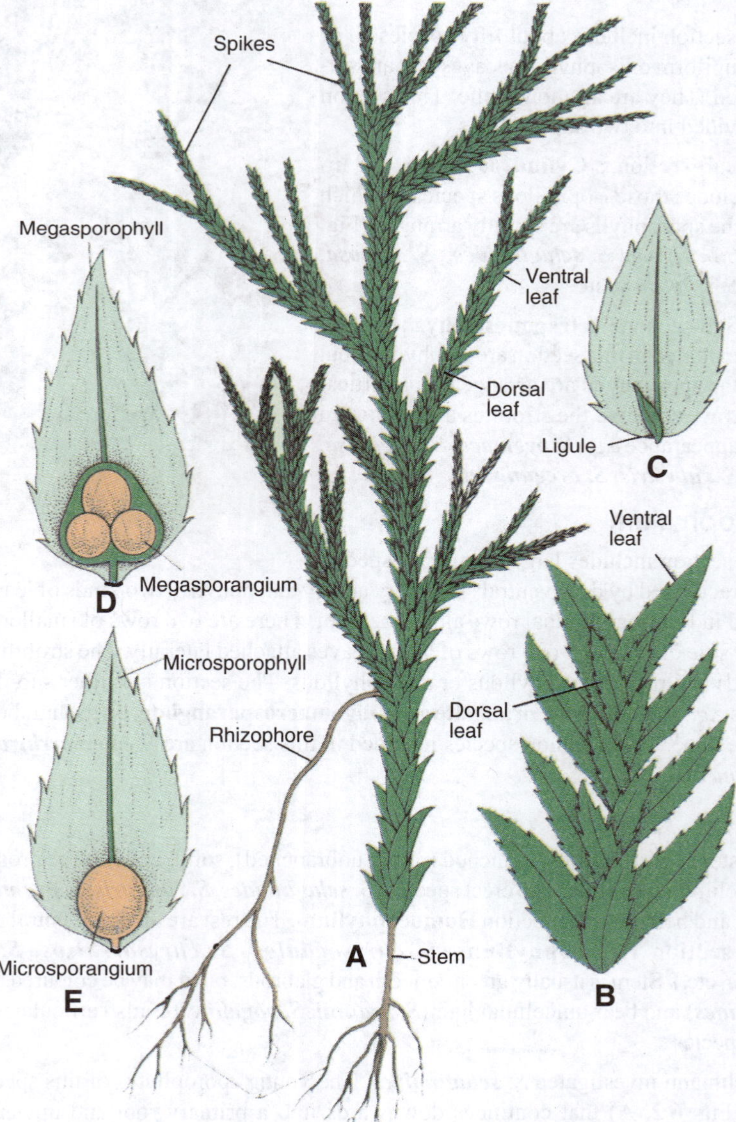

Fig. 6.1 (A—E). ***Selaginella chrysocaulos.*** A. Complete plant showing habit; B. Portion of a vegetative shoot showing leaf arrangement; C. A leaf with a ligule; D. A megasporophyll bearing a megasporangium; E. A microsporophyll bearing a microsporangium.

growing through and over the bushes (***S. wildenovi***). ***S. viridangula*** grows erect and is several feet tall. ***S. alligans*** is a climber and does so with the help of rhizophores that develop pad-like discs at their ends and attains a length of upto 2 to 8 metres. Some unbranched species are also known. The xerophytic species dry up and curl-like balls during dry seasons and again expand during favourable season or on wetting with water.

S. spinulosa

Hieronymus (1902) divided the genus into two sections:

1. Homoeophyllum

This section includes about fifty species all of which have uniform or isophyllous leaves that are spirally arranged. They are all monostelic. This section is further divided into two sub-sections :

(*a*) Sub-section : **Cylindrostachya :** It includes those isophyllous species in which the sporophylls are spirally arranged (Fig. 6.2, A) ***e.g.***, ***S. selaginoides***, ***S. spinosa***, ***S. spinulosa*** etc.

(*b*) Sub-section : **Tetragonostachya :** Species included in this section are isophyllous but the sporophylls are arranged in 4 vertical rows and give the strobilus a four-angled appearance ***e.g.***, ***S. pygmaea***, ***S. uliginosa***, ***S. rupestris***, ***S. oregana***, ***etc.***

2. Heterophyllum

This section includes large number of species that are characterised by dorsiventral symmetry and by anisophylly (two kinds of leaves). The leaves are arranged in four longitudinal rows along the stem. There are two rows of smaller leaves attached to the upper side of stem and two rows of larger leaves attached laterally. The strobili are tetragenous (four-angled) and may be isophyllous or anisophyllous. The section is further sub-divided into two sub-sections ***e.g.***, **Pleiomacrosporangiate** and **Oligomacrosporangiate,** on the number of megasporophylls in the cone. The common species included in this section are ***S. chrysorrhizos***, ***S. martensii***, ***S. kraussiana*** etc.

Stem

The stem is herbaceous, branched (rarely unbranched), solid, and may be prostrate, sub-erect, caulescent, climbing or erect. The erect species (***S. selaginoides***, ***S. rupestris***, ***S. pygmaea***) are radially constructed and belong to the section **Homoeophyllum.** The rest are all dorsiventral and are included under the section **Heterophyllum** (***S. chrysocaulos***, ***S. chrysorrhizos***, ***S. pallidissima***, ***S. krussiana***, etc.**)**. Stem is usually green, smooth and glabrous, but it may be coloured red (***S. umbrosa***, ***S. haematoides***) and bear unicellular hair (***S. braunii, S. vogelii***). It is also articulated in some North American species.

Bruchmann investigated ***S. selaginoides***. The young sporophytes of this species consist of a hypocotyl (Fig. 6.2, A) that continues downwards into a primary root and upwards it bears two cotyledons that surround the shoot apex. Later, pairs of **epicotylar leaves** make their appearance. After this the first stem develops and divides dichotomously into two equal limbs that bear spirally arranged isophyllous leaves. The two limbs of the first dichotomy again branch dichotomously and ultimately monopodially (Fig. 6.2, A). The monopodial lateral branches arise in one plane and are at right angles to the first dichotomy thus giving the entire shoot a fan-like appearanace. Strobili develop at the tips of stronger branches, whereas the rest remain sterile. The main axis and the hypocotyl remain as permanent structures (Fig. 6.2, A). The lower base of the hypocotyl swells up and gives out roots.

In the common and widely distributed ***S. chrysocaulos*** the prostrate and dorsiventral stem (Fig. 6.1, A) is repeatedly branched. The branching appears to be dichotomous in the younger portions of the stem and monopodial or lateral in the older portions. Smith, Scot and others hold that all the branches are really lateral but they appear to be dichotomous because they are formed so near the growing apex that the latter seems to give rise to two equal shoots. These shoots in the early formed portions of the plant grow with equal vigour so that it becomes difficult to distinguish between the main axis and the lateral branch. The equal growth of the main axis and the branch give the appearance of a dichotomy. The difference between the main axis and the branches becomes evident in later formed parts of the plant where the main axis grows vigorously and the branches remain small. Eames, Schoute, and others hold that branching in dorsiventral species of ***Selaginella*** is dichotomous or pseudomonopodial.

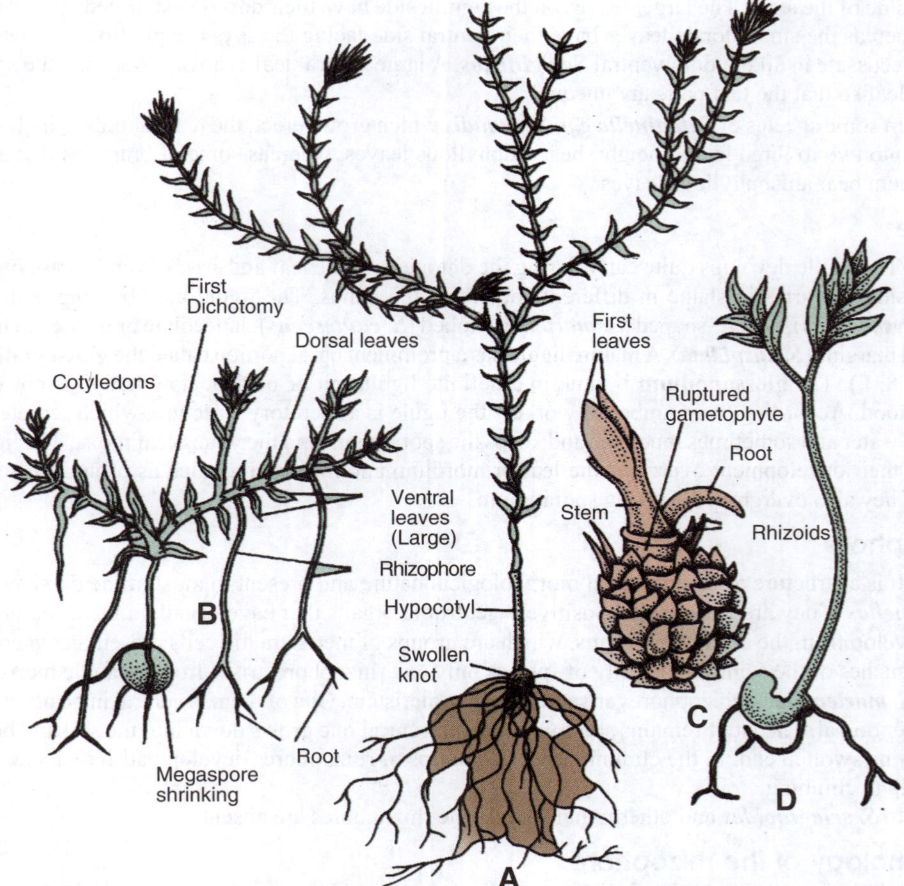

Fig. 6.2 (A—D). *Selaginella* A. Young plant of *S. selaginoides*; B. Young plant of *S. kraussiana*; C—D. Young sporophytes of *S. kraussiana*.

Leaves

The leaves are microphyllous. Each leaf is traversed by a single unbranched midrib. The leaf traces do not leave any leaf gaps in the stem stele. A ligule arises from the base of each leaf (ligulate). They are delicate, green, with entire or serrate margin and acute apex. In xerophytic species the leaves are rough and thick. They are variegated in ***S. picta.*** Mostly the leaves are glabrous but hairy in ***S. hispida*** and some other species. A periodic change in leaf colour occurs in ***S. serpens.***

The leaves are sessile and may be ovate, linear, lanceolate or cordate. Abnormal pinnatifid leaves have been reported in ***S. lyallii*** (Bruchmann, 1909).

There are about fifty species of ***Selaginella*** in which the leaves in a plant are all of one kind. Such a condition is called **isophylly** and the leaves are called **isophyllous.** They are included in section **Homoeophyllum.** The leaves in majority of species are of two kinds, ***i.e.***, small and large (Fig. 6.1, B). Such a condition is called **anisophylly** and the leaves are called **anisophyllous** or dimorphic (section **Heterophyllum**).

The phyllotaxy or the arrangement of the leaves on the stem is always spiral in the isophyllous species of Selaginella (Fig. 6.2, A).

In the anisophyllous species of ***Selaginella e.g., S. chrysocaulos*** (Fig. 6.1, A), the small and large leaves form four longitudinal rows. The smaller leaves are arranged in two rows along the dorsal side of the stem. The larger leaves on the ventral side have their dorsal side turned towards the sky, whereas the small dorsal leaves have their ventral side facing the sky. The phyllotaxy is essentially decussate in all the dorsiventral ***Selaginellas***. A large ventral leaf is always opposite the small dorsal leaf so that the leaf pairs are unequal.

In some species of ***Selaginella e.g. S. grandis*** which grows erect, the main shoots, which may grow upto two to three feet in height, bear isophyllous leaves, whereas some of branches that arise from them bear anisophyllous leaves.

Ligule

The ligule develops quite early during the ontogeny of the leaf and arises from its base on the upper side. It varies in shape in different species of the genus. The ligule may be tongueshaped (***S. chrysocaulos***), wedge-shaped (***S. martensii***), lobed (***S. caulescens***), lanceolate or may even have fringed margins (***S. cuspidata***). A mature ligule has a prominent basal portion called the **glossopodium** (Fig. 6.8, F). The **glossopodium** is sunk in a definite ligular pit or pocket. Its function is not well understood. According to a number of workers the ligule is a secretory structure, which secretes or exudes water and sometimes mucilage and keeps the sporangium and the young leaf moist. The ligule during their development overgrow the leaf primordium and may be regarded as protective structures. They also overgrow the young sporangium.

Rhizophore

It is a structure of controversial morphological nature and present in most of the dorsiventral ***Selaginellas***. They are leafless and positively geotropic organs that have usually a localised origin and develop from the **angle meristems**, which are groups of meristematic cells present between the two branches of the stem. In majority of species only one rhizophore arises from an angle meristem but in ***S. martensii*** two rhizophores arise from such a meristem. One of them is ventral in position and the other dorsal. The latter remains short whereas the ventral one grows down into the soil and bears roots at its swollen end. In the climbing species the tips of rhizophores develop pad-like thickening and help in climbing.

In ***S. selaginoides*** and other radial species the rhizophores are absent.

Morphology of the rhizophore

Three different types of morphological interpretations have been advanced to solve the morphological nature of the rhizophore. According to one view they have been regarded as modified stems that bear adventitious roots (Pfeffier, Treub, Bruchmann, Velenovsky, Worsdell, Troll, Schoute).

The supporters of the second view (Harvey-Gibson, Van-Tiegham, Uphoff) regard them as roots. Bower and Goebel consider the rhizophores as organs ***sui generis*** (neither roots nor shoots).

The rhizophores are regarded as root-like structures due to the following characters : 1. They are positively geotropic. 2. They bear no leaves. 3. Their internal structure resembles that of a root. 4. Their stelar organisation is always monostelic even if the stems are polystelic.

Webster and Steeves (1964) and Bierhorst (1971) have almost discarded the term "rhizophore" and consider it to be a root. Bierhorst regards this term as of only historical interest. According to him such stem-like features are also known in the well established roots of other taxa. It is now known (Webster and Steeves, 1964) that the tip of the rhizophore divides to form roots, and the older belief that roots arise endogenously from tip of rhizophore is not valid. These observations rule out any difference between the root and the rhizophore. Exogenous roots are also known in some angiosperms.

The stem-like characteristics of rhizophores are : 1. Their exogenous origin (roots of some angiosperms, ***e.g.***, ***Nasturtium austriacum*** arise exogenously from the stem). 2. They develop from special meristems called the angle meristems that are present between the two branches of the stem. 3. They lack root caps (capless roots are also known). 4. They have no root hair. 5. Experimental evidence that under certain experimental conditions the rhizophores develop into leaf-bearing shoots.

J.C. Schoute (1938) regards them as stem with root-bearing functions because they lack root caps and root hair and are exogenous in origin. Regarding its anatomy. Schoute points out that it is like stem in its organisation.

It has been experimentally demonstrated by Bruchmann (1905) in ***S. kraussiana*** and by Williams (1931, 1937) in ***S. grandis*** and ***S. martensii*** that cessation of vegetative activity by inducing strobili formation or by decapitating leaf-bearing shoots, the rhizophore rudiments can be made to grow into leafy shoots. Cusik (1954) concluded that some hormones produced by shoot apex induce angle meristems to grow into rhizophores. He proved it experimentally.

After these experimental evidences, Williams and others did not decide to regard the rhizophores as transformed shoots but agreed with Bower and Goebel in treating them as organs ***sui generis***. The term organs ***sui generis*** in this case implies that rhizophores are neither roots nor stems but are intermediate structures. According to Schoute an organ ***sui generis*** is the one that is a novelty and has not risen as a result of metamorphosis of any organ. He disagrees with Bower and others in regarding rhizophores as structure ***sui generis*** because they are not novel structures and possess features in common with both the stems and the roots.

S. umbrosa

S. braunii

Roots

The roots are adventitious and originate from the (*i*) tips of rhizophores, (ii) from the swollen bases of hypocotyl and (iii) directly from the stem. They arise endogenously and branch dichotomously. In ***S. chrysocaulos*** and other dorsiventral species they arise from the swollen tips of the rhizophores. In ***S. selaginoides*** they arise from the swollen bases of the hypocotyl (Fig. 6.2, A). In ***S. umbrosa***, ***S. braunii***, etc., they arise directly from the stem and have no relation to the branching ***i.e.,*** they can arise from anywhere all along the stem. In some cases the roots arise only from the places where the stems branch ***e.g.***, ***S. densa.***

The roots have root caps and bear root hair. In ***S. selaginoides*** an endophytic fungus has been reported in the cortical cells of the root.

ORGANIZATION OF THE STROBILUS

The strobilus (Fig. 6.3, A.B) is the sporangia-bearing region of the sporophyte. The sporangia arise in the axils of leaves called the **sporophylls.** The sporophylls are like ordinary vegetative leaves but may differ in shape and size. There are two kinds of sporangia in ***Selaginella*** : (*i*) the **microsporangia** and (*ii*) the **megasporangia.** The sporophylls bearing microsporangia may be called the **microsporophylls** and those bearing megasporangia as **megasporophylls.** There is no morphological or anatomical differentiation between the micro and the megasporophylls. The **strobilus** is always terminal in position. Its formation usually results in the cessation of apical growth. In some cases, however, the axis of the strobilus resumes growth, ***i.e.***, it poliferates and overgrows the strobilus to form a normal dorsiventral shoot, ***e.g.***, ***S. grandis.*** In ***S. erythropus*** the proliferating strobilus produces a sterile shoot which again bears another strobilus at its tip. Such a condition results in the formation of two strobili on the same branch.

In ***Selaginella selaginoides*** the storobilus is cylindrical. The sporophylls are spirally arranged and are isophyllous. It is clearly marked off from the vegetative region. In ***S. molliceps*** sporangia have also been found to develop in the axils of ordinary vegetative leaves. In a large number of species the strobilus is tetragonous or four-angled. The sporophylls in the tetragonous strobilus may be isophyllous or anisophyllous and may be spirally arranged or decussate. In ***S. pallidissima***, ***S. chrysocaulos***, ***S. chrysorrhizos***, etc., the vegetative region as well as the strobili (tetragonous) are anisophyllous. In all these cases the sporophylls are ligulate and the ligule is present between the sporangium and the base of the sporophylls.

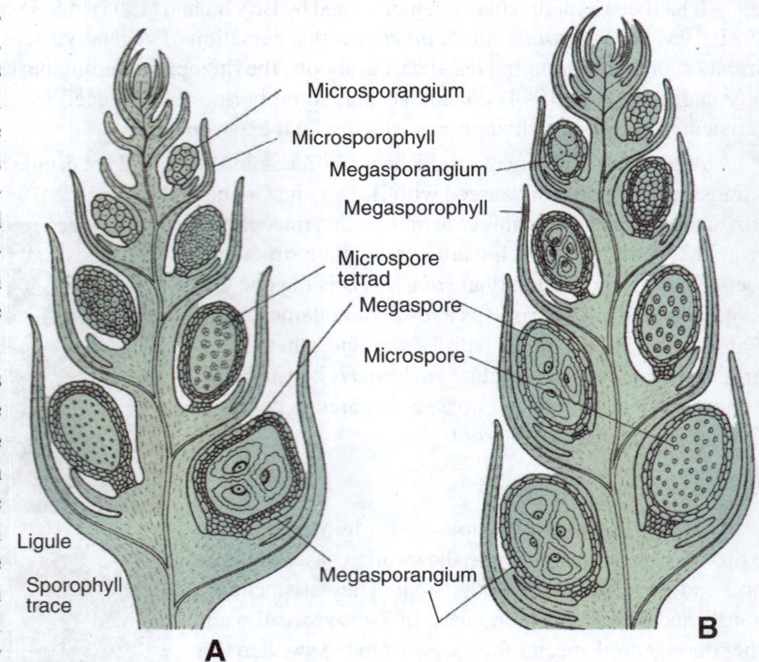

Fig. 6.3 (A—B). V. S. sporangiferous spikes of *Selaginella*. A. S. *kraussiana*; B. S. *oregana.*

The strobilus is also called the **sporangiferous spike** or the **cone.** Each sporophyll bears one sporangium in its axil (**axillary**) or a little upwards on the stem (**cauline**). The strobili are usually bisporangiate, *i.e.*, they bear both micro-and megasporangia in the axils of the respective sporo phylls. Such strobili are protandrous. The two kinds of sporophylls have no fixed position on the strobilus and they are variously positioned in different species of the genus. In ***S. chrysocaulos***, ***S. rupestris*** and many others the megasporophylls are restricted to the base of the strobilus and the microsporophylls to the upper region. In ***S. caulescens*** and ***S. martensii*** the two kinds of sporophylls are indiscriminately mingled along the entire length of the strobilus. In ***S. kraussiana*** (Fig. 6.3, A) there is only one megasporophyll at the base of the strobilus and the rest are all microsporophylls. In ***S. oregana*** (Fig. 6.3, B) the tetragonous stobilus has two rows of microsporophylls on one side and two rows of megasporophylls on the other side.

In ***S. atroviridis*** and ***S. gracilis*** the strobili are unisporangiate, ***i.e.***, a strobilus either bears microsporangia or it bears only megasporangia. Both the microsporangiate and megasporangiate strobili occur on the same plant.

ANATOMY

Stem

The vascular region or the ***stelar system*** in ***Selaginella*** exhibits considerable variation in the different species or even in different parts of the stem in the same species. Anatomy of ***S. chrysocaulos*** is given in detail. Variations in other species are, however, given in brief.

Selaginella chrysocaulos (Fig. 6.4, A).

It is a dorsiventral species and transverse section of a typical stem reveals the following structure :—

Epidermis

It is made up of a single layer of thin-walled rectangular or barrel-shaped cells and is covered by a thin layer of cuticle. The cells are colourless. There are no stomata.

Cortex

It consists of many layers of cells. The outermost layers (usually 2 to 4) of cells develop thick walls in the older regions of stem and form a **sclerenchymatous hypodermis.** The rest of the cortex is made up of thin-walled chlorenchymatous and polygonal cells. They enclose small intercellular spaces. In ***S. lepidophylla*** most of the cortex is sclerenchymatous. Sclerenchymatous hypodermis is absent in the younger parts of the stem in ***S. chrysocaulos***.

Air space

Next to the cortex there is large air space in the centre of which lies the stele suspended by means of **trabeculae.** The trabeculae are modified endodermal cells and possess casparian strips (Fig. 6.4, A). These trabeculae are actually cells of an endodermal layer. During the course of development (Fig. 6.4, C—E) the endodermal cells elongate radially (Fig. 6.4, D). As a result of this elongation the cells separate from each other (Fig. 6.4, D) and lead to the formation of air spaces. Further elongation results in the formation of a large air space traversed by elongated endodermal cells or the **trabeculae.** The trabeculae possess casparian strips. During the course of their radial elongation the trabeculae may divide and become two or more celled filaments.

Stele or the Vascular region

There is single central stele suspended in the air space by means of the trabeculae (Fig. 6.4, A). The stele in this species is flattened like a ribbon and is called a **protostele** because there is no pith in the centre. It consists of :—

(*a*) A single layer of thin-walled cells called the **pericycle.** The pericycle completely encircles the inner vascular tissues.

(*b*) Next to the pericycle is **phloem.** The phloem consists of a layer of phloem parenchyma cells next to the pericycle and a layer or two of **sieve cells.** The companion cells are absent. Phloem completely encircles the central xylem. The sieve cells appear as large polygonal cells in a cross section.

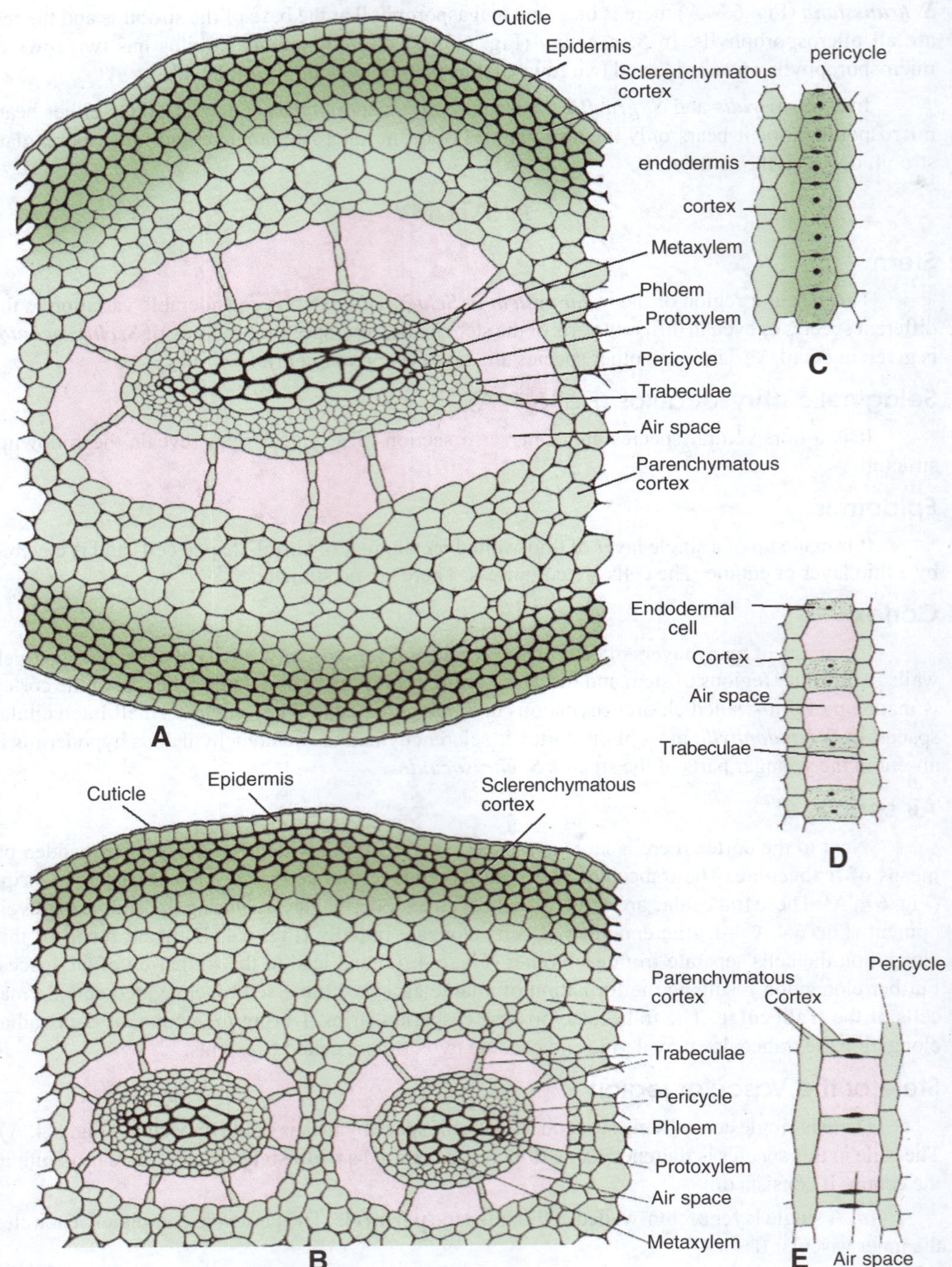

Fig. 6.4. (A—E). Anatomy of *Selaginella*. A. T.S. stem of *S. chrysocaulos*; B. T.S. stem of *S. kraussiana*. Note distelic condition; C—E. Development of trabeculae.

(*c*) The centre of the stele is occupied by the xylem tissue. The xylem, therefore, forms the core of the stele, there being no pith. The xylem consists of **metaxylem** and **protoxylem.** The metaxylem is composed of scalariform tracheids and xylem parenchyma cells. It forms the major portion of the flattened stele. The protoxylem occupies the two ends of the flattened metaxylem and is composed of annular and spiral tracheids. The xylem is, therefore, **exarch**. Since there are two groups of protoxylem at either end, the xylem is regarded as **diarch**. The xylem lacks vessels and fibres.

S. flabellata

The protostele is **monostelic** because the stem has only one such stele. Other species with such a stele are ***S. martensii***, ***S. chrysorrhizos***, ***S. flabellata***, etc.

In ***S. selaginoides*** a section of the stem from the upper region of the axis reveals an **actinostelic** (star-shaped) **protostele** with 7 groups of exarch-proto-xylem. A cross section through a portion of the stem in trailing region reveals a monostelic protostele with mesarch xylem. In this case protoxylem is in the centre and is surrounded on all sides by metaxylem. In ***S. kraussiana*** there are two protosteles suspended in the air space. They are exarch. Such a stele is called distelic (Fig. 6.4, B).

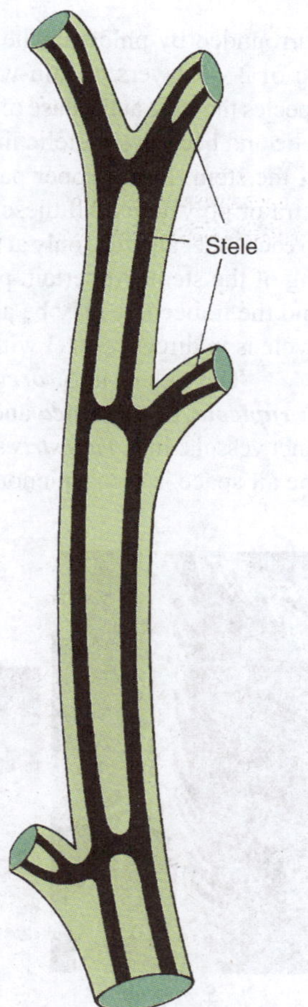

Fig. 6.5. *Selaginella kraussiana.* It shows longitudinal course of the two steles. They anastomose at the point of bifurcation of the stem.

The entire stem is traversed by two parallel steles that join only at the point of branching of the stem (Fig. 6.5) and again separate. Metaxylem consists of scalariform tracheids, whereas protoxylem is composed of annular and spiral tracheids. Bierhorst (1960) and Zamora (1960) reported that the helics of spiral tracheids in this species are variously wound in different parts of the same tracheid. The xylem is encircled by phloem which in turn is surrounded by pericycle. The phloem consists of a single layer of sieve cells and parenchyma cells. Esau (1953), and Cheadle and Gifford (1953), reported that the sieve cells have many sieve areas on their inclined end walls and lateral walls. ***Selaginella sulcata***, ***S. galeottei***, ***S. willdenovii***, etc., belong to this category.

Selaginella sulcata

In ***S. willdenovii*** (Fig. 6.6) there are three or even four ribbon-like protosteles. The xylem is exarch. There are two groups of protoxylem, one at either end, in the central stele whereas in the other two that lie on either side of it there are more than two groups of protoxylem (Fig. 6.6). The xylem is surrounded by phloem. The pericycle is made up of 2---4 layers of thin-walled cells. In this species the stele at the base of the stem is monostelic and becomes tristelic in the middle region of the stem. In the upper parts it may be even tetra or polystelic. All these steles run parallel to each other and fuse only at the point of branching of the stem. At certain places the air space and the trabeculae may be absent and the pericycle is in direct contact with the inner cortex.

The xylem in ***S. oregana***, ***S. rupestris***, ***S. rupicola***, ***S. arizonica*** and ***S. densa*** has distinct vessels. In ***S. rupestris*** and ***S. lepidophylla*** the air space is inconspicuous.

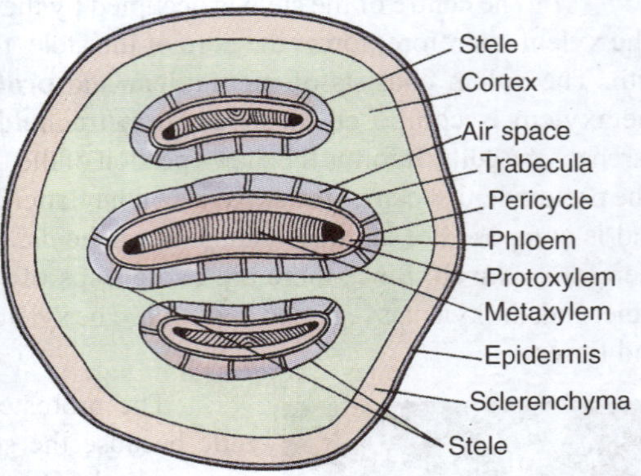

Fig. 6.6. T.S. (diagrammatic) stem of *S. willdenovii* showing three protosteles.

S. oregana *S. arizonica* *S. lepidophylla*

Rhizophore (Fig. 6.7, A)

In a transverse section the rhizophore reveals a singled-layered epidermis whose cells may be thick-walled. The cortex is extensive and is usually distinguished into an outer thick-walled or sclerenchymatous cortex and an inner thin-walled or parenchymatous cortex. Last layer of the cortex is called the **endodermis** and is followed by a single layer of parenchymatous cells called the **pericycle.** The endodermis is not very clear.

The stele is typically a **protostele** and shows variations in its form and arrangement of protoxylem in the different species of ***Selaginella***. In ***S. martensii*** (Fig. 6.7, A) the metaxylem occupies the centre and the protoxylem forms a group of smaller tracheids on one side, *i.e.*, ***it*** is **exarch.** In ***S. atroviridis*** the metaxylem is half moon-shaped or crescentic and the protoxylem occurs in the form of a few groups on the concave adaxial side. In ***S. kraussiana*** the condition is mesarch. The xylem is surrounded by phloem which consists of sieve cells and parenchyma.

Root. (Fig. 6.7, B)

The root epidermis is single-layered and is covered by a thin cuticle. Root hairs are present.

The cortex is wide and extensive and usually consists of an outer sclerenchymatous cortex of 3 to 5 or more layers in thickness. Webster and Steeves 1963) described the entire cortex of older root in ***S. densa*** as sclerenchymatous. In the younger roots the sclerenchyma is peripheral in position and is followed by thin-walled cortex. A distinct lacuna or air space has also been reported in the inner cortex of ***S. densa*** (Webster and Steeves, 1963). It is traversed by trabeculae that are not endodermal cells but cortical cells that elongate radially. This lacuna is absent in other species.

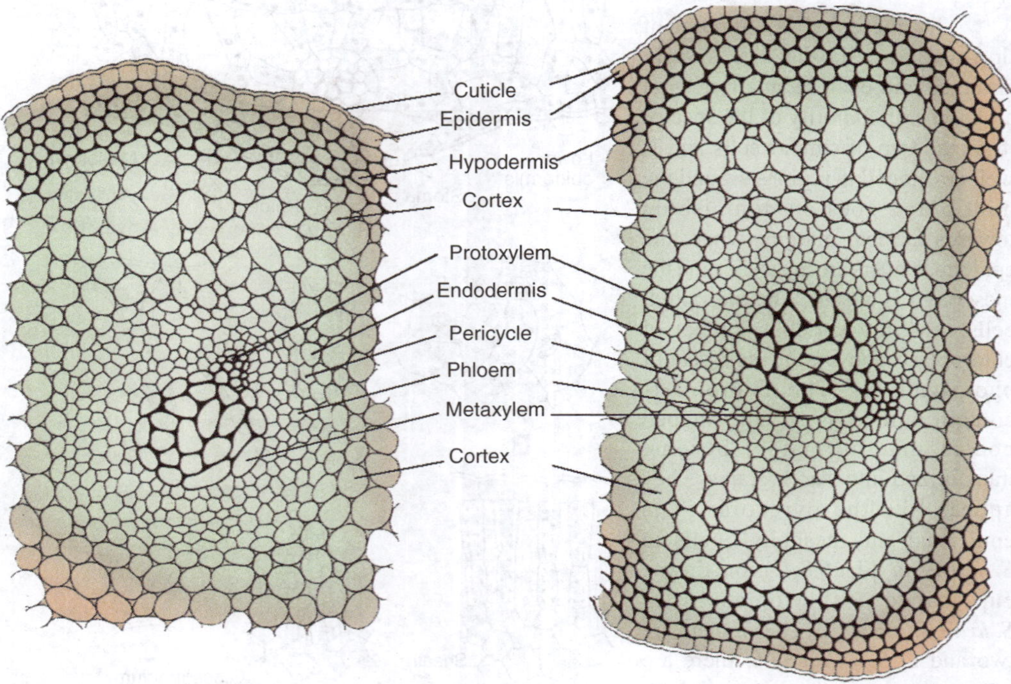

Fig. 6.7 (A—B). *Selaginella.* Anatomy of Root and Rhizophore. A. T. S. Rhizophore of *S. martensii*; B. T. S. Root of *S. chrysocaulos.*

The endodermis is usually indistinct but in some species it is distinguishable (***S. densa***, ***S. rubella***, etc.). Next to endodermis is a single layer of pericycle. The stele is a typical **protostele** with **exarch** and **monarch** xylem. There is only one group of protoxylem that is peripheral in position. The phloem almost encircles the xylem except that opposite the protoxylem group the sieve cells are absent or not properly developed.

S. densa

Leaf. (Fig. 6.8)

It is dorsiventral and consists of distinct upper and lower epidermises that are one cell in thickness. The stomata may be present on both the epidermal layers. In majority of the species the stomata are restricted to the lower epidermis in the vicinity of the mid-rib. The epidermal cells contain chloroplasts and the cells composing the two epidermal layers may be similar in shape and size or they may show some differences. The cells in the upper epidermis of ***S. martensii*** are large and conical and those of the lower epidermis are comparatively smaller. In this species the stomata are restricted to the lower epidermis. The epidermal cells in some species bear hair-like appendages.

The mesophyll is composed of thin-walled cells that are loosely arranged and enclose small or large air spaces. In majority of the species it is made up of similar cells and is well developed in regions around the midrib and goes on diminishing towards the leaf margins where the epidermal layers are in contact and the mesophyll is absent. The mesophyll cells contain a variable number of chloroplasts and are the main seat of photosynthesis. The chloroplasts contain numerous spindle-shaped bodies that ultimately become transformed into starch grains. They appear like the pyrenoids. The epidermal and mesophyll cells in ***S. martensii*** (Fig. 6.8, C) contain one cup-shaped chloroplast. In ***S. kraussiana*** and ***S. caesia*** there are two and in ***S. willdenovii*** there are upto eight chloroplasts in each cell. Ultrastructural studies on the chloroplasts of ***Selaginella*** (McHale, 1965) reveal that they are intermediate between the algal chloroplasts and the spermatophyta. The chloroplasts show grana-like structures called the **'granoids'.** In ***S. lyallii*** (Fig. 6.8, E) and ***S. concinna*** the mesophyll is distinguished into a distinct palisade layer and a spongy parenchyma.

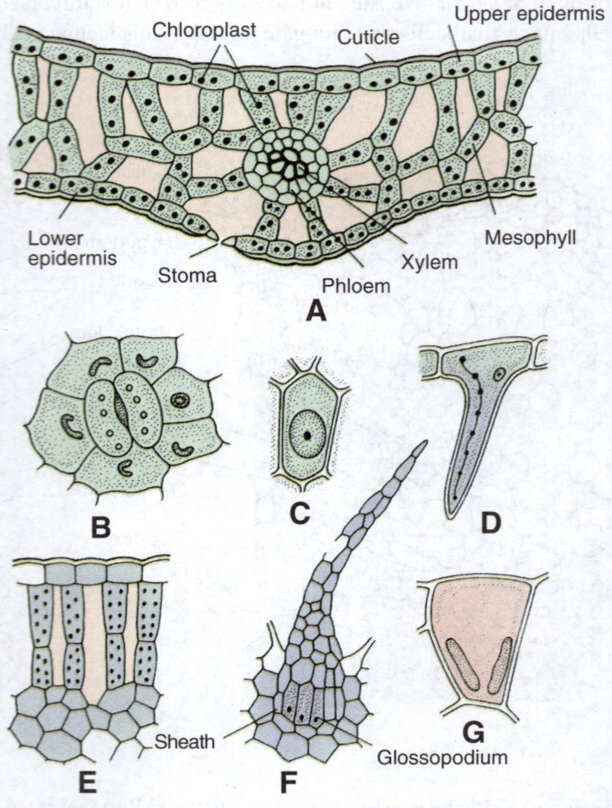

Fig. 6.8. *Selaginella.* A. V.S. leaf of *S. kraussiana*; B. A stoma from the living leaf of *S. martensii*; C. A cell from the mesophyll of *S. martensii* showing a single cup-shaped chloroplast; D. A tooth from the living leaf of *S. martensii*; E. V. S. portion of leaf of *S. lyallii* showing distinct palisade and spongy tissue; F. A ligule; G. A mesophyll cell from the leaf of *S. caesia* showing the two chloroplasts (B—D, after Emberger; G, after Haberlandt ; E-F, after Harvey-Gibson.)

The vascular bundle is very simple and surrounded by a layer of phloem. The xylem has tracheids only and there is no distinction into proto and metaxylem. The tracheids may be annular or spiral. The phloem has a few sieve cells and a number of elongated parenchyma cells. A single layer of cells completely encircles the phloem. It may be regarded as the **bundle sheath.**

Ligule

The ligule arises from several short rows of superficial cells. A fully developed ligule consists of a distinct and hemispherical basal region where cells are large and thin-walled and contain vacuolated cytoplasm. This region is the **glossopodium** (Fig. 6.8, F). It is surrounded by a sheath called the glossopodial sheath. In ***S. kraussiana*** and ***S. rupestris*** the sheath cells show casparian strips. The cells above the glossopodial region are large, polygonal and have densely granular cytoplasmic contents. The region is swollen and quite prominent. It is not surrounded by the sheath. Next to this region the ligule usually (in many species) narrows down, or broadens or is produced into finger-like processes. Depending upon the shape of the ligule the cells in the distal portion are likewise arranged. In tongueshaped and spindle-shaped ligules the cells in the distal region are narrow and elongated.

Apical Growth

De Bary (1884) and Barclay (1931) studied the shoot apices of certain species of ***Selaginella*** and concluded that there is an intergradation from a single apical cell to a general meristematic group. In ***S. willdenovii*** the young sporophyte has a tetrahedral apical cell which divides along three cutting faces and the first segments thus cut off undergo periclinal division to distinguish an outer layer that by further periclinal and anticlinal divisions gives rise to the epidermis and the cortex. The inner cells give rise to the endodermis, pericycle and the stele. In ***S. selaginoides***, Bruchmann (1897) reported the apical growth to take place by a group of cells. Strasburger (1891) reported the occurrence of two apical cells in ***S. wallichii***. Williams (1931) reported two apical cells in ***S. grandis***. The superficial cells very close to the stem apex give rise to the leaf primordia on either side of the apical region. During its early development procambial strands differentiate in the leaf and are connected with the vascular strand of the stem (Hsu, 1937). The procambial strands differentiate into primary xylem and phloem.

Bhambie (1960), and Bhambie and Puri (1963, 1985) report that in ***Selaginella*** the shoot apex, is "somewhat dorsiventrally flattered in mature plants and exhibits lateral symmetry. At the extreme tip is a laterally oriented band of 2-6, large wedge-shaped cells, larger in size; poor in cytoplasm and not showing division figures. These conspicuous cells are arranged either uniseriately or biseriately forming a group of laterally oriented cells at the extreme tip. Other cells subjacent to these apical initials, though homogenous and densely cytoplasmic, can be distinguished into an outer layer and an inner mass of cells. In vertical longitudinal sections, species having horizontally flattered apices show only a single or two large cells, giving the impression that the extreme tip has one or two apical initials. Earlier reports of one or two apical cells in this genus are probably based on such sections."

The rhizophore also grows in length by the activity of a single apical cell or a group of apical cells. In roots there is a single apical cell that has three cutting faces.

So far as leaves are concerned, Sifton (1944) observed that a group of cells is responsible for the formation of a leaf primordium.

REPRODUCTION

Vegetative Propagation

Vegetative propagation in ***Selaginella*** is effected by the following methods :

Fragmentation

It is effected only in species that grow under humid conditions. It has been reported in ***S. rupestris***. In this case the trailing branches of the stem develop adventitious branches and later sever their connection from the parent plant and grow into separate individual plants.

Tubers

Formation of tubers has been reported in ***S. chrysorrhizos*** and ***S. abyssinica***. In ***S. chrysorrhizos*** they are formed underground. The tubers bear rudimentary scales and appear towards the end of the growing season at the tips of underground branches that arise from the base of the stem (Fig. 6.9, A). During unfavourable circumstances the aerial parts of the plant die and tubers enable the plant to perennate. At the advent of favourable conditions the tubers germinate to produce new plants of ***Selaginella.***

Resting buds

(Fig. 6.9, B) have been reported to develop at the ends of some aerial branches in *S. chrysocaulos*. They develop at the close of the monsoon season and are very compact structures. The leaves in this region are closely arranged and overlap each other and cover the growing point. The buds give off **rhizophores** that bear roots at their tips and fix them to the soil. The resting buds survive the unfavourable periods when the rest of the plant dies. They grow into new individuals at the return of favourable conditions.

Reproduction by spores

Selaginella is heterosporous and the **spores** are of two kinds :

(*i*) microspores (*ii*) megaspores. They are produced in **microsporangia** and **megasporangia**, which arise in the axils of **microsporaphylls** and **megasporophylls** and may be **cauline** or **foliar** in nature. The sporophylls are arranged in definite loose or compact and terminal structures called the **strobili** or **spikes**. They develop at a certain stage of maturity.

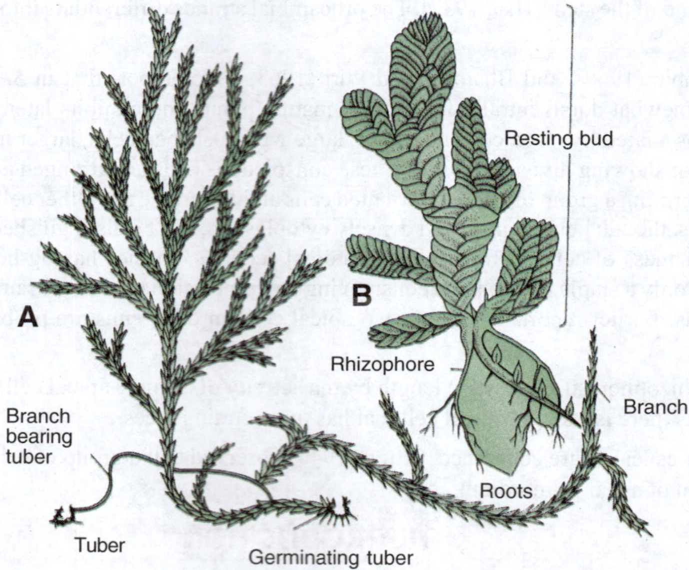

Fig. 6.9. (A-B). Vegetative propagation in *Selaginella*. A. Plant formed by the germination of a tuber. Also note that the same plant bears a new tuber at the tip of a specialised branch. This is in *S. chrysorrhizos*; B. Portion of plant of *S. chrysocaulos* bearing a resting bud. (Highly enlarged).

Sporangia

The two kinds of sporangia are variously coloured and have a variety of shapes and sizes. They are either seated strictly on sporophyll base (foliar) or on the stem (cone axis) just above the sporophyll axil (cauline).

Microsporangia

Structure

A microsporangium may be oval, obovoid, reniform or spherical in shape and varies in colour from red, yellow to brown. The outline of the sporangium is smooth and almost regular. It is smaller in size than the megasporangium and has a short multicellular stalk that varies in width in the different species. The wall of the microsporangium is two-layered thick. Next to the inner wall layer is the

tapetum which is nutritive in function and arises from the **archesporial tissue.** The wall layers and the tapetum enclose a large number of **microspore mother cells.** The mother cells are diploid and are the last cells of the diplophase or the sporophytic generation. These undergo meiosis and form numerous microspore **tetrads.** The microspores in a tetrad are tetrahedrally arranged and later on separate from each other. The cavity of the sporangium is now filled with numerous (1,500—2,000) microspores. At this stage the tapetum disorganises. In some cases the sporangium has a two-layered wall at maturity and in the latter case only one-layered. The cells of the outer layer develop thick walls. It is shortly stalked.

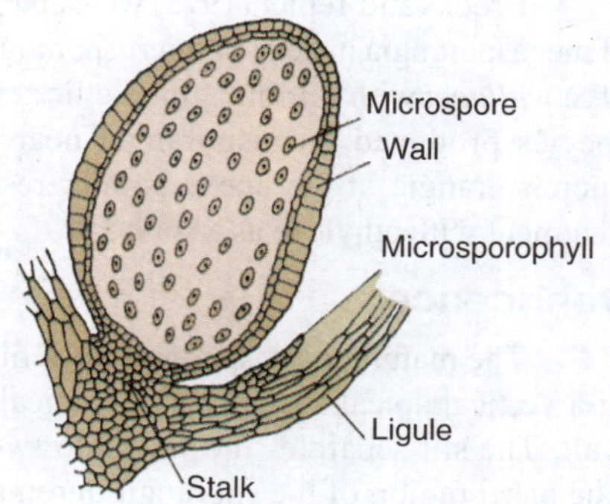

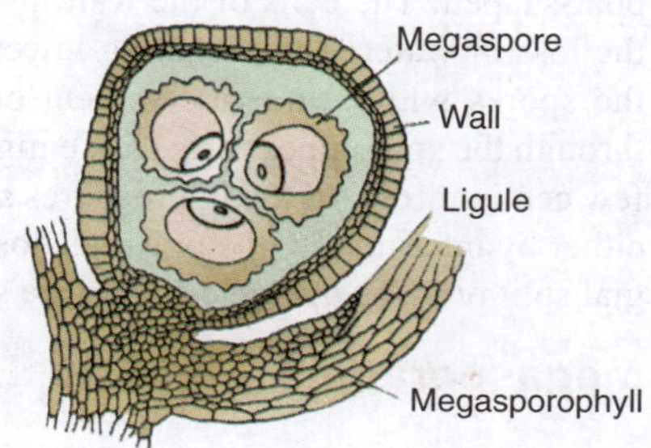

Fig. 6.10. *Selaginella.* A. Mature Microsporangium ; B.Mature Megasporangium.

Development

The developmental stages, in both the micro and mega-sporangia are similar upto the differentiation of mother cells. The development is essentially of **eusporangiate type.** In ***S. martensii*** and ***S. galeottii*** the sporangium arises from the axis of the cone and is, therefore, **cauline** (Fig. 6-12 A, B,C.). In ***S. selaginoides*** the sporangium is foliar in origin, because it arises from the base of the sporophyll (Fig. 6.12 D).

In these species the sporangium originates from a transverse row of cells (Fig. 6.12, A) called the **sporangial initials**. These initials divide by periclinal walls and cut off outer **primary jacket cells** or **wall cells** and inner **primary archesporial cells** (Fig. 6.12, B). The wall cells divide anticlinally and periclinally to form two-layered wall (Fig. 6.12, C—E). The archesporial cells divide repeatedly to form a group of **sporogenous cells** (Fig. 6.12, D,E). The outermost layer of sporogenous cells divides periclinally to cut off an outer layer of **tapetal cells** (Fig. 6.12, E). The tissue at the base of the sporangium divides to form the sporangial stalk (Fig. 6.12, F). The **tapetum** lies next to the inner layer of wall and is in direct contact with the sporogenous cells. It persists till the spores are formed and its cells may become papillate. The sporogenous cells divide repeatedly and their last division results in the formation of **spore mother cells**. Upto this stage the development of micro-and megasporangia is similar. After this stage they follow a different course.

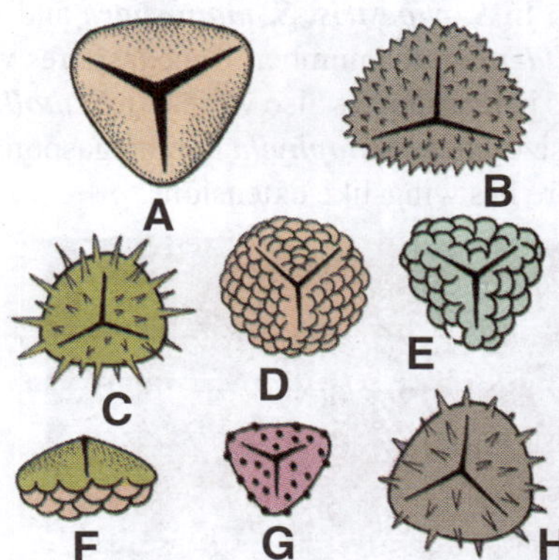

Fig. 6.11. Megaspores of various species of *Selaginella* illustrating spore coat morphology. A. ***S. repens*** with granulose ectine; B. Spinose ectine in *S. galcottei*; C. Spinose ectine in ***S. xiphophila***; D. Tuberculate ectine in ***S. chrysocaulos***; E. Verrucate ectine in ***S. mongolica***; F. same as *E*; G. Baculate ectine in *S.* haematodes; H. Spinose ectine in ***S. spinosa.*** (After Knox, 1931)

In the microsporangia usually all the **microspore mother cells** are functional and undergo **meiosis** to form **tetrahedral tetrads** of micrsopores. These, later, separate into individual microspores. The tapetal cells now disorganise and form a continuous **plasmodial fluid** in which the microspores float and receive nourishment.

Brooks and Tepfer (1972) were able to induce the formation of megasporangia instead of microsporangia by treating the plants of ***S. wallacei*** with 0.1 mm. gibberellic acid. Normally this radial species produced megasporangia near base of strobilus and microsporangia at the apex. Similar results were obtained by treatment with ethylene as a spray.

S. wallacei

Dehiscence

The mature microsporangium dehisces by the appearance of a vertical apical slit only in the apical part of the sporangial wall. The slit separates the wall into two valves that separate. The basal region of the sporangium remains intact and appears boat-shaped. The cells of the wall in this region shrink due to the loss of water. The shrinking lower part exerts pressure on the spores within and forces them out with a little violence through the split upper part. The semigerminated microspores fall at a little distance that may be a few centimetres. All the microspores are not thrown out at once but they are expelled at intervals either by intermittent opening and closing split upper part of the sporangial wall, or if the sporangial split remains open the spores are scattered, in small masses, by the wind.

Megasporangia (Fig. 6.10, B)

Structure

It is comparatively larger in size, four lobed, shortly stalked and may be green or cream-coloured in some species and white or dark brown in others. It has a two-layered wall and a third innermost layer is called the **tapetum**. Out of all spore mother cells only one remains functional and divides meiotically to form a tetrad of four haploid megaspores. By this time the tapetum usually disorganises (Fig. 6.10, B), but this is by no means true for all species. The megaspores separate and grow considerably in size so as to fill up the whole sporangium. The megasporangium becomes four-lobed due to extensive enlargement of the megaspores. In some species the developing megaspores undergo abortion so that their number varies from 1 to 4. In ***S. rupestris***, ***S. monospora*** and some other species only one megaspore is left behind. In ***S. willdenovii*** the number of megaspores varies between 4-24 or even 36 or 40. The size of the megaspores in this species also varies. In ***S. molliceps*** one megaspore is large and the other three are smaller in size. In ***S. stenophylla*** two megaspores are large and two small in size. In ***S. caulescens*** the megaspore has wing-like extensions.

S. willdenovii

S. caulescens

When fully mature the megasporangium has only one jacket layer whose walls become thick and radially elongated.

Development (Fig. 6.12)

It is similar to microsporangium up to the differentiation of spore mother cells. In case it is destined to become a megasporangium, only one **megaspore mother cell** remains functional and others start degenerating. The functional megaspore mother cell undergoes meiosis to form a tetrad of megaspores. The number of functional megaspore mother cells has been found to vary from one to ten. In majority of species only one megaspore mother cell is functional and produces 4 haploid megaspores. The number of megaspores in various species has been recorded to vary between 1 to 40. After the formation of tetrads there is sometimes abortion of the megaspores. This also leads to variations in the number of surviving megaspores.

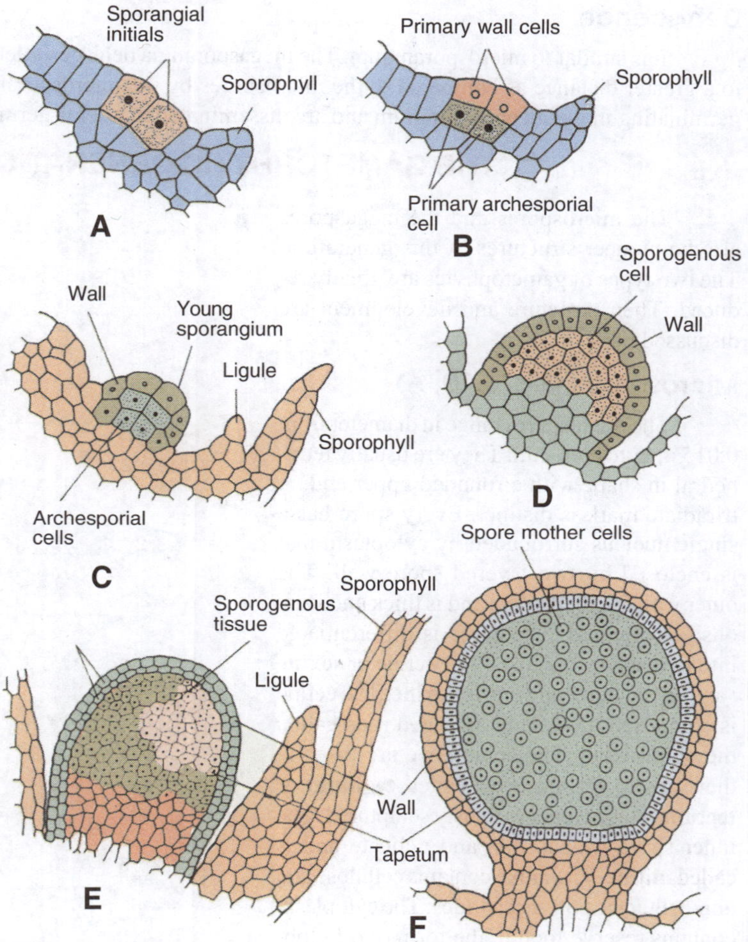

Fig. 6.12 (A—F). Various stages of development of sporangium in *Selaginella.* A— C. *Selaginella martensi* ; D— F. *S.selagionides.* (After Bower)

James French (1972) has observed that in ***S. bigelovii*** the megasporangia and microsporangia show growth differences before the sporocyte stage. He observed that in this species the developing megasporangia are larger than the microsporangia and grow more vigorously before sporocyte differentiation. Bierhorst (1971) is also of the same opinion.

S. bigelovii

Dehiscence

It is similar to microsporangium. The megasporangia dehisce violently and scatter megaspores to a greater distance as compared to the microspores by the microsporangia. The megaspores start germinating in the megasporangium and are disseminated in a semi-germinated condition.

THE GAMETOPHYTIC GENERATION

The microspores and the megaspores are the pioneer structures of this generation. The two types of gametophytes are greatly reduced. Their structure and development are discussed below :

Microspores (Fig. 6.10, A)

The microspores range in diameter from 0.015 mm. to 0.06 mm. They are usually tetrahedral in shape with a rounded upper end. A triradiate mark is distinct. Every spore has a single nucleus surrounded by cytoplasm that is enclosed by two layered spore wall. The outer wall is called **exine** and is thick and variously sculptured. The exine is differentiated into two layers : (*i*) the outer ectine or **sexine** ; and (*ii*) an inner **endine** or **nexine**. The ectine is composed of radially arranged rods called the **columellae**. The columellae are fused at their tips to form a layer called **tegillum** or **tectum**. This layer is variously sculptured. The inner spore wall is thin and delicate and is called **initine.** The intine contains cellulose and non-cellulosic polysaccharides. The cytoplasm contains reserve food in the form of oil globules and other fatty substances and even nitrogenous materials. It contains little water and no chloroplasts. The exine contains **sporopollenin**, waxy substances and a variety of carbohydrates.

Microgametophyte (Fig. 6.13)

Germination of the Microspore

A complete account of the germination of microspore has been given by Slagg (1932) for ***S. kraussiana***. The germination is **precocious**, ***i.e.***, it starts within the microsporangium. The microgametophyte is at 13-celled stage when the microsporangium dehisces. The stages of germination that take place within the microsporangium can be summarised as below:

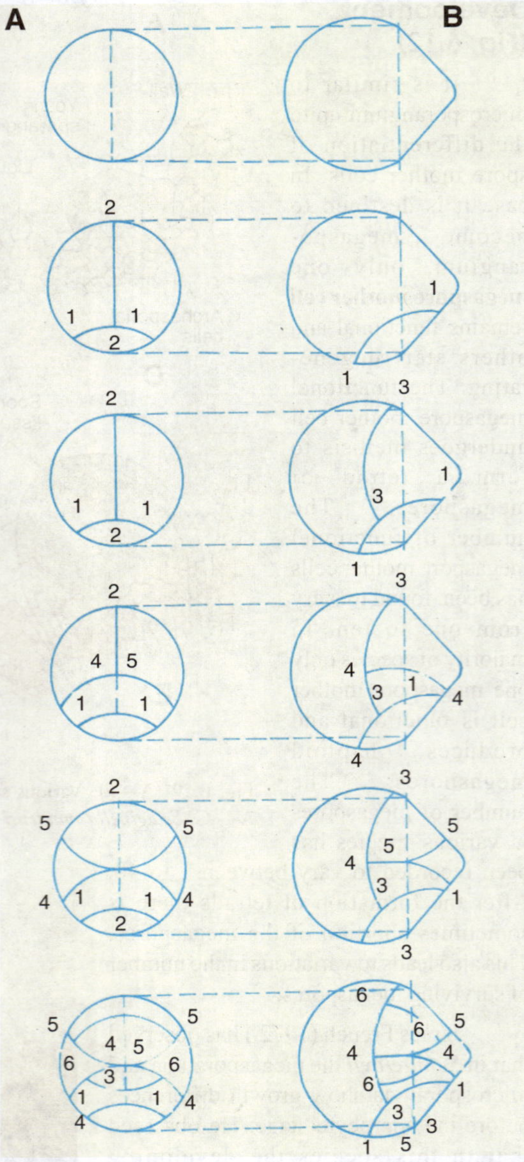

Fig. 6.13. *Selaginella kraussiana.* Development of Male gametophyte. In B the developing microspores have been shown in vertical sections. In A the spores are cut at right angles to those in B at the plane x—x (After Slagg)

The haploid nucleus of the microspore divides into two daughter nuclei. One of the two daughter nuclei migrates to one side of the spore. A cell wall is laid down separating a small, lens-shaped **prothallial cell** from a larger **antheridial cell** (Fig. 6.13, B). This smaller **prothallial cell** remains inactive and does not divide further. The antheridial cell divides along several planes and gives rise to an **antheridium.** The first division of the antheridial cell is by a vertical wall (Fig. 6.13. A, 2—2). The antheridial cell thus becomes two celled and each of these two cells divides further by a transverse wall (Fig. 6.13, B, 3-3) at right angles to the first plane of division. This wall can be seen only if we cut a vertical section of the spore. This is the five-celled stage : one prothallial cell plus four antheridial cells. Out of the four cells thus formed two basal cells do not divide further. The upper two cells divide further by a curved wall that meets the second wall (2-2) somewhere in the middle (Fig. 6.13, A, 4—4). The microgametophyte has seven cells at this stage, ***i.e.***, 6 + 1. Out of the four cells formed by the last division the two bigger ones divide again by curved walls (Fig. 6. 13, A, B, 5—5) and result in the formation of eight antheridial cells arranged in a manner that there are four cells in the middle quite separate from two cells on either side above and below. The middle four cells divide by periclinal walls (Fig. 6.13, 6—6) to cut off 4 **primary androgonial cells** from 8 jacket cells. This is the 13-celled stage. These cells are 8 **jacket cells, 4 primary androgonial cells** and one **prothallial cell**. At this stage the microgametophytes are shed from the microsporangium. The rest of the development proceeds when they fall on a suitable substratum. There are, however, example when the microgametophytes are liberated at an earlier stage. In this case they usually fall on the partially opened female gametophytes and proceed with further development in their close vicinity. In some species the male gametophytes are released at an advanced stage (more than 13-celled stage) of germination. In ***S. kraussiana*** microgametophytes are liberated only at the 13-celled stage and fall on moist soil.

During further development of the 13-celled stage only the four central **primary androgonial cells** divide. They give rise to 128-256 **androcytes** or antherozoid mother cells. During the formation of **androcytes** the eight jacket cells of the antheridium and the prothallial cell undergo degeneration so that the 256 androcytes float ***enmasse*** or in small groups in the mucilage filled cavity of the microspore (fig. 6.14 C). At this stage the androcytes are surrounded only by the microspore wall.

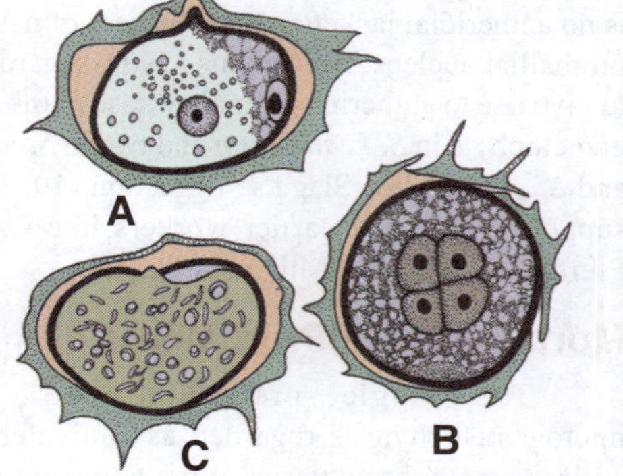

Fig. 6.14 (A-C). *Selaginella kraussiana.* **A.** T.S. spore showing a small prothallial cell; **B.** T.S. germinated spore showing 4 primary androgonial cells and jacket cells; **C.** T.S. mature male gametophyte containing spermatozoids.

Spermatozoid

The cytoplasm of each androcyte metamorphoses into a biflagellate spermatozoid (Fig. 6.15, A—C). The spermatozoid of ***Selaginella*** has the usual three parts ***viz.,*** the **nucleus,** the vesicle (the vesicle (the plasma part) and the **motor apparatus** that consists of cilia bearing band or the **stalk** of the **motor apparatus**. The spermatozoids are liberated by the cracking open of the spore wall along the triradiate ridge. The freshly released spermatozoids have their body coiled around the lens-shaped vesicle (Fig. 6.15. C). Later the vesicle absorbs water and becomes spherical and is soon set free. The nucleus of the spermatozoids is rod-shaped and is embedded in cytoplasm or the plasma part (Fig. 6.15, A,B). The stalk of the motor apparatus is located at the anterior end of the spermatozoid. It bears two flagella (Fig.6.15, A—C). One of them is at the extreme tip and the other a little below (Fig. 6.15, B). The former is smaller and the latter longer. Each one of these cilia is attached to a **basal granule**. Yuasa studied the spermatozoids of *S. involvens* and observed the two cilia arise at the same point and one of them adheres to the body of the spermatozoid for a certain distance (Fig. 6.15, A). The biciliate sperms of ***Selaginella*** resemble spermatozoids of ***Lycopodium***, liverworts ***and Chara***, in striking contrast to those of vascular cryptogams.

S. involvens

The male gametophyte of ***Selaginella*** is extremely reduced like that of ***Isoetes***, but in this case the wall of the antheridium degenerates before the spore wall cracks open and the androcytes are lying free within the spore wall. In ***Isoetes*** the antheridial wall persists and degenerates only after the spore wall cracks open. In ***Selaginella*** there are 8 wall cells as compared to 4 in ***Isoetes***. In ***Isoetes*** one antheridium produces only 4 multiflagellate spermatozoids. There is a greater reduction in the tissue of the microgametophyte in ***Isoetes***.

Morphology of Antheridium

Various morphological interpretations have been advanced regarding the nature of the antheridial cell in the microgametophyte of ***Selaginella***. Some authors regard the antheridial cells as the only antheridium of the reduced male gametophyte. This interpretation seems to be plausible as is clear from Slagg's account of the development of male gametophyte in ***S. kraussiana***. Lyon's (1901) work of ***S. apus*** led her to conclude that there is no antheridium in this species of ***Selaginella*** as there is no antheridial jacket and two masses of naked androcytes are not separated by any wall from the prothallial nucleus. Bruchmann (1909) regards the antheridial cell as an initial structure that divides to give rise to antheridial initials whose number varies in species. He interprets 4 antheridia per male gametophyte in ***S. kraussiana*** and two in ***S. martensii*** and ***S. caulescens***. Slagg's suggestion (1932) is also supported by other earlier workers like Campbell, Pfeffier, Belajeff and Millardet.

Morphology of Prothallial Cell

The single prothallial cell of the microgametophyte is regarded as equivalent to the entire gametophytic tissue of the homosporous vascular cryptogams, ***e.g.***, ***Lycopodium***, ***Equisetum***, ***Dryopteris***, etc. According to Bruchmann (1909), the eight jacket cells are also male prothallial cells.

Megaspores (Figs. 6.11, A—H, 6.16)

They range in diameter from one millimetre to five millimetres and are tetrahedral in shape. There is a distinct triradiate ridge on the outer wall. The megaspore is also unicellular and uninucleate. The cytoplasm is devoid of chloroplasts and contains fatty substances, oils, nitrogenous materials, water, etc. The contents of the megaspore are enveloped by a two-layered thick wall. The outer layer is thick and variously sculptured and called the **exine**. The exine is differentiated into two layers : (*i*) the outer heterogenous layer called the

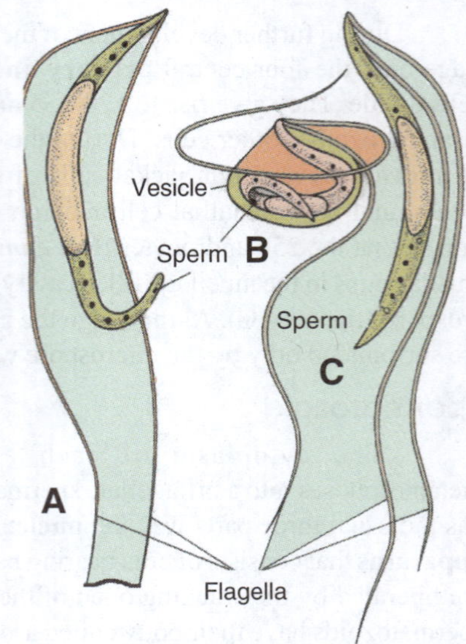

Fig. 6.15 (A—C). *Selaginella.* A. Liberated sperm of *S. involvens*; B. Liberated sperm of a species of *Selaginella*, in which vesicle has been dispensed with; C. A sperm with vesicle (A, after Yuasa; B—C, after Dracinschi).

ectine (or sexine) ; and (*ii*) an inner homogenous layer called the **endine** (or nexine). The ectine is composed of radially arranged rods called the **columellae** that are fused to form a layer called **tegillum** or **tectum** which is variously sculptured (Fig. 6.14). The endine is closely applied to the **intine**. The inner layer is thin and hyaline and is called the **intine.**

S. selaginoides

Megagametophyte (Fig. 6.16)

The megagametophytes develop as a result of the germination of megaspores. The germination is **precocious**, ***i.e.***, it starts within the megasporangium, in majority of the species, except in ***S. selaginoides*** and ***S. helvetica***. In these two species, the megaspore germinates after being shed. The species with precocious germination release the semi-germinated megaspores at various stages.

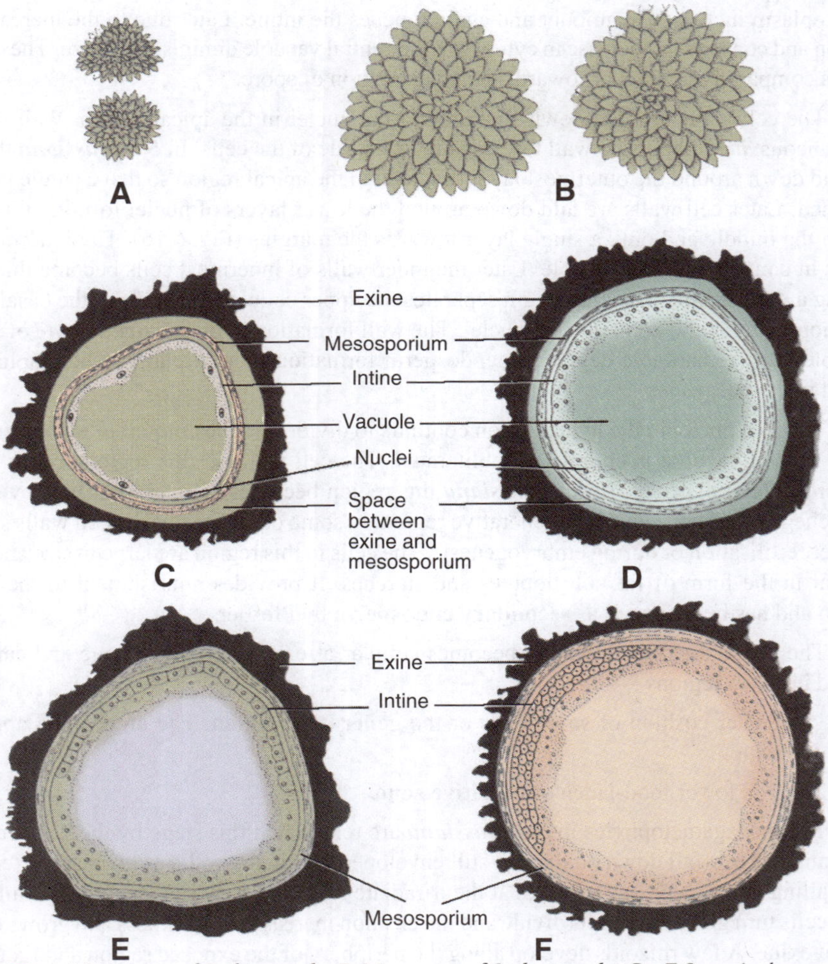

A B

Exine
Mesosporium
Intine
Vacuole
Nuclei
Space between exine and mesosporium

C D

Exine
Intine
Mesosporium

E F

Fig. 6.16 (A—F). A. Microspores of *S. chrysocaulos;* B. Megaspores of *S. chrysocaulos;* C—F. Stages in the germination of megaspore in *S. kraussiana* (C—F after Campbell).

S. kraussiana

Campbell (1902) studied the development of megagametophyte in ***S. kraussiana*** and the following account is based on his observations (Fig. 6.16). The growing megaspores obtain nourishment from the tapetal fluid and start germinating. The first visible change is the separation of **endine** from the **ectine** due to greater growth of the latter. The space thus formed is filled with a nonstainable fluid. The intine is closely applied to the endine. The cytoplasm forms a thin lining layer next to intine and encloses a big vacuole with nucleus embedded in the cytoplasmic layer. Before the commencement of free nuclear division the cytoplasm contracts slightly from the intine and forms a globular vesicle. As a consequence of free nuclear division the number of nuclei increases. At first the nuclei are flattened but they become rounded with increase in amount of the cytoplasm. The nuclei are evenly distributed throughout the cytoplasm. Later the distribution becomes uneven and the number of nuclei increases towards the apex of spore and are comparatively few towards the basal region. At this stage there is a big central vacuole and the cytoplasm increases in amount and again touches the intine. Later due to the increased nuclear division and consequent increase in cytoplasm the central vacuole diminishes in size. The cytoplasmic layer is comparatively thicker towards the apical region of spore.

The cell walls are laid down first around the nuclei in the apical region. Wall formation is simultaneous, ***i.e.***, there is no wall towards the inner side of the cells. In ***S. kraussiana*** the walls are first laid down around the outermost layer of nuclei in the apical region so that a single layer of cells is formed. Later cell walls are laid down around the lower layers of nuclei forming three layers of cells in the middle and only a single layer towards the margins (Fig. 6.16). The innermost layer of cells is in contact with the vacuole. Later the inner walls of innermost cells become thickened thus forming a **diaphragm**. The diaphragm separates the upper cellular region from the basal region that is yet non-cellular and contains free nuclei. The wall formation in megagametophyte of ***Selaginella*** resembles to a remarkable degree the endosperm formation in seed plants. The diaphragm is not formed in all the species.

The free nuclei in the lower region continue to divide and the amount of cytoplasm increases till the central vacuole is completely obliterated. The cell walls in this region appear at different stages in different species. In ***S. kraussiana*** this region becomes cellular after the development of first archegonium in the apical or generative region. In some other species the cell walls start appearing after fertilisation or during embryogenesis. The cells in this region are large in size and store food material in the form of fat, oil globules and starches. It provides nourishment to the developing embryo and has been termed as **secondary endosperm** by Pfeffier.

The entire megagametophyte becomes cellular at one or the other stage and can be distinguished into two regions :—

1. Upper cushion of small cells or the **generative region**. The archegonia appear in this region.
2. The lower food-laden or **nutritive zone.**

The megagametophytes in ***S. kraussiana*** are released at this stage by the dehiscence of megasporangia. They fall down on the soil still enveloped within the walls of megaspore (endosporic). After falling down the exine ruptures at the triradiate ridge and exposes the upper cellular cushion, whose cells turn green. The cells divide and the cushion increases in thickness and grows out through the split exine. A few rhizoids develop along the periphery of the exposed region and fix the gametophytes to the soil and absorb moisture.

In *S. galeottii* distinct mounds of tissue develop on the exposed region. These mounds develop rhizoids at their tips and surround the central archegonia bearing region. Besides absorption these mounds help in catching microspores with antheridia enclosed within them and thus help in fertilisation (Fig. 6.17, B).

Development of the archegonia. (Fig. 6.18)

The archegonia are produced in considerable numbers (***S. kraussiana***) and are not formed in any definite order. They usually occupy the central position on the apical cushion and are surrounded by the rhizoids. Each archegonium develops from a superficial cell of the apical cushions. This cell is called the **archegonial initial** (Fig. 6.18, A). It divides by a periclinal wall to form an upper **primary cover cell** and a lover central cell (Fig. 6.18, B). The central cell divides further by a periclinal wall to form an upper **primary canal cell** and a **lower primary ventral cell** (Fg. 6.18, D). The primary canal cell does not divide further and acts as a single **neck canal cell**. The primary ventral cell undergoes periclinal division to form a ventral **canal cell** and an **egg cell.** The former lies below neck canal cell. The ventral canal cell and the egg cell form an axial row of cells in the archegonium. When all these changes are going on, the primary cover cell divides into four **neck initials** by two vertically intersecting walls (Fig. 6.18, C). Each neck initial divides by a transverse wall to form eight-celled neck (Fig. 6.18, D) of the archegonium. These eight cells are arranged in two tiers of four cells each or four longitudinal rows of two cells each. The entire archegonium except the upper tier of neck cells is embedded in the apical cushion of the megagametophyte.

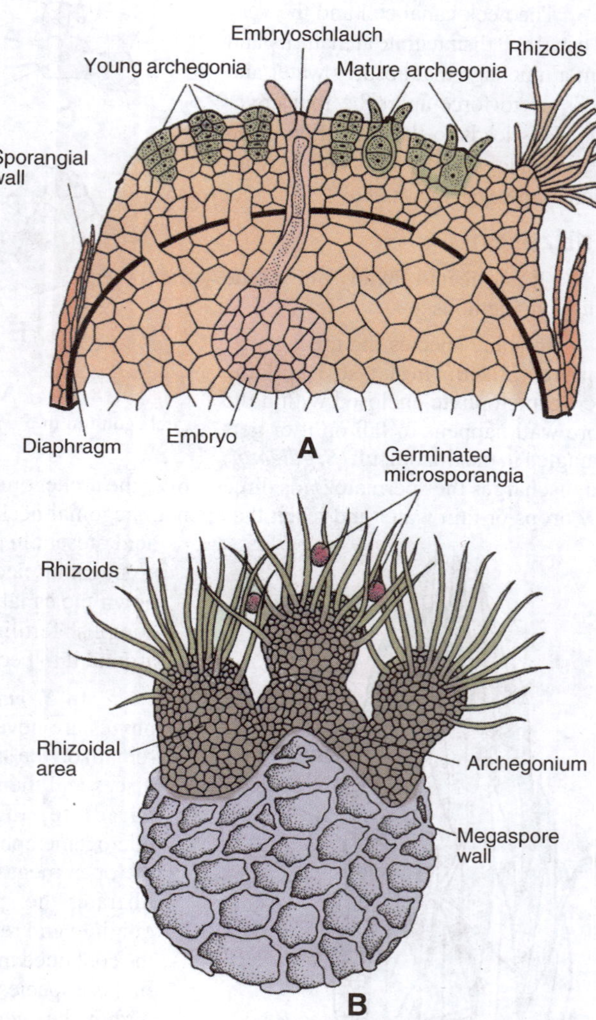

Fig. 6.17 (A-B). A. Median longitudinal section through mature female gametophyte of *S. kraussiana* showing the upper cushion with ; Also note the diaphragm and the lower tissue; B. Megagametophyte of *S. galcottii* showing rhizoidal mounds.

Structure

The mature archegonium (Fig. 6.18, E) consists of a distinct neck composed of two tiers of four cells each, the upper tier functioning as cover cells. The neck enclosed a single neck canal cell.

The venter portion consists of a single ventral canal cell and an egg. The venter is not surrounded by a sterile jacket because it is embedded in the tissue of the gametophyte. The only projecting portion is the upper tier of neck cells.

The neck canal cell and the ventral canal cell disintegrate at maturity and form a mucilage that absorbs water and swells up to force apart the four cover cells. This leads to the formation of an open canal for the spermatozoids to swim down to the egg (Fig. 6.18, F).

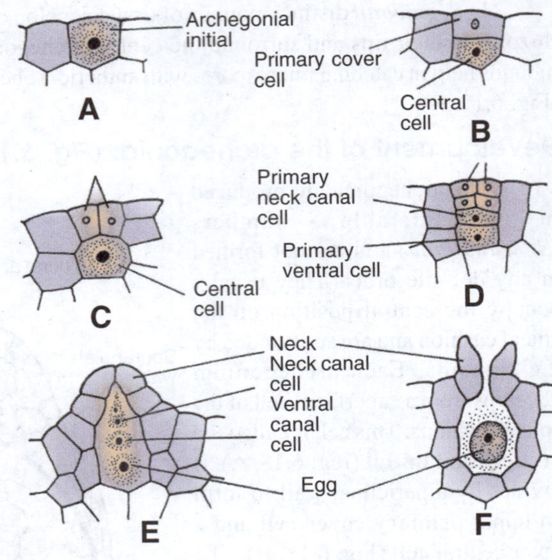

Fig. 6.18 (A-F). Various stages in the development of an archegonium in *S. kraussiana.*

Fertilization

It is effected by any one of the following methods :—

In some species the megagametophyte falls on moist soil and the microgametophyte enclosed within the spore wall happens to fall on it or gets entangled in the rhizoid tuft (***S. galeottii***) and discharges the spermatozoids directly over the archegonial surface. The spermatozoids swim in dew drops or rain water and reach the open archegonial necks, the source of attraction being malic acid present in the mucilage that oozes out of the open archegonial necks. More than one sperms may swim down the canal but only one penetrates the egg to accomplish fertilization. The fertilised egg secretes a wall around it to become a **zygote** or the **oospore**.

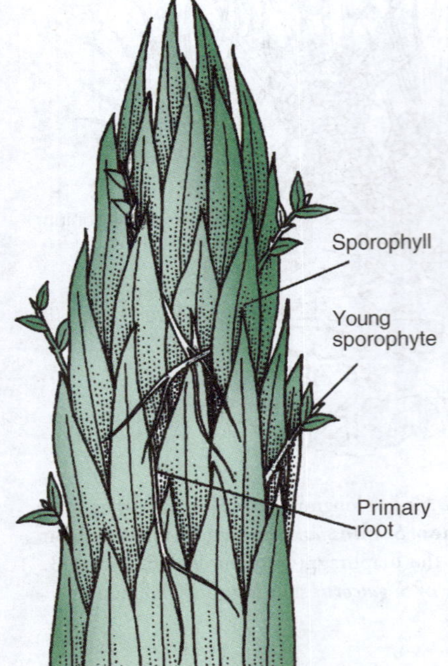

Fig. 6.19. *S. rupestris.* The young sporophytes have developed upon the strobilus of the parent plant. (After Lyon).

In ***S. rupestris*** and ***S. apus*** the megagametophytes are never shed from the megasporangium and remain on the strobilus. The megasporangium wall dehisces and the megaspore wall also splits open at the triradiate ridge to expose the archegonia. The microgametophytes enclosed within spore wall fall upon the open megasporangium, their wall bursts open and liberates the spermatozoids that enter the megasporangium and reach the exposed archegonial surface of the contained megagametophyte and effect fertilisation. In these species the oospore completes its development within the megasporangium and the young embryo grows into a seeding, develops primary root (Fig. 6.19) and then falls on the ground.

Campbell (1905) reported that in ***S. bigelovii*** which grows in the dry region of Southern California, the megaspores bearing young embryos become completely dried up after the embryo has attained some size. They remain in that state during dry season and resume growth when the favourable conditions come.

Parthenogenesis

This has been reported in ***S. atroviridis.*** In this species with unisporangiate strobili the egg develops into embryo without fertilisation. In some species with bi-sporangiate strobili the archegonial necks remain closed and do not permit the spermatozoids to enter. The egg, in these species, develops directly into embryo (***S. anacardia***, ***S. rubicaulis*** and ***S. helvetica***). ***This phenomenon where the egg develops into embryo without the act of fertilisation is called* parthenogenesis.**

S. bigelovii *S. helvetica* *S. martensii.*

Apogamy

Lyon (1904) reported the development of embryo directly from the archegonial initial. This is the case of apogamous development of embryo. Apogamy is not common in ***Selaginella.***

THE EMBRYO

The zygote is the pioneer structure of the sporophyte generation and as a consequence of a series of orderly changes develops into an embryo. The embryology in ***Selaginella*** varies slightly with species. We shall consider in detail the embryology of ***S. martensii***.

Embryology in Selaginella martensii (Fig. 6.20)

The zygote elongates in the axis of the archegonium. The first division (I—I) of the zygote is transverse (Fig. 6.20, A) forming an upper suspensor cell (hypobasal) and a lower embryonal cell (Epibasal). The embryonal cell gives rise to all the part of the embryo, e.g., stem, Leaf, hypocotyl, foot and primary rhizophore. The suspensor cell elongates and divides to form a multicellular suspensor that elongates downwards pushing the embryo deep into the food laden tissue of the gametophyte (Fig. 6.20, B). The embryonal cell divides into two equal cells by a vertical wall. (II—II). A second vertical wall at right angles to the first (Bruchmann, 1909) leads to the formation of four cells (only two are visible in one section) or the **quadrant stage**. One of the four cells divides by an oblique wall (III—III) resulting in the formation of the shoot apical cell (Fig. 6.20, B). The remaining four cells (3 + 1) divide by transverse walls to form eight cells arranged in two tiers of four cells each. The cells in both the tiers divide by vertical and transverse walls forming an undifferentiated mass of cells (Fig. 6.20 C). The cell divisions are more active in the quadrant nearer the suspensor and form a distinct mass of cells called the foot (Fig. 6.20, F). Greater growth in the foot region results in orientation of the embryonal mass at right angles to the suspensor (Fig. 6.20, E,F). Two superficial cells in two diagonally opposite quadrants (quandrants away from suspensor) differentiate as initials of the cotyledons (Fig. 6.20, D, E). They are on either side of the stem apex initial. The cotyledon initials, by their division give rise to two **cotyledons** each of which develops a **ligule** near to its base (Fig. 6.20, F). Meanwhile the apical cell of the stem has divided to form a distinct shoot apex enclosed by the cotyledons. The cells posterior to the cotyledons and the shoot apex give rise to the hypocotyl.

The root initial differentiaties quite late in the development of the embryo. A superficial cell lying on one side of the foot towards the suspensor functions as the **root initial**. So root initial in this species differentiates between the foot and the suspensor (Fig. 6.20, F). The root apical cell cuts off segments and gives rise to a structure called the primary rhizophore. The primary rhizophore comes out of the ruptured portion of the megaspore wall and grows into soil and later gives rise to the primary root.

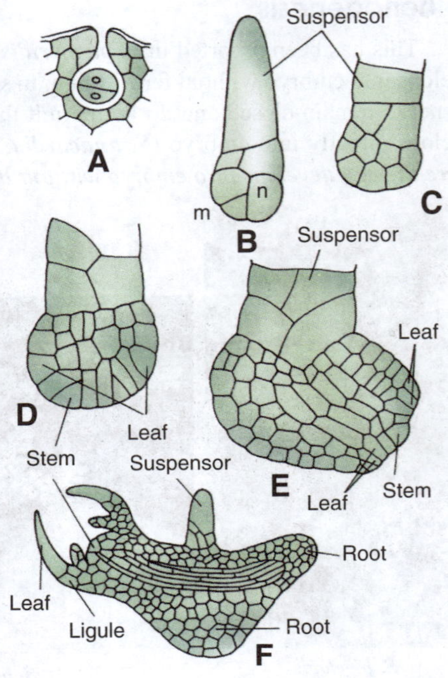

Fig. 6.20. Embryo-development in *S. martensii*

Up to this time the foot acted as a suctorial organ and absorbed food from the gametophytic tissue. After the formation of first root the young sporophyte becomes independent. By now the shoot apex has also grown out of the gametophytic tissue and the cotyledons have turned green. The young sporophyte with young stem, two cotyledons, and a primary root is now an independent individual. The stem may branch immediately or may grow unbranched for some time. The vascular elements or the stele appears first in the stem as narrow and elongated cells called the procambial strands. These cells differentiate into xylem and phloem elements. Later the stele extends into the rhizophore and the root.

In ***S. kraussiana*** the peculiar feature is the enlargement and elongation of the archegonial cavity deep into the gametophytic tissue. The archegonial cavity appears as a long tubular structure called the '**embryoschlauch**.' The two-celled embryo lies at the tip of the **embryoschlauch** where it undergoes further division. The embryonal cell gives rise to stem apex, leaf apices and hypocotyl. The foot, root, and suspensor develop from the suspensor cell. Root initial appears quite late between the foot and the suspensor. The suspensor is not well developed and is single-celled because its function is performed by the **embryoschlauch**. Embryoschlauch also develops in ***S. rubicaulis*** and ***S. galeottei*** but in this case the embryonal cell gives rise only to shoot apex and leaf initials, the root, foot and hypocotyl develop from suspensor cell. In these cases the first wall of the zygote is obliquely vertical and not transverse.

Taxonomic Position

Division	:	***Lycophyta***
Class	:	***Ligulopsida***
Order	:	***Selaginellales***
Family	:	***Selaginellaceae***
Genus	:	***Selaginella***

HETEROSPORY

Introduction

Heterospory is a condition that interprets the production of spores of two different sizes and two different developmental patterns. The two different sizes are the small and the larger spores. The former are called the **microspores** and the latter are termed as **megaspores**. They have differental development patterns because the microspores germinate to produce the male gametophytes or **microgametophytes** that bear male sex organs called the antheridia, and the megaspores germinate to produce the female gametophytes or the that bear archegonia or the female sex organs. The two kinds of spores are produced in two kinds of sporangia. The micro spores in **microsporangia** and the megaspores in **megasporangia**. The microspores are produced in large numbers and are smaller in size as compared to megaspores that are produced in comparatively much smaller number and are large in size. The large size of the megaspores is attributed to the availability of more nutrients. They are few in number per megasporangium and get more nutrition, as compared to many microspores per microsporangium. The reduction in the number of megaspores has been achieved either through reduction in number of spore mother cells, or by the disintegration of developing spores.

Importance of Heterospory

The most important aspect of heterospory is that it is an expression of sex determining process of the plant. It has brought about, along with its onset, a phenomenal shift of sex determining capacity from the gametophyte to the sporophyte. In all the homosporous individuals the sex determinants exert their effect in the gametophytes during the formation of the antheridia and the archegonia. In the heterosporous individuals the sex determinants exert their influence during sporogenesis. In ***Selaginella***, as we have seen, the stages of development of both the kinds of sporangia, 'are similar up to the formation of spore mother cells. Thereafter the differences start. In the microsporangia all the spore mother cells are functional and produce microspores, whereas in megasporangia only a few or one mother cell is functional. Heterospory has also freed, during the course of evolution, the sporophyte from aquatic habitats, and such an emancipation, has enabled the sporophyte to grow freely under varied environmental conditions. It is, therefore, a significantly evolutionary development during the struggle to conquer land.

Heterospory is the most important evolutionary development in the vascular plants because it has ultimately to lead to seed developments. It is rather a pre-requisite to seed habit. Besides bringing along with it the phenomenon of pollination, it has also resulted in the reduction of the gametophytic tissue, their endosporic nature, precocious germination of the spores, partial and ultimately complete retention of the megasporangium and female gametophyte on the sporophyte, reduction in the number of megaspores to one; reduction in the number of male gametes, and ultimately loss of mobility of the male gametes in angiosperms and organic union between the megaspore and the megasporangium. All these new developments are vital for seed habit. These changes have gradually developed in vascular plants and led to seed habit.

Origin of Heterospory

The heterospory can be better discussed if we examine evidences from **palaeobotany**, from **developmental studies** and from the **experimental results.**

(*a*) Evidence from Palaeobotany

The available fossil record suggests that there was an early and widespread occurrence of heterospory in almost all the major plant groups. A number of heterosporous genera belonging to the **Lycopsida**, **Sphenopsida** and **Pteropsida** were known in the late Devonian and early Carboniferous periods. Some of them exhibit stages where heterospory is not pronounced whereas in others heterospory was well developed and approached the seed condition. These various grades of heterospory in these early land plants advance strong suggestions as to the origin of the heterosporous condition. Williamson and Scott (1894) discovered and described two species of **Calamostachys** (**Sphenopsida**) that indicate the initial steps that have led to heterospory. These species are ***C. binneyana*** and ***C. casheana***. The former species is homosporous but in some sporangia there were a few spores of unequal sizes. ***C. casheana*** showed distinct heterospory as there were microspores and megaspores

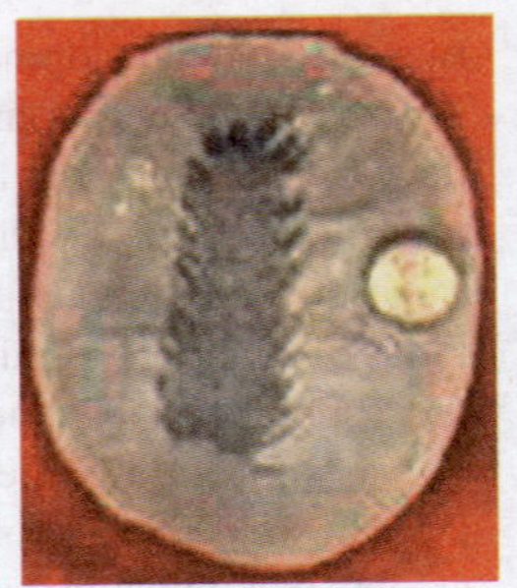

Lepidostrobus braidwoodensis

produced in separate sporangia. The megasporangia contained small and aborted spores. This shows that abortion of spores leads to the differences in size and number. Thoday (1906) described ***Sphenophyllum dawsoni*** (**sphenopsida**) which has usually been considered homosporous. Thoday found that some sporangia in this species showed abortion of some spore tetrads and the remaining spores grew larger in size than the spores usual for the species. In ***Stauropteris burntislandica***, Chaloner (1958) reported that the megasporangia contained tetrads of megaspores in which two spores were large and two small. This is also a case of reduction in the number of megaspores due to abortion. Scott (1901) reported that in ***Lepidocarpon*** (Lycopsida) three megaspores out of each tetrad had aborted and only one matured. Ramanujam and Stewart (1969) also reported a ***Lepidocarpon*** cone from Pennsylvanian of Illinois in which there are 3 abortive megaspores and one functional megaspore. In ***Lepidostrobus braidwoodensis***, Arnold (1938) reported that the megasporangia contained only one matured megaspore in addition to a number of aborted spores. In ***Miadesmia*** the seed like megasporangia contained only one megaspore and there was trace of aborted spores.

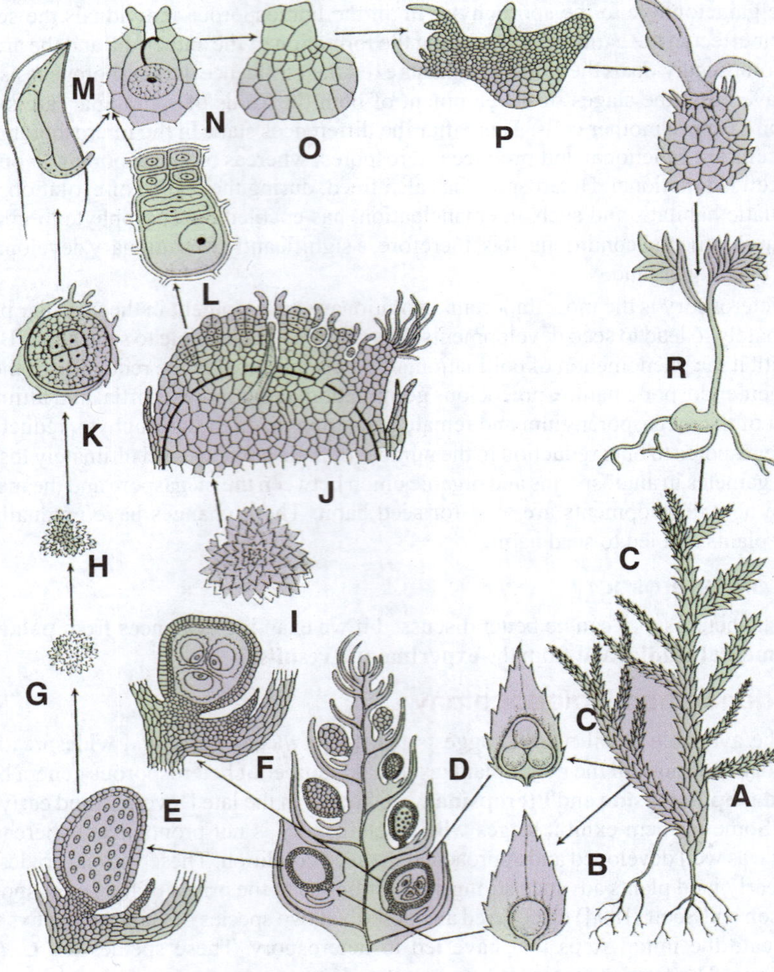

Fig. 6.21. *Selaginella.* A diagrammatic representation of life cycle.

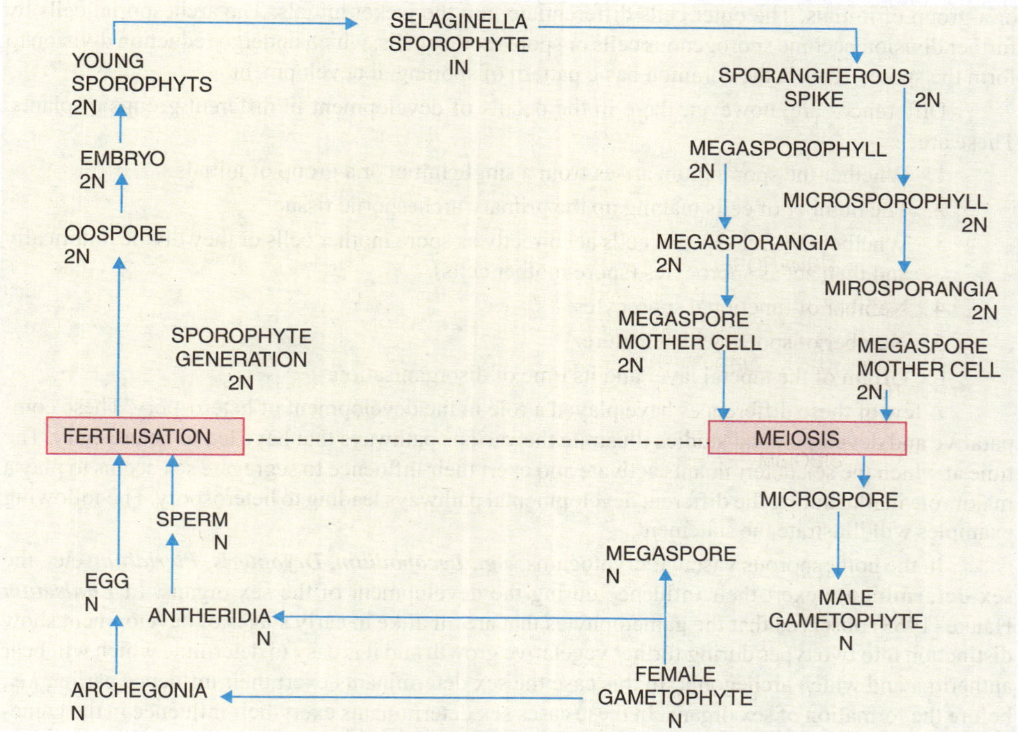

SELAGINELLA SPOROPHYTE IN

YOUNG SPOROPHYTS 2N

SPORANGIFEROUS SPIKE 2N

EMBRYO 2N

MEGASPOROPHYLL 2N

MICROSPOROPHYLL 2N

OOSPORE 2N

MEGASPORANGIA 2N

MIROSPORANGIA 2N

SPOROPHYTE GENERATION 2N

MEGASPORE MOTHER CELL 2N

MEGASPORE MOTHER CELL 2N

FERTILISATION

MEIOSIS

MICROSPORE

SPERM N

MEGASPORE N

EGG N

ANTHERIDIA N

MALE GAMETOPHYTE N

ARCHEGONIA N

FEMALE GAMETOPHYTE N

Fig. 6.22 Graphic life cycle of *Selaginella.*

These examples give us an idea of the reduction in the number of spores and differences in their sizes but whether it was achieved through abortion of the spores or through reduction in the number of spore mother cells, can not be determined as there is no information regarding the development of the sporangia and the spores. We have no developmental record to show that these spores produced heterothallic gametophytes. In the absence of such an information, it is not possible to shed light on the importance of the difference in the size of the spores, and whether these spores actually produced two kinds of gametophytes. Baxter (1972) discovered a heterosporous fern from a coal ball from Middle Pennsylvanian of Kansas. He found small pinnules that bear stalked pendant sporangia that appeared like blunt arrowheads in a median longitudinal section. Each sporangium has one megaspore and is attached to the lower surface of the pinnule. A marginal indusium covers the sporangia partially. The megaspore completely fills the sporangium.

(*b*) Developmental Evidence

The living representative of the major divisions of the vascular cryptogams (**Lycopsida**, **Sphenopsida**, **Pterospsida**) include a number of heterosporous genera (***Selaginella***, ***Isoetes***, ***Marsilea***, ***Pilularia***, ***Regnellidium***, ***Salvinia***, ***Azolla*** and ***Stylites***). They afford us the possibilities of extensive morphological, developmental and comparative studies that can throw much light on the origin of heterosporous condition in the plants. The development of the sporangium in all the living plants follows a common developmental pattern. In all cases archesporium is formed by the lower cell or cells cut off from the sporangial initial

Regnellidium

or a group of initials. The outer cells differentiate into the jacket initials. The archesporial cells by further division become sporogenous cells or spore mother cells, which undergo reduction division to form the spores. This is the common basic pattern of sporangial development.

Differences are, however, there in the details of development in different groups of plants. These are:

1. Whether the sporangium arises from a single initial or a group of initials.
2. The number of cells making up the primary archesporial tissue.
3. Whether the archesporial cells act directly as spore mother cells or they divide mitotically and then act as sporocytes (spore mother cells).
4. Number of functional sporocytes.
5. Number of spores which mature.
6. Origin of the tapetal layer and its time of disorganisation.

A few of these differences have played a role in the development of heterospory. These comparative and developmental studies illustrate the various pathways that have led to heterospory. The time at which the sex determinants activate and exert their influence to segregate sex seems to play a major role in illustrating the different developmental pathways leading to heterospory. The following examples will illustrate the statement.

In the homosporous vascular cryptogams, ***e.g.***, ***Lycopodium***, ***Dryopteris***, ***Pteridium***, etc., the **sex determinants** exert their influence during the development of the sex organs. In ***Equisetum*** Hauke (1963) observed that the gametophytes that are all alike in early stages of development show distinction into two types during further vegetative growth and it is easy to determine which will bear antheridia and which archegonia. In this case the sex determinants exert their influence earlier, ***i.e.***, before the formation of sex organs. In these cases sex determinants exert their influence in the gametophytes, their time, however, differs. In ***Selaginella***, which is a heterosporous vascular cryptogam the sex determinants exert their influence during the differentation of the sporocytes (spore mother cells). In the sporangia destined to develop into megasporangia only one megasporocyte is functional, whereas in the microsporangia all the sporocytes are functional. From this stage onwards there is clear differentiation between the two types of sporangia that will produce two types of spores that in turn develop into male and female gametophytes. The same is the case in ***Marsilea*** and ***Isoetes***. This is a significant drift in the time of activation of sex determinants. It has shifted from the gametophyte to the sporophyte. It is a small shift, but very significant, because it has brought along with it the condition of heterospory in the vascular cryptogams.

In the angiosperms the stamens (microsporophylls) and carpels (megasporophylls) appear quite different at maturity so that we can tell in advance that the stamens will bear microsporangia and the carpels will bear ovules that contain megasporangia. Here sex determinants exert their influence before the maturation of the sporophylls. In many gymnosperms and some angiosperms the reproductive bodies are borne on special branches which can be identified before they bear the reproductive bodies. This can be ascertained by their position and morphology. In these cases the sex determination occurs during the vegetative growth of the branch. So in the seed plants the sex determinants exert their influence much earlier than in the heterosporous pteridophytes.

These examples indicate that the phenomenal shift of sex determinants from gametophyte to sporophyte has accompanied heterospory.

A case of incipient heterospory has been discovered in a fern called ***Platyzoma microphylla*** (Tryon, 1964). In this fern the sori contain two types of sporangia that are intermingled. In one type there are 16 large spores and in the other 32 small spores. The smaller ones germinate to produce male prothalli whereas the larger ones produce female prothalli. But the female prothalli in case of failure of fertilization may bear antheridia at a later stage. Incipient heterospory is also observable in ***Ceratopteris thalictroides*** (Schedlbauer, 1972). Multispore cultures of this species produce two types of prothalli. These are spathulate male prothalli and cordate bisexual prothalli. Such a distinctness in the genders is due possibly to the effect of an antheridiogen.

(c) Experimental Evidence

Only a few experimental studies have been conducted to approach the problem of origin of heterospory. All these studies were conducted with the basis that differences in spores size are due to abortion of most of the spores, greater abortion leaving greater nutrition for the remaining and functional spores, that grow in size. Shattuck (1910) performed a series of experiments on ***Marsilea*** and tested the effect of nutrition on spore size. He was able to alter spore size by growing plants under variable conditions of light, temperature and nutrition. He found that in plants growing under favourable circumstances the microsporangia contained a number of aborted microspores. The functional spores enlarged, grew larger in size, and it was found that spore enlargement was proportional to spore abortion. He found that microsporangia showing microspore abortion developed spores that were 16 times larger than their original size. In extreme cases of abortion only a single spore survived and looked like a megaspore and showed all the structural characteristics of a megaspore. Under unfavourable conditions Shattuck (1910) was able to induce the formation of the large number of smaller spores in the megasporangia. He attributed these induced changes in the spore size to be due to change in the nutritional state of the sporangia. It was not possible for him to germinate these altered spores and therefore no conclusive results could be achieved. In the absence of such a conclusion these experimental studies cannot throw any light on the origin of heterospory.

Variations in the nutritional environments of the sporangia may alter the pattern of spore formation, but it appears that it is something else in the cellular controls system that regulates the differentiation of spores into two different sexual entities. It depends upon factors in the genetic system of the cell.

HETEROSPORY AND SEED HABIT

Heterospory is considered to be a pre-requisite to seed formation. In addition to heterospory the seed plants show the following characteristics that have led to the formation of the seed.

(*i*) Reduction in size of the male gametophyte.

(*ii*) Formation of one megaspore within a megasporangium or the nucellus, as we call it in the seed plants.

(*iii*) The megasporangium is surrounded by an additional protective coat called the integument.

(*iv*) The megaspore is never shed and is permanently retained within the megasporangium. Fertilization and formation of embryo takes place ***in situ***.

(*v*) Histological union between the megaspore and the megasporangium (nucellus). In the seed plants the megaspore is never free from the nucellar cells. They have a complete organic fusion, which is lacking in ***Selaginella*** and all other cryptogams.

Selaginella, no doubt, illustrates an example of heterosporous vascular cryptogams, that approach seed habit because of the following notable characteristics :—

1. It is heterosporous.
2. The megaspores start germination within the megasporangia and their time of release from the megasporangia varies with species.
3. The number of megaspores in ***S. rupestris*** and ***S. monospora*** is reduced to one.

S. rupestris

4. In. S. ***rupestris*** the megaspore is never shed and fertilisation and development of embryo up to the formation of rhizophore, stem and cotyledons takes place while the megaspore is enclosed within the megasporangium, which retains its connection with the parent plant. This condition can be linked to **Vivipary** in some angiosperms.

After considering the above listed points we can, however, reach conclusion that the heterosporous vascular cryptogams like ***Selaginella*** and ***Marsilea*** fail to develop seeds because :—

1. They have no protective structure like the integuments surrounding their megasporangia.
2. The retention of megaspores permanently within the megasporangia has not become established.
3. Histological union between the megaspore and the megasporangium is absent (Martens, 1966).
4. Lack of resting period after the development of embryo.

The megasporangium of ***Selaginella*** is comparable to the nucellus of ***Pinus*** and other seed plants, the contained megaspore to the megaspore or the embryo sac cell of ***Pinus***. The endosperm in ***Pinus*** is comparable to the female prothallus of ***Selaginella.*** Both have archegonia with eggs contained in their venters. In ***Selaginella*** only integuments are absent. The sporangial wall of ***Selaginella*** is two layered and could act as a protective structure. As a matter of fact the outermost wall of megasporangium in ***Selaginella*** becomes thick and sufficiently strong at maturity.

Regarding the retention of megaspore within the megasporangium, it has also been achieved by ***S. rupestris***. This feature could not become common to all the species because there is no histological union between the megaspore wall and the wall of the megasporangium. In the presence of such a union, the megaspore could never come out of the megasporangium. Such a union if present in ***Selaginella***, could also usher in an era of rest for the embryo after its development. In such a case the whole megasporangium could have fallen down from the plant and a megasporangium with contained embryo could easily be regarded as a primitive seed. In case we find a living example of a seed plant in which there are traces of separation of megaspore from the megasporangium we can regard it as a clue to the origin of seed habit. Martens (1966) has worked on these lines and has arrived at a more or less safe conclusion. Weterkeyn and Sloover (1962) conducted fluorescence microscopy on microspores of certain seed plants and megaspores of pteridophyta and found that a deposition of callose is responsible for isolating the spores from tetrads and from sporangial wall. This is called megaspore mother cell callose or **megasporocyte callose.** Such a callose formation is absent in the seed plants. They found it to be present in traces in one member of the cycadales (***Encephalartos***). In this case there is a little separation between megaspore and nucellus (megasporangium). Such a traces is lost in other groups of gymnosperms. As for the fossils no membrane analysis of the megasporocytes has been made. It can be presumed that such an isolation mechanism of megaspores from the megasporangia might have been present in the Pteridosperms.

Chromosome Number

Manton (1950) has reported n = 9 as the basic chromosome number for the genus. Such low basic number of chromosomes and their small size is quite characteristic for the genus. The chromosomes measure 1μ or even less in length.

Economic Importance

The genus has no economic importance except that some species are grown as ornamentals due to the lush green foliage that in some species gives beautiful tints such as bronze or blue. ***S. botryoides*** yields a medicine that has some curative values for the liver diseases.

QUESTIONS

ESSAY TYPE

1. What is heterospory and seed habit? Explain it with the help of *Selaginella.*
2. Give a comparative account of Reproductive structures of *Selaginella* and *Equisetum*
3. Draw neat well labelled diagrams of longitudinal sections of strobili of *Lycopodium*, *Selaginella*, and *Equisetum.*
4. Describe the life cycle of *Selaginella,* or *Equisetum* with the help of only neat and labelled diagrams.
5. (*a*) Compare the gametophytes of *Selaginella*, *Equisetum* and *Adiantum.*

   (*b*) Give a comparative account of the gametophytes of *Lycopodium* and *Selaginella.*
6. (*a*) Explain the life cycle of *Selaginella* with the help of neat and labelled diagrams. **(Rohilkhand, 2000)**

   (*b*) Describe various stages of the life cycle of *Selaginella.*
7. Draw neat and labelled diagrams only to show the life cycle of any heterosporous pteridophyte.
8. With the help of labelled diagrams, explain the structure of strobilus of *Selaginella* and *Equisetum.*
9. What is heterospory and seed habit? Name the plant having this condition and describe its life cycle briefly.
10. Give a comparative account of anatomical structure in the stem of *Selaginella* and *Equisetum.* **(Meerut, 2000; Allahabad, 2000)**
11. Describe development of sporophyte in *Selaginella.*
12. What is heterospory? Describe seed habit in *Selaginella.*
13. Draw labelled sketches of the spore bearing organs of *Selaginella*, *Equisetum* and *Adiantum* and point out the respective differences between them. **(Kerala, 2001)**
14. With labelled diagrams only compare the structure of spore bearing organs of *Lycopodium*, *Selaginella* and *Adiantum.* **(Lucknow, 2001)**
15. Give an illustrated account of the malegametophyte of *Selaginella.* **(Lucknow, 2001)**
16. Give an illustrated account of the development of female gametophyte of *Selaginella.*
17. What is heterospory? Illustrate your answer with reference to the life history of *Selaginella*, Discuss its significance. **(Poorvanchal, 2000)**
18. Draw labelled diagrams of the spore bearing organs of *Lycopodium*, *Selaginella*, *Equisetum* and *Adiantum* and point out the respective differences between them.
19. Give a detailed account of the gametophytic phase in *Selaginella.*

    or

    Give a brief account of gametophyte of *Lycopodium* or *Selaginella.*
20. Write an essay on Heterospory and origin of Seed habit.
21. (*a*) Write a comparative account of the structure of cones of *Lycopodium* and *Selaginella.*

    (*b*) Compare the structure or and organisation of strobili in *Lycopodium* and *Selaginella* with the aid of well labelled diagrams.
22. By means of labelled sketches only, illustrate the structure of the spore bearing organs of *Selaginella* and *Marsilea.*
23. Describe the structure of archegonia of *Lycopodium*, *Selaginella*, *Equisetum*, *Marsilea*, and *Pteridium* and discuss their evolution.
24. With labelled sketches only illustrate the stracture of T. S. of the stems of *Lycopodium, Selaginella*, *Equisetum*, and *Marsilea.*
25. Discuss the features of special interest found in *Selaginella.*
26. Define Heterospory? Illustrate you answer by describing the life history of a plant studied by you. In what way it is considered advanced our homospory?
27. What is heterospory and seed habit? Describe the life history of a plant showing such type of phenomenon. **(Poorvanchal, 2002)**
28. Give an account of internal structure and nature of Rhizophere. **(Poorvanchal, 2002)**
29. Describe the structure of strobilus of *Selaginella* with neat labelled diagrams. **(Bangalore, 2001)**

30. Describe the spore bearing organs of *Selaginella.* (Nagpur, 2002)

31. What is meant by heterospory? Discuss the phenomenon with reference to *Selaginella* and state how seeds might have developed from heterosporous condition. (Gauhati, 2000)

32. Give an account of the female gametophyte of *Selaginella* with the aid of diagrams.

33. Give an account of the stelar structure of stem in *Lycopodium* and *Selaginella.*

34. Describe the morphology of leaves in *Lycopodium*, *Selaginella*, *Equisetum* and *Pteris.*

35. What is heterospory? Describe sexual reproduction in *Selaginella.* (Kerala, 2000)

36. Discuss the modes of reproduction in *Selaginella*.

37. Explain the development of Seed habit in *Selaginella.*

38. Draw graphic life cycles of *Selaginella* and *Pteris.* List the major differences both in the sporophytic and gametophytic phases of the two.

39. With the help of labelled diagrams only, describe the life cycle of *Selaginella.*

40. With the help of diagrams describe the development of archegonia in *Selaginella*. (Kanpur, 2002)

41. Describe the characters of *Selaginella* which show its habit. (Kanpur, 2002)

42. Write an essay on *Heterospory.*

43. Describe the internal structure of *Selaginella* stem. (Mysore Univ., 2004)

44. Give an account of structure of strobilus in *Selaginella*. (Nagpur Univ., 2005)

45. Give an account of structure of strobilus in *Selaginella.* (Nagpur Univ., 2005)

46. Write an essay on structure of cones and reproduction in *Selaginella*. (M.S. Univ., Tirun., 2005)

47. (*a*) Explain the different types of vegetative reproduction in *Selaginella.*
    (*b*) Write the biological significance of heterospory in *Selaginella.* (M.S. Univ. Tirun., 2005)

SHORT ANSWER TYPE

48. Write short notes on
    (*i*) Rhizophore (UP College, 2000; Kumaon, 2000; Allahabad, 2000, Bangalore 2001, Kerala, 2000)
    (*ii*) Heterospory and Seed habit (Kumaon, 2000; Lucknow 2001; Himachal Pradesh, 2000).
    (*iii*) (*a*) Heterospory., (Meerut, 2000)
    (*b*) Importance of heterospory. (Garhwal 2000)
    (*iv*) Seed habit in *Selaginella.*
    (*v*) Ligule. (Bangalore, 2001)
    (*vi*) Morphological nature of Rhizophore of *Selaginella.*
    (*vii*) Importance of heterospory.
    (*viii*) Endodermis of *Selaginella.* (Lucknow, 2000)
    (*ix*) Male gametophyte of *Selaginella.*
    (*x*) Development of male gametophyte of *Selaginella* and its nature.
    (*xi*) Structure of strobilus in *Selaginella.*
    (*xii*) Cones of Selaginella.
    (*xiii*) Heterospory in *Selaginella.* (Lucknow, 2000)
    (*xiv*) Origin of heterospory, (Lucknow, 2000)

49. Why the gametophyte of *Selaginella* sp. is said to be reduced?

50. Give two reasons to prove that Rhizophore of *Selaginella* is a stem. (Poorvanchal, 2000)

51. Comment on Endodermis of *Selaginella* in only ***thirty words.***

52. Rhizophore of *Selaginella* represent roots as well as stems. Discuss.

53. Mention the seed habit of *Selaginella* by comparing it with *Cycas* or *Pinus*.

54. Mention at which stage the male gametophytes of *Selaginella* are released from the microsporangium?

55. In *Selaginella* how many celled male gametophyte the microspores have at the time of liberation from the microsporangia?

56. What changes are required to make *Selaginella* a seed plant? (Lucknow, 2001)

57. Draw well labelled diagrams of the following (No description needed):
(*i*) T. S. Stem of ***Selaginella.*** **(Poorvanchal, 2000; Garhwal, 2000; Allahabad, 2003)**
(*ii*) Antherozoids of ***Selaginella***, ***Equisetum*** and ***Marsilea.***
(*iii*) L. S. ***Selaginella*** Cone (Strobilus) **(UP, College, 2000; Allahabad, 2000, 2003, 2004)**
(*iv*) Rhizophore of ***Selaginella.***
(*v*) L. S. Spike of ***Selaginella.***
(*vi*) V. S. Megagametophyte of ***Selaginella.***
(*vii*) T. S. Rhizophore of ***Selaginella.***
(*viii*) Megasporangium of ***Selaginella.***
(*ix*) Female gametophyte of Selaginella.
(*x*) Strobilus of ***Selaginella.***

58. With the help of diagrams, differentiate between:
(*a*) Rhizophore and Root. **(Rohilkhand, 2000; Kerala, 2000)**
(*b*) Strobilus of ***Selaginella*** and ***Equisetum.***
(*c*) Gametophytes of ***Selaginella*** and ***Adiantum.***
(*d*) ***Selaginella*** root and ***Equisetum*** Root.
(*e*) Monostelic and polystelic stems of ***Selaginella.*** **(Allahabad, 2001)**
(*f*) Cones of ***Lycopodium*** and ***Selaginella.***
(*g*) Leaves of ***Lycopodium*** and ***Selaginella.***
(*h*) Spore producing organs in ***Isoetes*** and ***Selaginella.***

59. Which pteridophyte has trabeculated endodermis?

60. Write classification of ***Selaginella***.

61. In which plant ligule is found and where?

62. From where will you collect ***Selaginella***?

63. What changes would be necessary in the megasporangium of ***Selaginella*** to convert it into a possible seed?

64. (*a*) Name a pteridophyte which has trabeculated endodermis.
(*b*) In which plant and where do you find a ligule

65. Write special features of ***Selaginella*** Sporophyll.

66. How many prothallial cells are present in the male gametophyte of ***Selaginella*** and ***Pinus?*** **(Allahabad, 2001)**

67. Name two heterosporous forms in the pteridophytes studied by you.

68. What is ligule?

69. What are the two types of spores produced in ***Selaginella***?

70. Mention one difference between female gametophytes of ***Selaginella*** and ***Marsilea***.

71. Mention the name of pteridophyte in which rhizophore is present?

72. What is ligulate sporophyll?

73. Name the plant showing trabeculated endodermis **(Poorvanchal, 2002)**

74. Give a brief account of the male gametophyte of ***Selaginella.*** **(Lucknow, 2001)**

75. Describe the nature of Rhizophore. **(Kanpur, 2001)**

or

What is the morphological nature of Rhizophore?

76. What is parthenogenesis? Name one Species of ***Selaginella*** where it occurs. Write a short note on legule **(Himachal Pradesh, 2000)**

77. Comment on the occurrence and significance of pteridophytes. **(Calicut, 2001)**

78. Give an account of the progressive reduction in megasporangia of ***Selaginella***

79. Give two important points of ***Selaginella*** Stem and ligule.

80. Give the structure of rhizophore in ***Selaginella.***

81. Why ***Selaginella*** is said to be heterosporous?

82. Describe the stele found in ***Selaginella.***

83. Explain the evolutionary significance of heterospory.

84. Give two important features of leaves of ***Selaginella*** and rhizophore of ***Selaginella.***

85. At what stage microspores of ***Selaginella*** are liberated?

86. (*a*) Name any species of ***Selaginella*** where megaspores are never released from the megasporangium.

(*b*) Which cells are modified into trabeculae in the stem of ***Selaginella*** ?

(Allahabad, 2003)

87. Write short notes on leaf of ***Selaginella.***

(Garhwal Univ., 2005)

OBJECTIVE TYPE

88. Fill in the blanks :

(*i*) Trabeculated endodermis is found in ________ stem.

(*ii*) Male gametophyte of ***Selaginella*** has undergone extreme reduction and has only ________ cells.

(*iii*) Heterospory is noticed in ________ and ________.

(*iv*) The structure present in between microsporophyll and microsporangum of ***Selaginella*** is ________.

(*v*) Ligule is present in the leaf of ________.

(*vi*) Both micro and macrosporangia are produced in the ________ of sporophylls.

(*vii*) Sperms in ***Selaginella*** are ________.

(*viii*) In mature archegonium of ***Selaginella***, the neck consists of ________ cells each.

(*ix*) Heterospory means presence of ________ types of spores.

(*x*) A number of species of ***Selaginella*** have been reported from ________ Himalayas.

89. Select the correct answer:

(*i*) Stele in ***Selaginella*** is

(*a*) Protostele
(*b*) Siphonostele
(*c*) Solenostele
(*d*) Dictyostele.

(*ii*) Trabeculae in ***Selaginella*** is modification of

(*a*) Pericycle
(*b*) Endodermis
(*c*) Pericycle and Endodermis
(*d*) Cortex.

(*iii*) ***Selaginella*** is a member of

(*a*) Eligulopsida
(*b*) Ligulopsida
(*c*) Calamopsida
(*d*) Leptosporangiopsida.

(*iv*) In ***Selaginella*** the male gametes are

(*a*) Aflagellate
(*b*) Uniflagellate
(*c*) Biflagellate
(*d*) Multiflagellate.

(*v*) Endodermis is trabeculated in

(*a*) ***Selaginella***
(*b*) ***Equisetum***
(*c*) ***Marsilea***
(*d*) ***Lycopodium.***

(*vi*) In ***Selaginella***, the stem is

(*a*) herbaceous and branched
(*b*) herbaceous and unbranched
(*c*) Woady and branched
(*d*) None of the above.

(*vii*) The sporangia in the sporophylls of ***Selaginella*** are

(*a*) homosporous
(*b*) borne in the axil
(*c*) borne at the tip
(*d*) absent.

(*ix*) Reduction Division in the life cycle of ***Selaginella*** occurs during

(*a*) germination of zygote
(*b*) formation of spores
(*c*) germination of the spore
(*d*) both during germination of zygote and spore.

(*x*) Heterospory is found in

(*a*) ***Selaginella***
(*b*) ***Marsilea***
(*c*) ***Pinus***
(*d*) All of the above

(*xi*) Dimorphic leaves are present in

(*a*) ***Selaginella***
(*b*) ***Cycas***
(*c*) ***Pinus***
(*d*) All the above

(*xii*) Which of the following is a member of ***Ligulopsida*** ?

(*a*) ***Pteridium***

(*b*) ***Selaginella***

(*c*) ***Equisetum***

(*d*) ***Lycopodium***

(*xiii*) Which type of Antherozoids are present in ***Selaginella***?

(*a*) Aflagellate

(*b*) Uniflagellate

(*c*) Biflagellate

(*d*) Multi flagillate

(*xiv*) Ligule is found in

(*a*) Vegeatative leaves

(*b*) Sporophylls

(*c*) Rhizophore

(*d*) Root

Sphenophyta (Benson, 1957)

Equisetum hyemale

- Distinguishing features
- Classification
- Description of ***Equisetum*** Linn
- Classification of the genus
- Economic importance

Distinguishing features

1. Sporophyte is well differentiated into roots, stem and whorled leaves.

2. Stem jointed, solid or medullatid and protostelic. Few are with secondary thickening.

3. Plant homosporous or heterosporous, sporangia are thickwalled borne in reflexed position on spornagiophores.

4. Sporangia are arranged in whorls.

5. Multiflagellate antherozoids are present.

The division Sphenophyta is also known by two other names. One of these **Arthrophyta** is the oldest name. Smith (1955) suggested the name **Calamophyta**. The division has been classified as follows:

A. Class :—Sphenophyllopisda

It includes only one order. They are all extinct.

Order **Sphenophyllales.** The representatives of the order are distributed among two extinct families. They appeared on this earth in the Upper Devonian and persisted till the lower Triassic.

Family 1. **Sphenophyllaceae.** It includ the genus ***Sphenophyllum***, ***Sphenophyllostachys***, ***Bowmanites*** and ***Eviostachya.*** All of them are extinct.

Family 2. **Cheirostrobaceae.** It is represented by the genus ***Cheirostrobus*** that was discovered in the Lower Carboniferous of Scotland. It was a large cone measuring 3.5 cm. in diameter.

B. Class :—**Calamposida.** It includes three orders :

Order 1. **Calamitales.** This order was at its peak of development during the Upper Carboniferous along with the Lepidodendrales and were found in the coal measure and swamp forests. The order became extinct at the end of Permian. It includes two families.

Family 1. **Asterocalamitaceae.** It includes the genera ***Asterocalamites*** and ***Archaeocalmites.*** They are extinct.

Family 2. **Calamitaceae.** It includes the genera ***Calamites***, ***Protocalamites***, ***Annularia, Asterophyllites, Protocala-mostachys,*** and ***Palaeostachya.***

Order 2. **Hyeniales.** They are represented both in the Lower and Middle Devonian and include two families.

Family 1. **Protohyeniaceae.** It includes the genera ***Protohyenia*** that was discovered by Ananiev (1957) from the Lower Devonian rockes of Siberia.

Family 2. **Hyeniaceae.** It includes the genera ***Hyenia*** and ***Calamophyton.***

Equisetites - The extinct genus

Order 3. **Equisetales.** It includes the family **Equisetaceae** that is represented by one extinct and one living genus. ***Equisetites*** is the extinct genus and ***Equisetum*** is the only living genus of the division **Sphenophyta**.

EQUISETACEAE

The family includes a single living genus ***Equisetum***, which is the only living representative of the entire race, that has become extinct. The life history of ***Equisetum*** is discussed in detail in this chapter.

EQUISETUM Linn.

Distribution

Commonly known as the **horsetails,** the twenty-five species of ***Equisetum*** are widely distributed and reach luxuriant development in the North Temperate zone (N. America, Eurasia). Some occur in tropical regions (W. Indies, S. America, Chile); rarely in arctic and alpine zones. ***E. arvense*** is a cosmopolitan species and extends into S. Africa. It grows at higher altitudes on the Himalayas in India (Narkanda near Simla). ***E. arvense*** can assume many different growth forms

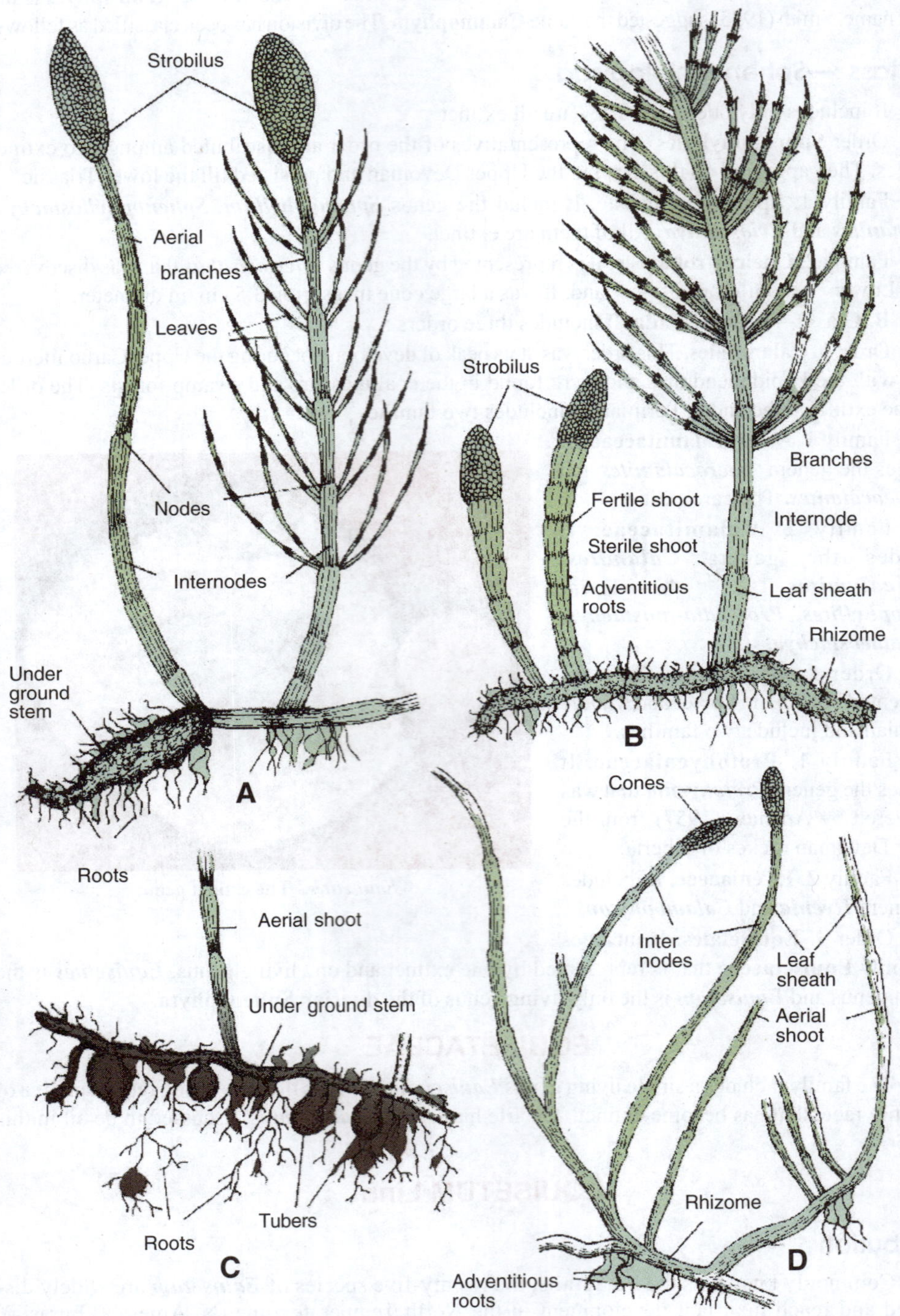

Fig. 7.1. (A – D) *Equisetum*. A. *Equisetum diffusum*; B. E. *arvense*; C. E. *arvense* rhizome with tubers; D. E. *debile*.

but according to Hauke (1966) these are only superficial modifications and do not deserve the status of varieties. The genus is absent in New Zealand and Australia. In India the common species include ***E. debile***, ***E. diffusum***, ***E. arvense***, ***E. robustum, E. palustre*** and ***E. ramosissimum.*** Clarke (1880) recorded ***E. diffusum, E. arvense, E. debile*** and ***E.elongatum*** (***E. ramosissimum***) from northern India. ***E. diffusum*** and ***E. ramosissimum*** are very widely distributed throughout the Himalayas from Simla eastwards to Darjeeling between 1,500-2100 metres above sea level. ***E. debile*** is of very wide distribution throughout India and is found in abundance in the plains of Punjab, U.P., Bengal, Madhya Pradesh, South India and extends to Ceylon and tropical Asia. ***E. arvense*** and ***E. palustre*** are restricted to western Himalayas. ***E. arvense*** is abundantly met with near the snow line and extends upto. Kumaon. In Simla hills it grows at Narkanda. Chowdhury (1973) recorded five species from the upper Gangetic plain. ***E. debile*** is also found in Malaya and some Pacific islands. According to Hauke (1963) there are only 15 species and five hybrids. A fossil form was known in Cretaceous times.

Taxonomy

The genus has traditionally been divided into two groups : **Eu-equisetum** (horsetails) and **Hippochaete** (Scouring rushes). The species belonging to the first group have stomata in line with the epidermis whereas the species in the second group have sunken stomata. Baker (1887) regarded **Hippochaete** as a sub-genus. Milde (1865, 1867), Farwell (1916), Rothmaler (1944), Love ***et al*** (1977) and Ching (1978) recognised Hippochaete as a genus. Manton (1950) clearly demonstrated the cytological differences between ***Equisetum*** and ***Hippochaete.***

Hauke (1963) in his monograph maintained a sub-generic status for ***Hippochaete.*** This status has also been supported by Indian workers (Bir, 1978) and Pichi-Sermolli (1977). So the family Equisetaceae is monogeneric and the single genus ***Equisetum*** consists of two sub-genera as ***Equisetum*** proper (***Eu-equisetum***) and ***Hippochaete.***

Taxonomy of the Indian horsetails is not yet fully understood, "and also it is not fully clear as to whether, ***E. robustum*** and ***E. maximum*** actually occur in India or not. Specific status of ***E. debile*** is still questionable". Hauke (1961, 1962) has opined that ***E. debile*** should better be recognised as a sub-species of ***E. ramosissmum*** because the 'typical' ***debile*** is differentiated from 'typical' ***ramosissimum*** by short, smooth and cylindrical truncated sheaths, individual endodermis around each vascular bundle and stomata in only one line. He recognised two names as ***E. ramosissimum*** sub-sp. ***ramosissimum*** and ***E. ramosissimum*** sub-sp. ***debile*** (Hauke, 1962, American Fern Journal 52:33).

E. robustum

Habitat

Equisetum grows under varied habitat, but thrives well under damp and amphibious conditions. ***E. debile*** grows under hydrophytic, mesophytic as well as xerophytic conditions. This capacity of the plants has enabled the genus to survive the test of time and live up to the present age. Commonly they are found along the banks of rivers, streams (***E. debile, E. pratense***) and swampy places (***E. palustris***). ***E. arvense*** grows in open grass lands, railway embankments, exposed sandy and dry places and at various altitudes on hills.

Silica is a common constituent of the epidermal cell walls of almost all the species Presence of silica gives a rough and abrasive touch to the stem.

THE SPOROPHYTE

The plant body of ***Equisetum*** represents the sporophyte generation and is differentiated into roots, stem and leaves. It will be described under the following headings and sub-headings.

Habit

The genus includes species that show a great variation in size. They are all herbaceous and possess an underground, perennial, and much branched rhizome from whose nodes arise a number of aerial branches (Fig. 7.1). Both the rhizome and the aerial branches present a jointed appearance due to the presence of distinct nodes and internodes. The aerial shoots usually die during unfavourable periods of the year and are hence annual but in some species they are perennial. Regarding size, the smallest species (***E. scripoides***) has a few inches high aerial shoot with 1 to 1.5 mm. across ; the largest (***E. giganteum***) has 6-13 metres tall aerial stems with 2.5 cm. across and are unable to stand erect. ***E. schaffneri*** is stout with 2.5 metres high stem and 10 cm. in diameter. The common Indian species ***E. debile*** is 1-2 feet high under xeric conditions but grows upto 2-5 metres tall under damp and shaped environs. In tall species the stem is weak (1/3 to 3/4 cm. across). ***E. arvense*** is medium sized.

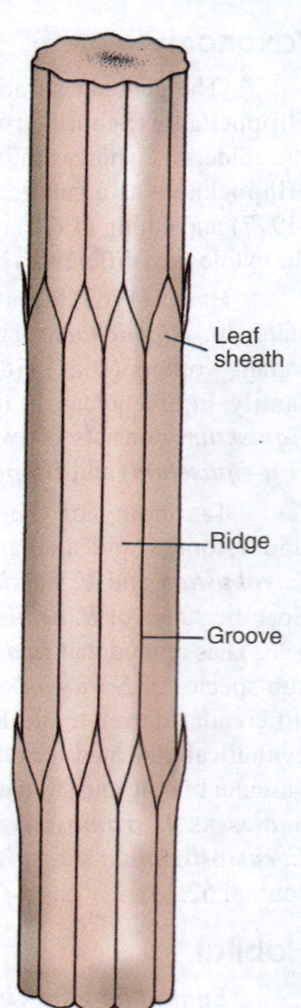

Fig. 7.2. ***Equisetum debile.*** Portion of an aerial stem showing ridges and groves and their position in the respective internodes.

Stem

The main stem is an extensively branched sympodial, dark brown or brown, rhizome that gives off branches from its nodes (Fig. 7.1). The rhizome penetrates deep into the soil and gives off aerial as well as subterranean branches that alternate with the brown scale leaves. The leaves are fused at their bases and along a portion of their length to form a leaf sheath. The branch primordia pierce the leaf sheath near its base and grow out as branches that may remain underground or become aerial, or may remain inactive for some period. In some species, e.g. ***E. arvense*** (Fig. 7.1, C) the branch primordia develop into tubers whose cells store food material and are capable of perennating. These tubers are, therefore, branches whose growth becomes arrested. They represent a single internode.

Aerial Branches

They are conspicuously jointed and are green, rough, abrasive (due to deposits of silica) and stiff, and erect in thick and short forms. The jointed appearance is due to the presence of distinct nodes and internodes. The nodes bear whorled leaves that are laterally fused at base. The internodes are long and hollow and are longitudinally ridged and grooved. The ridges alternate in successive internodes (Fig. 7.2) and so do the grooves. The number of ridges corresponds to the number of leaves at a node. The regular alternation of ridges at successive nodes is usually disturbed due to the variation in the number of leaves in the upper nodes of the stem (Bierhorst, 1959). ***Another characteristic feature of the stem is the presence of an*** **intercalary meristem** that is present at the base of the internodes and just above the node. This meristematic region is protected by the leaf sheath. The activity of this meristem is responsible for the increase in length of the internode. Its presence makes the stem structurally weak. If we pull the stem, the internodes are pulled apart at these weak spots.

The aerial stems may be branched or unbranched. In ***E. arvense*** the sterile aerial branches are profusely branched (Fig. 7.1, B). The branches arise in whorls at the nodes. They alternate with the leaves. The branch primordia pierce through the leaf sheath at its base and then grow into branches. The number of branch primordia usually corresponds to the number of leaves at a node. In ***E. sylvaticum*** the whorled branches on the main aerial stem may branch in turn. In ***E. diffusum*** (Fig. 7.1 A) the whorled branches do not branch in turn. In ***E. hyemale*** and ***E. debile*** the buds remain dormant throughout the life of the plant and the aerial stems are persistently unbranched. The dormant bud resume activity if the apical buds is injured and whorled branches may develop.

E. diffusum

In ***E. debile*** (Fig. 7.1, D) and ***E. diffusum*** (Fig. 7.1, A) the branches show no distinction into fertile and sterile shoots. All of them are green and may bear terminal strobili.

In ***E. arvense*** (Fig. 7.1, B) and ***E. telmateia*** there is segregation of aerial branches into separate fertile and sterile shoots. The fertile shoots are nongreen, unbranched and bear terminal strobili. They die after spore dispersal. The sterile shoots are large, profusely branched, green and do not bear strobili (Fig. 7.1).

In ***E. palustre*** the unbranched, reproductive shoots turn green and develop branches, after throwing off their mature strobili.

In ***E. pratense*** the rhizome bears three types of aerial branches on the same plant. These are the sterile, green and branched shoots ; the fertile unbranched and colourless shoots that die after spores dispersal, and fertile branches which behave like sterile branches after shedding their strobili.

A few examples of abnormal development of the shoot have been observed. These are : (*i*) internodes very small giving a stunted appearance : (*ii*) internodes flexuous and give a snake-like appearance to the plant ; (*iii*) dichotomous branching of the axis ; and (iv) growth of vegetative shoot beyond the terminal cone.

Leaves

E. schaffneri

Equisetum is microphyllous. The leaves are simple, eligulate, isophyllous and extremely reduced in size. They arise in whorls at each node and their number varies with species or even in the same species. In species with narrow stems the number is less (2 to 3 in ***E. scripoides***) and in those species bearing thick stems, e.g., ***E. schaffneri***, the number is large (up to 40). Sometimes the stems go on becoming thinner towards the apex and in such cases the number of leaves at the lower nodes is more as compared to the upper nodes (Bierhorst, 1959). All the leaves at a node are laterally fused to form a sheath around the base of the internodes. The sheath has free and pointed teeth like tips. It is brown in colour and is closely applied to the internode. This is mainly protective in function and in some species the leaves become dead and scale like at maturity. The leaves do not preform any photosynthetic function, which is, therefore confined to the green aerial stems. The members of a leaf whorl at one node alternate with those at the upper or lower nodes.

Roots

The roots are generally borne on the rhizome and except the first root of the embryo, the rest are all adventitious and arise endogenously. They arise from the bases of the branch primordia at each node and are, therefore, borne in whorls. The roots are extensively branched and have a root cap at their tips.

Apical growth

Stem

It is effected by means of a single pyramid-like and four -sided tetrahedral apical cell. The rounded base of the pyramid-like apical cell is towards the outer free end of the stem. The apical cell cuts off segments along its three lateral sides only. These are called cutting faces and are directed downwards. The apical cell cuts off three (one on each cutting face) cells and this process continues in a successive manner. Each of these cells or segments divides by an anticlinal wall into **segment cells** that are superimposed (i.e., one lies above and the other below). This division of the segments into segment cells differentiates two tiers, the upper segment cells develop into nodes (by further division) and the lower by a periclinal wall into two unequal cells. The outer larger cell divides further and gives rise to the **procambial cells, cortex and epidermis.** The inner smaller cell divides to form the pith. Like this the primary tissues are distinguished in the node and the internode. In the internode the central pith cells disintegrate by a process of mechanical tearing (Golub and Wetmore, 1948) and form a hollow cavity. This appears during the elongation of the internode. Some of the cortical cells also are unable to keep pace with the elongating internode and result in their separation from each other. This is how the air cavities appear in the cortex. These air cavities are called the **vallecular canals.**

The procambial strands lying at the periphery of the pith differentiate into **primary vascular tissue** ***i.e.*** **xylem** and **phloem**. The vertical direction of differentiation of these tissues has been studied by a number of workers. Golub and Wetmore (1948) studied it in ***E. arvense.*** They observed that **protoxylem** and **protophloem** elements start differentiating within the procambial strands at the fourth node of the young sporophyte. From here the differentiation proceeds basipetally into the internodes below and acropetally into the differentiating leaves above. As the internodes elongate the protoxylem and protophloem elements are subjected to a stretching force which results in their tearing apart. During the tearing the **carinal canals**, located in the protoxylem, are formed. The intercalary meristem located at the bases of internodes does not give rise to vascular elements.

Differentiation of **metaxylem** and **metaphloem** starts after considerable elongation of the internode has taken place. This differentiation begins from the node and proceeds basipetally to the lower internode and then continuity is established with the metaxylem and metaphloem differentiating in the lower node and so the process continues leading to the development of the entire and continuous vasculature of the stem.

In case we trace the course of one vascular bundle of the stem upwards through an internode we will find that at the point of insertion of the leaf, the vascular strand (bundle) divides into three branches. The middle ones enters the scale leaf whereas the two lateral branches diverge right and left and each of them joins laterally with another adjacent strand from one of the vascular bundles of the upper internode.

The lateral branches originate at about the sixth node from the shoot apex. The bud primordium grows with the help of a single pyramidal apical cell and the growth pattern is similar to that the stem.

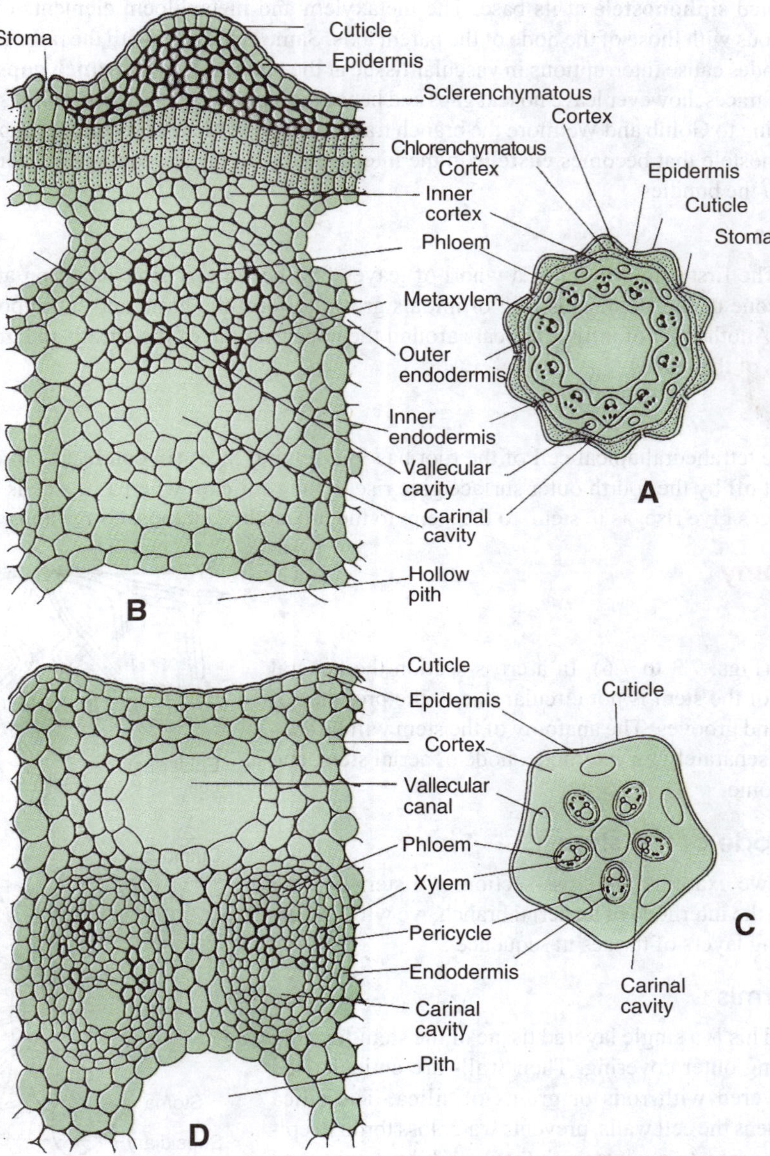

Fig. 7.3 (A–D). *Equisetum debile.* Anatomy of aerial stem and the rhizome A.T. S diagrammatic passing through the aerial internode of the stem; B. T. S. Portion in detail; C. T.S. diagrammatic passing through the rhizome; D. T. S. Portion in detail.

They arise exogenously (Campbell, 1918) but Taylor (1939) regards them as endogenous in origin. Regarding their position with respect to the left some authors (Campbell 1918) consider them axillary whereas others (Bower, 1908; Taylor 1939) regard them to be alternate in position with the leaves. The latter view is prevalent. When the branch grows and develop two or three nodes, a single endogenous **root initial** cell makes its appearance near the basal portion of each branch. The procambial strands develop into the vascular elements of the branch, which form an uninterrupted **siphonostele** at its base. The metaxylem and metaphloem elements of the branch are continuous with those of the node of the parent axis. Same is the case with the pith. The branch traces at the nodes cause interruptions in vascular tissue at the node and leave **branch gaps** (Jeffery, 1899). The leaf traces, however leave no leaf gaps and hence the siphonostele in ***Equisetum*** is **Cladosiphonic**. According to Golub and Wetmore the branch traces also do not leave any branch gaps and the stele is a siphonostele that becomes **eustelic** in the internode due to the appearance of ground parenchyma between the bundles.

Leaves

The first primordium of a whorl of leaves first becomes visible as a ring at the base of the apical cone of the stem. This ring of initials grows to form the basal and fused portion of the leaf sheath. Another set of initials appears around the upper margin of the sheath and grows to form the free tips of the leaves.

Roots

The tetrahedral apical cell of the root has four cutting faces (in contrast to three in stem). The cells cut off by the fourth outer surface give rise to the **root cap,** whereas the cells cut off by lower three faces give rise, as in stem, to the stem, to the procambial strands, cortex and epidermis.

Anatomy

Stem

(Figs. 7.3 to 7.6). In a cross-section the general outline of the stem is not circular due to the presence of ridges and grooves. The anatomy of the stem will be considered separately for internode, node of aerial stem, and the rhizome.

Internode of the stem

If we examine the cross-section of a stem passing through the internode of an aerial branch, we will find the following layers of tissues in sequence :

Epidermis

This is a single layered tissue of the stem forming the strong outer covering. Their walls are cuticularised and covered with rods or grains of silica. The silica strengthens the cell walls, prevents water loss through epidermis, protects against pathogens and predators, and helps in keeping the plants erect. Kaufman and his co-workers (1972) have shown that in *E. scripoides* the silica (SiO_2) deposits first in the stomatal apparatuses and later in the long epidermal cell walls, progressing through the top of the internode to the basal intercalary meristematic region in a basipetalous sequence. They are of the opinion that silicifications must play a major role in *Equisetum* shoot development in breaking internodal extension, strengthening the shoot system and triggering "senescence" in last stages of internodal development. Here and there the epidermis is perforated by the stomata which are present in the furrows. The stomata (Fig. 7.4, A, B) are peculiar in having a double set of cells, lying one above the other. The two innermost cells constitut-

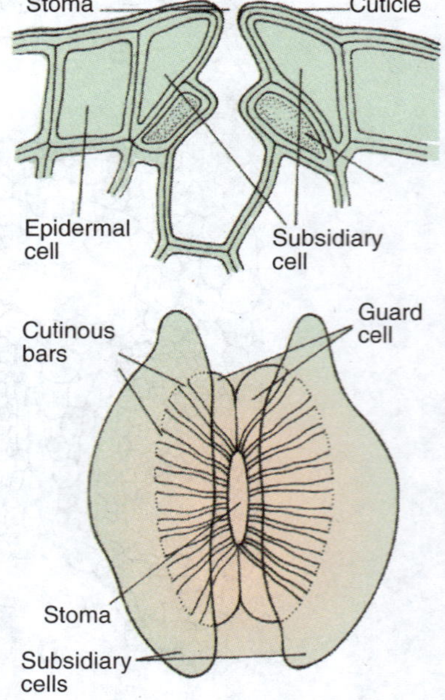

Fig. 7.4. Stomata from the stem internode of ***Equisetum***. **A.** T.S. epidermis of ***E. arvense*** *showing* sunken stomata; **B.** stoma of ***E. telmatia*** showing siliceous bars.

ing a pair of guard cells and the two uppermost a pair of accessory or subsidiary cells (Fig. 7.4). The subsidiary cells completely cover the guard cells and hide them from view. A number of transverse bands of siliceous matter radiate from the stomatal pore and traverse the guard cells (Fig. 7.4, B). The stomata in some species are sunken, each lying in a pit. In some species the stomata are on the surface. The stomatum develops from a single superficial cell called the stomatal initial. It divides by two longitudinal walls into three cells. The middle cell acts as the guard mother cell whereas the two lateral cells act as subsidiary cells. The guard mother cell divides again by a longitudinal wall to form two guard cells. A pore appears between the guard cells. Later the subsidiary cells grow over the guard cells and overarch them. Such a development pattern of the stomata recalls syndetocheilic type of stomatal ontogeny found in some fossil groups of the gymnosperms (Bennettitales). Such an affinity has been pointed out by Pant and Mehra (1964) while discussing the development of stoma in the vascular cryptogams.

The siliceous bands of thickening have been regarded by some authors (Campbell 1911; Johnson 1933; Copeland 1936; Kersimo 1954) as belonging to the guard cells. Others (Hauke, 1957; Jyotsana, 1964) consider them to arise from the inner walls of the subsidiary cells.

Cortex

The cortex (Fig. 7.3) can be divided into two parts (*i*) outer cortex and (*ii*) inner cortex. The ***outer cortex*** is differentiated into two types of cells :—

(*i*) Sclerenchymatous Cells (Fig. 7.3). These are thick walled cells that give mechanical strength to the stem and are sub-epidermal in position. Below the ridges, they occur in large and heavy groups. There is an equal number of smaller groups of sclerenchyma beneath the epidermis of the grooves or the furrows. They are absent beneath the stomata. They are very well developed in ***E. giganteum*** and project to sufficient depth in cortex. They have been designated as fibre like collenchyma by Hauke (1963).

(*ii*) Chlorenchymatous Cells. These are lying lateral to and below the sclerenchyma (forming a curved band) and contain a large number of chloroplasts in their cells and form the assimilatory region of the stem (Fig. 7.3). From below the sclerenchyma of the ridge it reaches the surface on either side between the support tissue of the ridge and that of the furrow. Here lie the stomata, therefore communicating directly with the green tissue. ***E. debile*** (Fig 7.3) reveals a continuous band of chlorenchyma composed of elongated cells arranged in 3 or 4 layers below the sclerenchyma.

Inner Cortex

It consists of a few layers of larger celled parenchyma traversed by a ring of large air space (Fig. 7.3) forming an aeriating system. They are called the **vallecular canals.** Each of these lies below the furrow of the external surface and is thus close beneath the photosynthetic tissue.

Deep seated rhizome, thick cuticle, reduced leaves, presence of sclerenchyma, a well developed chlorenchyma, and sunken stomata are the xerophytic characteristics of the stem. Presence of distinct ridges and grooves and stiff epidermis may also be regarded as xerophytic characters. Presence of vallecular canals is hydrophytic character.

Endodermis

A variety of conditions is found as to the position of endodermis in various species. The most common condition as is found in ***E. arvense, E. palustre, E. scripoides*** and a few other species is that a common ring of endodermis is present surrounding the whole ring of the vascular bundles on their outside (Fig. 7.5). In other species a common ring of endodermis is present both outside as well as inside the vascular bundles as in ***E. debile*** and ***E. hymale*** (Fig. 7.3). The outer endodermis dips in between the vascular bundles, in some cases, coming to lie against the inner endodermis. In ***E. limosum,***

E. giganteum and a few other species there is a separate endodermal layer encircling each vascular bundle. Some botanists consider the first condition to be the most primitive and the last the most specialised. Evidences for this are derived from comparative studies of stelar structures in the genus and from the structure of the ***Calamites,*** an ancient group related to ***Equisetum.***

Stele

Within the endodermis there is a single layer of pericycle forming the outermost layer of the stele. The stele consists of a well defined ring of small vascular bundles surrounding a large pith cavity. The bundles are conjoint, collateral and endarch. They alternate in position with the **vallecular canals** and so lie beneath the ridges. The individual bundles in a ring are separated from one another by a large zone of parenchyma. Internal to the vascular bundles and so beneath the ridges are smaller **carinal cavities.** One of these belongs to each vascular bundle and is filled with water. The carinal canals are formed by the tearing apart of the protoxylem elements. Since protoxylem distinguishes itself quite early during the ontogeny of shoot, the elongating internodes exert a tearing effect on the protoxylem elements. This leads to the formation of water filled carinal canals. Bierhorst (1958), however, experimentally proved that the carinal canal is not a cavity but a conducting channel. The vascular bundle consists of xylem and phloem. There is no cambium hence the bundles are closed.

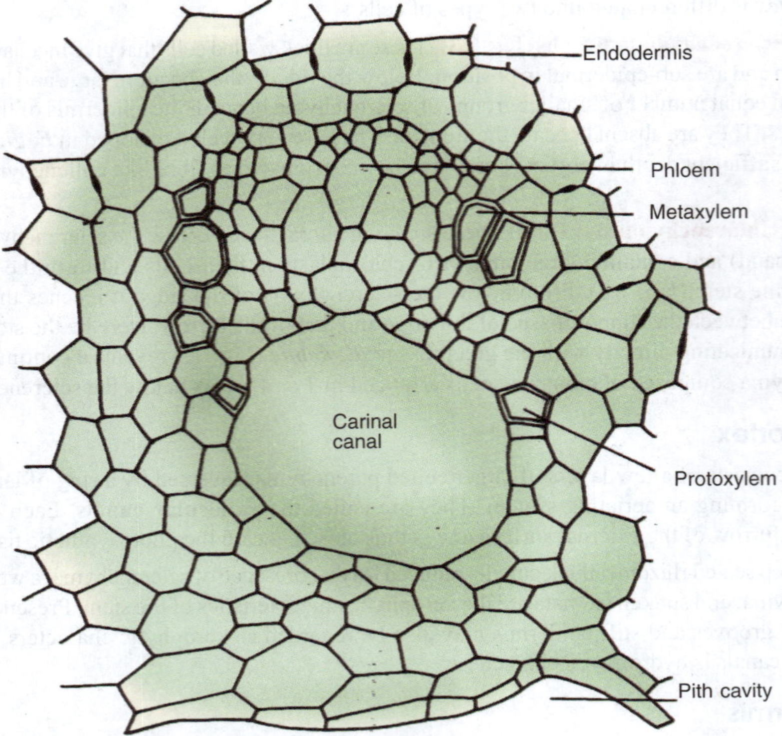

Fig. 7.5. A vascular bundle from the aerial stem of ***E. arvense.*** The endodermis is present only on the outer side of the bundle.

Xylem is poorly developed. It has the shape of a "V". The point of "V" lies towards the centre and the ends diverge towards the outside. The metaxylem which consists of scalariform tracheids forms the two limbs of the "V". Bierhorst (1958) studied in detail the xylem elements in some species of ***Equisetum*** and reported the presence of two types of vessel elements in the metaxylem of internodes alone. One type of vessels has simple perforation plates and the other type has reticulate. He reported that the vessels occur only three in a row and do not form long conducting channels as in seed plants. The protoxylem which occupies the point of "V" consists of a few annular and spiral tracheids. Later on, its position is occupied by the carinal canals as a result of its disorganisation. The meta and protoxylem are separated from each other by a little parenchyma.

Bierhorst (1958) has, however, reached a different conclusion as a result of his studies on xylem of 10 species of ***Equisetum***. He observed that the metaxylem groups occupying the lateral arms of "V" are quite independent of the few xylem elements and carinal cavity. He regards it as carinal groups of xylem because according to him carinal canal is a conducting channel. He found that the carinal group has its own metaxylem and protoxylem so do the xylem groups at the lateral ends of 'V'. The lateral xylem group may in some cases be entirely made up of protoxylem (***E. fluviatile***), or of metaxylem alone (***E. hyemale***, etc.) or of both proto and metaxylem (***E. sylvaticum***).

The phloem consists of phloem parenchyma and the sieve tubes. It lies between the two limbs of the "V" but a little to the outer side.

E. fluviatile

Pith

The pith is present in young specimens but is absent in the mature stems and is represented by a large central cavity. Hollow pith, presence of carinal cavities in the xylem, and less developed xylem are hydrophytic characters.

Node

The anatomy of the node shows some difference from that of the internode. Vallecular canals are present in some species (***E. debile***), and may be absent in others. The pith is present at the nodes and forms a diaphragm separating the internodes above and below. The vascular bundles, that are separate in the internodes, are connected by a continuous cylinder of xylem in nodal region. So the vasculature in the node is in the form of a continuous cylinder. Eames (1909) reported that carinal canals are absent in the nodal region. There are two views regarding the presence of protoxylem groups in place of carinal canals. Eames (1909) and Bierhorst (1958) are of the view that the protoxylem elements are present. Some, however, regard it to be absent. The continuous vascular cylinder of the nodes gives rise to the leaf traces and the branch traces. Some workers regard the separate vascular strands of the internodes as leaf traces that extend down the internodes of the node below. This view is, however, not held to be correct. Another view regarded the stelar system to be a **dictyostele** and the parenchymatous gaps between the bundles as leaf gaps. This is also untrue as the leaf traces in microphyllous pteridophytes do not leave any leaf gaps. Moreover this view is not supported by the studies on the extinct relatives of ***Equisetum***. This view also overlooks the presence of intercalary meristem whose activity is responsible for growth in length of the internode. The nodal vascular cylinder must be looked upon as a siphonostele. The appearance of the parenchymatous gaps in the internode may be due to some activity of the intercalary meristem.

Rhizome (Fig. 7.3, C, D and 7.6)

It differs from the aerial stems in certain respects : (*i*) There are no stomata in the epidermis; (*ii*) the sclerenchyma is not very well developed and forms a continuous layer ; (*iii*) Chlorenchyma is absent ; (iv) the pith is present in some species but is absent in others. In ***E. debile*** the endodermis encircles every vascular bundle, whereas in ***E. arvense*** a common endodermal layer encircles the stelar region externally (Fig. 7.6).

The fertile aerial shoots of ***E. arvense*** are colourless and a transverse section reveals the absence of stomata in the epidermis. The sclerenchyma and chlorenchyma are also absent. The rest of the structure is similar to that of the rhizome. In ***E. sylvaticum*** the colourless fertile shoots become green and act as ordinary sterile shoots after the dispersal of spores. In this case the stomata, the sclerenchymatous tissue and the chlorenchyma appear after the spores are shed.

The anatomical features of ***Equisetum*** show both xerophytic and hydrophytic characters. The **xerophytic characters** of the plant are :

1. Presence of thick coat of cuticle all over the epidermal layer of stem.
2. Photosynthetic stem having chlorenchymatous cortex.
3. Stem behaving as leaves.
4. Presence of small, scale like leaves with limited number of chloroplast.
5. Presence of sunken stomata in groove of stem.
6. Stiff Epidermis and presence of ridges and furrows.
7. Deep seated rhizome.

Similarly the hydrophytic characters of the plant are :

1. Vascular tissues specially xylem is poorly developed.
2. Vallecular canals are present in the inner cortex region, which always contains air thus forming an aerating system.
3. Presence of large pith cavity surrounded by a few remaining pith parenchyma cells. Pith cavity is lacking in ***E. seripoides.***
4. The central pith cavity and the carinal canals usually contain water in them.

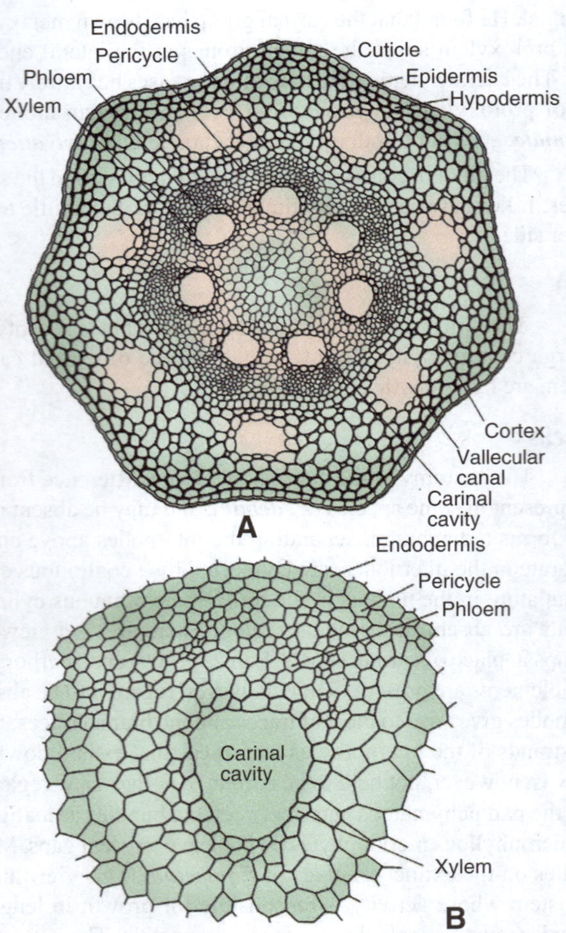

Fig. 7.6 (A–B). ***Equisetum arvense.*** A. T. S. portion of rhizome showing detailed internal structure; B. A portion of above showing a structure of a bundle in detail.

Nature of the stele and gaps

Studies on the developmental anatomy of the axis of ***Equisetum*** (Golub and Wetmore, 1948) reveal a complete absence of the foliar and branch gaps in the stelar organization. The parenchymatous areas between the vascular bundles in the internode have been regarded by these authors to be ground tissue parenchyma. It might have developed due to the activity of the intercalary meristem. The observations suggest that the stelar organization in the stem of ***Equisetum*** is purely a **siphonostele** which forms a complete and uninterrupted cylinder at the nodes. At the internodes its continuity is interrupted by the appearance of parenchymatous areas that are neither foliar gaps nor branch gaps.

Earlier workers like Browne (1912, 1915, 1920, 1921, 1933) suggest the presence of leaf gaps in region of the strobilus and she regards the sporangiophores as modified leaves. Moore (1914) also agrees with Browne. Jeffery (1917) regards the parenchymatous areas between the vascular bundles in the internode as branch gaps. Barratt (1920) conducted a detailed study on the development of vascular tissues in ***E. arvense*** and ***E. maximum*** and regarded these gaps to be due to the encroachment of parenchyma on the xylem tracheids. Studies by Barratt reveal that stele in the first embryonic shoot is a typical **protostele.** It changes to a **siphonostele** at the point where a secondary branch originates. All branches that arise later have an uninterrupted siphonostele at their bases. The same is the case at the nodes. In the internodal region alone the continuity is interrupted by the appearance of parenchymatous gaps, which according to Barratt are due to encroachment of the ground tissue parenchyma. These studies were later confirmed by Walton (1944) on the basis of his studies on ***E. limosum.*** These detailed ontogenetic studies, therefore, support the view that there are no **leaf gaps and branch gaps** in stelar set-up of ***Equisetum*** and the stelar organization is essentially a **siphonostele.**

Sporne (1966) suggests that these parenchymatous gaps are of a peculiar "intercalary" origin. Namboodri and Beck (1968) also suggest that during the evolution of the stele, parenchymatous areas may arise having no association with leaf gaps. White (1970) has suggested that in ***Tectaria decurrens*** and ***T. devesa*** such gaps are directly related to characteristics of the apical meristem when the pattern is established. Slade (1971) has supported Namboodri and Beck in her recent paper on the "stelar evolution in vascular plants. "Such parenchyma gaps that are not leaf gaps or branch gaps have now been reported in many gymnosperms, and angiosperms (Devadas and Beck, 1972) and ferns (White, 1970). They are also known in ***Tmesipteris*** (Sykes, 1908 ; Sahni, 1925). These observations have suggested a new concept of evolution of stele (see Chapter 8).

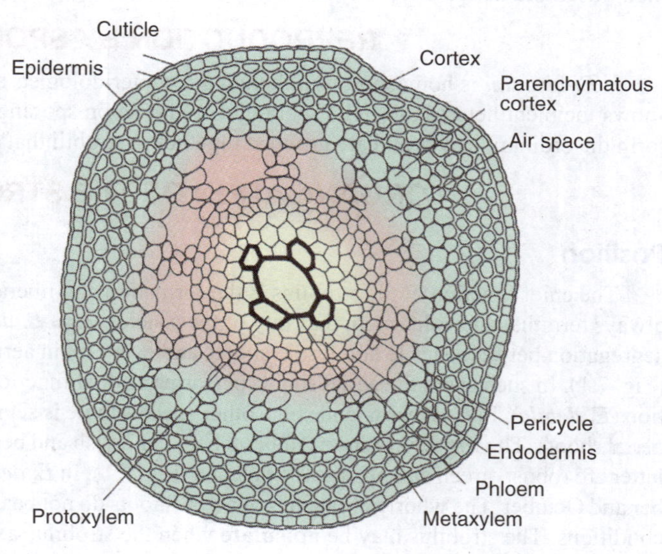

Fig. 7.7. T. S. root of *E. debile.*

Root (Fig. 7.7).

The outermost layer of the root is the single layered epidermis or the piliferous layer. It bears root hair in the root hair zone. It is made up of a single layer of thinwalled cells.

Underneath it, is the many layered cortex consisting of parenchymatous cells. The larger roots in ***E. limosum*** and ***E. debile*** possess a row of large air space in the inner cortex (Fig. 7.7). These spaces develop schizogenously. In some species the cortical cells beneath the epidermis are thick-walled and lignified. They constitute the tissue usually referred to as **exodermis.** Cortex is limited on the inside by a well defined endodermis followed by a pericycle which is one cell thick.

Since the cells of the pericycle fit accurately to those of the endodermis, some botanists like Smith consider the endodermis to be two cells in thickness. Since there are no casparian strips in the pericycle cells, it cannot, therefore, be regarded as a second layer of endodermis. Casparian strips are, however, clearly visible in the outer layer. Lateral roots originate from the pericycle. The stele is triarch or tetrarch. The three or four protoxylem groups meet and surround a central metaxylem vessel. There is no pith. The phloem elements alternate with the protoxylem groups. Both the xylem and the phloem are exarch because both the protoxylem and the proto-phloem lie towards the periphery.

VEGETATIVE PROPAGATION

In some species the branch primordia on the rhizome remain short and develop into round or ovoid tubers (Fig. 7.1, C). When detached from the parent rhizome each tuber grows into a new plant and thus serves as a means of vegetative propagation of the sporophyte. The cells of the tuber are filled with starch and every tuber has an outer thick-walled protective layer of cells.

Every branch of the rhizome bears preformed branch primordia which can develop into new subterranean and aerial branches at any time. The older portions of the deep seated rhizome may decay setting free the branches. Each branch with the help of preformed branch primordia, which now become active (from the hormonal inhibition) can grow into a new plant. The aerial shoots can also behave in a similar manner. Each node-internode segment of the aerial stem has the potentially of giving rise to a new plant. The root primordia are always associated with branch primordia at their bases and hence there is no difficulty in establishing the new vegetatively developing plants.

REPRODUCTION BY SPORES

Equisetum is homosporous although a few pteridologists are of the opinion that ***Equisetum*** shows incipient heterospory. The sporangia are borne on sporangiophores which are very different form the ordinary leaves and are grouped together in strobili that are terminal in position.

ORGANISATION OF THE STROBILUS

Position

The chief function of the strobilus is the formation and liberation of the spores. The strobili are always terminal and solitary. In majority of the species (***e.g., E. debile, E.diffusum,*** etc.) there is no segregation between sterile and fertile shoots and every main aerial shoot bears a terminal strobilus (Fig. 7.1). In such cases the aerial shoots perform the dual function of photosynthesis and reproduction (***E. debile***). In ***E. arvense*** and some other species there is segregation between fertile and sterile aerial shoots. The former are short , unbranched, yellowish and bear a terminal strobilus, whereas the latter are robust, green and profusely branched (Fig. 7.1). In ***E. debile*** the strobili mature in September and October. The whorled branches of aerial shoots do not bear strobili, except under exceptional conditions. The strobilus may be **apiculate** when the strobilus axis grows a little beyond the upper whorl of sporangiophores and becomes pointed. In others the strobilus has a rounded apex and bears sporangiophores at its apex.

Abnormalities

A few abnormalities have been recorded in ***E. debile*** regarding the strobilus. In some cases it has been noticed that in addition to the terminal cone borne on the main axis there are present lateral branches arising from the nodes and each bearing a terminal cone. In some cases two separate cones are borne terminally whereas in certain plants bifurcated cones have also been reported. In ***E.pratense*** proliferation of the strobilus has also been recorded. In this case the apex of the strobilus continues its growth as a vegetative shoot ; the effect is thus gained of a fertile zone bearing sporangiophores, threaded upon an axis ; or of a succession of such zones separated by leaf sheaths.

Structure of Strobilus

The cone of ***Equisetum*** is peculiar. It consists (Fig. 7.8, A) of a central thick axis which bears a number of densely crowded T-shaped peltate scales called the sporangiophores. The sporangiophores are arranged on the axis in successive whorls alternating, though somewhat irregularly, with one another. The number of sporangiophores in each whorl varies from few to many. Unlike ***Lycopodium*** and ***Selaginella*** they are not leaf-like. The morphology of the sporangiophores is an uncertain question. Some botanists regard them as reduced spore bearing leaves. Others regard them as specialised structures more nearly related to stems than to the leaves.

Structure of Sporangiophore

The sporangiophore (Fig. 7.8, A—D) has two regions : (*i*) a small proximal cylindrical and stalk-like portion attached at right angles to the cone axis, and (*ii*) a shield-shaped peltate disc attached to the distal or outer end of the stalk. The peltate disc bears on its under-surface close to its edge, a ring of several usually 5-10 sac-like pendent sporangia (Fig. 7.8). They extend inwards towards the axis of the cone and fill up all the space that is left between it and the sporangiophores. The peltate heads of the sporangiophores fit closely together so that the sporangia are concealed. Due to mutual pressure the peltate disc usually becomes hexagonal in outline. At the base of the cone, the central axis bears a ring-like outgrowth the **annulus.** Occasionally there are two such rings. Sometimes the **annulus** is toothed and apparently represents reduced leaf whorl. It seems to some extent a protective structure. Some botanists consider it to be sporangiophoric in nature because the annuli of some species regularly bear small sporangia (***E. cryptophora***).

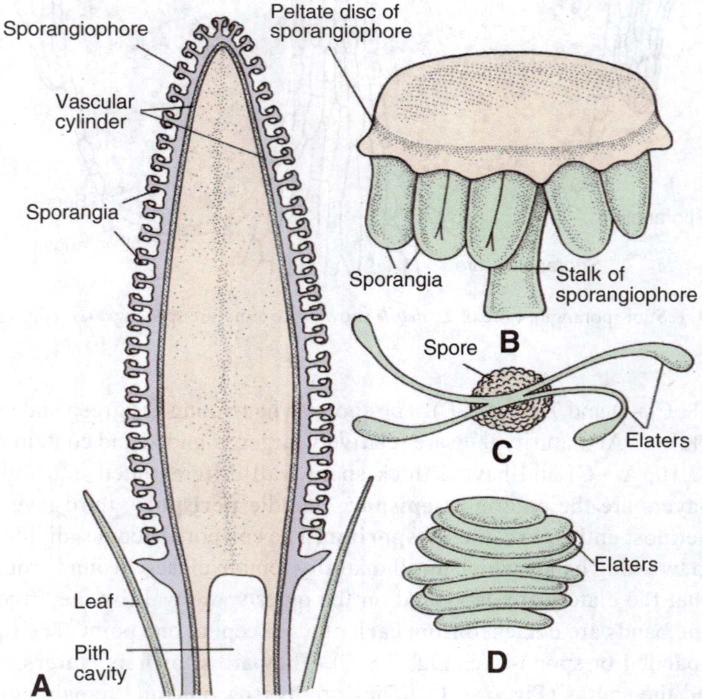

Fig. 7.8 (A–D) Organisation and structure of strobilus of *E. debile*.A.V. S. strobilus; B. A. Peltate sporangiophore bearing pendent sporangia; C. A. spore with elaters stretched; D. A spore with elaters wound spirally around it.

Sporangia

A mature sporangium is an elongated sac-like structure (Fig. 7.9) full of large number of haploid spores. The spores are of equal size (homosporous). The mature sporangium has a single layer of wall cells. The cell walls of this layer are spirally thickened. A young sporangium has 2—4 layered thick wall. The innermost wall layer is derived from the outermost sporogenous cells and is called the **tapetum.** The inner wall layers and the tapetum disorganise as the spores are formed.

Dehiscence

At maturity the strobilus axis elongates slightly and thus loosens its compactness. The tightly fitting sporangiophores separate from each other. Later due to the loss of water the sporangiophores shrink and fall apart. The sporangia are now exposed. The sporangia open by longitudinal slits, down the side next to the sporangiophore stalk and the spores sift out. They are light and fluffy and are carried away singly or in clusters by the air currents.

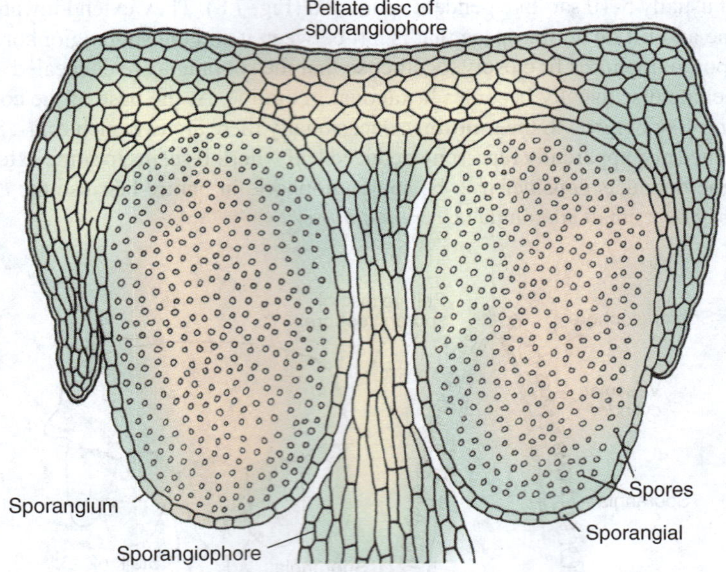

Fig. 7.9. L.S. of sporangiophore of ***E. debile*** showing two mature sporangia containing spores.

Spores

(Figs. 7.8, C—D and 7.10, A—E). The spores when young are green and are covered by a thin wall of cellulose. At maturity they are relatively larger, rounded and contain numerous chloroplasts (Fig. 7.10, A—C) and have a thick spore wall differentiated into four layers (Beer, 1909). These layers are the outermost **epispore**, middle **perispore**, third layer called the **exospore** and innermost **endospore** or **endosporium.** The epispore becomes divided along several spiral lines into two long bands which untill maturity remain closely wound around it. Bierhorst (1971) states that the elaters are deposited on the outer spore wall by the "free liquid" in the sporangium. The bands are detached from each other except at one point. The tips of the bands are slightly expanded or spoon-like (Fig. 7.8 C). They are known as **elaters** and are spirally wrapped around the spores (Fig. 7.8, C). They are hygroscopic and remain coiled around the spores in moist air. When dry the elaters stretch themselves out crosswise, remaining attached only in the middle of their length and at one point so that they appear as four distinct appendages

(7.8 C). The elaters are different from those of the bryophytes as they do not differentiate from complete cells. The elaters in bryophytes are diploid, spindle-like and have spiral thickenings, whereas in ***Equisetum*** they are haploid and have no spiral thickenings. The spores have no triradiate ridge. The cytoplasm is filled with lipoproteins and lipid granules. The nucleus usually lies in the centre of the cytoplasm.

Functions of Elaters

The elaters help in the dispersal of spores and dehiscence of the sporangium. At maturity the sporangiophores wither and the sporangia lose water. The spores inside dry up and begin to uncoil their elaters. This loosens the whole mass of spores and makes them take more room and so to press on the wall of the sporangium which gives in and opens by a longitudinal slit. It has also been suggested that as the sporangium loses moisture the stretched out elaters become entangled with those of other spores. Consequently the spores are dispersed in masses and obliged to germinate in company. This is an advantage as it increase the chances of fertilization in dioecious prothalli. Large number of spores entangled by elaters offer a greater surface to the wind and thus act as a parachute mechanism.

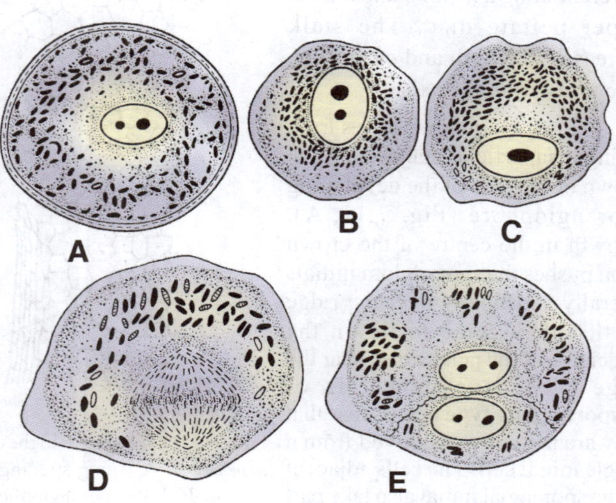

Fig. 7.10 (A–E). ***Equisetum.*** Early stages in spore germination.

A. Mature spore of *E. arvense* containing large plastids (After Py); B. Chloroplasts radially arranged around the nucleus (After Nienburg); C. Chloroplasts have migrated towards lighted side (After Nienburg); D. First division (After Nienburg); E. First wall has appeared cutting off a rhizoid cell with few plastids and a prothallia cell with many plastids (After Nienburg).

Incipient Heterospory

The spores ***E. arvense*** are apparently all the alike and the species is said, therefore, to be homosporous. McClean and Cook (1951) reported the occurrence of two kinds of spores in ***E. arvense.*** They examined a large number of spores and measured their diameter and plotted them on square paper. It was found that two curves are produced with separate maxima and very little overlapping. The smaller series are much pale green than the larger ones and can be picked out by eye in a fresh sporangium. The two kinds are not equal in number nor is the number in each kind constant in a tetrad. This is not perfect heterospory, because though the small spores produce small male prothalli, the large spores produce female prothalli but if fertilisation fails to occur these prothalli may also develop antheridia (Schratz, 1928). This is kind of incipient heterothallism. The case is interesting, however, as showing apparently an early stage in the development of heterospory. Duckett (1970) examined the spores of several species and did not observe any trace of size difference.

Development of Sporangium

The aerial shoot destined to give rise to the strobilus at its tip exhibits very slow apical growth and assumes a conical shape. It soon shows small conical protuberances that arise in an acropetalous succession (younger near the apex and the older below). These outgrowths assume a hemispherical shape and are the primordia of the sporangiophores (Fig. 7.11). Further growth results in the formation of a constriction near the base of these outgrowths thus differentiating a lower stalk and the upper peltate disc. The stalk increases in length and carries the disc away from the axis.

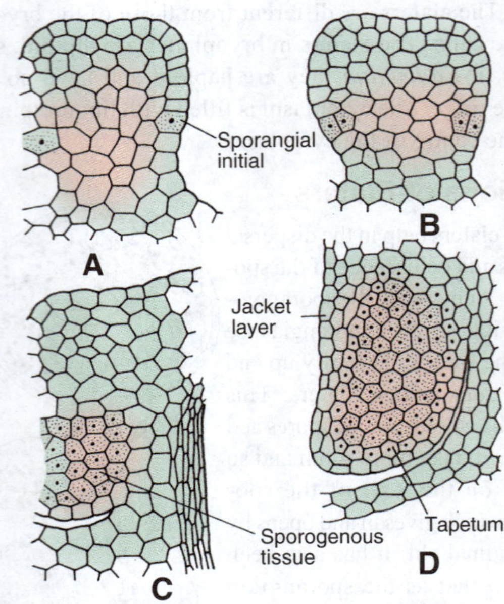

Fig. 7.11. (A–D) Stages in the development of a sporangium in *Equisteum arvense*. A. A young sporangiophore in L.S. showing two sporangial initials; B. L. S. sporangiophore with sporangial initial divided into an outer and inner cells; C. A later stage of development; D. L. S. sporangiophore with young sporangium.

Each sporangium arises from a single superficial cell around the crown or the disc of the developing sporangiophore (Fig. 7.11, A). Growth in the centre of the crown soon pushes the sporangium initials laterally and finally over the ridge so that they come to lie on the underside of the peltate disc near the edge. The development is of **eusporangiate type** in the sense that they are not entirely derived from a single initial cell. The cells adjacent to the sporangial initial also take part in development. However, all the sporogenous tissues in a sporangium can be traced back to a single superficial cell of the young sporangiophore.

The sporangial initial divides into an outer and an inner cell by a periclinal wall. The inner or the **primary sporogenous cell** divides anticlinally as well as periclinally to give rise to a mass of sporogenous cells. The outer cell divides periclinally as well as anticlinally into a number of cell layers. The outer most two or four layers act as wall layers of the sporangium. The inner layers constitute the sporogenous tissue along with the sporogenous tissue formed by the primary sporogenous cell. The cell of the sporangiophore lateral to the outer wall cells also divide and contribute to the sporangial wall. It, therefore, becomes clear that the sporangial wall is not derived from one initial cell. The outermost sporogenous cells divide periclinally and cut off a **tapetal layer** internal to the wall layers. The sporogenous cells divide many times and the cells of the last cells of the last cell generation act as **spore mother cells.** About 1/3 of them degenerate along with the tapetal cells. The surviving or functional spore mother cells round off and float in the **periplasmodial fluid** formed by the degenerated tapetal cells and the spore mother cells. The inner wall layers, except the outermost layer, also degenerate. The rounded spore mother cells undergo meiosis and form tetrads of haploid spores. Later the spores secrete their own wall layers and separate as individual spores. The spores contain chloroplasts and are surrounded by 4 layered wall. The sporangial wall cells become spirally thickened.

MORPHOLOGICAL NATURE OF THE STROBILUS AND THE SPORANGIOPHORES

The strobilus of ***Equisetum*** has been regarded by some author (Browne, 1912, 1915, 1920, 1923, 1927, 1933) to be a shortened axis differentiated into nodes and internodes. The sporangiophores have been regarded to arise from the nodes. Barratt (1920) on the other hand regards it to be a shortened axis without any differentiation into nodes and internodes. The former view is supported by many botanists.

The morphology of sporangiophores is a debatable question and a number of explanations have been forwarded. Milde (1867) regarded the sporangiophores to be modified leaves. From his extensive and excellent observations on the transitional forms between the sheath leaves and the annulus teeth, and between these teeth and the sporophylls, Milde arrived at the firm conviction that leaf, annulus teeth and sporangiophores are all homologous organs. Goebel also came to the same conclusion. His experiments on ***E. telmateia*** prove this view. He was able to convert the fertile shoots into sterile shoots. The sporangiophores exhibited a regular change to ordinary leaves during his experiments. Tschudy (1939) also reached the same conclusion as a result of his studies on ***E. telmateia***.

Bower (1908) and Browne (1927) regarded the sporangiophore as a modified branch and considered it to be axial in nature.

Hirmer (1933), Zeller (1893), Seward (1898) regard the sporangiophore to be a ventral lobe of a dorsiventrally split sporophyll. The sterile dorsal lobe has disappeared.

The consensus of opinion, however, favours its foliar nature, ***i.e.***, the sporangiophore is a modified leaf.

GAMETOPHYTE

Germination of the Spore

The haploid spores germinate into a gametophyte. They are green and remain viable from 5—20 days and are unable to germinate after this period. Medium light, sufficient oxygen and moisture are most suitable for germination. Wollersheim (1953) attributed their brief viability to the thick spore wall. Hauke (1963) attributed it due to high rate of respiration caused by the chloroplasts and abundant mitochondria present in the spores.

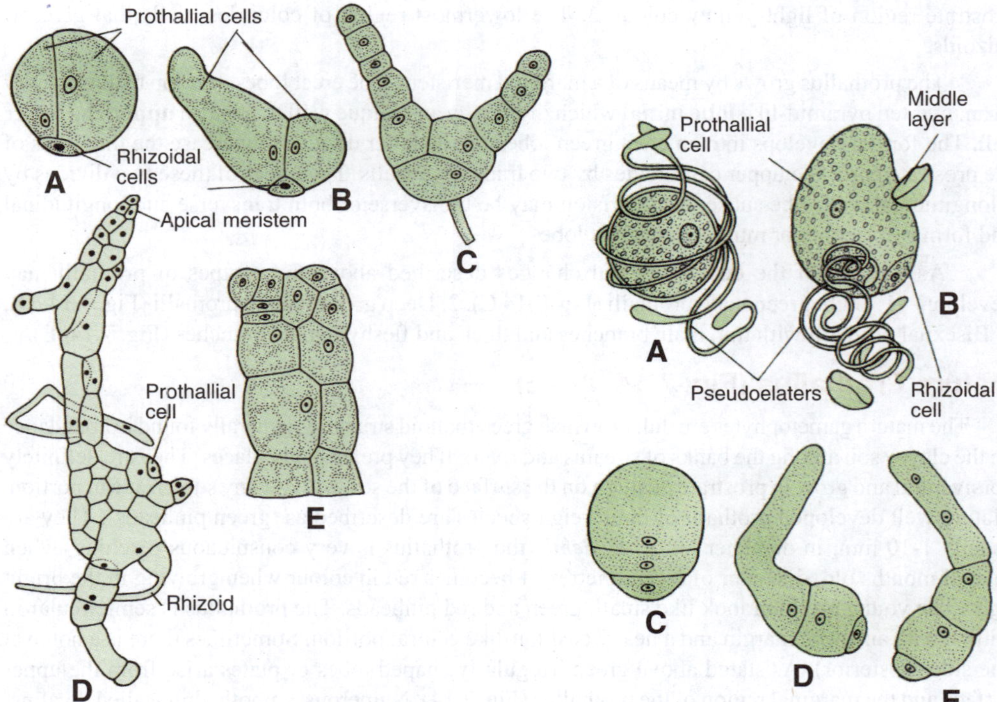

Fig. 7.12. ***Equisetum debile.*** Early stages in spore germination. (After Jyotsna and Mohan Ram).

Fig. 7.13. E. ***debile.*** Stages in the development of gametophyte. (After Jyotsna and Mohan Ram.)

Jyotsna and Mohan Ram (1964) studied the germination of spores in ***E. debile*** (now called ***E. ramosissimum*** sub. ***sp. ramosissimum***). The following account is based on their observations (Figs. 7.12, 7.13). The spores germinated within 2-3 days.

Prior to germination the large vacuole is replaced by several small vaculoes; the chloroplasts surround the nucelus. The spore absorbs water, swells up and the middle spore layer ruptures. The **pseudoelaters** are cast off (Fig. 7.12). The spore contents divide by a wall into two unequal cells, the small lenticular **primary rhizoidal cell** and a large **prothallial** or **gametophytic cell.** The rhizoid cell has only few or no chloroplasts whereas the prothallial cell has many chloroplasts and oil droplets. The rhizoidal cell elongates into the first or **primary rhizoid**. The prothallial cell divides in various planes to give rise to the rest of the prothallus. It usually divides by a transverse wall (Fig. 7.12, C—E) that may lie close to or away from the rhizoidal cell. In the former case the upper cell is larger whereas in the latter case it is smaller. In any case it divides further by transverse walls to form a filament of green cells (Fig. 7.12). The prothallial cell usually faces light whereas the rhizoidal cell is away from light. The filament formed by transverse division of prothallial cell divides in all directions to form a flat green and leaf-like expanse of tissue or to an elongated and branched thallus. Any superficial cell of the thallus at this stage may divide unequally into a small **secondary rhizoidal cell** and larger cell. The former (Fig. 7.13) develops into a secondary rhizoid. Many such rhizoids may arise. Later the thalloid tissue increases in thickness by further anticlinal and periclinal walls into a bulky cell mass gorged with starch. Cell division later takes place in all planes to form a cushion-shaped massive thallus that is several cells in thickness and bears numerous rhizoids on its lower surface. Earlier growth of the thallus is by single apical cell, which is later on replaced by a group of meristematic cells (Fig. 7.13). Further development leads to the formation of a prothallus with three distinct regions :— 1. The spongy green upper part of erect green lobes. 2. The middle prostrate region of light-yellow colour. 3. The lowermost region of colourless cells that give off rhizoids.

The prothallus grows by means of a marginal meristem. The erect lobes develop by the activity of an inverted pyramid-like **lobe initial** which divides by an oblique wall to form an **upper** and **lower cell.** The former develops into an erect green lobe and the latter divides to increase the thickness of the prostrate part. The upper cell divides by two transverse walls. The upper of these two divides by a longitudinal wall. The subsequent division may be transverse or both transverse and longitudinal and form a uniseriate or multiseriate erect lobe.

As a result of the developmental changes described above three types of prothalli may develop:— 1. Light green male prothalli (Fig. 7.14 C). 2. Deep green female prothalli (Fig. 7.14, D). 3. Bisexual prothalli with thin male branches and thick and fleshy female branches (Fig. 7.14, E).

Mature Prothallus (Fig. 7.14, A—E)

The mature gametophytes are dull brownish green thalloid structures generally found in abundance on the clayey soil among the banks of streams and rivers. They prefer shady places. They are definitely dorsiventral and grow in prostrate position on the surface of the soil without any subterranean portion. Mature well developed prothalli of the foreign species are described as 'green pinheads'. They are usually 1-10 mm. in diameter but in ***E. debile*** the prothallus is very conspicuous reaching, when several months old, diameter of 3 centimetres. It becomes red in colour when growing in the bright light. The young prothalli look like small, green and red pinheads. The prothallus is semicircular in outline with an entire margin and a heavy cushion-like central portion. Sometimes there is a notch at one side (posterior). As stated above green irregularly shaped lobes or plates arise from the upper surface and the marginal region of the prothallus (Fig. 7.14). Numerous, smooth, thin walled, hyaline, unicellular and unbranched rhizoids grow from the underside of the central region of the cushion and attach the gametophyte to the substratum. The entire margin of the cushion is meristematic in nature.

The prothallus is often attacked by a fungus in the upper cells of the lobes. Generally it is dorsiventral but in ***E. debile*** it is radial (Kashyap, 1924), heart-shaped and has meristem all around. It differs from ***L. cernuum*** in the absence of a cylindrical body.

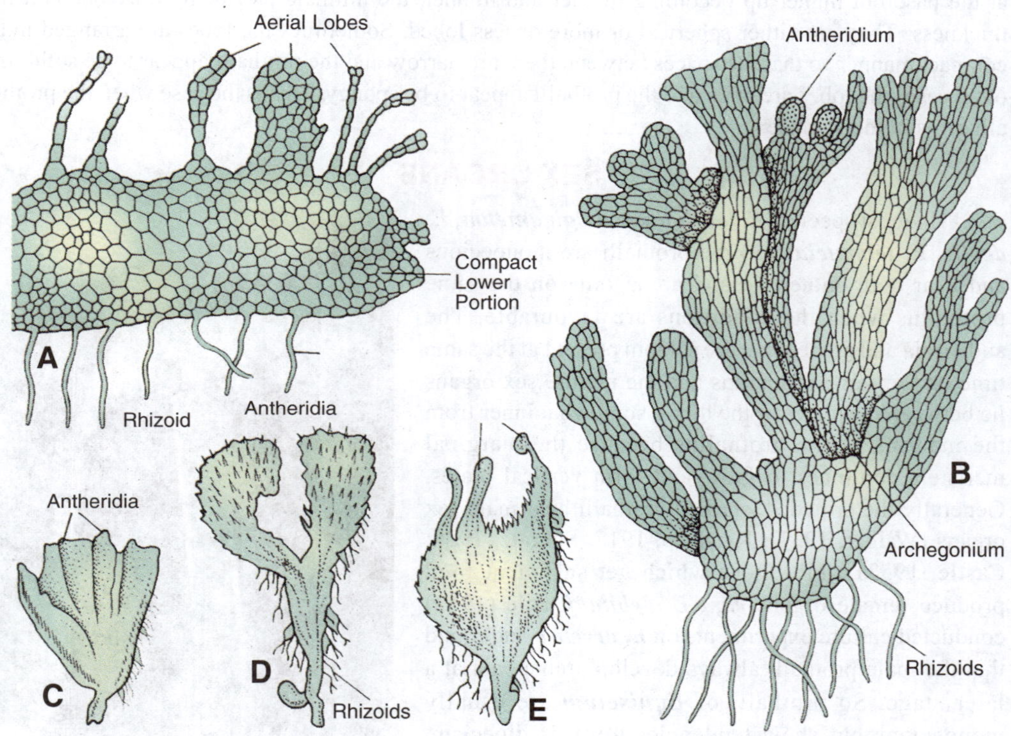

Fig. 7.14. (A–E). Prothallus of *E. debile.* A. L.S. of young prothallus; B. A bisexual prothallus showing a lower compact region bearing rhizoids and archegonia and upper spongy region bearing antheridia; C. A male prothallus; D. A female prothallus; E. A bisexualprothallus.(C–E. After Jyotsna and Mohan Ram)

ANATOMY OF THE PROTHALLUS

Internally the prothallus is differentiated into two well marked regions:

1. Lower or basal compact, rounded, parenchymatous portion forming the disc (Fig. 7.14, A).
2. An upper spongy portion (Fig. 7.14, B).

Disc

It consists of a large celled parenchymatous tissue without any air spaces between them. The cells are usually all alike. They lack chlorophyll and are full of starch grains. The chloroplasts which were present in the whole body of the prothallus in the earlier stages have all been transformed into leucoplasts. This transition is gradual and starts from the cells in the lower region whereas the cells in the upper region still contain chloroplasts and are green. The disc has an outer meristematic region which increases the diameter of the disc radially. It produces the parenchymatous cells of the compact portion and the spongy lobes of the upper portion and also the sex organs.

Spongy Upper Portion

In consists of densely crowded green vertical lobes which completely cover the disc delow (Fig. 7.14). The lobes are irregular, plate-like expansions of photosynthetic tissue several cells thick at the base but higher up becoming thinner and thinner, the ultimate part being only one cell in thickness. They are either spherical or more or less lobed. Sometimes the lobes are arranged in a compact manner so that the spaces between them are narrow and the prothalli appear to be solid. In other cases the lobes are open and the prothalli appear to be spongy. This is the case when the prothalli grow in moist place.

SEX ORGANS

In most species (***E. laevigatum***, ***E. giganteum***, ***E. debile***, ***E. telmateia*** etc.) the prothalli are monoecious and bear both antheridia and archegonia on the same prothallus when the conditions are favourable. The antheridia and archegonia are seldom present at the same time on the same prothallus and the mature sex organs lie between the plates or the lobes, somewhat inner from the margins of the prothallus because the marginal meristem continues to grow and form vertical plates. Generally the crowded, starved prothalli bear male sex organs (Walker, 1931 ; Kashyap, 1917; Schratz, 1928; Castle, 1953) while those which get sufficient food produce female organs ***e.g.***, ***E. debile***. Castle (1953) conducted culture experiments on ***E. arvense*** and found that the male prothalli always develop archegonia at a later stage. So prothalli of ***Equisetum*** are usually monoecious but show tendencies towards dioecism. Hauke (1963) also found that in some species of ***Equisetum*** the male and the female prothalli can be recognised quite early during their development, ***i.e.***, even before they bear sex organs. The male prothallus is small and the part bearing the antheridia is more or less folded. The female prothallus is larger of the two. The male prothalli are yellowish to reddish as contrasted to the green female prothalli. The colour of male prothalli is due to the presence of orange-red pigment which develops in the chloroplasts of the antheridial jacket cells. Joyet—Lavergne (1926-34) found that in some species of ***Equisetum*** (***E. arvense***, ***E. maximum***, ***E. limosum***) there are two categories of spores based on their physico-chemical reactions. These spores are alike so far as their size and structure are concerned but they show different chemical reactions. The spores of category 'A' germinate to produce female prothalli and those of category 'B' produce male prothalli. The former stain more heavily with neutral dyes than the latter. Gullvag (1968) and Duckett (1970) do not agree with Joyet-Lavergne and regard his observations to be faulty.

E. giganteum

Archegonia (Fig. 7.15)

They develop on young prothalli in the meristematic rim when the upright lobes are beginning to develop. They mature with the lobes and are usually found near their base and between them. In the older prothalli they become numerous and are distributed over the upper surface of the prothallus. The prothallus ceases to grow after fertilisation of the first formed archegonia.

Structure

The mature archegonia (Fig. 7.15,I, J) have the base sunken in the prothallus tissue with only its neck protruding. The neck is short and consists of four rows with usually 3 or 4 cells in each row. The neck cells of the uppermost tier are divaricate (bent back) at maturity thus leaving a wide opening for the entry of the sperms. The axial row consists of the egg cell, the ventral canal cell, and one or two neck canal cells. If the number of the neck canal cells is two, they lie side by side and are "bootshaped" (Fig. 7.15, I). At maturity there is the usual gelatinisation of all axial cells but the egg.

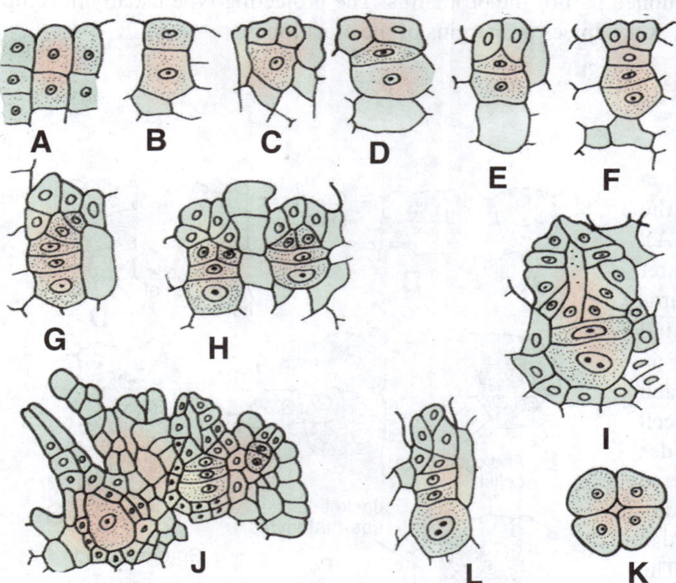

Fig. 7.15. (A–L) Various stages in the development of archegonium in *E. debile* (After Jyostsna and Mohan Ram)

Development (Fig. 7.15)

Archegonium initial (Fig. 7.15, A) is a superficial cell of the marginal meristem. It divides by a periclinal wall. The outer daughter cell is the primary **cover cell** (Fig. 7.15, B). The inner daughter cell is the **central cell** (Fig. 7.15, B). The primary cover cell divides quadrately to form four primary neck cells (Fig. 7.15, C-D). Each of these divides transversely to form an archegonial neck 3 or 4 cells in height (Fig. 7.15, G). It projects vertically above the thallus. The central cell divides transversely into a **primary canal cell** and a **primary ventral cell** (Fig. 7.15, E). In some species there is rarely a division of the **primary canal cell** whereas in others it divides regularly into two 'boot-shaped' neck canal cells (Fig. 7.15, H,I) by a vertical division (***E. debile***, ***E. arvense***, ***E. hyemale*** and ***E. limosum***). Kashyap (1914) reported only one canal cell in ***E. debile*** but Sethi (1928) reported two boot-shaped cells lying side by side. The primary ventral cell divides by a transverse division into a **ventral canal cell** and **an egg cell**. Jyotsna and Mohan Ram (1968) also observed the division of the primary neck canal cell, in ***E. ramosissimum***, into two boot-shaped neck canal cells. They also observed archegonia with two ventral canal cells (Fig. 7.15, K).

E. ramosissimum

The last step in the archegonial development is the usual gelatinisation of the neck canal cells and the ventral canal cell, thus forming a passage for the entry of spermatozoids.

Antheridia (Fig. 7.16)

They develop later when the prothallus is several months old and is larger in size with a red colour and folded margins. They are produced in large numbers, occurring chiefly on that part of the prothallus which does not develop lobes and lacks chlorophyll. In ***E. debile***, the margins bearing them have bright colour and smooth shining surface. They develop in an acropetalous succession and are of two types. These are the **embedded type** and the **projecting type.** The embedded type develop on the lower massive and cushioned part of the prothallus. The projecting type usually develop in starved prothalli and are found at the apices or margins of the erect lobes.

Development of Antheridia

A. Embedded type

It develops from a superficial cell cut off by the marginal meristem (Fig. 7.16, A). The younger antheridia lie nearer to the margin and the older towards the centre. The antheridia initial divides by a periclinal wall into an outer **primary jacket initial** and an inner **primary androgonial cell** (Fig. 7.16, B). The former divides anticlinally to form a single layered jacket or wall of the antheridium. Sometimes a triangular cell also differentiates in the wall cell. The primary androgonial cell divides several times to form a group of androgonial cells (Fig. 7.16, E—G) which divide further and form **androcyte mother cells** (Fig. 7.16, H). There are usually 256 such cells and each one of them has a centrally placed nucleus surrounded by chloroplasts which later on disintegrate. Two **blepharoplast granules** appear on either side of the nucleus. The nucleus divides and a wall appears between the two daughter nuclei thus forming two androcytes (Fig. 7.16, D, E). Like this the 256 androcyte mother cells form 512 androcytes. The protoplast of each androcyte metamorphoses into a multiflagellate spermatozoid (Fig. 7.17, A—J).

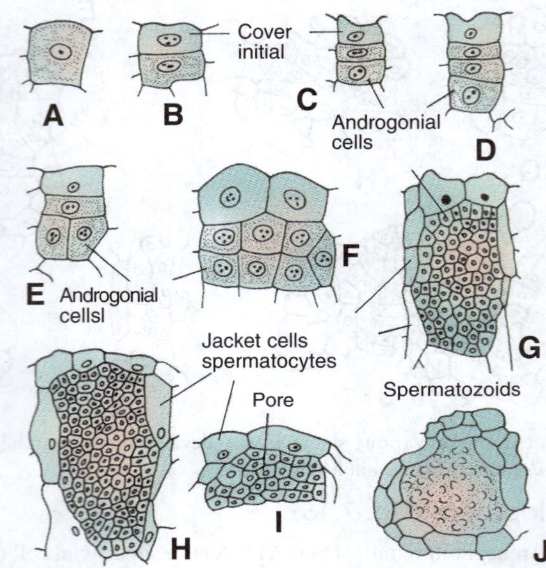

Fig. 7.16. (A–J) Stages in the development of antheridium (after Jyotsna & Mohan Ram)

B. Projecting type (Fig. 7.14, B)

In this type a marginal cell or a terminal cell of the erect lobe undergoes division by three walls laid in three directions. This results in cutting off of a central tetrahedral cell called the **antheridial initial**. The initial divides by a periclinal wall to form an outer **primary wall cell** and an inner **primary androgonial cell.** The rest of stages are similar to the embedded type. The number of androcytes formed in this case is smaller than in the embedded type.

Spermatozoid

The androcyte metamorphoses into a spermatozoid. Its contents (Fig. 7.17, A) are a distinct nucleus, a **blepharoplast, mitochondria**, a few degenerating **chloroplasts** and cytoplasm. The blepharoplast is regarded by Sharp (1912) to be composed of a number of granules. These granules fuse to form a continuous thread (Fig. 7.17, G,H). The nucleus shifts to one side of the androcytes and elongates and becomes curved (Fig. 7.17, G, I). One end of the nucleus is next to the blepharoplast (Fig. 7.17, G, H). The nucleus now elongates considerably and becomes spirally coiled (Fig. 7.17, I).

The blepharoplast thread also coils spirally. Later numerous flagella appear on the anterior part of the broad and spirally coiled protoplast. The coiled nucelus is surrounded by cytoplasm. The posterior part of antherozoid has vacuolate protoplasm (Fig. 7.17, J). A vesicle is also visible at the posterior end. It contains remains of the androcyte cytoplasm.

The mature spermatozoid has (Fig. 7.17, J) the usual three parts, the nucleus, the motor apparatus and the vesicle. The motor apparatus is the cilia bearing band, also called the **stalk.** The vesicle becomes visible only after the liberation of the spermatozoid. It absorbs water and increases in size. Major portion of the spermatozoid is made up of nucleus and blepharoplast. Cytoplasm also envelops the entire spermatozoid.

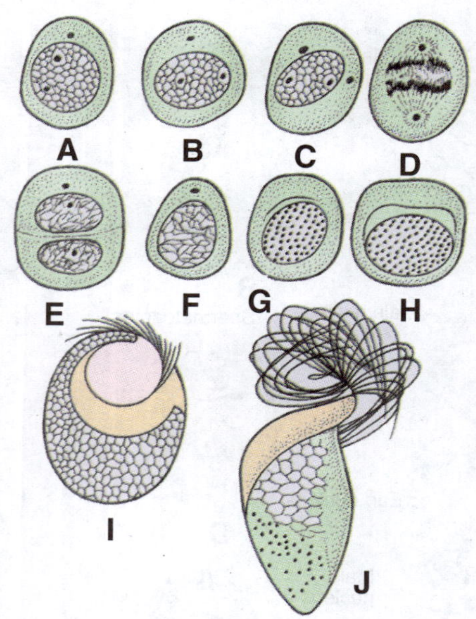

Fig.7.17. (A–J) Stages in the development of spermatozoid in *Equisetum.* (After Sharp)

Mature antheridium

The mature antheridium is a more or less globular structure. It is sessile and may be embedded in the lower massive cushion-like part of the prothallus or it may be borne on the margin or apex of the erect lobes. The body of the antheridium consists of a single layered outer wall or the jacket. The jacket may consist of 3—5 cells. It encloses a mass of up to 582 spermatozoids.

Hauke (1963) studied the methods of dehiscence of the antheridium and release of spermatozoids. Presence of water is essential for dehiscence. The cells around the antheridium absorb water by osmosis and enlarge. Water is also absorbed by a hydrophilic substance surrounding the androcytes. The pressure caused by the swelling of hydrophilic substance and the cells around the antheridium force the mass of spermatozoids out through an opening in the antheridial wall. This opening or a slit-like aperture is formed in the antheridial wall by the separation of its cells.

Fertilization (Fig. 7.18)

Water is essential for fertilization. The spermatozoids are attracted by the smell of malic acid present in the mucilage around the open archegonial necks. A number of spermatozoids enter the neck and reach the venter but only one is able to effect fertilization (Fig. 7.18, C—F) and convert the haploid egg into a diploid structure called the **zygote** or the **oospore**. Sethi (1928) noted that the entire spermatozoid penetrates into the egg cell in ***E. debile***. Many archegonia are fertilized on one prothallus. Kashyap (1914) observed as many as 15 embryos on the prothallus in ***E. debile***. This is quite contrary to other vascular cryptogams where only one embryo develops on one prothallus. Fertilization is usually effected in the months of December.

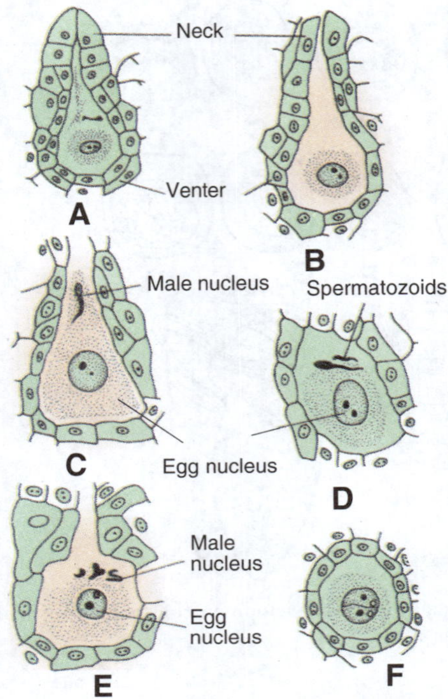

Fig. 7.18. (A–F) *Equisetum debile.* Stages in fertilization. (After Jyotsna & Mohan Ram)

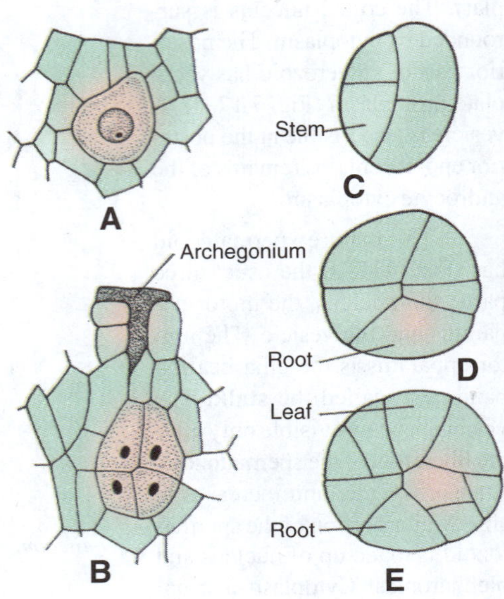

Fig. 7.19. (A–E) Earlier stages in the development of embryo in *E. arvense.*

DEVELOPMENT OF EMBRYO

The oospore is the pioneer structure of the sporophyte. It develops into an embryo within the archegonial venter, which grows into a two or more layered **calyptra**. The archegonial neck is also seen to persist in some species (***E. arvense***). The detailed embryology of ***E. arvense*** is described below :—

E. arvense

The oospore divides by a transverse wall into an upper **epibasal** cell and a lower **hypobasal** cell. There is no suspensor. The epibasal cell lies nearest to the archegonial neck. Both the cells divide by a longitudinal wall to form **quadrants** (Fig. 7.19). The epibasal quadrants are larger than the hypobasal ones. The next wall in each quadrant is at right angles to the second wall thus forming an eight celled or an **octant stage.** Out of the epibasal octants the largest cell functions as the **shoot apical cell.** According to Sadebeck (1905), this cell divides by three interesecting walls enclosing a central tetrahedral cell which functions as shoot apical cell. The peripheral three cells act as the **leaf initials**, which remain fused to form the leaf sheath (Fig. 7.20). According to Kashyap (1914), Sethi (1928), Campbell (1928) and Bower (1935), the shoot apical cell does not divide in a regular manner to cut off leaf initials but the cells lying lateral to them grow out as a first sheath of three scalelike leaves. The hypobasal octants by now have divided to form a massive foot. One of the large cells lying at the base of the hypobasal half acts as the **root initial** and cuts off a **cap initial** in

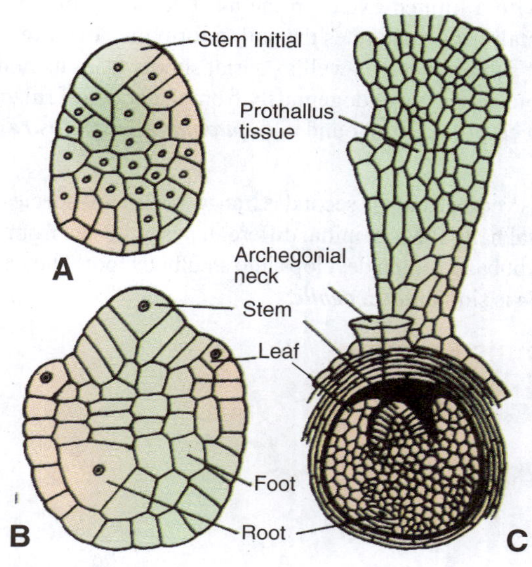

Fig.7.20. Embryology in *E. arvense.* (after Sadebeck)

the beginning. Since the stem and the leaves differentiate in the epibasal region lying close to the archegonial neck and there is no suspensor, the embryology in ***Equisetum*** is **exoscopic.**

The shoot initial is tetrahedral in form and cuts off segments towards its three faces and leads to the formation of a conical meristem that constitutes the stem apex of the adult shoot. The three leaf initials form the first leaf sheath. The root grows slowly and pierces through the underlying prothallus tissue and grows into a small and short-lived first or **primary root.** The stem apex divides actively and grows out after piercing the calyptra. Second and third leaf sheaths arise from the nodes of the young stem and give it a characteristic jointed appearance. Other leaf whorls arise at regular intervals on the stem. The later formed leaf sheaths have 4—7 leaves. The primary stem grows in air, turns green and stops growing after forming a few nodes and internodes (Fig. 7.21, A—F). Later a secondary branch arises from its base (Fig. 7.21, A—F). It grows to form 10—15 nodes and internodes and stops growing. Like this

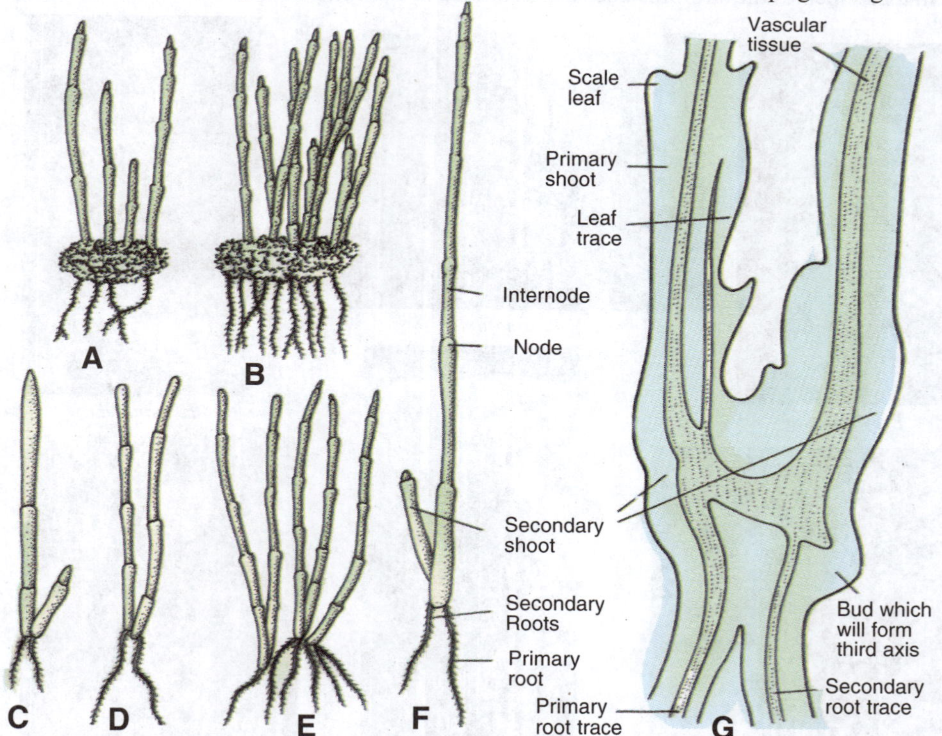

Fig.7.21. Sporophytes of ***Equisetum debile*** (A – E); and ***E. arvense*** (F–G); F. A young sporophyte of ***E. arvense*** with primary and secondary; branches; G. L.S. of F.

tertiary and quaternary shoots arise and grow to a limited extent in the air. Ultimately one of the shoots instead of growing erect grows horizontally and penetrates the soil to form the **rhizome.** The rhizome then grows indefinitely and gives off subterranean as well as aerial shoots from its nodes. The primary root dies and whorls of adventitious root arise endogenously from the nodes of rhizome and establish the young plant. This sequence of embryology is found in ***E. maximum***, ***E. palustre*** and ***E. telmateia.***

In ***E. debile*** the first wall of the oospore is vertical and the second is transverse. In this species the shoot, leaves and root initials arise in the epibasal half. The root initial differentiates quite late from one side of the lower half of epibasal region. The hypobasal region develops into a bulbous foot. The rest of the stages are similar to ***E. arvense. E. hyemale*** is similar to ***E. debile.***

Taxonomic Position

Division	:	*Sphenophyta*
Class	:	*Calamopsida*
Order	:	*Equisetales*
Family	:	*Equisetaceae*
Genus	:	***Equisetum***

Classification of the Genus

The genus has further been classified into two subgenera. These are: (*i*) subgenus ***Equisetum*** and (***ii***) sub-genus ***Hippochaete. Equisetum*** includes species in which the stomata are not sunken and are in level with epidermal cells, ***e.g.***, ***E. sylvaticum***, ***E. pratense***, ***E. limosum***, ***E. palustre***, etc. In these species the strobilus does not terminate in a pointed or apiculate tip of the axis, but it

E. sylvaticum — *E. limosum* — *E. telmateia.* — *E. maximum* — *E. scripoides* — *E. palustre*

bears sporangiophores at its apex also. The apex of the strobilus is, therefore, rounded. The stomata are not scattered or grouped in broad bands. In ***Hippochaete*** the stomata are sunken below the epidermis and are not scattered. They are always arranged in narrow bands. The cone axes project slightly beyond the sporangiophores and are apiculate. Some of the species included in this subgenus are ***E. scripoides***, ***E. giganteum***, ***E. hyemale***, ***E. ramosissimum***, ***E. laevigatum*** etc.

Chromosome Number

The basic chromosome number established as a result of cytological studies on some species is reported to be n = 108 (Manton, 1950; Ninan, 1955; Mehra and Bir, 1959).

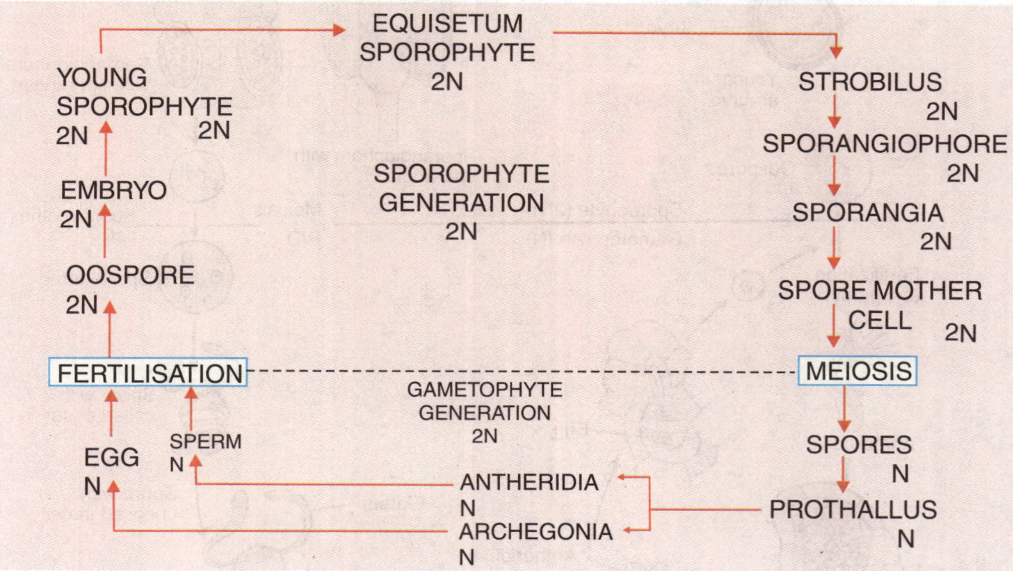

Fig. 7.22. A graphic life sketch of *Equisetum*.

Economic Importance

Presence of silica in the stem gives a rough and abrasive touch and bundles of stems are used to clean utensils and for polishing wood. Some species are of great medicinal value. The plant of ***E. arvense*** is used as a diuretic in Germany. The ashes of this plant are used to relieve acidity and dyspepsia. This plant yields silica in a therapeutically active form. ***E. debile*** yields a cooling medicine that relieves gonorrhoea. The stems of ***E. hyemale*** are used for polishing. Benedict (1914)

E. debile

E. hyemale

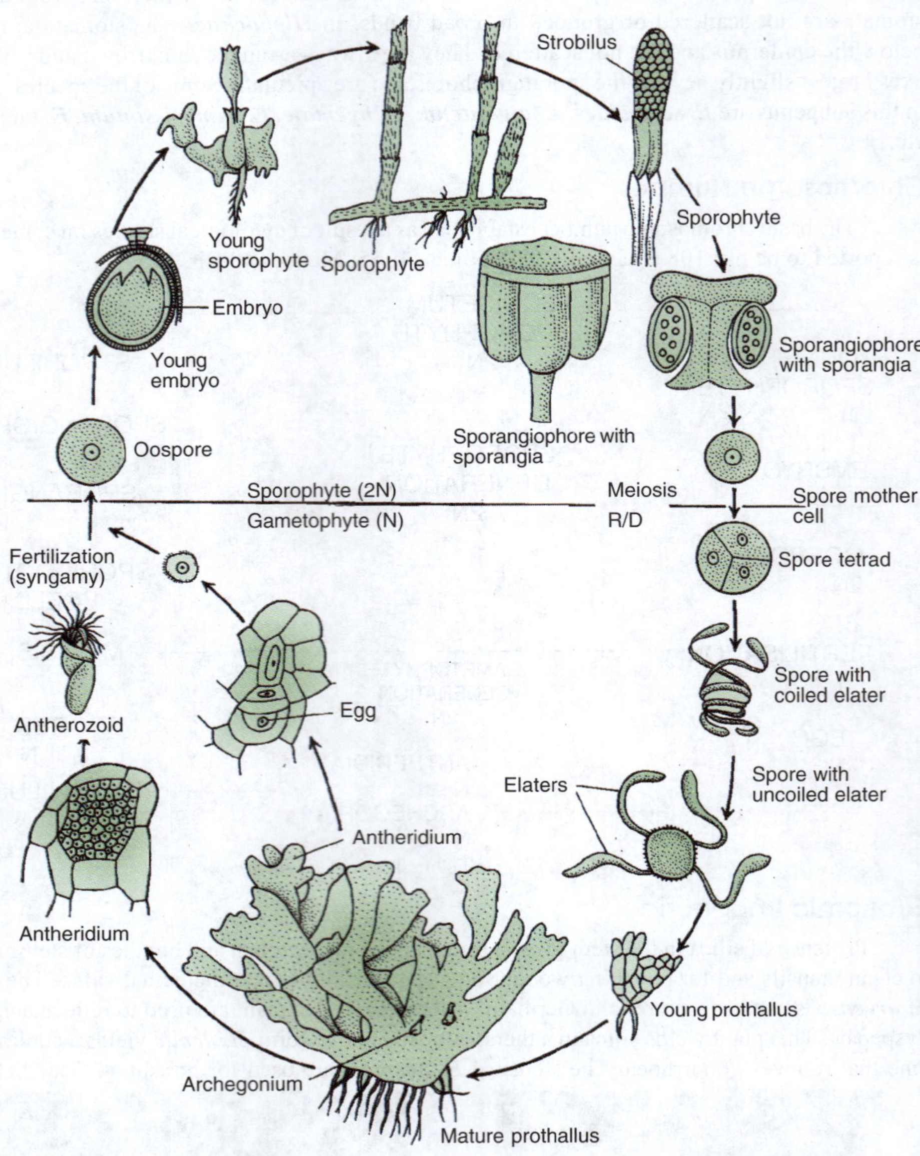

Fig. 7.23. ***Equisetum sp.*** Digrammatic life cycle

reported the presence of gold (4½ oz. per ton) in some species. Some species are also used as indicators of mineral contents in the soil.

Graphical and diagrammatic life cycle of Equisetum sp. are represented in Fig. 7.22 & 7.23.

QUESTIONS

ESSAY TYPE

1. Describe the structure of sporangiophore of ***Equisetum.***
2. Describe the distribution and structure of ***Equisetum*** and point out the features of special interest in its anatomy.
3. Describe in detail the gametophytic generation in ***Equisetum.***
4. Describe the development and structure of gametophyte of ***Equisetum.***
5. ***Equisetum*** has both hydrophytic and gametophytic characters. Discuss this statement.
6. Give an illustrated account of life-history of ***Equisetum.***
7. The sporangiophore of ***Equisetum*** is not a sporophyte. Discuss.

   or

   Describe the general features of ***Equisetum.*** In what respects do they differ from ***Marsilea***? Discuss.
8. Give a comparative account of the prothallus of ***Equisetum*** and ***Adiantum.***
9. With the help of labelled diagrams only describe the life-cycle of ***Equisetum.*** **(Kanpur, 2001; Poorvanchal, 2002)**
10. Describe the internal structure of internode of aerial stem of ***Equisetum.***
11. Explain in brief the life cycle of ***Equisetum*** indicating the stages where changes in number of chromosomes take place.
12. Give in brief a comparative account of internal structure of rhizomes of ***Equisetum*** and ***Marsilea***.
13. Describe the structure of ***Equisetum*** plant, cone and prothallus.
14. Give a detailed and illustrated account of the spore bearing organ (strobilus) in ***Equisetum.***
15. Describe the structural organisation of the Cone of ***Equisetum*** with suitable diagrams.
16. Give an account of habit, habitat and internal structure of ***Equisetum*** sporophyte. Give the organisation of its cone also.
17. Describe briefly the life history of ***Equisetum.***
18. Describe the external and internal structure of sporophyte of ***Equisetum.*** **(Bangalore, 2001)**
19. Describe the ecology of external and internal structure of ***Equisetum*** stem. **(Himachal Pradesh, 2000)**
20. Draw a well labelled diagram of T. S. Stem of ***Equisetum***. Point out features of anatomical and ecological interest.
21. Describe the organisation of strobilus, structure of sporangiophore and dehiscence of sporangium in ***Equisetum.***
22. Describe the distribution and external and internal anatomy of ***Psilotum*** and ***Equisetum.***
23. How do the stems of ***Selaginella*** and ***Equisetum*** differ on their internal/anatomical characteristic features? Give suitable diagrams only for your answer.
24. Describe the spore bearing organ of ***Equisetum***. In what way it is different from its fossically?
25. (*a*) With a diagram, give an account of internodal anatomy of ***Equisetum*** stem.

    (*b*) Describe the morphology and anatomy of aerial stem of ***Equisetum.*** **(Awadh, 2003)**

    (*c*) Draw a labelled diagram of ***Equisetum*** plant and mention the xerophytic and hydrophytic characters which are present in it. **(Allahabad, 2003)**
26. Give the comparative account of strobilus of ***Selaginella, Lycopodium*** and ***Equisetum***. **(Garhwal Univ., 2005)**
27. Describe the L/S of ***Equisetum*** cone **(Kuvempu Univ., 2005)**
28. Describe the morphology of sporophyte of ***Equisetum***. **(Mysore Univ., 2004)**
29. Discuss the cone of ***Equisetum*** with suitable diagrams. **(Garhwal Univ., 2005)**

SHORT ANSWER TYPE

30. Describe the development and structure of archegonium of ***Equisetum***.
31. Draw labelled sketch of the following.

    (*i*) L. S. of strobilus of ***Equisetum.***

    (*ii*) T. S. and L. S. of the cone of ***Equisetum.*** **(Lucknow, 2000, 2004)**

    (*iii*) T. S. Rhizome of ***Equisetum.***

    (*iv*) T. S. of aerial stem of ***Equisetum.*** **(Lucknow 2001; Awadh, 2000; UP College, 2000; Rohilkhand, 2000)**

    (*v*) T. S. of aerial stem of ***Equisetum*** through internode. **(Kanpur, 2000; Kerala, 2001; Nagpur, 2002)** (*vi*) T. S. Cone of ***Equisetum.***

(*vii*) T. S. aerial stem (through node) of ***Equisetum.***

(*viii*) Vertical section of Prothallus of ***Equisetum***.

(*ix*) L. S. Cone of ***Equisetum.*** **(Allahabad, 2004; Awadh, 2003)**

(*x*) Internal structure of (T.S.) root of ***Equisetum.*** **(Poorvanchal, 2000; Kanpur, 2004)**

(*xi*) Mature cone.

(*xii*) Dry spore of ***Equisetum.***

(*xiii*) L. S. prothallus of ***Equisetum.***

(*xiv*) Cone of ***Equisetum.*** **(Kerala, 2001)**

(*xv*) Development of ***Equisetum*** Prothallus. **(Garhwal, 2000)**

(*xvi*) (*a*) L.S. of ***Equisetum*** cone.
(*b*) T.S. Stem of ***Equisetum*** (Internode). **(Nagpur Univ., 2005)**

32. Write briefly on the following :

(*a*) Cone of ***Equisetum.***

(*b*) Sporangiophore of ***Equisetum.*** **(Awadh, 2003)**

(*c*) Strobilus of ***Equisetum.*** **(Nagpur, 2002)**

(*d*) Spore of ***Equisetum*** **(Kanpur, 2005; Poorvanchal, 2000; Meerut, 2000; Lucknow, 2003)**

(*e*) Prothallus of ***Equisetum.*** **(Awadh, 2000)**

(*f*) Gametophytes of ***Equisetum.***

(*g*) Xerophytic and hydrophytic characters of ***Equisetum.***

(*h*) Ecological features in the anatomy of aerial stem of ***Equisetum.***

(*i*) Root of ***Equisetum.***

(*j*) Internal structure of ***Equisetum*** stem. **(Poorvanchal, 2002)**

(*k*) Xerophytic characters of ***Equisetum.*** **(Gauhati, 2000)**

(*l*) Stem of ***Equisetum.***

(*m*) Stem anatomy of ***Equisetum.***

33. In which plant and where do you find vallecular canal.

34. Describe in 30 words antherozoids of ***Equisetum.***

35. Name a pteridophyte whose spores have elaters.

36. Which pteridophyte has a spore with elaters?

37. Name the type of stele found in ***Equisetum.***

38. Where and in which plant vallecular Canal is found?

39. How is epidermis of the stem of ***Equisetum*** different from that of ***Cycas*** leaflet?

40. How is spermatozoid of ***Equisetum*** different from that of ***Cycas***?

41. What is the function of elaters of spores of ***Equisetum***?

42. What is the function of camial canal?

43. Mention the name of the pteridophyte in which vessels are found :

44. Describe the hydrophytic and xerophytic characters of stewes by ***Equisetum.***

45. Draw a diagram of ***Equisetum*** spore.

46. Name the ***cavity*** which is formed by the disintegration of protoxylem tracheids in ***Equisetum***.

47. (*a*) Name a pteridophyte in which both xero phytic and hydrophyte characters are present.
(*b*) Name the lysigenous cavities found in the stem of ***Equisetum.*** **(Allahabad, 2003)**

48. How is epidermis of ***Equisetum*** different from that of other pteridophytes? **(Allahabad, 2002)**

49. Explain with suitable diagram the structure of mature dry spore of ***Equisetum.***

50. Describe briefly the internal structure of root of ***Equisetum.*** **(Lucknow, 2001)**

51. Draw a labelled diagram of the L. S. of ***Equisetum*** strobilus. **(Kanpur, 2001)**

52. Though ***Equisetum*** is homosporous yet it produces dioecious thalli. How do you account for it? Explain.

53. What changes do you expect, if the spores of ***Equisetum*** are subjected to moist and dry conditions. **(Poorvanchal, 2000; Awadh, 2003)**

54. Draw labelled diagrams of dry and wet spores of ***Equisetum.***

55. Discuss the nature of gametophytes of ***Equisetum.***

56. Mention the hydrophytic characters of ***Equisetum.***

57. Name the two canals found in ***Equisetum.***

58. Name a pteridophyte which shows both xerophytic and hydrophytic characters.

59. How is epidermis of ***Equisetum*** different from that of other pteridophytes? **(Poorvanchal, 2000)**

60. Name the type of development of sporangia in ***Equisetum.*** **(Poorvanchal, 2002)**

61. Differentiate the following pairs (by ***diagrams only***):

(*a*) Stomata of ***Equisetum*** and ***Cycas.***

(*b*) Carinal canal and vallecular canal.

(*c*) Sperms of ***Lycopodium*** and ***Equisetum.***

(*d*) Sperms of ***Equisetum*** and ***Marsilea.***

(*e*) Spores of ***Equisetum*** and ***Pteridium.***

(*f*) Roots of ***Marsilea*** and ***Equisetum.***

(*g*) Autherozoids of ***Equisetum*** and ***Marsilea.***

(*h*) Morphology of aerial stem of ***Lycopodium*** and ***Equisetum.***

(*i*) (1) Strobilus of ***Selaginella*** and ***Equisetum.***

(2) Prothallus of ***Lycopodium*** and ***Equisetum.***

(*j*) ***Selaginella*** root and ***Equisetum*** root.

(*k*) Prothallus and sex organs of ***Equisetum*** and ***Adiantum.***

(*l*) Prothallus of ***Lycopodium*** and ***Equisetum.***

(Lucknow, 2003; Kanpur, 2005)

62. Draw an outline diagram showing aerial stem of ***Equisetum*** in a T. S. List its ecological features.

63. Draw a labelled diagram showing aerial stem of ***Equisetum*** in a T. S. Name the type of stele present in it.

64. (*a*) Describe the structure of antherozoid of ***Equisetum***.

(*b*) Describe the occurrence and external morphology of the vegetative parts of ***Equisetum.***

65. Describe the organisation of strobilus in ***Equisetum.***

66. Mention the xerophytic characters in ***Equisetum.***

67. How are sporangia arranged in ***Equisetum.***

68. Mention any two features of the following :

(*i*) Cone of ***Equisetum***

(*ii*) Spore of ***Equisetum***

(*iii*) Stem of ***Equisetum***

(*iv*) Root of ***Equisetum***

(*v*) Elaters of ***Equisetum***

69. Define Elaters of ***Equisetum.***

70. Critically review the views on the morphology of the sporangiophore of ***Equisetum.***

71. Describe the structure and development of ***Equisetum*** antherozooid.

(Kanpur, 2003)

OBJECTIVE TYPE

72. Fill in the blanks:

(*a*) The ***Equisetum*** and ***Ephedra*** stem have similarity in having ---------------------- and ----------------------.

(*b*) Spores of ***Equisetum*** are unique in having ----------------------.

(*c*) Cone of ***Equisetum*** is known as ---------- -------------.

(*d*) Multiflagellate sperm are found in -------- ---------------.

(*e*) Vallecular and carinal canals are found in ----------------------.

(*f*) The cavities in the cortex of ***Equisetum*** stem are called ----------------------.

(*g*) ***Equisetum*** is commonly known as ------- ----------------.

(*h*) A cavity formed by the disintegration of protoxylem elements in ***Equisetum*** is called ----------------------.

(*i*) Simple collateral vascular bundles are found in ----------------------.

(*j*) Endodermis in the root of ***Equisetum*** is ---------------------- layered.

(*k*) ***Equisetum*** is ---------------------- heterosporous.

(*l*) Intercalary meristem is present in -------- ---------------.

(*m*) ***Equisetum*** is characterised by the presence of ----------------------.

73. Select the correct answer :

(*i*) The spores with elaters are found in

(*a*) ***Equisetum***

(*b*) ***Adiantum***

(*c*) ***Lycopodium***

(*d*) ***Marsilea.***

(*ii*) Vallecular canal in ***Equisetum*** is situated.

(*a*) below the ridge

(*b*) Below the furrows

(*c*) Between the ridges and furrows

(*d*) Between pith and endodermis.

(*iii*) Sporangia are developed in ***Equisetum*** on

(*a*) upper surface of sporophylls

(*b*) Sporangiophores

(*c*) Scale leaves
(*d*) Lower surface of sporophyll.

(*iv*) Horsetail is the common name of
(*a*) ***Lycopodium***
(*b*) ***Selaginella***
(*c*) ***Equisetum***
(*d*) ***Marsilea.***

(*v*) Simple collateral vascular bundes are found in
(*a*) ***Pleopeltis***
(*b*) ***Equisetum***
(*c*) ***Selaginella***
(*d*) ***Lycopodium.***

(*vi*) In ***Equisetum*** the sporangia are developed on
(*a*) upper surface of sporophylls
(*b*) sporangiophores
(*c*) Vegetative leaves
(*d*) lower surface of sporophylls.

(*vii*) In the root of ***Equisetum***, the endodermis is
(*a*) absent
(*b*) one layered
(*c*) two layered
(*d*) three layered.

(*viii*) Vallecular Canal in ***Equisetum*** is situated
(*a*) below the ridges
(*b*) below the furrows
(*c*) between ridges and furrows
(*d*) between pith and endodermis.

(*ix*) In which of the following the carinal canal is present in the stem?
(*a*) ***Selaginella***
(*b*) ***Equisetum***
(*c*) ***Marsilea***
(*d*) ***Adiantum.***

(*x*) Hygroscopic spoon shaped elaters are found on the spores of
(*a*) ***Equisetum***
(*b*) ***Lycopodium***
(*c*) ***Pteris***
(*d*) ***Cycas***

Matteucia struthiopteris

CHAPTER 8

Filicophyta or Pterophyta

- Introduction
- Fossil Records from India
- Distinctive characters
- Vegetative propagation
- Stelar system in vascular cryptogams
- Spore producing members
- Prothallus/the gametophyte
- The sex organs
- Embryology
- Evolutionary trend among Filicophyta
- Classification

Introduction

The Filicophyta (ferns) are an assemblage of vascular cryptogams that have established themselves most successfully to life on land. Their successful invasion of the varied habitats, their diverse habit, their supremacy in vegetative propagation and their remarkable success in competition with the modern seed plants stamp the modern Filicophyta as having reached the highest point in the homosporous evolution. The living Filicophyta comprise about 305 genera and 10,000 species (Holttum, 1949) that have spread right from the equator to the arctic regions. The Indian hills especially the Himalayan ranges afford a luxuriant growth of these plants and more than 500 species have been reported. Dixit (1984) raised the number of species to more than one thousand. Bir, S.S. (1987) has given at detailed account of the work done in various fields of Indian ferns and fern allies. It is a comprehensive account of Pteridology in India and almost all aspects on which work has been done in India have been discussed. The account is followed by an exhaustive bibliography which is complete upto 1986. Kaur and Raza (1983) compiled bibliography of Indian Pteridology till the end of 1980. B. D. Sharma (1994) gave a comprehensive account of morphology, anatomy and reproductive biology of pteridophyta while literature on morphogenetic studies on pteridophyta was brilliantly reviewed by Harvinder K. Cheema (1994). Both the review papers were published in "Botany in India" edited by B. M. Johri (1994). They have cited nearly 1,500 references.

Geological records reveal that the ferns first appeared in the Post-Carboniferous period of the Palaeozoic times. They dominated the earth's crust in Carboniferous period which has been described as "the age of the ferns." Most of those fern-like plants have now been regarded as the seed plants and it is, therefore, not justifiable to call that age as the age of 'ferns'. Rather ferns were in minority in that age as compared to the lycopods and the seed plants. They are represented in much greater number in the modern age. Many of the modern families of the Filicophyta had their representatives in the Mesozoic times especially the Triassic and Jurrasic periods. The Polypodiaceous Filicophyta that are regarded as recent ferns, also had their representative in the early Coenozoic. They had their origin somewhere in the Middle Devonian period and have successfully braced the stress and strain of the adverse climate of the late Permian with the result that they competed well and are represented in the modern period by a varied and a dominant flora.

Fossil Records from India

The fossil record of the Indian filicophytes dates back to Lower Carboniferous of India. Hoeg, Bose and Shukla (1956) described some fern-like leaves from the Spiti Valley. These are compressions of ***Adiantites***, ***Sphenopteris***, ***Sphenopteridium***, ***Rhacopteris*** and ***Rhodea***. Structural details of these fragmentary compressions have not been worked out.

A large number of ferns and fern-like leaves have been reported from the Lower Gondwana of India. These include ***Alethopteris***, ***Asterotheca***, ***Belemnopteris***, ***Merianopteris***, ***Pecopteris***, ***Ptychocarpus***, ***Neuropteridium*** and ***Sphenopteris***. With our increasing knowledge of the fossil remains of Lower Gondwana, the names of above genera have undergone changes and Pant and Misra (1977) recognise only two fossil families of Lower Gondwana ferns in India. These are the families ***Damundopteridaceae*** and ***Asterothecaceae***. The former family includes the genera ***Damundopteris*** and ***Damundosorus*** and eusporangiate sporangia that are free. The family ***Asterothecaceae*** includes the genera ***Asterotheca***, ***Asansolia***, ***Dizengotheca*** and ***Trithecopteris***. They are characterised by synangiate type of sporangia found in the living family ***Marattiaceae***. The relationships of ***Damundopteridaceae*** are obscure. According to Pant and Misra (1977), "It would thus appear that out of the two contemporaneous lines of Lower Gondwana ferns, the damundopteridaceous stock became extinct but apparently the modern ***Marattiaceae*** are the remnants of the asterothecaceous line."

In a monograph on Indian Pteridophytes, Surange (1966) reported 67 members of the filicophyta from India. In a latest review on the Indian fossil Pteridophytes Sharma ***et al.*** (1986) gave a vivid account of the fossil ferns belonging to the families Osmundaceae, Gleicheniaceae, Dipteridaceae, Matoniaceae, Dicksoniaceae, Schizaeaceae, Cyatheaceae and Polypodiaceae. According to them the pteridophytes occur throughout the Gondwana system but more frequently during Mesozoic era and

Fossils of Pterophyta (Ferns)

the ferns make the major bulk of the assemblage. Upper Gondawana rocks are comparatively very rich in fossil ferns. Luthar and Sharma (1986) described ***Solenostelopteris jurassica*** of Schizaeaceae from Jurassic flora of Rajmahal hills, and presented a complete restoration of the plant with anatomical details of different organs.

Distinctive Characters

The filicophyta differ from lycophyta in possessing large and megaphyllous leaves. The leaf lamina is traversed by prominent and branched veins. The leaf traces and branch traces leave distinct gaps in the vascular organisation of the stem. The sporophyte reproduces extensively by vegetative means. The leaves have unique property of performing a dual function of photosynthesis and reproduction. They range in size from 2—5 cm. in ***Anogramma leptophyla*** to large tree ferns of ***Cyatheacea*** and some other families. Many love shaded and humid environments, a few like ***Ceratopteris***, ***Marselia***, ***Salvinia***, ***Azolla*** are aquatic; some are xerophytic, ***e.g.***, ***Woodsia.*** Hundreds of species are epiphytic (***Polypodiaceae***, some ***Davalliaceae***, some ***Aspleniaceae***). The rhizomes may be creeping upright or grow above the soil into columnar aerial stems. In many ferns, ***e.g.***, ***Stenochlaena*** the stem is weak and climbing, attached usually by adventitious roots. Branching may be dichotomous (***Pellaea***, ***Dryopteris***, ***Polypodium***, ***Pteridium***) or lateral with no definite relation to leaf insertion (***Adiantum***, ***Pteris***). Epipetiolar buds grow into branches in ***Pteris*** and ***Pteridium***. Leaves may be simple entire or lobed, pinnate compound or palmate compound or highly dissected in many ferns; pinnae may be dichotomously branched in some. Leaf dimorphism is quite common in many genera ***e.g.***, ***Pellaea***, ***Onoclea***, ***Vittaria*** etc. Venation may be dichtomous or reticulate. Dickson (1946) described various patterns of venation in the ferns. The sporangia are grouped in distinct **sori** and arise in groups from a swollen cushion of cells called the **placenta.** The sori are protected by **indusia** that may be true or false. The sporangia in most cases possess a distinct **annulus** and a **stomium.** The sporangial development may be **eusporangiate** or **leptosporangiate.** Majority are **homosporous**, but a few like ***Salvinia***, ***Azolla***, ***Marselia***, ***Pilularia*** and ***Regnelidium*** are **heterosporous.** The gametophytes may be **endosporic** or **exosporic.** In the homosporous types they are usually monoecious but dioecious condition is not rare. The antheridia and the archegonia are usually embedded or partly embedded in the gametophytic tissue and have

Anogramma leptophyla *Ceratopteris* *Woodsia* *Stenochlaena* *Salvinia* and *Azolla* *Pilularia*

Living Pterophyta (Ferns)

Cyathea gigantea

Marselia minuta

a structure resembling the other vascular cryptogams. The embryos may or may not have suspensor.

Anatomically the stem exhibits complex structure and ranges from a simple or a highly dissected protostele, or a siphonostele or a dictyostele to polycyclic conditions (see under stelar system). The development of leaf traces and leaf gaps makes the stem anatomy complex. The xylem lacks vessels except in ***Pteridium aquilinum*** and ***Marselia.*** The vessels have now been reported in larger number taxa ***e.g.***, ***Ophioglossum***, ***Helminthostachys***, ***Pteris vittata***; ***Nephrolepis cordifolia*** etc. Presence of sieve tubes with distinct sieve plates have been reported in ***Cyathea gigantea*** (Shah and Fotedar 1974) and in ***Marselia minuta*** (Loyal and Maheshwari, 1979). Branched tracheids (Bir ***et. al.*** 1984) and sieve elements (Shah et. al. 1976) have also been reported in some ferns. The roots exhibit a similar and an almost a constant pattern of tissues among the ferns. The leaf lamina and the petiole show a great amount of variation in their internal structure.

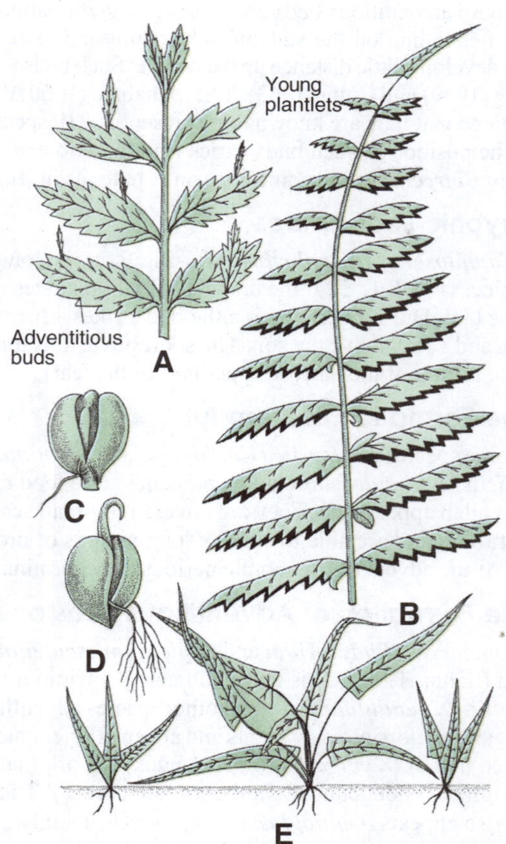

Fig. 8.1. Methods of vegetative propagation in ferns. A. *Asplenium viviparum* showing marginal adventitious buds that have grown into young plantlets on the pinnae; B. *Cystopteris bulbifera* : Portion of a leaf showing pinnae bearing adventitious buds on their bases; C—D. Germinating buds as in B; E. *Comptosorus rhizophyllus* : Vegetative propagation by means of drawn out leaf spices.

VEGETATIVE PROPAGATION

The ferns reproduce vegetatively by the following methods :

1. Fragmentation

It is of common occurrence among ferns with creeping axis or the rhizome. The older parts of the rhizome decay. When the progressive rotting of the older portions reaches beyond the branches, the latter separate and act as independent plants. Each branch has its own apical cell and is secured to the soil by adventitious roots. It grows into a new plant. The best examples are ***Pteridium aquilinum***, ***Dryopteris rigida***, ***Pteris*** and ***Adiantum.***

2. By the Formation of Adventitious Buds

Some ferns develop adventitious buds on the leaf lamina. They are source of vegetative propagation. In ***Asplenium bulbiferum*** (Fig. 8.1, A) and ***Diplazium celtidifolium*** develop them on the upper surface of the leaf lamina. In both cases the buds appear to be marginal in the first instance. In ***Cystopteris bulbifera*** they develop at the bases of the pinnae (Fig. 8.1, B—D). The adventitious buds in ***Woodwardia radicans*** are found on the upper side of the leaf at points immediately above the sori that have stopped growing. These adventitious buds are capable of germinating into new sporophyte plants. They may germinate after falling on the suitable substratum or ***in situ*** (***Asplenium viviparum***). In ***Dryopteris rigida*** buds develop a little distance up the petiole. Such buds were experimentally induced in ***D. dilatata*** by Wardlaw (1949) and Cutter 1963). T.S. Mahabale (1960) discovered the presence of buds in Indian ferns and stated that they are known to be present in 219 species belonging to different families of the Filicales. The position of such buds varies from base to apex of a leaf in different species. A few are axillary as in ***Botrychium*** and some arise on stipules as in ***Angiopteris evecta.***"

3. Embryonic Leaf Apices

In ***Camptosorus rhizophyllus*** (Walking fern) the long and simple leaves have drawn out and tapering apices (Fig. 8.1, E). If the drawn out leaf tip comes in contact with moist soil, it develops an adventitious bud. During germination the bud strikes adventitious roots in the soil and gives rise to young stem and a leaf above the soil. The sporophyte thus formed is separated from the parent sporophyte by the rotting of the narrow upper part of the leaf.

4. By the Formation of Stem Tubers

In some species of ***Marsilea*** (***M. hirsuta*** and ***M. minuta***) irregular tuberous bodies are formed on the stem. In ***M. hirsuta***, short lateral branches store food material in the form of starch and assume irregular swollen appearance. They are covered by small scale-like leaves. These swollen structures are called **tubers** and are able to survive long periods of drought. They persist when the rest of the plant dies. At the advent of favourable periods they germinate into new plants.

5. By the Formation of Adventitious Buds on Roots

In some ferns (***Platycerium*** and ***Asplenium esculentum***), the root apex develops directly into a leafy bud (Sahni, 1917). This leafy bud can grow into a new plant. In ***Ophioglossum vulgatum***, ***O. aitchisonii***, ***O. pendulum*** and some other species adventitious buds (Fig. 9.1) appear on the roots. Such buds can develop into new plants and are an efficient means of vegetative propagation. Wardlaw (1953) stated that in ***O. vulgatum*** the root buds may originate from segments of the apical meristem "or they originate endogenously in the middle cortex." They form spherical or ellipsoidal masses. Later root also appears. ***Ophioglossum*** reproduces mainly by this method.

The phenomenon of apogamy has been dealt in detail in Chapter I.

STELAR SYSTEM IN VASCULAR CRYPTOGAMS

Stelar Theory

Sachs (1875) formed an idea that the vascular system of the plant body is a continuous system. The idea was introduced with a greater emphasis by Van Tiegham and Douliot in 1886. They proposed

Pteridium aquilinum

Dryopteris rigida

Pteris

Adiantum

Diplazium celtidifolium

Cystopteris bulbifera

Woodwardia radicans

Asplenium viviparum

Platycerium

and developed the **Stelar theory** according to which the root and the stem are fundamentally similar in gross anatomy, because in both the cortex encloses the central part of the axis. The central part or the core of the axis is called the stele. So a ***stele (Greek word meaning column) is the core of the axis which includes the vascular system, the interfascicular portion, the pith (if present) and some surrounding portions of the fundamental tissue in the vicinity of vascular bundles*** (pericycle). This concept of the stele as proposed by Van Tiegham and Douliot was widely accepted by the plant morphologists and plant anatomists. The term stele is restricted only to the primary vascular tissue.

Nature of the Stele and Effect of Leaf Gaps

A debatable question is the nature of the stele, or the vascular cylinder in the filicophyta (ferns) and the seed plants. In these plants the large leaf traces seem to play a prominent role in determining the vascular organisation of the stem. Wardlaw and Wetmore (1951) and also Wardlaw (1946) consider these large leaf traces in the megaphyllous vascular plants as distinct units of the structure and regard the stele or the vascular cylinder of the stem as a composite structure. According to them the stem stele is made up of both foliar (leaf) and cauline (stem) vascular elements. The ratio between the two types of elements constituting the stem stele no doubt varies with plants. Wetmore (1953), however, regards that in some ferns the entire vascular organisation of the stem is foliar in nature. It has no doubt been shown (Wardlaw, 1944) that in ***Dryopteris dilatata*** "the incipient vascular tissue originates in continuity with the active meristems of shoots, buds, leaves and roots." The distinguishing vascular strand of the young leaf primordium, at the shoot apex, becomes united with the complete vascular cylinder of the stem. There are no leaf gaps at this stage. But lower down, ***i.e.*** as we proceed towards the mature parts of the stem, small leaf gaps begin to appear at the points of insertion of the leaf trace on the stele of the stem. These small **leaf gaps** enlarge in size with the enlargement of leaf trace bundles. Since the leaves are arranged in close spirals on the mature stem, the stele of the stem becomes broken by the overlapping leaf gaps and appears dissected. These observations indicate that there is a close relationship between the stelar organisation of the stem and the enlargement of the leaves. The formation of the leaf gaps has been explained by Wardlaw (1968) in his own words : "An anatomical analysis also shows that in growing leaf bases, there is a very considerable tangential enlargement, due chiefly to an increase in the volume of the cortical and medullary parenchyma. Concomitant with these developments, the leaf trace which is initially a crescentic mass in cross section, is pulled apart into four or five separate strands, with parenchyma in between. In brief the distribution of growth in the leaf base is such that the initially crescentic vascular trace is disrupted by being subject to tensile stress. The cylindrical, still undifferentiated stele of the shoot is also affected by this tensile stress and hence the formation of the leaf gaps."

It is evident from the above description that changes in the stelar organisation of the shoot are brought about by the appearance of leaf traces and leaf gaps. Wardlaw suggested that these changes can be stopped by defoliation of the shoot apices. He (1944, 1949) conducted experiments on certain ferns (***Dryopteris dilatata***, ***D. filix-mas***, ***Angiopteris evecta***, ***Onoclea sensibilis***, ***species of Osmunda*** and ***Todea***) by removing the young leaf primordia at the apices of the shoots. He obtained the desired results with ***Dryopteris dilatata*** and ***D. filix-mas.*** In both these cases the young leaf primordia were removed from the shoot apices. Normally the stems show a dictyostelic condition (Fig. 8.5, E). Under experimental conditions, ***i.e.***, when leaf primordia were removed the stem in that region showed solenostelic stelar organisation. Even protostelic condition was also obtained by such experiments. These experiments show that the stelar organisation of the stem in the filicophyta is dependent, to a greater degree on the presence or absence of leaves.

In the microphyllous vascular cryptogams (***Selaginella***, ***Lycopodium***, ***Isoetes***) the leaf traces do not leave any leaf gaps in the stem stele, which therefore remains undisturbed. In these cases the vascular system of the leaf has no bearing on the stem stele.

Dryopteris dilatata

D. filix-mas

Angiopteris evecta

Onoclea sensibilis

Types of steles

Easu (1953) and Smith (1955) recognise two principal types of stelar organisation among the vascular plants. These are : (A) **the Protostele, and** (B) **Siphonostele.**

A. The Protostele

It is simplest and the primitive type of stele. In this case the vascular cylinder consists of the solid core of xylem surrounded by phloem, pericycle, and endodermis (Fig. 8.2). There is no pith. The name **protostele** was suggested by Jeffrey (1897,1899, 1902) and is regarded as a fundamental stelar type of the vascular plants from which the other type originated in the course of evolution. Brebner (1902) classified the protostele into two :—

(*a*) Haplostele (Fig. 8.2.)

In this type of protostele the central core of xylem is smooth and is surrounded by a uniform layer of phloem. It was found in the extinct psilophytales like ***Rhynia*** (Fig. 2.2) and ***Horneophyton***. In the living genera it is characteristic of many species of ***Selaginella***, ***e.g.***, ***S. chrysocaulos***, ***S. kraussiana***, etc. In ***S. chrysocaulos*** the ribbon-shaped (Fig. 6.6, A) **haplostele** has two protoxylem groups, ***i.e.***, it is diarch and exarch. In ***S. kraussiana*** there are two haplosteles (Fig. 6.6, B). Such a condition is called **distelic** as compared to **monostelic** condition in ***S. chrysocaulos***. In ***S. willdenovii*** (Fig. 6.8) the condition is tristelic or even tetrastelic because in this case there are three or four haplosteles. In ***S. selaginoides*** the monostelic haplostele has metaxylem surrounding completely the protoxylem. This condition of xylem is called **mesarch.** Some ferns like ***Gleichenia dichotoma***, ***Cheiropleuria*** and ***Lygodium*** have stems that maintain protostelic condition throughout the life of the plant. In the former case the xylem is composed of tracheids mixed with parenchyma cells.

(*b*) Actinostele

In this case the xylem core is stellate or starshaped (Fig. 8.2) as in ***Lycopodium serratum*** (Fig. 4.6, C) and in the upper portions of the stem in ***Selaginella selaginoides***. It also occurs in ***Psilotum***

S. kraussiana *S. selaginoides* *Gleichenia dichotoma* *Lygodium*

(Fig. 3.2). The extinct psilophyte named ***Asteroxylon*** (Fig. 2.8) also possessed a star-shaped actinostele. In ***Lycopodium volubile*** and ***L. clavatum*** the xylem occurs in the form of parallel plates alternating with phloem plates (Fig. 4.6). Such an actinostele has been named as **plectostele** by Zimmermann (1930, 1938). In ***L. cernuum*** the actinostele consists of irregular groups of xylem embedded in mass of phloem (Fig. 5.6). Such actinostele is called **mixed protostele.**

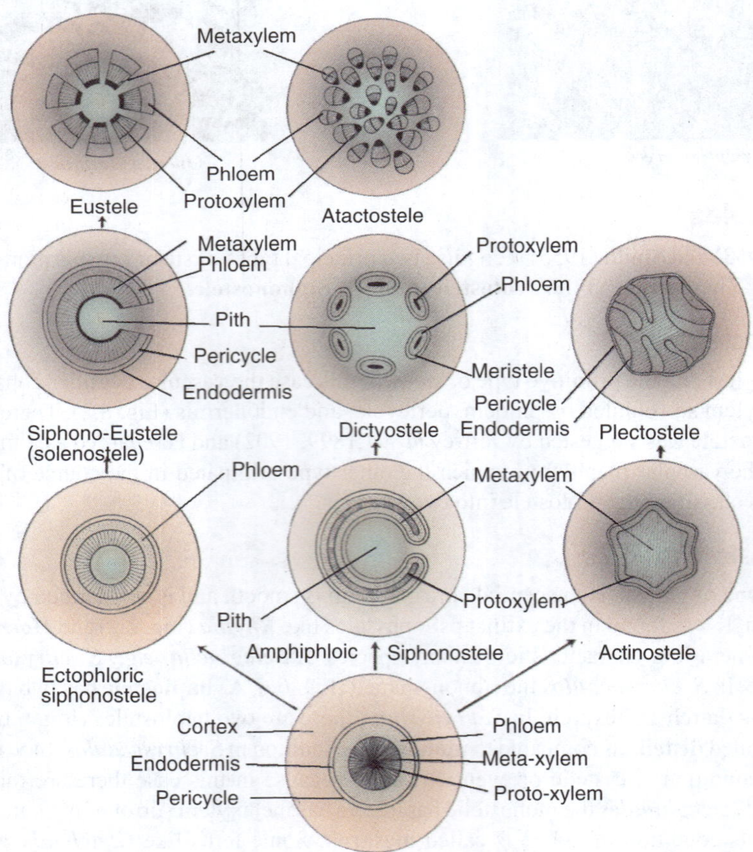

Fig. 8.2. Various types of stelar organisation in the vascular plants. (After Zimmermann)

A very interesting sequence of variation in the protostele is illustrated by members of the family ***Hymenophyllaceae***. These can be listed as below :—

1. In *Trichomanes scandens* the xylem forms a central core with a little parenchyma in the centre. The protoxylem forms peripheral groups (exarch).

2. In ***Hymenophyllum dilatatum*** and ***H. demissum*** the metaxylem forms a ring around the central parenchyma in which lie embedded a few protoxylem tracheids.

Hymenophyllum dilatatum

H. demissum

3. In ***H. scabrum*** the metaxylem ring is interrupted by parenchyma gap thus forming two arcs. The protoxylem lies embedded in the central parenchyma.

4. In ***Trichomanes muscoides*** the lower xylem arc and phloem are absent thus making a collateral bundle.

5. In ***T. microphyllum*** xylem is represented by a single tracheid.

6. In ***T. motleyi*** there is no xylem at all.

H. scabrum

B. The Siphonostele (Fig. 8.2)

(*a*) Origin

Medullated protostele is called **siphonostele.** It is characteristic of the filicophyta. During the development of siphonostele the central core of xylem is replaced by parenchymatous cells so that a definite pith surrounded by xylem appears in the centre. Series of transverse sections cut from the base upwards of rhizomes of certain ferns (***Aneimia***, ***Schizaea***) clearly reveal the development of siphonostele. In ***Aneimia phyllitidis*** (Bower 1923) the stele at the base of the rhizome is a

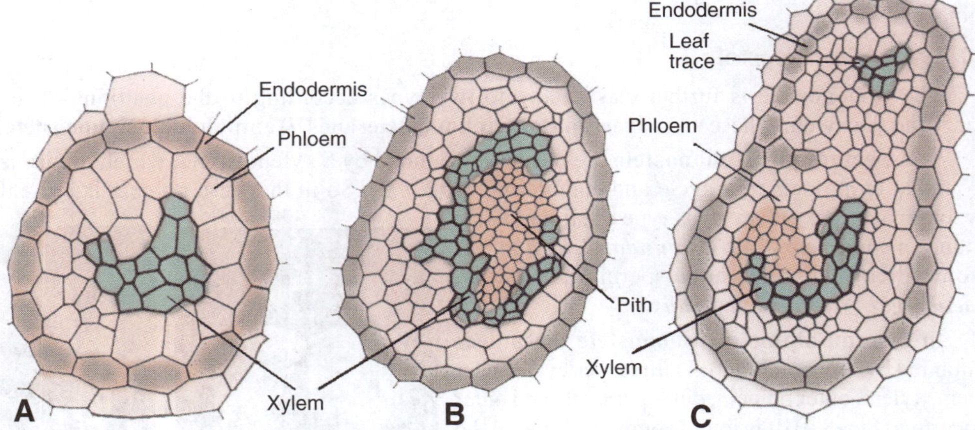

Fig. 8.3. (A-C). *Aneimia phyllitidis.* **A.** T.S. near the base of rhizome showing protostele; **B.** T.S. at a higher P-level showing the appearance of pith; **C.** T.S. showing siphonostele giving off a leaf trace without breaking the endodermis. (After Bower)

typical protostele (Fig. 8.3, A). The leaf traces given out at this stage do not disturb the protostele and also do not interrupt the endodermis. Higher up the sections reveal the presence of a central pith (Fig. 8.3, B) enclosed completely by a ring of tracheids. In this case the whole vascular system is surrounded by endodermis during the development of pith. Similar developmental studies in ***Schizaea***, and in the sporelings of ***Botrychium***, ***Helminthostachys***, ***Osmunda*** and ***Gleichenia pectinata***, reveal that the vascular cylinder is completely surrounded by endodermis during the development of pith. Gewirtz and Fahn (1960) conducted similar studies on developmental anatomy of ***Ophioglossum lusitanicum*** (see chapter 9). These observations support the view that the pith is **intraxylary** or **intrastelar** in origin. Boodle (1901), Gwynne Vaughan (1908), Bower (1911), Petry (1914), Thompson (1920), and Gewirtz and Fahn (1960) support this theory of intrastelar origin of pith. Presence of isolated tracheids in the pith in ***Botrychium virginianum***, ***B. lunaria***, ***B. ternatum*** and ***Osmunda regails*** support the intrastelar origin of pith.

Botrychium virginianum

B. lunaria

B. ternatum

Jeffrey (1897, 1899, 1902, 1917) put forth his view that the pith is extrastelar in origin. He believed that pith originated as a result of invasion of the parenchymatous cells of the cortex into the stele. This he believed took place through the leaf gaps and branch gaps. He used the presence of inner endodermis between the pith and the vascular tissue as a proof of extrastelar origin of pith. According to him endodermis penetrated inwards together with the parenchyma of the cortex. This argument has been challenged by certain observations on ***Selaginella*** and ***Pteridium***. In both these genera the endodermis has been proved to be stelar in origin. In ***Pteridium aquilinum*** Chang (1927) observed that endodermis pericycle and protophloem arise from a common layer of procambial cells.

(*b*) Types of Siphonostele

The siphonostele is further classified into two types according to the positions of the phloem and the xylem. These are : (*i*) **ectophloic siphonostele**, and (*ii*) **amphiphloic siphonostele.**

In the **ectophloic siphonostele** the pith is surrounded by a xylem strand, which in turn is surrounded by phloem, pericycle and endodermis (Fig. 8.2). So in this case phloem is present only external to the xylem. This type of siphonostele is found in some ferns like ***Osmunda***, ***Schizaea*** and in some dicotyledonous angiosperms like ***Phlox***, ***Lindenbergia***, ***Nicotiana*** and ***Salix.***

In the **amphiphloic siphonostele** the pith is surrounded by inner endodermis, inner pericycle, inner phloem, xylem, outer phloem, outer pericycle and outer endodermis (Fig. 8.2). It is found commonly among the fern, ***e.g.***, ***Adiantum***, ***Marsilea***, ***Dipteris***, ***Platyzoma*** etc. So in amphiphloic siphonostele the pholem surrounds the xylem internally as well as externally.

In ***Todea hymenophylloides*** the inner endodermis appears as a discontinuous sheath. There is no internal phloem.

Dipteris

Platyzoma

Both the ectophloic and amphiphloic siphonosteles may occur in the form of a continuous cylinder of vascular tissue. In these cases the leaf traces do not leave leaf gaps and therefore the continuity of the stele is uninterrupted as in some species of ***Selaginella***. Such a siphonostele is termed by Jeffrey (1910) as **Cladosiphonic siphonostele.** In the filicophya the siphonostele may be perforated by smaller or larger leaf gaps. Such a perforated siphonostele is named by Jeffrey as **Phyllosiphonic.** Both these types can be distinguished among ectophloic and amphiphloic siphonosteles. Appearance of leaf gaps in the siphonostele (Phyllosiphonic) leads to a number of modifications.

If the stem or the rhizome bears leaves at distance the leaf gaps also appear at a considerable distance form each other. At such places the stele is interrupted and appears horse-shoe-shaped. The leaf trace appears as a small crescentic bundle (Fig. 8.2). ***e.g. Adiantum pedatum***. The leaf gap closes at a higher level and the stele again becomes complete and circular. So the stele is perforated at places corresponding to the origin of leaf traces and is complete in between. Such a siphonostele is called **solenostele** (Fig. 8.2). The solenostele may be ectophloic or amphiphloic depending upon the type of siphonostele from which it is derived.

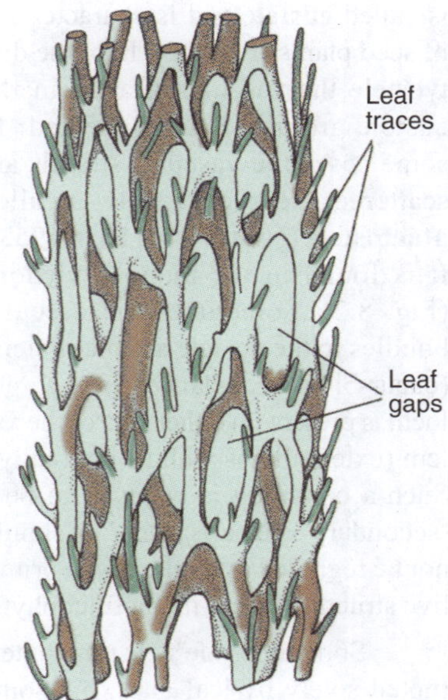

Fig. 8.4. *Dryopteris filix-mas.* Stelar organisation of rhizome showing a cylindrical network of interconnected vascular strands. It is a dictyostele.

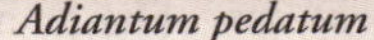

Adiantum pedatum

Pellaea rotundifolia

In a number of species of ***Dryopteris***, ***Pteris***, ***Pellaea rotundifolia*** and a number of other filicophyta, the rhizome is short and the leaves overlap each other. This leads to the overlapping of leaf gaps in the stele so that the lower part of one leaf gap is parallel with the upper part of another gap. Such an arrangement of the leaf gaps results in the formation of a cylindrical network of (Fig. 8.4) interconnected vascular strands. A cross-section of such a stele reveals the presence of separate vascular bundles (Fig. 8.5, E) separated by parenchymatous patches. Each such vascular bundle is of concentric type and consists of a central core of xylem surrounded by phloem, pericycle and endodermis (Fig. 8.5, E) Individually these bundles are called the **meristeles.** Anatomically the meristeles are amphicribral bundles with exarch, mesarch or endarch, xylem. Such a siphonostele is termed as s **dictyostele** (Fig. 8.5, A,C,D,E). It is derived from the amphiphloic siphonostele

Appearance of overlapping leaf gaps in an ectophloic siphonostele results in the splitting of such a stele into separate collateral vascular bundles. Such a modification of ectophloic siphonostele is called **eustele** and is characteristic of seed plants (Fig. 8.2). Like the dictyostele the vascular strands in the eustele are also interconnected. In some cases the vascular strands are scattered and such a stele is called **atactostele** (Nast, 1944 ; Esan 1953). It is found in the monocotyledons (Fig. 8.2). Sometimes the vascular bundles in the eustele are bicollateral (cucur-bitaceae, solanaceae), ***i.e.***, phloem is present on either side of the xylem (externally as well as internally). Such a condition is believed to be a 'secondary specialisation.' It should not be regraded as a relic of the primitive structure found in the filicophyta.

Sometimes the stele is not interrupted solely by leaf gaps, but some interruptions appear as a result of secondary reductions of the vascular tissue and form interfascicular parenchyma termed as **medullary rays.** Such parenchymatous interruptions are called **perforations.**

In some filicophyta ***e.g.***, ***Matonia***, ***Pteridium***, ***Pteris vittata*** and ***Marattia***, two or more concentric cylinders of vascular tissue are present. Such a stele is called **polycyclic stele.** In ***Pteridium aquilinum*** (Fig. 8.5, D) there are two concentric circles of vascular tissue. The outer circle is composed of a number of meristeles, ***i.e.***, it is dictyostelic. The inner circle consists of only two meristeles. Two patches of sclerenchyma of the ground tissue are present between the two circles of meristeles. Polycylic stele in ***Pteris elata*** var. ***karsteniana*** consists of larger number of meristeles

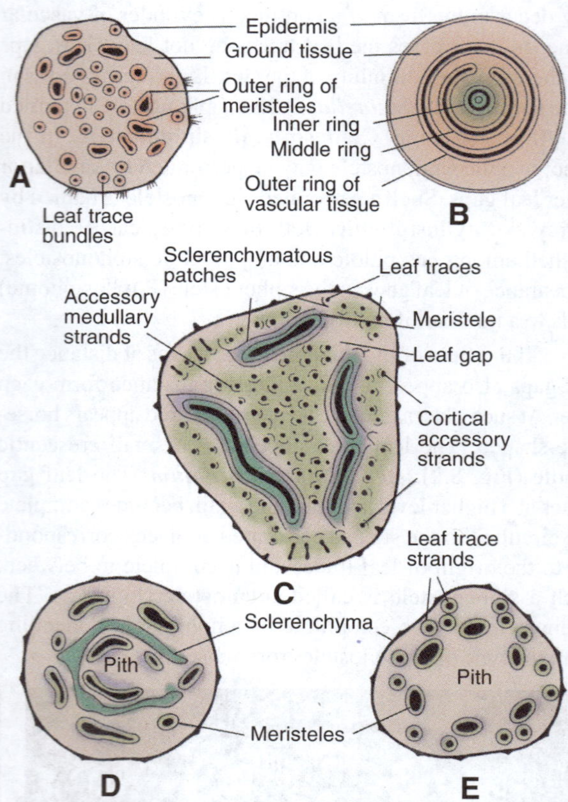

Fig. 8.5 (A–E). *Stelar types in ferns.* A. Polycyclic dictyostele in *Platycerium aethiopicum*; B. Polycyclic stele in *Matonia pectinata;* C. Dictyostele in *Cyathea imrayana;* Note the sclerenchymatous strands in the cortex and pith; D. T.S. rhizome of *Pteridium aquilinum* showing dicyclic dictyostele; E. T.S. rhizome of *Dryopteris chrysocoma* showing dictyostele.

Matonia

Pteris vittata

in the inner ring than in the outer ring. In this case there are no sclerenchyma patches. There are three concentric rings of vascular tissue in ***Matonia pectinata*** (Fig. 8.5, B). Each ring has a typical solenostelic structure (Fig. 8.5). Four concentric cylinders of vascular tissue have been recorded in ***Pteris podophylla*** (Bower, 1923). All the four are siphonostelic in nature. In ***Platycerium aethiopicum*** there are four or more irregular concentric circles of meristeles (Fig. 8.5, A).

Matonia pectinata

The anatomical study of the stems in dendroid ***Cyatheaceae*** reveals a still more complex vasculature. Transverse section through the stem of ***Cyathea imrayana*** (Fig. 8.5, C) reveals the presence of broad meristeles that are enclosed within broad plates of sclerenchyma. The leaf traces arise as number of strands that run obliquely through the cortex (Fig. 8.5, C). Each leaf trace arises from the lower margin of each leaf gap. In addition to this vasculature a number of other vascular strands are present in the pith and the cortex. Each one of them is surrounded by endodermis and is often accompanied by a sclerenchymatous band. These cortical and medullary strands may anastomose freely and end blindly during their course downwards. These vascular strands are considered to be the **accessory strands** that have originated in the pith or the cortex without any connection with the general vascular system.

Cyathea imrayana

Sporne (1962) uses a different nomenclature for the stelar organisation in the vascular cryptogams. He regards the ectophloic siphonostele as **medullated protostele.** The term **solenostele** has been used for amphiphloic siphonostele and such siphonosteles that are perforated by leaf gaps at considerable distances. Sporne has not used the term siphonostele for the pteridophytes. The term ***dicytostele*** is used when leaf gaps overlap and the solenostele is much dissected. So according to Sporne the steles are of the following types: (*i*) protostele; (*ii*) medullated protostele; (*iii*) solenostele; (*iv*) dictyostele; and (*v*) polycyclic stele.

Evolution of the Stelar System (Fig. 8.2)

The consensus of opinion now favours Jeffrey's belief (1898) that protostele is the primitive type of stele. It is considered to be a fundamental stelar organisation that was present in the earliest vascular plants and is now retained by some living vascular cyptogams like ***Psilotum*** (Fig. 3.2), ***Tmesipteris***, ***Selaginella***, ***Lycopodium***, ***Gleichenia***, ***etc.*** The primitive vascular plants like the extinct psilophytales also possessed protostelic vascular organisation ***e.g.***, ***Rhynia*** (Fig. 2.2), ***Horneophyton***, ***Asteroxylon*** (Fig. 2.8). Presence of exclusively protostelic stems in the earliest vascular plants and their retention in some of living vascular cryptogams lends a strong support to Jeffrey's view that protostele is phylogenetically a primitive type of vasculature. In its simplest form the protostele is **haplostelic** (Fig. 8.2). During further elaboration the central core of xylem became irregular and assumed an almost (Fig. 2.8) star-like shape (***Asteroxylon***, some species of ***Selaginella and Lycopodium***). Such a modification of protostele was termed as **actinostele** by ***Brebner*** (Fig. 8.2). As a result of further elaboration, the xylem splits up into a number of parallel plates alternating with phloem (Fig. 8.2). Such a modification was called **plectostele** by Zimmermann (1930). It is found in some species of ***Lycopodium***. Haplostele to actinostele and then to plectostele is considered to be one line of evolution of the protostele (Zimmermann, 1930). It can be regarded as a **Lycopsid line of evolution.**

Another very important evolutionary change that occurred in the protostele was the appearance of the central pith. This step led to many important changes in the protostelic organisation and gave rise, on further elaboration, to complicated stelar types. The origin of pith has been explained variously by various authors. Two theories have been put forth to accord for its origin. These are the **intrastelar theory** and the **extrastelar** or **invasion theory.** Appearance of pith led to the conversion of the protostele into a new type of stele called the **siphonostele.** Elaboration of **siphonostele** also followed two courses of evolution :

(*a*) The appearance of pith resulted in the formation of a stele consisting of a central pith surrounded by a complete ring of xylem, which in turn was surrounded by a complete ring of outer phloem, pericycle and endodermis (Fig. 8.2). Such a stele was designated as **ectophloic siphonostele** (medullated protostele). In its simplest form such a siphonostele is uninterrupted by leaf gaps and is also called **Cladosiphonic.** In the megaphyllous vascular plants the complete and uninterrupted cylinder of ectophloic siphonostele becomes interrupted by the appearance of leaf gaps. It is now called **phyllosiphonic.** In case the leaves do not overlap the stele is interrupted only at considerable distances (nodes) by one leaf gap (Fig. 8.2). In between the two gaps the vascular cylinder remains complete. Such a stele is also called solenostele or **siphono-eustele** (Zimmermann, 1930, 1938). In case the leaves on the stem overlap the leaf gaps also overlap and lead to the formation of a much dissected stelar organisation (Fig. 8.2). It is made up of a number of separate and collateral vascular bundles. This condition is called **eustele** and is met with in the seed plants. In case the vascular bundles are scattered, as in monocotyledons, the stele is termed as **atactostele** (Fig. 8.2).

(*b*) During another line of evolution the medullation of the protostele was followed by the appearance of phloem on either side of the xylem and likewise internal pericycle and endodermis also appeared. As a result the siphonostele consisted of a central pith surrounded by internal endodermis, internal periycle, internal phloem, xylem, external phloem, external pericycle and external endodermis (Fig. 8.2). Such a siphonostele is called **amphiphloic siphonostele** (***Marsilea***, ***Adiantum***). It may be **cladosiphonic** or **phyllosiphonic.** The phyllosiphonic amphiphloic siphonostele with only one leaf gap at the node is called **amphiphloic solenostele.** In case the leaf gaps overlap the resultant stele is called the **dictyostele.** Dictyostele is very common in the filicophyta. In many eusporangiate and leptosoporangiate ferns the dictyostelic stems are protostelic at their bases. One such example is afforded by ***Ophioglossum lusitanicum*** (Gewirtz and Fahn, 1960). Recent experimental studies (Wardlaw, 1944, 1968) also reveal that dictyostelic condition can be changed to solonostelic or even protostelic condition by removing the young leaf primordia from shoot apices (***Dryopteris dilatata***). All these observations prove that protostele is the basic or the fundamental stelar type from which the complicated steles or vascular systems arose by elaboration.

Ophioglossum lusitanicum

Polycyclic condition exhibited by some ferns like ***Marattia***, ***Matonia***, ***Pterdium***, ***Cyathea***, etc., also originated from the protostelic condition by further elaboration. This is borne out by the fact that in ***Matonia pectinata*** there is a regular transition from protostelic condition to solenostele and then to a polycyclic condition. The stem is protostelic at the base, then becomes solenostelic and ultimately polycyclic. (Three concentric circles of siphonosteles). This developmental phenomenon is termed as **recapitulation.** Occurrence of such a developmental phenomenon lends further support to Jeffery's view that protostele is a primitive condition.

A number of workers have presented a different interpretation of the evolution of stelar system in the seed plants. They believe that the leaf gaps of the seed plants (Gymnosperms and Angiosperms) are not morphologically equivalent to those of ferns. Namboodri and Beck (1968) and Beck (1970) have put forth an explanation that the **eustele** in gymnosperms is not derived from filicean type of **siphonostele** nor are there any filicean type of leaf gaps. These authors studied the stelar organisation in primitive gymnosperms like ***Calamopitys foerstei*** and ***Lygenopteris oldhamia*** and found that the **eustele** in these originated direct from the **protostele** through longitudinal dissection, without any intervening siphonostelic stage. Such a mode of organisation of the primary vasculature in the woody angiosperms that are now living has been studied by Dormer (1945, 1946, 1954); Devadas (1970), Devadas and Beck (1971, 1972), Sporne (1958) and Slade (1971). According to their view filicean leaf gaps are absent in the eustele of primitive gymnosperms like ***Lygenopteris***. It has now been held that angiosperms originated from the Pteridosperms (Takhtajan, 1969; Delevoryas, 1962; Beck, 1962, 1970), and if this view is correct then there are no fern type leaf gaps in the angiosperms also. Such a view has been supported by Sporne (1958) and Slade (1971). So according to this interpretation the **eustele** originated directly from the **protostele** without the intervention of Siphonostele.

Namboodri and Beck (1968) have put forth a new concept that parenchymatous areas may arise in the stele without having any connection with the leaf gaps. Such a conclusion is based on detailed study of the primary vasculature in gymnosperms and angiosperms (Slade, 1971) and is against Jeffery's leaf gap concept for ferns.

The above view supports another new interpretation that the primary vasculature of the stem is **Cauline** and not **foliar.** The protostele is clearly axial, not a foliar structure. "That this is true becomes very apparent when one considers that the ancestral psilophytes from which the protostelic **progymnosperms** must have evolved were leafless" (Devadas and Beck, 1972).

Usually primary vasculature of stem is considered to be foliar in nature (Esau, 1965 b; Philipson and Balfour, 1963; O'Neil, 1961; Nast 1944); but the above view of its cauline nature is at variance with this.

Leaf Traces and Leaf Gaps

A leaf trace is that part of the vascular cylinder that extends between the leaf base and the point where it merges with the vascular system of the stem. The number of the leaf traces associated with one leaf may be more than one. So leaf traces connect the vascular system of the leaf with that of the stem.

In ferns and the seed plants the portion, lying immediately above the point of divergence of leaf trace from the vascular cylinder of the stem, becomes parenchymatous. In this parenchymatous region the vascular elements are absent, but this region is limited to only a short distance up, beyond which the vascular tissue is present. These parenchymatous regions in the vascular cylinder are described as **Leaf gaps** or **lacunae.** The vascular plants in which the leaf traces leave leaf gaps are called **megaphyllous.** There are no leaf gaps in the lower vascular plants such as ***Selaginella*** and ***Lycopodium.*** In these cases the leaves are small and their comparatively thin leaf traces do not leave leaf gaps. Such plants are called **microphyllous.** The effect of the leaf gaps on the vasculature of vascular cryptogams has been discussed in detail in the preceding pages.

Branch Traces and Branch Gaps

The branches develop from buds and have vascular connections with the main axis. These vascular connections that connect the vascular system of a branch with that of the stem are called branch traces. Similarly there are **branch gaps** which are the parenchymatous areas in stem vascular system just above the origin of branch trace. The leaf gaps and the branch gaps are not actually breaks in the continuity of the vascular cylinder because the vascular tissue maintains lateral connection just above and below the gap.

SPORE PRODUCING MEMBERS

The spores are produced within specialised structures called the **sporangia.** The sporangia, in most of the living filicophyta, occur in small or large groups called the **sori** (singular = sorus). The sori are variously arranged on the margins or ventral surfaces of leaves or leaflets. Such leaves are called the **sporophylls.**

The leaves, in most of the filicophyta, serve a dual purpose of photosynthesis and reproduction. In the common ferns like ***Adiantum***, ***Pteris*** and ***Dryopteris*** and leaf or leaflet can bear sori on its under surface. There is no distinction between fertile and sterile leaves. There are, however, some ferns which exhibit segregation of photosynthetic and reproductive function. The genus ***Osmunda*** affords a good example of cases where a few pinnae or leaflets of a leaf are set aside for bearing sporangia whereas the others are sterile and photosynthetic. ***Osmunda regalis*** has a pinnate compound leaf in which the upper or distal pinnae are fertile whereas the lower or proximal pinnae are sterile. In ***Osmunda claytonniana*** and ***O. javanica*** certain pinnae, anywhere along the leaf have smaller size and bear sporangia. These examples illustrate restriction of sporangiferous area in the same leaf. In ***O. cinnamomea*** the leaves are dimorphic, *i.e.*, there are two types of leaves. There are: (***i***) sterile leaves that have well developed leaf blades and appear late in season, and (***ii***) fertile leaves that lack lamina and appear earlier in season. Other examples of this type are ***Matteuccia struthiopteris*** and ***Blechnum spicant*** in which the rhizome bears : (***i***) sterile leaves that are large and broadly expanded and (***ii***) fertile leaves or sporophylls that are attenuated and have reduced lamina. These two types of leaves occur in distinct alternating zones. In ***Pellaea altropurpurea*** the sterile leaves have oval pinnae and fertile fronds have narrow and elongated pinnae.

Osmunda regalis

Osmunda claytonniana

Matteuccia struthiopteris

O. cinnamomea

Blechnum spicant

The Sorus

The term sorus has a Greek origin. It actually means a heap or a group. The group of sporangia is, therefore, called a **sorus.** A sorus may have from two to many sporangia. The sorus may be protected by a revolute margin or by a special outgrowth called the **indusium** or it may be unprotected and naked. Besides the sporangia, the sorus also includes the **receptacle** or the **placenta** on which the sporangia arise. The indusium if present may also be regarded as a part of the sorus.

There are some ferns in which the sporangia are naked (without indusium) and occur scattered along the veinlets. They do not form sori, ***e.g.***, ***Leptopteris hymenophylloides***. In ***Todea barbara*** the sporangia are densely scattered and are not covered by an indusium. The sori in this case are ill defined. The sporangia in both these cases are superficial. In the living genera of the Marattiales the superficial sporangia are arranged in definite sori. In ***Christensenia*** the sori are naked, circular and are irregularly arranged between the lateral veins. In ***Marattia***, ***Angiopteris***, ***Danaea***, etc., the sori are elongated and are situated below the lateral veins. Among the polyopoidaceae the sori are naked in ***Polypodium*** and some other genera.

Leptopteris hymenophylloides *Danaea*

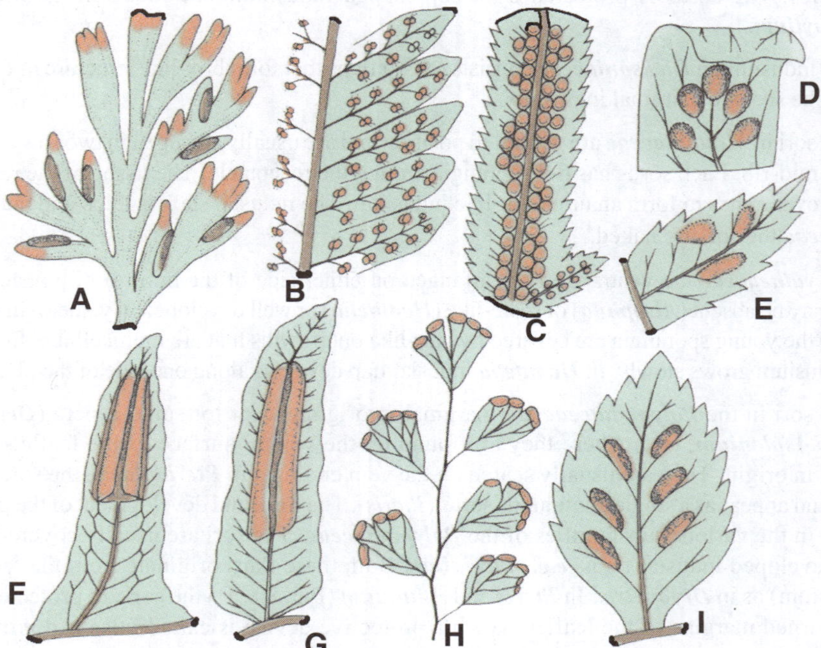

Fig. 8.6. (A–I) Fern Sori. A. Sori of *Asplenium* with unilateral indusia; B. Sori of *Polystichum*; C. Sori of *Dryopteris* covered by reniform indusia; D. Sori of *Mattecucia struthiopteris* is covered by peltate indusia; E. *Aethyrium filix-foemina.* A pinna with three reniform indusia with lacerated margins; F. Elongate sorus of *Blechnum occidentale*; G. Elongate indusium of *Lomaria spicant*; H. Sori of *Adiantum*; I. Sori with unilateral indusia in *Asplenium lanceolatum.*

The Ophioglossales are a group of interesting ferns in which the sporangia are marginal in position and are borne on fertile spikes that arise on ventral surface of a sterile lobe. The leaf, in the ophioglossales, consists of a petiole which bears a simple (Fig. 9.1) or a dissected sterile lamina from whose ventral surface arises a simple or a branched fertile spike. The fertile spike in ***Ophioglossum*** bears sporangia arranged in two rows (Fig. 9.1). The sporangia are embedded and fused together. The spike terminates in a small sterile process. Every sporangium is supplied with a vascular trace. The fertile spike in ***Helminthostachys*** bears groups of sporangiophores arranged in many rows. Each sporangiophore bears groups of sporangia and a few green scale-like lobes at the tip. The spike in ***Botrychium*** is branched once, twice or many times. Each branch bears two rows of marginal sporangia. The sporangia in ***Osmunda***, ***Davallia***, ***Trichomanes*** and many other ferns are also marginal in position (Fig. 8.6, A). In ***Osmunda*** the fertile pinnae are bladeless, ***i.e.***, have either reduced or no lamina. In ***Schizaeaceae*** the sporangia occur isolated and are protected by a false indusium. They originate from initials that arise marginally, but get displaced to ventral side during development.

Blechnum occidentale

In ***Glecheniaceae*** the sporangia form distinct sori on the ventral surface of the leaves. The sori are seated at the vein ends or rarely along the middle of the vein. The sori are covered only by hair and scales. There are no indusial flaps. The sorus consists of a single circle of sporangia.

In ***Hymenophyllaceae*** the sori are marginal and are characteristic in that the vein supplying the receptacle grows through it (by the activity of a basal meristem) and forms a long thin bristle, ***e.g.***, ***Trichomanes***. The sorus is protected by a cup-shaped indusium. The indusium is bilipped in ***Hymenophyllum***.

The indusium in ***Dicksoniaceae*** consists of two flaps that form box-like structure in ***Cibotium baromet.*** The sori are marginal in position.

The sori in ***Mattoniaceae*** are ventral in position and are usually arranged in two rows on either side of the mid-rib. Each sorus has 6-9 sporangia around the receptacle. The tissue of the receptacle grows and overarches to form an umbrella-like indusium. The indusial stalk is thick and massive. In ***Dipteridaceae*** the sori are naked.

In ***Cyatheaceae*** the ventral sori are arranged on either side of the mib-rib of pinnae and the indusium may be absent (***Alsophila***) or scale-like (***Hemitelia***) or well developed (Cyathea). In ***Cyathea medullaris*** the young sporangia are covered by hair-like outgrowths that are multicellular. In this species the indusium grows slowly. In ***Hemitelia*** indusial flap develops along one side of the placenta.

The sori in the ***Polypodiaceae*** are marginal in origin except for some genera (***Dryopteris***, ***Athyrium***, ***Asplenium***, etc.) where they are found on the ventral surface of the leaflets and are superficial in origin. They are usually seated on the vein endings. In ***Pteridoideae*** the sori become confluent and appear as a single continuous sorus (***Pteris***). The form and development of the indusium is variable in the various sub-families of the ***Polypodiaceae.*** The sori are usually covered by two laterally developed indusial flaps. ***e.g.***, ***Pteridium.*** Both these flaps originate from the receptacle (true indusium) as in ***Dryopetris.*** In ***Pteris*** and ***Adiantum*** (Fig. 8.6, H) the sori are protected by the inwardly turned margins of the leaflets. Such a protective device is called **false indusium.** The indusium is reniform in ***Dryopteris*** (Fig. 8.6, C); circular in ***Polystichum lobatum*** (Fig. 8.6 D), funnel-shaped in ***Davallia***; elongated and curved in ***Asplenium lanceolatum*** (Fig. 8.6, I), ***Lomaria spicant*** (Fig. 8.6, G) and ***Blechnum occidentale*** (Fig. 8.6, F). In the last named genera the sori become confluent and are covered by a common indusium. In ***Lomaria,*** the lamina is

Davallia

Cibotium baromet

Alsophila

Lomaria spicant

Athyrium filixfoemina

Asplenicum lanceolatum

Hemitelia

greatly reduced. In ***Athyrium filixfoemina*** the indusium is reniform with lacerated margins (Fig. 8.6 E). In ***Matteuccia struthiopteris***, the indusium is cup-shaped with dentate margins (Fig. 8.6, D) and is thin and papery. In this case the leaflet margins become strongly, inrolled and afford additional protection to the sori. The same is the case in ***Onoclea***. In ***Adiantum*** the sporangia develop on the underside of special marginal flaps of lamina that become reflexed (Fig. 8.6, I) and protect the sorus.

In marsileales and the salviniales the sporangia develop in sori that are borne within distinct structure called the sporocarps. The sporocarps in ***Marsilea*** enclose sori that contain both micro and megasporangia. In Salviniales the smaller sporocarps contain many microsporangia each and the larger ones contain one or more megasporangia per sporocarp.

According to the mode of development of sporangia in a sorus Bower (1935) classified the sori in ferns, into the following three types:

1. The simple sorus. The sporangia in such a sorus develop simultaneously and all of them mature together, ***e.g.***, ***Ophioglossum*** and ***Osmunda.***

Adiantum

2. The Gradate or Basipetal Sorus. The placentae or the receptacles are long and almost cylindrical. They bear mature or older sporangia at their distal ends and younger sporangia near the proximal or basal part. Such a sorus is found in ***Dicksonia***, ***Loxosoma***, ***Trichomanes***, ***Cyathea***, ***Alsophila***, etc.

3. The Mixed Sorus : (Fig. 8.7). Such a sorus is an aggregation of old and young sporangia that occur mixed and show no regular arrangement in a sorus. The young and old sporangia are indiscriminately mixed. The sporangia usually have long stalks and vertical annulus. The number of spores varies between 12—64. Such a sorus is found in majority of the living ferns, ***e.g.***, polypodiaceae (***Adiantum***, ***Pteris***, ***Pteridium***, ***Davallia***, etc.)

The Sporangium

The sporangium in the filicophyta exhibits a greater range in shape, form and structure. Regarding the development of the sporangia, it is of two types (*i*) **Eusporangiate development** and (*ii*) **Leptosporangiate development.** The sporangium is essentially a sporeproducing organ. Bower (1935) defined sporangium as "***the essential feature of a sorus, stripped of all its variants and accessories, is the sporangium.***"

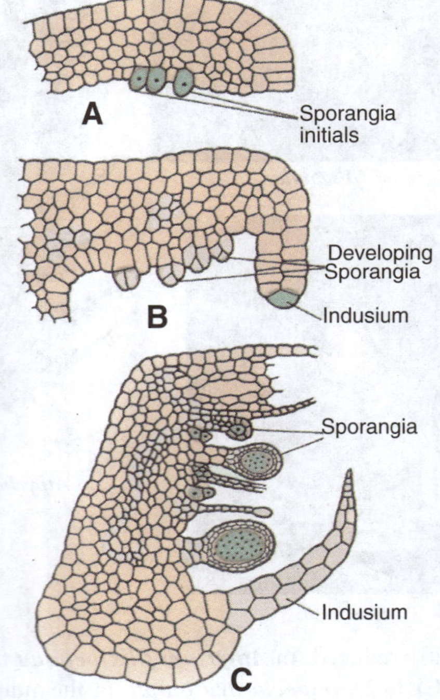

Fig. 8.7. (A–C) **Mixed sori in *Pteris*..** A–B. V.S. young sori of *Pteris serrulata*; A. Showing sporangial initials arising from the ventral surface of leaf; B. Early stages of sporangial development; C. V. S. mature sorus of *Pteris cretica* showing indusium, placenta and sporangia (After Bower)

The sporangia with massive and short stalks; many layered wall provided with stomata; absence of annulus, presence of annulus on one side of the sporangium; production of large number of spores; and dehiscence of sporangium by vertical or distal slits are the primitive characters. Eusporangiate mode of development is also considered to be a primitive feature. Such primitive sporangia were present in most of the fossil ferns (***Stauropteris***). The living filicophyta also possess such sporangia. The Marattiales and the Ophioglossales are good examples that possess such primitive types of sporangia. The sporangia in both these orders develop in a **eusporangiate** manner. The sporangium originates from a single initial cell, differentiated in the epidermis. This cell divides into two. The outer cell gives rise to a part of the sporangial wall, the rest of the jacket develops from the surrounding adjacent cells. The inner cell alone gives rise to the sporogenous tissue. The entire wall of the sporangium is, therefore, not derived from a single cell but from a group of cells. Such a development, where the sporangium develops from a group of cells, is called eusporangiate development. The wall becomes many layered thick (eusporangiate feature). The innermost wall layer functions as **tapetum**, which later forms a nourishing plasmodial fluid. Numerous stomata occur in the wall, perhaps due to its thickness. The stalk is short and massive and develops from the tissue lying at the base of sporangial head or capsule and is not derived from sporangial initial or its derivatives. The spore output is large. In the Ophioglossales the sporangia produced more than 2,000 spores in ***Botrychium*** and up to 15,000 spores in ***Ophioglossum vulgatum***. In Marattiales it varies from 1,440 in ***Angiopteris***, 2,500 in ***Marattia*** to 7,000 in ***Christensenia.***

The Marattiales except for ***Angiopteris*** are characteristic in possessing fused sporangia or synangia. Annulus is absent in the sporangia of ***Ophioglossum***, ***Botrychium,*** and Marattiales except ***Angiopteris***. ***Angiopteris*** has free sporangia with a slightly differentiated annulus. Dehiscence is transverse in ***Ophioglossum***, apical in ***Botrychium***, longitudinal in ***Helminthostachys***. In Synangiate ***Marattiaceae*** the whole sorus dehisces by a vertical slit.

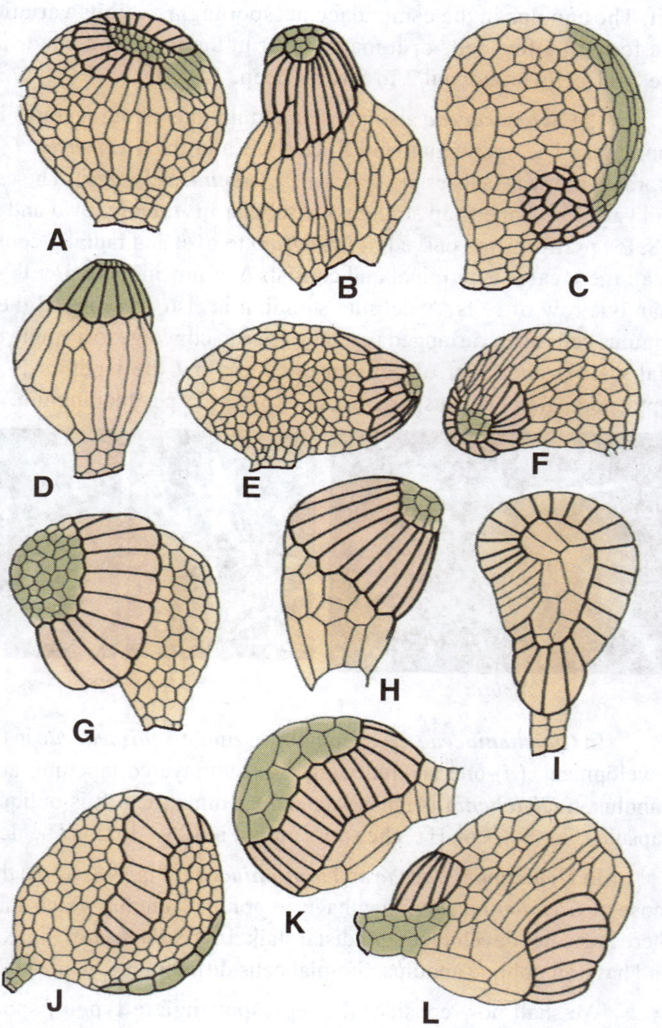

Fig. 8.8. (A–L) Various types of sporangia in ferns. A. Sporangium of *Mohria calfforum*; B. Sporangium of *Schizaea dichotoma*; C. Sporangium of *Osmunda cinnamomea*; D. Sporangium of *Actinostachys oligostachys* with terminal annulus; E. Sporangium of *Lygodium reticulatum*; F. Sporangium of *Anemia phyllilidis*; G. *Gleichenia caudata*; H. *Gleichenia bracknirdgi*; I. *Cyathea capensis*; J. *Matonia pectinata*; K–L. *Hymenophyllum dilatatum*.

The sporangium in the ***Osmundaceae*** shows both eusporangiate and leptosporangiate characteristics. The eusporangiate characteristics are : ***(i)*** several cells are concerned in its initiation ; ***(ii)*** the mature sporangium has a short and thick stalk; ***(iii)*** the sporogenous cells in some cases are cubical; ***(iv)*** the annulus is less developed and is present only on one side of the sporangium; ***(v)*** the number of the spores is comparatively large (***Leptopteris*** 128-236; ***Osmunda*** 256—521) and ***(vi)*** presence of chlorophyll in the spores. The leptosporangiate characters are: ***(i)*** conical shape of sporangial initial; ***(ii)*** the wall is one or two layered; ***(iii)*** the tapetum originates from the outermost sporogenous cells and ***(iv)*** presence of a thin walled stomium. The shape of the sporangium is generally pear—shaped and it dehisces by longitudinal slit. The annulus in the osmundaceous sporangia exhibits variations in structure. It may occur in the form of a few cells or it may appear in the form of a band-like group of indurated cells that are 8-11 cells broad and 2 to 3 rows deep.

The ***Schizaeaceae*** also present certain interesting features in the form and structure of their sporangia. The sporangial stalks are short and thick (***Mohria, Lygodium*** : Fig. 8.8, A, E) or the sporangia may be subsessile or sessile (***Anemia***, ***Schizaea***). The capsule of the sporangium is large and varies in shape from straight and tubular in ***Anemia***, oval and slightly curved in ***Schizaea*** (Fig. 8.8, B), pyriform and curved in ***Lygodium*** to oval and radially constructed in ***Mohria. The annulus*** in all these cases is terminal and consists of a few indurated or thick-walled cells arranged in more than one row of cells. A definite stomium is also present and the dehiscence is longitudinal. The annulus cells are so arranged that they arise to converge to a single or a few central cells. The sporangial wall is one layer of cells in thickness but the tapetum is two layered. The sporangia are leptosporangiate with a spore output of 64—128 per sporangium.

Mohria

Lygodium

Anemia

In ***Gleicheniaceae*** the sporangia resemble ***Schizaeaceae*** in their : (***i***) mode of leptosporangiate development; (***ii***) one layered wall; (***iii***) two layered tapetum; and (***iv***) short and thick stalk. The annulus in ***Gleichenia*** is uniseriate and asymmetrical. It is obliquely placed around the sporangial capsule (Fig. 8.8, G—H). The spore output is 128—1,024. The dehiscence is vertical.

In ***Hymenophyllaceae*** and ***Mattoniaceae*** (Fig. 8.8, J—L) the sporangia are similar in form to those of ***Gleicheniaceae.*** They have an oblique annulus which shifts more towards the stalk so that there is greater development of distal stalk. In ***Cyatheaceae*** (Fig. 8.10, 1) also the sporangia are large and have an oblique annulus. Stomial cells differentiate.

We shall now consider the **leptosporangiate type** of sporangium (Fig. 8.9, A—H). These sporangia are characterised by the following features:

1. The sporangium originates from a single superficial sporangial initial (Fig. 8.9, A), which divides into two (Fig. 8.9, B) by a transverse or oblique transverse wall. The lower or basal cell takes no further part in sporangial development.

2. The outer cell along gives rise to the entire sporangium (stalk, jacket, tapetum and sporogenous tissue). The outer cell functions as a tetrahedral apical cell (Fig. 8.9, C) which cuts off segments along its three sides. The lower segments give rise to the stalk that consists of three rows of cells. The upper three segments give rise to a portion of the sporangial wall.

3. The apical cell stops growing by the appearance of a periclinal curved wall that separates an outer jacket initial or cap cell. Now four jacket initials surround a single central cell or the **primary sporogenous cell** (Fig. 8.9, D).

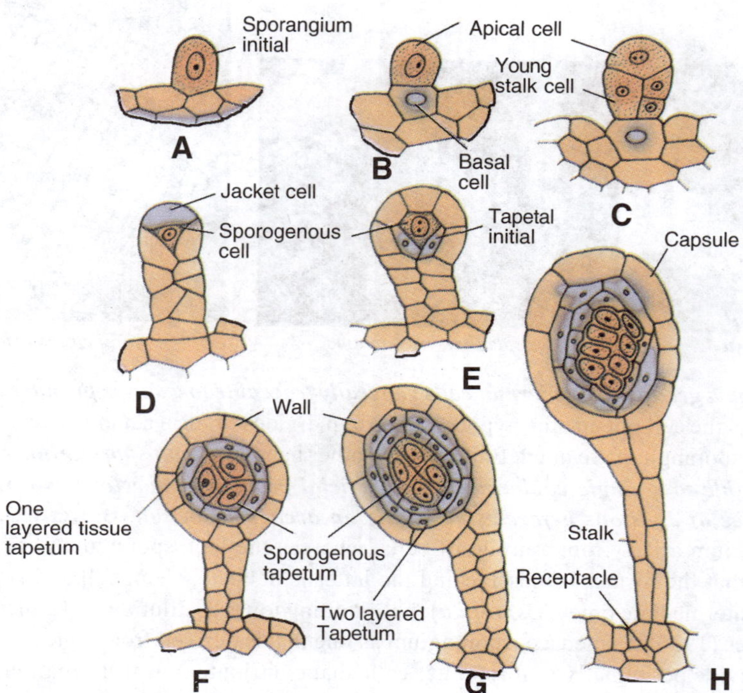

Fig. 8.9. (A–H). Various stages in the development of leptoporangiate sporangium in *Polypodium.*

4. The jacket cells divide anticlinally and give rise to a single layered sporangial wall (Fig. 8.9. E)

5. The primary sporogenous cell cuts off 4 tapetal initials (Fig. 8.9, E, F) which divide both periclinally and anticlinally to form a two layered tapetum.

6. The sporogenous cell later divides to form 8—32 spore mother cells (Fig. 8.9, G, H) which undergo reduction division to produce 32—128 spores.

7. The sporangial stalk is long and thin (Fig. 8.9, H).

8. The annulus is made up of a single layer of thick-walled cells. The annulus is usually vertical in position. The dehiscence is transverse. The annulus is thickened only along its three inner walls, the outer wall being thin. The annulus is attached at one end with the stalk. As its other end the annulus is attached to the **epistomium** which is a pair of thin-walled cells adjacent to the **stomium.**

9. The stomium or the lip is made of a group of 4 transversely elongated and parenchymatous cells. It is capped above by two thin-walled cells called the **epistomium** and below by two thin-walled cells called the **hypostomium**. The **hypostomium** is connected with the sporangial stalk.

The stalks of leptosporangiate sporangia are usually three rowed but variations are there. In ***Asplenium trichomanes*** the stalk consists of a single row of cells. In ***Metaxya*** (***Cyatheaceae***) and ***Cheiropleuria*** (***Dipteridaceae***) the stalk consists of 4 rows of cells.

Bierhorst (1971) considers that the terms **eusporangium** and **leptosporangium** are vague and inconsistent and should be discarded. The two types have been variously defined. D.H. Campbell (1928) defined the ***eusporangiate sporangium as the type in which sporogenous***

Metaxya *Asplenium trichomanes* *Cheiropleuria*

tissue arises from a group of hypodermal cells generally traceable to a single primary archesporial cell, whereas the cell outside the hypodermal group divides to form a many layered wall and the limits of the sporangium are not definite. On the other hand, ***the leptosporangium can on the whole be traceable to a single epidermal or superficial cell and it develops as a result of a regular sequence of divisions before establishing an archesporial cell***. Bower (1923) stated that a eusporangium arises from a group of cells, whereas the leptosporangium arises from a single cell and that the former is massive and the latter is delicate. Campbell and Bower have also described intermediate types (***Osmunda***) thus pointing towards dilution in the distinction of two types. Eames (1936) defined a eusporangium as one that is derived from a group of superficial cells that divide periclinally to form outer cells that contribute to wall formation and inner cells that give rise only to the sporogenous cells. Eames defined the leptosporangium as "Following the first division of the initial, the outer cell becomes a pyramidal apical cell which cuts off basal cells to form a stalk until a distal, arched, periclinal division limits growth in length. There is thus formed a central cell, usually tetrahedral in form surrounded by a one layered wall. Anticlinal divisions in the wall divide this layer into many cells. The central cell represents the primary sporogenous tissue. From it are cut off by periclinal divisions one or two layers of thin cells which become tapetum." Such a development is considered as characteristic of the **Filicales.**

Studies in sporangial development in various pteridophytes suggest that these two types are not consistent and various grades of differences met with in each type lead to their apparent mergence. Bower described eusporangiate development in the lycopods and later in Ophioglossales and Marattiales and found many differences from the usual eusporangiate type. Similarly in many leptosporangiate filicales some types have many layered walls that are traceable to the inner cell—a eusporangiate character. Such instances have ultimately led Bierhorst to suggest the terms eusporangiate and leptosporangiate obsolete.

Further Bierhorst (1971) regards that the ancestral filicalean sporangium must be an elongated structure situated at the tip of an aerial axis and lacks annulus and dehisces longitudinally. Such sporangium has been reported in an extinct genus ***Trimerophyton*** (Trimerophytaceae) which is now regarded as ancestral to the ferns and fern-like plants. It has been established that such a sporangium developed from a single initial which was also the apical cell of the axis. The development was leptosporangiate as the initial produced lateral segments and then stopped producing them after dividing by a dome-shaped periclinal wall (leptosporangiate character). The outer cell produced outer wall layer and the inner one produced tapetal layers, sporogenous tissue and some inner wall layers. Such a discovery contradicts the previously held view that eusporangium and massive sporangia are primitive. It indicates that leptosporangium is primitive.

The Spores

They are the pioneer structures of the gameophytic generation and are differentiated as a result of sporogenesis. The spore mother cells undergo reduction division and form tetrads of haploid spores or meiospores. They young spores are nourished by the plasmodial fluid formed by the disorganisation of the tapetal layers. In the primitive ferns like Ophioglossales, the plasmodial fluid is consumed by the developing spores. In some of the advanced ferns the plasmodial fluid remains as a deposit on the outermost wall of the spore. This deposited plasmodial fluid is often designated as the **"perispore"** or **perine.** The perispore is present in Asplenieae and Aspidieae. In the megaspore of ***Salvinia*** the perispore forms a characteristic layer with pollen-chamber like depression at its apex. The spores may be tetrahedral (Fig. 8.10) in shape or bilateral with two flattened sides (Fig. 8.10). At the **proximal end** of the spore, there is a triradiate or a monolete aperture, a portion where the exine is fissured. The spores are usually shed after separation from the tetrad, *i.e.*, they are liberated in a one-celled condition. In ***Hymenophyllaceae*** and ***Vittariaceae*** the spores start germinating before they are shed. The period of spore viability is also variable. In ***Osmunda*** they remain viable for a few days, whereas in ***Onychium*** they remain viable for over a year. The spore has usually a centrally placed nucleus surrounded by vacuolate cytoplasm that has chloroplastids or leucoplastids and oil globules suspended in it. Majority of ferns are homosporous but ***Marsileales*** and ***Salviniales*** are heterosporous. The plasmodial fluid in the ***sporangia of Azolla*** forms characteristic structures called the **massulae.**

Onychium

Azolla

Regarding polarity of the spore, it had **proximal pole** (the end directed towards the centre of the tetrad) and a **distal pole** (end directed towards outside). The hypothetical line connecting the two poles is called **polar axis**, and the one perpendicular to it is the **equatorial axis.** If the equatorial axis divides the spore into two equal polar faces, the spore is isopolar, otherwise it is heteropolar.

The spore wall encloses the spore protoplasm and usually consists of two layers : **exine** and **intine**. Intine is colourless. The exine or the outer covering is further made up of two layers :—

(*i*) the outer layer or the **ectine** (***Syn-Sexine*** of Erdtman, 1952); and (*ii*) the inner layer or **endine (Syn-Nesine** of Erdtman, 1952). The ectine is made up of elongaed or columnar and radially placed rods called the **columellae.** The columellae may be free at their tips or may be united to form a continuous layer called the **tegillum** or **tectum.** Perine, as stated above, may be present or absent. It is usually absent in trilete spores (except in ***Adiantum*** and ***Cheilanthes***), and may be present or absent in monolete spores.

Cheilanthes

The **ectine** or the outer layer of exine is variously ornamented. The ornamentation is due to the arrangement of **columellae** or due to some outgrowths on the ectine. These ornamentation patterns are considered to be of great taxonomic importance. In case the columellae are free at their ends the pattern is called **pilate.** In many pteridophytes the free ends of columellae are fused and the columellae are arranged in distinct patterns *e.g.*, **reticulate** when they from a network (***Schizaea laevigata*** and some ***Osmundaceae***), **striate** when they are parallel (***e.g.***, ***Schizaea***) **rugulate** when they anastomose (***e.g.***, ***Cheilanthes farinosa***). In many cases the upper ends of columellae are fused to form a layer called the **tegillum** or **tectum** and this layer bears many types of outgrowths or excrescences which are usually of following types :—

(*i*) **Granulose** (Fig. 8.10) when they form a granular or grainlike pattern, *e.g.*, ***Tmesipteris***, ***Mohria***, ***Schizaea bifida***, ***Osmunda***, ***Ophioglossum***, ***Lygodium***, ***Pteridium***, ***Adiantum capillus-veneris*** (Fig. 8.10) etc.

(*ii*) **Spinulose** when outgrowths are spine-like with pointed or blunt end *e.g.*, ***Psilotum***, ***Selaginella fruticulosa***, ***Lonchitis*** (Fig. 8.10), ***Asplenium abysinicum***).

(*iii*) **Verrucate** when there are wart-like unconstricted outgrowths, *e.g.*, ***Pyrossia beddomiana***, ***Selaginella mongolica***, etc.

(*iv*) **Tuberculate** when in the form of finger-like tubercles, *e.g.*, ***Todea, Pteris comeroonianum*** (Fig. 8.10), ***P. togoensis*** etc.

(*v*) **Baculate** when the outgrowths are rod-shaped, *e.g.*, ***Selaginella haematodes.***

(*vi*) **Saccate** when vesicular outgrowths arise *e.g.*, ***Selaginella parkei***, ***S. rupestris*** etc.

(*vii*) **Psilate** when there are fine dots on the surface, *e.g.*, ***Cheilanthes chrysophila***, ***Loxogramma involuta***, ***Pyrossia fissa.***

Schizaea bifida

Lonchitis

Selaginella haematodes

Pyrossia fissa

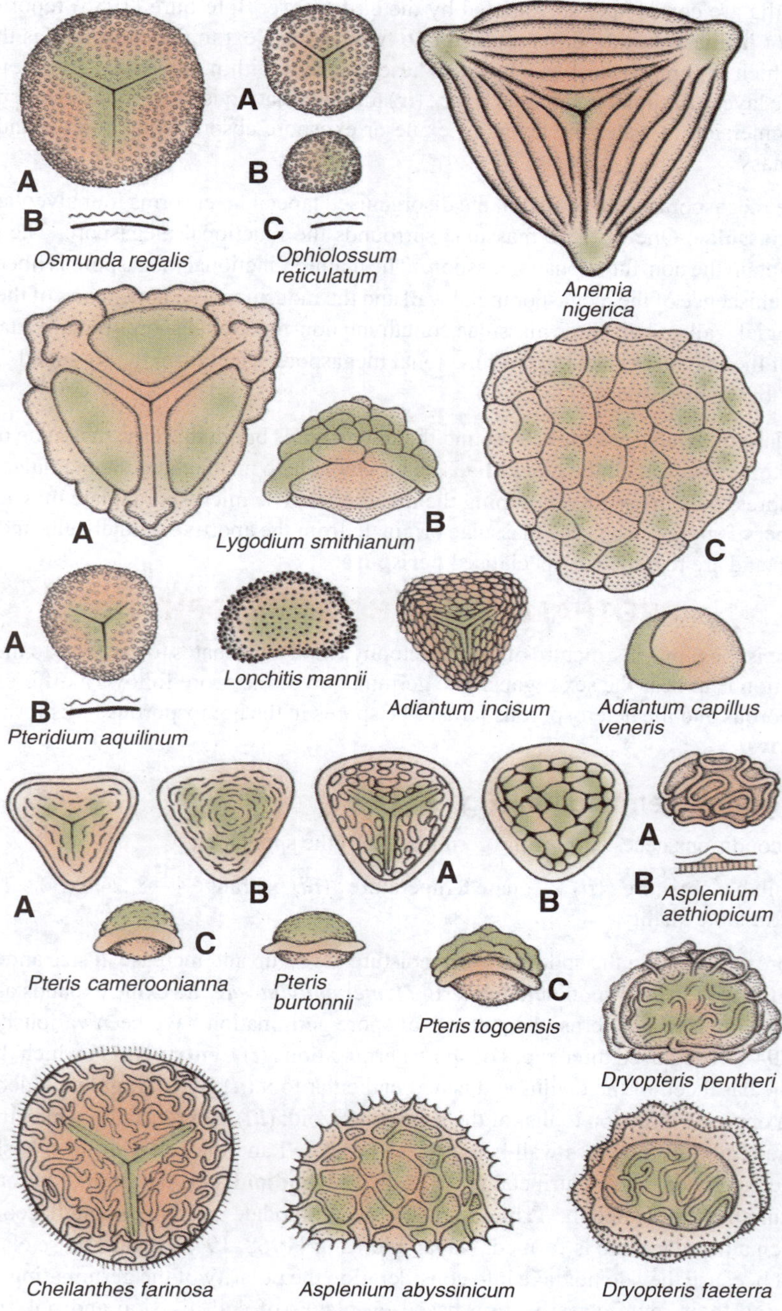

Fig. 8.10. Fillicophyta. Spore types

In some heterosporous ferns like ***Marsilea*** the megaspore while within the megasporangium, is surrounded by an acellular mass formed by the tapetum (which is two to three cell layers thick) and the aborted megaspores. This acellular mass differentiates into a number of layers. The megaspore has an almost oval shape with a small papilla at its apex. Both the megaspore protoplast and the papilla are completely surrounded by the **endospore.** Boterburg (1956) reported that the **endospore** or **intine** is further differentiated into two layers. Next to the endospore is the **exine** or **exospore** which is composed of five layers. These are from within outwards : (*i*) reticulate layer; (*ii*) prismatic layer, (*iii*) inner gelatinous layer ; (*iv*) middle gelatinous layer ; and (*v*) outer gelatinous layer. On immersion in water the layers of exine or exospore absorb water and expand to form a gelatinous mass.

In the megasporangium of ***Azolla*** the disorganised tapetal layer forms four alveolar structures called the **massulae.** One of these massulae surrounds the functional megaspore. The other three massulae contain the non-functional megaspore. The mature functional megaspore is liberated by the transverse dehiscence of the megasporangial wall and the indusium. The upper parts of the indusium, megasporangial wall and the three massulae containing non-functional spores remain attached to the upper end of the massula enclosing the functional megaspore. They form the so-called "**swimming apparatus**" of the megaspore.

Similarly massulae are formed around the microspores but in this case massulae bear numerous pin-like processes called the **glochidia.** The glochidia help in attaching the massulae containing the microspores to the massula enveloping the megaspore. The microsporangiate massulae enclose larger number of spores, each. The massulae originate from the abortive mother cells and the tapetal plasmodium and are regarded as specialised **perispore.**

PROTHALLUS OR THE GAMETOPHYTE

Spore is the pioneer structure of the gametophyte and germinates to give rise to the prothallus whose function is to bear the sex organs. The germination of the spore follows a different course in the homosporous and the heterosporous ferns. The spores in the homosporous ferns form **exosporic gametophytes.**

Early Stages of Germination (Fig. 8.11)

The conditions necessary for the germination of the spore are :

(*i*) suitable moisture ; (*ii*) adequate temperature ; (*iii*) pH range 4—8 ; and (*iv*) sufficient light intensity of suitable quality.

Before germination the spores absorb moisture, swell up and increase in size and the exine is ruptured at the triradiate or monolete mark. In ***Hymenophyllaceae*** the exine expands as the spores swell. In the Homosporous ferns, the patterns of spore germination have been variously classified. Momose (1942) recognised three types of spore germination : (*i*) **Centripetal** in which the rhizoid is seated on the basal cell of the thallus and is perpendicular to it (*ii*) **Centrifugal** in which the thallus grows in an opposite direction to that of the primary rhizoid; (*iii*) **Tangential** in which the rhizoid is seated at the region of the cross wall between the basal cell and the cell next to it. Nishida (1965) renamed the Tangential and Centripetal types as the **Polypodioid** and **Aspidioid** types, respectively and added another type called the **Tripolar-**type to accommodate the Hymenophyllaceous germination in which a tripolar plate is formed. Nayar and Kaur (1968, 1971) regard these classifications as incorrect because they do not take into consideration the polarity of the germinating spore. They have proposed their own classification based on planes of cell division and polarity of spore (Fig. 8.11). They recognised three distinct categories :—

1. Polar Germination (Fig. 8.11)

In this type of germination the first wall of the spores is parallel to its equatorial plane. The growth and elongation of primary rhizoid and the young thallus is parallel to the polar axis of the spore. This type is further sub-divided into three sub-types :—

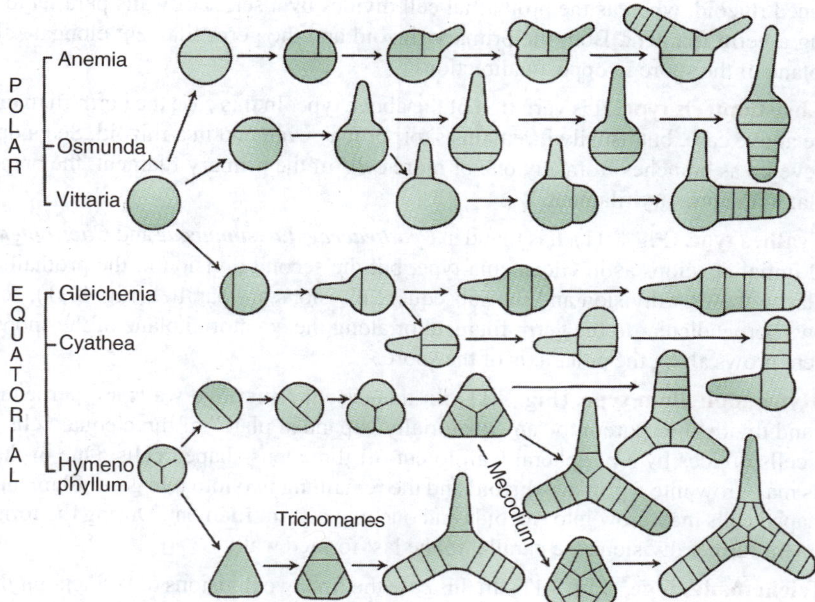

Fig. 8.11. Earlier stages of spore germination in homosporous ferns. The proximal pole of the spore is indicated by the tri-radiate mark. (After Nayar and Kaur, 1971)

(*a*) Osmunda type. It is found in ***Osmundaceae*** and is the simplest type of polar germination. The first wall cuts off a small **primary rhizoidal initial** that grows parallel to the polar axis, into a primary rhizoid. The other cell divides by one or two or even more walls parallel to the first wall (Fig. 8. 11) into a short uniseriate **germ filament.** The germ filament and the rhizoid grow in opposite directions.

(*b*) Anemia type (Fig. 8.11). In this type the spore divides into two equal cells of which the distal one remains inactive throughout. The proximal cell divides by wall perpendicular to the first to cut off a small lateral **rhizoid initial** and a large **prothallial initial.** The prothallial initial divides by a series of walls parallel to the first wall and grows into a germ filament that elongates in a direction parallel to the polar axis of the spores (facing the proximal pole). The primary rhiziod elongates parallel to the equatorial axis of the spore. This type is characteristic of ***Anemiaceae*** and ***Lygodiaceae*** or Nayar.

(*c*) Vittaria type (Fig. 8.11). It is the commonest type of polar germination. The first division is similar to Osmunda type. The smaller cell grows into a primary rhizoid that elongated parallel to the polar axis. The larger or the distal cell acts as the prothallial initial and divides by a wall perpendicular to the first wall into two equal daughter cells. One of them remains inactive and the other divides by series of walls parallel to the second wall into a germ filament that is perpendicular to the primary rhizoid.

2. Equatorial Germination

In this type of germination (Fig. 8.11) the first wall of the spore is formed parallel to the polar axis of spore. The prothallus elongates in a plane parallel to the equatorial plane of the spore. It is further sub-divided into six types :—

(*a*) Gleichenia type (Fig. 8.11). It is characteristic of ***Gleicheniaceae***, ***Dipteridaceae***, ***Loxogrammaceae*** and most of the ***Polypodiaceae***. The first wall is parallel to the polar axis cutting off a small primary rhizoidal and a large prothallial cell. The primary rhizoidal cell elongates into a laterally placed rhizoid, whereas the prothallial cell divides by a series of walls parallel to the first wall forming a germ filament. Both the primary rhizoid and the germ filament elongate along the equatorial plane of the spore in opposite directions.

(*b*) Christiopteris type. It is variation of the above type. In this case the germ filament develops as in the above case, but usually it remains short or may even end in a rhizoid. Secondary germ filaments develop as branches from any one or more cells of the primary filament; the branches are perpendicular to the primary filament.

(*c*) Cyathea type. (Fig. 8.11). It is found in ***Cyatheaceae***, ***Loxsomaceae*** and ***Cheiropleuriaceae***. The rhizoid initial develops as in Gleichenia type, but the second division in the prothallial cell is perpendicular to the first division and the subsequent divisions are parallel (Fig. 8.11), so that the rhizoid grows perpendicular to the germ filament or along the equatorial plane of the spore and the germ filament grows along the polar axis of the spore.

(*d*) Hymenophyllum type. (Fig. 8.11). In this case the first two walls are perpendicular to each other and divide the spore into "an equatorially expanded plate" of three equal cells. Each of these three cells divides by a peripheral wall to cut off three lens-shaped cells. One of these lens-shaped cells may grow into a primary rhizoid and the remaining two into two germ filaments, or two of the lenshaped cells may grow into rhizoids and one into a germ filament. During the formation of a germ filament further divisions are parallel to the last formed wall.

(*e*) Trichomanes type. (Fig. 8.11). In this case the first two divisions of Hymenophyllum type are omitted with the result that three lens-shaped cells are formed simultaneously at the periphery of the triangular spore. Further development is like the Hymenophyllaceous type.

(*f*) Mecodium type. (Fig. 8.11). It is like Hymenophyllaceous type up to the formation of three cells. Extra divisions occur in these three cells resulting in the formation of a 9 to 12 celled triangular plate of cells, one cell in thickness. It is expanded along the equatorial plane of the germinating spore. Three lens-shaped peripheral cells are cut off in this plate, at each corner, and develop further as in the Hymenophyllaceous type.

3. Amorphous Germination

It is characteristic of some primitive ferns, *e.g.*, ***Marattiaceae***, ***Matoniaceae***, ***Ophioglossaceae***, a few members of ***Schizaeaceae*** (***Actinostachys*** and ***Lophidium***) and ***Gleicheniaceae*** (***Stromatopteris***). The spore divides without any polarity in cell divisions and direction of growth. It results in a plate of cells in which a meristematic cell is differentiated in one of the peripheral marginal cells. Further growth of the prothallus is in the direction of the meristematic cell.

Stromatopteris

Later Stages of Development (Fig. 8.12)

These stages lead to the development of a characteristic adult prothallus or the gametophyte. Nayar and Kaur (1969, 1971) recognised seven different patterns among the homosporous ferns. These are briefly described below :—

1. Osmunda type (Fig. 8.12)

It is characteristic of ***Osmundaceae***. The two celled long germ filament turns into a plate of 4 cells by the longitudinal division of each cell. In case the germ filament is longer then only the anterior pair of cells will divide to form 4 cells. These four cells divide further to form a circular plate of prothallial cells. A meristematic cell is established in a marginal cell. The quadrant in which the meristematic cell appears grows more and the prothallus assumes an asymmetrical shape (Fig. 8.12). The young thallus elongates and later becomes notched at the growing region. It later on assumes a symmetrical cordate shape as in Adiantum type.

2. Marattia Type (Fig. 8.12)

In this case spore germination is of amorphous type resulting in a circular plate of cells, that may later become two or more cells thick. A meristematic cells is established, notch appears and ultimately a cordate prothallus develops. Later growth is like the Adiantum type.

3. Adiantum Type (Fig. 8.12)

In this case the terminal cell of the germ filament divides first by an oblique wall followed by a wall perpendicular to it. This establishes a meristematic cell flanked on the either side by two cells. Sometimes the terminal cell of the filament divides by a wall parallel to the long axis of the filament followed by a second oblique wall cutting off a median cell, which in turn divides by a wall perpendicular to the second oblique wall. This cuts off a median meristematic cell. The meristematic cell divides by walls parallel to the oblique wall, each wall being perpendicular to the one preceding it. These divisions and further divisions of the daughter cells lead to the formation of an obovate plate of cells which later becomes notched in the region of the meristematic cell (Fig. 8.12). The meristematic cell then divides by a transverse wall (Fig. 8.12). The anterior daughter cell divides by two or three walls parallel to each other but perpendicular to the original transverse wall, forming a row of two or three cells. These cells are elongated in the direction parallel to the long axis of the thallus and constitute the 3-celled meristem. These cells divide by walls parallel either to their lateral or basal walls. These divisions in two planes lead to the formation of a mid-rib behind the meristematic region in the median plane of the prothallus and a cordate prothallus with semicircular lateral wings.

4. Drynaria Type (Fig. 8.12)

In this type the formation of apical cell (meristematic cell) is delayed and the terminal cell of the germ filament divides by longitudinal and transverse walls into a spathulate plate of cells (Fig. 8.12). At a later stage the meristematic cells is distinguished in one of the marginal cells of this plate by two oblique divisions (Fig. 8.12). Later divisions are exactly similar to the Adiantum type and lead to a cordate prothallus with a terminal notch.

5. Kaulinia Type (Fig. 8.12)

It is like ***Drynaria*** type up to the formation of a spathulate plate. In this case no meristematic cell is formed. All the cells of the prothallus divide actively. The prothallus elongates and becomes ribbon-like with rounded anterior ends (Fig. 8.12). Branches may arise from the marginal cells. There is no midrib. At maturity irregular cushions of two to four cells in thickness may develop along the median plane of the prothallus.

Drynaria

6. Ceratopteris Type (Fig. 8.12)

It is also similar to ***Drynaria*** type upto the formation of a spathulate plate of cells. Now in some cases the anterior cells of the germ plate may not grow further or grow very slow and take no further part in the development of the prothallus.

The inter-calary cells, instead, divide actively and gradually, the meristematic activity becomes restricted only to some marginal cells on one side and away from the apical portion (Fig. 8.12).

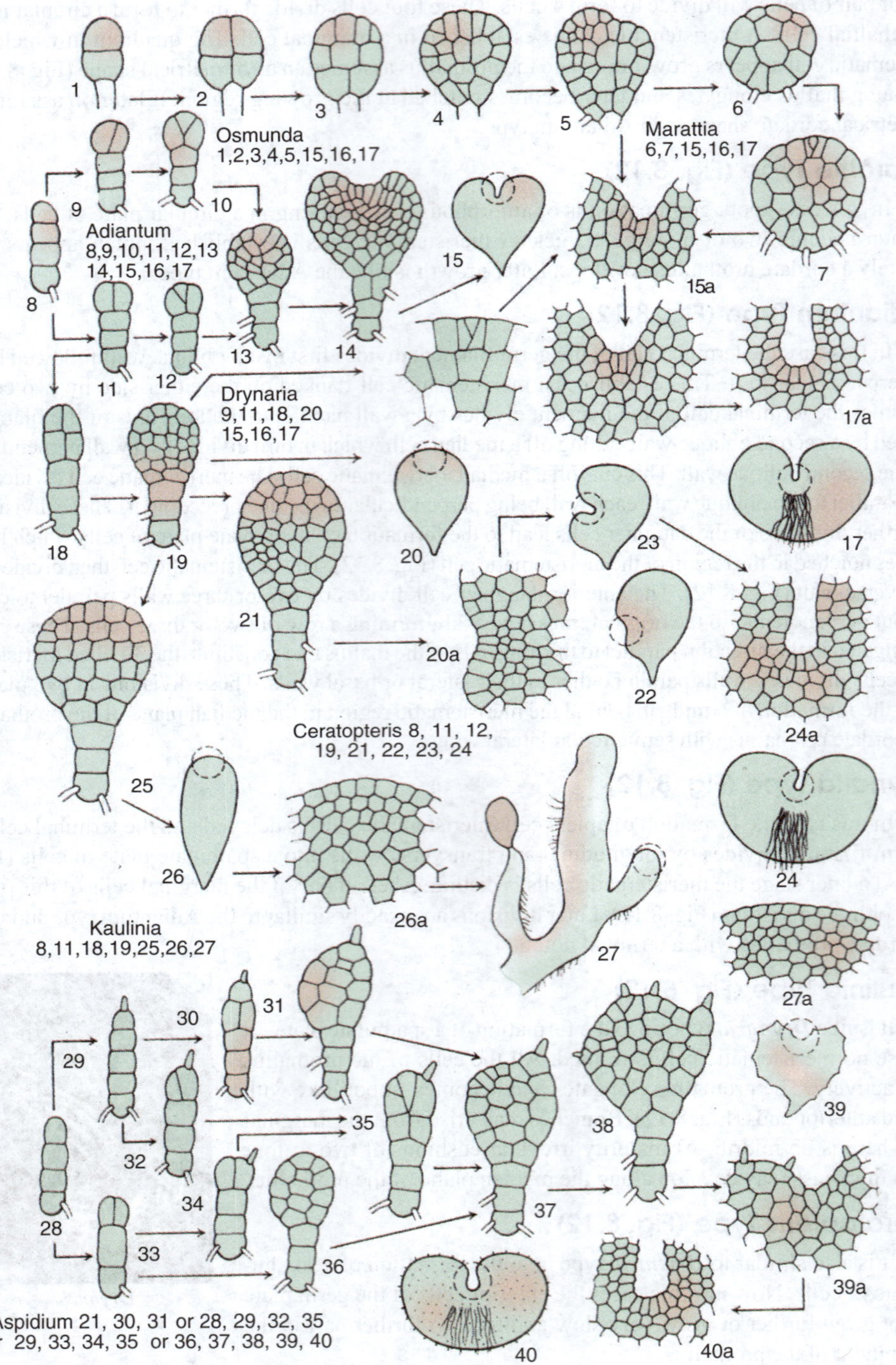

Fig. 8.12. Later stages in spore germination in homosporous ferns. (After Nayar and Kaur, 1971). The various types are indicated by numbers. Each specified type with regard to its numbers is given within squares and rectangles.

Later a lateral notch appears. The meristematic cells lie at the base of this notch, which goes on becoming pronounced. Later a mid-rib develops behind the meristematic region. The young prothallus is cordate but asymmetrical with one wing larger than the other or in extreme cases only one wing is developed. In some cases the marginal meristem becomes apical by unilateral growth of both the wings, thus making the prothallus cordate.

7. Aspidium Type (Fig. 8.12)

It shows great variations in development. These variations are conditioned by the development of a hair. Usually the terminal cell of the germ filament grows into (Fig. 8.12). a unicellular hair. This cell becomes sluggish and takes no part in prothallus development. Sometimes a few cells below the terminal cell also remain sluggish and the posterior cells of the filament divide actively to form a plate of cells (Fig. 8.12). This plate is lopsided ***i.e.***, one side is broader than the other. The larger side of the plate distinguishes a meristematic cell by two oblique divisions of one of its marginal cells (Fig. 8.12). In some cases the terminal cell of the germ filament divides longitudinally into a smaller cell and a larger cell that bears the unicellular hair. The smaller cell divides actively and produces the prothallial plate (Fig. 8.12). The plate is slightly lop-sided at its anterior end. A meristematic cell appears in the margin. Marginal unicellular hair always develop from the prothallial plate. In another variation hair development is delayed until the terminal cell of the filament has divided longitudinally. One of the two cells grows slowly and divides, ultimately bearing a hair. The other cell divides actively to form a plate which develops a marginal meristematic cell. Later development is like the Adiantum type (Fig. 8.12).

All these variations described above may occur in the same species. The young thalli are asymmetrical, cordate and lopsided due to marginal meristematic cell being lateral. Later growth, no doubt, results in the loss of asymmetry and finally a symmetrical, cordate prothallus is produced (Fig. 8.12). Marginal hair are profuse so, are the superficial ones.

The net result of spore germination is the prothallus that also varies in shape, size and structure. Cases are not rare where the prothallus in the same genus or even in the same species shows variations in form and size (***Dryopteris***, ***Schizaea***, etc).

The Adult Prothallus (Fig. 8.13)

Nayar and Kaur (1971) recognised 5 types of adult prothalli in the homosporous ferns. These are; (*i*) **cordate;** (*ii*) **filamentous ;** (*iii*) **strap shaped ;** (*iv*) **ribbobn-shaped;** and (*v*) **tuberous.** These are described below.

The Cordate Type

It is the most common type of prothallus (Fig. 8.13) found among the ***Polypodiaceae*** (***Dryopteris***, ***Nephrolepis***, ***Pteridium***, ***Pteris***, ***Adiantum,*** etc.)., ***Dicksoniaceae***, ***Gleicheniaceae***, ***Dipteridaceae***, ***Matoniaceae***, ***Osmundaceae***, ***Marattiaceae*** and ***Cyatheaceae.*** In the ***Polypodiaceae*** the spore germinates to give rise to a green plate of cells with a apical notch. Later the apical cell is replaced by a row of meristematic cells. Under suitable conditions of growth and nourishment this notched plate develops into a heart-shaped or a cordate prothallus. The gametophyte is one cell in thickness except in the region posterior to the notch, where it is many celled thick. The sex organs and rhizoids develop on the ventral side. Under crowed and undernourished conditions the prothallus becomes filamentous, *e.g.*, ***Dryopteris*** and bears only antheridia. It can develop into a more or less cordate type if transferred to suitable environmental conditions. Normally the prothallus is **monoecious.** The prothallus in ***Cyatheaceae*** also resembles that of ***Polypodiaceae*** in its shape and structure, but differs in the earlier mode of development. In this case the short filament of cells divides longitudinally to form a plate of green cells. The apical cell later appears in this thin plate of cells. Later it is replaced by a group of apical cells. Further development is similar to that of polypodiaceous prothallus. In ***Dipteridaceae*** (***Dipteris conjugata***) the spore germinates to form a plate of cells which later develops into a cordate prothallus with sinuate or ruffled margins. The sex organs usually develop on the ventral surface but they can also develop on the dorsal side.

The Filamentous Type

The filamentous type of prothallus is characteristic of ***Schizaea*** (Fig. 8.13, C) and some members of the family ***Hymenophyllaceae*** (***Trichomanes***). Filamentous prothalli have, no doubt, been seen to develop in several ferns but they are the result of abnormal environmental conditions. In ***Schizaea*** and ***Trichomanes*** filamentous condition persists under all circumstances. The prothalli in both these genera resemble some branched algal filaments or moss protonemata. Some of the branches penetrate the soil and act as rhizoidal branches, whereas other are green and transversely septate. In ***Schizaea pusilla*** the rhizoids are unicellular and develop from specialised short branches whose cells are infected with a mycorrhizic fungus. The sex organs are seated on unicellular lateral branches. In ***Trichomanes*** the antheridia occur on any branch of the filamentous prothallus but archegoina appear in clusters on specialised multicellular structures called the archegoniophores.

Elaphoglossum

Strap-shaped Thallus

This is found in the family Grammitidaceae and some Lomariopsidaceae and Polypodiaceae. In ***Rhipidopteris peltata*** and ***Elaphoglossum stenophyllum*** the thalli are long and strap-like with fimbriated and hairy margins. The breadth of the thallus is intermediate between cordate type and ribbon-shaped prothalli. They are much longer than broad (Fig. 8.13, F) and are slow growing and unbranched with cordate apex. The apical meristem is pluricellular, the midrib is thin and interrupted. Sex organs and rhizoids are borne on the midrib. The rhizoids may occur along the margins too.

Ribbon-shaped Thallus (Fig. 8.13, E)

It is found in ***Loxogrammaceae, Vittariaceae*** and some ***Hymenophyllaceae*** and ***Polypodiaceae***. The thalli are thin, one cell in thickness, flat, dorsiventral, perennial and slow growing. It is devoid of a mid-rib and has rounded apex that has no well defined meristem. It is usually profusely branched. The branching is lateral. The sex organs are borne on irregularly scattered 2-4 celled thick cushions. In ***Hymenophyllaceae*** these cushions are marginal and in others they are superficial. The rhizoids arise in marginal clusters.

The Tuberous Type

This type of prothallus is prevalent in the ***Ophioglossaceae***. ***Helminthostachys*** affords a very good example of such a type of prothallus (Fig. 8.13, A). In this case the prothallus is saporophytic in nutrition and is mostly underground or subterranean. The underground portion of the prothallus is lobed and is attached to the soil by means of rhizoids. It is also called the vegetative region of the prothallus. From the vegetative region arises a cylindrical branch that grows vertically upwards and bears the sex organs. This cylindrical and generative region may remain underground or it may slightly project above ground. The cylindrical region grows by means of 4-sided apical cell. The endophytic fungus is largely confined to the lower vegetative region but it may extend up to the cylindrical region. The thalli are perennial and very slow growing.

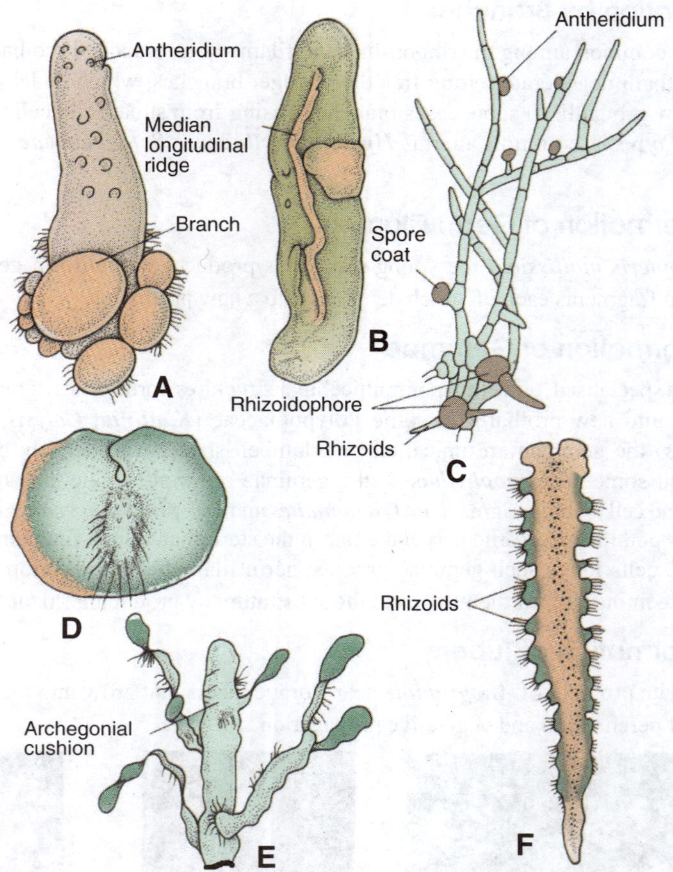

Fig. 8.13 (A–F). Mature prothallus types of homosporous ferns. A. Tuberous prothallus of ***Helminthostachys***; B. Tuberous prothallus of ***Botrychium***; C. Filamentous prothallus of ***Schizaea pusilla***; D. Cordate prothallus of ***Adiantum caudatum***; E. Ribbon-shaped prothallus of ***Paraleptochilus decurrens*** (Polypodiaceae); F. Strapshaped prothallus of ***Elaphoglossum*** **sp.**

The mature gametophytes of ***Ophioglossum*** may be cylindrical, conical or irregular in shape. They may be branched or unbranched. They are usually wholly subterranean or in some cases partly above ground and partly underground. In ***O. pedunculosum*** the exposed portion of the prothallus assumes green colour and may become flattened and irregularly lobed.

The gametophytes of ***Botrychium*** are also mycorrhizic and may be cylindrical or have a tendency towards flattened and dorsiventral form (***Botrychium virginianum***). There is no well defined meristem in ***Helminthostachys*** and ***Botrychium.***

Such prothalli are also found in ***Actinostachys*** and ***Lophidium*** of ***Schizaeaceae***, and ***Stromatopteris*** of ***Gleicheniaceae.***

Vegetative Propagation of the Prothallus

Vegetative propagation of mature prothalli of homosporous ferns is quite common. It is effected by following methods :—

1. Regeneration by Branches

It is very common among the ribbon-like and filamentous branched prothalli. The older portions of the prothalli degenerate setting free the younger branches, which under certain conditions develop into new prothalli. In some cases branches arising from superficial cells develop into new prothalli. This type is common among ***Hymenophyllaceae***, ***Schizaeaceae***, ***Vittariaceae*** and ***Polypodiaceae.***

2. By the Formation of Germ Filaments

In ***Ctenopteris mollissima*** the young prothallus produces monoliform germ filaments that easily break into fragments each of which develops into a new prothallus.

3. By the Formation of Gemmae

These are specialised, unicellular or multicellular structures borne on sterigmata and are capable of germinating into new prothalli. In some Polypodiaceae (***Kaulinia***, ***Colysis***, ***Leptochilus*** and ***Paraleptochilus***) the gemmae are unicellular or dumbell-shaped and densely chlorophyllous. In ***Vittariaceae*** and some ***Hymenophyllaceae*** the gemmae are multiseriate and spindle-shaped are attached by an end cell to the sterigmata. In ***Trichomanes*** and ***Polyphlebium venosum*** the multicellular and filamentous gemmae are borne perpendicular to the sterigmata. Such filamentous gemmae may become 10—12 cells long. Such gemmae produce germ filaments that develop into prothalli. On detachment the gemmae attach themselves to the substratum by producing rhizoids.

4. By the Formation of Tubers

The cordate prothalli of ***Anogramme*** bear storage tubers that grow into the soil and serve the dual function of perennation and vegetative propagation.

Colysis

Leptochilus

Anogramme

THE SEX ORGANS

The sex organs develop on the gametophyte and are always multicellular and jacketed. The male sex organs are called **antheridia** and the female are known as **archegonia**. Both may arise on the same prothallus (bisexual or monoecious) or they may be borne on different (unisexual or dioecious) prothalli. Bisexual prothalli may be **protandrous** or **protogynous.** The antheridia produce male gametes or the sperms or antherozoids that are flattened structures and require a liquid medium to swim to the female sex organs containing eggs. The number of sperms produced per antheridium varies considerably. It may be from more than one hundred (***Ophioglossum***) to 16 to 32 sperms per antheridium as in ***Dryopteris***, ***Pteris***, ***Pteridium***, ***Adiantum***, etc. The female gamete or the egg is non-motile and is present in the venter of the archegonium. The archegonia show a remarkable uniformity in their structure and development and in possessing a single egg.

The hermaphrodite prothalli usually produce antheridia earlier than the archegonia. Antheridial development often begins before the differentiation of the mid-rib or even before the organisation of the plate-like stage. The archegonial development starts after the formation of the mid-rib or the cushion or cushions, as the case may be. It has been noticed that unfavourable growth conditions lead to the production of male prothalli and exceptionally favourable conditions tend to produced only female prothalli whereas less favourable conditions favour the production of bisexual prothalli. Nutritional conditions of the substratum and availability of light affect the sex of the prothallus (Sossuntzov, 1954, 1957; Hurel-Py 1950 ; Schraudolf, 1967). Recently it has been shown that antheridia formation in the ferns is controlled by chemical substances called the **antheridiogens.** Antheridiogens are specific for certain groups of plants, *i.e.*, an antheridiogen effective in ***Pteridium*** may not be effective in ***Ceratopteris*** and ***vice versa*** (Dopp, 1950, 1959; Naf 1956, 59, 60, 65; Voeller, 1964; Schedlbauer, 1972). The antheridiogens are produced by the prothallus at specific periods and then diffuse into the substratum. Their small concentrations induce antheridial formations in species sensitive to the particular antheridiogen.

Position of the sex organs

As has been stated before, the gametophytes may be **exosporic** or **endosporic.** The former are typical of the homosporous forms whereas the latter are restricted to the heterosporous pteridophytes. The exosporic prothalli are primitive and are usually monoecious or may become dioecious due to certain changed environmental conditions.

Ceratopteris

The sex organs in the exosporic gametophytes may be embedded in the gametophyte tissue or they may be of the projecting or emergent type. The former is the case in the eusoporangiate ferns whereas the latter type is commonly found in the leptosporongiate ferns. In ***Ophioglossum vulgatum*** and other species of this genus the sex organs are sunken in the prothallus tissue. In ***Botrychium virginianum***, a member of the same family, the antheridia are sunken, but the archegonia have projecting necks. The sex organs develop on the upright cylindrical branches in ***Ophioglossum***, that arise from the lobed basal part. They develop in an acropetalous manner. The antheridia are numerous and develop in an irregular sequence all around the cylindrical body. In ***Marattiaceae*** (***Marattia***, ***Angiopteris***), the dorsiventral prothallus is monoecious and bears antheridia on both the dorsal and the ventral surfaces. In ***Ceratopteris*** the antheridia are partially embedded in the prothallus tissue (Fig. 8.14, 12) and are borne along the margins. The archegonia are confined only to the ventral surface and occur along the central cushion along with the rhizoids. The antheridia and the archegonia are both sunken in the prothallus tissue. The sex organs in Osmundales arise on the ventral surface of the prothallus surface of the prothallus and project above the surface (emergent type). The antheridia in ***Osmunda***, arise along the margins or on the ventral side of the lateral wing like expansions of the prothallus. The archegonia arise on either side of the distinct mid-rib and have projecting necks. In some of the ***Osmundaceae*** young archegonia are also found mixed with older ones. Majority of the leptosporangiate ferns have their sex organs distributed on the ventral surface of the prothallus. They are of emergent type. The antheridia occur among the rhizoids whereas the archegonia are restricted to the cushion behind the apical notch (***Dryopteris***, ***Pteris***, ***Pteridium***, ***Adiantum***, ***Nephrolepis***). In the filamentous prothalli the sex organs arise laterally on the filaments, but in some cases archegonia are borne on special multicellular cushions.

Nephrolepis

The endosporic female gametophytes of ***Azolla*** and ***Salvinia*** develop several archegonia, whereas those of ***Marsilea***, ***Pilularia*** and ***Regnellidium*** bear only one archegonium. The endosporic male gametophytes in ***Marsilea*** and ***Azolla*** bear two antheridia and produce 32 and 8 antherozoids, respectively, The endosporic male gametophytes in ***Salvinia*** produce two antheridia with a total of 8 antherozoids from both.

Structure of sex organs

(*a*) Antheridium. In the homosporous ferns the antheridium is a simple object containing a variable number of spermatozoids enclosed by a single layer of wall cells. The number of wall cells varies from 3 to many. In the primitive homosporous groups Ophloglossales and the Marattiales the antheridia are embedded in the prothallus tissue. In ***Ceratopteris*** which belongs to the advanced group of ferns the antheridia are partially embedded (Fig. 8.14, 6 to 12). There are two types of antheridia within the homosporous ferns. These are **(*i*)** the **eusporangiate antheridia** that are characteristic of the primitive groups, and **(*ii*)** the **leptosporangiate antheridia** that are found mostly in the advanced groups.

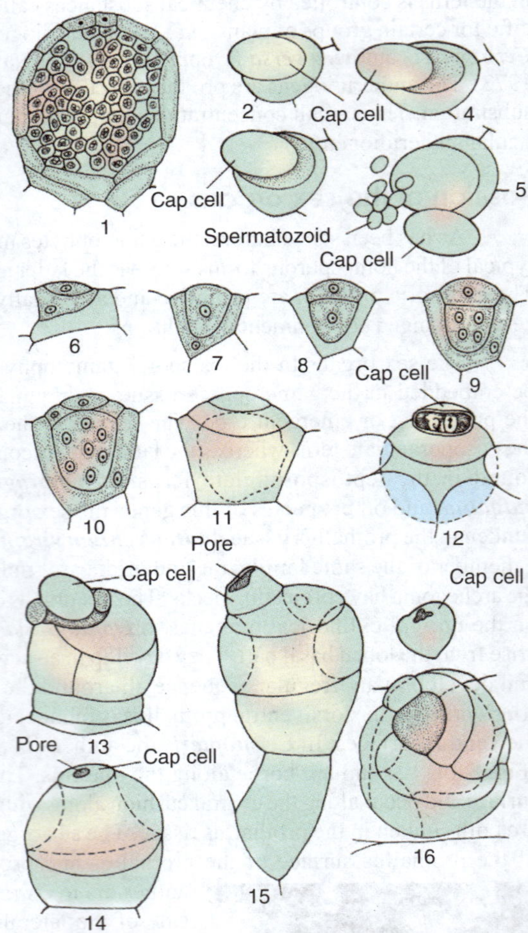

Fig. 8.14. **Antheridia of homosporous ferns.**
1. L.S. antheridium of *Todea barbara* (After Stokey); 2—5. Stages in antheridial dehiscence of *Anemia* by slipping of cap cell; 6—12. Stages in the development of antheridium of *Ceratopteris thallicroides*; 11. Lateral view of mature antheridum; 12. Showing dehiscence (After Nayar & Kaur, 1971); 13. Dehisence of antheridium in *Alsophila excelsa* (After Stokey); 14. Dehiscence of antheridium by apical pore in *Stenochlaena tenuifolia* (After Nayar and Kaur. 1971); 15. Dehiscence of antheridium in *Ctenopteris suspensa* by apical pore (After Stokey); 16. Antheridium of *Chieropleuria bicuspis* showing dehiscence by throwing off cap cell (After Stokey and Atkinson).

In the eusporangiate type the antheridia are large and massive and produce over 100 spermatozoids. They are embedded and are surrounded by a wall consisting of 10—30 narrow and elongated cells. In Ophioglossales the wall may become 2—3 cells thick. From one of the wall cells in the anterior part of the antheridium, one or two small triangular or ovate cap cells are cut off. They constitute the **operculum.** The antheridium is attached to prothallus by two or three wedge-shaped or flat basal cells. During dehiscence the cap cell is thrown off intact to release the spermatozoids.

In the leptosporangiate type of antheridium the wall is made up the three cells : (*i*) the basal cell which may be funnel-shaped (***Dryopteris, Adiantum***, etc.), barrel-shaped (***Ctenopteris***) or discoid (***Stenochlaena***) ; (*ii*) the annular or the ring cell ; and (*iii*) the apical cap cell or the cover cell. It encloses 16—32 spermatozoids. During dehiscence the cap cell behaves in the following manner :—

(*a*) In some leptosporangiopsida the cap cell is thrown off (Polypodiaceae) and it often collapses during the process (Hartman, 1931), (Fig. 8.14, 16).

(*b*) In ***Cheilanthaceae*** and ***Parkeriaceae*** the cap cell is lifted like a hinged lid (Fig. 8.14, 12, 13).

(*c*) In ***Ctenopteris*** and ***Stenochlaena*** a pore is formed in the cap cell (Fig. 8.14, 14—15) and the sperms are released through it.

(*d*) In ***Anemiaceae*** (Fig. 8.14, 2—5), the cap cell is detached all around and slips off sideways, thus opening up the antheridium.

The heterosporous ferns have antheridia enclosed completely within the microspore wall. The antheridial wall is single layered. Some regard that in ***Marsilea*** a single male gametophyte has two antheridia because there are two distinct groups of androcyte mother cells. There is no cap or opercular cell in the heterosporous ferns. The entire jacket disintegrates. This is followed by the dissolution of the microspore wall thus liberating the sperm mass.

(*b*) Archegonium. The archegonium in all the ferns is made up of two distinct parts : (*i*) the **neck** and (*ii*) the **venter.** The neck is composed of four longitudinal rows of cells, each row having an equal number of cells that vary among the ferns. The neck is closed by four cells called the **cover cells** and encloses a neck canal which is filled with one or more **neck canal cells.** The **venter** is the lower swollen region of the archegonium and contains a single **egg** and a **ventral canal cell.** The neck canal cells, the ventral canal cell and the egg form an axial row of cells. The venter is surrounded by a single layer of wall cells that may remain distinct from the surrounding prothallial cell or may become indistinguishable. The archegonia in ***Ophioglossum*** are embedded in the tissue of the prothallus with only the upper portion of the neck slighty projecting. In ***Botrychium*** the archegonial neck projects sufficiently above the prothallus. The neck in ***Ophioglossum*** is made up or 4 longitudinal rows of cells, each row 3 to 4 cells in height. There is a single binucleate neck canal cell. The same is the case in Marattiales. The archegonial neck is up to 8 cells high in ***Osmunda cinnamomea*** and projects above the prothallus surface. The archegonial neck in ***Gleichenia*** (***Gleicheniaceae***) may be 7—14 cells in height and is always directed towards the growing apex. The neck may be curved or straight. The necks of archegonia in the polypodiaceae are strongly curved away from the apex of the prothallus and may be 3—6 cells in height. ***Anemiaceae***, ***Matoniaceae***, ***Gleicheniaceae*** and ***Cheiropleuriaceae*** the archegonial neck is curved towards the apex of the prothallus whereas in others it is curved away from the apex. The former positions along with long neck is a primitive condition. The single neck canal cell is binucleate. The archegonia are emergent in ferns with filamentous prothalli. The neck canal cell is binucleate and generally swollen towards the apex at maturity. This is an advanced condition. In ***Gleicheniaceae***, ***Cyatheaceae***, ***Schizaeaceae*** and ***Dicksoniaceae***, the neck canal cells are sometimes 4-nucleate. In ***Plagiogyraceae*** the neck canal has 4 cells. Nishida and Sakuma (1961) have reported abnormal archegonia with multinucleate and multicellular neck canals and with two ventral canal cells in some primitive families.

The axial row of cells, except the egg or the oosphere, degenerate thus making a passage for the sperms to swim through. The four terminal cells or the cap cells separate from one another and diverge or are thrown away thus opening the archegonial neck. It is probable that the degenerated mass of axial cells absorbs water and swells up thus exerting a pressure on the cover cells. As a result a drop of liquid oozes out through the opening. The exudate contains an organic acid (usually malic acid) or some other substance whose smell attracts the spermatozoids. Under this chaemotactic response the spermatozoids reach the neck, enter it and one of them is able to effect fertilization.

Development of sex organs

The mature sex organs look strikingly different but they show a remarkable similarity, at least in their early stages of development. Both originate as one-celled protuberances, from the superficial layer of the prothallus. It is very difficult to say at this stage as to which will develop into antheridium or an archegonium. At this one-celled stage these initials seem to be uncommitted. It appears that their nutritional biochemical milieu might be the determining factor guiding their developmental destiny, whether antheridium or archegonium. It seems probable that the genetic system controlling the processes involved can invoke different biochemical path-ways and produce different phenotypic end products.

Archegonium

The superficial archegonial initial divides first by periclinal wall. This wall separates an outer sterile **primary cover cell** and an **inner cell.** The former divides by two intersecting walls to form four cover cells, which further by transverse divisions give rise to 4 longitudinal rows of neck wall. The inner cell may divide transversely into a lower **basal cell** and an upper **central cell** or may function directly as the **central cell.** The central cell is the potential fertile cell which by further transverse divisions gives rise to the neck canal cell, ventral canal cell and the egg. The central cell divides transversely into an upper **primary neck canal cell** and a lower **primary ventral cell.** The former may divide into two or more neck canal cells, but in the ferns it functions directly as the single canal cell that is usually binucleate. The primary ventral cell divides into an upper smaller **ventral canal cell** and lower **egg cell.**

Nishida and Sakuma (1961) studied archegonial development in a number of homosporous ferns and pointed out that there are variations in the nuclear division in the axial row of cells. In ***Ophioglossales***, ***Adiantaceae***, ***Cheilanthaceae***, ***Dryopteridaceae***, ***Landseaceae***, ***Pteridaceae***, and ***Dennstaedtiaceae*** the ventral canal cell is differentiated only after the neck canal cell has completed its nuclear division. In ***Anemiaceae***, ***Cyathaeaceae***, ***Gleicheniaceae***, ***Grammitidaceae***, ***Hymenophyllaceae***, ***Loxosomaceae***, ***Marattiaceae***, ***Osmundaceae, Parkeriaceae***, ***Plagiogyriaceae***, and ***Polypodiaceae*** the ventral canal cell is formed before the nucleus of neck canal cell divides . In ***Blechnaceae*** and ***Mattoniaceae*** both the conditions are found. They consider this character as of significance in the phylogeny of the ferns.

Antheridium

The eusporangiate type of antheridium develops from a superficial wedgeshaped initial that differentiates from a superficial cell of the prothallus. The divisions of the antheridial initial do not follow any regular sequence of cell divisions. Earlier divisions distinguish a central androgonial cell surrounded by many wall cells. The androgonial cell divides repeatedly to form androcytes whose protoplasts metamorphose into spermatozoids.

In the embedded type of antheridium in Ophioglossales and Marattiales the embedded antheridium is of massive eusporangiate type. It originates as a single antheridial cells that is superficial but does not project above the prothallus surfaces. It divides by a transverse wall into a lower **androgonial cell** and an upper **primary cover cell.** The cover cell divides by 3 or 4 vertically intersecting walls to form a row of cover cells. Out of these the central cell acts as an **opercular cell** (or cap cell). Sometimes the cover cells divide by a transerse wall so that, this layer becomes two celled thick. The androgonial cell enlarges and divides repeatedly to form a mass of spermatocytes whose protoplasts metamorphose into spermatozoids. As the spermatozoids are developing the surrounding cells of the prothallus divide to cut off an antheridial wall of 2 or 3 cells thick.

The embedded antheridium of ***Ceratopteris*** (Fig. 8.14, 6—12) is basically of leptosporangiate type but differs in some respects. In this case a prothallial cell acts as a mother cell and divides by an oblique wall to cut off a wedge-shaped **antheridial initial**. The first division in the antheridial initial is by a curved wall that deepens sufficiently (Fig. 8.14, 6—7). It cuts off a peripheral cup-shaped cell surrounding an anterior cell on all sides except the upper flat side. It has a dome-shaped base and a flat upper end. The second wall appears in the anterior cell near its upper end. This wall is flat (Fig. 8.14, 8) and touches the first wall all around. It separates a large central **andogonial cell** and a flat or discoid peripheral cell. The third wall appears in the flat peripheral cell and is concave touching the upper wall of the androgonial cell (Fig. 8.14, 10) It cuts off a circular cap cell. This sequence of development was studied by Nayar and Kaur (1969).

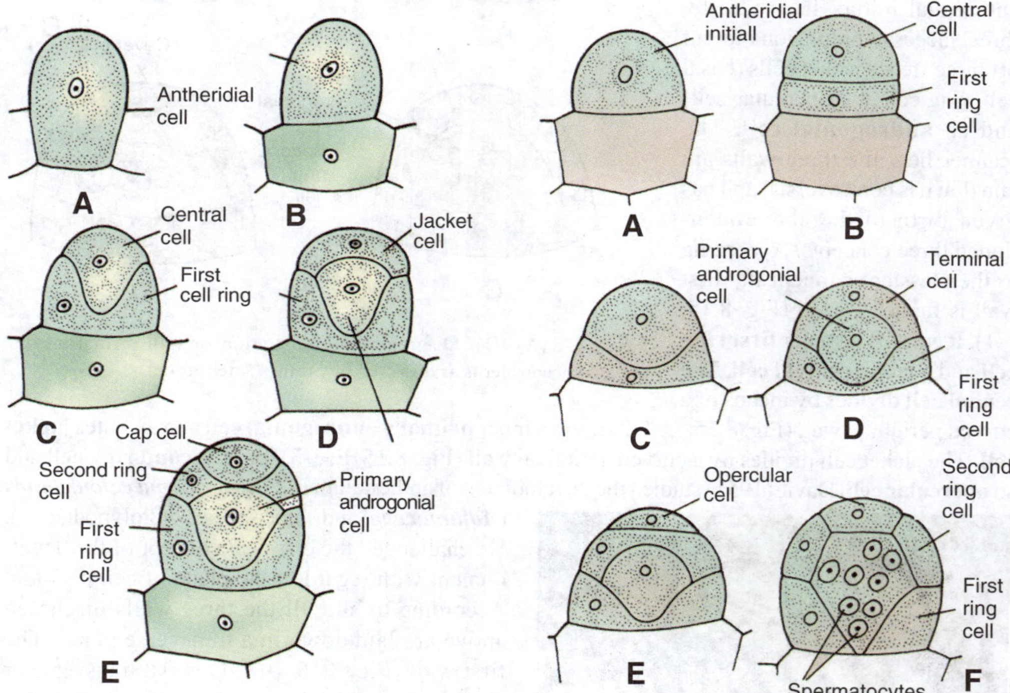

Fig. 8.15. (A—E). Illustrating the classical concept of antheridial development in ferns. (Diagrammatic)

Fig. 8.16. (A—F). Illustrating Davie's concept of wall formation during antheridial development in the polypodiaceae. (*Diagrammatic. Adapted from Davie*)

Development of the emergent or projecting type of antheridium is leptosporangiate and differs considerably from the embedded type. In this type of antheridium the first wall does not set apart a sterile jacket cell and a fertile primary androgonial cell. The formation of the latter is delayed until two or more sterile jacket cells have been cut off. The details of development of the emergent type of antheridium have been studied in the polypodiaceous ferns by many workers. The studies reveal certain variations in the laying down of the three walls that lead to the separation of three jacket cells and an androgonial cell. They fall into three categories :

(*i*) The classical concept which was upheld by Strassburger (1869), Campbell (1886, 1895, 1905). Atkinson (1894) and Laderberg (1907).

(*ii*) Davie's concept that was put forth by Davie in 1951. It is widely accepted and has been quoted in many texts that appeared after 1951 (Smith, 1955; Sporne, 1962 ; Bell and Coombe 1965).

(*iii*) Stone's concept was put forth by Stone (1958, 1961, 1962) as a result of her studies on ***Blechnaceae***, ***Polypodiaceae*** (Fig. 8.17) and ***Hymenophyllaceae.***

The antheridium arises from a superficial cell of the prothallus. This cell enlarges in size and forms a distinct swelling (Fig. 8.15, A) that separates from the prothallial cell by a transverse wall (Fig. 81.5; B, a—a). This projecting or emergent cell is called the antheridial initial. It divides by three successive divisions to cut off three sterile jacket cells (basal cell, ring cell and opercular cell) and an **androgonial cell.** The manner how the three walls are laid down is controversial and has given birth to the above-mentioned three concepts. According to the **classical concept** the first wall is funnel-shaped (Fig. 8.15, 1-1). It cuts off an outer **first ring cell** and a larger **central cell.** The central cell divides by means of a curved periclinal wall (Fig. 8.15, D, 2-2) into an inner **primary androgonial** cell and an outer **jacket cell.** The jacket cell divides by a curved anticlinal wall (Fig. 8.15, E, 3-3) into a **second ring cell** and an **opercular cell.** Davie (1951) studied the development of antheridium in ***Pityrogramma calomelanos*** (***Adiantaceae***) and some species of Polypodiaceae. He challenged the classical concept of the development with regard to the plane of cell division. According to him all the three walls discussed above are laid down in a transverse plane. The first wall (Fig. 8.16, B, 1-1) is flat transverse or slightly curved and never funnel-shaped. It becomes bent downward (Fig. 8.16, C) till it touches the basal wall (*a-a*), due to the increasing turgor within the upper cell. According to Davie the primary androgonial cell enlarges and forces the second wall (2-2) into contact with the third wall (Fig. 8.16, F). Stone (1958, 1961, 1962) agrees with Davie (1951) that first wall is transverse and not funnel-shaped (Fig. 8.17 A—D). Regarding the second wall she disagrees with Davie (1951) and agrees with the classical concept. According to her the second wall is hemispherical and parallel to the outer wall of antheridium (Fig. 8.17, C, 2.2). Verma and Khullar (1966) regard that the third wall is transverse and flat or slightly concave and "is usually in contact with the convexity of the second wall."

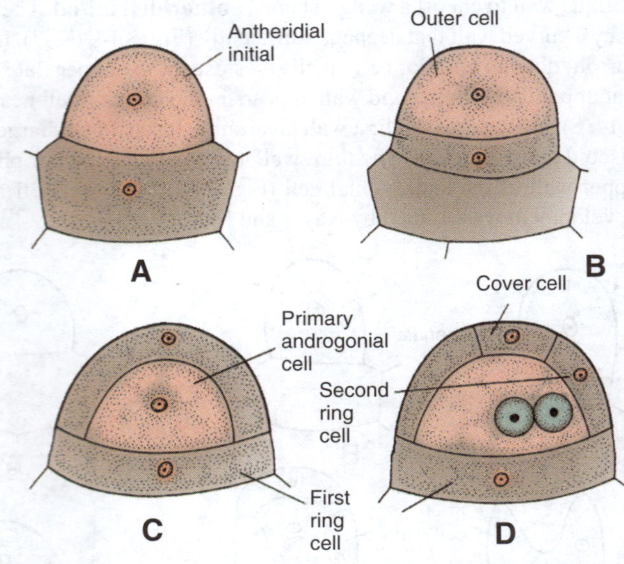

Fig. 8.17. (A—D). Diagrammatic illustration of wall formation in polypodiaceous ferns as observed by Stone. (After Stone)

Pityrogramma

An excellent account of the morphology of the gametophytes of homosporous ferns was given by Atkinson and Stokey in 1964. They have referred to the nature of the first wall in the antheridium of the polypodiaceous ferns. They agree with Davie (1951) in that the first wall is always flat in the beginning. They have not discussed the second and the third walls. The consensus of opinion now is in favour of the first wall being transverse.

In ***Adiantum lunulatum*** (Verma and Khullar, 1966) the first wall is transverse and may remain flat or become variously curved. The second wall is always hemispherical and periclinal. The third wall may be flat transverse and usually touches the second wall. In some cases the third wall was not touching the second wall. In this species there is lot of variation in the curvature of the first and the third walls.

EMBRYOLOGY

In Filicophyta, rather in all the archegoniate plants, fertilisation or the fusion of the sperm nucleus and the egg nucleus takes place within the venter of the archegonium. Prior to fertilisation certain remarkable biochemical changes have been reported to take place in ferns. Bell (1963) investigated the unfertilized egg or ovum of ***Pteridium aquilinum.*** By using some special staining techniques and autoradiography, Bell demonstrated that the ovum is quite different in cytoplasmic organisation from the surrounding cells of the prothallus. The DNA was determined to be in stable state in the nucleus and the cytoplasm, except around the nucleus, where it was in higher concentration. The basis proteins and the RNA were present in large quantities in the ovum cytoplasm than those of the surrounding prothallial cells. During nutrition the nucleus of the ovum was seen to give out small outgrowths or 'blebs' from its periphery into the cytoplasm. This occurred during the outward movement of the nuclear DNA. This study of the ovum reveals that certain significant cytoplasmic transformations occur in the egg, before fertilisation.

As a result of fertilisation a diploid cell called the **zygote** or the **oospore** is established. It is protected by the venter of the archegonium and the surrounding prothallus tissue. The earlier stages of development of the zygote into a young embryo are effected in this protected condition, *i.e.*, within the venter. The protoplast of the zygote represents a complicated and a specialised diffusion reaction system. Before it undergoes divisions it is now known that special biochemical reactions are initiated in its cytoplasm. These include a patternized distribution of enzymes and metabolites. These biochemical changes are said to guide the future developmental course of the embryo (Rondet, 1962). The entire organisation of the future sporophyte plant is designed in the developing zygote. This small structure presents a microcosm of interactions of a complexity equal to or even greater than that of the mature growing plant.

The first prerequisite of the development of zygote into an embryo is the determination of **polarity**, *i.e.*, the establishment of the base and apex of the embryo. The plane of first division wall in the zygote decides the polarity of the embryo. Three types of polarity have been recognised in the vascular cryptogams. These are :—

A. Exoscopic (Fig. 8.18, A)

The first division wall in the zygote is at right angles to the long axis of the archegonium (it may be said to be transverse). The cell nearer to the neck of the archegonium acts at the apical cell and gives rise to the shoot apex. The lower cell gives rise to the foot. No suspensor is differentiated. It is characteristic of Ophioglossum, Botrychium simplex, B. lunaria, Azolla and Salvinia. Tmesipteris, Psilotum, Isoetes and Equisetum are the other pteridophytes in which the polarity is **exoscopic.**

B. Endoscopic (Fig. 8.18, B, C)

The first division wall in the zygote is at right angles to the long axis of the archegonium but the cell nearer to the neck of the archegonium acts either as suspensor or it gives rise to the foot. The inner cell acts as the apical cell and gives rise to the shoot apex. Endoscopic polarity is therefore of two types (Fig. 8.18, B).

(a) Endoscopic without suspensor. Such embryos are endoscopic but have no suspensor, *e.g.*, ***Marattia*** and some related genera, ***Kaulfussia***, ***Angiopteris.*** The cell nearer to the archegonial neck gives rise to the foot whereas the lower cell gives rise to the shoot apex.

(b) Endoscopic with suspensor (Fig. 8.18, C). Suspensor and foot are formed near to the archegonial neck, *e.g.*, ***B. obliquum***, ***Helminthostachys***, ***Danaea***, ***Macroglossum.*** In ***Angiopteris evecta*** sometimes the suspensor is present and sometimes it is absent.

C. Lateral (Fig. 8.18 D)

This type of polarity is characteristic of the leptosporangiate ferns (***Dryopteris***, ***Adiantum***, ***Pteris***, etc). The first division wall of the zygote is parallel to the longitudinal axis of the archegonium (Fig. 8.18, D). No suspensor is differentiated. One of them (nearer the prothallus apex) is called the **epibasal cell** and forms the shoot apex and the first leaf. The other cell is the **hypobasal cell** and gives rise to the foot and the root. The second wall is at right angles to the first and results in the formation of the quadrant stage (Fig. 8.18, E).

The first division wall of the zygote may divide it into two equal or unequal cells. In ***Pteris serrulata*** the epibasal cell is smaller than the hypobasal whereas in ***Marsilea*** the hypobasal

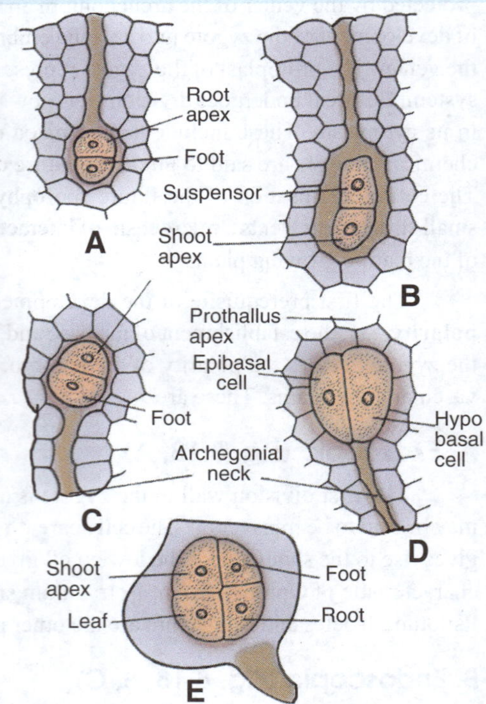

Fig. 8.18. (A–E) Kind of polarity in Embryo development in filicophyta. A. Exoscopic; B. Endoscopic with suspensor; C. Endoscopic without suspensor; D. Lateral; E. Quadrant of D.

cell is smaller. This difference in the size of the two cells has been attributed to the cytoplasmic and metabolic differences at the two poles of the zygote (Wardlaw, 1968).

Early Embryology

The early embryology includes the determination of polarity and early segmentation leading to the establishment of apices of stem, first leaves and root and also the development of foot. Further segmentation of the two celled zygote will be discussed separately among the eusporangiate, proto-leptosporangiate and the leptosporangiate ferns.

Pteris serrulata

A. Eusporangiate ferns (*Eusporangiopsida*)

The eusporangiopsida include two orders, *viz*, the Ophioglossales and the Marattiales. A comparative account of earlier stages of embryo development will be given below.

There are slight variations in the earlier and later stages of development of the embryo in various species of ***Ophioglossum***. Based on these variations, Campbell (1911, 1940) recognised three distinct patterns of embryo development in this genus. These are represented by ***Ophioglossum vulgatum***, ***O. moluccanum*** and ***O. pendulum***. In ***O. vulgatum*** the first wall of the zygote differentiates an **epibasal** segment nearer to the archegonial neck and a **hypobasal** segment. The next division leads to the quadrant stage. One of the hypobasal quadrants gives rise to the foot, whereas the endogenous root apex differentiates by the activity of the other hypobasal quadrant. The epibasal quadrants divide into a number of undifferentiated meristematic cells. It remains undifferentiated for several seasons. The first root develops rapidly and grows into the soil. The foot also becomes conspicuous but the epibasal mass of cells remains undifferentiated. At a very late stage of development the shoot apex, the apex of the first leaf and the leaf sheath make their appearance in epibasal half. Later the apex of second root also becomes evident. The second leaf makes its appearance in the epibasal half. The vascular strands appear first in root and even in the leaves and become joined. In ***O. moluccanum*** the hypobasal quadrants of the embryo give rise only to the foot. The epibasal segment gives rise only to the first leaf. The primary root arises endogenously from the middle of the embryo and is not referable in origin to either of two halves. The shoot apex originates endogenously in the primary root and consists of a group of meristematic cells. The first leaf and the first root almost elongate rapidly at the same time and come out of the prothallial tissue. The second leaf appears close to the stem apex but has no connection with it. The vascular strands of the two leaves become cojoined with that of the primary root. In ***O. pendulum*** the things are quite different. The epibasal segment gives rise to the root or sometimes to both the first and second roots. The hypobasal segment gives rise to a large and massive foot. The shoot apex makes its appearance very late and appears at the base of the primary root. It is believed that shoot apex becomes organised in the epibasal half but due to slow development it is displaced from its original position by the rapidly developing root.

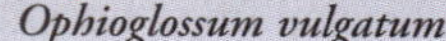

Ophioglossum vulgatum

O. pendulum

B. Protolepto-sporangiate Ferns

The protoleptosporangiopsida, that are represented by various species of ***Osmunda***, exhibit lateral polarity in their embryos. The development is typically leptosporangiate. The first wall of the zygote is parallel to the long axis of the archegonium. The second wall is at right angles to the long axis of the archegonium and forms a **quadrant stage.** The four cells of the quadrant stage divide by a transverse wall to form an **Octant stage.** Further divisions do not follow a regular pattern and lead to the formation of a globular embryo. The primary organs (shoot apex, leaf, root apex and foot) differentiate at a later stage. Gross (1931) studied the embryology in ***Osmunda cinnamomea*** and described that first leaf, stem and the root are formed from the half of the octant adjacent to the archegonial neck. The foot develops from the entire other half. The first root has an endogenous origin. The foot is massive and bulky (eusporangiate character).

(*c*) **Leptosporangiate Ferns.** The archegonia in the **leptosporangiate** ferns ***Dryopteris***, ***Pteris***, ***Adiantum***, ***Pteridium***, etc., are formed on the lower or the ventral surface of the prothallus. Their necks are curved. The fertilized egg or the zygote is almost spherical in form. The first division wall is parallel to the longitudinal axis of the archegonium (Fig. 8.18, D). ***Cardiomanes reniforme*** (***Trichomanes reniforme***) is an exception. In this fern the first partition wall is at right angles to the long axis of the archegonium . The second wall in majority of the Laptosporangiate ferns is at right angles to the axis of the archegonium and forms a **quadrant stage.** In ***Cyathea*** (Stokey, 1930) the first two partition walls in the zygote are vertical. The third one is also supposed to be vertical in ***Cyathea.*** The third wall in majority of the Leptosporangiate ferns is transverse and forms the **octant stage**. The first leaf, the shoot apex, root and foot are referable in origin from definite portions of the octant stage. The first division wall of the zygote in most of the ferns produces two unequal cells (Fig. 8.18, D). The cell towards the apex of the prothallus is known as the **epibasal cell** and is smaller in size (in ***Marsilea*** it is larger in size) whereas the other one is larger and is called the **hypobasal cell** (Fig. 8.18, D). The earlier workers believed that the primary organs like stem, leaf and root and foot can be definitely traced back to these two segments or to the individual cell of the quadrant stage. According to them the shoot apex and the first leaf develop from the epibasal cell. The foot and the root can be traced back to the hypobasal cell. If we consider the octant state with respect to the origin of these four organs, the first leaf originates from the two epibasal octants the lie nearer to the archegonial neck (**superior epibasal octants**). The shoot originates from the other two epibasal octants (**inferior epibasal octants**). The two hypobasal octants nearer to the archegonial neck give rise to the primary root (**superior hypobasal octants)** and the other two to the foot **(inferior hypobasal octants).** The octant stage is followed by 16-celled and 32-celled stages. According to Vladesco (1935) the divisions leading to these stages follow a regular sequence. The plane of these dividing walls is not determined by external factor (light and gravity) but by certain internal factors operating with the zygote and the surrounding prothallial cells. The preciseness regarding the origin of the primary organs from definite segments to the quadrant stage or the octant stage has been challenged by some workers (Vladesco, 1935; Ward 1954). Vladesco (1935) reported that in ***Gymnogramme sulphurea*** the foot is derived partly from the superior epibasal octants and party from superior hypobasal octants. The shoot, in this fern, does not develop from the inferior epibasal

Trichomanes reniforme

Gymnogramme sulphurea

octants but from the equatorial region of the embryo. The first leaf originates from both the inferior epibasal octants. Ward (1954) stated that in ***Phlebodium aureum*** the shoot cannot be traced back to any definite position in the embryo. The foot develops from both the inferior quadrants, the leaf from the superior epibasal quadrant and root from superior hypobasal quadrant. Lot of variations have been reported in the leptosporangiopsida regarding the origin of primary organs, but fuller details of many of them remain to be worked out. Vladesco (1935) noted some irregularities in the embryology of ***Dryopteris filixmas***. The embryology of this species was worked out by Hofmeister, Becquerel (1931) and Vladesco (1935). The first wall, which is usually parallel to the axis of the archegonium, divides the zygote into a smaller epibasal (towards the apex of the prothallus) cell and larger hypobasal cell. The second wall is at right angles to the archegonial axis and produced a quadrant stage with unequal cells. The third wall is transverse and leads to the octant stage. The organs make their appearance in the post-octant stage. There is some difference in their inception as compared to other leptosporangiopsida. The foot develops from the superior epibasal and hypobasl octants. The root arises from the equatorial region of the embryo. The root initial cannot be assigned definite position in the octants. The first leaf arise from the inferior epibasal octants. Such irregularities, as described above, have also been noted in ***Pteridium aquilinum.*** Vladesco is of the opinion that octants have no organogenic significance. He reported two embryos developing in one archegonium of ***Dryopteris parasitica.***

Phlebodium aureum

EVOLUTIONARY TREND AMONG FILICOPHYTA

Recently after considering a number of morphological features of the sporophyte and the gametophyte, Holttum (1949, 1964), Atkinson and Stokey (1964), Bierhorst (1971), Nayar and Kaur (1968, 1969, 1971) and others pointed out a number of characters that can decide the phylogenetic position of a taxon. These have been discussed in detail in the preceding pages. Here only a complete list of such factors is given. A list of such factors was also proposed by Bower (1923), but it has been greatly modified by these authors.

Primitive Features

1. Terrestrial habitat ; plants small in size and with horizontal or upright short stems. 2. Leaves are poorly differentiated from the stem. 3. Leaves pinnate and limited in growth. 4. Dichotomously veined leaves. 5. The leaf has apical cell with three cutting faces. 6. The leaves have discontinuous marginal meristematic patches that are situated at vein ends. 7. Leaf arrangement is irregular. 8. The leaves are monomorphic, *i.e.,* of one type. 9. Stem branching is dichotomous. 10. Stem bears simple trichomes or hair. 11. Roots either absent or poorly developed. 12. Stem is protostelic. 13. Xylem is mesarch. 14. Large simple and adaxially curved leaf traces. 15. Sori are distinct, terminal or marginal in position. 16. Sori elongate. 17. Sporangial maturation in a sorus is gradate or simultaneous. 18. The sori are exindusiate. 19 Sporangia are large and bulky. 20. Sporangia produce numerous spores. 21. Sporangial walls many layered. 22. Spores are monolete, tetrahedral, are without perispore and are homosporous. 23. Sporangia have massive stalks. 24. Annulus absent or apical or sub-apical in position. 25. Dehiscence is longitudinal. 26. The line of dehiscence is not regular and anatomically specialised. 27. The spore germination is amorphous. 28. The growth of the prothallus is by a distinct apical cell or meristem. 29. The prothallus is tuberous, large and subterranean. 30. The prothallus is endophytic and slow in development. 31. The prothallus is devoid of hair

and trichomes. 32. The prothallus bears septate rhizoids. 33. The antheridia are large and produce many spermatozoids. 34. The antheridial wall is many celled. 35. The opercular cell of antheridium is lateral. 36. The archegonia have long necks with many neck canal cells. 37. The necks of archegonia are bent towards the apex of the thallus. 38. The zygote divides first by a transverse wall. 39. Embryo is exoscopic. 40. The embryo is devoid of roots and leaves. 41. The embryo has a massive foot.

Advanced Features

1. The plants have long and creeping stems. 2. Epiphytic habit. 3. Leaves are quite distinct from the stem. 4. The leaves are simple and indeterminate. 5. The leaves have reticulate venation. 6. The leaf apical cell has two cutting faces. 7. The leaves have a continuos and uniform marginal meristem. 8. The leaves have a regular arrangement. 9. The leaves may be dimorphic. 10. The stem branching is lateral. 11. The stem bears multicellular, branched and complicated trichomes 12. The stem has well developed roots. 13. The stele is siphonostelic or solenostelic or dictyostelic. 14. Xylem is exarch. 15. Leaf traces may be multiple or double or treble. 16. Sori are superficial and extend along the veins; they may form cenosori or may be acrosticoid. 17. Sori indusiate. 18. Sporangia in sorus may not mature simultaneously, sori are mixed. 19. Sporangia small with less massive sporangial stalks. 20. Sporangial wall one or two cells thick. 21. Sporangia have thin stalks. 22. They produce smaller number of spores. 23. They have an oblique annulus interrupted by a stomium. 24. Dehiscence is transverse. 25. Dehiscence lines anatomically distinct. 26. Spores are trilete, bilateral, have a perispore and are heterosporous. 27. Spores germinate to produce a filamentous protonema. 28. Prothallus growth is diffuse or by lateral or intercalary meristem. 29. The mature thallus is either cordate or strap-shaped or ribbon-like. 30. The mature thallus is covered with hair or trichomes of diverse types. 31. The prothallus is not endophytic and its growth is quick. 32. The prothallial rhizoids are nonseptate. 33. The antheridia are small and produce only a few spermatozoids. 34. The antheridial wall is 3-celled. 35. The opercular cell is terminal. 36. The archegonia have short, curved necks. 37. The necks are bent towards posterior side of the prothallus. 38. The zygote divides first by a longitudinal wall. 39. Embryology is endoscopic. 40. The embryo has distinct roots and leaves. 41. The foot is poorly developed.

CLASSIFICATION

History : Linnaeus in his ***Species plantarum*** (1753) and ***Genera plantarum*** (1754) grouped the Pteridophytes under the order **filices** of class **cryptogamia**. He described 17 genera and 213 species of pteridophytes with ferns having 14 genera and 182 species. He treated ***Equisetum***, ***Marsilea***, ***Pilularia*** and ***Isoetes*** as "fern allies". ***Lycopodium*** was put under the order **Musci.** Linnaeus based his classification on the position and structure of sori. He did not consider the presence or absence of indusium as of any importance.

Adanson (1793) regarded indusium as an important character and formed two series :— (*i*) **Indusiatae** which included all those genera in which the sori were covered with indusium and (*ii*) **Ex-indusiatae** which included genera with naked sori.

J.E. Smith (1793) regarded the presence or absence of annulus as an important character and divided the Pteridophytes into two series:— (*i*) **Annulatae** which include those pteridophytes whose sporangia possessed annulus and (*ii*) **Ex-annulatae** with sporangia without an annulus. **Bernhardi** (1799, 1801, 1805) also regarded the presence and absence of annulus as an important character.

Swartz (1806) in his ***Synopsis Filicum*** divided the ferns into three orders:— (*i*) **Order Gyratae** (annulatae of Smith) which was further divided into two sub-orders :— (*a*) **Nudae** with naked sporangia and (*b*) **Indusiatae** with a distinct indusium; (*ii*) order **Spurie Gyratae** which included the families now named as ***Schiziaceae***, ***Osmundaceae***, ***Gleicheniaceae*** and the genus ***Angiopteris*** and (*iii*) Order **Agyratae** which included the families ***Marattiaceae*** and ***Ophioglossaceae***. He described 38 genera and 710 species of ferns.

Later **Willdenow (1810)** in his ***Species Plantarum*** and after that **Kaulfuss** (1824), **Blume** (1828), **Kunze** (1840), **Gray** (1821), **Bary** (1824) and **Desvaux** (1827) applied the characters of position, shape and structure of sori, indusia and sporangia for arranging the ferns in a systematic order. **Willdenow** included the water ferns (***Marselia***, ***Pilularia***, ***Salvinia***) alongwith ***Isoetes*** under a distinct group the **Hydropterids.** Mirbel (1803) recognised the four major groups of Pteridophyta (the horsetails, lycopods, homosporous and heterosporous ferns). He included ***Psilotum*** and ***Tmesipteris*** among lycopods and separated ***Selaginella*** from ***Lycopodium***. De Candolle (1813) segregated the pteridophytes as vascular cryptogams. Endlicher (1836-41) recognised pteridophyta as a distinct class for the first time.

Gandichaud (1826-30) and Schott (1834) recognised the importance of vegetative characters in classifying the Pteridophytes.

C.B. Presl (1836, 1845, 1847) published his classification in '**Tentamen Pteridographie**', created many new genera and species based on the types of leaf venation and number of vascular bundles entering the petiole. He recognised 112 genera and 1,500 species. Later Hooker wrote ***Genera Filicum*** (1838-42), ***Species Filicum*** (1844-64) and ***Synopsis Filicum*** (1865-68). He recognised for the first time 2,400 species and illustrated 135 genera of ferns.

J. Smith (1875) in his ***Historia Filicum*** recognised 275 genera of ferns and divided the Polypodiaceae into two divisions (*i*) **Eremobrya** and (*ii*) **Desmobrya. Fee** (1852) in his ***Genera Filicum*** listed 181 genera of ferns excluding ***Cyathea***, ***Dicksonia*** etc. **George Mettenius** (1856) also proposed a detailed classification of ferns but could not complete it due to his death is 1866. He seggregated various fern families some of which are recognised even now. **Moor** (1857) in ***Index Filicum*** and **Frank (1877)** also proposed their classification of the ferns. Diels in the 4th volume of of **Engler and Prantl's *Die naturlichen Pflanzenfamilien*** (1899) gave an enumeration of all known ferns. **Engler and Prantl (1898)** in Diels *Nat. Pfl.* vol 2. listed 141 genera of the ferns. They distinguished and characterised the major groups of ferns except the leptosporangiates. All these authors took into consideration both the vegetative and reproductive features of the ferns but they differed in the concept of families, genera and the number of species recognised. **Fee** (1852) and **John Smith** (1875) introduced such characters as form and shape of spores, venation of leaves, articulated and non-articulated nature of the fronds, mode of growth of the stem, anatomy of the stem in separating genera and other groups of pteridophyta. Many of Smith taxonomic groups are even recognised today. **Goebel** (1880) as a result of his morphological studies on ferns recognised the **Eusporangiate** and **Leptosporangiate** ferns. **Campbell** established that Eusporangiate ferns are primitive than the leptosporangiate ferns. The phenomenon of alternation of generations was explained in 1894 by the distinguished botanist **Strasburger.** This discovery threw a new light in understanding some important aspects of the Pteridophytes. The major task of employing anatomical characters (especially of the vascular region or stele) as taxomoic aids was due to the work of **Sachs** (1875), **Van Tiegham** (1886), **Jeffrey** (1897, 1899, 1902) and **Bower** (1911).

Carl Christensen (1905-1906) published a complete enumeration of fern genera and species (149 genera and 6000 species) in his ***Index Filicum***. In 1934 he enumerated 213 genera.

Bower (1923-1928) published his three monumental volumes entitled **Filicales** and gave an excellent account of the ferns. He considered that besides soral characters, the structure and ontogeny of the dermal appendages, the venation, the stelar organisation etc., were also important in classifying the ferns. Bower, however, did not propose any new classification. In the year 1928 Bower proposed the evolutionary trends in the present day ferns and also erected a phylogenetic scheme of the ferns. The evolutionary trends proposed by him have been greatly modified by recent workers (For details see pages 241, 242). In his phylogenetic scheme Bower pointed out that all the modern ferns evolved from the ***Botryopteridaceae*** (extinct) along five distinct evolutionary lines. These lines are represented by (*i*) Ophioglossaceae, (*ii*) Marattiaceae, (*iii*) Osmundaceae, (*iv*) Schizaeaceae and (*v*) Gleicheniaceae. **Copeland, E.B.** (1929) tried to arrange, in a phyletic sequence, most of the polypodiaceous genera, but his work was incomplete because he had left many African and American genera. Later (1947) he published his detailed classification.

Engler and Diels (1936) recognised two sub-classes of ferns (*i*) Osmundidae and (*ii*) Leptosporangiatae. The Osmundidae included the only order Osmundales whereas the Leptosporangiatae included orders Filicales, Marsileales and Salviniales. They included 12 families in the Filicales and divided the Polypodiaceae into 12 sub-familes. They considered Ophioglossaceous and Marattiaceous ferns as excluded from the above scheme.

Christensen's System of Classification (1938)

Carl Christensen (1938) published his new phyletic scheme in "Manual of Pteridology" and has tried to include most of the fern genera (247) and species that now number 10,000. He included all the ferns in class **Filicinae** and divided it into two series (*i*) Series Filices **Eusporangiatae** and (*ii*) Series Filices **Leptosporangiatae.** The outline of his system is given below :—

Class : FILICINAE

Series I. **Filices Eusporangiopsida**: It included two orders and three families.

Order 1. **Ophioglossales:** It included only one family with three living genera.

Family **Ophioglossaceae:** It included the genera ***Ophioglossum***, ***Botrychium*** and ***Helminthostachys***. The family is characterised by :—

1. The sporophytes are terrestrial, herbaceous and short in size.
2. The rhizome is short and fleshy.
3. The roots are adventitious and bear adventitious buds in ***Ophioglossum.***
4. The leaves are not circinately coiled and are exstipulate.
5. The fertile leaves have a common petiole that bears a sterile undissected or dissected blade and fertile spike.
6. The sporangia are marginal.
7. Spore output is large and may be upto 2,000 spores in a sporangium.
8. They are homosporous.
9. The prothalli are underground and mycorrhizic.
10. The prothalli are monoecious.

2. Order **Marattiales**. It included two families.

Family 1. **Angiopteridaceae**. It includes the genera ***Angiopteris***, ***Macroglossum*** and ***Pseudomarattia.*** It is characterised by linear and biseriate sori that are protected by a false indusium. Sporangia have a rudimentary annulus at the top. Leaves are pinnately compound with free veins.

Family 2. **Marattiaceae.** It includes seven genera and 80 species. The important genera include ***Marattia***, ***Christensenia*** and ***Danaea.*** These genera and species are grouped in three sub-families. In this family the sporangia fuse to form synangia and annulus is rudimentary.

Series II. Filices **Leptosporangiatae.** It includes two orders:—

Order 1. **Filicales**. It includes all the homosporous ferns with leptosporangiate development of sporangia and sex organs with the exceptionof ***Marsileaceae*** which is heterosporous. The leaves are circinately coiled in bud condition. It includes the families:—

Osmundaceae, Schizaeaceae, Marsileaceae, Gleicheniaceae, Matoniaceae, Hymenophyllaceac, Loxsomaceae, Hymenophyllopsidaceae, Plagiogyraceae, Dicksoniaceae, Cyatheaceae and Polypodiaceae.

The family Polypodiaceae has been divided into fifteen sub-families *e.g.* Dennstaedtioideae, Lindsayoideae, Davalloideae (***Nephrolepis***), Oleandroideae (***Oleandrea***), Pteridoideae (***Pteridium***, ***Pteris***), Gymnogrammeoideae (***Adiantum***, ***Chilanthes***), Vittarioideae (***Vittaria***), Onocleoideae, Blechnoideae, Asplenoideae, Woodsioideae, Dryopteridoideae (***Dryopteris***, ***Polystichum***, ***Cystopteris***),

Dipteridoideae, Woodsioideae, Dryopteridoideae (***Dryopteris***, ***Polystichum***, ***Crystopteris***), Dipteridoideae, Polypodioideae (***Polypodium***, ***Drynaria***, ***Pleopeltis***) and Elaphoglossoideae (***Elaphoglossum***).

Order 2. **Salviniales.** It includes the heterosporous water ferns that have sporangia arranged in a basipetalous manner; have globose and indusiate sori. There is no annulus in sporangia. Leaves are not circinate in bud condition. The microsporangia have numerous microspores. The megasporangia have usually one megaspore. It includes the families : Salviniaceae (***Salvinia***) and Azollaceae (***Azolla***).

In the above classification Christensen (1938) grouped together the protoleptosporangiates and the leptosporangiates and included them under the order filicales. The protoleptosporangiates include the Osmundaceous ferns that are intermediate in character regarding the morphology of their sporangia *i.e.* they resemble both the Eusporangiate and Leptosporangiate ferns in some characters. The name Protoleptosporangiata was suggested by Hirmer (1936). In classifying the ferns Christensen has taken into consideration both vegetative and reproductive characters *e.g.* 1. Structure of dermal appendages (hairs, trichomes, scales) 2. The stelar structure and organisation. 3. Posture of rhizome. 4. Spore morphology. 5. Structure, shape and position of sorus. 6. Prothallial morphology. He, however, depended more upon soral morphology in recognising several sub-families of Polypodiaceae and also in grouping together some un-related genera.

Ching's (1940) System of Classification

R.C. Ching (1940), a Chinese botanist dealt with the classification of the Polypodiaceous ferns and divided them into 33 families that were grouped in six phyletic groups. He suggested polyphyletic origin for these ferns. Ching recognised five hypothetical ancestral groups out of which three were named as (*i*) The **Dipteroid stock** (*ii*) **Cyatheoid stock** and (*iii*) **Dicksnoid stock.** He recognised two un-named extinct stocks one of which gave rise to the family **Didymochlaenaceae** and the other to the families **Aspleniaceae** and **Blechnaceae.** His classification was mainly based on the following taxonomical criteria:—

1. The shape, position and nature of sorus.
2. Stelar structure of rhizome.
3. Morphology of the dermal appendages (hair, paleae) on the rhizome.
4. Nature of venation.
5. Dimorphism of the fronds.
6. Articulated or otherwise nature of the leaves to the rhizome.
7. Tetrahedral and bilateral nature of the spores.

Out of the 33 families recognised by him 13 are monotypic *e.g.* Acrostichaceae, Adiantaceae, Ceratopteridaceae, Cheiropleuriaceae, Didymochlaenaceae, Dipteridaceae, Hydoderriaceae; Hypolepidaceae, Loxogrammaceae, Oleandraceae, Platyceriaceae, Sphaerostephanaceae and Woodsiaceae. The rest of the families included more than one genera *e.g.* the families Aspleniaceae, Vittariaceae, Lindsaeaceae, Grammetidaceae, Polypodiaceae, Thelypteridaceae, Blechnaceae, Aspidiaceae, Elaphoglossaceae, Pteridaceae, Culcitaceae, Onocleaceae, Gymmnogramaceae, Monachosoaceae, Matoniaceae, Davalliaceae, Dictyosphaceae, Sinopteridaceae, Cyatheaceae, and Dicksoniaceae.

In the year 1978 (a and b) Ching classified the Pteridophyta into five orders of fern allies (**Lycopodiales, Selaginellales, Isoetales, Equisetales** and **Psilotales**) and the ferns were put under **Filicophytina** which were further divided into **Eusporangiopsida** (Ophioglossales and Marattiales), **Protoleptosporangiopsida** (Osmundales) and **Leptosporangiopsida** (Filicales or Polypodiales, Marsileales, and Salviniales). The **Filicales** or the **Polypodiales** was further divided into 46 families with Polypodiaceae subdivided into sub-families ***Lepisorioideae***, ***Pyrrosioideae***, ***Crypsinoideae*** and ***Microsorioideae.*** He recognised three new families *e.g.* ***Stenochlaenaceae***, ***Bolbitidaceae***, and ***Drynariaceae***. Earlier he also recognised families ***Dryopteridaceae*** (Ching 1974). Ching recognised 197 genera (1978 a) of Polypodiales or Filicales from China. This number includes eight new genera described by him (Ching 1978 b) singly or jointy with S.K. Wu and W.M. Chu.

Dickson's System of Classification (1946)

Dickson (1949) also attempted a phylogenetic classification of the ferns. His system is based on his studies of the ferns of Burma. He regarded the position of the sorus with regard to the veins as the most significant character in classifying the leptosporangiate ferns. Besides he also took into consideration the comparative morphology, venation patterns and the trends in the evolution of the dermal appendages. He gave importance to the position of the sorus on the vein endings or superficially on the veins rather their marginal or superficial position on the leaf lamina. He was more or less guided by Ching's views. He recognised five orders in the ferns—**Ophioglossales** (***Ophioglossaceae***), **Marattiales** (***Marattiaceae*** and ***Angiopteridaceae***), **Marsileales** (***Marsileaceae***), **Salviniales** (***Salviniaceae***) and **Filicales** (including the bulk of modern living ferns), which were divided into 42 families that were grouped under three groups:—

1. Sorus terminal on veins.
2. Sorus superficial on veins.
3. Sporangia scattered on elongate receptacles positioned on the veins or extend between the veins.

Dickson recognised 12 independent lines or stocks A—L along which the filicales evolved. Four out of these stocks were most active in evolution. These were:—

(*i*) D-Stock evolving into Woodsiaceae, Peranemaceae, Hypoderriaceae, Aspidiaceae, Onocleaceae, Thelipteridaceae, Aspleniaceae, Oleandraceae and Blechnaceae.

(*ii*) E-Stock evolving into Cheiropleuriaceae, Dipteridaceae, Platyceriaceae, Grammitaceae, Polypodiaceae, Loxogrammaceae, Antrophyceae, Vittariaceae and Elaphoglossaceae.

(*iii*) H-stock giving rise to Hypolepidaceae, Monochosoraceae, Sinopteridaceae, Pteridaceae and Adiantaceae.

(*iv*) J-Stock leading to Culcitaceae, Dennstaedtiaceae, Lindsayaceae, Dictyoxiphaceae and Didymochaenaceae.

Copeland's System Of Classification (1947)

E.B. Copeland, (1947) proposed his system of classification in his ***Genera Filicum*** with complete generic descriptions of 308 genera grouped under 21 families and three orders. He has classified all the living ferns including the eusporangiate families and the hydrophytic ferns. He has clubbed all the ferns under class Filicinae with three orders and 21 families. The outline of his system is given below:—

Class : FILICINAE

Order 1. **Ophioglossales.** It includes family ***Ophioglossaceae*** (4 genera)

Order 2. **Marattiales.** It includes family ***Marattiaceae*** (4 genera)

Order 3. **Filicales.** It includes 19 families and 300 genera. The families are listed below:—

Osmundaceae, Plagiogyriaceae, Schizaeaceae, Parkeriaceae, Vittariaceae, Pteridaceae, Cyatheaceae, Loxomaceae, Hymenophyllaceae, Blechnaceae, Davalliaceae, Aspleniaceae, Aspidiaceae, Grammitidaceae, Polypodiaceae, Gleicheniaceae, Matoniaceae, Marsileaceae and Salviniaceae.

Copeland's system of classification has received wide acclaim and popularity and is followed uptodate in many parts of the world and in some herbaria. He has given a very clear concept of the delimitation and characterisation of majority of the fern genera and species. The only draw back in his system is the grouping of these genera into families as for example out of the 300 genera of the ferns he has included 200 in the families Pteridaceae, Aspidiaceae and Polypodiaceae. Each one of these families includes more than 60 genera each. This large size of these families has raised certain doubts in the minds of most of the pteridologists. In the family Pteridaceae Copeland has included the Dicksonoid group of true ferns only on the assumption that they are ancestral to the Pteris group of genera. Along with Dicksonoid ferns he included the Dennstaedtioid ferns also in the Pteridaceae because the Dennstaedtioids ferns originated from the Dicksonoid stock. On the basis of the nature of indusium the Chaetopterid genera ***Paesia***, ***Pteridium*** etc. were regarded as intermediate between Dennstaedtioids and Pteroids and were hence included in Pteridaceae. Further, due to the similarity of structure of indusium of ***Cheilanthes*** and ***Hypolepis***, he merged the Gymnogrammeoid ferns (***Cheilanthes***, ***Pellaea***, ***Onychium*** and others) with the Pteridaceae. The genus ***Adiantum*** is also regarded as a member of the Pteridaceae. There is not even a single morphological character common to all the genera included in Copeland's Pteridaceae. The Pteridologists do not regard it as a natural family. Manton (1958) has remarked that the genera 31-63 in Copeland's Pteridaceae should have been included in a separate family Adiantaceae. Her conclusions were based on the cytological studies of the group. The genera 31-63 included Pteroids, Chielanthoids and Adiantum. They all have a basic chromosome number x = 29 and 30. The other genera also (apart from 31-63) have wide difference in their basic chromosome numbers e.g. the Pteridium group has x = 26, Dennstaeoids and Lindsaeoids x = 41-44 and Dicksonoids x = 65-68. Based on these observations Dicksoniaceae, Dennstaediaceae and Lindsaeaceae can be easily regarded as separate families instead of being merged with Pteridaceae. Copeland depended much on common ancestory and supposed relationships in composing his family Pteridaceae which seems to be ill defined and un-natural.

The same is the case with the family Aspidiaceae of Copeland (1947). He has merged the sub-families Onocleoideae, Woodsioideae, Drypteridioideae, Elaphoglossoideae and some genera of Asplenioideae of Christensen (1938) with his family Aspidiaceae. Apart from these he has also included the Thelypteris group, the Athyrium group and the Lomariopsis group in Aspidiaceae. Morphologically all these groups are distinctly different and do not deserve to be lumped together in a single family. In circumscribing his family Aspidiaceae Copeland has again overdepended on similarities in soral morphology and on the hypothesis of soral evolution as proposed by Prof Bower. As compared to Pteridaceae and Aspidiaceae the Polypodiaceae of Copeland appears to be morphologically a more compact and coherent group except for the inclusion of some genera of doubtful relationship *e.g.* ***Dipteris*** and ***Cheiropleuria.*** He followed Christensen (1938) in including Grammitid ferns in the Polypodiaceae.

Copeland has merged Christensen's sub-families Davallioideae and Oleandrioideae into one family Davalliaceae and has raised the status of his subfamilies Vittarioideae and Blechnoideae to the family rank (Vittariaceae and Blechnaceae + the genus ***Stenochlaena***). He has separated the genus ***Ceratopteris*** from Christensen's sub-family Gymmnogrameoideae and has included it in a new family Parkeriaceae. Christensen's tribe Asplenieae (Sub-family Gymmnogrameoideae) has been raised to the rank of Family Aspleniaceae and his tribe Athyrieae (Gymmnogrameoideae) was included in Aspidiaceae. His concept of the families Osmundaceae, Schizaeaceae, Gleicheniaceae, Mattoniaceae, Hymenophyllaceae, Loxomaceae, Plagiogyriaceae and Cyatheaceae is almost the same as that of Christensen.

Van Cotthem (1973) based on the structure distribution and development of Stomata in the higher ferns suggested that Copeland's families Ophioglossaceae, Marattiaceae, Schizaeaceae, Pteridaceac, Davalliaceae, Aspidiaceae, Cyatheaceae, Polypodiaceae and Salviniaceae need to be splitted as they do not show uniformity in stomatal-structure and distribution, whereas Osmundaceae, Gleicheniaceae, Loxomaceae, Parkeriaceae, Blechnaceae, Aspleniaceae, Matoniaceae, Vittariaceae and Marseliaceae do not need any splitting because they show uniformity in stomatal structure. According to Van Cotthem (*i*) Loxomaceae is related to Dicksoniaceae, Cyatheaceae, and Dipteridaceae, (*ii*) Plagiogyriaceae shows similarities with Osmundaceae and (*iii*) There are no stomata in ***Leptopteris.*** The views are also supported to a great extent by cyto-taxonomic studies.

Copeland's overdependence on the soral morphology has lead to the grouping together of unrelated genera. He has not given due importance to other morphological characters in determining the natural relationship and phylogeny of the modern ferns especially the larger groups like Pteridaceae, Aspidiaceae and the Polypodiaceae. He has not given multicharacter approach to his system of classification. Copeland has, however, given a new approach in introducing the idea of geographical distribution in unfolding the evolutionary trends of the ferns. Copeland has discussed in detail the probable antarctic origin of most of the living fern flora. In his own words "Practically the whole fern world of the tropics is descended from the ferns of Old Antarctica". He considered that most pantropic groups owe their present distribution to separate migrations northwords from the antarctic, "and the most primitive, even archaic, representatives of many groups are to be found as near as is now possible to their antarctic source". While drawing and determining the affinities of the modern ferns he has attributed an immediate Antarctic ancestry to most of the modern ferns *e.g.* Hymenophyllaceae, Cyatheaceae, Gleicheniaceae, Aspleniaceae, and the Grammitidaceae. Apart from the Ophioglossaceae (which he considers to have directly descended from the Anthocerotales), Marattiaceae and Osmundaceae (both of which he regards as isolated families without any relationship with any past or present groups), Copeland regards the above named five families to represent five separate lines of evolution, which gave rise to the modern homosporous ferns. These five families are regarded to have originated from some unknown fossil forms.

R.E. Holttum's System of Classification

Holttum proposed a system of classification of leptosporangiate ferns in 1947 and elaborated the same in 1949. He based his system of classification on his detailed and extensive studies of the ferns in Malaya and its neighbouring regions. He later proposed certain changes in the years 1954, 1959 ad in 1973. The system proposed in later years will also be discussed after considering his earlier system (1949). In the year 1949 Holttum included 305 genera and 10,000 species of ferns. He recognised the division of ferns into **eusporangiates** and **leptosporangiates.** In the eusporangiates he recognised the orders **Ophioglossales** and **Marattiales** (Page 268) and regarded, "neither of them very closely related to the leptosporangiate ferns." He did not discuss the eusporangiate ferns in 1949. The **leptosporangiate** ferns were included in 14 families excluding the heterosporous ferns (Marseliaceae and Salviniaceae). The list of these families with their characters is given below.

Family 1. Osmundaceae

It exhibits the following characters:—

1. The sporophytes are terrestrial with a radially symmetrical and erect rhizome (root stock).
2. The rhizome bears a close rosette of leaves or fronds that are devoid of scales.
3. The leaves may be pinnate, bipinnate or deeply dissected.
4. The veins are free.
5. The petiole (stipe) bears stipule like organs at its base.
6. The stele in the petiole is C-shaped in cross section.
7. The sporangia are large and have rudimentary annulus and are borne either on specialised reduced leaflets devoid of laminae or may be closely arranged along the veins of ultimate leaflets.
8. The spores are green in colour in ***Osmunda***. Holttum included three genera in this family *e.g.* ***Osmunda***, ***Todea*** and ***Leptopteris.***

Bower regarded this family as intermediate between the eusporangiate and leptosporangiate ferns. Hirmer (1936) included this family under **protoleptosporangiopsida**, which resemble the eusporangiopsida in certain respects and the leptosporangiopsida in some respects. Holttum considers it to be an ancient family represented in the Palaeozoic by some fossil members. He agreed with Bower (1928) in regarding primitive Osmundaceae to be the possible ancestors of ***Plagiogyria*** and some gymmnogrammoid ferns.

Family 2. Schizaeaceae

It is characterised by:—

1. Terrestrial habit.

2. The rhizome may be dorsiventral (***Lygodium***) or erect.

3. The rhizome is covered with scales in ***Mohria*** and hairy in all other genera.

4. The rhizome is protostelic in ***Lygodium*** and either solenostelic or dictyostelic in other members.

5. The fronds may be pinnate (***Anemia*** and ***Mohria***), unequally dichotomous (***Lygodium***) or equally dichotomous (***Schizaea***).

6. The sporangia may be borne singly and marginal in position, each protected by a small flap. They arise in close rows on pinnately arranged lobes in ***Schizaea*** and ***Lygodium.*** They are borne on reduced lamina of two specialised basal pinnae in ***Anemia***. The sporangia are borne at vein endings throughout the frond in ***Mohria.***

7. The sporangia are large in size.

8. The annulus is apical.

9. The dehiscence of sporangia is lateral.

10. The spores are tetrahedral or bilateral in ***Schizaea.***

Holttum included four genera in the family namely ***Schizaea***, ***Lygodium***, ***Anemia*** and ***Mohria***. It is also regarded as an ancient family with fossils known in the Carboniferous.

Family 3. Gleicheniaceae

This family is characterised by:—

1. The members are terrestrial in habit with a creeping, slender and dorsiventral rhizome.

2. The rhizome is covered with hair or scales with stiff marginal hair.

3. The rhizome is characterised by protostelic or solenostelic vascular system. Mixed protostele is found in some species of ***Gleichenia.***

4. The leaves usually large in size and 'deeply divided'. In some species of ***Gleichenia*** they are simple pinnate. The ultimate leaflets have free veins. The leaves grow in stages by means of 'resting' of the apical bud.

5. The sori contain few sporangia and develop on the ventral side of the leaflets.

6. The soral development is simultaneous.

7. The sori arise on veins.

8. The sporangia have complete oblique annulus and are large in size.

9. The sporangial dehiscence is longitudinal.

10. The spores are tetrahedral.

The family includes only a single genus ***Gleichenia.***

Family 4. Matoniaceae

It shows the following characters:—

1. The plants are terrestrial with creeping rhizomes that are covered with hair.
2. The rhizomes have concentric siphonosteles.
3. The leaves are erect and dichotomously branched.
4. The veins may be free or may anastomose near the costae.
5. The sori are covered by umbrella—shaped indusium and may arise superficially at a junction of several veins (***Matonia***) or terminal on small veins (***Phanerosorus***).
6. The sporangi are few in a sorus, large in size with an oblique and incomplete annulus.
7. The spores are tetrahedral.

The family includes the genera ***Matonia*** and ***Phanerosorus***, and is restricted in distribution found only in some localities in Malaya.

Family 5. Hymenophyllaceae

The family is characterised by:—

1. The rhizome is usually long, slender and creeping.
2. The rhizome is covered with hair in younger regions.
3. In some species the rhizome is short and devoid of roots.
4. In species with long and creeping rhizomes, the leaves arise at greater distances.
5. The leaves may be simple and small or even large and finely dissected upto 60 cm. or more in some species of ***Trichomanes***. The leaves are mostly unicostate.
6. The lamina is one cell in thickness except in the region of the vein.
7. The sori are indusiate and may be terminal or marginal in position. The receptacle on which the sporangia develop is more or less elongated and is protected at its base by the indusium. The indusium is conical or tubular near the base and dilated near the apex with two distinct lips.
8. The sporangia have an oblique annulus.
9. The spores are tetrahedral.

The family includes the genera ***Trichomanes*** and ***Hymenophyllum***. There are 650 species.

Family 6. Plagiogyriaceae

This monotypic family includes the genus ***Plagiogyria*** and is characterised by :—

1. The rhizomes are stout, erect and short, without scales but with scanty hair.
2. The rhizomes are dictyostelic.
3. The leaves are once pinnate (pinnae simple) and dimorphic. The leaves arise in a rosette and have petioles that are wide and flattened near the base with prominent ridge on the back. The ridge bears 'wart-like aerophores' on either side and bear mucilage secreting hair.
4. The fertile leaves have longer petioles and bear comparatively narrow pinnae. The pinnae are distinct and winged at the base. The veins are free.
5. The sporangia develop on the ventral side and are attached only to the once forked veins. The indusium is false formed by the reflexed margins of the pinnae only during earlier stages of sporangial development.
6. The annulus is complete and oblique. The dehiscence is lateral.

Family 7. Dicksoniaceae

This family includes four genera and thirty species. It is characterised by :—

1. It includes tree ferns that are mostly tropical and southern in distribution and possess tall and stout trunks or in some cases short and prostrate rhizomes (stocks) bearing large leaves at their spices.

2. The trunk and the stock are profusely covered with hair.

3. The leaves are large tripinnatifid to tripinnate. The veins are free.

4. The indusiate sori are marginal and terminal on veins.

5. The indusium is cup-shaped with distinct upper and lower flaps. The flaps are concave.

6. The development of Sporangia is gradate. The sporangia possess an oblique annulus.

7. The spores are tetrahedral.

Genera :— Placed in this family are ***Dicksonia***, ***Cibotium***, ***Cystodium*** and ***Thyrsopteris***. Christensen (1938) included ***Culcita*** in this family but Holttum included it in the family Dennstaediaceae.

Family 8. Protocyatheaceae

The family is represented by only two genera ***Lophosoria*** and ***Metaxya*** (each including one species). Both these genera resemble each other in possessing superficial sori and hairy rhizomes. Holttum regards them as being intermediate between Cyatheaceae and Gleicheniaceae. The characters of both the genera are listed separately:—

1. ***Lophosoria*** :— It has erect, short and stumpy rhizome called stock. The leaves large, bipinnate and spirally arranged, each possessing an abaxial bud at the base. The bud develops into runner with a solenostelic vascular region. Hair are present on the apex of the stock and all over the leaves. The sori contain a few large sporangia. There is no indusium. The sporangia dehisce simultaneously and are superficial. The annulus is oblique with a lateral stomium.

2. ***Metaxya*** :— It differs from ***Lophosoria*** in possessing:—

(*i*) Creeping rhizomes bearing uni-pinnate leaves with free veins; (*ii*) the sori are irregularly arranged; (*iii*) paraphyses are present among the sporangia and (*iv*) the annulus is vertical.

Family 9. Cyatheaceae

The family includes only a single genus ***Cyathea*** and is characterised by :—

1. The stem may be in the form of a massive trunk or an erect stock. The old stem is usually covered with a thick mass of interlacing roots. The apex of the trunk is protected by dense scales.

2. The leaves are large and may be simple or compound. The compound leaves may be once pinnate or bipinnate with tripinnatifid leaflets.

3. The rachises and costae are covered with scales on the ventral surface.

4. The veins arise from the costae at right angles. The veinlets are simple or sometimes forked.

5. The petioles are scaly at the base. The bases of the scales may be raised or even spinous.

6. The sori are never terminal and are found on the veins ***i.e.*** they are superficial.

7. The sporangial receptacle is spherical and the sporangia are mixed with hair.

8. The indusium may be absent or even present. The indusium when present is thin and cup-shaped and completely surrounds the young sorus.

9. The annulus is oblique.

10. The spores are tetrahedral.

Holttum included eight hundred species (of ***Cyathea*** alone) in this family. These species are mostly southern and pantropic in distribution.

Holttum and Sen (1961) included some more genera in this family. These are ***Thyrsopteris*** (1 species), ***Dicksonia*** (nearly 25 sp), ***Cystodium*** (1), ***Cibotium*** (10), ***Culcita*** (9).

Family 10. Polypodiaceae

The family includes twenty one genera that have been grouped into five distinct groups:—

(*a*) This group includes two genera ***Dipteris*** and ***Cheiropleuria*** that are regarded to be primitive. They are terrestrial with creeping and bristly rhizome having solenostelic organisation of the vascular region. The leaves are dichotomously branched, large and with reticulate small veins.

(*b*) This group includes the genera ***Platycerium***, ***Pyrrosia*** (***Cyclophorus***), and ***Drymoglossum***. The leaves are covered with stellate hair. They are dichotomously branched in ***Platycerium.***

(*c*) It includes five genera namely ***Neocheiropteris***, ***Pleopeltis***, ***Lepisorus***, ***Lemmaphyllum*** and ***Hymenolepis***. In these genera the sori are covered with peltate paraphyses. The fronds may be branched (***Neocheiropteris***) or simple.

(*d*) It includes the genera ***Microsorium***, ***Colysis***, ***Leptochilus*** and ***Drynaria***. In ***Microsorium*** the leaves are simple or pinnately lobed and the numerous small sori lack peltate scales. In ***Colysis*** and ***Leptochilus*** the sori are laterally fused and acrostichoid. The base of the frond is specially well developed in ***Drynaria***.

(*e*) The genera ***Phymatodes*** (with evenly spaced round sori; fronds coriaceous); ***Selliguea*** (sori laterally fused); ***Pycnoloma*** (acrostichoid); ***Grammatopteridium*** (acrostichoid); ***Lecanopteris*** (the rhizome is inhabited by ants); and ***Polypodium*** (the leaves are pinnate).

The genus ***Loxogramme*** is not placed in any group. It resembles the polypodioid ferns in the vascular anatomy of the rhizome and petiole, in possessing scales on the rhizome. In its venation pattern it resembles ***Colysis.*** It resembles ***Dipteris*** and ***Cheiropleuria*** in having non articulated fronds. It differs from all polypodioid ferns in possessing tetrahedral and spherical spores.

The family is generally characterised by:—

1. The rhizomes are creeping and may be terrestrial or epiphytic. In some cases infested with ants. It is covered with peltate scales. Hairs are present in general included in Group (a).
2. The leaves are generally arranged in two rows on the upper side of rhizome and are always jointed to it (except in ***Loxogramme***, ***Dipteris*** and ***Cheiropleuria***).
3. The leaf lamina may be simple and entire, lobed or even pinnate.
4. Venation is reticulate. The veins are free in aeroles or rarely entirely free.
5. The vascular region in the rhizome either solenostelic or dictyostelic with a number of meristeles.
6. Sori without indusium. In some members they are covered (when young) by peltate scales (Genera in group C). The sori may be large or small and usually round in shape. They may be:— (***i***) sunk in cavities in the surface of the pinnae; (***ii***) may be placed on free or connected veins; (***iii***) on a plexus of veins; (***iv***) may be elongated and parallel to the main veins; (***v***) may be elongated and parallel to the margin and (***vi***) acrostichoid (***Leptochilus***).
7. The spores lack perispore and are bilateral except in ***Loxogramme*** where the spores are spherical in some species.

Family 11. Grammitidaceae

The family includes nine genera and 250 species. The characteristic features are listed below:—

1. It includes mostly the epiphytic ferns with dorsiventral, creeping rhizomes (more or less ascending). The vascular system may be a simple or dissected solenostele. It is covered with brown peltate scales with stiff unicellular bristles on the edges.

2. The leaves are arranged in two alternating rows on the dorsal surface of the rhizome. The leaves are petiolate, the petioles may or may not be jointed to the rhizome. The entire leaf is covered with unicellular or multicellular hair, the latter with unicellular out growths. The hair may be long or short.

3. The leaf lamina may be (*i*) simple, (*ii*) pinnate or pinnatifid or (*iii*) in rare cases bipinnatifid.

4. The veins may be free or anastomosing.

5. The petioles have a single Vascular strand.

6. The sori are naked or sometimes sunk in cavities which may extend into grooves. They are mostly superficial. The shape of the sorus may be round or sometimes elongate along the veins.

7. The sporangia may be naked or covered with hair.

8. The spores are tetrahedral.

Genera :— ***Grammitis***, ***Ctenopteris***, ***Prosaptia***, ***Calymmodon***, ***Acrosorus***, ***Xiphopteris***, ***Cochlidium***, ***Scleroglossum***, ***Nematopteris.***

Family 12. Thelypteridaceae

The family includes the genera ***Thelypteris***, ***Cyclosorus***, ***Goniopteris***, ***Meniscium***, ***Glaphyropteris and Steiropteris.*** In 1971 Holttum recognised the genera ***Ampelopteris***, ***Phegopteris***, ***Psedophegopteris***, ***Metathelypteris***, ***Macrothelypteris***, ***Glaphyropteridopsis***, ***Parathelypteris***, ***Pronephrium***, ***Stenogramma***, ***Sphaerostephanos***, ***Pneumatopteris***, ***Christella***, ***Trigonspora***, ***Coryphopteris and Amphinemon***. The last three are new genera described by Holttum. The family shows the following characters:—

1. The rhizome is creeping or sub-erect, not dorsiventral.

2. The Vascular region is a dictyostele or a modified solenostele.

3. The rhizome is covered with narrow scales that bear stiff or glandular unicellular hair on their surfaces or margins.

4. The leaves and roots arise from all the sides of the rhizome.

5. The leaves may be pinnate or bipinnate and are much longer than broad. The pinnae near the base of leaf are smaller than the upper ones. They subtend a broad angle with the rachis and are usually symmetrical, rarely unequal. The veins in each lobe are pinnate with simple or forked veinlets.

6. A translucent membrane is always present across the sinus between the two adjacent pinnae.

7. The petioles show two separate vascular strands near the base. Higher up these veins unite to form a single U-shaped strand (as it appears in a cross-section of the petiole).

8. The surfaces of the rachis and lamina are covered with unicellular stiff hair or sometimes with glandular hair.

9. Sori are situated dorsally on the veins and round in shape. They may or may not be protected by an indusium. The indusium when present is reniform (Kidney-shaped).

10. The sporangia are often setose and often bear glandular hair on their stalks. Glandular cells or setae have also been reported on their surfaces.

11. The spores are usually bilateral or rarely tetrahedral with a winged perispore. In some cases the wings are more or less divided into small laminae or spines.

Separation of thelepteridoids as a distinct family from the Dryopteridoid-Tectaroid assemblage (Charistensen, 1938) is uncontroversial. The differences, in the morphology of the dermal appendages, form of frond, sinus membrane, sori and spores clearly support Holttum's idea of treating Thelypteridaceae as a distinct family. His conclusion that Cyatheoid and Thelypteridoid ferns have originated from the Glechiniaceae or some primitive allied source is open to question.

Family 13. Dennstaedtiaceae

Holttum has divided this family into eleven sub-families as below:—

Sub-family (*i*) Dennstaedtioideae

It includes the genera ***Dennstaedtia***, ***Microlepia***, ***Leptolepia***, ***Oenotrichia***, ***Saccoloma***, ***Orthiopteris*** (including ***Ithycaulon***), ***Hypolepis***, ***Culcita and Monachosorum***. The subfamily is characterised by:—

1. The rhizomes are creeping, slender and covered with hair with the exception of ***Orthiopteris*** in which the stock is erect and covered with scales at the apex. The vascular region is solenostelic.

2. The leaves are rarely simply pinnate and usually finely dissected with the ultimate pinnae having an asymmetrical base.

3. The potiole is traversed by a more or less U-shaped vascular strand.

4. The Sori are indusiate and terminal in position on the veins.

5. The indusium is cup-shaped or two lipped. In some cases the lower lip is lacking or one of the lips is fused with the lamina of the frond and the other forming a more or less pouch shaped indusium.

6. The sporangia may be gradate (in primitive genera) or mixed.

7. The annulus in the gradate sporangia is oblique whereas in others it is vertical.

8. The spores are tetrahedral.

This sub-family has been regarded to be the most primitive of all the other subfamilies recognised in the family Dennstaedtiaceae. Within this family Holttum believes that the various sub-families represent separate evolutionary lines, all originating from the Dennstaedtioideae, viz.,

(*a*) From ***Microlepia to Davallia*** (s.f. iii) and so to ***Asplenium*** (s.f. vi) and to ***Lomariopsis*** (s.f. viii) and possibly to ***Blechnum*** (s.f. vii); probably also to ***Nephrolepis*** (s.f. iv).

(*b*) from ***Hypolepis*** to the ***Pteris*** sub-family (v).

(*c*) To ***Dryopteris*** (s.f. ix) and ***Athyrium*** (s.f. xi)

(*d*) To the ***Tectaria*** sub-family (x).

Sub-family (*ii*) Lindsayoideae

It includes the genera ***Lindsaya***, ***Tapeinidium***, ***Sphenomeris*** (or ***Stenolana***), ***Odontosoria***, ***Isoloma and Schizoloma.*** It is characterised by:

1. Habitat terrestrial with some genera epiphytic and a few climbers.

2. The rhizome slender and covered with bristles or narrow scales, that are not peltate.

3. The vascular region is a modified protostele.

4. The petioles or the stipes are not joined to the rhizome.

5. The leaf lamina may be simple or divided.

6. The veins are free in the deeply divided forms and anastomosing in less divided forms. The midribs of the leaflets sunk forming a groove which runs into the rachis - groove.

7. The margin of leaflets is thickened and decurrent with the margin of the rachis - groove.

8. The sori are situated terminally on the vein and are nearly marginal. They may occur singly or those occuring on a adjacent veins may join laterally.

9. The sori are covered with indusia that open towards the leaf margin.

10. The spores lack perispore and are tetrahedral.

This sub-family is regarded as nearly of equal status to the first sub-family. Both of them are regarded as representing independent parallel lines of evolution from a common primitive stock. Holttum has suggested "the Lindsayoid ferns seem not be on main evolutionary line, and it might be better to place them in a separate family of thier own."

Sub-family (*iii*) Davallioideae

This sub-family includes six genera (***Davallia***, ***Humata***, ***Araiostegia***, ***Davallodes***, ***Leucostegia*** and ***Rumohra***) that are characterised by:—

1. Majority are epiphytic with creeping and scaly rhizomes. The scales are peltate at base.

2. The rhizome in cross section shows two broad dorsal and ventral steles in addition to a variable number of smaller strands on the sides.

3. The petioles or the stipes are joined to the rhizome and arise in two rows. Several vascular strands enter the stipe at the base with two larger ones uniting together during their upward course.

4. The leaves may be simple, pinnate or finely dissected, deltoid, with terminal leaflets unequal at the base.

5. The veins are free with raised midribs. The rachis of the pinna is winged and decurrent with the margin of the leaflets.

6. The sori are nearly marginal and situated terminally on the veins, and are distinct *i.e.* they do not fuse laterally.

7. The indusium is fused at the base and sometimes along the sides too.

8. The spores are bilateral and without a perispore.

Sub-family (*iv*) Oleandroideae

The sub-family includes three genera that are separately characterised as below:—

(*a*) *Nephrolepis*:— Stem erect, short, covered with peltate scales and dictyostelic, reproducing vegetatively by means of short or long runners that develop apical buds which develop into new stems bearing rosettes of leaves. The fronds are not jointed at the base with the stem. The leaves are simple and pinnate. The pinnae are articulate with free veins. The sori are terminal at vein ends and are covered with reniform or semicircular indusia.

(*b*) *Oleandra*:— The stem may be erect and much branched or creeping and less branched, covered with scales, with leaves arising in close groups. The leaves are jointed near the base with the stem. The leaves are simple with free veins. The sori are arranged on either side of the mid rib and are seated on the veins (not terminal). The shape of the sorus may be semicircular or reniform.

(c) *Arthropteris*:— The genus has slender, creeping stems covered with scales. The leaves are arranged in two rows on the stem and are articulated. The sori may be naked or covered with reniform indusia and are situated at vein ends (terminal).

Sub-family (*v*) Pteridoideae

The sub-family includes nine genera (***Pteridium***, ***Paesia***, ***Lonchitis***, ***Anisosorus***, ***Histiopteris***, ***Pteris*** and three nearly related genera ***Neurocallis***, ***Acrostichium*** and ***Stenochlaena***) and is characterised as below:—

1. The primitive genera possess creeping and hairy rhizomes whereas the other genera possess rhizomes with radial symmetry and are covered with scales. In ***Stenochlaena*** the rhizomes are climbers.

2. The primitive members bear highly dissected fronds with distinctly grooved rachises and rachis-branches. The mid-ribs of leaflets are also grooved *i.e.* sunk in the lamina. In the advanced members within the family the leaves are less divided or even simply pinnate (***Pteris sp.***)

3. The veins are mostly free, rarely anastomosing without included free veinlets.

4. The sori are usually confluent (fusion sori) with or without the abortion of the inner indusium. Acrostichoid in ***Acrostichium*** and ***Stenochlaena.***

Sub-family (*vi*) Asplenioideae

It includes seven genera (***Asplenium***, ***Phyllitis***, ***Diplora***, ***Comptosorus***, ***Ceterach***, ***Pleurosorus and Pleurosoriopsis***) and is characterised as:—

1. The rhizomes are mostly sub-erect or erect and radial in organisation. In some genera it may be creeping or climbing and dorsiventral. The rhizome is covered with scales that are peltate at báse. The scales are usually clathrate. The vascular system is a dictyostele.

2. The leaves possess finely dissected laminae, sometimes simple, rarely dimorphous. The veins are mostly fee, rarely fused or reticulate.

3. Two separate vascular strands enter the petiole and later join together to form a single strand with four arms (X-shaped).

4. The sori are always linear, usually on one side of the vein, rarely on both the sides of the vein. The indusium is narrow, opening towards the vein or veinlet.

5. The spores are bilateral.

6. Perispore is present.

Sub-family (*vii*) Blechnoideae

The sub-family includes 5 genera (***Blechnum***, ***Sadleria***, ***Woodwardia***, ***Doodia*** and ***Brainea***) with the following characteristic features:—

1. They are terrestrial in habitat.

2. The rhizomes are erect, stout and scaly.

3. The leaves may be pinnate (***Blechnum indicum***) or bipinnate (***Woodwardia***).

4. The veins are usually free, sometimes anastomose.

5. Several vascular strands enter the petiole.

6. The sori are indusiate. The indusium may be marginal, superficial, or intramarginal to sub-costae, exterior and introrse membranous type.

7. The sori may be short or long situated in a single row on veins or vascular commissures parallel to the midrib of a pinna. In ***Doodia*** the sori may occur in one to three lines on each side of the mid-rib.

8. The spores are bilateral.

9. Perispore may be present or absent.

Sub-family (*viii*) Lomariopsidoideae

The genera included in this family are : ***Egenolfia***, ***Bolbitis***, ***Lomariopsis***, ***Thysanosoria***, ***Teratophyllum***, ***Lomagramma*** and ***Elaphoglossum.*** It shows the following characters :—

1. The habit of the rhizome is different in the genera listed above and Holttum has divided the genera into three groups depending upon the characters of rhizome alone. These are (***a***) rhizome short

ad creeping, usually growing on rocks (lithophytes) in shady forest or short climbing (never high climbing) on adjacent trees *e.g.*, ***Egenolfia and Bolbitis*** ; (***b***) rhizome high climbing, 'usually starting life on rocks or the bases of trees or forest flora, and always rooted in the ground' *e.g.* ***Lomariopsis***, ***Thysanosoria***, ***Teratophyllum*** and ***Lamagrama*** ; and (***c***) rhizome epiphytic or lithophytic, short and never high climbing *e.g.* ***Elaphoglossum.***

2. Anatomically the rhizome in all the genera has a broad basal vascular strand and one or more dorsal strands (probably all are meristelic). The basal strand supplies the roots where-as the dorsal strands supply the leaves. The symmetry of the rhizome is always dorsiventral.

3. The leaves may arise in two or more rows on the dorsal side depending upon the number of dorsal vascular strands.

4. The scales on the rhizome and other parts (stipe and frond) are peltate. They are larger in the rhizome. There are no unicellular hair.

5. The petioles or the stipes are usually not jointed to the rhizome except in ***Elaphoglossum*** where they are jointed.

6. Fronds simple (***Elaphoglossum***), simply pinnate or bipinnate (***Teratophyllum***). The pinnae in climbing and epiphytic genera are jointed to the rachis which is always winged. The terminal unjointed pinna may be present or absent.

7. The stipe is traversed by several separate vascular strands.

8. The climibing genera (***Teratophyllum***) have distinctive bathyphylls that are more dissected than the acrophylls.

9. The veins may be free or uniting near the margins.

10. The fertile fronds are usually distinct with reduced lamina bearing sporangia on the ventral side.

11. Sporangia may or may not be supplied by a special vascular strand.

12. The spores have a perispore except in ***Lomagramma***.

Sub-family (*ix*) Dryopteridoideae

This is one of the largest sub-families and includes 14 genera namely ***Woodsia***, ***Stenolepia***, ***Acrophorus***, ***Diacalpe***, ***Peranema***, ***Lithostegia***, ***Dryopteris***, ***Polystichum***, ***Polystichopsis***, ***Cyrtomium***, ***Didymochlaena***, ***Polybotrya***, ***Maxonia*** and ***Gymnocarpium***. The subfamily is characterised by:—

1. These ferns are mostly terrestrial in habitat.

2. The rhizomes are usually erect or sub-erect and are radial in symmetry.

3. The stelar organisation is a dictyostele.

4. Numerous vascular strands enter the petiole.

5. The rhizome is covered with scales that are devoid of stiff unicellular hair. Marginal teeth when present develop from two adjacent cells.

6. The leaves are amply divided with leaflets unequal at base. The mid-ribs of the pinnae and pinnules are sunk in the lamina forming a groove whose edge is decurrent with the groove of rachis.

7. The margins of lamina of each pinna is also decurrent with the margin of the rachis.

8. The veins are free (except ***Cyrtomium***).

9. Sori spherical and covered with a true indusium that may be round or reniform or peltate or globose in shape.

10. The spores are bilateral and have a perispore.

Gymnocarpium differs from the rest of genera in possessing a creeping rhizome with dorsiventral symmetry and in bearing two rows of leaves on its upper side.

Sub-family (x) Tectarioideae

This sub-family also includes 14 genera namely ***Ctenitis***, ***Lastreopsis***, ***Tectaria***, ***Heterogonium***, ***Quercifilix***, ***Stenosemia***, ***Cyclopeltis***, ***Hemigramma***, ***Amphiblestra***, ***Pleuroderris***, ***Dictoxyphium***, ***Pteridys***, ***Pleocnemia*** and ***Alcypteris***. The characters of this sub-family can be listed as follows :—

1. The members are mortly terrestrial with erect or sub-erect and almost radial rhizomes. The stelar system is dictyostelic. The petioles of leaves are not jointed to the rhizome.

2. The petioles are supplied by several vascular strands.

3. The leaves may be simple or much divided and narrow (***Heterogonium***) deltoid. The basal pinnae may be basi-scopically enlarged.

4. The midribs of the pinnae and pinnules are raised and not grooved.

5. The rachis, midribs and other parts of fronds are covered with multicellular hair on the dorsal surface. The veins may be free or anastomosing (***Heterogonium***).

6. The sori are spherical and covered with indusia that are reniform in shape. The sori are either terminal or dorsal on the veins. In some cases the sori are without indusia and are spread along the veins. In others arostichoid (***Heterogonium***). In ***Dictyoxiphium*** the marginal sori are continuous and indusiate:

Sub-family (xi) Athyrioideae

It includes two genera (***Athyrium*** and ***Cystopteris***) and shows the following characters :—

1. Terrestrial in habitat with erect or sub-erect and radial rhizomes.

2. The rhizomes are covered with dentate scales, the teeth are formed by two adjacent cells.

3. The petioles are supplied by two vascular strands at base, uniting during their upward course to form a single U—or V-shaped single strand.

4. The fronds may be simple (***Cystopteris***) or amply dissected (***Athyrium***), small, or very large in some species of ***Athyrium*** which look like trees ferns.

5. The veins are usually free but in some species of ***Athyrium*** there is a distinct anastomoses of veins which form parallelogram-shaped areoles.

6. The mid-ribs and the rachises are covered with scales. They are hairless but in some cases the upper surface is covered with multicellular hair.

7. The mid-ribs are grooved that are decurrent with the grooves of the rachis.

8. The sori are usually linear on the vein, with linear indusia. In some cases the indusia are reniform or even J-shaped. In ***Cystopteris*** the sori are round with a basal indusium.

9. Spores possess a perisporium.

The genus ***Athyrium*** (***including Diplazium***) is composed of 294 species whereas ***Cystopteris*** includes 13 species.

Family 14. Adiantaceae

Holtum (1949) has united the Gymnogrammoid and Vittarioid ferns of Bower in forming the family **Adiantaceae.** He has followed Bower (1923-1925) in doing so. Since ***Adiantum*** is the largest genus (184 Species) of this assemblage, the family is named after it.

Regarding the characteristic features of **Gymnogrammoid ferns** Holttum (1949) states, "These are an extremely varied and specialised group, mostly of dry climates and more or less xerophytic and reduced vegetatively." He did not discuss any characters and made a reference to Bower (1928, chapter 39).

Regarding **Vittarioid ferns** he gives the characters as :

"Epiphytes with densely hairy roots, short rhizome covered with dark clathrate scales and tufted, usually limply pendulus simple fronds (dichotomously lobed in ***Hecistopteris***). ***Veins*** reticulate except in ***Hecistopteris*** and in the very reduced genus ***Monogramme***; in ***Vittaria*** with a single row of areoles each side of the costa and continuous sub-marginal vein formed of the outer margins of the areoles; sori along the veins, more or less immersed; paraphysis present."

For informations sake the list of genera of the Gymnogrammoid ferns and Vittarioid ferns (included by Holttum in Adiantaceae, but treated by Bower separately) is given below.

Bower divided the Gymnogrammoid ferns into four groups:—

Group I. Relatively Primitive Genera:—

(*i*) ***Llavea*** (1 species)
(*ii*) ***Gryptogramme*** (4 species)
(*iii*) ***Onychium*** (6 species)
(*iv*) ***Jamesonia*** (14 Species)
(*v*) ***Ceratopteris*** (1 species)

Group II. Central Group of Genera:

(*vi*) ***Pterozonium*** (2 species)
(*vii*) ***Syngramme*** (16 species)
(*viii*) ***Anogramme*** (9 species)
(*ix*) ***Gymnogramme*** (50 species)
(*x*) ***Coniogramme*** (2 species)
(*xi*) ***Hemionitis*** (8 species)
(*xii*) ***Gymnopteris*** (12 species)
(*xiii*) ***Ceropteris*** (7 species)
(*xiv*) ***Trismeria*** (2 species)

Group III Adiantoid Ferns:—

(*xv*) ***Adiantum*** (184 species)

Group IV. Cheilanthoid Ferns:

(*xvi*) ***Pellaea*** (68 species)
(*xvii*) ***Doryopteris*** (39 Species)
(*xviii*) ***Saffordia*** (1 Species)
(*xix*) ***Trachypteris*** (1 species)
(*xx*) ***Cheilanthes*** (101 species)
(*xxi*) ***Notholaena*** (47 species)

In the Vittarioid ferns Bower included five genera:

(*i*) ***Hecisopteris*** (open venation) one species
(*ii*) ***Monogramme*** (occasional areolae) 15 species
(*iii*) ***Vittaria*** (closed venation) 46 species
(*iv*) ***Antrophyum*** (closed venation 27 species
(*v*) ***Anetium*** (closed venation and acrostichoid sorus) 1 Species.

The total number of genera in Holttum's Adiantaceae comes to 26 genera and 655 species.

In the year 1954 Holttum (in revised Flora of Malaya, Vol II) discussed the heterosporous ferns and included ***Salvinia*** and ***Azolla*** in different families (***Azollaceae*** and ***Salviniaceae***). In the year 1959 ("Pteridophyta. ferns and Fern allies". Flora Malesiana) he revised his previous (1947-1949) system of classification whose outline is given below. In this System Holttum recognised ***Gleichenia*** and ***Dicranopteris*** as two distirct genera (earlier they were merged into one in ***Gleichenia***) and ***Azolla*** and ***Salvinia*** were treated under one family Salviniaceae.

Holttum (1959) *revised his system which is as follows :*

Class : PTEROPSIDA

Order 1. ***Ophioglossales***	Ophioglossaceae
Order 2. ***Marattiales***	Marattiaceae
Order 3. ***Filicales***	Osmundaceae, Schizaeaceae, Gleicheniaceae, Hymenophyllaceae, Matoniaceae, Cyatheaceae, Polypodiaceae (S.st.), Grammitidaceae.

The heterosporus ferns were included in the families Marsileaceae and Salviniaceae (***Salvinia***, ***Azolla***).

The remaining genera of ferns that could not be included in the eight families listed above were arranged in sixteen separate groups namely ***Adiantum*** group, ***Asplenium*** group, ***Athyrium*** group, ***Blechnum*** group, ***Davallia*** group, ***Dennstaedtia*** group, ***Dicksonia*** group, ***Dryopteris*** group, ***Lindsaea*** group, ***Lomariopsis*** group, ***Nephrolepis*** group, ***Plagiogyria*** group, ***Pteris*** group, ***Tectaria*** group, ***Thelypteris*** group, and ***Vittaria*** group.

Holttum and Sen (1961) projected the use of anatomical features in understanding the taxonomy and classification of Cyatheaceae.

Holttum (1973 a, b) introduced significant modifications in his earlier treatment of the evolution of the various fern taxa and proposed a new scheme of phyletic relationships of the various groups. Holttum (1938-1982) has worked for about five decades on the taxonomical aspects of fens and his contributions are significant in strengthening various problems in fern taxonomy and classification. In the year 1982, in a "Dedication" note to the ***Flora Malesiana***, Holttum has dealt in detail the subject of taxonomy and classification of ferns and has rightly remarked," the objectives on any scheme of biological classification are to show natural relationships and to provide a means for the identification of individual organisms". His valuable suggestions in this article provide a useful guidence to the future ferns taxomonists. Before concluding Holttum's system of classification (1947-1949) the merits and demerits of his system should also be discussed.

Merits

These can be listed as below:—

1. His approach towards classification of ferns is nearly of the multicharacter type. He has taken into consideration both the Vegetative as well as the reproductive features of the ferns. Even some minor characters like the nature of the grooves on the rachises and stalks of pinnae; submerged and raised costae; sinus membranes etc. have been taken into consideration. He has laid great stress on anatomical features of the rhizone and petioles of leaves in addition to the dermal appendages and other morphological features.

2. His arguments in tracing the lines of evolution are also sound.

3. He has recognised Grammitidaceae as a distinct family phyletically removed from Polypodiaceae.

4. He has treated Polypodiaceae *s. str.* as a family representing a separate line of evolution quite distinct from other higher ferns.

5. He has given a rank of a separate family to the Thelypteridoid ferns (Thelypteridaceae) after removing them from the Dryopteridoid ferns.

6. He has separated the Gymnogrammeoid ferns from the Pteridoids.

7. The separation of Asplenoid ferns from the Athyrioid ferns and treating them as separate sub-families.

8. In treating ***Nephrolepis*** as primitive and ancestoral to ***Oleandra*** and regarding the Oleandroid ferns representing a separate line of evolution apart from ***Davallia.***

9. The inclusion of ***Loxogramme*** in the family Polypodiaceae (rather than Grammitidoid group) and ***Elaphoglossum*** in Lomariopsidoideae.

10. He opposed Bower in placing section Lomaria of ***Blechnum*** as primitive in the subfamily Blechnoideae and in deriving from it the section Eu-blechnum. He regarded Eu-blechnum to be more primitive than Lomaria section.

De-Merits

The major draw back in his system are few and can be listed as below:—

1. His Dennstaedtiaceae is an overlarge assemblage of ferns that have not been properly charactersied and are clubbed together merely on the basis of certain supposed lines of evolution. Some of these sub-families need revision and proper characterisation. Holttum in his revised system of classification in 1959 has done so.

2. Gleicheniaceous ancestory for both Cyatheaceae and Thelypteridaceae is also open to question.

3. The evidences presented by him to support the Dennstaedtioid origin of both the Athyrioid and the Asplenoid ferns also need further elaboration and revision.

In the year 1973a Holttum greatly revised his phyletic scheme and regarded Cyatheaceae and Thelypteridaceae to have evolved as a separate line of evolution from the Dicksonioid ferns. He has further suggested that the ***Athyrium*** and ***Diplazium*** have evolved from the Dryopteris group, which also gave rise as an offshort to ***Dennstaedtia, Microlepia***, ***Hypolepis***, ***Paesia***, ***Pteridium*** and ***Pteris*** (?). Regarding ***Asplenium*** Holttum (1973a) regarded that it had ancestors with superficial sori but its phyletic position is doubtful along with those of Onocleioids and Blechnoids.

4. The clubbing together of Gymnogrammoid and Viltarioid ferns under Adiantaceae. These deserve the rank of separate families. The Adiantaceae should be treated as a monotoypic family with only ***Adiantum*** as its sole representative. This genus alone has 184 or more species showing distinctive characters. In 1959 Holttum grouped together the Adiantoid ferns and Gymnogrammoid ferns under the heading **Adiantium group**. He separated (1959) the Vittarioid ferns and put them under Vittaria group. In 1973 Holttum treated Vittariaceae as a separate family and considered it to have evolved from some ancestral ferns with exindusiate marginal sori or sporangia along with ***Ceratopteris***, ***Negripteris***, ***Platyzoma***, ***Eriosorus***, ***Cheilanthes***, ***Gymnogramma***, ***Adiantum*** and ***Pteris. Schizaea***, ***Mohria***, ***Anemia***, and ***Lygodium*** are also considered to have evolved from the same source but along separate lines.

In 1973 a Holttum recognised three major evolutionary groups:—

Group I. Ancestral Ferns with Marginal Sori and Double Indusium.

Group II. Ancestral Ferns with Marginal Sori without Indusium.

Group III. Ancestral Ferns with Superficial Sori.

Group I gave rise to three lines of evolution:—

(*a*) Hymenophyllaceae, Loxosomaceae, ***Cibotium***, ***Metaxya***

(*b*) ***Dicksonia*** line

(*c*) ***Culcita—Thyrsopteris*** line

The first line is believed to have ended blindly whereas ***Dicksonia*** gave rise to ***Cyathea*** and ***Thelypteridaceae*** on one hand and ***Tectaria***, ***Ctenitis*** and Lomariopsis group, ***Oleandria - Nephrolepis***, ***Davallia*** group and Polypodiaceae on the other hand. ***Culcita - Thyrsopteris*** represent the end points of an other active line from where developed at different levels of evolution, ***Lindsaea*** group, ***Dryopteris*** group which further gave rise ***Athyrium*** and ***Diplazium*** and an offshoot resulting into ***Dennstaedtia***, ***Microlepia***, ***Hypolepis***, ***Paesia***, ***Pteridium*** and ***Pteris.***

The second group gave rise to ***Schizaeaceae*** (***Schizaea***, ***Mohir'a***, ***Anemia etc.***) and some other general listed above including Vittariaceae.

The third group evolved into (*i*) Osmundaceae and ***Plagiogyria***, (*ii*) Matoniaceae; and (*iii*) Glecheniaceae. Grammitidaceae, polypodiaceae and ***Dipteris*** having evolved from Gleicheniaceae at different levels. The position of ***Asplenium***, Onocleoids and Blechnaeceae is regarded as doubtful.

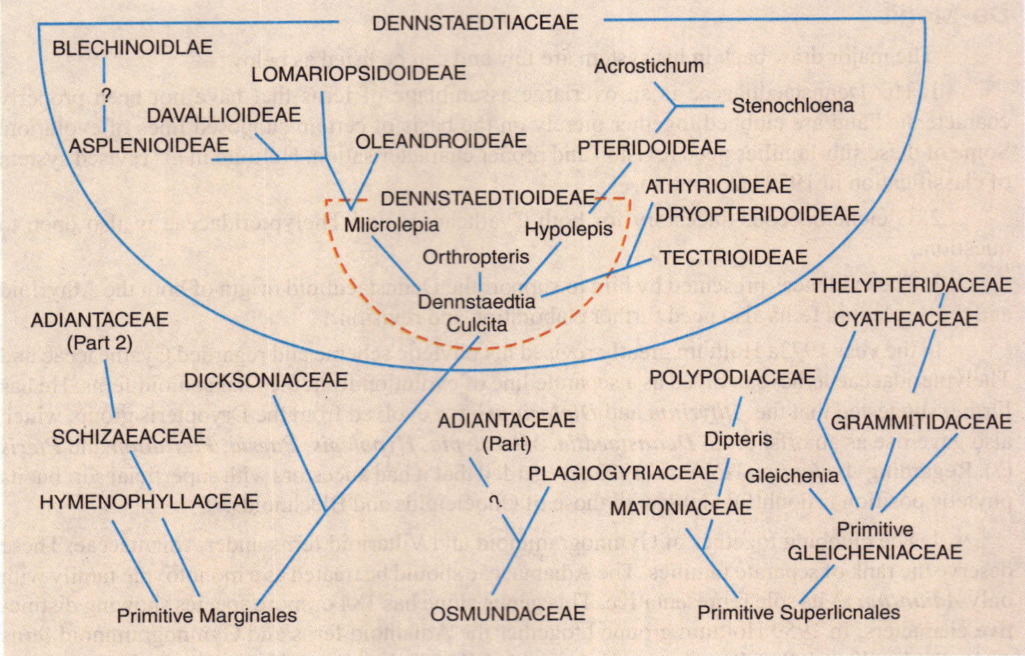

Fig. 8.19. Phylogenetic scheme of the homosporous ferns as presented by Holttum (1949)

After Holttum (1949), **Reimer** published his classification in 1954. He followed Christensen (1935) and grouped together most of leptosporangiate homosporous ferns in family Polypodiaceae and sub-divided it into 14 sub-familiers. He divided the class **Filicineae** into four sub-classes:—

1. Sub-class **Primofilices** in which he included the fossil members of the class.

2. Sub-class **Eusporangiatae** in which he included the order Ophioglossales with a single family Ophioglossaceae.

3. Sub-class **Osmundidae** which included a single order (***Osmundales***) with a single family (Osmundaceae).

4. Sub-class **Leptosporangiatae.** It was divided into three orders (Filicales, Marsileales and Salvineales). Under the filicales he included all the homosporous ferns. He recognised seven humosporous families namely Gleicheniaceae, Schizaeaceae, Matoniaceae, Cyatheaceae, Hymenophyllaceae, Dicksoniaceae and Polypodiaceae (with 14 sub-families). **Marseliales** Included family ***Marseliaceae***, (***with three genera***) whereas, **Salviniales** included Salviniaceae and Azollaceae (with one genus in each family).

Alston (1956) gave a multicharacter approach to fern taxonomy and proposed outlines of his system based on the following characters :

(*i*) Spore morphology (bilateral or tetrahedral nature).

(*ii*) Chromosome numbers.

(*iii*) Stipe anatomy.

(iv) Creeping or erect habit, the arrangement of leaves, their nature and venation.

(v) Presence or abscence of hair and scales on the rhizome. Peltate or simple nature of the scales and unicellular and mutlicellular hair.

(vi) Spores trilete or monolete.

(vii) Structure of sporangium, whether sporangia are sessile, shortly stalked or stalks long; whether annulus is absent, slightly developed or fully developed, oblique or vertical. Kinds of sorus and presence or absence of indusium etc.

(viii) Presence or absence of perispore.

Alston did not live long to elaborate his system but it is based mainly on Holttum's system with the following main differences:—

1. He separated the Vittarioid ferns from adiantaceae and treated them as a separate family Vittariaceae (following Copeland, 1947).

2. He recognised Cheiropleuriaceae as a distinct family by separating ***Cheiropleuria*** from Polypodiaceae.

3. He made certain drastic changes in Holttum's Dennestaedtiaceae by raising some of the sub-families to the family level e.g. Lindsaydaceae, Davalliaceae, Aspleniaceae, Blechnaceae, Lomariopsidaceae and Athyriaceae.

4. He merged ***Pteridium***, ***Histiopteris*** and ***Hypolepis*** with ***Dennstaedtiaceae,*** and ***Acrostichum*** with Adiantaceae. The genus ***Stenochlaena*** was merged with the Polypodiaceae.

5. He constituted the family Aspidiaceae by removing some genera from Dryopteridoideae and Tectarioideae.

6. The genera ***Matteucia***, ***Wordsia***, ***Stenolipia***, ***Cheilanthopis*** and ***Cystopteris*** were removed from Dryopteridoideae and merged with Athyriaceae.

Lyman Benson (1957) included all the ferns under the division ***Pteridophyta***, which included only one class ***Pteropsida*** (fern class). The Pteropsida were further divided into three orders of living ferns as follows:— Order 1. **Filicales**—It included eight families e.g. ***Osmundaceae***, ***Gleicheniaceae***, ***Cyatheaceae***, ***Schizaeaceae***, ***Hymenophyllaceae***, ***Polypodiaceae***, ***Salviniaceae*** and ***Marseliaceae.***

Order 2. **Marattiales,** which included family ***Marattiaceae.***

Order 3. **Ophioglossales,** which included family ***Ophioglossaceae.***

He included all the fossil ferns under the order ***Coenopteridales.***

Pichi-Sermolli's Classification (1958, 1959)

Pichi-Sermolli divided the Pteridophyta into six classes:—

Class I **Lycopsida**

Class II **Sphenopsida**

Class III **Noeggerathiopsida**

Class IV **Psilotopsida**

Class V **Psilophytopsida**

Class VI **Filicopsida:**— It is further divided into seven sub-classes:—

Sub-class	Order	Family
1. Primofilicidae:— It includes all the fossil ferns.		
2. Ophioglossidae	***Ophioglossales***	Ophioglossaceae
3. Marattidae	***Marattiales***	Marattiaceae
4. Osmundidae	***Osmundales***	Osmundaceae
5. Filicidae	***Gleicheniales***	Gleicheniaceae
	Schizaeales	Schizaeaceae
	Hymenophyllales	Hymenophyllaceae
	Matoniales	Matoniaceae
	Platyzomales	Platyzomaceae
	Polypodiales	Dipteridaceae
		Cheiropleuriaceae
		Grammitidaceae
		Polypodiaceae
	Cyatheales	Cyatheaceae
		Lophosoriaceae
	Dicksoniales	Dicksoniaceae
		Dennstaedtiaceae
		Lindsaeaceae
	Pteridales	Pteridaceae
		Negripteridaceae
		Sinopteridaceae
		Cryptogrammaceae
		Gymnogrammaceae
		Actinopteridaceae
		Adiantaceae
		Parkeriaceae
		Vittariaceae
	Davalliales	Davalliaceae
		Oleandraceae
	Aspidiales	Aspleniaceae
		Thelypteridaceae
		Athyriaceae
		Aspidiaceae
		Lomariopsidaceae
	Blechnales	Blechnaceae
	Loxomales	Loxomaceae
	Plagiogyriales	Plagiogyriaceae
	Hymenophyllopsidales	Hymenophyllopsidaceae
6. Marsileidae	***Marsileales***	Marsileaceae
7. Salviniidae	***Salviniales***	Salviniaceae
		Azollaceae

In 1977, **Pichi-Sermolli** proposed system of arrangement of genera of Pteridophyta and recognised 58 families of ferns, thus believing in splitting rather than lumping of taxa. In proposing this classification he emphasized his preference for classification of various groups according to their presumed origin and relationship rather than degree of evolution ("treating the primitive groups, even if not allied to each other, first and the advanced ones afterwards" Pichi-Sermolli 1977, P. 315). In this system he has dealt in detail the phyletic schemes of the presumed relationship of genera of quite a large number of taxa. This arrangement is the best treatment so far attempted by any taxonomist and it is beyond the scope of this text to give details (For details consult Pichi-Sermolli, R.E.G. 1977.

In his scheme (Pichi-Sermolli 1958, 59) ***Ophioglossidae*** and ***Marattidae*** include only one order and one family each. The same is the case with **Osmundidae** (one order and one family), which according to him, "differ from all other ferns in several important characteristics of the Sporophyte and the gametophyte." He considers the **Osmundidae** to be a group of primitive ferns that form an isolated group equal in rank to the other sub-classes (Ophioglossidae, Marattidae and Filicidae). He further stated that the Osmundidae are intermediate between the eusporangiates and the leptosporangiates, only morphologically. In his own words, "Its (Osmundidae) few living species are the last survivors of this conservative and primitive evolutionary line. No other living fern is regarded as derived from the Osmundidae, they being relicts of a group which appears to have lost to evolutionary potentialities." His system differs from other systems discussed before in the nuember of families recognised, abolition of sub-families and transference of some genera from one group to the other group e.g. Christensen (1938) included all the homosporous genera into 11 families (with Polypodiaceae divided into 15 Sub-families); Holttum (1949) into 14 families (with Dennstaedtiaceae divided into 11 sub-families); Copeland (1947) into 16 families; Ching (1940) into 33 families; Dickason (1946) into 42 families; and Pichi Sermolli (1959) into 32 families. The main differences between Holttum (1949) and Pichi-Sermolli (1958, 59) are listed below:—

1. The polypodiaceous genera ***Dipteris*** and ***Cheiropleuria*** of Holttum (1949) are elevated to the rank of families Dipteridaceae and Cheiropleuriaceae.

2. Pichi Sermolli has raised the various Sub-families of Holttum's Dennstaedtiaceae to the level of families e.g. the sub-families Dryopteridoideae and Tectarioideae are merged to form one family Aspidiaceae; the sub-families Lomariopsidoideae, Athyrioideae, Asplenioideae, Blechnoideae, Davalloideae, Oleandroideae, Lindsaeoideae are regarded as separate families; the Dennstaedtoideae and Pteridoideae are merged into one family Dennstaedtiaceae.

3. Apart from the Chaetopterid genera of Pteridoideae (which have been included in the family Dennstaedtiaceae) the rest have been included in a separate family Pteridaceae.

4. The various Gymnogrammeoid genera of Adiantaceae (of Holttum) have been grouped under families Parkeriaceae, Gymnogrammaceae, Adiantaceae, Cryptogammaceae, Sinopteridaceae and Negripteridaceae.

5. The Vittarioideae of Holttum's Adiantaceae have been included in a separate family Vittariaceae.

6. These two systems differ widely so far as the concepts of phylogeny and relationships of evolutionary processes among the homosporous ferns are concerned.

In his system of classification Pichi-Sermolli (1959) has taken into account most of the fern characters, including vegetative characters, comparative morphology, reproductive characters, chromosome numbers and even the gametophyte morphology. It appears that, like most of the earlier taxonomists, he also based his systems fundamentally on the morphology of the sorus, structure of sporangia and the types of spores. It is a highly improved, and extremely dependable system of classification of the ferns and fern allies. Van Cotthem (1973) utilising the cumulative data on stomatal types gave information about their distribution in Filicopsida. His conclusions nearly fall in conformity with Pichi-Sermolli's familial concepts of recognising small families.

A large number of systems of classification of the ferns have been proposed after Pichi Semolli (1958, 59) but a few of them will be considered in this text.

P.N. Mehra, (1961) in his Presidential address to the Botanical section of the 48th Indian Science Congress discussed in detail the cytological evolution in ferns and proposed a system of classification of leptosporangiate ferns based mainly on the chromosome numbers. He has also taken into account other morphological factors and even the morphology of the ganetophyte and the sex organs. Based on his observations and a very long research experience in the field of pteridology he envisaged the evolution of ferns to have taken place from six stocks or series :

Series 1. **Ophioglossaceous Stock** which gave rise to Ophioglossaceae

Series 2. **Marattiaceous Stock** which gave rise to the Marattiaceae

Series 3. **Osmundaceous Stock** which gave rise to the Osmundaceae and the Plagiogyriaceae

Series 4. **Schizaeaceous Stock** which gave rise to Schizaeaceae, Marsileaceae, Parkeriaceae, Vittariaceae, Adiantaceae, Cryptogrammaceae, Pteridaceae, Gymnogrammaceae and Sinopteridaceae.

Series 5. **Hymennophyllaceous Stock** which evolved into the families Salviniaceae, Hymenophyllaceae, Lindsaeaceae, Oleandraceae, Davalliaceae, Hypolepidaceae, Dennstaedtiaceae, and Dicksoniaceae.

Series 6. **Gleicheniaceous Stock** which evolved into Gleicheniaceae, Blechnaceae, Aspleniaceae, Thelypteridaceae, Athyriaceae, Aspidiaceae, Woodsiaceae, and Cyatheaceae on one hand and Matoniaceae, Dipteridaceae, Cheiro pleuriaceae, Polypodiaceae and Grammitidaceae on the other hand.

Ito (1962) arranged the families and sub-families of leptosporangiate ferns on a graph according to their inter-relationships based on spore forms, spore germination patterns, types of vascular organisation and Vegetative characters. He listed twenty two families with Polypodiaceae, pteridaceae, Vittariaceae, Davalliaceae and Aspidiaceae divided into several sub-families as below:—

(*i*) Polypodiaceae divided into sub-families Polypodioideae, Loxogrammoideae and Grammitoideae.

(*ii*) Vittariaceae into subfamilies Antrophyoideae and Vittarioideae

(*iii*) Pteridaceae into sub-families Lindsaeideae, Dennistaedtioideae, Adiantoideae, Pteridoideae, Chelanthoideae and Gymnogrammoideae.

(*iv*) Davalliaceae into sub-families Davallioideae and Oleandrioideae.

(*v*) Aspidiaceae into Sub-families Woodsioideae, Aspidioideae, Onocleoideae, Elaphoglossoideae, Blechnoideae, Thelypteridoideae and Asplenioideae.

The grouping of Pteridaceae and Aspidiaceae is most un-natural.

Nayar's System of Classification (1970, 74). **Nayar (1970, 1974) proposed a system of classification, the details of which are as follows :**

	CLASS	FILICOPSIDA	
Sub-class	**Order**	**Family**	**Sub-family**
1. Ophioglossidae	(*i*) ***Ophioglossales***	Ophioglossaceae	—
	(Only one family)		
2. Marattidae	(*ii*) ***Marattiales***	***Danaeaceae***	
	(with four families)	Marattiaceae	—
		Kaulfussiaceae	
		Angiopteridaceae	
3. Filicidae	(*iii*) ***Osmundales***	Osmundaceae	—
	(with two families)	Plagiogyriaceae	
	(*iv*) ***Schizaeales***	Schizaeaceae	
	(with eight families)	Anemiaceae	
		Lygodiaceae	
		Parkeriaceae	
		Vittariaceae	
		Adiantaceae	
		Pteridaceae	
		Cheilanthaceae	
	(*v*) ***Cyatheales***	Dennstaedtiaceae	
	(with fourteen	Hypolepidaceae	
	families)	Lindsaeaceae	
		Cyatheaceae	Thysopteridoideae
		(with five	Dicksonioideae
		sub-families)	Cibotioideae
			Metaxyoideae
			Cyatheoideae
		Loxomaceae	
		Hymenophyllaceae	
		Blechnaceae	
		Davalliaceae	
		Oleandraceae	
		Aspleniaceae	
		Dryopteridaceae	Onocleoideae,
			— Athyroideae,
			Hypodematioideae,
			Dryopteridioideae, and
			Tectarioideae.
		Lomariopsidaceae	
		Thelypteridaceae	
		Grammitidaceae	
	(*vi*) ***Gleicheniales***	Loxogrammaceae	
	(with six families)	Polypodiaceae	— Platycerioideae,
		Dipteridaceae	Microsorioideae,
		Cheiropleuriaceae	Pleopeltidioideae,
		Gleicheniaceae	Polypodioideae and
		Matoniaceae	Crypsinioideae

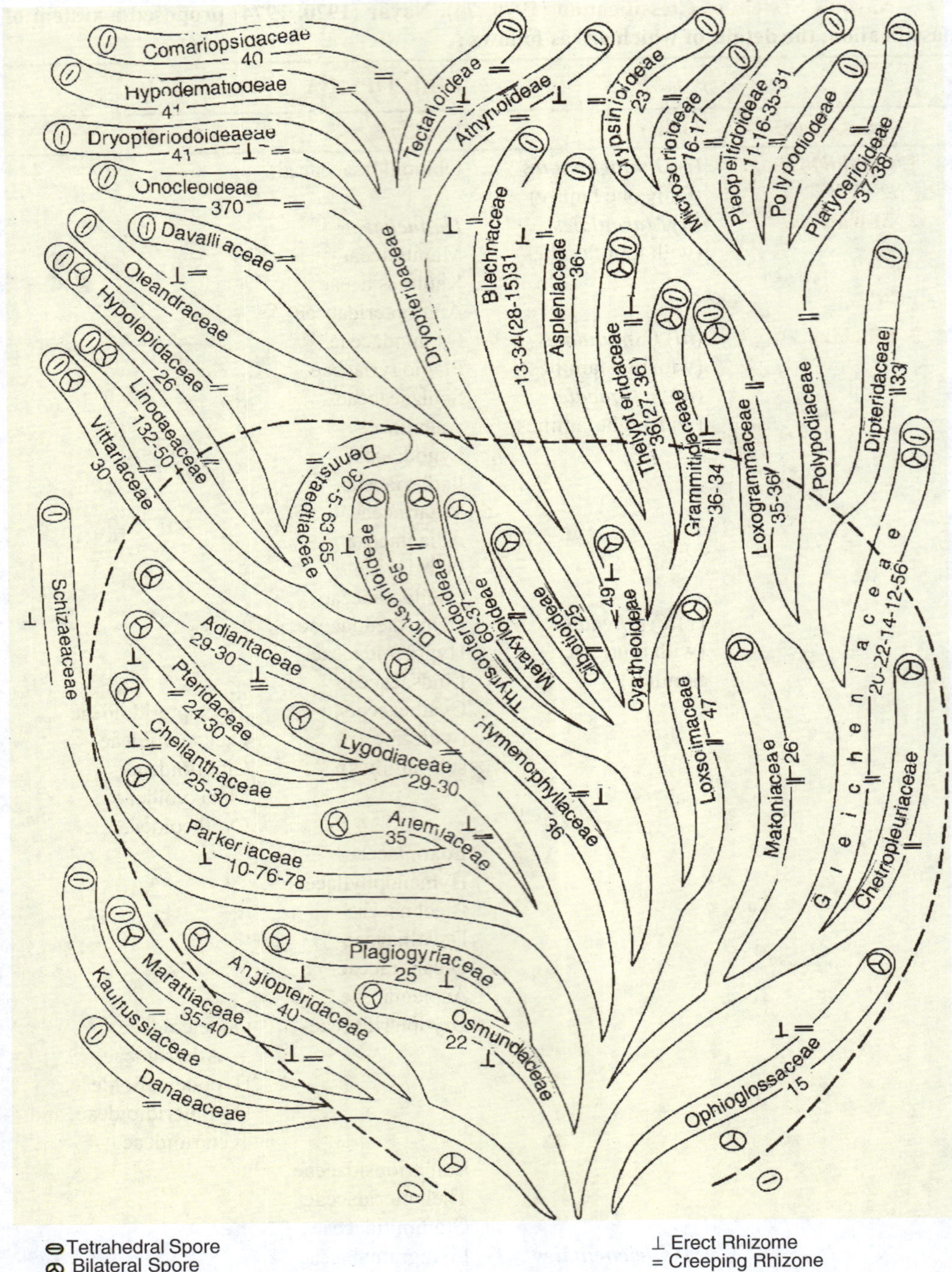

Fig. 8.20 : Phylogeny of Filocopsida as presented by Nayar (1974). The dotted line represents the region of transition from tetrahedral to bilateral spore. The families within the circle possess tertrahedral spores whereas those outside the circle possess bilateral spores. The families which cross the circle possess both types of spores.

In all there are three sub-classes six orders and thirty five families with Cyatheaceae as well as Polypodiaceae sub-divided into five sib-families each. Nayar has not considered the heterosporous ferns in his system. Nayar's system is broad based as he has taken into consideration almost all the characters about whom information was available upto the time of publication of his classification. It is in reality a multicharacter approach to fern classification. He deals with only the homosporus ferns. In his own words, "In addition to the classical taxonomical criterion, viz., the structure and morphology of the fern sorus, Nayar bases his system on choromosome numbers, prothallial morphology, spore morphology, details regarding vascular organisation of the rhizome as well as on comparative criteria like morphology of the dermal appendages as enumerated by Bower."

His system of classification differs in many respects from the systems of classification proposed earlier. The major differences are listed below:

1. Citing instances from vascular morphology, soral morphology, spore morphology and prothallial morphology Nayar made drastic changes in the grouping of some Gymmnogrammeoid and Pteridoid genera into distinct families e.g. (*i*) he included the Gymmnogrammeoid genera (of other authors) ***Actinopteris***, ***Pityogramma***, ***Oncychium***, ***Eriosorus***, ***Jamesonia*** in his family Pteridaceae because of their close alliance to ***Pteris***; (***ii***) he included ***Acrostichium*** (Pteridoideae of Holttum and Christensen), ***Coniogramma***, ***Syngramma and Anogramma*** (Gymnogrammeoideae of Holttum and Christensen) in Adiantaceae because of their similarities with ***Adiantum*** and (***iii***) the rest of the Gymnogrammeoid genera were included in a separate family called Cheilanthaceae.

2. Another major difference from other systems is the creation of a family Hypolepidaceae in which the Chaetopterid genera of Christensen's Pteridoideae and Pichi-sermolli's Dennstaedtiaceae were included. These genera were separated because of their peculiar kind of stelar organisation and chromsome number x = 26.

3. The family Cyatheaceae of Nayar is a bigger group because he has clubed together the families Dicksoniaceae, Lophosoriaceae and Protocyatheaceae of other authors like Holttum, Christensen and Pichi-Sermolli. He has sub-divided the Cyatheaceae into five sub-families.

4. He has merged Pichi-sermolli's Aspidiaceae and Athyriaceae into his Dryopteridaceae.

5. He has splitted Copeland's large family Aspidiaceae into three separate families Dryopteridaceae, Lomaropsidaceae and Thelypteridaceae.

6. The Polypodiaceae of Copeland has been split up into three separate families Polypodiaceae, Dipteridaceae and Cheiropleuriaceae. He has sub-divided his Polypodiaceae into five sub-families.

7. The genus ***Loxogramme***, which was included by Copeland in Grammitidaceae and by Holttum in Polypodiaceae, has been included in a separate family Loxogrammaceae included under the order Gleicheniales.

8. The family Schizaeaceae has been split up into three separate families Schizaeaceae, Anemiaceae and Lygodiaceae.

9. The Marattiaceae has also been split up into four separate families namely Danaeaceae, Marattiaceae, Kaulfussiaceae and Angiopteridaceae.

10. As opposed to Bower (1923), Nayar has regarded the creeping and elongated condition of rhizome as an advanced feature whereas the short, erect or sub-erect and condensed rhizomes bearing spirally aranged leaves as a primitive condition.

11. He has suggested that in some fern families the solenostetic condition of the vascular region is derived from the dictyostelic condition.

12. The tetrahedral spores (trilete) are regarded to be primitive condition whereas the bilateral or monolete spores are regarded as advanced. Nayar regards this change as irreversible.

Nayar's circumvention of Cyatheaceae (into 5 sub-families), Dryopteridaceae (5 sub-families) and Polypodiaceae (5 sub-families) is open to question.

Bierhorst's System of Classification (1971)

Bierhorst (1971) presented his system of classification for Pteridophytes and the seed plants. The outline is given at the end of first chapter. Here the details of his system concerning the ferns is given :

Class	Order	Family	Number of Genera
Filicopsida	*Noeggerathiales*	Psilotaceae	Two
		Noeggerathiaceae	One
	Filicales	Stromatopteridaceae	One
		Gleicheniaceae	One (Gleichenia with 130 Species)
		Schizaeaceae	Two (41 Species)
		Anemiaceae	Two (93 species)
		Lygodiaceae	One (39 species)
		Osmundaceae	Three (13 species)
		Hymenophyllaceae	Two (650 species)
		Matoniaceae	Two (4 species)
		Platyzomaceae	One
		Dipteridaceae	One
		Cheiropleuriaceae	One
		Polypodiaceae	Fifty five
		Vittariaceae	Nine
		Cyatheaceae	Thirteen (with 996 species)
		Pteridaceae	Fifty five
		Parkeriaceae	One
		Davalliaceae	Twelve
		Aspleniaceae	Nine
		Aspidiaceae	Sixty six
		Blechnaceae	Eight
		Hymenophyllopsidaceae	One
		Loxomaceae	Two (with four species)
		Plagiogyriaceae	One (with 36 species)
	Marsileales	Marsileaceae	Three (with 67 species)
	Salviniales	Salviniaceae	One (with 10 species)
		Azollaceae	One (with 6 species)
Ophioglossopsida	*Ophioglossales*	Ophioglossaceae	Three
Marattiopsida	*Marattiales*	Marattiaceae	Four (with 166 species)
		Danaeaceae	Two (with 31 species)

Disputable points in the above system of classification can be listed as below:—

1. Bierhorst's placing of Psilotaceous genera in the Filicales and in regarding them allied to Stromatopteridaceae, Gleicheniaceae, and Schizaeaceae.

2. His inclusion of the families Thelypteridaceae, Athyriaceae, Aspidiaceae and Lomariopsidaceae into one family Aspidiaceae with sixty six genera.

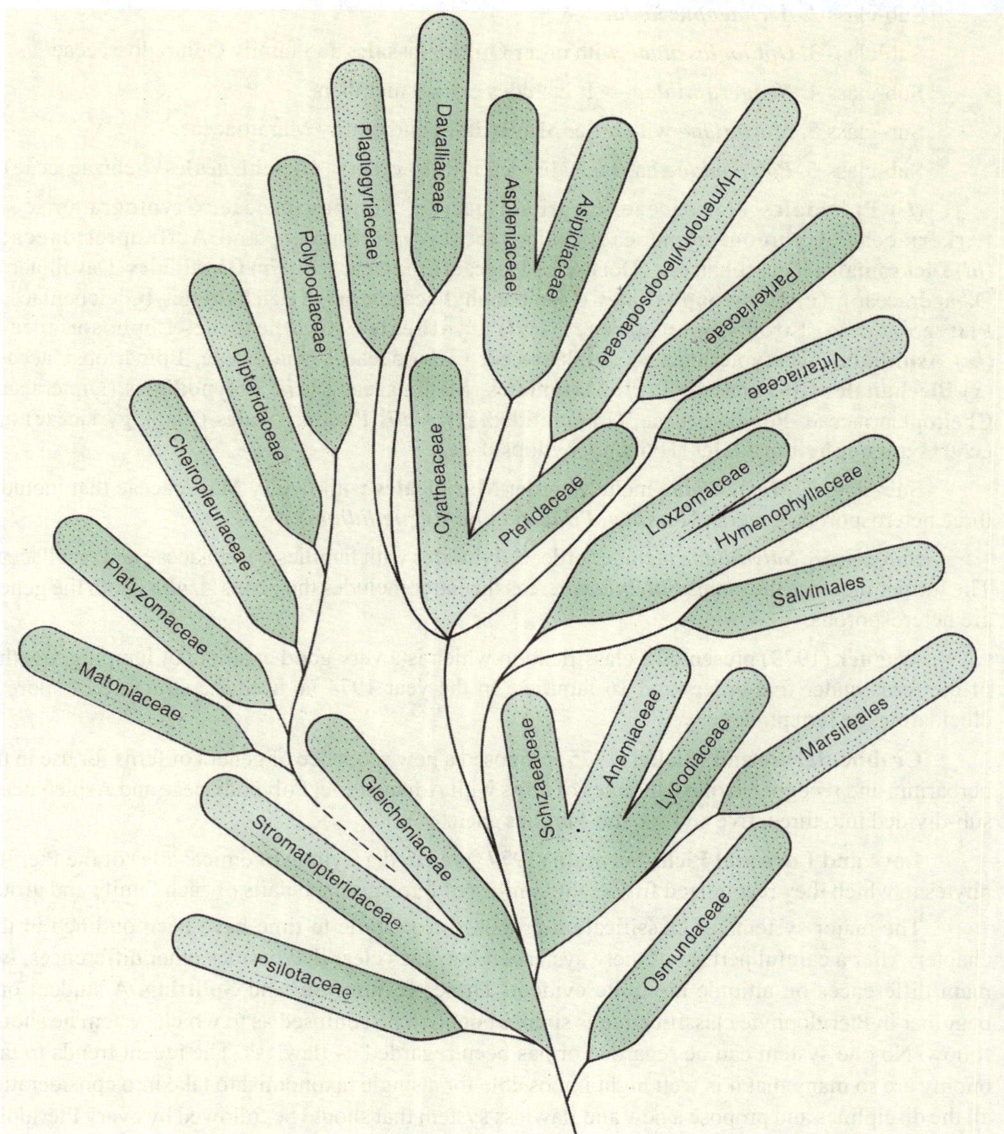

Fig. 8.21 : Phylogeny of ferns as presented by Bierhorst (1971)

3. His clubbing together of the families Dennstaedtiaceae, Lindsaeaceae, Pteridaceae, Sinopteridaceae, Cryptogrammaceae, Gymnogrammaceae, Adiantaceae etc., into one family Pteridaceae with about fifty five genera.

Cronquist et. al. (1966) included the ferns in division ***Polypodiophyta*** with one class ***Polypodiatae***. The brief outline is given below:—

Division:— **Polypodiophyta**

Class:— **Polypodiatae**

Sub-class 1. ***Propterididae***] Fossils

Sub-class 2. ***Archaeopterididae***]

Sub-class 3. ***Ophioglossiidae*** with order **Ophioglossales** and family Ophioglossaceae

Sub-class 4. ***Neogerathiidae***:— It includes extinct members.

Sub-class 5. ***Marattidae*** with order **Marattiales** and family Marattiaceae.

Sub-class 6. ***Polypodidae*** has been divided into 14 orders : (*i*) **Schizaeales** (Schizaeaceae);

(***ii***) **Pteridales** Pteridaceae, Negripteridaceae, Sompteridaceae, Cryptogrammaceae, Parkeriaceae, Gymnogrammaceae, Adiantaceae, Vittariaceae, and Actinopteridaceae); (***iii***) **Dicksoniales** (Dicksoniaceae,. Dennstaeadtiaceae, Lindsaeaceae); (***iv***) **Davalliales** (Davalliaceae, Oleandraceae); (***v***) **Hymenophyllales** (Hymenophyllaceae); (***vi***) **Gleicheniales** (Gleicheniaceae, Platyzomaceae, Stromatopteridaceae), (***vii***) **Cyatheales** (Cyatheaceae, Lophosoriaceae); (***ix***) **Aspidiales** (Therypteridaceae, Aspleniaceae, Thyriaceae, Aspidiaceae, Lomariopsidaceae), (***x***) **Blechanales** (Blechnaceae); (***xi***) **Matoniales** (Matoniaceae); (***xii***) **Polypodiales** (Dipteraceae, Cheiropleuriaceae, Polypodiaceae, Grammitidaceae); (***xiii***) **Plagiogyriales** (Plagiogyriaceae) and (***xiv***) **Hymenophyllopsidales** (Hymenophyllopsidaceae),

Sub-class 7. ***Marselliidae*** includes order **Marsileales** with family Masileaceae that includes three heterosporous genera (***Marsilea, Pilularia and Regnellidium***)

Sub-class 8. ***Salviniidae*** includes order **Salviniales** with families Salviniaceae and Azollaceae. The former includes the genus ***Salvinia*** whereas the latter includes the genus ***Azolla***. Both the genera are heterosporous.

Wagner (1973) presented a classification which is a very good example of lumping together of the ferns under five orders and 16 families. In the year 1974 he used the structure of spore in elucidating the fern phylogeny.

Crabbe, Jermy and Mickel (1975) proposed a new sequence of genera of ferns for use in the herbarium and recognized thirty families of ferns with Adiantaceae, Polypodiaceae and Aspleniaceae sub-divided into three, five and six sub-families each.

Love and Love and Pichi-Sermolli (1977) prepared a Cytotaxonomical atlas of the Pteridophytes in which they recognized fifty eight families of ferns giving delails of each family and group.

The major systems of classification proposed from time to time have been outlined in this chapter. After a careful perusal of these systems it becomes clear that besides other differences, two main differences on attitude are quite evident. These are **lumping** and **Splitting.** A student or a begginer in Pteridophyte classification is simply bound to be confused as to which system he should follow. No one system can be regarded or has been regarded as flawless. The recent trends in taxonomy are so many that it is well night impossible for a single taxonomist to take into consideration all the disciplines and propose a new and flawless system that should be followed by every Pteridologist or should be acceptable to all. The areas of disagreement regarding the composition, relationships and contents of families, the generic level differences and the taxonomic difficulties at specific level are so many that it is an enigma in itself for the students of fern taxonomy. The excellent taxonomic work done by Christensen, Copeland, Holttum, Ching, Pichi-Sermolli, Bierhorst, Wagner and others abroad and by Mehra, Bir, Nayar and their students in India have straightened many problems, but still lot needs to be done. Taxonomy is an unending synthesis and refinements in taxonomic system have been due to multi-character approach, combining, as it does, external morphology with internal micromorphology like anatomy, palynology, epidermology, embryology, spore morphology, physiology, ecology, cytology, genetics, population biology, serodiagnois, biochemistry and chemistry, DNA hybridisation, ultrastructure, mathematics and electronics. It is evident, from the recent trends listed above, that a student of systematics in the coming decades will have to arm himself with an adequate knowledge of all those disciplines, before venturing at a scheme of classification. As pointed out by Wagner (1968) the modern systematist should be a '***Chemo-cyto-histo-morpho-taxono-metrician***'. If one can master all these disciplines in his life time or even be capable of utilising the collected data to erect a flawless system of classification, then certainly the new systematics will be one of the most indispensable, exciting, rewarding and challenging field of study. Will it be possible for taxonomists to identify a plant in the field by utilising all these characters or disciplines? Certainly it is not possible. "No one mind can adequately survey the whole field covered by modern research" (Holttum 1973). "Let us be clear in our mind that floristic taxonomy for encyclopaedic purposes has to be utilitarian, general purpose and practical so that plants can be identified in the field with ease and one should not have to determine chromosome numbers, pollen or spore type, iozyme pattern or the like of it, to identify a plant. The most important consideration has to be morphological discontinuity," (T.N. Khoshro, 1980). Holttum (1967) has very aptly remarked, "Comparative morphology is the basis of taxonomy", and "clearer understanding of morphology, whether of whole organs or of microscopic parts of organs, makes possible the improvement of classification". So it is not essential that a taxonomist must necessarily be a chemo-cyto-histomorpho-taxono-metrician. These refinements can be taken up at the level of monographs on critical taxa where one can have synthesis of as many parameters as possible, which would emanate from collective studies of groups of specialists.

The system of classification of ferns followed in this text is outlined below:—

Division Filicophyta or Pterophyta

1. Class: **Primofilicopsida** that includes all extinct genera.
2. Class: **Eusporangiopsida** that includes orders Ophioglossales and Marattiales.
3. Class: **Protoleptosporangiopsida** that includes the order Osmundales.
4. Class: **Leptosporangiopsida** that includes the orders Filicales, Marsileales and Salviniales.

QUESTIONS

ESSAY TYPE

1. Write a detailed note on the stelar system in pteridophytes.
2. With the help of suitable diagrams describe the evolution of Stelar system in pteridophytes. **(Kumaon, 2000)**
3. Describe the range of Sporangial Structure in pteridophytes studied by you.
4. Write an essay on phylogeny and evolutionary tendencies in pteridophytes.
5. What is stele? Describe Stelar system of pteridophytes.
6. With the help of diagrams, describe the structure and development of Leptosporangiate and Eusporangiate pteridophytes.
7. Give an illustrated account of Stelar organisation found in pteridophytes studied by you.
8. Write classification of Pteropsida giving suitable examples.

9. Give an account of Stelar system in pteridophytes by means of labelled diagrams only.

10. Give a brief account of evolution of sorus in ferns (filicales) giving suitable examples. **(Awadh M.Sc. 2000)**

11. Give the characteristic features and classification of filicinae.

12. Define the tern stele. Give labelled diagrams of steles found in pteridophytes studied by you. **(Allahabad, 2000)**

13. Give an illustrated account of the different types of steles found in pteridophytes studied by you. **(Lucknow, 1993, 2000; Kerala, 2002; Poorvanchal, 2002)**

14. (*a*) What is stele? Describe its various types in pteridophytic plants **(Kanpur, 2005)**

(*b*) What is stele? What are its types? Explain protostele in detail. **(Kanpur, 2000)**

15. Explain how the complex types of stele evolved from the simplest type, Give an example for each.

16. Give an illustrated account of different types of spore germination leading to prothallus.

17. Describe different types of prothallus in ferns.

18. Briefly describe the evolutionary trend in filicophyta.

19. Giving suitable examples discuss / explain.

(*a*) Development of eusporangiate and leptosporangiate types of sporangia.

(*b*) Apogamy and Apospory in Pteridosphytes.

(*c*) Apogamy and Apospory in pteridophytes.

(*d*) Polyploidy in ferns.

(*e*) Origin of Sporophylls in Pteropsida. **(Garhwal, 2000)**

20. Give different systems of classification of pterophyta or filicophyta. State which system of classification is the best and why?

21. Describe briefly the system of classification of pteropsida proposed by Holtum. Also Mention merits and demerits of this classification.

22. Give an outline of the classification of ferns proposed by Cronquist ***et. al.*** and Bierhorst.

SHORT ANSWER TYPE

23. Write short notes on:

(*i*) Difference between protostele and siphonostele. **(Rohilkhand, 2000)**

(*ii*) Protostele and its types. **(Meerut, 2000)**

(*iii*) Stelar System in pteridophytes. **(Kanpur, M.Sc., 2000)**

(*iv*) Actinostele. **(Himachal Pradesh, 2000)**

(*v*) Siphonostele. **(Nagpur 2001; Garhwal 2002)**

(*vi*) Solenostele.

(*vii*) Dictyostele.

(*viii*) Protostele and dictyostele in pteridophytes.

(*ix*) Types of steles in Pteridophytes.

(*x*) Alternation of generation in pteridophytes.

(*xi*) Types of sporangia found in ferns

(*xii*) Amphiphloic Siphonostele. **(Nagpur, 2002; Himachal Pradesh, 2000)**

(*xiii*) Mixed protostele.

(*xiv*) Classification of ferns. **(Rohilkhand, 2000)**

(*xv*) Types of steles in Pteridophytes.

(*xvi*) Eusporangiate and Leptusporangiate sporangium **(Lucknow, 2005)**

24. Differentiate between:

(*i*) Eusporangiate and leptosporangiate development of sporangia.

(*ii*) Ectophloic and Endophloic siphonostele.

(*iii*) Exosporic and Endosporic gametophytes.

(*iv*) Actinostele and plectostele.

(*v*) Solenostele and dictyostele.

(*vi*) Sorus and synangium in pteridophytes.

(*vii*) Siphonostele and Solenostele.

25. Write briefly on:

(*a*) Evolution of Sorus in ferns.

(*b*) Stelar types in pteridophytes.

(*c*) Evolution of stelar structure.

(*d*) Vegetative propagation in ferns. **(Awadh M.Sc. 2000)**

(*e*) Mechanism of sporangial dehiscence and spore dispersal in terns.

(*f*) Methods of vegetative propagation in ferns.

(*g*) Exosporic and Endosporic sex organs.

(*h*) Exoscopic and Endoscopic embryo development.

(*i*) Indusium **(Lucknow, 2004)**

26. Name the stele in which xylem is star shaped. Define a polystele.

27. Name the type of stele in pteridophytes where the vascular cylinder is medullated.

28. Name the scientist who proposed stelar theory.

29. Name most primitive stele.

30. What is leptosporangiate development?

31. Define Plectostele.

32. Rearrange the following types of steles in the order of their development and evolution. Siphonostele, Plectostele, Protostele, Actinostele, Dictyostele.

33. What is the name given to each component of a dictyostele?

34. What do you call a siphonostele which has developed a few leaf gaps?

35. Do vessels occur in pteridophytes? If yes, cite examples.

36. Draw labelled diagrams of :

(*a*) Mixed protostele

(*b*) Haplostele

(*c*) Dictyostele **(Kanpur 2003)**

(*d*) Protostele

(*e*) Actinostele

(*f*) Plectostele

37. Short notes on– Various kinds of stelar systems in pteridophytes **(Garhwal Univ., 2005)**

38. Short notes on :

(*a*) Ligule

(*b*) Dictyoxylic Stele

(Mysore Univ., 2004)

(*c*) Amphiphloic Siphonostele

(Nagpur Univ., 2005)

(*d*) Plectostele **(Nagpur Univ., 2005)**

OBJECTIVE TYPE

39. State whether the following statement is True or False :

(*i*) Ferns differ from Mosses in having alternation of generations.

(*ii*) A temporary organ of attachment and nutrition of embryo of Pteridophytes is called foot.

(*iii*) The sporangia developing from a single initial giving rise to the entire sporangium is called leptosporangiate.

(*iv*) Circinate Vernation is characteristic of ferns.

(*v*) Ferns also reproduce vegetatively.

40. Select the correct answer :

(*i*) Stelar theory was proposed by

(*a*) Sachs

(*b*) Van Teghem and Douliot

(*c*) Foster and Gifford

(*d*) DD Pant.

(*ii*) In a stele the xylem core is stellate or star shaped. Such a stele will be called

(*a*) Protostele

(*b*) Haplostele

(*c*) Actinostele

(*d*) Plectostele.

(*iii*) In ***Gleichenia dichotoma*** the stele is protostele but the xylem is

(*a*) Exarch

(*b*) Mesarch

(*c*) Endarch

(*d*) Diarch.

(*iv*) Medullated protostele is called

(*a*) Haplostele

(*b*) Actinostele

(*c*) Siphonostele

(*d*) Dictyostele.

(*v*) In Dictyostele each component is called

(*a*) Meristele

(*b*) Plectostele

(*c*) Actinostele

(*d*) Protostele.

(*vi*) Sori bearing leaves are known as

(*a*) Prophylls

(*b*) Sporophylls

(*c*) Both

(*d*) None.

(*vii*) When the sporangium develops from a group of cells, this type of development is regarded as

(*a*) Simple

(*b*) Complex
(*c*) Eusporangiate
(*d*) Leptosporangiate.

(*viii*) In leptosporangiate type of Sporangial development the sporangium develops from
(*a*) Single superficial cells
(*b*) two cells
(*c*) three cells
(*d*) a group of cells.

(*ix*) Which of the following is an advanced feature in pteridophytes?
(*a*) small plants with upright short branches
(*b*) Presence of protostele
(*c*) Large sporangia
(*d*) Dimorphic leaves.

(*x*) Fern gametophyte is
(*a*) Homothallic
(*b*) Heterothallic
(*c*) Heteroccious
(*d*) Autoccious.

(*xi*) The spores of the fern plant germinate to form
(*a*) thickwalled zygospore
(*b*) an embryo
(*c*) a new fern plant
(*d*) prothallus.

Ophioglossum petiolatum

CHAPTER 9

Eusporangiopsida Ophioglossales

- Characteristic features
- Description of ***Ophioglossum L.***
- Description of ***Botrychium*** Swartz.
- Description of ***Helminthostachys*** Kaulf.

The Eusporangiopsida include the primitive ferns that are characterised by the typical eusporangiate development of their spore producing organs and the antheridia. The sporangia originate from a group of initials and possess a several layered thick wall. In the Ophioglossales the fertile leaves have a common petiole that bears a sterile dissected or simple lamina and a stalked sporangiferous **spoke** or the **sporophyll**. The spike may be simple or branched and beats large sessile or shortly stalked sporangia in two marginal rows. Dehiscence may be by the vertical or transverse silts. They are all homosporous and the spores are produced in enormous numbers. The spores do not have a perispore. The antheridia are always embedded. The first division wall of the antheridial initial sets apart an outer sterile **jacket cell** and an inner fertile primary **androgonial cell.** The antheridia produce large number of multiflagellate spermatozoids. Embryogeny may be **exoscopic** or **endoscopic** with or without suspensor. The gametophytes are long lived and are attacked by an endophytic fungus. They may be colourless; subterranean and saprophytic (***Ophioglossum, Botrychium***) or green aerial and photosynthetic (***Marattia***). The leaves are not circinately coiled in bud conditions (Ophioglossales) and may be stipulate or exstipulate.

The class includes two orders : (***i***) Ophioglossales and (***ii***) Marattiales.

Order : **Ophioglossales**

Family : **Ophioglossaceae**

The family includes 4 genera (***Ophioglossum***, ***Botrychium***, ***Helminthostachys*** and ***Rhizoglossum***) and seventy species (Willis, 1966) of terrestrial and herbaceous plants that have no fossil history.

The family is characterised by:

(*i*) The sporophytes are terrestrial, herbaceous and short in size;
(*ii*) The rhizome is short and fleshy ;
(*iii*) The roots are adventitious and bear adventitious buds in ***Ophioglossum*** ;
(*iv*) Leaves are not circinately coiled and exstipulate;
(*v*) The fertile leaves have a common petiole that bears a sterile undissected or dissected blade and fertile spike;
(*vi*) Sporangia are marginal;
(*vii*) Spore output is large and may be upto 2000 spores in a sporangium;
(*viii*) They are homosporous;
(*ix*) The prothalli are under ground and mycorrhizic
(*x*) The prothalli are monoecious.

Genera ***Ophioglossum***, ***Botrychium*** and ***Helminthostachys*** are discussed in some detail.

Marattia douglassi

Helminthostachys

Ophioglossum

Botrychium

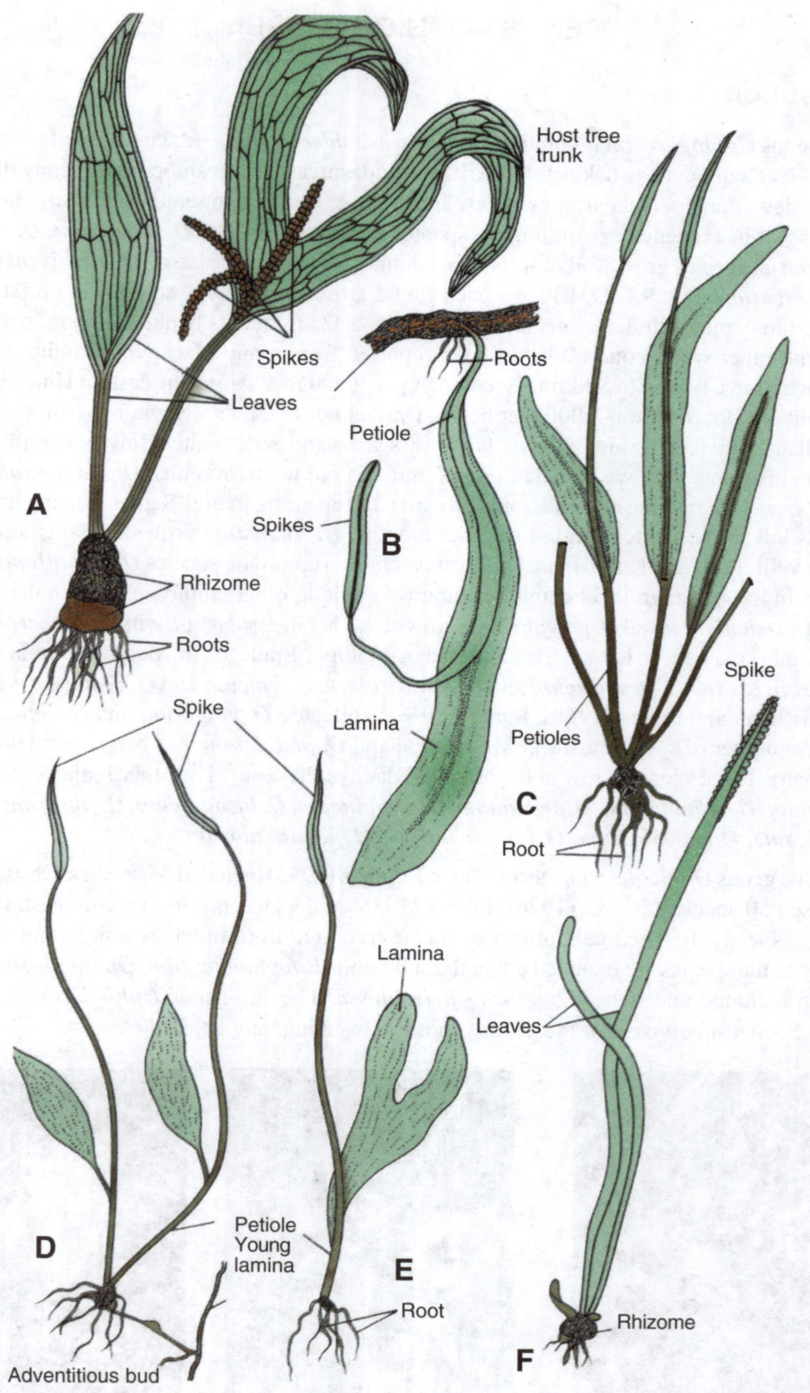

Fig. 9.1 (A–F). ***Ophioglossum.*** A. *O. palmatum*; B. *O. pendulum*; C. *O. aitchisini*; D. *O. vulgatum* bearing a germinated root bud; E. *O. vulgatum* with a divided leaf lamina; F. *O. simplex* with reduced rhjzome and two leaves that have extremely reduced lamina.

GENUS –OPHIOGLOSSUM L.

Distribution

Genus ***Ophioglossum*** is commonly known as ***Adder's tongue*** fern because of the resemblance of fertile leaf segment to snakes tongue. It is a widespread and cosmopolitan genus that achieves luxuriant development in the tropics. A few species occur in the temperate zone also. The sprophytes are terrestrial in existence except for two species that are epiphytic (***O. palmatum***, ***O. pendulum***.). The terrestrial species grow well in soils rich in humus. A few species, *e.g.*, ***O. aitchisoni*** (Fig. 9.1, C) and ***O. vulgatum*** (Fig. 9.1, D, E) have been found growing in clayey soils with a mixture of sand. They are the common Indian species. The latter species frequents banks of rivers, streams, pools, ponds, and other water courses. It has been reported from many places in N. India and occurs in abundance from Chamba to Sikkim, ascending up to 9,000 feet. Along the eastern Himalayas it grows luxuriantly at Mount Hattu, 2,000 feet below Darjeeling. In Chhota Nagpur it grows well on Paras Nath Hill at 2,500 feet. It flourishes well in rainy season and early winter (July-November) and again reappears in spring . It slows down in growth during cold winter months. ***Ophioglossum aitchisoni*** also grows well during the rainy season, and starts disappearing in mid-September and no trace of the plants is left in the winter and hot summer months. ***O. nudicaule*** grows in Bangalore, Belgaum, Palghat, Velli, Palai and Kumbanad. Of the two common epiphytic species ***O. pendulum*** (Fig. 9.1, B) grows in India. It is high level epiphyte and grows among other epiphytic ferns in the tropical rain forests. ***O. reticulatum*** and ***O. polyphyllum*** grows together in association with ***Anthoceros*** and mosses under bamboo shade in Keskal Bastar district, Madhya Pradesh. In Madras state at Courtallam, Tiruneleveli district ***O. gramineum*** and ***O. nudicaule*** var . ***typicum*** and ***O. costatum*** were collected by Balkrishnan and others (1960). Banerji (1961) collected ***O. vulgatum*** and ***O. reticulatum*** from Nepal. A number of species grow at Mount Abu and ***O. petiolatum***, has been recorded from Ajmer (Rajasthan). Three species grow at Pachmarhi (Madhya Pradesh). The other Indian species included ***O. costatum***, ***O. reticulatum***, ***O. petiolatum***, ***O. gramineum***, ***O. lusitanicum***, ***O. fibrosum***, ***O. capense***, ***O. nudicaule***, ***O. polyphyllum***, ***O. japonicum*** and ***O. pedunculosum.***

The genus ***Ophioglossum***, according to Reimers (1954) included 45 species. Chrisensen (1938) recognised 50 species. Bower (1926), Eames (1936) and Clausen (1938) recognised 43, 30 and 28 species, respectively. The total number of species reported from India are a little more than twelve. Majority of the species are included under the sub-genus ***Euophioglossum. Ophioglossum palmatum*** has been included under the sub-genus ***Cheiroglossa.*** The sub-genus ***Ophioderma*** includes three species. Such a subdivision of the genus is not followed in many texts.

Ophioglossum nudicaule

Ophioglossum vulgatum

Ophioglossum.-Adder's tongue fern

Ophioglossum polyphyllum · Ophioglossum · Ophioglossum reticulatum · Cheiroglossa · Ophioglossum costatum · Ophioderma

External features of the sporophyte

Rhizome

The rhizome (Fig. 9.1) is usually short, cylindrical, radially symmetrical and upright organ. In ***O. aitchisoni*** it is covered with a tuft of darkish brown persistent sheaths which are the remains of the bases of the preceding year. Such sheaths are also present in ***O. lusitanicum***, but are absent in ***O. vulgatum*** and ***O. moluccanum*** (Fig. 9.1, B). The rhizome is short and thick in ***O. palmatum.*** It is very much reduced in ***O. simplex*** (Fig. 9.1, F). The epiphytic ***O. pendulum*** has a creeping or a dorsiventral rhizome. The rhizome is generally unbranched but Petry (1914) has reported dichotomous branching.

Leaf

The sporophytes may be monophyllous or polyphyllous. ***Ophioglossum vulgatum*** (Fig. 9.2) and ***O. molucannum*** are ordinarily monophyllous, *i.e.*, only one leaf unfolds during one growing season, but two or even three leaves have also been reported to spring up in one season (Fig. 9.1, C, D, E). ***Ophioglossum aitchisonii*** and ***O. lusitanicum*** are polyphyllous (Fig. 9.1) and the number of leaves unfolding in a single season varies from two to five or six. The leaves are not circinately coiled in the bud condition. The slow growing temperate species unfold a single leaf after three years. The terminal bud of the rhizome, in these species is seen to enclose the developing leaf primordia of a few succeeding years.

The sterile unfolded leaf is divisible into a lower petiole and an upper green lamina. The lamina is simple and undivided in majority of the species and may vary in form from linear, lanceolate, oval, ovate, oblong to elliptical. The leaf apices may be acute, obtuse, emarginate or acute mucronate. It is herbaceous, leathery and sometimes fleshy in texture. The surface is smooth. The fertile leaves bear a short or a long and an unbranched spike that arises from the base of the lamina or a little below it (Fig. 9.1). In ***O. simplex*** the sterile lamina is lacking and the fertile spike is the only conspicuous part of the above ground portion of the leaf (Fig. 9.1, F). The leaf is also called the frond. The lamina is well developed and palmetly lobed in ***O. palmatum*** (Fig. 9.1, A). This species is characterised by the presence of several spikes that are arranged in two rows along the margins of the basal portion of the palmate lamina and the distal portion of the petiole. The leaf reaches its maximum size in the epiphytic ***O. pendulum*** (about 1.5 metres), where it has short petiole that merges insensibly into a branching and narrow, strap-shaped lamina (Fig. 9.1, B). The venation is reticulate but a distinct midrib (Fig. 9.1, B) is lacking.

Ophioglossum palmatum

Ophioglossum pendulum

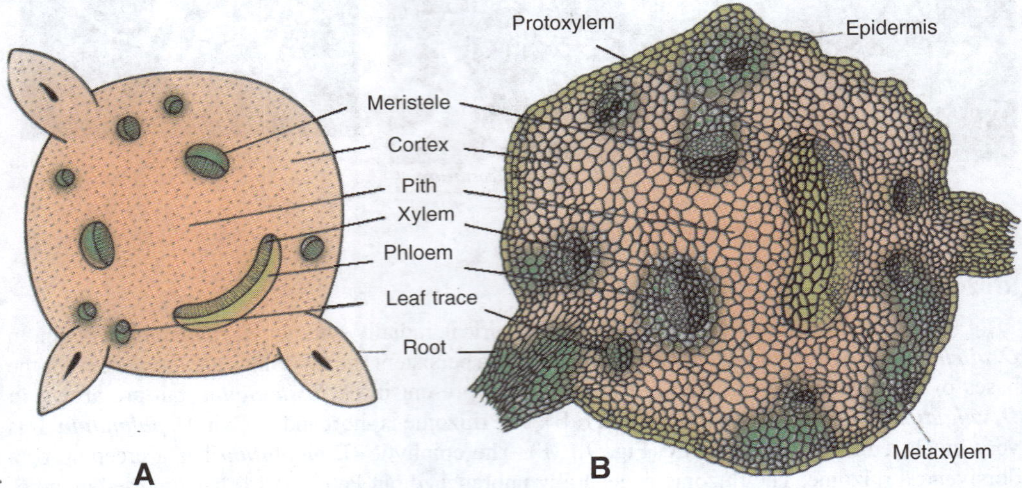

Fig. 9.2 (A–B); *T.S. rhizome of* ***O. vulgatum*** A. *Diagrammatic*; B. *Cellular*

Root

The adventitious roots arise from the rhizome and may be branched or unbranched. They do not bear root hair and are heavily infected with an endophytic fungus. It has been observed that only one root arises opposite the point of origin of a leaf or opposite the scars left by the old and fallen leaves. The roots of many plants bear adventitious buds that grow into new plants and serve as means of vegetative propagation (Fig. 9.1, D).

ANATOMY

A. Rhizome

The internal structure of the rhizome has been studied by a number of workers in different species (Vasishta, 1927; Petry, 1910; Lang, 1912; Maheshwari and Singh, 1934; Campbell,1911; Bower,1911; Boodle, 1899; and Gewritz and Fahn, 1960). A transverse section of the rhizome reveals an irregular outer contour. This is due to the presence of numerous leaf bases and adventitious root bases. There is a distinct outer layer called the epidermis (Fig. 9.2). The cortex is composed of many layers of thin walled cells that store abundant starch (Fig. 9.2). The cells are mostly oval or elliposidal in outline and enclose small intercellular spaces. The older parts of the rhizome in some species reveal the presence of periderm which is not referable in origin to any cork cambium. Endodermis and pericycle are absent in majority of the species. Vasishta (1927) reported the presence of a distinct outer endodermis in the basal portions of the rhizomes in ***O. aitchisoni*** and ***O. vulgatum***. Petry (1910) reported it in ***O. pendulum***. Both outer and inner endodermis layers have been observed on the first formed portions of the stem in ***Ophioglossum*** from Ceylon (now Sri Lanka) (Lang 1912).

Fig. 9.3. Stereo-diagram of stele of *O. lusitanicum.* (After Gewirtz and Fahn)

Stele

In ***Ophioglossum vulgatum*** and ***O. lusitanicum*** the young rhizome reveals a protostele. There is a central xylem patch surrounded by two to four layers of xylem parenchyma, a single layer of phloem elements and a definite endodermis which is present in ***O. vulgatum*** and absent in ***O. lusitanicum***. The endodermis possess distinct casparian strips. In the mature rhizomes the same structure protostele is shown by the basal portion (first formed) of the rhizome. A little above, the protostele becomes medullated, *i.e.*, a pith appears in the centre. This is also called **ectophloic siphonostele.** At this stage the stele in ***O. vulgatum,*** consists of an outer endodermis, a single layer of phloem, xylem and a central pith. The endodermis, in this region of the rhizome, is considered to be an extension of the root endodermis. The continuity of the stele is not broken at this stage, by the gaps. Sections cut a little above the basal part reveal the absence of endodermis and perforation of the siphonostele by one or more leaf gaps . This perforated ectophloic siphonostele is called the **ectophloic solenostele.** There are two unequal bundles. The large one is semicircular and the smaller may be oval or ellipsoidal in shape. Each one of them has phloem facing the epidermis and the xylem facing the pith (Collateral). The xylem is **endarch.** Higher up the ectophloic solenostele changes to a dictyostele (Fig. 9.2) due to the appearance of overlapping leaf gaps. The individual bundles composing a **dictyostele** are called the **meristeles.** Gewritz and Fahn (1900) studied the vasculature of ***O. lusitanicum*** (Fig. 9.3) and revealed the same sequence of protostele, siphonostele, ectophloic solenostele and dictyostele from below upwards. They also reported that pith is intrastelar in origin. The mature rhizomes of ***O. reticulatum*** and ***O. aitchisoni*** also reveal the same variations in their stelar organisations from base upwards. In the latter species two leaf traces depart from the leaf gap (Vasishta, 1927). In this respect it resembles ***O. palmatum***, ***O. pendulum*** and ***O. fibrosum***.

Ophioglossum lusitanicum

The mature rhizome ultimately has a dictyostele composed of a few meristeles. Each meristele is characteristic in lacking an endodermal layer. It has an outer patch of phloem which does not surround the xylem on all sides. So the organisation of the meristele is collateral and not concentric. The phloem consists of a single layer of sieve elements in ***O. aitchisonii*** and ***O. pendulum***. In other species it may be four or five layers thick. Sharma and Singh (1984) reported the presence of cauline stele in ***O. costatum***. Next to the phloem are a few layers of xylem parenchyma. Xylem faces the pith and is endarch. It is composed of tracheids the metaxylen trachieds possess scalariform pits. The protoxylem shows variations in the lignification of its walls. Early formed protoxylem trachieds are usually annular. Later the thickening pattern may change to reticulate type. Circular bordered pits may also be present in the mature protoxylem tracheids. There is no secondary growth in ***Ophioglossum***. Khandewal and Goswami (1977) reported the presence of periderm in the roots and rhizome of ***Ophioglossum***. A functional cork cambium or phellogen has been reported.

The pith is made up of starch filled, thin-walled cells. The cells are smaller in size in the vicinity of the xylem elements and go on increasing in size towards the centre. Scattered tracheids have also been observed in the pith of ***O. aitchisonii*** (Vasishta, 1927) and ***O. pendulum*** (Petry, 1914). Interxylary origin of pith has been reported in ***O. reticulatum*** (Sharma and Singh 1984).

Growing point

The rhizome grows by means of an apical cell. A median longitudinal section through the apical region shows the apex to be hidden among the bases of the young leaves. There is a cavity which encloses the sunken apex, in the centre of which lies the apical cell. In longitudinal section the apical cell, in ***O. aitchisonii*** and ***O. vulgatum*** is approximately tetrahedral in form with four cutting faces and apex slightly truncate. The segmentation, as followed by Vasishta (1927) in ***O. aitchisonii***, is regularly along the three lateral sides. It is doubtful whether the segments on the basal side are so regular cut off as on the three other sides. The large lateral segments retain their identity for a considerable time. The first division in each segment is transverse. Thus the segment is divided into the lower smaller and the upper larger cell. The lower cell divides again by a longitudinal wall into two. After that the divisions are irregular and difficult to follow.

B. The leaf

(*a*) Petiole (Fig. 9.4). The petiole has a distinct epidermis composed of a single layer of slightly thick-walled cells. The ground tissue of the petiole has a few layers of chlorenchymatous cells below the epidermis. The rest of it is made up of parenchymatous cells, that may enclose small intercellular spaces. The vascular bundles are embedded in the ground tissue.

The course of vascular bundles in the petiole of ***O. vulgatum*** is described below :—

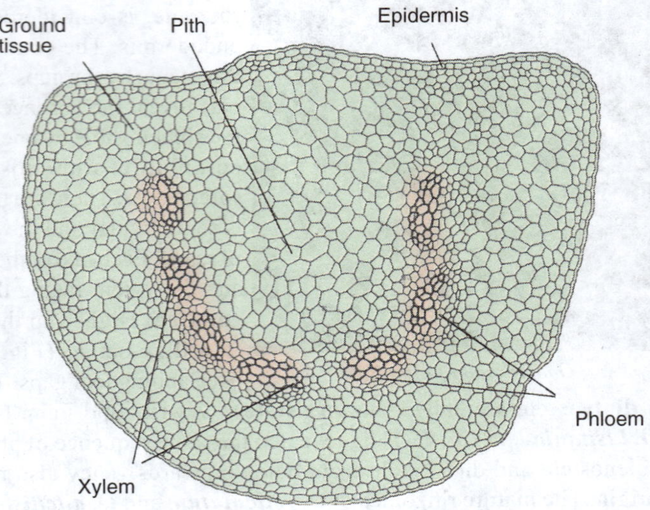

Fig. 9.4. T.S. petiole of *O. vulgatum*.

At the base of the petiole there is a single collateral and endarch vascular bundle. It soon divides into three bundles which arrange themselves as the angles of a triangle. In the upper region of the petiole there are 5 to 7 bundles arranged in a semicircle in the sterile leaf (Fig. 9.4). In the fertile leaf there are 7–10 bundles in the upper portion on the petiole. They are arranged in the form of a semi-circle or an ellipse. Three adaxial bundles go to form the vascular system of the spike, and the rest enter the lamina. Each vascular bundle of the petiole is more or less circular in outline. The xylem mass is flattened. The protoxylem elements are situated on the inner surface (endarch). The phloem is on the outer face and is separated from the xylem tracheids by xylem parenchyma. Thus the bundles are collateral and endarch. There is no distinct bundle sheath.

In ***O. aichisoni*** there are two vascular bundles at the base of the petiole. Their xylem faces the adaxial surface. Both the bundles divide into four and as we go upwards the bundles divides repeatedly to form large number of vascular bundles in the upper region of the petiole. In the sterile leaves the bundles arrange themselves in the form of a 'U' open on the adaxial side. In the fertile leaves the bundles in the upper part of the petiole, are arranged in a circle. At the base of the lamina they arrange themselves into two series, the inner and the outer. The inner series, consisting of three or four bundles, constitutes the vascular system of spike. The inner series, consisting of about 7–9 bundles forms the vascular system of the lamina, where they are arranged in a flattened arc.

(*b*) Lamina (Fig. 9.5). It consists of a distinct epidermal layer which is present on both the dorsal and the ventral surfaces of leaf. Its continuity is interrupted at intervals by large stomata which are more numerous on the lower epidermis than on the upper epidermis in ***O. aitchisoni*** and ***O. reticulatum***. In ***O. vulgatum*** the stomata are almost equal in number on both the epidermal layers (Fig. 9.5). The mesophyll is not distinguished into palisade and spongy parenchyma (Fig. 9.5). It consists of loosely arranged thin-walled and chlorenchymatous cells. The cells are almost rounded or oval in outline and enclose small or large intercellular spaces. The **mesophyll** is traversed by numerous delicate veins, which do not form any projections on the surfaces of the lamina. The structure and outline of each bundle is similar. They are collateral and endarch. A more or less distinct bundle sheath is recognisable around each vascular bundle. The phloem faces lower epidermis.

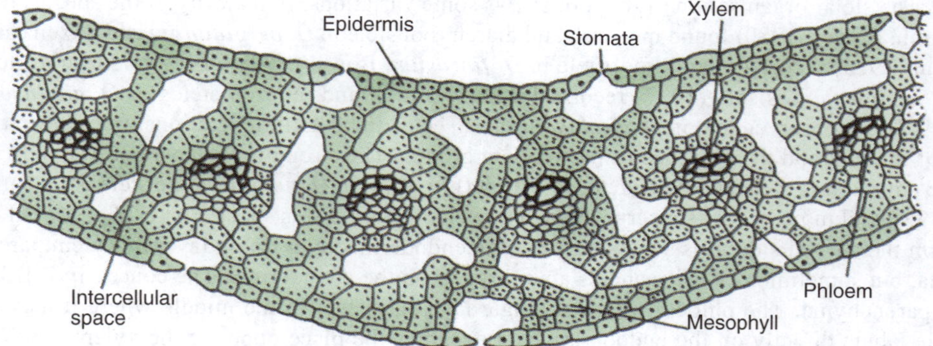

Fig. 9.5. V. S. lamina of *O. vulgatum*.

(c) The Root (Fig. 9.6, 9.7). In majority of the species a distinct epidermal layer is descernible throughout the life of the root. In such cases the outer walls of the epidermal cells become thickened and suberised. In ***O. fibrosum*** the epidermis, in the older roots is replaced by the next cortical layer or the **exodermis.** The cortex is wide. Two distinct regions can be recognised : (*i*) outer cortex composed of 3—6 layers or more of closely packed and thin-walled cells filled with the hyphae of an endophytic and mycorrhizic fungus and (*ii*) the inner cortex made up of several layers of large, thin-walled, starch-filled and oval or rounded cells that enclose small intercellular spaces (Fig. 9.6). There is a distinct endodermis. Pericycle is absent.

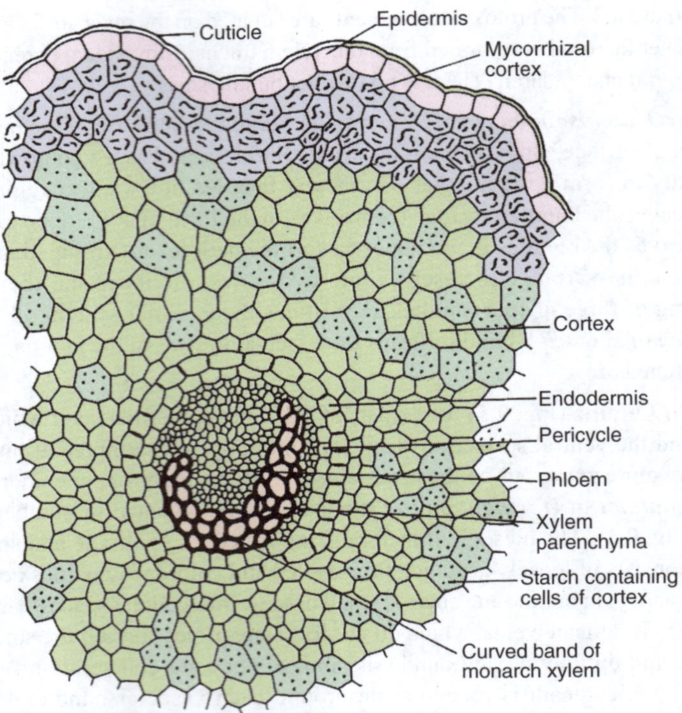

Fig. 9.6. T. S. portion of root of *O. vulgatum* showing internal structure. The stele is monarch.

The stelar organisation of the root shows some variations. In majority of the species it is monarch. Bower (1980) found monarch and diarch root stele in ***O. berginanum*** and Saxena and Mathur (1925) reported diarch root stele in ***O. fibrosum***. Bower (1908) also found triarch stele in ***O. decipiens***. Vasishta (1927) recorded diarch, triarch and tetrarch steles in ***O. aitchisoni*** (Fig. 9.7). He observed that in ***O. aitchisoni*** there seems to be definite relationship between the size of the root and the presence of protoxylem groups. The strong roots have got triarch (Fig. 9.7, B) or tetrarch stele and delicate ones diarch (Fig. 9.7). In ***O. vulgatum*** the stele is monarch (Fig. 9.6) and more or less saucer-shaped. The xylem mass in this species takes the form of an arc. On the curved side, it is separated from the endodermis by a single layer of xylem parenchyma, but sometimes a few tracheids abut directly on the endodermis. The concavity is filled with parenchyma. The phloem which is single layered except in the middle where it may be double, abuts directly on the endodermis and occupies the place opposite the xylem arc. The roots lack secondary growth.

Growth of the root

The root grows by means of a tetrahedral apical cell. It cuts off segments on all the three sides, the outer of which forms the root cap. The root cap is not formed only from the outer segment, but cells are also contributed to it by the lateral segments. The divisions in the lateral segment, in ***O. aitchisoni***, are irregular and the tissue here consists of very irregularly arranged cells.

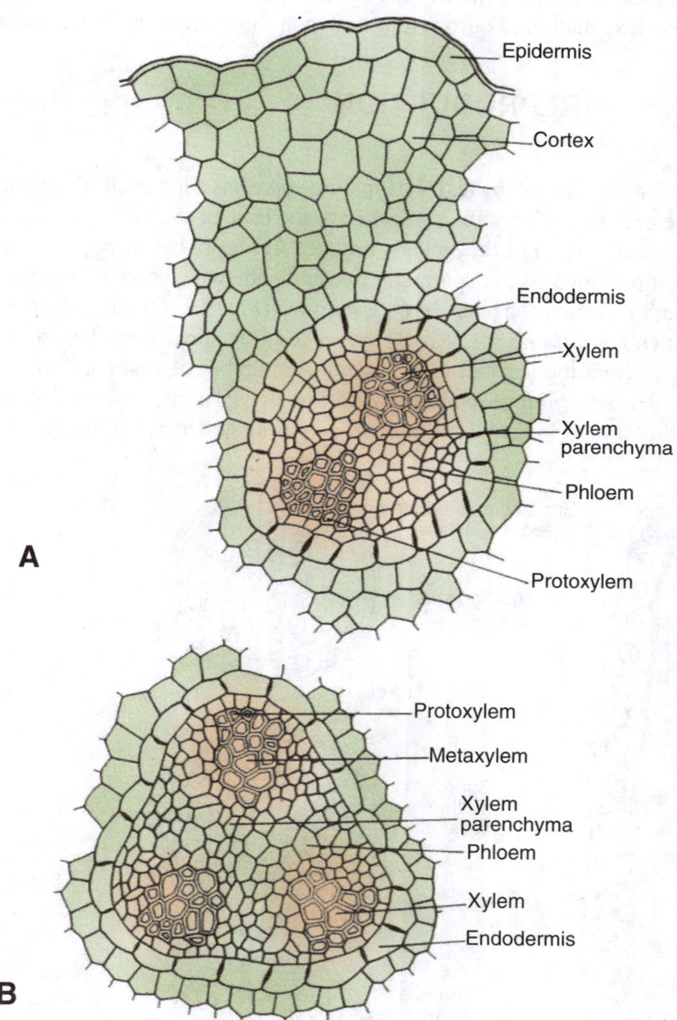

Fig. 9.7. (A–B). *O. aitchisoni.* **A.** T. S. portion of a diarch root; **B.** T.S. portion of a triarch root.

Vegetative propagation

Most of the species of ***Ophioglossum*** reproduce vegetatively by the formation of adventitious buds on the roots. Vasishta (1927) reported their presence in ***O. aitchisoni*** and ***O. vulgatum***. Later Wardlaw (1953) induced their formation and development under natural and experimental conditions, in ***O. vulgatum***. Wardlaw (1953) induced shoot buds on the entire and decapitated roots of ***O. vulgatum***. The majority of such buds develop from the cortical cells of the roots.

Goebel (1902) suggested that these adventitious buds are modified lateral roots. Lateral roots are absent in ***O. vulgatum*** and hence according to Wardlaw, there is justification of such a hypothesis. Peterson and Cutter (1967) have demonstrated that decapitated roots of ***O. petiolatum*** bear larger number of adventitious shoot buds.

Rostowzew (1891) noted the appearance of lateral buds on the rhizome of ***O. vulgatum***. Such buds are of rare occurrence under natural conditions. Later (1953) Wardlaw induced the formation of buds on the decapitated shoots of ***O. vulgatum***. They have an endogenous origin either from the pith or the cortex. Such buds grow into new plant and are therefore a means of vegetative propagation.

REPRODUCTION

The fertile spike

The spike is a specialised segment of the leaf. The mature spike has a lower sterile cylindrical portion called the stalk or the peduncle. The stalk is attached to the base of the lamina of the sterile leaf on its ventral side. The upper portion of the spike is broader and bears two rows of embedded sporangia on either side. The tip of the spike is sporangia free and somewhat conical in shape (Fig. 9.8). The spike varies in length in various species. In ***O. aitchisoni*** (Fig. 9.1, C) the spike is longer than the sterile leaf blade. In ***O. pendulum*** (Fig. 9.1, B) the spikes are shorter than the leaf blades. Usually one fertile spike arises from the leaf but in ***O. palmatum*** number of small spikes arise on either side of the petiole (Fig. 9.1, A). ***Ophioglossum simplex*** (Fig. 9.1, F) is characteristic in lacking a sterile lamina. The leaf, in this case, is long and cylindrical and terminated in a sporangiferous area.

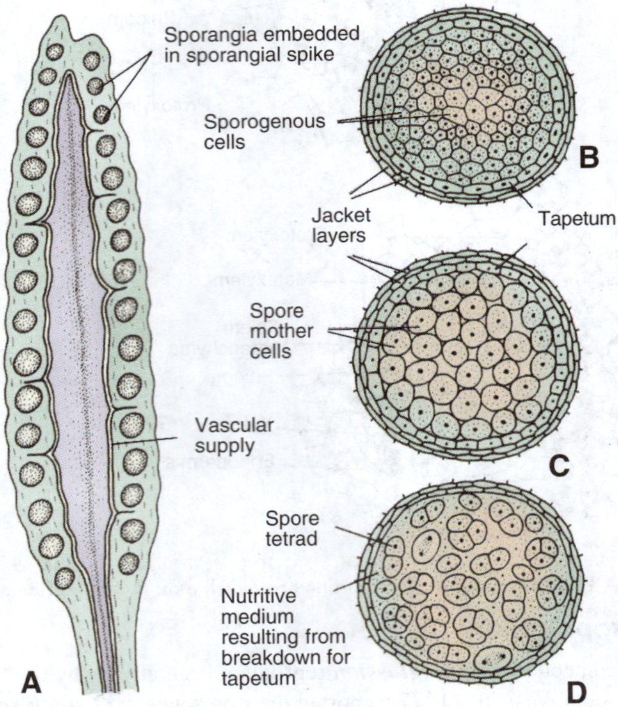

Fig. 9.8 (A–D). ***Ophioglossum vulgatum.*** A. Diagrammatic V.S. through a young spike showing sporangia free tip, the sporangia, the penduncle and the vascular supply; B—D. Diagrammatic cross sections at various levels of the spike.

The sporangia free tip of the spike consists of parenchyma and vascular tissue. It no longer contributes to the growth in length of the spike. The sporangia originate during the early ontogeny of the spike when it is very small in size. Peterson and Elizabeth Cutter (1967) attribute the growth in length of the spike due to an intercalary meristem present below the sporangial area. They removed the sporangia free tip of the spike and found that it did not affect the growth of the stalk or the peduncle.

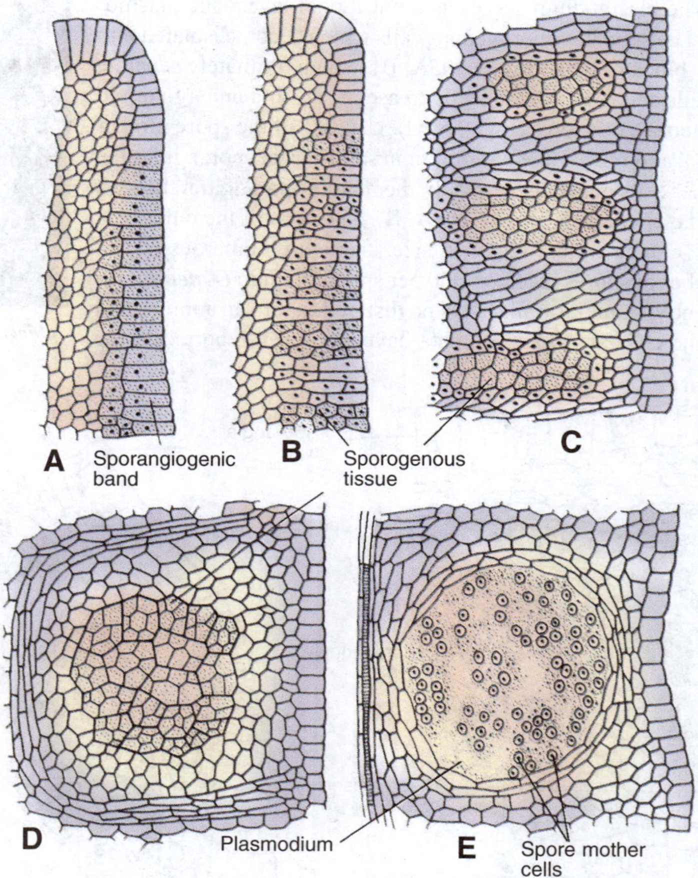

Fig. 9.9. (A—E). Various stages in the development of sporangium in ***Ophioglossum vulgatum.*** **A.** Section of a young spike showing the sporangiogenic band; **B.** Section of a spike showing fertile sporogenous cells; C—E. Later stages of development. (A,D, E, after Bower)

The peduncle continues in the sporangial region as a central axis and terminates in a small sterile and conical tip. The sporangia are borne in two-distinct and longitudinal rows on either side of the axis (Fig. 9.8). The sporangia are completely embedded in the tissue of the spike and their position is indicated by distinct transverse furrows of fertile tissue. These furrows alternate with the sporangia .The spike receives its vascular supply from the petiole. The vascular strands run upward and anastomose during their upward course. They give out lateral strands that enter the sterile mass between the sporangia (Fig. 9.8). These lateral strands then bend towards the base of the sporangia and constitute their vascular supply. The number of sporangia embedded in each spike varies from six to twenty or even more.

Structure of the sporangium

The completely embedded sporangium may by spherical or oval in shape. It has a several layered thick wall. The innermost wall layer functions as a **tapetum.** The cavity of the sporangium is filled with a compact mass of **spore mother cells** (Fig. 9.9, D). The tapetal cells disorganise and their cytoplasm forms a continuous mass of tapetal plasmodium with the several tapetal nuclei embedded in it (Fig. 9.9, E). These free nuclei have been reported to increase in number by fragmentation

(Bower, 1908). The plasmodium serves as a nutritional layer. The plasmodium penetrates in between the sporogenous cells which become isolated into large and smaller block of cells (Fig. 9.10, A, B), which ultimately separate into individual cells and round off. They increase in size and undergo tetrad division to form tetrads of spores (Fig. 9.8, D). Usually all the spore mother cells are functional and form spore tetrads but in some larger sporangia some of the spore mother cells may degenerate. In the mature sporangium the wall consists of 3—6 layers of cells (Fig. 9.10, A, B). The cells of the outermost layer are radially elongated and larger in size. There are numerous spores within . The number of spores reach 15,000 per sporangium in ***O. pendulum*** (Bower, 1935). The sporangial wall shows no distinction into an annulus and a stomium. All the cells are alike. Stomata have also been reported in the sporangial wall.

Ophioglossum pendulum

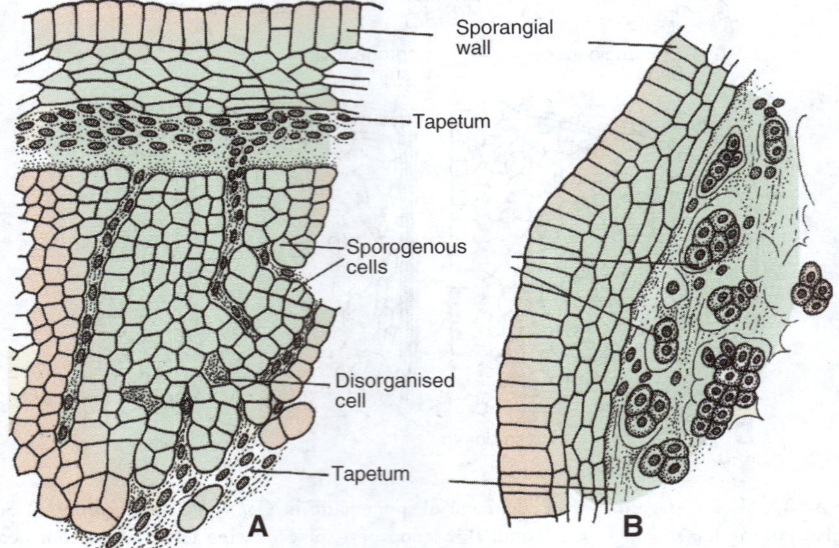

Fig. 9.10. (A–B) Section through mature sporangia of *O. vulgatum* showing penetration of plasmodium in between the spore mother cells thus breaking them into larger (A) and smaller groups or blocks (B). (After Bower)

Dehiscence

Due to the absence of any specialisation in the sporangial wall there is no special mechanism concerned in the dehiscence of sporangial wall. A transverse slit appears in the sporangial wall and causes its dehiscence. The shrinking of the masses of sterile tissue in the spike may cause the dehiscence of the sporangium. After the maturation of the spores the sterile cells of the spike lose water and shrink and set up a strain that causes some weaker cells of sporangial wall to separate and form a transverse slit .

Development of the spike and sporangia

Bower (1896) studied the developments of spike and sporangia in ***Ophioglossum***. The spike originates as a small and a conical outgrowth of meristematic tissue. This outgrowth arises at the base of the lamina on the ventral side. Soon a pyramid-shaped apical cells appears in this meristematic tissue. This apical cell cuts off segments along its four sides and its activity leads to the formation of four quardrants, two of which lie parallel to the surface of the sterile lamina and the other two in a plane perpendicular to it . Further growth and division of the segments of the apical cell lead to the formation of a mound of tissue with a well defined epidermal layer. The epidermal layer formed along the two quadrants that ate oriented parallel to the surface of the leaf lamina shows differentiation of cells into two vertical strips of cells. These strips are called the **sporangiogenic bands** (Fig. 9.9,A) and are several cells in height and two to three cells in breadth. The sporangiogenic band does not extend throughout the length of the young spike. It occupies only the portion which is destined to bear the sporangia. The apex of the spike and its basal portions remain sterile. The latter forms the peduncle.

The cells of the sporangiogenic band act as initials of sporangia. So the two bands produce two marginal rows of sporangia. Each cell of the sporangiogenic band undergoes periclinal divisions so that the band becomes up to four cells thick. Some outer cells also divide anticlinally and contribute to the portion of sporangial wall. The inner cells, at certain places along the length of the band, divide in more than one plane and form groups of isolated cells called the **archesporial cells** (Fig. 9.9, B, C). Their cytoplasmic contents are denser and they possess big nuclei and stain darker than other cells. Like this alternate bands of sterile and archesporial cells become distinguished in the hypodermal region of each sporangiogenic band (Fig. 9.9, B, C). The positions of groups of archesporial cell mark the positions of future-sporangia. The cells of the sporangiogenic band external to each archesporial group divide preclinically and anticlinally and contribute to the portion of sporangial wall. The archesporial cells divide repeatedly to form a large group of sporogenous cells (Fig. 9.9, C). The cells of the spike surrounding each such group also contribute to the sporangial wall. They join the external wall layers (formed by the sporangiogenic band) and complete the sporangial wall. So the jacket or wall layers of the sporangia are derived partly from a sporangiogenic band and partly from the surrounding steriler cells of the spike. The wall is several cells in thickness. Its innermost layers form the ill-defined tapetal layers, which later disorganise to form the plasmodial fluid. The sporogenous cells divide further and form the spore mother cells. The spore mother cells later isolate into groups (Fig. 9.10, B, C) and finally separate as individual round cells which undergo meiosis and form tetrahedral tetrads of haploid spores. (Fig. 9.8, D).

Morphological nature of the spike

The following views have been expressed by various authors to explain the morphological nature of the spike in ***Ophioglossum***.

1. Bower (1896,1935) suggested that the simple spike of ***Ophioglossum*** originated as a result of condensation and lateral concrescence of the branched spikes of genera like ***Botrychium***. He believed that the sporangia of the fertile spike are actually synangia derived by the partition or "fission" of the continuous sporangia. He believed the two marginal rows of sporangia to be two continuous sporangia which have become partitioned to form two rows of sporangia. Such an assumption is not supported by the course of vascular supply from the petiole to the sporangia. Bower later (1926) abandoned his own hypothesis and agreed to a greater extent with Zimmermann (1930) in considering the spike to be a fertile lobe of the leaf.

2. Roeper (1859) considered the leaf of ***Ophioglossum*** to be derived from a compound leaf whose two basal pinnae or leaflets became laterally fused to form the spike. The remaining pinnae fused to form a sterile blade. Such an assumption is supported by the nature of the vascular supply to the spike. Goebel (1915) regarded the spike to be a modified single pinna of a leaf. Roeper's hypothesis was later upheld by Chrysler (1910) and Bower (1926).

3. Zimmermann (1930) suggested that the fertile spike and the sterile lamina are the two dichotomies of a shoot. The former is meant for reproduction and the latter for photosynthetic functions.

The vascular supply from the petiole to the spike and the sterile leaf blade, no doubt, suggest that the spike is a modified segment of the leaf. A noteworthy feature in this connection is the growth of the spike. After the inception of the sporangiferous region the spike ceased to grow by means of an apical cell. It seems to elongate by the activity of an intercalary meristem (Peterson and Cutter, 1967) situated at the base of the sporangiferous region.

GAMETOPHYTIC GENERATION

Ophioglossum is homosporous and the spores germinate to give rise to bisexual prothalli or the gametophytes.

The spores

They are produced in large numbers and are visible as small, yellow, dust-like particles. They are tetrahedral in shape and possess a two layered thick wall. The outer wall is called the **exine** or the **exosporium** and is slightly thick and sculptured. The inner wall is thin and delicate and is called the **endosporium** or the **intine**. The cytoplasm is granular and stores food in the form of oil droplets. The nucleus is centrally located. The young spores are green, but they lose this colour at maturity. Devi (1980) studied the spores of eleven species of ***Ophioglossum***. Khandelwel (1986) studied the spores of eleven species of ***Ophioglossum*** and reported the presence of three types of surface sculptures i.e. reticulate, verrulate and scabrate. In ***O. eliminatum*** the presence of both alete and trilete spores has been reported. Kremp (1967) reported the presence of monolete and trilete spores in the same sporangium.

Germination of the spore

The germination of spores of some species of ***Ophioglossum*** has been studied under experimental conditions. The full details of germination leading

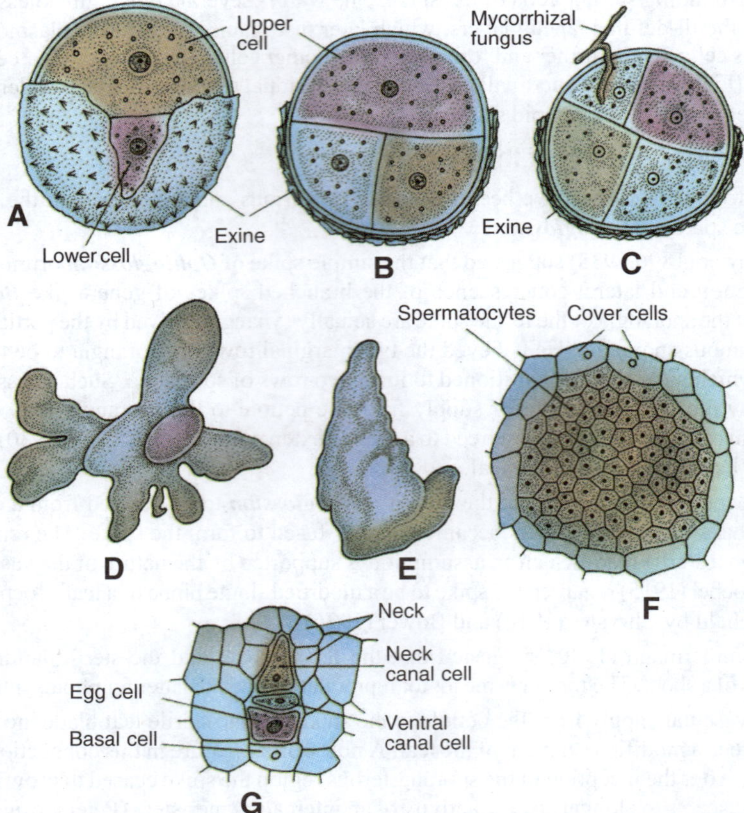

Fig. 9.11 (A—G) ***Ophioglossum pendulum.*** A. Two-celled stage; B. Three-celled stage; C. Four-celled stage. Also note the entering mycorrhizal fungus; D. Mature prothallus of *O. pendulum* as figured by Lang (1902); E. Mature prothallus of *O. pendulum* as figured by Smith (1955); F. A mature antheridium; G. A mature archegonium of *O. pendulum.* (A—C after Campbell).

to the formation of a mature prothallus are lacking. The time taken by the spores to germinate varies from a few days to a few months or even a year or two. Campbell (1907, 1911) studied the earlier stages of spore germination in ***O. pendulum*** (Fig. 9.11, A—C). During germination the spores absorb water and swell The exine splits along the triradiate mark. The spore contents enclosed within the intine emerge slightly from ruptured exine. The spore divides into two nearly equal cells by a transverse wall (Fig. 9.11, A). The cell below the triradiate mark is the upper cell. The lower or the basal cell divides next by a vertical wall (Fig. 9.11, B) forming two basal cells. The divisions beyond the three-celled stage have not been properly studied. After the three-celled stage, further divisions were observed mostly in the lower cell. The upper cell divided only occasionally. Campbell (1907) observed that the mycorrhizal fungus enters the young prothallus in the four-celled stage (Fig. 9.11, C). The entry is effected through the lower cell. The thirteen-celled gametophytes of ***O. pendulum*** were more or less spherical or globose in shape and did not bear rhizoids. Nothing is known about the establishment of an apical cell. Lang (1902) collected the prothalli of ***O. pendulum***. She found that the young prothalli were almost spherical in shape and unbranched. A symmetrical **apical meristem** is reported at later stages of development (Bierhorst, 1971). The older prothalli were, however, much branched (Fig. 9.11, D).

Structure of the mature prothallus

The prothalli of ***Ophioglossum*** are subterranean and may be hyaline, brownish, light greyish or yellowish in colour. They are essentially saprophytic and are always infected partly or wholly by a mycorrhizic fungus. They usually lack rhizoids. The size of the prothallus varies with species. It usually ranges between 4 to 12 millimetres. The prothalli in most species are perennial, but in ***O. moluccanum*** they are annual. The shape of the prothallus also varies. It may be roughly cylindrical or sometimes irregularly lobed and in some cases conical. The prothallus in the common ***O. vulgatum*** is more or less cylindrical (Fig. 9.12, A, B) and is usually unbranched but branched and massive prothalli have also been reported by Bruchmann (Fig,. 9.12, B). The lower portion of the prothallus is massive and almost bulbous and rounded. From this lower part arises an upright and cylindrical region that grows straight by means of a tetrahedral apical cell. This region may bear adventitious branches. The apical region or the upright cylindrical portion is free from the mycorrhizic fungus. It is also absent in the outer layers of cells in the lower portions. The endophytic fungus is restricted to the inner layers of cells in the lower and older portion of the prothallus. The prothallus also does not exhibit any internal cellular differentiation except that the inner cells are slightly elongated and filled with starch grains. A peripheral **thickening meristem** has been reported near the older end of gametophytes in ***O. vulgatum*** (Bierhorst, 1971). The sex organs are distributed all over the upright, cylindrical portion. Their arrangement is acropetalous but have no regular sequence of distribution. They are scattered all over the upper portion and both the antheridia and the archegonia are intermingled.

Ophioglossum palmatum

M.R. Mesler, (1972) studied the gametophytes of ***O. palmatum*** and found that the mature gametophytes are cylindrical but much branched. The youngest portions of the prothalli are circular in cross section whereas the older ones are flattened. He attributed it to the activity of a **thickening meristem.** The branches are oriented in three dimensions. He discovered that older portions of the prothallus die leaving the number of branches free and radiating in several directions. Each branch grows into a full fledged prothallus. This is a case of vegetative propagation of the prothallus by fragmentation. The prothallus is infected by endophytic fungus. Unicellular and papilla-like rhizoids sparsely cover the prothalli of this species.

Pant and Misra (1973) have studied in detail the gametophytes of six Indian species (***O. vulgatum***, ***O. reticulatum***, ***O. petiolatum***, ***O. nudicaule***, ***O. costatum*** and ***O. gramineum***). In all species they found the prothalli to be cylindrical and in some cases branched. They are subterranean but their apices may project above soil and turn green. They did not observe stomata but found rhizoids in all the species. Some prothalli were monoecious whereas others were dioecious. In all of them sex organs develop acropetally from superficial initials. They reported the presence of multi-cellular in ***O. nudicaule.***

Sex organs

The antheridia and archegonia are found on the same prothallus (monoecious) and usually occur intermingled (Fig. 9.12, A). They develop a little below the growing apex and arise in an acropetalous manner.

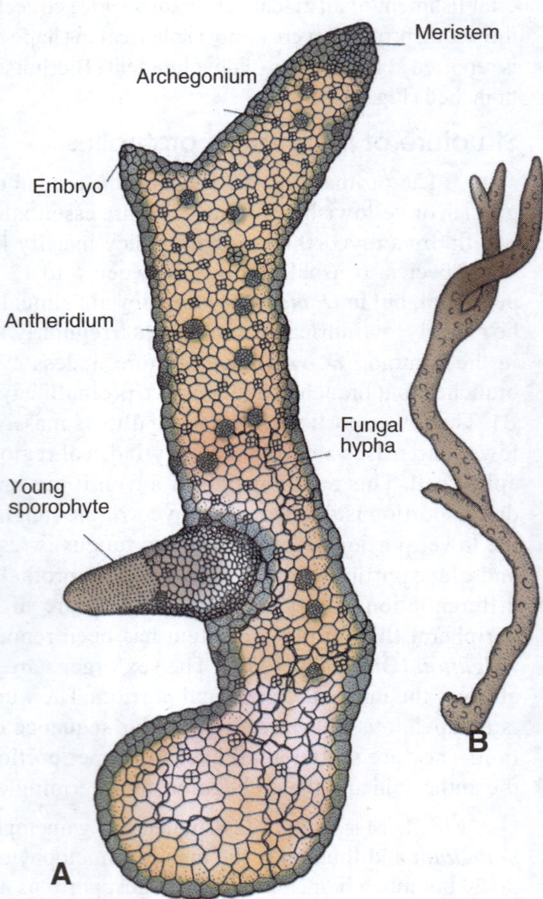

Fig. 9.12. (A—B). *Ophioglossum vulgatum*. A. A mature gametophyte bearing sex organs. A young sporophyte is also seen on the side; B. The same as A, showing branching. (After Bruchman)

Antheridium

(a) **Structure** (Fig. 9.13, F). The antheridia are generally completely embedded in the prothallus tissue or may be sightly projecting (***O. nudicaule***). They are usually spherical or more or less ovoid in shape at maturity. The antheridial wall is one layer of cells in thickness. Nozu (1961) described a two-layered thick wall in ***O. vulgatum***. Such a condition is not common. The wall encloses a large number of androcytes which later metamorphose into spermatozoids. The number of spermatozoids in each antheridium may be up to thousands. A triangular cap cell or opercular cell is visible in the antheridial wall when viewed from surface. This cell disintegrates at maturity and leaves a small pore through which the spermatozoids escape.

O. nudicaule

Presence of mycorrhizal fungal hyphae in the wall cells of antheridia has been reported in ***O. nudicaule*** (Pant, *et al*, 1984).

The spermatozoids are multiflagellate (Fig. 9.13, G) and those of ***O. pendulum*** are considered to be the largest known among the living vascular cryptogams. The anterior narrow and coiled region of the spermatozoid is wholly composed of the blepharoplast and bears flagella. The posterior portion is broad and loosely coiled. It is nuclear in origin. A distinct vesicle is also discernible at the posterior region.

(*b*) Development (Fig. 9.13, A—E). A superficial cell of the prothallus, a little distance behind the apex, functions as the antheridial initial. It divides by a periclinal wall into an outer cell called **primary cover cell** or **primary wall cell.** The outer cell gives rise to a portion of the antheridial wall and an **opercular cell.** The inner cell divides to give rise to the **androgonial cells.**

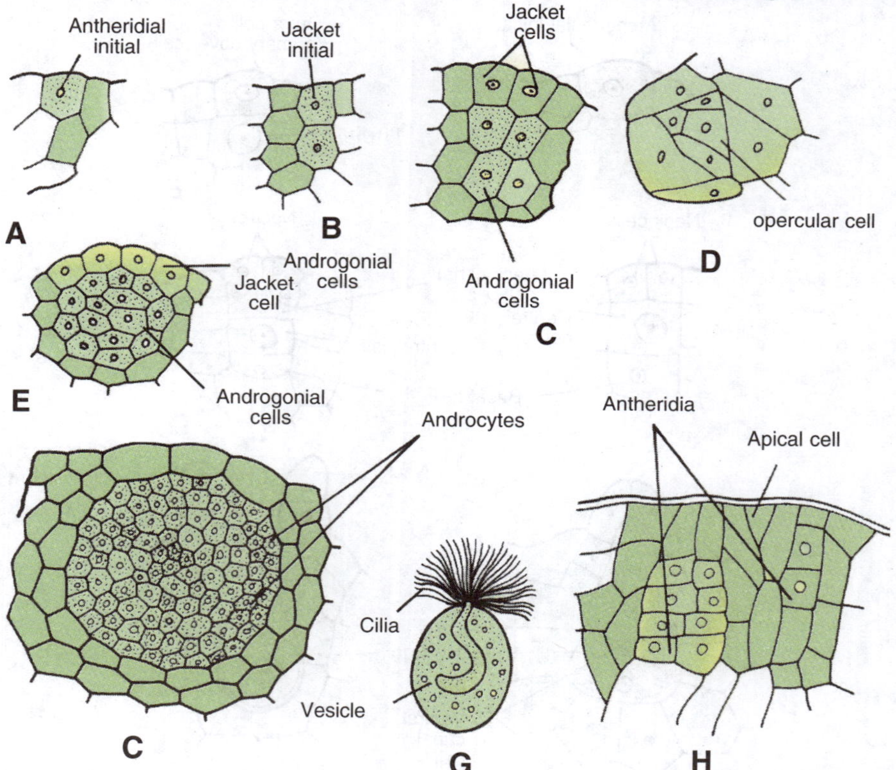

Fig. 9.13 (A–H). *Ophioglossum vulgatum.* (A-C; E-G). Stages in the development of antheridium; *Ophioglossum pendulum* D; surface view of a mature antheridium showing opercular cell H; L.S. of gametophyte apex showing two young antheridia and an apical cell.

The outer cell divides into equal cells by means of an anticlinal wall. A second wall is laid down in such a manner so as to establish a **triangular apical** cell with three cutting faces (Fig. 9.13, D). This cell is capable of dividing along its three faces. The segments thus cut off form a portion of the antheridial wall. This triangular cell, after stopping to divide, functions as the opercula cell. The rest of the antheridial jacket is derived from the surrounding prothallial cells.

The inner cell divides by transverse and vertical walls to form a quadrant of androgonial cells. These cells divide further by vertical walls and then by both vertical and transverse walls to form a large number of androgonial cells that divide further to form several thousand **androcyte mother cells.** The androcyte mother cells divide once to form **androcytes.** The protoplast of each androcyte metamorphoses into a single multiflagellate antherozoid.

Each androcyte mother cell has two blepharoplast granules, one either side of its nucleus. The resulting two androcytes have a blepharoplast each. During the metamorphoses of the spermatozoid, the blepharoplast elongates into a coiled thread-like structure that touches the nucleus at one end. The nucleus also becomes loosely coiled and forms the posterior portion of the spermatozoid. The coiled blepharoplast forms the anterior end of the spermatozoid and bears numerous flagella. The unused cytoplasm forms a small vesicle at the posterior end.

Archegonium

***(a)* Structure** (Fig. 9.14, E—F). The mature archegonium has the usual flask-shaped organisation. It has a short neck made up of four longitudinal rows of cells. Each row is 3 to 4 cells in height. The neck encloses a neck canal that lodges a single binucleate **neck canal cell.** Sometimes

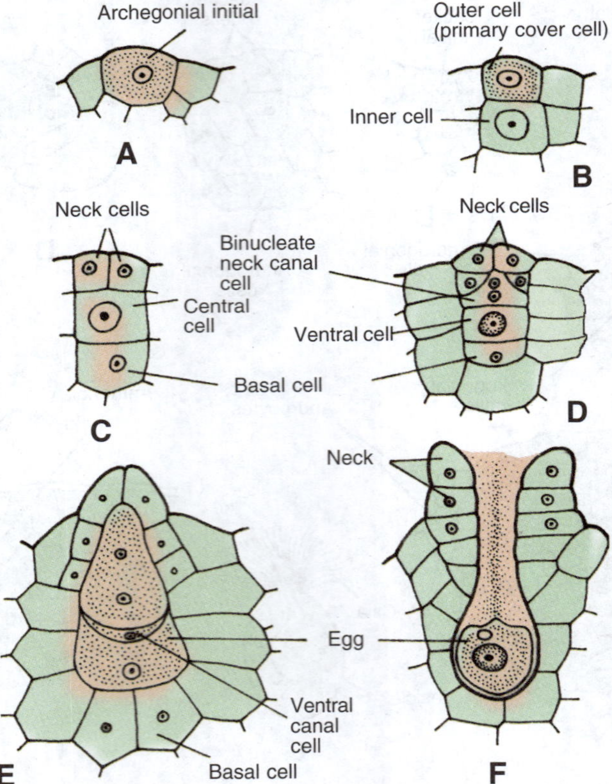

Fig. 9.14 (A–F). A-E. Stages in the development of an archegonium in *O. vulgatum*; F. A mature archegonium after dehiscence.

the two nuclei are separated by a cell wall thus leading to two uninucleate neck canal cells. The venter consists of a single **ventral canal cell,** which soon disorganises, and an **egg cell** that contains the female nucleus. The neck may project slightly above the prothallus tissue or may be completely embedded. The neck canal cell and the ventral canal cell disorganise in a mature archegonium. The neck cells at the top also separate thus making an open channel for the sperms to enter (Fig. 9.14, F).

***(b)* Development** (Fig. 9.14, A—F). The archegonial initial is also superficial cell of the prothallus. It divides into an outer and an inner cell by a periclinal wall. The former is called **the primary neck cell** or the **primary cover cell.** The inner cell divides by a periclinal wall into an upper

central cell (lying below the primary neck cell) and a lower basal cell. The primary cover cell divides by two intersecting anticlinal walls to form four quadrately arranged **neck-cells.** The cells by further transverse divisions give rise to four longitudinal rows of neck cells, each row having 3 to 4 cells. Meanwhile the central cell divides into an outer **primary neck canal cell** and an inner **primary venter cell.** The former enlarges in size and pushes in between the neck cells. Its nucleus divides into two daughter nuclei forming a binucleate neck canal cell. The primary ventral cell divides into an upper **ventral canal cell** and a lower **egg cell.** The ventral canal cell is ephemeral and short-lived. It soon degenerates. The **basal cell** does not divide further. Nozu (1961) observed that in ***O. vulgatum*** the two nuclei of the neck canal cell become separated by a wall, thus forming two uninucleate neck canal cells.

The process of fertilisation has not been observed in any species of ***Ophioglossum***.

EMBRYOGENY

The oospore is the pioneer structure of the sporophyte generation and develops first into an embryo.

Campbell (1911, 1940) reviewed the embryogeny of the genus ***Ophioglossum*** and recognised the following three types :

1. ***Ophioglossum vulgatum*** type as described by Bruchmann in 1904 (Fig. 9.15, A—E).
2. ***O. moluccanum*** type as described by Mettenius in 1856.
3. ***O. pendulum*** type as described by Lang (1902) and by Campbell (1911, 1921 and 1940).

Embryogeny in *O. vulgatum*

Bruchmann (1904) gave a detailed account of the development of embryo in ***O. vulgatum***. After fertilisation the surrounding prothallial tisue forms a calyptra-like envelope around the oospore. The oospore divides by a transverse wall at right angles to the long axis of the archegonium (Fig. 9.15, A) into an upper (nearer to the archegonial neck) **epibasal** and a lower **hypobasal cell**. There is no suspensor formed and embryogeny is **exoscopic.** The epibasal cell gives rise to the shoot and the first leaf. The hypobasal cell gives rise to the foot and the root. The quadrant and octant stages have not been observed. The next stage figured by Bruchmann (Fig. 9.15, b) is an oval mass of cells as visible in a median section. The first wall (1—1) is indicated and it is clearly shown that both the foot and the root develop from the hypobasal region. The epibasal part remains an undifferentiated mass of meristematic cells for a considerable length of time. During further development the root initial grows rapidly and the primary root soon emerges out of the prothallial tissue (Fig. 9.15). A vascular strand of the root is also traceable along its entire length extending from apex backwards to the region from where root originated (Fig. 9.15, C).

There is no endophytic fungus. During further development, the stem apex, the first leaf and its surrounding leaf sheath develop from the epibasal half (Fig. 9.15, D). A few elongated cells between the first leaf and root strand represent the incipient vascular strand of the first leaf. Later it becomes joined with root strand. Similarly a second leaf along with leaf sheath makes its appearance in the apical region (Fig. 9.15, E). It also develops its vascular strand that cojoins the vascular strand of the root. Meanwhile the apex of the second root also becomes organised (Fig. 9.15, D). The first leaf cannot grow beyond a certain height and remains below ground. It is the second leaf that overgrows the first leaf and comes above ground as the first green leaf. It is sterile and bears no spike. Development up to this stage took 8—10 years time .The third leaf that appears in the next season may be fertile and may bear a spike. The shoot apex is very slow growing and appears as a small bud at this stage. The young sporophyte at this stage has three or four unbranched roots. Bruchmann also observed a small adventitious bud on one the roots . Endoscopic embryo have been recorded in ***O. nudicaule*** (Pant, 1984).

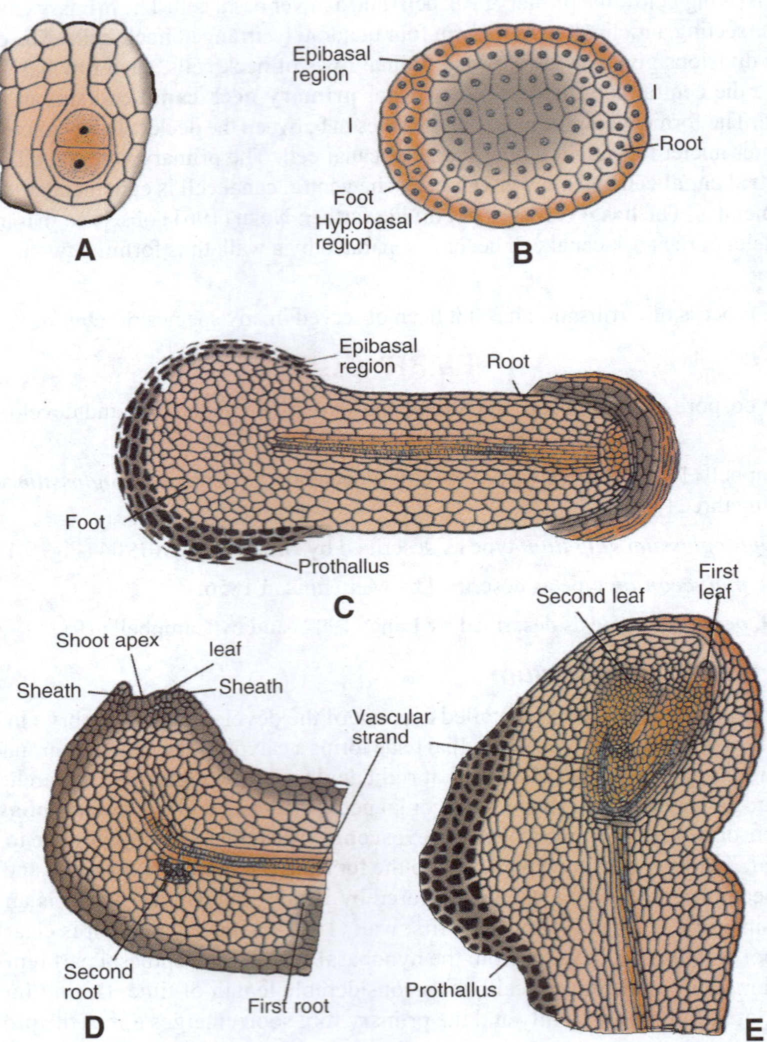

Fig. 9.15 (A—E) ***Ophioglossum vulgatum*. Stages in the development of embryo.** **A.** L.S. archegonium showing the first division of the zygote; **B.** Post-octant stage of embryo; **C.** L.S. older embryo showing well developed first root; **D.** L. S. of mature embryo showing shoot apex and first leaf; **E.** L. S. older embryo showing shoot apex and first two leaves; (After Bruchmann)

For variations in embryo development in other species please see chapter 8 (Early embryo in Eusporangiate Ferns).

Chromosome number

The number of chromosomes has been ascertained in a number of species. Abraham and Ninan (1954) counted 2n = 1,260 in ***O. reticulatum.*** In ***O. petiolatum*** the diploid number is about 1,000. The haploid number in ***O. vulgatum*** is n = 250-260 and in ***O. lusitanicum*** it ranges between 125-130 (Manton, 1950 : Verma, 1957). Verma (1957) observed both univalent and multivalent chromosomes in a population of ***O. vulgatum.*** He also observed the formation of diploid spores in

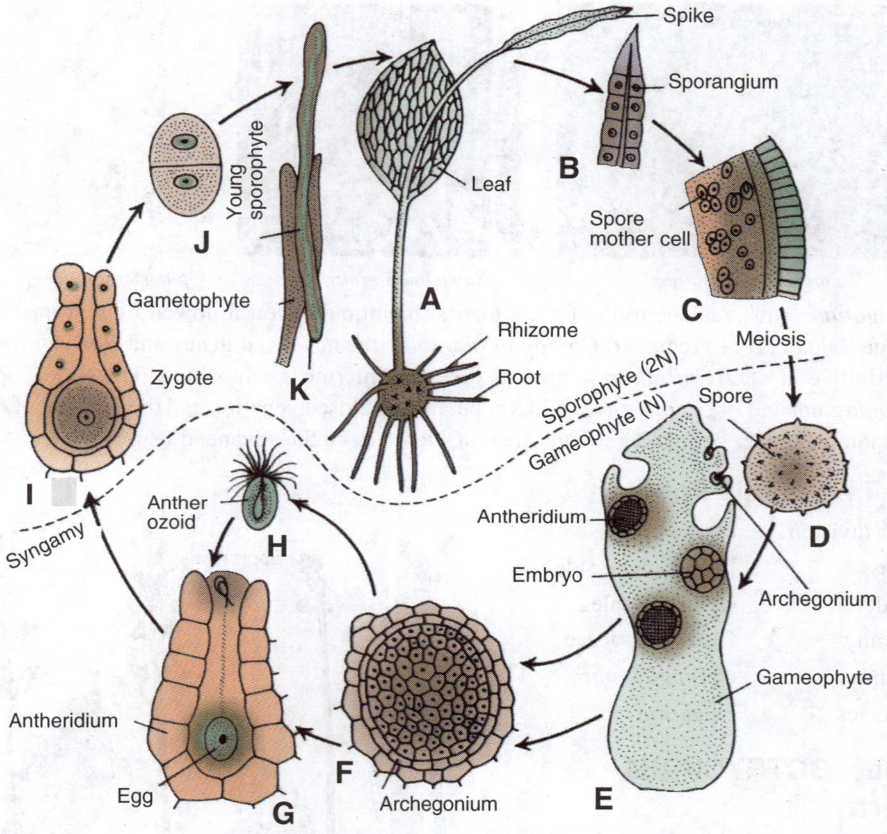

Spike
Sporangium
B
Leaf
Young sporophyte
J
Spore mother cell
C
Gametophyte
A
Meiosis
Rhizome
Root
Zygote
K
Sporophyte (2N)
Gameophyte (N)
Spore
I
Anther ozoid
Antheridium
D
Syngamy
H
Embryo
Archegonium
Gameophyte
Antheridium
F
Egg
G
Archegonium
E

Fig. 9.16 *Ophioglossum sp.* Diagrammatic life cycle.

Oospre
2x
Embryo
2x
Ophioglossum Sporophyte
2x
Fertile Spike
2x
Fertilisation
Sporangium
2x
Oosphore
x
Antherozoid
x
Sporophytic generation
2x
Gametophytic generation
2x
Archegonium
x
Spore mother cells
2x
Antheridium
x
R/D
Prothallus
x
Spore
x

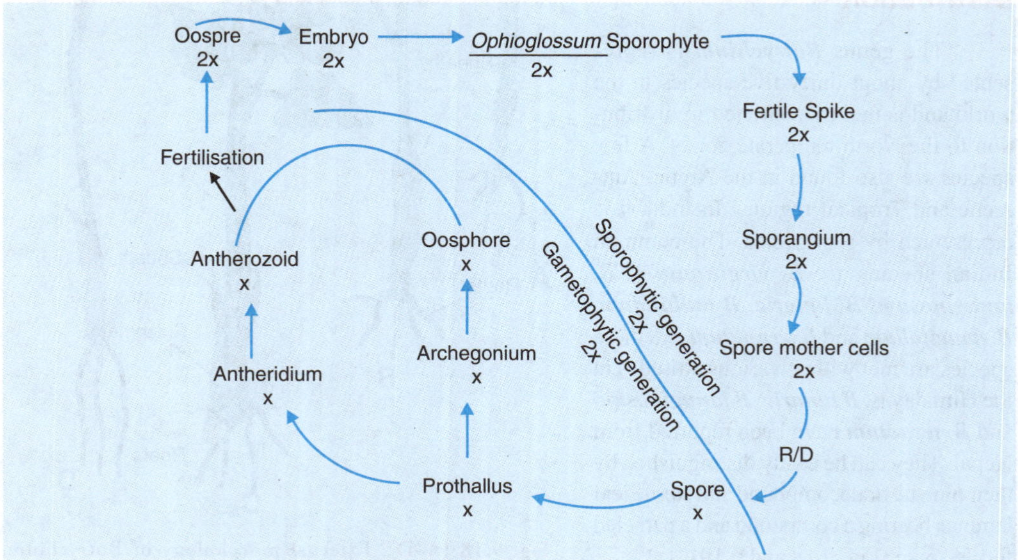

Fig. 9.17. *Ophioglossum sp.* Graphic life Cycle.

Botrychium ternatum

Botrychium multifidum

Ophioglossum daucifolium

O. vulgatum. This was due to the failure of first meiotic division followed by a normal second division. Ninan (1958) regarded ***Ophioglossum*** to be the advanced genus and ***Botrychium*** to be the primitive most. ***Helminthostachys*** was given an intermediate position. The simple spike of ***Ophioglossum*** and its simple leaves with reticulate and closed venation and the division of the leaf traces into a number of smaller strands are considered to be the advanced features.

Taxonomic Position

Sub division	:	Filicophyta
Class	:	Eusporangiopsida
Order	:	Ophioglossales
Family	:	Ophioglossaceae
Genus	:	***Ophioglossum***
Species	:	***vulgatum***

Genus–BOTRYCHIUM Swartz.

Distribution

The genus ***Botrychium*** is represented by about thirty-five species in the world and is mainly restricted in distribution to the North temperate zones. A few species are also found in the Arctic, Antarctic and Tropical regions. In India it is represented by six species. The common Indian species are ***B. virginianum***, ***B. lanuginosum***, ***B. lunaria***, ***B multifidum***, ***B. daucifolium*** and ***B ternnatum***. All these species are met with at various altitudes in the Himalayas. ***B lunaria***, ***B lanuginosum*** and ***B. ternatum*** have been reported from Nepal. They can be easily distinguished by their pinnate or decompound and sterile leaf laminae bearing a compound and a panicled fertile spike (Fig. 9.18 and 9.19)

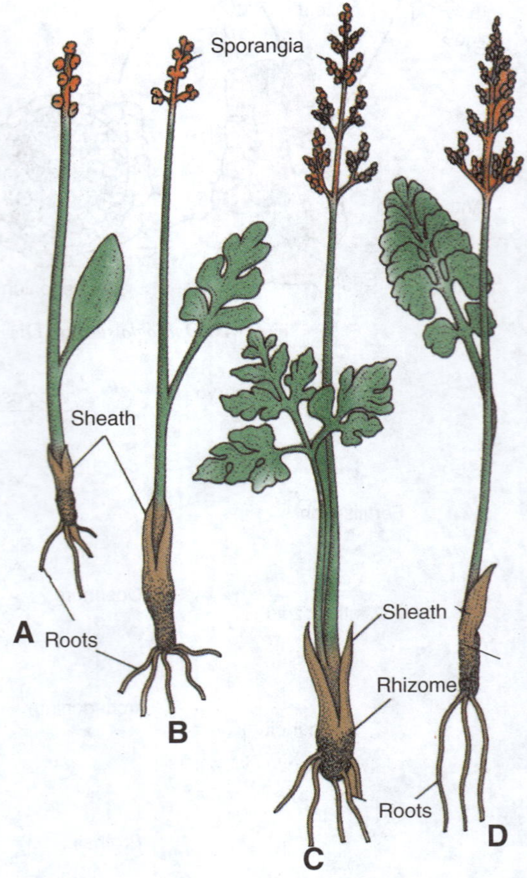

Fig. 9.18 (A–D). External morphology of Botrychium. A, B. *B. simplex*; C. *B. ternatum*; D. *B. lunaria*.

External Morphology of the Sporophyte

Regarding its external morphology the various species of ***Botrychium*** can be serially arranged from simplest types like ***B. simplex*** (Fig. 9.18, A, B) to complicated types like ***B. virginianum*** (Fig. 9.19, B) through intermediate forms exemplified by ***B. lunaria*** (Fig. 9.18, D) , ***B. ternatum*** (Fig. 9.18) and ***B. dissectum*** (Fig. 9.19, A). The simplest species ***B. simplex*** has a short unbranched rhizome bearing a number of unbranched adventitious roots . The sterile leaf has a petiole ending in a simple (Fig. 9.18, A) or a once pinnate (Fig. 9.18, B), lamina. The spike is also simple and unbranched. In ***B. lunaria*** (Fig. 9.18. D) the leaf is always once pinnate, but the spike is bipinnate. The leaflet of ***B. lunaria*** resembles the pinnule of ***Adiantum***. The leaf lamina in ***B. ternatum*** (Fig. 9.18, C) is twice pinnate. The fertile spike in this species is also bipinnate. The leaf lamina in ***B. dissectum*** and ***B. virginianum*** is decompound and the same is the condition in the fertile spike (Fig, 9.19, A, B). The venation is dichotomous in the simpler forms and approaches the pinnate type in ***B. virginianum.***

Botrychium virginianum is the largest species and the entire plant reaches a height of about 50 to 60 centimetres. The rhizome is well developed and may be up to six centimetres in length. It is upright and unbranched. A number of thick and fleshy roots arise from the rhizome. Each root arises opposite a leaf. The roots branch and form a well developed root system. The roots have no root hairs. The branches arise laterally. In some species the roots are unbranched, in others sparingly branched and the branching may be dichotomous or monopodial. The roots have an endophytic and mycorrhizal fungus. The leaf is not circinately coiled. At the base of each old leaf is large and more or less triangular in outline. The sterile lamina is three to four times pinnate and may be as broad as 30 to 40 centimetres. Only one leaf unfolds in a year. The young leaf is a sheath which encloses the young leaf of the next season. The sheath arises from the leaf base and is open on one side in ***B. virginianum*** so it gives the appearance of a pair of stipules (Fig. 9.19, A). The fertile segment of the leaf or the sporangiferous spike arises from the inner surface of the petiole and is twice or thrice pinnate (Fig. 9.19 A). The apical bud when cut open longitudinally shows in addition to the mature unfolded leaf of the present year, the younger leaves of the next two years. The stem apex is evident at the base of the youngest leaf. The apical buds in ***B. virginianum*** and ***B. ternatum*** are covered with numerous short hair that are absent in other species *e.g.* ***B. simplex***.

Fig. 9.19 (A–B). External morphology of *Botrychium*. A. *B. virginianum*; B. *B. dissectum*.

Anatomy of the Sporophyte

(*a*) **Rhizome:** A transverse section of the stem before the commencement of secondary growth reveals a central stele that is essentially an ectophloic siphonostele (Fig. 9.20). The leaf gaps do not overlap so that the stele is in the form of a continuous vascular cylinder interrupted here and there by a single leaf gap (Fig, 9.20). The leaf trace is semi-lunar in outline. The entire stele is surrounded by a single layer of endodermis.

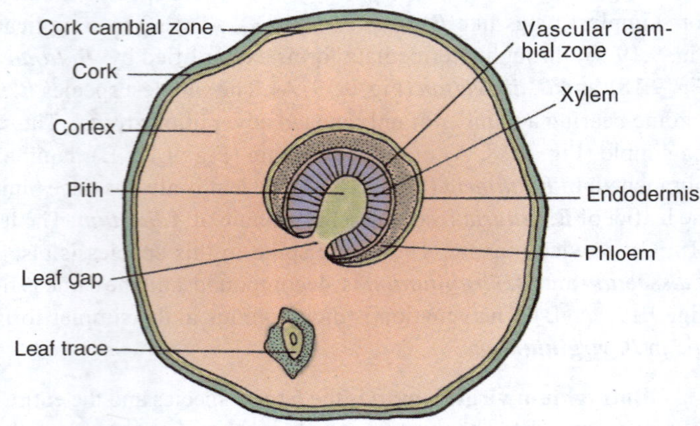

Fig. 9.20 T.S. rhizome of ***B. lunaria*** (outline sketch)

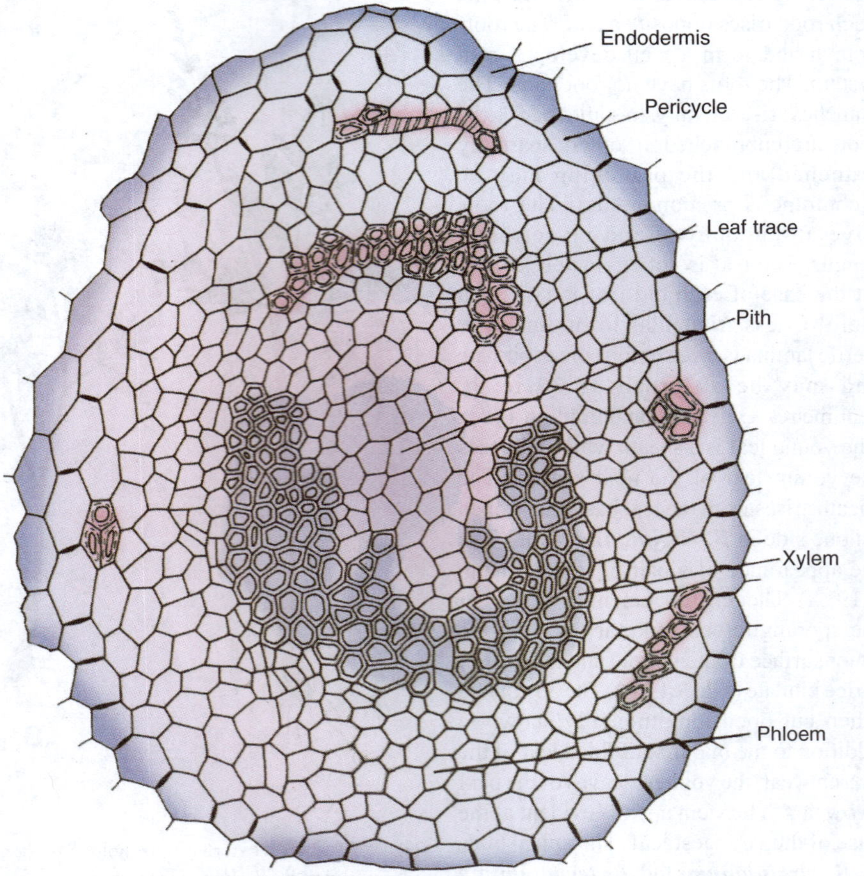

Fig. 9.21. ***Botrychium lunaria.*** T.S. young stem showing the departure of leaf trace and the formation of leaf gap. There is a distinct pith in the centre. Some thick-walled tracheids are also present in the pith. Signs of cambial cavity are also visible external to the xylem. (After Bower, 1926)

The endodermis remains unbroken even at the exit of the leaf trace (Fig. 9.21). At this stage there are two groups of steles in the stem. The large one represents the stem stele and the smaller one is the leaf trace. There is a distinct pith in the centre. Isolated tracheidal cells have been recorded near the periphery of the pith in the stems of ***B. virginianum***, ***B. ternatum*** and ***B. lunaria*** (Fig. 9.21). Their presence points towards the intrastelar origin of pith. Encircling the pith is the xylem tissue which in turn is surrounded by phloem (Fig. 9.21). Between the xylem and the phloem is present a zone of cambium (Fig. 9.20). The xylem is composed of tracheids and has endarch development. The tracheids have bordered pits that may be circular or oval in outline. Characteristic elongated type of pits present in the ferns are absent in ***Botrychium***. The pits along the walls of adjacent trachieds are separated by delicate membranes. In mature portions of the stem the continuity of the xylem is interrupted by the presence of medullary rats that are composed of radially elongated and thin-walled cells (Fig. 9.22 a, b). The cells of the medullary rays are filled with granular contents.

The phloem is made up of sieve tubes and parenchyma cells. The phloem fibres are absent. The sieve tubes in the primary phloem have thickened walls except at the sieve areas. Esau, Cheadle, and Gifford (1953) reported the absence or callose over the sieve areas.

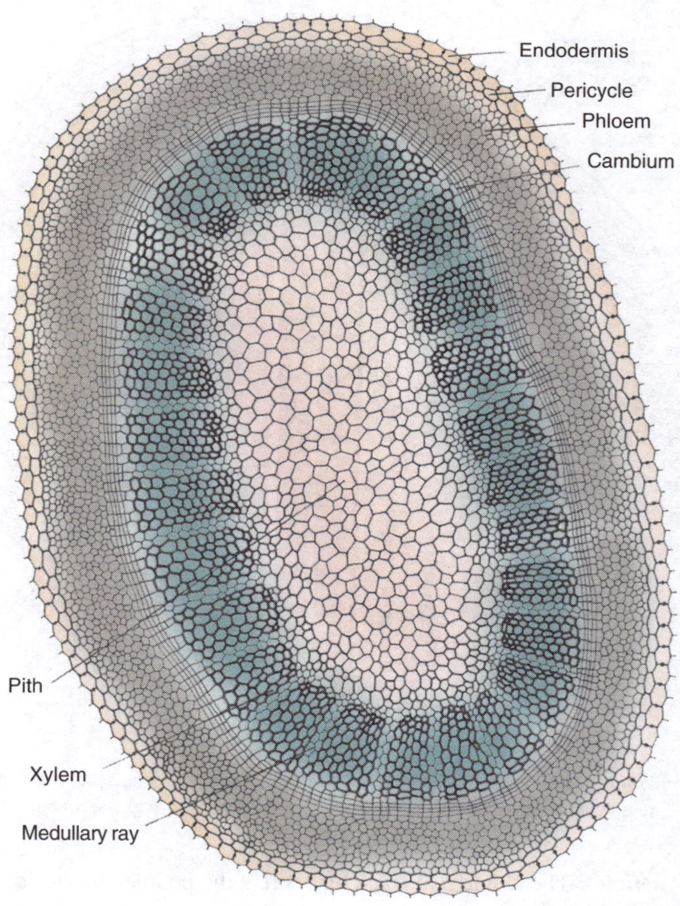

Fig. 9.22a. T.S. stem of *B. lunaria* showing secondary growth.

The cortex is made up of several layers of parenchymatous cells. The cells are filled with reserve food . The periphery of the cortex is occupied by a few layers of compactly arranged cork cells (Fig. 9.20). Cork is not formed by the activity of any definite cork cambium. They are formed by the repeated tangential division of the cortical cells. There is no distinct epidermal layer. It has now been observed that in almost all the species of ***Botrychium*** a definite **cork cambium** or **phellogen** originates in the cortex and gives rise to the cork layers externally and phelloderm internally.

The activity of the cambium adds to the primary vascular cylinder (Fig. 9.22 a, and 9.22 b). The cambial cells divide in a tangential plane and cut off cells towards their inner face as well as outer face. The former act as secondary xylem mother cells and mature into secondary xylem tracheids. The cells cut off towards the outer face develop into secondary phloem elements.

Lang reported the presence of internal endodermis in ***Botrychium lunaria***. She reported the presence of internal endodermis in that region of the stem which is intermediate between the juvenile and adult portions.

The stem grows by means of a single apical cell which is pyramidal in form and cuts off lateral segments in a regular manner.

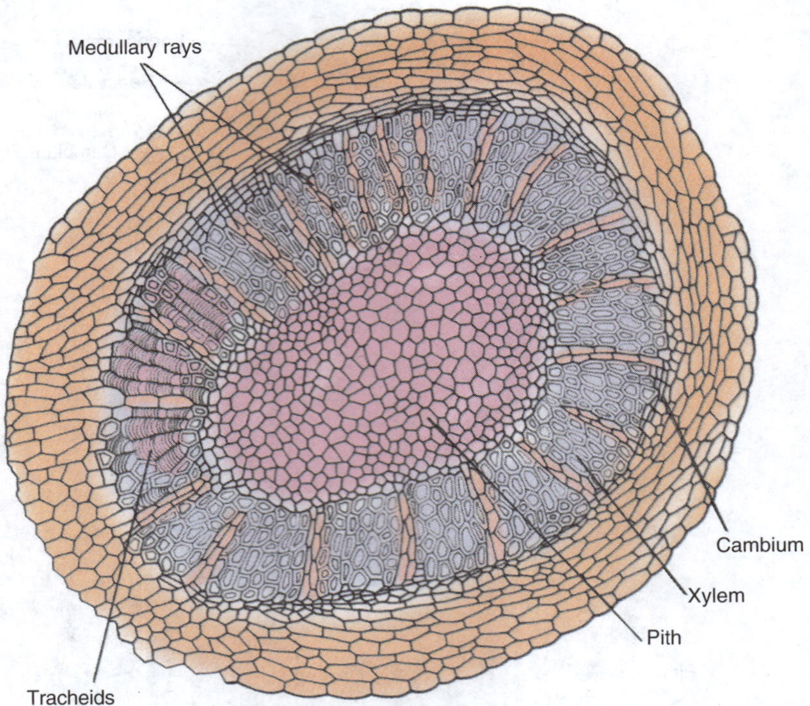

Fig. 9.22 b. *Botrychium virginianum.* T.S. older portion of rhizome showing secondary growth, only the stelar region is shown. The outermost layer of cells is the endodermis. Next to it is the secondary phloem (not labelled). (After Atkinson)

(*b*) The petiole : The single leaf trace that enters the petiole consists of only one vascular bundle. It usually divides into two vascular bundles while entering the petiole or after entering the petiole. In some cases it may not divide till it has traversed a little distance up the petiole. The petiole in the youngest leaves has only one central vascular bundle. In the mature leaves the number of vascular bundles in the petiole may be two near the base and it may increase to four in the distal portions of the petiole especially at the point of origin of the fertile spike. Out of these two are large and two small. The two larger vascular bundles are generally semilunar or crescent-shaped. The smaller ones are almost oval. In ***B. virginianum*** the larger bundles are concentric. The xylem occupies the centre and is completely surround by phloem which in turn is encircled by endodermis. In ***B. lunaria*** and ***B. ternatum*** the concentric bundles are collateral. The phloem is present only on the outer shape. The phloem is similar in composition to that of the stem. The end walls of the sieve tubes are swollen and are perforated by pores.

B. ternatum

Near the distal ends of the petiole the large crescentic vascular bundles give off a small branch so that there are four vascular bundles, two large and crescentic and two small and oval. These smaller vascular bundles extend into the rachis of the spike. The larger bundles continue upwards into the rachis of the sterile leaf.

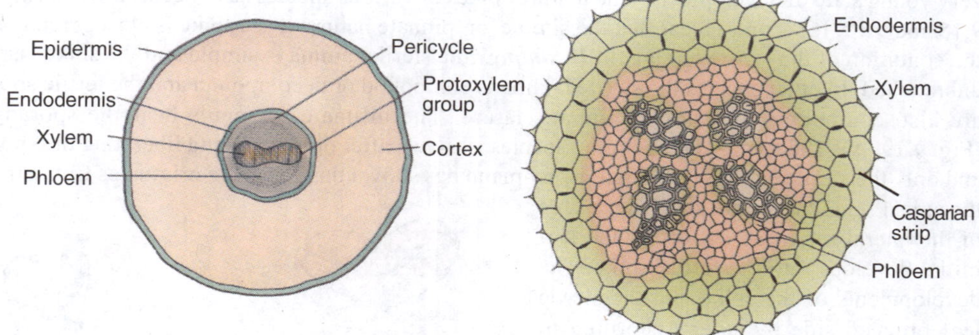

Fig. 9.23. *B. lunaria.* T.S. root (outline sketch).

Fig. 9.24. **Portion of T.S. of root of *B. virginianum.***

(*c*) The Sterile Leaflet : It has the usual upper and the lower epidermal layers. Both the epidermal layers are made up of a single layer of cells. Stomata are restricted only to the lower epidermis. A thin layer of cuticle is present over both the surfaces. The mesophyll is undifferentiated. It consists of spongy parenchyma cells that are more compact below the upper epidermis and more or less loosely arranged towards the lower surface. The vascular bundles may be concentric (***B. virginianum***) or collateral.

(*d*) The root (Fig. 9.23, 9.24). The outermost boundary of the mature root is occupied by a few layers of compactly arranged cork cells. Next to them is the wide cortex which is made up of several layers of loosely arranged thin-walled cells. The cells of the cortex increase in size as the root matures.

The endodermis (Fig. 9.24) is faint and encloses the entire stelar region. Its cells are thin-walled and possess the characteristic casparian strips. The xylem and phloem masses alternate with each other. Their number varies in different species or even in the same species. The mature roots in ***B. virginianum*** are tetrarch (Fig. 9.24). The xylem is exarch. In ***B. ternatum*** the stele is triarch and in ***B. lunaria*** it is diarch (Fig. 9.23).

The roots increase in diameter as they grow in size. This increase is due to the enlargement of the cortical tissue. The cortical cells increase considerably in size (Campbell, 1918). There is no secondary growth in the roots. The root grows by means of a three sided apical cell.

REPRODUCTION

Vegetative propagation, by any means, has not been reported in ***Botrychium***. A sexual reproduction takes place by means of meiospores that are homosporous and are produced within sac-like sporangia. The sporangia are borne on branched (***B. virginianum***, ***B. dissectum***, ***B. lunaria***, etc.) or unbranched (***B. simplex***) spikes or fertile segments (Fig. 9.18 and 9.19).

The Fertile Spike or Fertile Segment

The spike is attached to the sterile segment or the leaf near its central or upper portion. In some species it appears to arise lower down from the base of the petiole. The spike is attached on the ventral surface of the sterile segment or the leaf. The spike appears when the leaf is quite young. Small and reduced fertile spikes have been seen even in the bud condition in ***B. virginianum***. At the same time the spikes of various ages can be seen on a single plant. The mature unfolded leaf of the current year bears a mature and a well developed spike, the leaf destined to appear next year and enclosed still in the sheath of the current year's leaf bears a young fertile spike bearing sporangia in earlier stages of development, and a leaf that will appear in the third year shows the spike in a very young state of development. The mature spikes of various species have been shown in figures 9.18 and 9.19. It becomes clear that the simple or pinnate nature of the spike is closely related to the condition of the sterile lamina. In ***B. simplex*** the sterile lamina is simple and the spike is also unbranched. In other species the sterile lamina is compound or decompound and the fertile spikes are also compound and show a pinnate nature. The ultimate segments bear the sporangia (Fig. 9.19, A). The fertile spike in some species, shrivels after producing and liberating the spores and only the sterile leaf is left behind. Bruchmann has shown that the spike originates from one of the apical segments of the sterile leaf. One of the lateral segments cut off by the leaf initial develops into the fertile spike. The developments of the spike and the sterile leaf proceed side by side. According to Bierhorst (1971), " The ***fertile spike in Botrychium*** is seen when the leaf like organ is about four years old as a conical protuberance overarched slightly by the apex of the vegetative part of the complex." Petry (1915) reported that the spike arises from an apical cell which appears on the adaxial side of the apical cell of the vegetative structure.

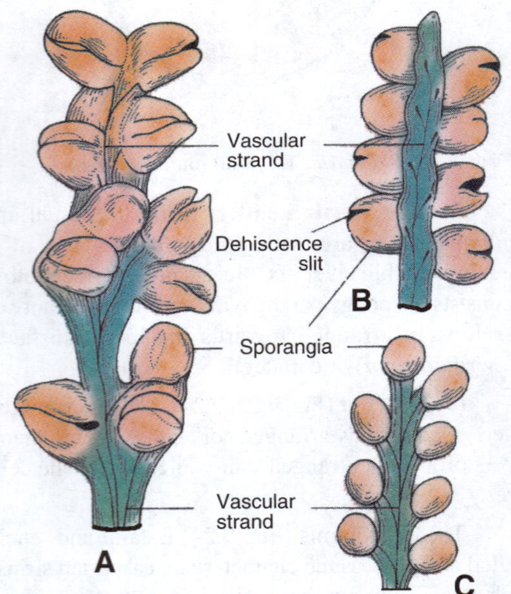

Fig. 9.25 (A–C). Vascular supply to the sporangia of *Ophioglossaceae*. A. *Botrychium simplex*; B. *B. dissectum*; C. *B. virginianum* (A—B. after Goebel; C. after Eames)

The sporangia are borne in an acropetalous order along the margins of the ultimate branches of the fertile spike (Fig. 9.25, A—C). Each ultimate branch terminates in a sporangium. So in ***Botrychium*** the sporangia are marginal as well as terminal in position. On the unbranched spikes of ***B. simplex*** the sporangia develop directly on the rachis (Fig. 9.25, A) and in this species they are arranged in an alternate manner. The sporangia of ***B. simplex*** are larger in size and form two rows (Fig. 9.25, A) very much similar to ***Ophioglossum***.

Vascular Supply to the Spike

Chrysler (1925) described the vascular supply to the leaf (both sterile and fertile segments) in ***B. virginianum*** and ***B. lanuginosum***. As described earlier, the petiole possesses two crescent-shaped vascular strands. At the base of the fertile spike each crescent-shaped strand cuts off a small oval vascular bundle. These two lateral vascular strands or the spike traces traverse the fertile spike and supply the branches and the sporangia of their respective sides (Fig. 9.25, A–C).

Nature of the Fertile Spike

Chrysler (1925) after studying the vasculature of the spike and the sterile segment concluded that the fertile spike or the segment is formed by the fusion of two pinnae that stand erect. Such a conclusion is derived from the occurrence of two separate vascular bundles in the spike.

Bower (1935), Zimmermann (1942), Chrysler (1945) and Nishida (1952) regarded that the leaf of ***Botrychium*** represents a modification of a reduced and a dichotomously branching axis. Such an interpretation is supported by the origin of the spike from a lateral segment quite close to the apical cell of the leaf (Bower, 1935). The alternate insertion of the sporangia also points towards this direction (Bower, 1935). The frequent bifurcation of the vascular tissue lends further support to this hypothesis.

Sporangium

(*a*) Structure (Fig. 9.26, C). The position and insertion of the sporangia have been discussed above. The sporangium has a short and massive, multicellular stalk (Fig. 9.26, C). The stalk id traversed by a vascular strand of its own. Borne on the stalk is a capsule. The vascular strand of the stalk terminates below the capsule. The wall of the capsule is composed of several layers of cells (4—6 layers). Bierhorst (1971) reported that the sporangial wall is made up of 4—7 inner layers of cells with an outer layer of cells where inner and anticlinal walls are thickened. The cells of the outer layer are "anticlinally" short at the site of dehiscence and form a distinct "groove". The tapetum is believed to be derived from the outermost cells of the sporogenous tissue (Campbell, 1918). Eames (1936) mentions that tapetum is derived "from the cells surrounding the sporogenous tissue". He does not clearly state whether it originates from the outer cells of sporogenous tissue or from the inner wall layer. The tapetum is not well defined and appears to be two to four layers of cells in thickness. The tapetal cells disorganise and form a nutritive plasmodial fluid in which the spore mother cells float freely and derive nourishment. The sporogenous cells enlarge and have distinct nuclei and dense cytoplasmic contents. These are the spore mother cells. They undergo meiosis and form haploid spores. The spore output of a sporangium, in ***Botrychium*** is large and varies between 1000—2000 spores per

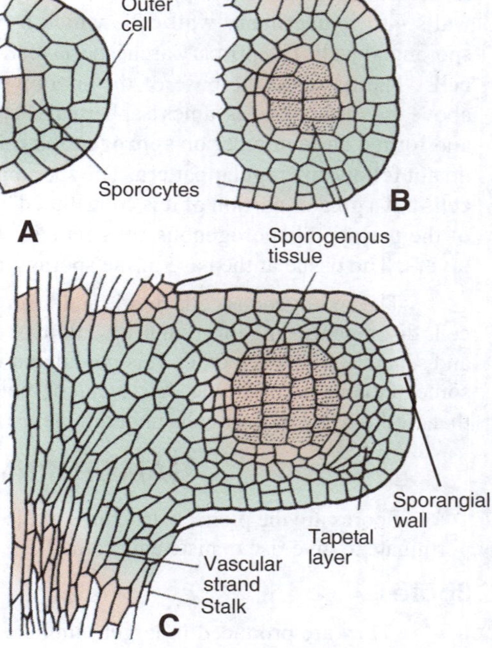

Fig. 9.26 (A—C). Stages in the development of sporangium in *B. daucifolium* (*B. lunaria*) (After Bower)

sporangium. The massive stalk, vascular supply reaching the base of the capsule, the thick and massive sporangial wall, many-layered tapetum and large spore output are the primitive features of the sporangium. The mature sporangia are yellow or yellowish brown in colour.

(*b*) Dehiscence. Campbell (1918) states, "The wall of the sporangium has 4—6 layers and sometimes the place of dehiscence is indicated, as in ***Ophioglossum***, by two rows of smaller cells". In Campbell's figure (148 C, page 269; 1918) these two rows are terminal in position and hence the dehiscence, according to him is terminal. Eames (1936) also states that the dehiscence in all the Ophioglossaceae is terminal. He states that it appears to be lateral in ***Ophioglossum*** because of the position of the spoprangium with respect to the fertile spike. Bower (1935) also states that the dehiscence is by means of a "distal" slit. Due to the absence of any specialisation of the spoprangial wall (annulus and stomium) there is no well defined mechanism of dehiscence. It may be caused by a shrinkage of the walls due to the loss of water.

(*c*) Development (Fig. 9.26, A—C). The development is eusporangiate. Bower (1896) and Goebel (1904) studied it in ***B. lunaria*** (= ***B. daucifolium)*** and Campbell (1918) and Holtzman (1892) followed it in ***B. virginianum.*** The sporangia develop in an acropetalous manner. In ***B. virginianum,*** Campbell observed that the young sporangia appear as small protuberances on the spike. These protuberances go on becoming smaller and smaller towards the tip of the fertile segment. A median longitudinal section passing through the tip of the spike reveals that the sporangium develops from a group of cells with a distinct pyramidal cell in their centre. He described that the central pyramidal cell divides into an outer and an inner cell by a periclinal wall (Fig. 9.26, A). The inner cell functions as the **primary archesporial cell.** The outer cell divides by two intersecting walls into four primary wall cells which by further periclinal walls forms a 4—6 layered thick sporangial wall. The primary archesporial cell also divides by cross walls to form four **archesporial cells.** The tissue at the base of the sporangium grows actively so that the sporangium projects above the surface and assumes a spherical shape. Meanwhile the archesporial cells divide repeatedly and form a large number on **sporogenous cells** (Fig. 9.26, B). The divisions of archesporial cells do not follow any regular pattern. The sporangial wall is not derived wholly from the primary walls cells but a greater portion of it is contributed by the tissue surrounding the sporogenous cells. Some of the peripheral sporogenous cells remain sterile and contribute to the formation of the **tapetal layers.** The tissue at the base of the sporangium forms a thick and massive, but short stalk.

The sporogenous cells divide and form the **spore mother cells** or the **sporocytes.** The tapetal cells disoraganise and form a nutritive fluid or the **plasmodial fluid.** The spore mother cells round off and separate from each other. It is not known with certainty whether all of them undergo meiosis or some of them remain sterile. As a result of meiosis tetrads of spores are formed. Eames (1936) stated that all the spore mother cells undergo meiosis and form spore tetrads.

GAMETOPHYTIC GENERATION

Spores are the pioneer structures of the gametophytic generation. They are homosporous and germinate to give rise to bisexual prothalli.

Spores

They are produced in larger numbers per sporangium. A single sporangium of ***B. lunaria*** produces 1,500—2,000 spores. The mature spores are more or less spherical structures with distinct triradiate mark on the **exosporium.** The spore wall is well developed and thick and is usually colourless or pale yellow. The outer wall or the **exosporium** is thick and resistant. It is variously sculptured and may show shallow pit-like depressions or small projections or tubercles. The **intine** or the **endosporium** is thin. The spore contains scanty cytoplasm filled with reserve food and a single nucleus.

Germination of the Spore

The details of spore germination are not available. The earlier stages seem to be similar to ***Ophioglossum***. The smallest prothalli of ***B. lanceolatum*** and ***B. matricariaefolium***, obtained by Foster (1967) were egg-shaped and consisted of a mass of cells with a "dome-shaped" apex. A pyramidal apical cell was distinct in the centre of the apex. The segments cut off by this apical cell on the side show increased meristematic activity and push the initial to the opposite side. Later, one apical cell is replaced by several apical cells which form an apical meristem. The sex organs originate in this "polarised" meristematic zone. The first formed region (oldest) of the prothallus is conical in shape and stands vertically upright. The later formed or the younger portion is horizontally placed and bears sex organs. Foster calls the former as "primary structure" and the latter as a "secondary structure".

Mature Prothallus

The prothalli are subterranean, brownish in colour and mycorrhizic. They are dorsiventral and hence differ from the cylindrical and radically constructed gametophytes of ***Ophioglossum*** (Fig. 9.26 a. A–E). Regarding size, the prothalli of ***B. virginianum*** are the largest (up to 20 mm.

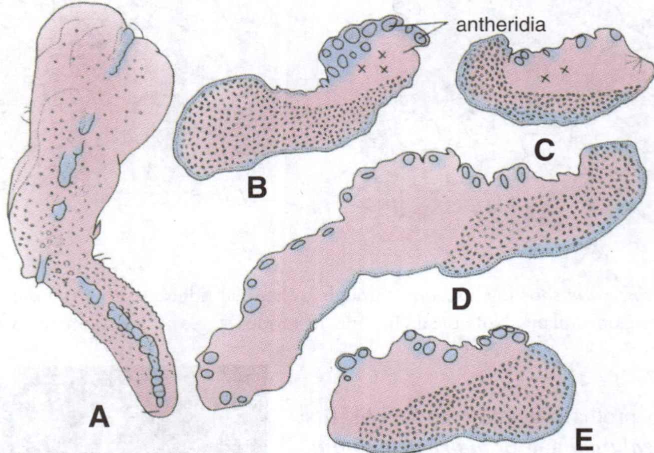

Fig. 9.26a (A—E) *Botrychium* A. Mature prothallus of B. *virginianum*; B—E. Median longitudinal sections through the prothalli of; *B. dissectum* showing antheridia and archegonia (X) (After Bierhorst, 1957).

long), those of B. ***dissectum*** (= ***B. obliquum***) comes next and measure 3-6 mm. In length and 1½ to 3 mm. in breadth. Those of ***B. lunaria***, ***B. simplex***, ***B. lanceolatum*** and ***B. matricariaefolium*** are the smallest The mature prothalli of most of the species (***B. virginianum***, ***B. dissectum***, ***B. lunaria***) are tuberous and dorsiventrally flattened structures. The dorsal side of the prothallus is marked by the presence of a distinct ridge. It is called the antheridial ridge because antheridia develop only in this region. The archegonia develop along the base of the antheridial ridge. The prothalli of ***B. virginainum*** (Campbell 1918). The prothalli of this species live for several years and may become irregular in shape. The older prothalli also show indications of dichotomy at the apex (Campbell, 1918). The growth of the prothalli is by means of an apical meristem. The same is the case in ***B. dissectum*** (Whittier, 1972) and ***B. obliquum***. Both have rhizoids.

B. matricariaefolium

The endophytic fungus is restricted to the lower older regions of the prothallus. The meristematic region is devoid of endophytic fungus. The inner calls of the prothallus in the ventral part are also filled with abundant reserve food material. Whittier (1972) cultured the prothalli of ***B. dissectum*** in auxenic culture and obtained non-chlorophyllous prothalli that lacked an endophytic fungus and grew well. He believes that endophytic fungus is not essential for the growth of ***Botrychium*** gametophytes in culture.

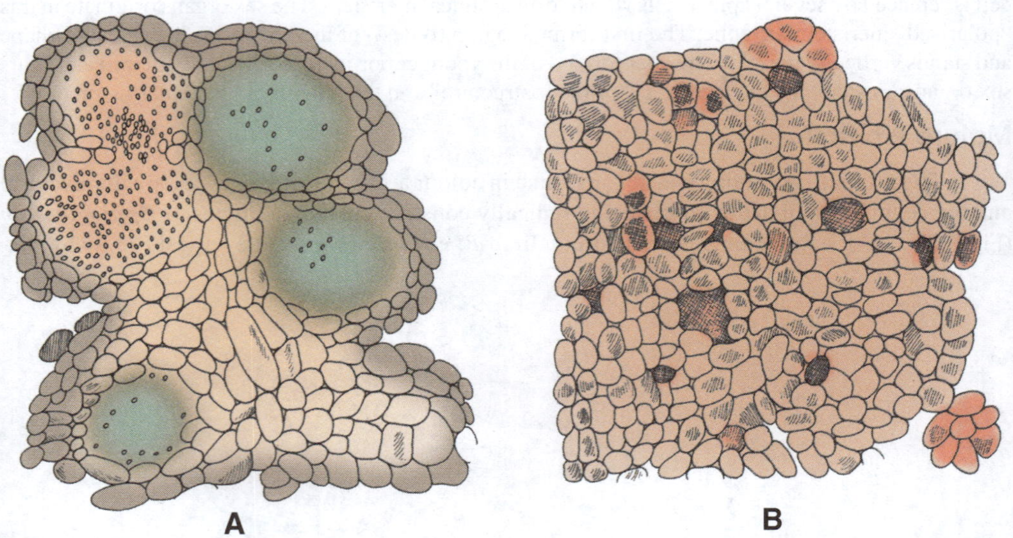

Fig. 9.26 b. A. *B. dissectum* showing a section through antheridial ridge; B. *B. virginianum* showing a section tangential to the archegonial plane. Note the archegonia in various stages of development (After Bierhorst).

Sex Organs

The young prothalli develop antheridia first except in ***B. lanceolatum*** and ***B. matricariaefolium*** where the two types of sex organs develop simultaneously on their dorsal side. Later the antheridia are restricted only to the prominent antheridia ridge. The archegonia develop along the base of the antheridial ridge and usually appear late. The structure and development of the sex organs (Fig. 9.27& 9.28) are almost similar to those of ***Ophioglossum*** except for some minor details. The antheridia are embedded in the tissue of the prothallus. The archegonial necks are longer and projects above the surface of the prothallus. Structurally the antheridium in ***B. virginianum*** differs from that of ***Ophioglossum*** in possessing a double layer of cover cells (Fig. 9.27, D—G). The spermatozoids of ***Botrychuim*** are smaller in size as compared to that of ***Ophioglossum. Each*** spermatozoid has numerous flagella and the body has 1½ coils.

B. lanceolatum

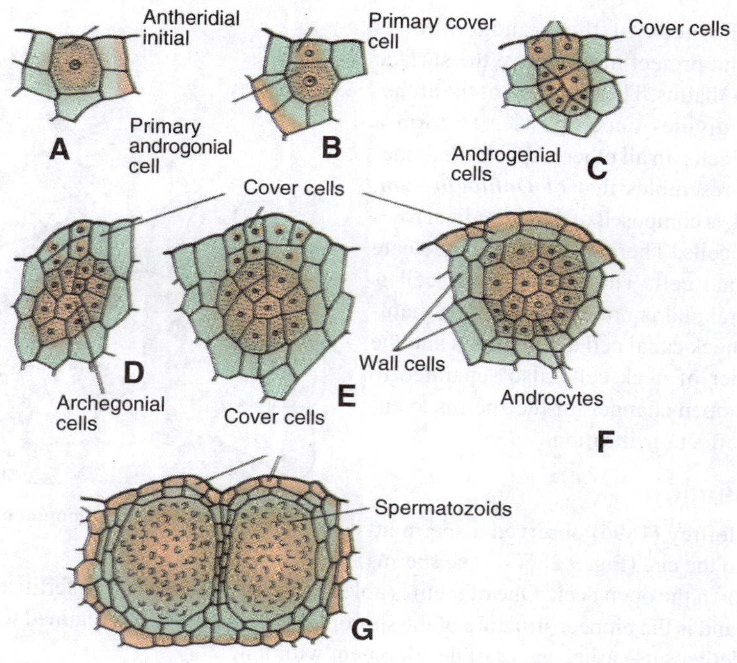

Fig. 9.27. Stages in the development of antheridium in *B. virginianum* (After Jeffrey).

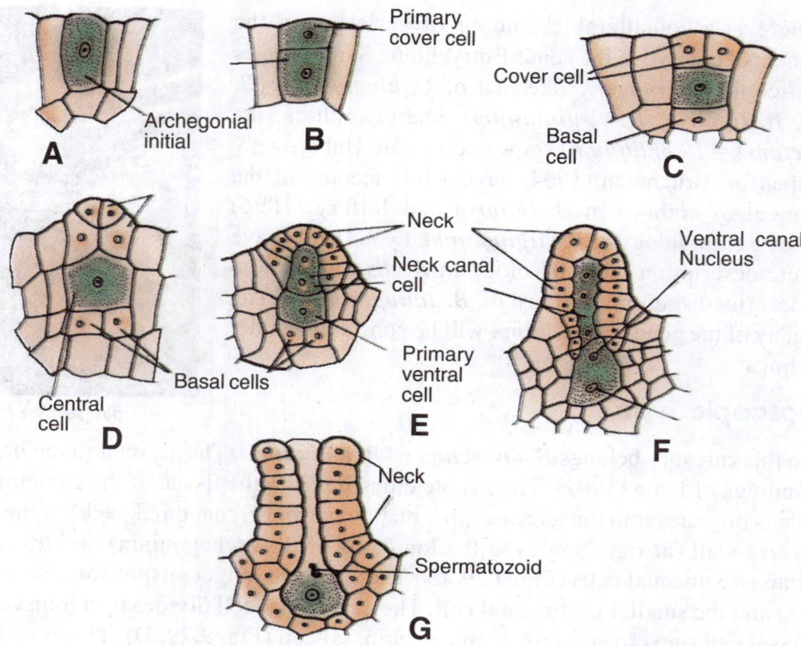

Fig. 9.28. Stages in the development of archegonium in *B. virginianum* (After Jeffrey).

The neck of the archegonium is longer and project much above the surface of the prothallus. The basal cell of the archegonium divides once or twice to form a group of cells. In all other respects the archegonium resembles that of ***Ophioglossum***. The neck is composed of 4 longitudinal rows of neck cells. There is a single binucleate neck canal cell. The ventral canal cell is ephemeral and is not easily seen. At maturity the neck canal cell disorganises and the apical tier of neck cells also separated to form an open channel for the sperms to enter and effect fertilisation.

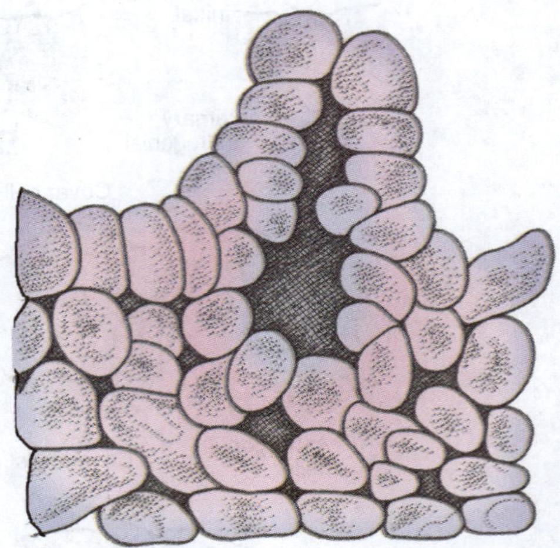

Fig. 9.28a. L.S. mature archegonium of *B. virginianum*. (After Bierhorst)

Fertilisation

Jeffrey (1896) observed a sperm attached to the egg (Fig. 9.28, G). The sperms swim down the open neck. One of them is able to effect fertilisation. The fertilised egg or Oospore is diploid and is the pioneer structure of the sporophytic generation. It is retained within the venter wall and undergoes its earlier stages of development within it.

EMBRYOLOGY

There is a considerable amount of variation in the development of embryo in the genus Botrychium. Some species have exoscopic embryogeny like that of ***Ophioglossum* (*B. simplex*, *B. lunaria*, *B. virginianum*)** whereas others like ***B. dissectum*** (= ***B. obliquum***) show endoscopic embryogeny with suspensor. Bruchman (1904) gave a full account of the development of embryo in ***B. lunaria*** and Jeffery (1896) described the embryology of ***B. virginianum***. Lyon (1905) gave a complete description of embryology in ***B. dissectum.*** Rao (1962) described the embryology of ***B. lanuginosum.*** The embryogeny of the genus ***Botrychium*** will be considered under two headings:

Botrychium lunaria

A. Endoscopic Type

To this category belongs ***B. dissectum*** (= ***B. obliquum***). The account given below is based on the findings of Lyon (1905). The zygote elongates into the tissue of the gametophyte below. The nucleus migrates into the terminal portion (away from archegonial neck) of the zygote tube. A transverse wall (at right angles to the long axis of the archegonium) divides the elongated zygote into two unequal cells (Fig. 9.29, C). These are the longer **suspensor cell** (near archegonial neck) and the smaller **embryonal cell.** The embryonal cell divides by a transverse wall into a **hypobasal cell** (next to suspensor) and an epibasal cell (Fig. 9.29, D). This wall is sometimes laid down in an oblique manner. The suspensor cell may also divide by a transverse wall to from a two-celled suspensor (Fig. 9.29, E). The epibasal cell gives rise to the first leaf and the stem. The hypobasal half develops into a root. The first root arises "near the center of the embryo very near the basal wall", which divides the embryonal cell into hypobasal cells. As the embryo, grows the epibasal cell divides to form an apical meristem (Fig. 9.29, E) from which the stem tip and the leaf initial originate. The hypobasal cell undergoes divisions to form a large foot (Fig. 9.29, E). The sequence of division in both the cells is not well understood. The hypobasal cell seems to divide by a number of oblique transverse and vertical walls to form a group of cells that develop into a foot. Most of the meristematic cells in the epibasal part contribute to the development of the first leaf. The stem tip lies at the base of the leaf and consists of a few cells, one out of which functions as the stem apical cell. The embryo at this stage appears to be an oval structure. The root becomes evident at a later stage in the median portion of the embryo in the form of a group of actively dividing cells. One of these differentiates as an apical cell of the root. As the root grows it penetrates the foot and elongates rapidly. The embryo now appears bipolar (Fig. 9.29, F). The first leaf also grows rapidly. The embryo undergoes a strong curvature so that the leaf and the stem face upwards (Fig. 9.29, G) and the root downwards towards the soil.

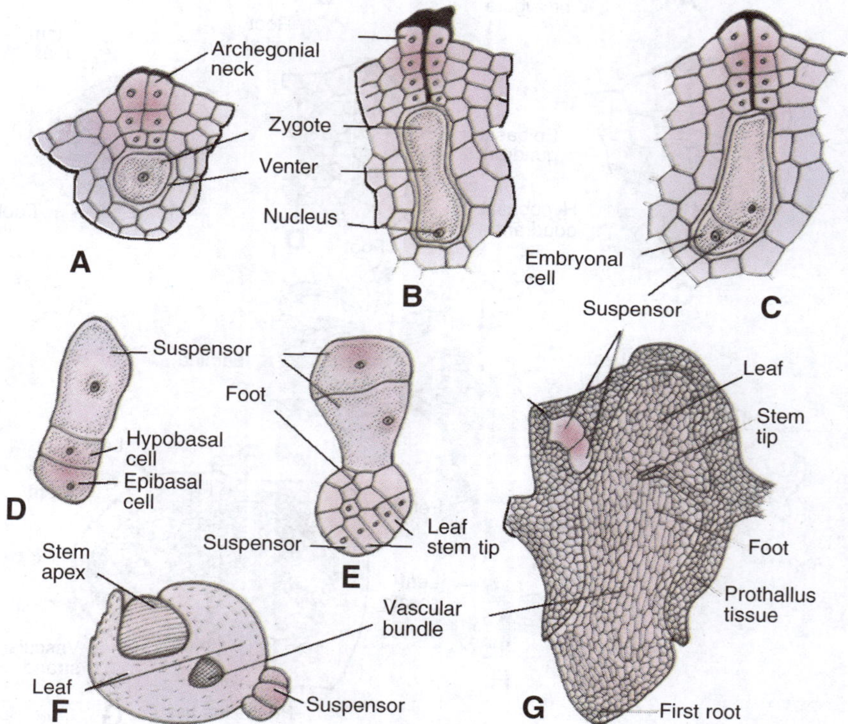

Fig. 9.29. (A—G). ***Botrychium obliquum*** **(=** ***B dissectum*****).** Various stages in the development of embryo. (After Lyon)

The root soon grows and pierces through the prothallus tissue to enter the soil (Fig. 9.29, G). A primary vascular strand is also evident at this stage. It runs uninterrupted through the primary root and the first leaf. It does not extend into the stem tip. A distinct and well developed sheath encloses the base of the first leaf. It also embraces the stem tip and second leaf. The vascular strand of the second leaf joins with that of the first. The first leaf when it comes above ground has well developed lamina and a long stalk of the petiole.

B. Exoscopic Type

To this category belong those species of ***Botrychium*** in which there is no suspensor and the stem and the leaf initials originate from the quadrant nearer to the archegonical neck (epibasal). The foot and the root arise from the hypobasal quadrant. ***Botrychium virginianum*** (Fig. 9.30), ***B. simplex***, ***B. lunaria*** and ***B. lanuginosum*** belong to this category. It is similar to ***Ophioglossum*** in all respects.

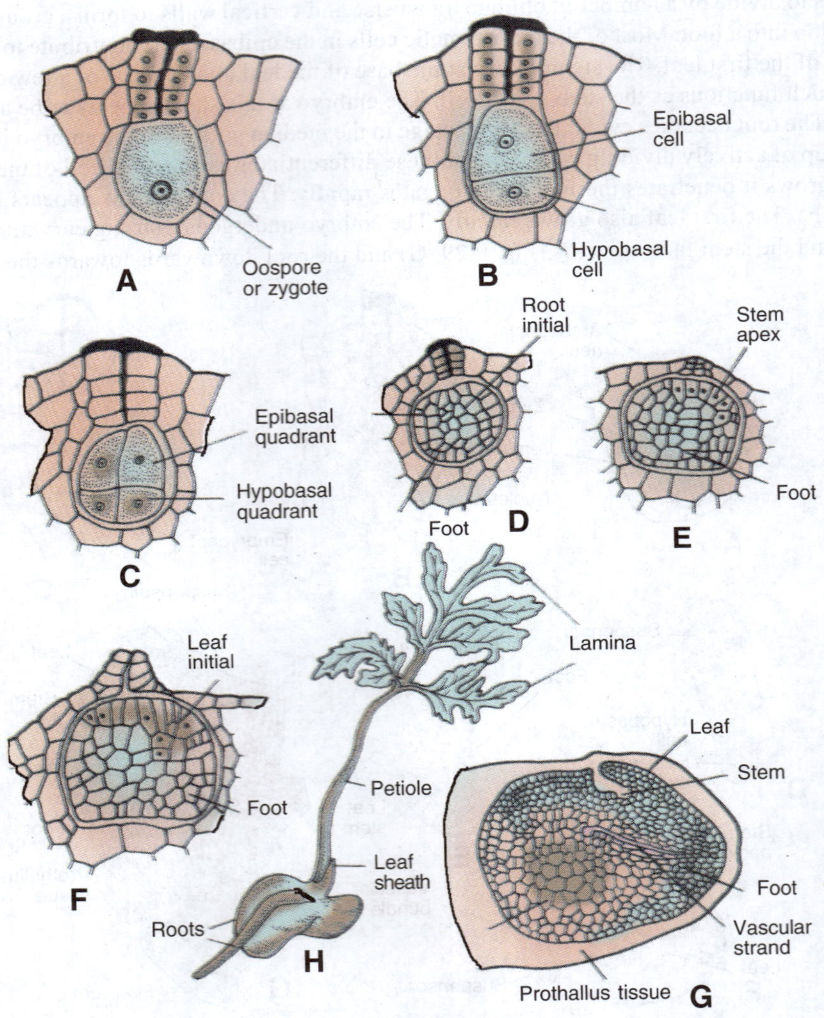

Fig. 9.30. (A—H). Various stages in the development of embryo of *B. virginianum*. (After Campbell)

A comparative study of embryogeny of the ***Botrychium*** reveals an outstanding difference of the presence and absence of suspensor in some species. Besides this there are many other differences in the details of development. These are the extent of development of the root which is well developed in ***B. viriginianum***, ***B. simplex*** and ***B. lunaria*** and comparatively less developed in ***B. dissectum***. The other differences include the initiation of the primary vascular strand and the development of first green leaf. In some species the first formed leaf comes (Fig. 9.31) above ground (***B. dissectum*** and ***B. virginianum***), in others the second leaf comes above ground (***B. simplex***) and in ***B. lunaria*** the third or fourth leaf comes above ground. The rate of growth of the embryonic stem tip also varies in different species. The origin of root and the first leaf is also variable. Wardlaw considered that these variations are due to variations in the biochemical pattern and the reaction system operating in the zygote or the egg. He attributed the presence or absence of suspensor to be due to some hereditary differences within the various species of the genus.

B. dissectum B. simplex

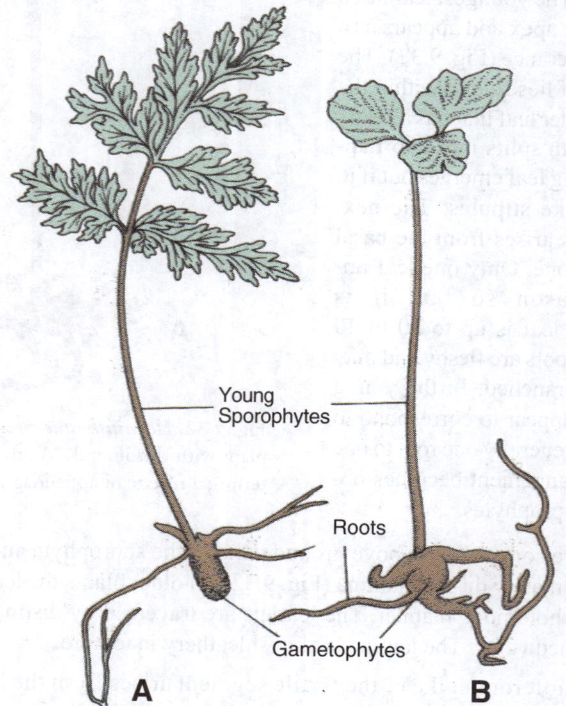

Fig. 9.31 (A—B). Young sporophytes of *Botrychium virginianum* (A) and *B. dissectum* (B).

Taxonomic Position

Sub division	:	Filicophyta
Class	:	Eusporangiopsida
Order	:	Ophioglossales
Family	:	Ophioglossaceae
Genus	:	***Botrychium***
Species	:	***lunaria***

Genus : HELMINTHOSTACHYS Kaulf

External Features

This monotypic genus is represented by ***H. zeylanica,*** which grows abundantly in North Australia and along Indo-Malayan region. In India it is found in Eastern U.P, Bengal plains, Assam and South India. The plant body consists of a prostrate rhizome that is much elongated and conspicuous than the outer two genera of the ***Ophiglossaceae***. From the rhizome arise two rows of erect and palmate leaves. The youngest leaf lies a little behind rhizome apex and appears as a small conical protuberance (Fig. 9.32). The second youngest leaf lies folded within the sheath of the next older leaf that has already expanded. The sheath splits into two flap-like parts as the young leaf emerges out if it. These flaps look like stipules. The next younger leaf always arises from the basal sheath of the older one. Only one leaf unfolds in one season so that it is monophyllous. The leaf is up to 30 to 40 cms. in height. The roots are fleshy and may be branched or unbranched. In the young rhizomes the roots appear to correspond to the leaves. There is generally one root to one leaf but such an arrangement becomes obscure in the older sporophytes.

Fig. 9.32. *Helminthostachys zeylanica.* **A.** A complete plant with rhizome; **B.** A group of sporangia with a terminal rosette of appendages. (B. after Bower)

The leaves are conspicuous above ground parts of the sporophyte and possess long and slender petioles with palmately divided lamina (Fig. 9.32). In older plants the leaflets are further divided in an apparently dichotomous manner. The leaflets are traversed by distinct veins that give off dichotomously branched veins. The lamina is almost leathery in texture.

The sporangioferous spike or the fertile segment arises from the base of the lamina (Fig. 9.32) and is further divided into short lateral branches which bear groups of terminal sporangia (Fig. 9.30). They do not form distinct marginal rows as in ***Ophioglossum.*** In some cases the tips of branches bearing sporangia give rise to green and leaf-like appendages. They are arranged in a rosette-like manner at the apices of branches. The sporangia appear to arise below them and there is a close resemblance with the sporangiophores of ***Equisetum***. But closer examination clearly reveals that the sporangia have no connection with the leaf-like appendages. The sporangia arise directly on the short branch. These green appendages can be compared to the indusia of the ferns.

Anatomy

Rhizome

The rhizome has a central stele that is medullated (Fig. 9.33, A, B). The xylem is surrounded on the outside by phloem and outer endodermis. Indistinct patches of internal endodermis have also been recorded in ***H. zeylanica***. The xylem is mesarch ***i.e.*** the protoxylem is surrounded by metaxylem on either side. The stele is perforated by leaf gaps on its dorsal face. There is no cambium and thus no secondary growth.

The tracheids of the xylem have bordered pits. The phloem consists of sieve tubes and phloem parenchyma. The end walls are also perforated. The rhizome grows by means of a single apical cell.

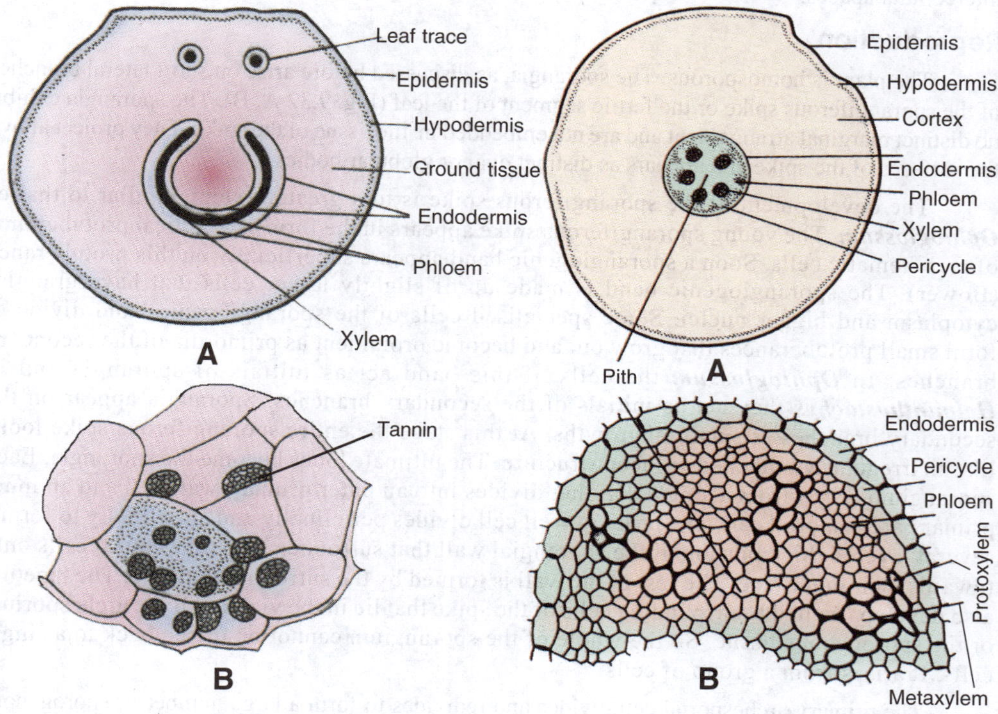

Fig. 9.33 (A–B). *Rhizome of Helminthostachys zeylanica.* A. Diagrammatic cross section; B. Tannin filled cells (After Campbell)

Fig. 9.34 (A–B). *Helminthostachys zeylanica.* A. Diagrammatic cross section of root; B. Portion of stele in detail. (After Campbell)

Roots

The roots may be triarch, pentarch or even have seven xylem masses alternating with an equal number of phloem groups. There is a distinct endodermis surrounding the stele. Pith is also distinct. The cortex is wide and composed of thin-walled cells filled with reserved food. (Fig. 9.34 A, B).

The root grows by means of a single apical cell which may later be replaced by a group of apical initials.

Leaf

(*a*) Petiole : The petiole has an outer layer of epidermal cells followed by a layer or two of thick-walled hypodermal cells. The cortex or the ground tissue is parenchymatous and in it lie embedded a ring of concentric vascular bundles. On the adaxial side of the petiole there are two small vascular bundles lying internal to the usual ring of vascular bundles. These adaxial bundles enter the sporangiferous spike or the fertile segment and constitute its vasculature. It appears that a single leaf trace on entering the petiole divides into two petiolar bundles. Subsequently these two divide into a number of separate vascular strands. The two vascular strands of the spike arise from the adaxial bundles of the petiolar ring.

(*b*) Lamina : The sterile leaflets have the two epidermal layers each made up of a single layer or compactly arranged cells. The stomata are restricted to the abaxial epidermis only. Around each stomata the epidermal cells are arranged in concentric layers. The mesophyll is distinguished into palisade and spongy tissue. The palisade is situated below the upper epidermis and consists of a single row of columnar cells. The cells of the spongy tissue are variously lobed and enclose large intercellular spaces.

Reproduction

The plant is homosporous. The sporangia, as discussed before arise on short lateral branches of the sporangiferous spike or the fertile segment of the leaf (Fig. 9.32 A, B). The sporangia exhibit no distinct marginal arrangement and are not embedded in the tissue of the spike. They project above the surfaces of the spike and appears as distinct oval or globular bodies.

The development of the sporangiferous spike is to a greater extent, similar to that of ***Ophioglossum***. The young sporangiferous spike appears in the form of a conical protuberance of meristematic cells. Soon a sporangiogenic band appears superficially on this protuberance (Bower). The sporangiogenic band is made up of slightly larger cells that have granular cytoplasm and bigger nuclei. Some specialised cells of the sporangiogenic band divide to form small protuberances that grow out and become prominent as primordia of the secondary branches. In ***Ophioglossum*** the cells of this band act as initials of sporangia and in ***Helminthostachys*** they act as initials of the secondary branches. Sporangia appear on the secondary branches as small outgrowths. At this stage the entire sporangiferous spike looks like an irregularly branched or lobed structure. The ultimate lobes become the sporangia. Each sporangium arises from a single cell that divides into an outer primary wall cell and an inner primary archesporial cell. The primary well cell divides periclinally and anticlinally to form a several layered thick portion of the sporangial wall that surrounds the archesporial cells only towards their outer face. The rest of the wall is formed by the surrounding tissue. The tapetum also develops from the surrounding cells of the spike that lie in the vicinity of the archesporium or the sporogenous tissue. So the whole of the sporangium cannot be traced back to a single cells. It arises from a group of cells.

The primary archesporial cell divides and redivides to form a large number of sporogenous cells. The cells surrounding them form a many layered thick **tapetum** and the lower and the middle portion of the sporangial wall. The stalk originates in the same manner as in ***Botrychium***. The sporogenous cells later function as spore mother cells and round off and separate. By this time the tapetal cells had degenerated to form the plasmodial fluid which nourishes the spore mother cells.

The mature sporangium has a structure exactly similar to that of ***Botrychium***.

GAMETOPHYTIC GENERATION

The gametophytes of ***Helminthostachys zeylanica*** (Fig. 9.35, A, B) live underground and are infected with a mycorrhizic fungus. They are thick and short and have an irregular shape. There is a distinct basal portion and an upright cylindrical portion. The former is tuberous and bears rhizoids. It is also infected with mycorrhizal fungus which is present in the central zone. The cylindrical portion bears the sex organs (Fig. 9.35, A, B). Its tip is free of mycorrhiza. The prothalli are bisexual ***i.e.*** monoecious. A tendency towards unisexuality is also noticeable in the prothalli. The bisexual prothalli are protandrous. The unisexual prothalli are chiefly antheridial.

Helminthostachys zeylanica

The sex organs resemble those of ***Botrychium***. The antheridia (Fig. 9.35, B, C) are completely embedded in the prothallus. The antheridial jacket is two layered and enclosed large number of androcytes which later on metamorphose into spermatozoids. The spermatozoids are multiflagellate.

The archegonia (Fig. 9.35, D) have long necks that project above the surface of the prothallus. Their structure resembles those of ***Botrychium***.

EMBRYOGENY

The embryogeny is endoscopic and has been studied by Lang (1902, 1910, 1914). The oospore enlarge considerably in size and penetrates into the prothallus tissue. It divides by two transverse walls into three cells. The upper two that are next to the archegonial neck constitute the two-celled suspensor (Fig. 9.35a B—C). The inner or the outermost cell is the primary embryonal cell (Fig. 9.35, E). Both the suspensor cells may divide repeatedly into two tiers of cells. The embryonal cell divides into a hypobasal cell and an epibasal cell (Fig. 9.35a C). The former gives rise only to foot whereas the latter gives rise to shoot apex, first leaf and also the primary root (Fig. 9.35a D). As the embryo grows further it undergoes a strong curvature so that the shoot apex assumes an upright position (Fig. 9.35, F). At a later stage the embryo possesses an elongated hypocotyl (Fig. 9.35, G), a foot embedded in the prothallial tissue, a well developed primary root, conspicuous apex and a scale-like first leaf. The second leaf comes above ground and is functional. The shoot apex develops first into a small upright axis, which later becomes horizontal and develops into a prostrate rhizome. The second leaf has a conspicuous basal sheath which also encloses the third leaf.

The vascular strand runs from below the shoot apex to the primary root. Just below the primordium of the first leaf the vascular strands give off a small branch that constitutes its stele. The vasculature of the young sporophyte appears to be partly cauline in origin.

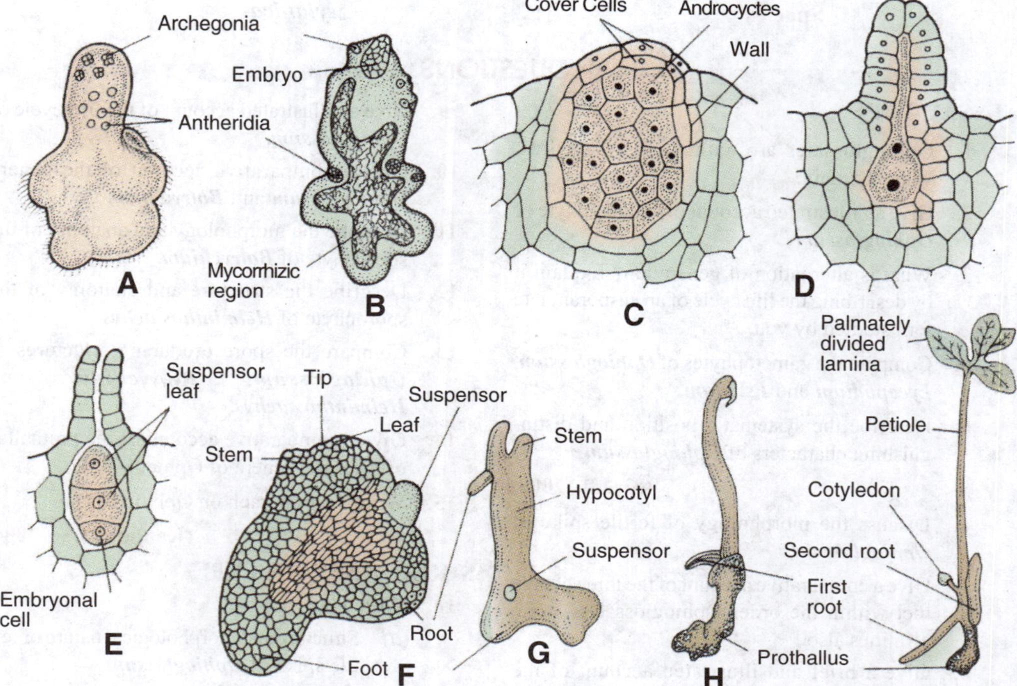

Fig. 9.35. Gametophyte (A, B). sex organs (C, D) embryo (F, G) and young sporophytes (H, I) of *Helminthostachys zeylanica*. (After Lang)

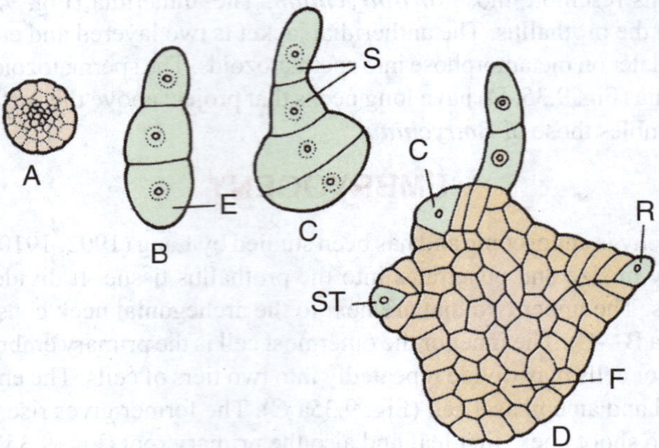

Fig. 9.35 a. Helminthostachys (A-D) Early stages in the development of embryo (E embryo cell; S, suspensor; ST. stem tip; L, leaf tip; R, root tip; F, foot) (After tang 1914)

Taxonomic Position

Sub division	:	Filicophyta
Class	:	Eusporangiopsida
Order	:	Ophioglossales
Family	:	Ophioglossaceae
Genus	:	***Helminthostachys***
Species	:	***zeylanica***

QUESTIONS

ESSAY TYPE

1. Ophioglossales are without fossil records. Why? Explain.
2. Give an illustrated account of Gametophyte of ***Ophioglossum***.
3. What is alternation of generation? Explain it by describing the life cycle of an eusporangiate fern studied by you.
4. Compare the gametophytes of ***Ophioglossum***, ***Lycopodium*** and ***Psilotum***.
5. Describe the systematic position and distinguishing characters of ***Ophioglossum***. **(Awadh, 2000)**
6. Discuss the morphology of fertile spike of ***Ophioglossum.***
7. Give a comparative account of the three genera met within the order Ophioglossales. Write affinities also.
8. Give a brief and illustrated account of the morphology and anatomy of the sporophyte of ***Ophioglossum***.
9. Give an illustrated account of the life-cycle of ***Ophioglossum.***
10. Give a comparative account of the genera ***Ophioglossum*** and ***Botrychium.***
11. Describe the morphology and anatomy of the sporophyte of ***Botrychium.***
12. Describe the structure and anatomy of the sporophyte of ***Helminthostachys***.
13. Compare the spore producing structures of ***Ophioglossum***, ***Botrychium*** and ***Helminthostachys.***
14. Give a comparative account of the prothallus of the three genera of Ophioglossales.
15. Write a monograph on Ophioglossum. **(Kanpur M.Sc. 2004)**

SHORT ANSWER TYPE

16. Write notes on:
    (*a*) Structure and morphological nature of fertile spike of ***Ophioglossum***.
    (*b*) ***Botrychium.***

(*c*) Spore producing organs in ***Ophioglossum.***

(*d*) Eusporangiate and leptosporangiate ferns. **(Kanpur M.Sc. 2000)**

(*e*) Fertile spike of ***Ophioglossum.*** **(Kanpur M.Sc. 2000; Rohilkhand, 2000)**

(*f*) Spike of ***Ophioglossum.***

(*g*) Sporophyte of ***Ophioglossum.***

(*h*) Morphological nature of spore producing structure in ***Ophioglossum.*** **(Poorvanchal, 2001)**

(*i*) Spore producing structures of ***Botrychium.***

(*j*) Spore producing structure of ***Helminthostachys.***

(*k*) Gametophyte of ***Ophioglossum.***

(*l*) Describe Meorphological nature of spike of ***Ophioglossum.*** **(Kanpur M.Sc. 2002, 2003)**

16. How will you distinguish between:

(*a*) Fertile fronds of ***Botrychium*** and ***Osmunda***

(*b*) Leaf of ***Ophioglossum*** and ***Botrychium***

(*c*) Eusporangiate and Leptosporangiate.

(*d*) External features of ***Ophioglossum*** and ***Botrychium.***

(*e*) External features of ***Ophioglossum*** and ***Botrychium.***

17. What is the botanical name of Adder's tongue fern? **(Kanpur 2000)**

18. Give the name of an epiphytic pteridophyte. **(Poorvanchal, 2003)**

19. Compare the Systematic position of ***Ophioglossum*** and ***Psilotum.***

20. Draw well labelled diagrams of the L. S. of spore producing organs of ***Ophioglossum***, ***Botrychium*** and ***Helminthostachys.***

21. Draw a labelled diagram of the T. S. of rhizome of ***Botrychium*** showing secondary growth.

22. Describe the development of spike and sporangia in ***Ophioglossum.***

OBJECTIVE TYPE

23. Select the correct answer :

(i) Secondary growth is observed in the stem of

(*a*) ***Botrychium*** (*b*) ***Lycopodium***
(*c*) ***Equisetum*** (*d*) ***Marselia.***

(ii) Development of Sporangium in ***Ophioglossum*** is

(*a*) Leptosporangiate
(*b*) Eusporangiate
(*c*) Both of the above
(*d*) None of the above.

(iii) Sporangia of ***Ophioglossum*** are seen in the form of a

(*a*) Strobilus (*b*) Sporophylls
(*c*) Spike (*d*) Tassel.

(iv) The stele in stem of ***Ophioglossum*** is

(*a*) Protostele
(*b*) Solenostele
(*c*) Dictyostele
(*d*) Siphonostele.

(v) Medullated protostele is observed in the

(*a*) lower portion of the stem of ***Ophioglossum***
(*b*) little above the base of stem of ***Ophioglossum.***
(*c*) upper portion of the stem of ***Ophioglossum***
(*d*) root of the ***Ophioglossum.***

(vi) Which of the following is commonly known as Adder's Tongue fern?

(*a*) ***Ophioglossum***
(*b*) ***Osmunda***
(*c*) ***Marattia***
(*d*) ***Dryopteris.***

(vii) Mycorrhizal fungus enters the spore germinating into a young prothallus in

(*a*) ***Ophioglossum*** (*b*) ***Lycopodium***
(*c*) ***Equisetum*** (*d*) ***Pteris.***

(viii) In most species of *ophioglossum* the prothalli are usually unbranched and

(*a*) Cylindrical (*b*) Oval
(*c*) Rectangular (*d*) Epiphytic.

(ix) Leaf gap and Leaf trace are usually seen in the rhizome of

(*a*) ***Lycopodium*** (*b*) ***Selaginella***
(*c*) ***Botrychium*** (*d*) ***Equisetum.***

(x) Which of the following is monotypic?

(*a*) ***Ophioglossum***
(*b*) ***Botrychium***
(*c*) ***Helminthostachys***
(*d*) ***Equisetum.***

Angiopteris-Fertile Leaf

CHAPTER 10

Eusporangiopsida (Contd.) - Marattiales

- Characteristic features
- Description of ***Marratia*** Swartz

CHARACTERISTIC FEATURES

The Marattiales are characterised by the follow ing important features :—

1. The sporophytes are terrestrial and have well developed subterranean stems or rhizomes that are massive (Fig. 10.1), tuberous (***Angiopteris, Marattia***) or sometimes horizontal (***Kaulfussia***). They range in size from a few inches (***Danaea***) to 5 to 6 metres in length (***Angiopteris, Marattia***).

2. The leaves show circinate ptyxis in a bud condition (Fig. 10.2).

3. The leaves have a swollen and stipulate leaf base (Figs. 10.1, 10.2).

4. The leaves (Figs. 10.1, 10.2) are pinnately compound (***Marattia, Angiopteris, Danaea***). In ***Christensenia*** and ***Kaulifussia*** the leaves are palmate compound. In ***Danaea simplicifolia*** the leaves are simple and lanceolate.

5. Mucilage canals and tannin cells are present in the tissues of the stem.

6. The sporangia occur in sori on the dorsal surface of the sporophylls. They are situated on the backs of the veins and on vein endings.

7. The sori may be long and linear, oval or even circular in outline.

8. The sporangia may be fused to form synangia (***Marattia***) or they may be free (***Angiopteris***). Rudimentary scales with fringed margins cover the sporangia in Angiopteridaceae and ***Marattia.***

Fig. 10.1. Habit of *Angiopteris evecta.*

Angiopteris

Marattia

Christensenia

Kaulfussia

9. The sporangia are sessile and do not possess an annulus. In ***Angiopteris*** (Fig. 10.3) there is a rudimentary annulus. Stomata may occur in the sporangial or synangial walls.

10. The sporangia are emergent and develop simultaneously and dehisce by means of vertical slits. Their development is eusporangiate. Each sporangium produces numerous (400—700) tetrahedral or bilateral spores.

11. The prothalli are above ground and green in colour. They vary in shape and may be cordate or oblong and live long. The central portion of the prothallus is cushion-like and thick. This part bears sex organs. The prothallus is mycorrhizic and monoecious.

12. The sex organs are borne on the ventral side of the prothallus and are embedded in the prothallus tissue. The necks of archegonia may project only slightly. In some species the archegonia may be borne on the dorsal side of the prothallus.

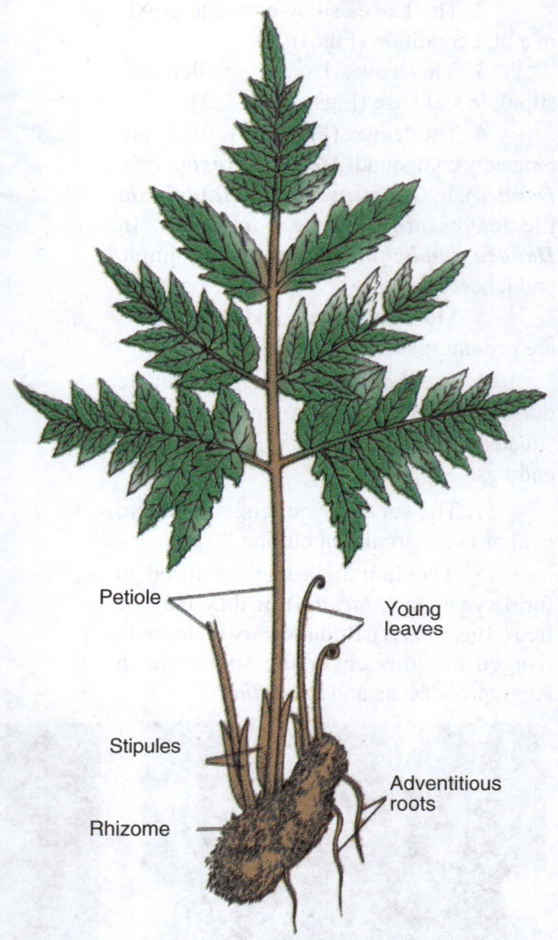

Fig. 10.2 *Marattia sp.* A young Sporophyte.

Christensen (1938) divided the order into three families : (1) ***Marattiaceae,*** (2) ***Angiopteridaceae*** and (3) ***Asterothecaceae.*** He placed ***Angiopteris*** (with about 100 species), ***Macroglossum*** (with two species) and ***Archangepteris*** (with 4 species) in the family ***Angiopteridaceae.*** In the family ***Marattiaceae*** he included the genera ***Marattia*** (with 50 species), ***Christensenia*** (with 2 species) and ***Danaea*** (with 30 species). The fossil genera are included in family ***Asterothecaceae*** (***Asterotheca***, ***Acitheca***, ***Eoangiopteris***, ***Psaronius*** and ***Scolocopteris***. Smith places all the known genera in a single family ***Marattiaceae***.

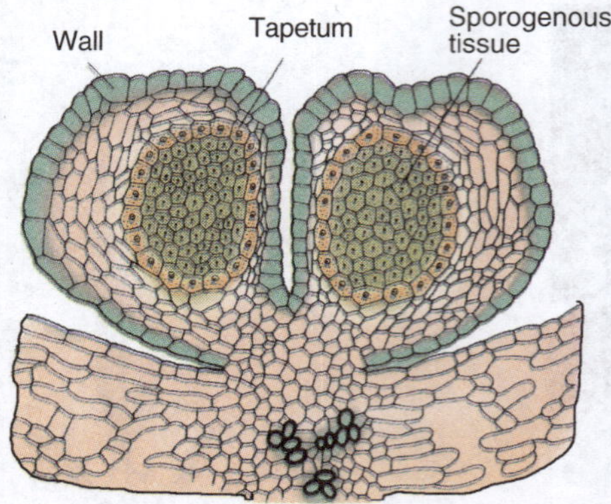

Fig. 10.3 *Angiopteris evecta.* A section through the sorus.

Spore (1968) recognised the following 5 families : (1) ***Asterothecaceae*** which includes all the fossil genera, (2) ***Angiopteridaceae*** (***Angiopteris***) ; (3) ***Marattiaceae*** (***Marattia***) ; (4) ***Danaeaceae*** (***Danaea***) and (5) ***Christenseniaceae*** (***Christensenia***). Due to close similarities between the various genera it is reasonable to follow Christensen's classification in this text. Biehorst (1971) includes only families in this order, which

Acitheca
Psaronius
Macroglossum
Archangiopteris

he treats as an order with "obscure affinities." He has included ***Marattia*** (60 sp.), ***Angiopteris*** (100 sp), ***Macroglossum*** (2 sp.) and ***Archangeopteris*** (1-4 sp.) in family ***Marattiaceae. Danaea*** (30 sp.) and ***Christensenia*** (1 sp.) are included under the family ***Danaeaceae***.

Family MARATTIACEAE

The family is characterised by the coalescence of sporangia into synangia. The annulus and the recurrent veins are absent. The indusial scales may or may not be present. The synangia are oval or linear in ***Marattia*** (Fig. 10.4 A) and ***Danaea*** (Fig. 10.4, E) and circular in ***Christensenia*** (Fig. 10.4, C). The leaves (Fig. 10.2) may be unipinnate, bipinnate or even tripinnate. They are rarely entire. Venation is dichotomous in ***Marattia*** and reticulate in ***Christensenia***. The rhizomes may be tuberous, conical and upright as in ***Marattia*** (Fig. 10.2) or it may be creeping as in ***Danaea*** and ***Christensenia***. The sporangia open by an internal slit in ***Marattia*** and ***Christensenia*** and by a terminal pore in ***Danaea*** (Fig. 10.4, F). The indusrial scales are absent in ***Christensenia*** and ***Danaea*** and may be present in ***Marattia*** (Fig. 10.4, B). The sori are scattered in between the veins of the leaflets in ***Christensenia***. They are situated below the veinlet endings. In ***Marattia*** they are near the margin. In ***Danaea*** the sporangial sori extend from the midrib to the margin of the leaflets. The sporangia in this genus are slightly sunk in the fleshy tissue of the leaftlet so that the fleshy flanges of tissue project in between the adjacent synangia (Fig. 10.4, B). The synangia form circular sori or patches (Fig. 10.4, C,D) in ***Christensenia*** and open by internal slits into a central furrow. They are closely united. In ***Marattia*** the synangia form linear or oval sori (Fig. 10.4, A, B) and the individual sporangia are fused in their basal portions (Fig. 10.4, B). In ***Danaea elliptica*** (Fig. 10.4, E, F) the synangia form linear sori and the sporangia are so intimately fused that their individuality is lost (Fig. 10.4, F). The number of sporangia fused to form a synangial sorus varies from 16-40 in ***Marattia***, 10-20 in ***Christensenia*** and up to 40 in ***Danaea***.

GENUS– MARATTIA Swartz

The genus is widely distributed in the tropical regions of the world and is represented by about 60 species (Sporne, 1966). According to Christensen (1938) the genus includes about 50 species. None of the known species have been reported from India and the United States of America. A few species are known from West Indies, Mexico, Central America and Hawaii islands. The common species are ***Marattia douglasii***, ***M. alata***, and ***M. fraxinea***.

Marratia fraxinia

SPOROPHYTE

External Features

A full grown plant of ***Marattia*** consists of a massive, tuberous and an upright subterraneant rhizome (Fig. 10.2). The rhizome may reach a height of about one metre and may be up to a metre in diametre. The leaves arise in

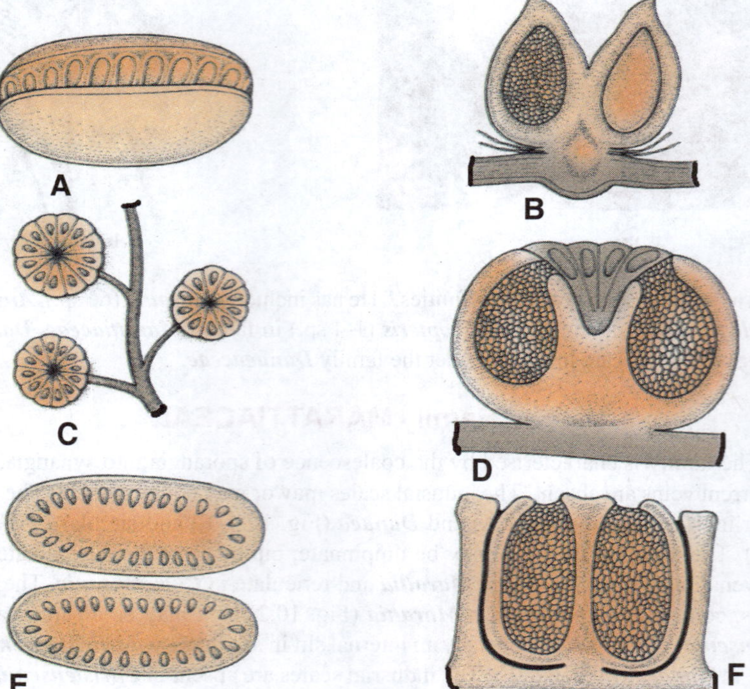

Fig. 10.4 (A–F). Structure and arrangement of sporangia and synangia in the Marattiaceae. A–B. The synangium of ***Marattia fraxinea*** in surface view (A) and in section (B); **C,D.** The synangia in ***Christensentia.*** In surface view (C) and a synangium in transverse section (D); **E,F.** The synangia in ***Danaea elliptica*** in surface view (E) and one as it appears in a section (F) (A, after Bitter; B, after Hooker-Baker; C, after Bitter, D, after Hooker-Baker; E, after Bitter and F, after Hooker-Baker).

close spirals and form a tuft of large fronds that may achieve a height of 5-6 metres. The leaves are pinnately compound (Fig. 10.2) and may be bipinnate or tripnnate. The margins of the leaftlets are serrate in ***M. fraxinea*** and ***M. alata***. They may be slightly or deeply incised in ***M. douglasii***. The leaves have open furcate venation. Every leaf has a stipulate leaf base (Fig. 10.2). Two stipules are persistent and rise on either side of the pulvinate base of the petiole and almost cover it. The long and branched adventitious roots from the rhizome and each root arises opposite the leaf base so that there is a regular insertion of one root to one leaf (Fig. 10.2.) The younger parts of the plant are covered with short simple hair. The young leaves are circinately coiled (Fig. 10.2) and they gradually uncoil as they grow in size and age and ultimately unfold as pinnately compound fronds. The leaves of ***M. alata*** may be 3—5 times pinnate and 2—3 metres in length with petiole reaching a diameter 4—6 centimetres. Each pinnule is traversed by a distinct midrib that gives off lateral veins that may be simple or dichotomously branched.

Anatomy

Stem

A transverse section of the stem reveals an irregular outline. There is a distinct epidermal layer whose cells are thick-walled in mature rhizomes. The entire cortex is parenchymatous. A ring of mucilage canals is present in the outer layers of the cortex (Fig. 10.5, A-C). These canals arise lysigenously. Tannin filled cells are also present in the cortex. Endodermis is distinct in the younger stem. Its identity is lost in the older rhizomes.

The stelar organisation in the younger or earlier formed portions of the rhizome is **protostelic** (Fig. 10.5, A). The **protostele** becomes medullated and later at a level where the third or fourth leaf arises the medullated protostele or the uninterrupted **siphonostele** becomes interrupted by the departure of a leaf trace that leaves a gap. At first only one leaf trace departs from the siphonostele and enters the leaf. Later two leaf traces depart and enter a leaf. A characteristic feature of the stelar organisation of ***Maratiaceae*** is presence of a central commissural strand (Fig. 10.5, B) in the stem. This strand traverses across the pith at a level where the protostele changes to a siphonostele. In the older rhizomes of ***Marattia*** the central commissural strand breaks up into a number of separate strands in the centre of the stem (Fig. 10.5, C). These strands appear as a ring of vascular bundles internal to the siphonostele of the rhizome. The siphonostelic vascular strand becomes broken up into a number of separate strands in older portions of the rhizome. This is due to the appearance of overlapping leaf gaps left by the departing leaf traces. So in the older portions of the stem the vasculature becomes a dictyostele. The meristeles of the **dictyostele,** the inner ring of commissural strands and the scattered leaf and root traces give the appearance of a **polycyclic dictyostele** (Fig. 10.5, C).

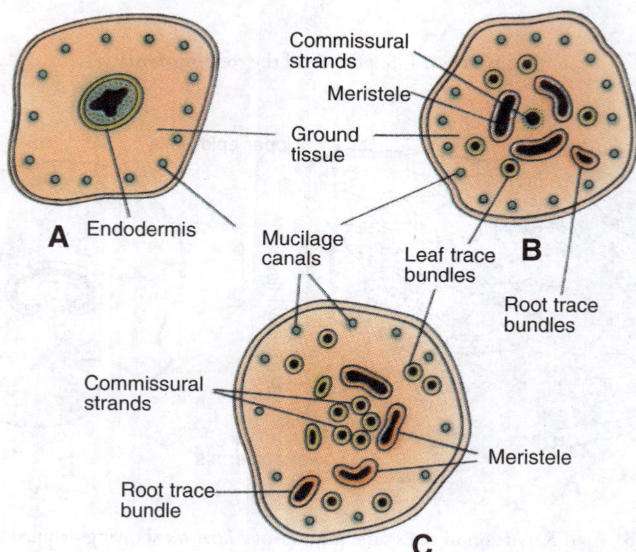

Fig. 10.5 (A–C). Anatomy of the stem of *Marattia.* Outline sketch showing the stelar organisation at various levels of the stem

Root

The roots have a many layered and parenchymatous cortex. The cortex contains numbers mucilage canals. An endophytic fungus has been reported in the younger roots. It is absent in the older roots. The stele consists of up to twelve xylem bundles alternating with the same number of phloem bundles (Fig. 10.6). Tanin cells are also present in the parenchymatous bands that separate the xylem bands. The xylem is exarch.

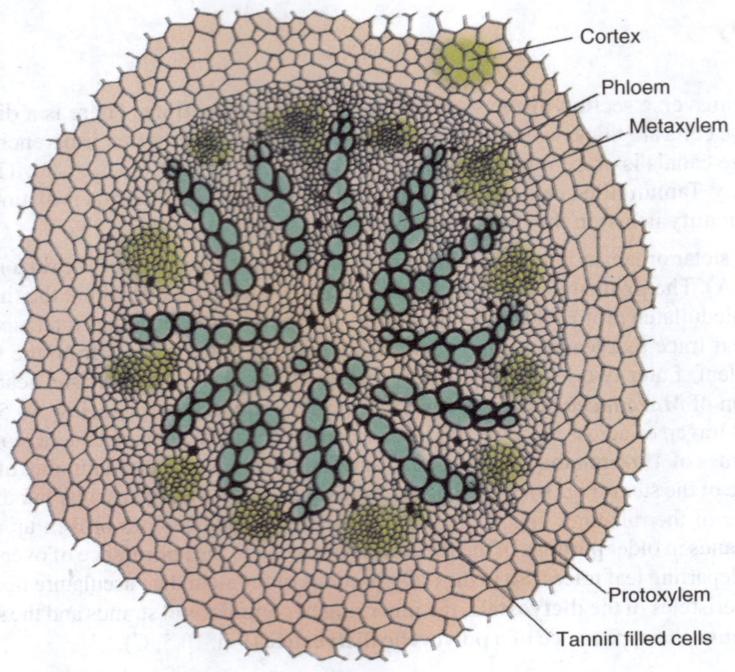

Fig. 10.6. T. S. portion of the root of *Marattia.*

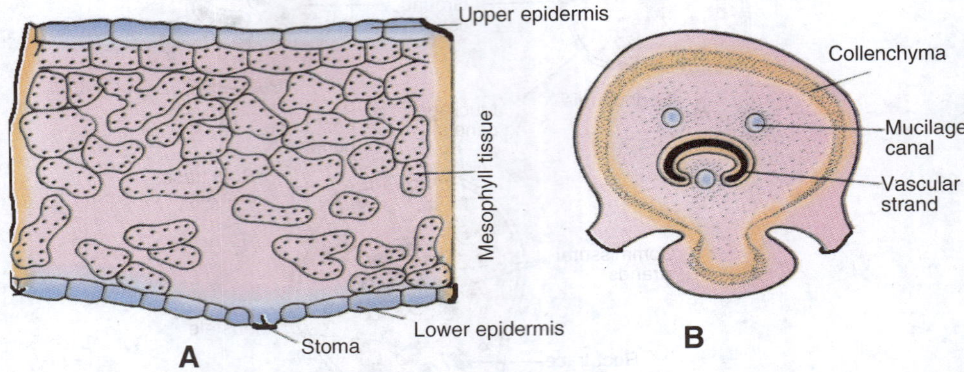

Fig. 10.7. (A—B). A. T. S. portion of the sterile pinnule of *Marattia* showing detailed internal structure; B. Outline sketch of a cross section through the mid-rib region of the pinnule of Marattia. (B. after Campbell).

Leaf

The structure of petiole is similar to that of the rhizome except that a distinct collenchymatous hypodermis is present below the epidermis. The rest of the ground tissue is parenchymatous. There are two or more concentric circles of vascular bundles. Endodermal boundaries are lacking around the meristeles.

The sterile lamina of the pinnules has distinct upper and lower epidermal layers with stoma restricted to the lower epidermis (Fig. 10.7, A). The mesophyll is distinguished into a well defined palisade below the upper epidermis and a spongy parenchyma. The palisade cells are more or less compactly arranged and have only small intercellular spaces. The spongy parenchyma cells are loosely arranged and form a conspicuous aerating tissue. The cells in the midrib region are compactly arranged. A well developed collenchymatous band of tissues is also seen in this region (Fig. 10.7, B). A few mucilage canals are also present in this region. There is a single crescentic vascular bundle in the midrib. The bundle is concentric but phloem is well developed on the side facing lower epidermis. A distinct endodermis is present around the bundle.

Reproduction

Marattia is homosporous. The spores are produced within thick and massive sporangia that coalesce together to form synangia. The sporangia are arranged in two marginal rows in a linear or oval sorus (Fig. 10.8, A). All the sporangia in a sorus are fused to form a single synangium (Fig. 10.8, B, C). Each sours contains 16—40 fused sporangia. The sori lie on the lateral veins a little below their ends (Fig. 10.8, A) and are marginal in position. They lie a little distance away from the margin of pinnule and are present on their dorsal sides. A few scale-like outgrowths arise from the margins of sori (Fig. 10.8, B) and cover them partially. The pinnules bearing the sori are called the sporophylls and are alike in size and shape to the sterile pinnules.

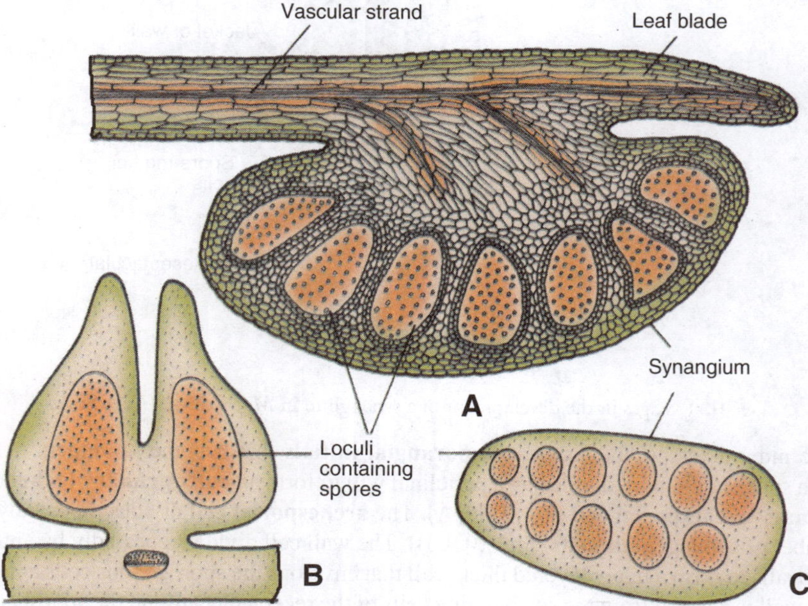

Fig. 10.8. (A—C). Synangia of *Marattia alata*. A. Longitudinal vertical section through a synangium showing vascular supply; B. T.S. synangium; C. A mature synangium cut in a horizontal section. (B, after Hooker-baker)

Development of the synangium starts when the sporophyll is very young. Some superficial or the epidermis cells become distinct by their denser cytoplasmic contents and prominent nuclei. These groups of cells are at first in line with the epidermis surface of the sporophylls. Later they divide and grow to form an elevated receptacle of the sorus (Fig. 10.9, A). The receptacle is now visible as a distinctly elevated structure above the epidermis. A few sterile hair grow from the cells of the epidermis surrounding the receptacle. These hairs are simple and multicellular (Fig. 10.9, A). The elevated receptacle has a distinct median groove or cleft so that the receptacular area appears in the form of two parallel ridges. These ridges grow and meet over the median cleft so that the latter is hidden from view in the mature sorus. On these ridges develop the sporangia whose lateral walls become confluent to form a synangium. A horizontal section of the sorus or the synangium (Fig. 10.8, C) at this stage shows two marginal rows of fused sporangia.

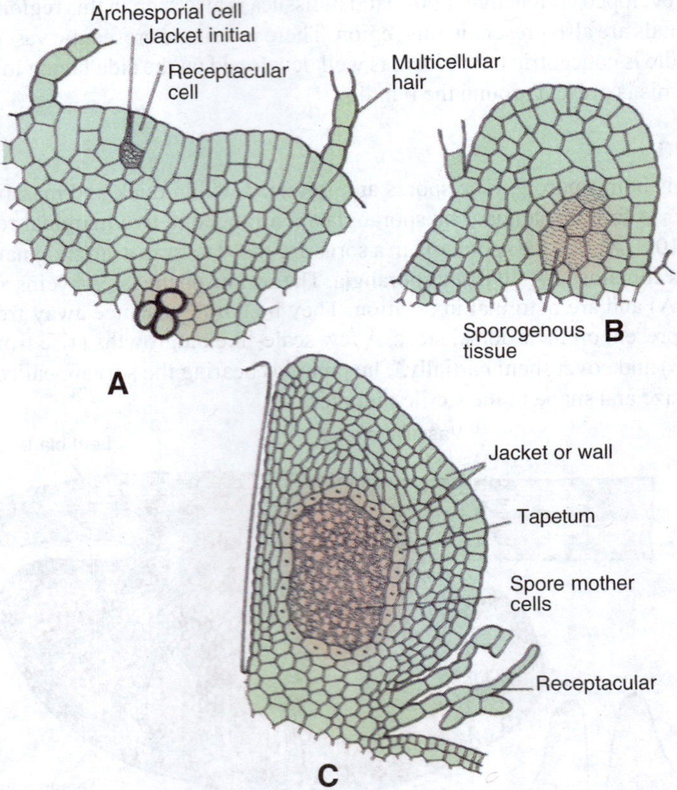

Fig. 10.9. Stages in the development of a synangium in *M. fraxinea*. (After Bower)

The individual sporangia develop **sporangial initials** that become prominent in the receptacle. Each sporangial initial divides by a periclinal will to form an outer **primary jacket** cell and an inner **primary archesporial cell** (Fig. 10.9, A). The **archesporial cell** divides repeatedly to form a large number of **sporogenous cells** (Fig. 10.9, B). The wall cell divides repeatedly by anticlinal and periclinal walls to form a many layered thick wall that envelops the sporogenous cells at the top. The rest of the wall is derived from the surrounding cells of the receptacle, around the sporogenous mass. So a sporangium is derived from a group of cells and is, therefore, **eusporangiate** in development. The cells of the jacket immediately surrounding the sporogenous mass form the **tapetal layers.** All the 500 or 512 spore mother cells are functional and undergo meiosis to form 2,000 or 2,048 spores. The spores are arranged in **tetrahedral tetrads.**

At maturity the outer walls of the fused sporangia become thickened. During dehiscence of the sporangium these walls contract due to dehydration and cause the separation of two fused ridges of the receptacle. This process can be linked to the opening of a closed book. The enclosed cleft is now exposed. Each sporangium which looks like a loculus of the sporangium now dehisces by an inner vertical slit. The spores sift out into the cleft and are then dispersed by the wind.

GAMETOPHYTE

The spores germinate to give rise to the prothalli or the gametophytes. The spores are minute tetrahedral structures. They are non-green and uninuculate. The cytoplasm is filled with starch grains, oil and albuminous substances. Jonkmann (1878) studied the germination of the spores in ***Marattia fraxina*** and ***Angiopteris.*** The details of germination are not clear. On reaching a suitable subtratum the spores absorb water and grow in size. The spore wall ruptures exposing the contents enclosed within the inner wall. The spore increases many times its original diameter before it sets to divide. The first division wall divides it into a smaller rhizoidal cell and a slightly large prothallial cell (Fig. 10.10 A). Both the cells develop abundant chloroplasts. Distinction into a rhizoidal cell and a prothallial cell is not always clear. In the former case the upper cell develops into the main body of the prothallus. In the latter case the first two cells are equal in size and then divide by walls at right angles to the first wall (Fig. 10.10, C). This results in the formation of a quadrant. One of these cells divides by two oblique walls to form a two-sided apical cell whose activity leads to the formation of a young prothallus. As the prothallus grows older the single apical cell is replaced by a row of apical cells. Further growth of the prothallus takes place by the activity of this group of apical cells.

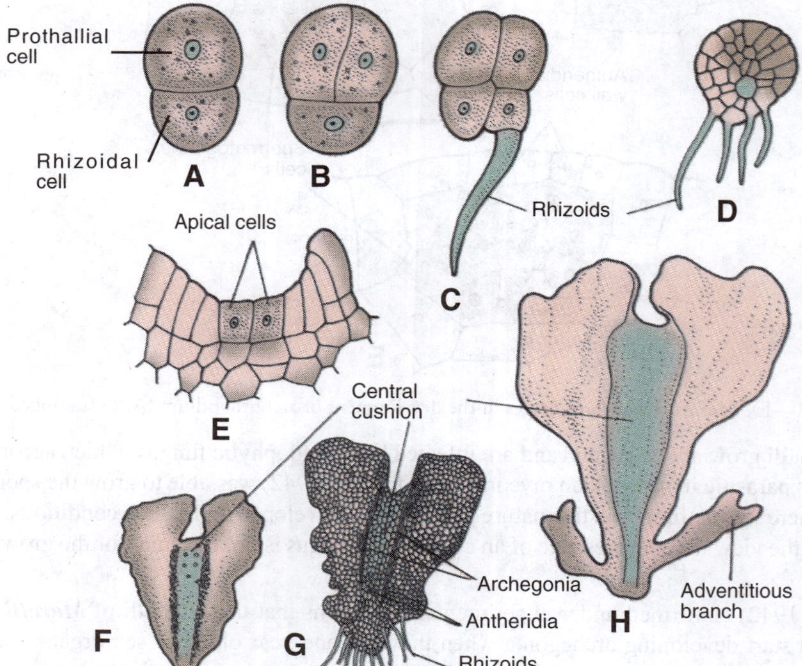

Fig. 10.10. Stages in the germination of spores in *M. fraxina* (A—D) and mature prothali of ***Marattia***; E. Horizontal section of the apex of prothallus of ***M. douglasii*** showing two apical cells; F. A young prothallium of ***M. douglasii***; G. A mature prothallus of ***M. douglasii*** showing position of sex organs; H. A mature prothallus bearing adventitious branches. (A—D, after Jonkman; E—H, after Campbell)

Before the inception of a group of apical cells the prothallus is more or less spathulate in form. Soon the segments cut off by the apical cell grow rapidly so that the prothallus develops a notch at the apex. The apical cell now lies in the notch. The prothallus assumes a heart-shaped appearance (Fig. 10.10, F, G). A cushion-like hemispherical mass of tissue becomes prominent in the middle region of the prothallus. Towards the margins the thickness of the prothallus goes on decreasing.

Numerous rhizoids arise from the ventral surface of the prothallus and penetrate the soil. The rhizoids are unicellular. The prothallus is covered by cuticle on both the surfaces and is green and photosynthetic. Dichotomous branching of the mature prothallus has also been recorded in ***M. douglasii*** (Campbell, 1911). In the same species Campbell recorded the formation of adventitious branches from the margins of the prothallus (Fig. 10.10, H). These branches grow from small buds and later get detached and grow into new prothalli.

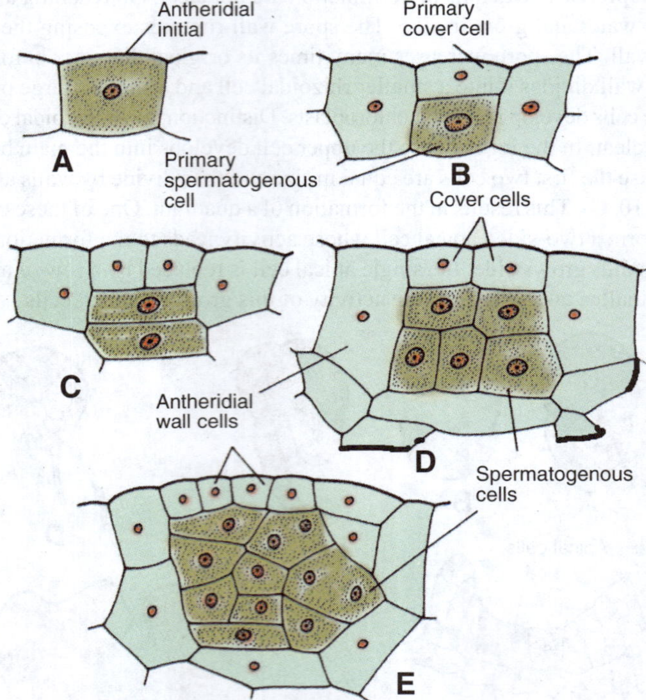

Fig. 10.11. (A—E). ***Marattia douglasii.*** Stages in the development of an antheridium.(After Campbell)

The prothalli grow above ground and are infected by an endophytic fungus, which according to West (1917) is parasitic in nature than mycorrhizal. Stockey (1942) was able to grow the spore on sterile media. There was no fungus in the mature prothalli that developed under such conditions. This lends support to the view that the presence of an endophytic fungus is not essential for the growth of the prothallus.

Stockey (1942) confirmed under experimental conditions that the prothalli of ***Marattia*** are protandrous and start developing archegonia when they are one year old. The sex organs usually appear on the ventral side in the region of the cushion. The archegonia develop near the notch and the antheridia below. Stockey (1942) observed that after the formation of archegonia had commenced the antheridia started appearing on the dorsal side of the prothallus. So the prothalli bear archegonia on the lower surface and the antheridia on both the upper and the lower surfaces. The sex organs are sunk in the tissue of the prothallus.

The structure and development of antheridium (Fig. 10.11) and archegonium (Fig. 10.12) are similar to that of ***Ophioglossum.*** (refer to Chapter IX). Mature spermatozoids of ***Marattia*** sp. are coiled and uniflagellate (Fig. 10.13) In ***Marattia*** the ventral canal cell is distinct and not transitory and obscure as in ***Ophioglossum***.

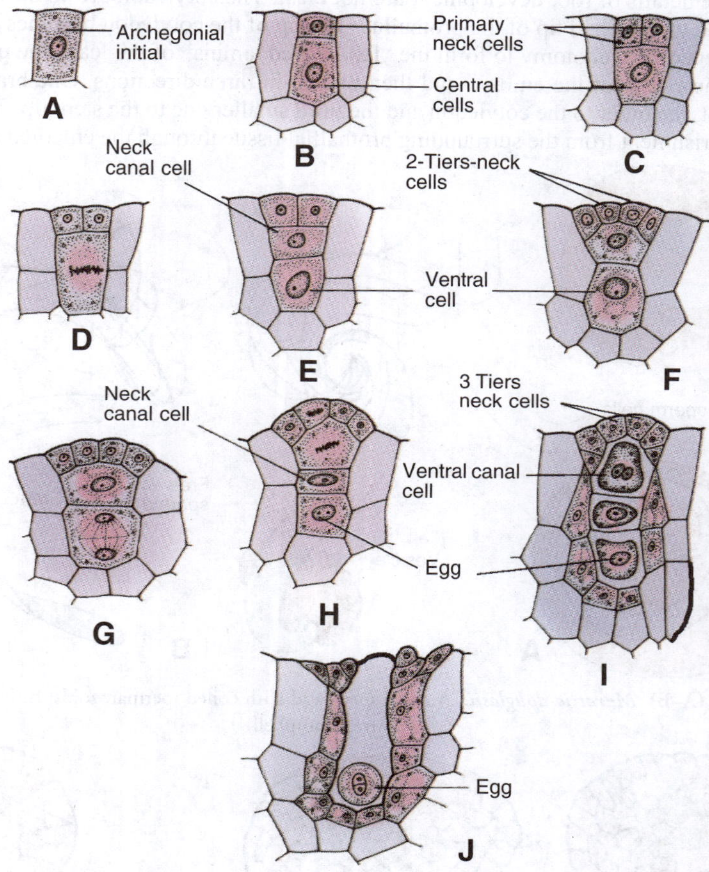

Fig. 10.12. (A—J). Various stages in the development of archegonium in *Marattia douglasii.*

(After Campbell)

EMBRYOGENY

The embryogeny of ***Marattia*** (Fig. 10.14, A—F) is not perfectly known. Jonkman (1879), Farmer (1892) and Campbell (1894) studied the embryogeny of ***Angiopteris*** and ***Marattia.*** In ***M. douglasii*** (Campbell, 1894) the zygote divides by a transverse wall into a hypobasal cell (towards archegonial neck) and an epibasal cell. The former gives rise to the foot alone. The stem, the leaf and the root arise from the epibasal cell. The second and third divisions give rise to a globular eight-celled embryo. The subsequent divisions are not laid in any regular sequence. The stem apex and foot become evident when the embryo becomes massive and consists of many cells (Fig. 10.14, A). The cotyledon and the primary leaf makes its appearance at a later stage from the epibasal half of the embryo. The anterior pair of epibasal octants is considered to give rise to the primordium of first leaf. It is believed that the apical cell might be present in the earlier stages of growth of the cotyledon, because the distinct apical cell becomes evident only at a later stage when it has become quite conspicuous. The foot also becomes conspicuous by this time. The stem apex remains overshadowed by the fast developing cotyledon. The root initial appears endogenously in the middle portion of the embryo where the epibasal and hypobasal halves meet (Campbell, 1894). This apical cell, according to Campbell (1894) appears to be four sided. The segments are cut off from all the four sides (Campbell, 1894). It develops into a primary root that grows downwards through the prothallial cells and enters the soil. The details of root development are not clear. The cotyledon grows upwards and does not curve around the notched tip of the prothallus. The tip of the cotyledon becomes flattened and then undergoes unequal dichotomy to form the "fan-shaped lamina" of the leaf. The procambial strands appear in the centre of the embryo and then branch in three directions. One branch goes into the primary root, the other to the cotyledon and the third smaller one to the stem tip. The young embryo derives nourishment from the surrounding prothallial tissue through the enlarged cells of the foot.

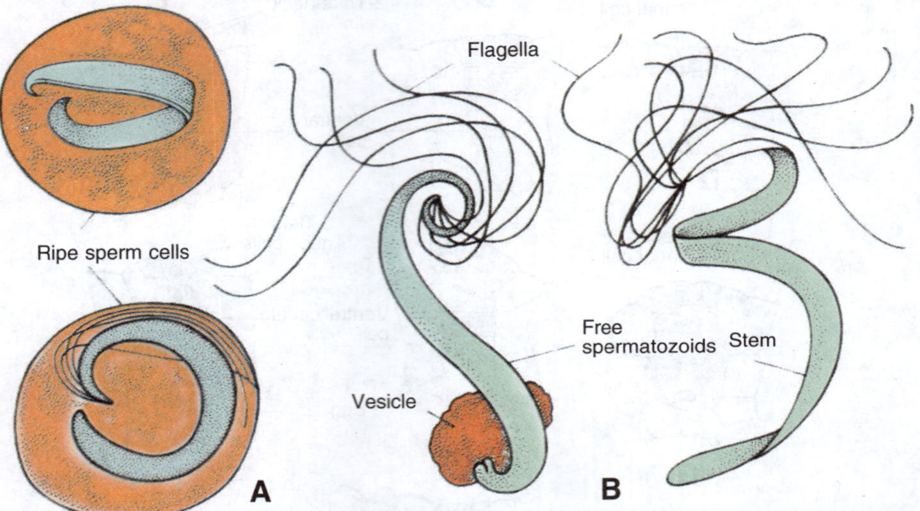

Fig. 10.13 (A–B). ***Marattia douglasii.*** **A.** Ripe spermatid with coiled spermatozoids; **B.** Free spermatozoids. (After Campbell)

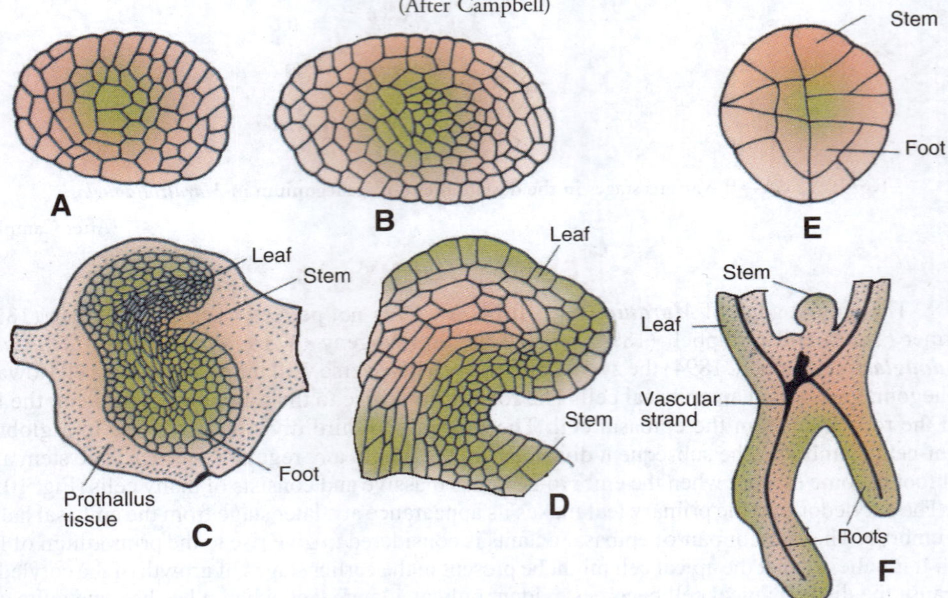

Fig. 10.14. ***M. douglasii*** Stages in the development of embryo in (A—E); F. L.S. Young sporophyte showing stem tip, part of leaf and two roots. (After Campbell)

Fig. 10.15. A young sporophyte of *M. douglasii* (After Campbell)

The prothallus continues to live for many months even after the young sporophyte has become independent. Tracheidal elements with reticulate thickenings start appearing in the procambial tissue lying at the junctions of the primary root, cotyledon and the stem. Later the differentiation of tracheids and other tracheary elements proceed towards the growing points of these organs. Like this the primary vasculature of the young sporophyte is laid down and it becomes independent.

Taxonomic Position

Sub-division	:	Filicophyte
Class	:	Eusporangiopsida
Order	:	Marattiales
Family	:	Marattiaceae
Genus	:	***Marattia***
Species	:	***douglasii***

QUESTIONS

ESSAY TYPE

1. What is alternation of generation? Explain it by describing the life-cycle of an eusporangiate form studied by you.
2. Describe the habit of ***Marattia*** with the help of labelled diagrams. Describe the structure of the sporophyte of ***Marattia.***
3. Give a brief account of the reproduction in the genus ***Marattia.***
4. How will you differentreate the spore producing structures of ***Marattia*** and ***Ophioglossum***? Explain with labelled diagrams.
5. Give an account of the gametophyte of ***Marattia***.
6. With labelled diagrams describe the development of antheridium and archegonium in ***Marattia.***
7. Give salient features of Marattiales. Classify the order into families with the names of the genus.

SHORT ANSWER TYPE

8. Ophioglossales are without fossil records. Explain.
9. Write short notes on
   (*a*) ***Marattia***
   (*b*) Sorus of ***Angiopteris*** and ***Marattia*** **(Awadh M.Sc., 2000)**
   (*c*) Synangia
   (*d*) Gametophyte of ***Marattia.***
   (*e*) Structure of mature antheridium and archegonium in ***Marattia.***
10. Describe the systematic position and distinguishing features of ***Marattia.***
11. Draw labelled diagrams of the following :
   (*i*) A young Sporophyte of ***Marattia.***
   (*ii*) Habit of ***Angiopteris.***
   (*iii*) Sporangia and Synangia in different genus of Maratheaceae.
   (*iv*) Gametophyte of ***Marattia.***
   (*v*) A Spermatozoid of ***Marattia.***
12. Briefly describe the embryogeny in ***Marattia.***
13. Name the genus of ophioglossales in which synangia are present.

OBJECTIVE TYPE

14. Select the correct answer :
   (*i*) Which of the following possesses a synangium as spore production structure?
   (*a*) ***Ophioglossum***
   (*b*) ***Marattia***
   (*c*) ***Botrychium***
   (*d*) ***Lycopodium***.
   (*ii*) The rhizome is massive and tuberous in
   (*a*) ***Equisetum***
   (*b*) ***Ophioglossum***
   (*c*) ***Marattia***
   (*d*) ***Lycopodium***.
   (*iii*) Mucliage canals and tannin cells are present in the tissues of the stem in
   (*a*) ***Selaginella***
   (*b*) ***Marattia***
   (*c*) ***Lycopodium***
   (*d*) ***Equisetum***.
   (*iv*) In ***Marattia*** the sporangia are
   (*a*) borne separately
   (*b*) fused to form synangia
   (*c*) present in pairs
   (*d*) absent.
   (*v*) Sex organs are borne on ventral side of prothallus in
   (*a*) ***Marattia***
   (*b*) ***Ophioglossum***
   (*c*) ***Lycopodium***
   (*d*) ***Equisetum***.

Osmunda cinnamomea

CHAPTER 11

Protoleptosporangiopsida

- Characteristic features
- Description of *Osmunda* L.

The class Protoleptosporangiopsida is an assemblage of interesting ferns. The class includes both living and fossil members all of which are included in a single family ***Osmundaceae*** which is placed under the order ***Osmundales***. The members of this class can be regarded as intermediate between the Eusporangiopsida and the Leptosporangiopsida, but they should in no case be regarded as phylogenetic links between them. They have a long evolutionary history and were represented in the Permian by such fossil genera as ***Zalesskya*** and ***Thamnopteris***. The living members of this class can be regarded as the 'living fossils'. The living Osmundales resemble the Eusporangiate ferns in the following repects :

1. Eusporangiate development of the sporangium.
2. The sporangia are massive and produce large number of spores. The sporangia are not arranged in distinct sori.
3. The structure and development of the archegonium.
4. The antheridia are largest in size and have many wall cells and produce many spermatozoids.
5. The prothallus is thick and massive and long lived like that of ***Marattia***.
6. Internal structure of the petiole. The endodermis is not very distinct and there are also present mucilage canals near the vascular bundle. In this respect they resemble the ***Marattiales***.
7. Presence of stipule-like expansion at the base of the petiole.

They resemble the leptosporangiate ferns in the following characteristics :

1. Origin of tapetum from the archesporial cells.
2. Presence of a primitive type of annulus in the sporangia.
3. Presence of a thin-walled stomium.
4. The wall of the sporangium is single layered.
5. The antheridia and archegonia are of projecting type.
6. The prothallus lacks endophytic fungus and is of cordate type.
7. Embryo development is of prone type, *i.e.*, the first division of zygote is vertical.

Leptopteris *Osmunda* *O. japonica*

T. barbara *O. cinnamomea* *O. claytoniana* *Todea*

Hirmer (1936) named this class as **Protoleptoporangiata.** Reimer (1934) and Pichi-Sermolli (1959) named it as **Osmundidae.** The class includes a single order **Osmundales** which includes the family **Osmundaceae.** The family includes three living genera (***Osmunda***, ***Todea*** and ***Leptopteris***) and about three fossil genera (***Zalesskya*** , ***Thamnopteris*** and ***Osmundites***). The genus ***Osmunda*** is described as a type in this text. ***Todea*** is a monotypic genus and is represented by ***T. barbara*** which is met with in Australia, New Zealand and South Africa. ***Leptopteris*** is represented by six species that are retricted to New Zealand and Polynesia. It is characterised by thin leaves that lack stomata.

GENUS–OSMUNDA L.

It is a widely distributed genus and is found in both tropical and temperate regions of the world. It is represented by twelve species (Christensen, 1938). Smith mentions 12 to 14 species as belonging to this genus. In India it is represented by five species (***O. regalis***, ***O. claytoniana***, ***O. javanica***, ***O. japonica*** and ***O. cinnamomea***). ***O. claytoniana*** (Fig. 11.1) is very common at higher altitudes in the north and north-western Himalayas. The natives use its young sterile leaves as a vegetable. It grows in abundance along the slopes of hills and also in exposed places at Kedar Nath at altitudes as high as 10,000 feet. It grows at a height of 3,000 meters or even more in the Simla Hills (Hutto Peak). ***O. regalis*** also grows in Simla Hills at an altitude of 1,500 maters at a place known as Chadwick falls. The plants growing here are usually sterile. They seem to prefer shaded and cool localities and are very much restricted in distribution. They occur in isolated patches. ***O. japonica*** is known from Garhwal to Assam whereas ***O. japonica*** is cultivated in South India. ***O. cinnamomea*** was reported from East India by Panigrahi (1960).

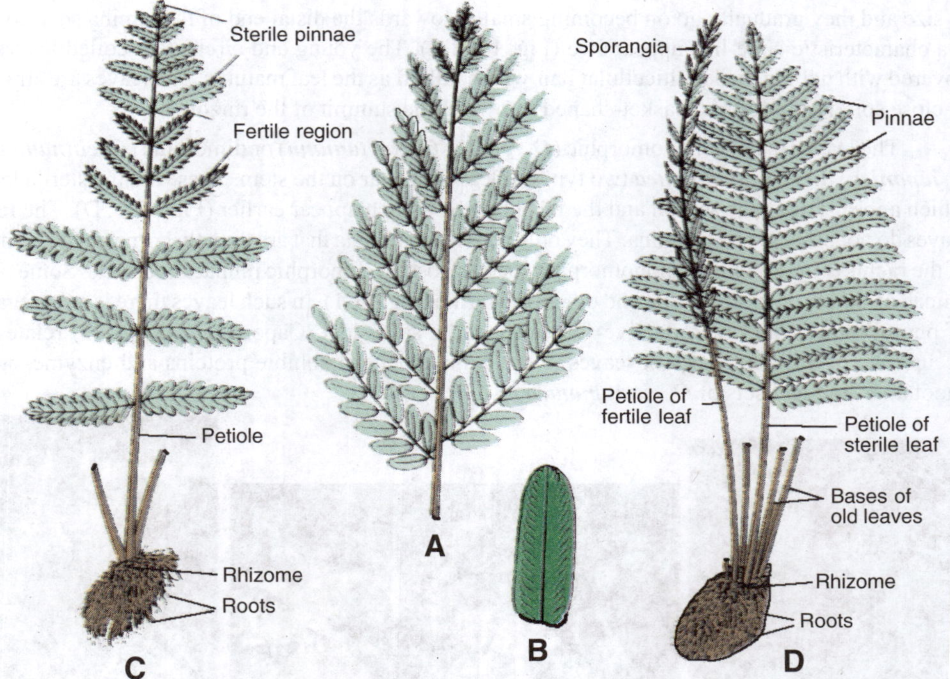

Fig. 11.1. (A-D) *Osmunda regalis.* A. shows a frond bearing sterile and fertile pinnae; B. A pinnule showing venation; C. *O. claytoniana* (complete plant); D. A complete plant of *O. cinnamomea*, bearing sterile and fertile fronds.

External Features

All the species of ***Osmunda*** are medium sized ferns, a few like ***O. cinnamomea*** (Fig. 11.1, D) achieve a height of two to three meters. The rhizomes are wholly subterranean and upright in pose. The rhizomes appear as short-sized, hard and stumpy structures that are sparingly branched. The branching is dichotomous. Faul (1901) recorded that one of the branches may again bifurcate. The rhizome is profusely beset with the bases of old leaves that have developed sclerenchymatous tissue. They give additional strength to the rhizome and are arranged in close spirals so that they cover the whole of its surface. Numerous endogenously developed adventitious roots arise from the rhizome near the bases of leaves and fix it to the substratum (Fig. 11.1, C, D). Usually two roots arise near each leaf base. Due to closely appressed leaf bases the roots flatten at the points of their origin and wind their way out. The roots are hard, rough to touch and black or dark brown in colour. They are profusely branched and bear brown root hair just below threir tips. The rhizomes with their leaf bases and tufts of roots may reach a diameter of a foot or so.

The leaves are circinately coiled when young. An unfolded leaf is pinnately compound and may be unipinnate (***O. claytoniana***) or bipinnate (***O. regalis***; Fig. 11.1 B). The leaves of ***O. cinnamomea*** and a few other species may reach a length of two to three meters. The leaf bases are conspicuously flattened to form stipule like expansions or wings which are covered with minute glandular hair. The leaves live only for one season (annual) and leave their persistent leaf bases on the stem. The proximal portions of the petiole in the older leaves are hard and stout, whereas in the young leaves they are herbaceous (***O. claytoniana***). The leaflets of the lamina are usually leathery in texture and may be entirely or variously incised. The unipinnate leaves of ***O. claytoniana*** are imparipinnate. The number of pinnules in the bipinnate leaves of ***O. regalis*** is also odd. The lower leaflets are larger in size and they gradually go on becoming smaller towards the distal end of the lamina so as to give it a characteristic cone-like appearance (Fig. 11.1, A). The young and circinately coiled leaves are covered with unbranched multicellular hair which fall off as the leaf matures. The leaves are arranged in close spirals and form a 'basket-shaped' crown at the summit of the rhizome.

The leaves may be monomorphic (***O. regalis***, ***O. claytoniana***) or dimorphic (***O. cinnamomea***, ***O. japonica***). In ***O. cinnamomea*** two types of leaves appear on the stem. These are the sterile leaves which appear late in the season and the fertile leaves which appear earlier (Fig. 11.1, D). The fertile leaves do not possess green lamina. They only possess sporangia that are directly borne on the branches of the rachis (Fig. 11.1). The monomorphic species possess dimorphic pinnae or leaflets. Some of the pinnae or the pinnules are sterile and others are fertile (Fig. 11.1). In such leaves there is a segregation of photosynthetic and reproductive functions in the same leaf. Caponetti, J.D. (1972) relates the occurrence of different types of leaves to the complement of soluble proteins and enzymes in the shoot apex and leaf sets of ***O. cinnamomea***.

O. cinnamomea

O. japonica

O. regalis

ANATOMY

Rhizome

The rhizome grows by means of tetrahedral apical cell. A transverse section (Fig. 11.2.) of the rhizome reveals the following arrangement of the tissue from outside within.

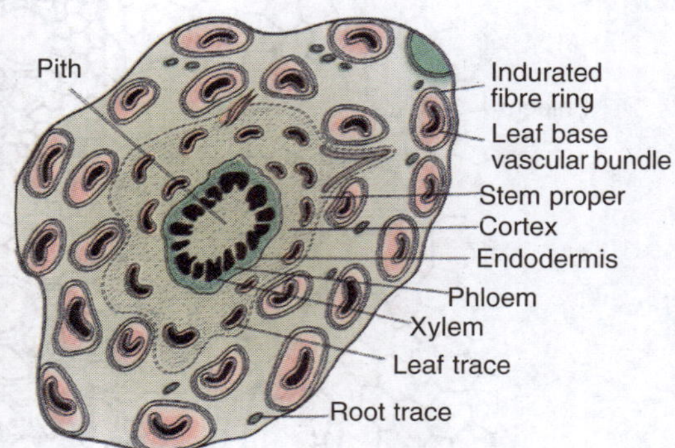

Fig. 11.2. T. S. (diagrammatic) of rhizome of *O. regalis.*

Cortex

It is distinguishable into two well defined regions : (***a***) the outer cortex and (***b***) the inner cortex. The outer is composed of several layers of thick-walled cells. The boundary of the outer cortex is irregular due to the presence of the persistent leaf bases. There is no definite layer of epidermis bounding it as all the cells are sclerenchymatous. The cells are all dead and black in colour. In this region lie embedded numerous spirally disposed leaf traces (Fig. 11.2). This region gives mechanical strength to the rhizome. The inner cortex is made up of a few layers of thin-walled cells. The cells are filled with starch grains. Some leaf trace bundles are also present in this region (Fig. 11.2).

Each leaf trace bundle is C-shaped (Fig. 11.2) and consists of an outer endodermal layer followed by 2 to 3 layers of pericycle. The phloem comes next and surrounds completely the central xylem strand. The xylem is horse-shoe shaped (Fig. 11.3). The protoxylem is situated on the concave side as a small mass of spirally thickened tracheids. The rest of the xylem is made up of larger scalariform tracheids. The convex side of the xylem faces the periphery of the stem. The phloem consists of sieve tubes with sieve plates on their lateral walls.

Endodermis

Next to the inner cortex is the endodermis. The endodermal cells possess distinct casparian bands (Fig. 11.3) on their radial walls. In ***O. cinnamomea***, Faul (1901) recorded the presence of an inner endodermis. Endodermis forms a complete boundary around the stelar region and runs uninterrupted. It is, no doubt, absent at places where there is a branch gap.

Pericycle

It forms 2—4 layers of parenchymatous cells next to the endodermis.

Stele

The stele consists of a varying number of C-shaped or horse shoe-shaped xylem bundles surrounded by a continuous ring of phloem elements (Fig. 11.3). The outer cells of phloem constitute the protophloen. The phloem is composed of distinct sieve tubes with sieve plates on their lateral walls. The phloem cylinder is thicker at places opposite the leaf gaps or the so-called medullary rays. The sieve tubes are characteristic in having their long axis tangential to that of the stele. Next to the phloem sheath is the several layered xylem sheath which is made up of parenchyma cells (Fig. 11.3). It completely encircles the xylem strands. There is every possibility that it is composed of xylem parenchyma and phloem parenchyma cells.

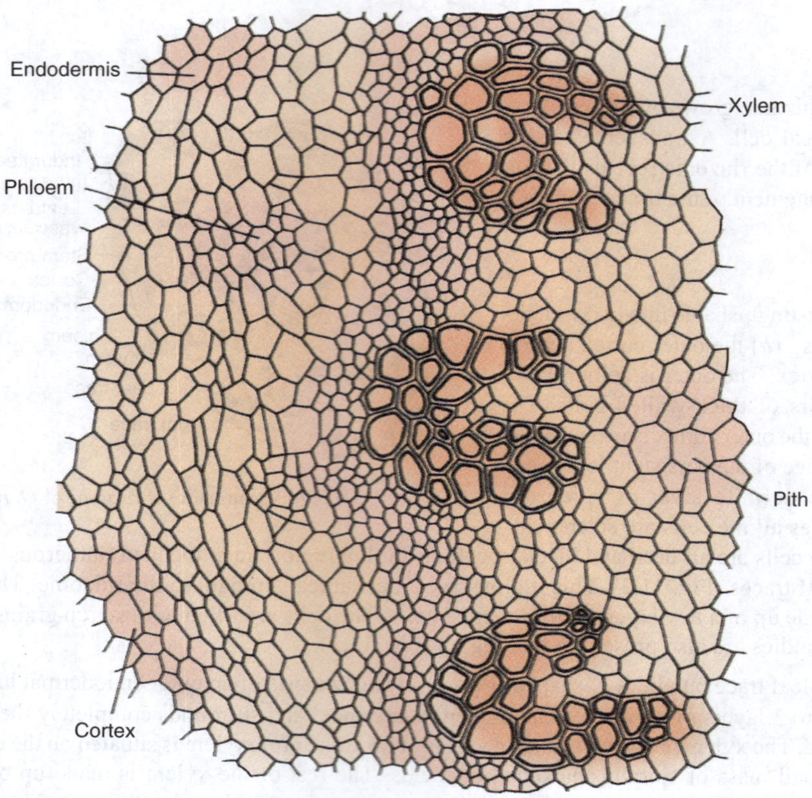

Fig. 11.3. T. S. portion of stelar region of *O. claytoniana*

The xylem appears in the form of discrete strands of variable shapes and sizes. Most of them are horse-shoe shaped, some are oval and others irregular in shape. These strands are separated by the narrow parenchymatous bands that connect the central pith with the xylem sheath. These are the leaf gaps. Each xylem strand consists of scalariform metaxylem tracheids and annular and spiral protoxylem tracheids. The former are larger in diameter than the latter and constitute most of the xylem tissue. In the horse-shoe shaped strands the protoxylem forms a small group of tracheids lying in the concavity of the metaxylem (Fig. 11.3). In the oval or irregular strands the protoxylem is surrounded on all sides by the metaxylem (mesarch). The number of xylem strands may be as many as forty in ***O. claytoniana*** and about fifteen in ***O. regalis***. Bierhorst (1960) recorded the presence of pits in the walls of metaxylem tracheids.

Faul (1901) reported the presence of both external and internal endodermis and phloem in ***O. cinnamomea***.

Pith is quite conspicuous and may be entirely parenchymatous (***O. claytoniana***) or a few sclerenchymatous strands may be present (***O. regalis***, ***O. cinnamomea***).

The stele in ***Osmunda*** has been designated as **dictyostele** by some workers. Sporne (1966) regards it as **dictyoxylic.** Due to the presence of continuous endodermis, pericycle and phloem layers external to the xylem, it does not fulfil the conditions met with in the dictyostele. The individual xylem strands can also not be regarded as meristeles. A meristele in its essentials has its own separate phloem, pericycle, and an endodermis. This condition is not obtained in any species of ***Osmunda***. The stelar organisation in the younger and first formed parts of the rhizome is **protostic** (Gwynne-Vaughan, 1911).

The Leaf

Petiole

A transverse section of the petiole reveals a distinct layer of epidermis, which in younger leaves is covered with numerous simple and multicellular hair. Next to the epidermis is a few layers thick sclerenchymatous hypodermis. The hyodermis encircles a broad central ground tissue. The ground tissue is composed of thin-walled cells. In it is embedded a single petiolar bundle. The bundle is crescentic or horse-shoe shaped and has a central xylem core surrounded on all sides by phloem. Endodermis is not very clear. The xylem has several protoxylem groups along the concave side. The metaxylem consists of large tracheids that have scalariform thickenings. It makes up most of the xylem tissue. A number of mucilage canals are present towards concave side of the vascular bundle—a feature which it shares with the Marattiales.

Leaflet

A cross section of the leaflet reveals the presence of two epidermal layers. The stomata are present on the lower epidermis. In between the two epidermal layers is the undifferentiated mesophyll region. The cells of the mesophyll enclose intercellular spaces of variable dimensions. The cells contain chloroplasts and are the main seat of photosynthesis. They are compactly arranged below the upper epidermis and go on becoming loosely arranged towards the lower epidermis. The mesophyll cells are compactly arranged in the region of the midrib and the veins. There is a single vescular bundle in the midrib. It is concentric.

Root

The roots arise endogenously from the rhizome. There is a distinct outer layer or the epidermis in the young roots (Fig. 11.4). It is later replaced by the outermost layer of the cortex. It is called the exodermis. Next to the epidermis is a few-layered hypodermis which is made up of thick-walled cells. Its cells are smaller in diameter than those of the cortex. The rest of the cortex is parenchymatous. There is a distinct endodermis surrounding the central stele. The endodermis of the root is continuous with that of the rhizome and the petiole. Next to the endodermis are two layers of thin-walled cells. They constitute the pericycle.

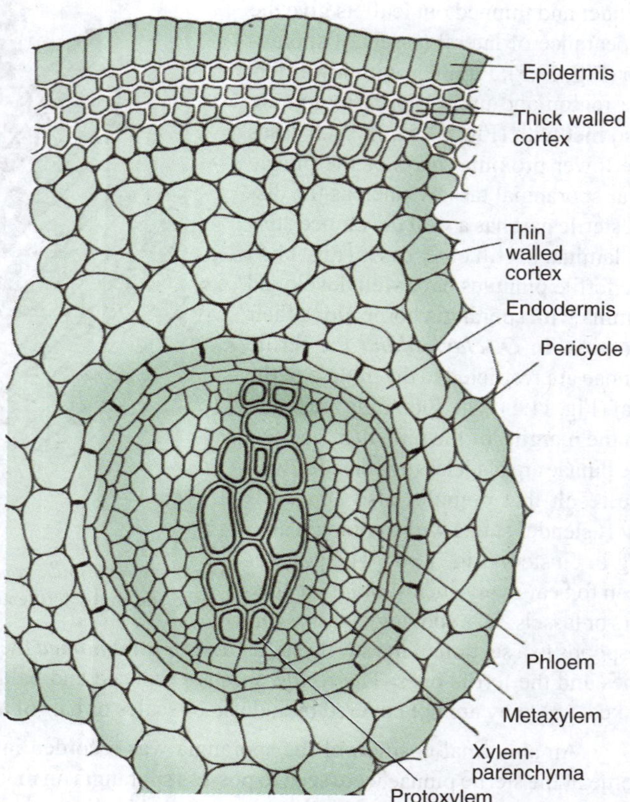

Fig. 11.4. T.S. part of root of *O. regalis* showing detailed internal structure.

The stele is usually diarch but Faul (1901) reported triarch stele in some species. There are two xylem bundles that meet in the centre and appear as an elliptical central core of xylem. The protoxylem is exarch in position.

It is represented by two groups on either side of the ellipsoidal xylem mass. There is no pith. Phloem forms two patches on either side of the xylem (Fig. 11.4). There are a few layers of thin-walled cells between the xylem and the phloem bundles.

Miller (1971) has beautifully elaborated the evolution of family Osmundaceae on the basis of anatomical studies. According to Miller the family has 16 living genera and nearly 100 fossil species, of which the modern species evolved from the protostelic ancestors by intrastelar origin of pith and subsequent or concurrent dissection of xylem cylinder by leaf gaps. ***Leptopteris*** and ***Todea*** are generically distinct from one another and from ***Osmunda***, whose three sub-genera namely ***Osmunda***, ***Osmundastrum*** and ***Plenasium*** represented three distinct lines of evolution in Paleoceone. Miller advocated that these taxa must be treated at generic rank. Interestingly, he emphasised that Osmundaceae arose in the Permian and do not represent ancestors of any other fern families.

REPRODUCTION

The sporophyte is homosporous. The spores are produced within sporangia that are not arranged in definite sori and are usually marginal in position. In ***Osmunda regalis*** the fertile pinnules are restricted to the distal region of the leaf. In some cases the leaves developed late in season are entirely fertile. The fertile pinnae are very much reduced. They have no sterile lamina and the sporangia appear to develop on the veins arising from the midrib. Clusters of sporangia borne on these slender and thinned out leaflets give the appearance of lateral tassels of sporangia (Fig. 11.5). Pinnules with lower fertile region and upper sterile region are also met with (Fig. 11.5). In such cases the lower proximal parts of the pinnae bear sporangial tassels whereas the distal sterile part has a well developed sterile lamina. In still other cases (Fig. 11.5) the fertile pinnules have well developed lamina with sporangia borne along their margins. In ***O. claytoniana*** the fertile pinnae are restricted to the middle of the leaf (Fig. 11.1). The sporangia are borne on the margins of the pinnules. The fertile pinnae are bladeless and slender, with the result that pinnules also appear as short slender stalks with sporangia arising in clusters (Fig. 11.6). The pinna is seen to bear many such sporangial clusters or tassels. So a sporangial tassel corresponds to a section of the sterile laeflet. In ***O. cinnamomea*** there are two kinds of leaves, the sterile ones and the fertile ones. The fertile ones are reduced and baldeless (Fig. 11.1). The sporangia are naked, *i.e.*, they are not covered by indusia or scales or hair of any type.

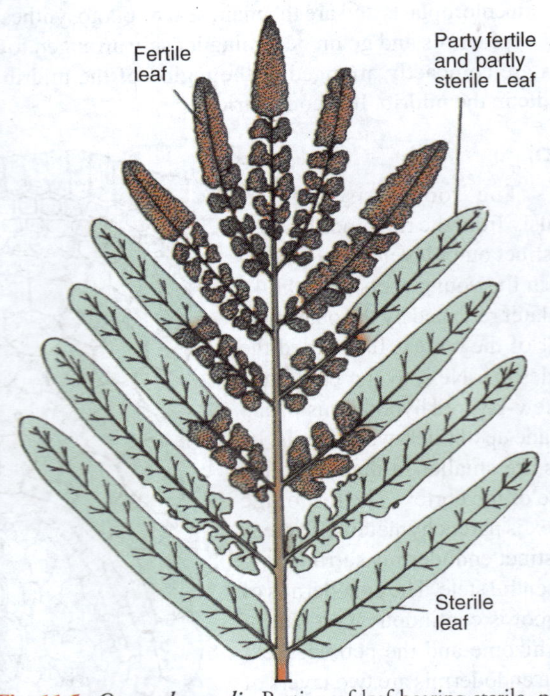

Fig. 11.5. ***Osmunda regalis.*** Portion of leaf bearing sterile, partly sterile and partly fertile and wholly fertile pinnules.

An abnormal position of the sporangia was recorded in ***O. claytoniana*** (Chowdhry, 1932). Some of the sterile pinnae were seen to possess sporangia on their abaxial suface. The sporangia were arranged in definite sori. Such pinnae were found in the transtition zone between the fertile and the sterile pinnae.

SPORANGIA

Structure and Dehiscence

A mature sporangium of ***O. regalis*** and most other species of ***Osmunda*** is a massive, short stalked and a pyriform structure. The stalk is many-celled thick and varies in length in different species. The capsule has a single layer of sterile jacket cells. Next to it there are two layers or rarely three layers of tapetal cells surrounding a central mass of 32 to 128 spore mother cells. The tapetal cells do not disorganise to form the nourishing plasmodial fluid. In ***O. claytoniana*** there are two layers of tapetal cells. The inner layer is derived from the sterile sporogenous cells and later disorganises to form the plasmodial fluid (Fig. 11.7, G). They elongate and push their way in between the spore mother cells. Their walls disintegrate and their protoplasts fuse with each other to form a network like plasmodial fluid. It disappears during the maturation of spores. The outer layer of tapetal cells is derived from the primary archesporial cell. It is made up of a single layer of flattened cells and persists throughout the life of the sporangium. It may be regarded as a functionless second wall layer. It plays no role in protection and dehiscence of the sporangium nor it nourishes the spores. The spore mother cells undergo meiosis and produce 128—512 spores.

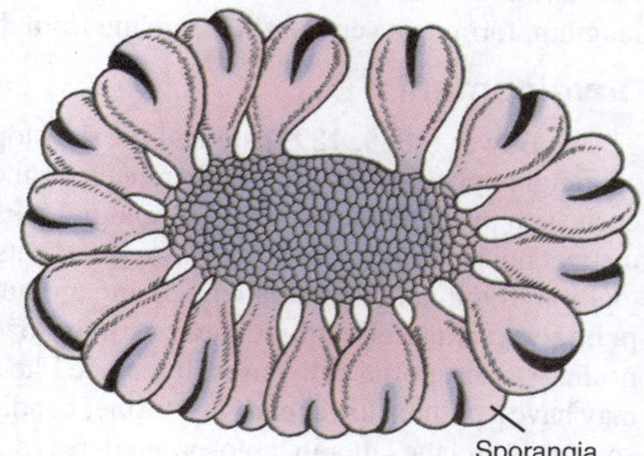

Fig. 11.6. *Osmunda regalis*. Transverse section of a tassel of sporangia. Note the arrangement of sporangia. (After Bower)

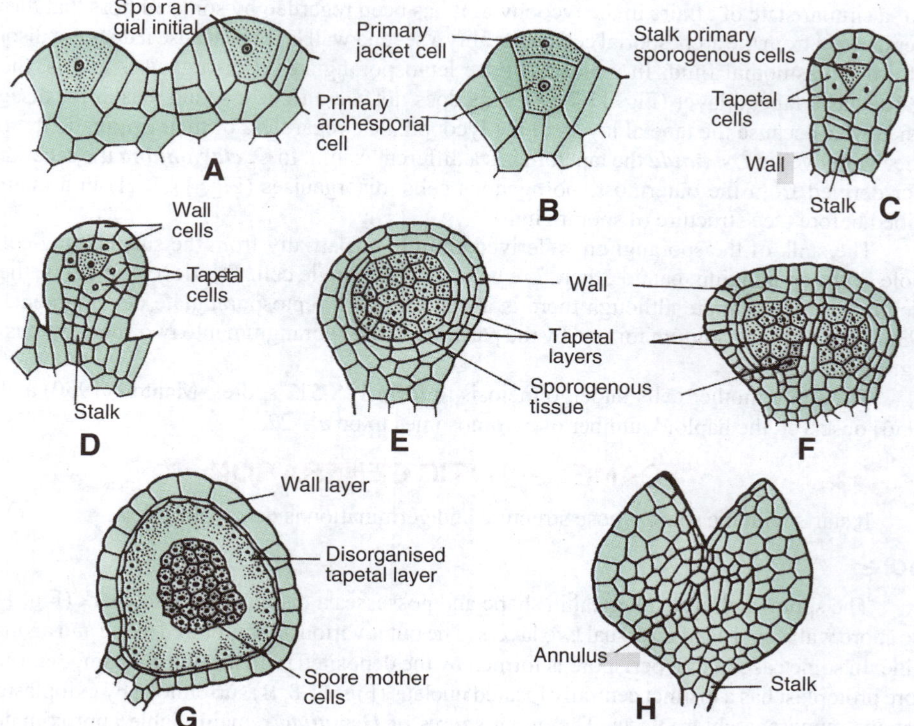

Fig. 11.7. (A—H). Stages in the development of sporangium. A—E. *O. regalis*; F. A synangium of *O. regalis;* G—H. *O. claytoniana* (A—F, after Bower, G—H, after Smith)

The mature sporangium develops a transverse group of thick-walled cells on one side. (Fig. 11.7, H). This is the annulas. In some sporangia the annulus covers the entire circumference of the apical region of the sporangium and may be many cells in breadth and about two or three cells in height (Williams, 1928). A vertical row of elongated cells on the side opposite to the annulus represents the future line of dehiscence of the sporangium (Fig. 11.7, H). These cells separate and open the sporangium, forming a vertical slit extending from the annulus to the outer side of the sporangium.

Development

Bower (1889, 1899) studied the development of sporangium in ***O. regalis***. The developing sporangia appear as small protuberances on the fertile and bladeless leaflets or their segments. In each such protuberance a single sporangial initial becomes distinct by its comparatively large size, distinct nucleus and denser cytoplasmic contents (Fig. 11.7, A). A transverse wall divides this sporangial initial into an outer smaller cell and an inner larger cell (Fig. 11.7 B). The former is called the **primary jacket cell** and the latter as **primary archesporial cell** (Fig. 11.7, B). The shape of the primary archesporial cell varies. It may be like an inverted pyramid with a truncated lower end or it may have a pointed lower end. The former condition is considered to be prevalent in the eusporangiate sporangia and the latter in leptosporangiate types. The archesporial cell cuts off three peripheral cells (Fig. 11.7, C) enclosing a **central primary sporogenous cell.** The peripheral cells were designated as **primary tapetal cells** by Bower. Their mode of origin is exactly similar to the tapetal cells in the leptosporangiate ferns. They divide anticlinally and form a complete layer of tapetum internal to the wall layer (Fig. 11.7, D). The tapetum may become two or even three-layered thick. This may happen by the periclinal division of the primary tapetal cells or the second layer of tapetum is derived from the outermost cells of the sporogenous mass. The latter condition is achieved in ***O. claytoniana*** (Smith, 1938). The primary sporogenous cell divides repeatedly to form 32—128 spore mother cells. The tapetal layers in ***O. regalis*** do not disorganise to form a plasmodial fluid. This condition reminds us of a similar state of affairs in the lycophyta. It has been regarded by some authors that the tapetal layer derived from the archesporial cell is in reality a second wall layer beacause it does not disorganise to form a plasmodial fluid. In majority of the leptosporangiates it disorganises and is, thersfore, regarded as a tapetal layer (Fig. 11.7, G). This does not seem to be a sound reason for disagreeing with Bower because the tapetal layers in the lycopodiales, regardless of their origin, do not disintegrate. Moreover in ***Osmunda*** the tapetum has a different origin. In ***O. claytoniana*** the second tapetal layer derived from the outermost sporogenous cells, disorganises (Fig. 11.7, H) in a manner described before (see structure of sporangium).

The stalk of the sporangium is derived entirely or partially from the surrounding cells. The whole of the sporangium cannot, therefore be traced to a single cell. The development is, therefore, essentially eusporangiate although there is a trend towards leptosporangiate development. Bower (1935) reported a synangium formed by the partition of the sporangium into two sporogenous groups (Fig. 11.7, F).

The spore mother cells undergo meiosis to form 128-512 spores. Manton (1950) and Ninan (1956) observed the haploid number of chromosomes to be $n = 22$.

GAMETOPHYTIC GENERATION

It starts with the spore whose structure and germination is described below.

Spore

The spore is almost spherical in shape and possesses a distinct triradiate mark (Fig. 11.8, A). The spore wall consists of the usual two layers. The outer variously sculptured exine and an inner thin intine. In some cases a thin perispore is formed by the deposited remains of the plasmodial fluid. The spore protoplast has a distinct centrally located nucleus (Fig. 11.8, B) surrounded by cytoplasm which contains numerous chloroplasts. The green spores of ***Osmunda*** remain viable approximately one week under field conditions (Lloyd and Klekowsky, 1970).

Germination of the spore

Campbell (1892) studieds the germination of spores in ***O. claytoniana*** and ***O. cinnamomea***. Under adequate conditions of moisture the exine ruptures at the tri-radiate mark. The intine grows out in the form of a small conical papilla (Fig. 11.8, C). The protruding spore protoplast divides by a transverse wall into a small **primary rhizoidal cell** and a large **primary prothallial cell.** The former comes out first in the form of a conical protuberance (Fig. 11.8, C) and the latter is still enclosed within the rupturing exine. The rhizoidal cell contains a few chloroplasts which later on, as it pushes its way down into the soil as a first rhizoid, disappear (Fig. 11.8, D, E). The larger prothallial cell contains numerous chlorplasts which move towards the periphery of the cytoplasm. The prothallial cell grows exactly opposite to the first rhizoid (Bipolar germination). There are some variations in the pattern of the laying down of the first walls in the prothallial cell. In ***O. claytoniana*** (Campbell, 1892) the prothallial cell divides by a few transverse walls (Fig. 11.8, D) to form a filament of green cells. The terminal cell of the filament divides by two intersecting oblique walls (Fig. 11.8, F) to include a triangular apical cell with two cutting faces. The lower cells of the filament may now divide by longitudinal walls (Fig. 11.8, F). The apical cell cuts off segment towards its right and left in a regular, alternate manner. The segments cut off by the apical cell divide each by a transverse wall to form an outer and an inner cell. The inner cells divide by horizontal walls. This adds to the thickness of the prothallus in the middle portion which becomes distinct as a midrib. The outer cells divide rapidly and overgrow the apical cell thus giving the young prothallus a heart-shaped appearance. The apical cell may remain distinct and functional for a long period or it may be replaced by a row of few marginal cells. These divide in such a manner so as to cut off segment laterally as well as towards dorsal and ventral sides. This type of growth leads to the formation of an elongated heart-shaped prothallus with distinct cushion below the notch (Fig. 11.9, A). The cushion projects prominently on ventral side. The prothallus may continue to grow by means of this marginal meristem for a considerable length of time (sometimes for a few years). The full grown prothallus of ***O. claytoniana*** may be as long as 3—6 cms. Campbell (1892) reported dichotomous branching of the prothallus in ***O. claytaniana***. The prothallus in the same species was also reported by him to produce adventitious branches (Fig. 11.9, B), which on separation are capable of developing into independent prothalli. The adventitious branches also grow

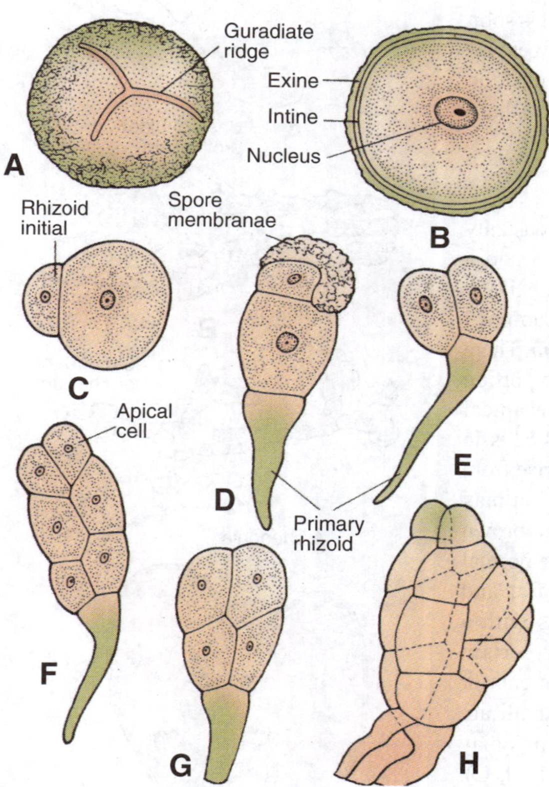

Fig. 11.8. (A–H). Germination of the spore and early stage in the development of the prothallus in *Osmunda*; A—F. *O. claytoniana*; G—H. *O. cinamomea.* (After Campbell)

by means of a two-sided apical cell. The rhizoids arise from swollen mid-rib portion on the ventral side and are unicellular in young prothalli, but become septate in older ones.

Sometimes the spores (***O. claytoniana***) develop into irregularly branched filamentous prothalli (Fig. 11.9, C) that are either wholly male or are protandrous and produce archegonia very late. The male prothalli remain small and may be produced in larger numbers.

In ***O. regalis*** and ***O. cinnamomea*** prothallial cell divides first by a longitudinal wall to form a quadrant stage (Fig. 11.8, G). In this case no filament is produced as in ***O. claytoniana***. Cases are not rare when the primary prothallus divides along three planes to form a mass of cells. An apical cell is soon established in this mass and by its activity gives rise to a flat and cordate porthallus. The central cushion is not very prominent in these species. The prothalli are dark green in colour.

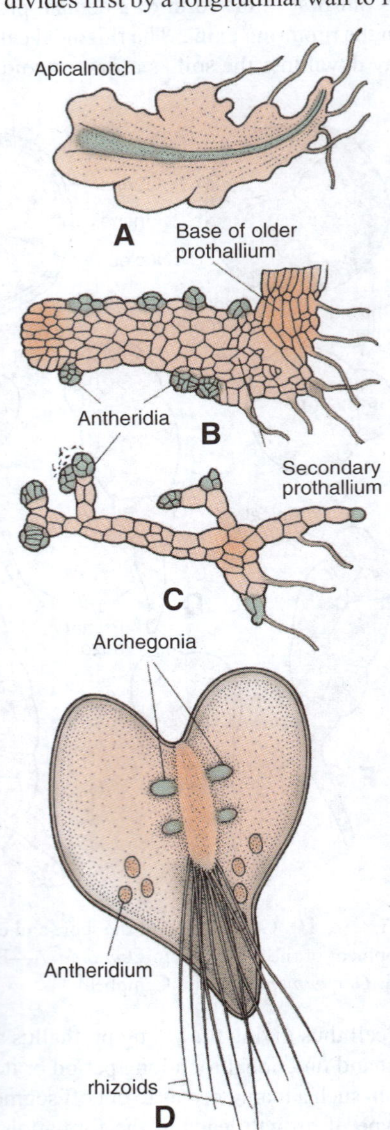

Fig. 11.9. (A—C). Prothalli of **O. claytoniana.** A. A young prothallus with a distinct apical notch; B. Old prothallus with secondary prothallus developd from an adventitious branch; C. A filamentous and a branching prothallus with antheridia; D. Mature prothallus. (After Campbell)

Structure of the prothallus

The mature prothallus of ***Osmunda*** is usually elongated, unbranched, dorsiventral, green and a cordate or heart-shaped structure (Fig. 11.9, A,D). It varies in length from 2 to 3 centimetres. Campbell (1892) reported a dichotomy in ***O. claytoniana***. There is a distinct midrib in the centre. The midrib portion is usually thick and cushioned below the apical notch.This cushion may be up to 16 to 18 cells in thickness in ***O. claytoniana***. The rhizoids arise from the ventral side and are usually unicellular but may become septate. They are light brown or dark brown in colour. The prothalli that develop under normal condition are monoecious, protandous, flat and ribbon-shaped and unbranched structures with a smooth or slighly wavy margin. Under certain conditions the prothalli develop into filaments or irregularly branched structures. Such prothalli are usually unisexual and bear only antheridia that may be marginal or terminal in position (Fig. 11.9, C).

Osmunda claytoniana

The growth of such prothallus is not by any definite apical cell. Goebel (1893) reported that the unfertilised prothalli of ***O. regalis*** live long and reach a length of 4 centimetres. He also recorded dichotomy of the prothallus in this species.

SEX ORGANS

The monoecious prothalli are protandrous and bear antheridia after about two weeks (***O. cinnamomea***) to one month (***O. claytoniana***) of their development from the spore. The antheridia usually appear along the margins or on the ventral surface along the wings. The archegonia appear late and develop either along the margins of the thick midrib (***O. cinnamoea***) or are scattered along over its surface (***O. claytoniana***).

Antheridia

Structure

The antheridia are of projecting or emergent type. They are large and globular structures and vary in position from terminal to marginal or even to the ventral side of the prothallus. The antheridia may or may not have a distinct stalk. The body of the antheridium has a single-layered wall made up of many curved cells that contain chloroplasts (Fig. 11.10, I, J). A triangular cell at the apes of the antheridium on one side, is the opercular cell (Fig. 11.10, F, G). It is thrown off during dehiscence thus permitting the spermatozoids to come out and get dispersed (Fig. 11.10, J). The spermatozoids are minute, coiled and multicilate structures (Fig. 11.10, K). It has two coils. The posterior coil is slightly broader than the anterior one.

Development

The antheridium develops from a single cell. The antheridial initial is cut off from a large superficial cell of the prothallus. In terminal antheridial the terminal cell of a branch becomes enlarged and divides by an oblique wall into an outer **antheridial initial** and a basal cell (Fig. 11.10, A). The division is similar to the leptosporangiate antheridial development. Subsequent divisions are, however, different. The antheridial initials divide by two intersecting oblique walls to cut off a tetrahedral cell surrounded on either side by two flattened cells (Fig. 11.10, B). The tetrahedral cell has its broad end facing outwards and the narrow end facing inwards. It cuts off segments along its three lower sides. These segments form the stalk of the antheridium. The length and thickness of the stalk depends upon the number of segments cut off by the apical cell. Usually one or only a few segments are cut off by the apical cell so that the stalk is either indistinct or very short. After the apical cell stops cutting segments from the three lower faces, it divides by a curved periclinal wall towards its outer broad face thus distinguishing an outer curved **primary jacket cell** and an inner **primary androgonial cell** (Fig. 11.10, B). The primary jacket cell undergoes anticlinal division to form two cells (Fig. 11.10, C). Both the jacket cells divide further by anticlinal and oblique walls to form the antheridial jacket. The antheridial jacket thus formed consists of a single layer of variously curved, chlorenchymatous cells (Fig. 11.10, D, E). A triangular cell or the **opercular cell** is formed by the division of the last formed jacket cells. The opercular cell is either apical or lateral in position. The androgonial cell has by now divided by repeated cell divisions to form a large number (100 or more) of androcyes or spermatocytes (Fig. 11.10, E). The protoplasts of each androcyte matamorphoses into a coiled and multiflagellate spermetozoid (Fig. 11-10, K). The spermatozoids remain within the androcyte walls and escape by their gelatinisation after the dehiscence of the antheridium.

The position and mode of formation of the triangular cell is quite similar to the eusporangipsida. The emergent position of the antherdium and its origin from a single cell are leptosporangiate characters. The mode of wall formation and the formation of the stalk are characters quite peculiar to

Osmunda. The number of androcytes is neither eusporangiate nor leptosporangiate. In the former they far outnumber ***Osmunda*** and in the latter they are few. The necks of the archegonia are also longer than leptosporangiates.

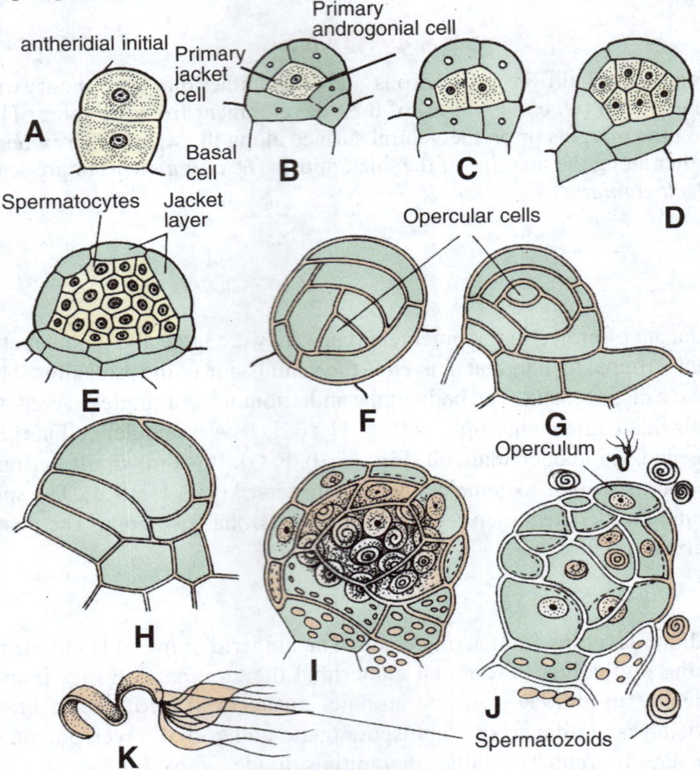

Fig. 11.10. (A—K). Stages in the development of antheridium in **Osmunda.** A—H. *O. cinnamomea*; I—K. *O. claytoniana.* (After Campbell)

ARCHEGONIUM

Structure

The archegonia are always produced on the ventral surface and have necks that project in a horizontal direction from the median cushion. The neck consists of 4 vertical rows of cells, each row 6 cells in height *i.e.*, it consits of 6 tiers of 4 cells each. The neck is, therefore, longer than the leptosporangiates. The neck encloses a single binucleate neck canal cell. The venter is embedded in the prothallus tissue and is not surrounded by its own wall. It contains a single ventral canal cell and a large egg cell (Fig. 11.11).

Before fertilisation, the neck canal cells and the ventral canal cell disorganise and become mucilaginous. The mucilage absorbs water, swells up and forces the apical tier of cells apart, thus making an open passage for the sperms to enter. The mucilage drop oozes out and attracts the spermatozids.

Development

The archegonium develops from a single superficial lying on the ventral side of the prothallus somewhere along the flanks of the central cushion. This cell is called the **archegonial initial** and can be distinguished by its large size (Fig. 11.11, A). The archegonial initial undergoes periclinal division and cuts off an outer smaller **primary cover cell** (Fig. 11.11, B) and an inner large **mother cell of the central cell.** The former divides by two vertically intersecting walls to form four diagonally arranged neck initials (Fig. 11.11, C). The mother cell of the central and lower basal cell. (Fig. 11.11, C). The latter takes no futher part in the development of the archegonium. It may divide once or twice to form cells that are distinguishable from the surrounding prothallial cells. The neck initial divides transversely to form 6 to 8 tiers of neck cells (Fig. 11.11, D—F). The neck is straight or curved like most of the polypodiaceae. The central cell divides into an upper **neck canal cell** (Fig. 11.11, D) and a lower **primary ventral cell.** The former forms a single binucleate **neck canal cell** and the latter divides into an upper smaller ventral canal cell and a lower large **egg cell.** The neck canal cell pushes in between the neck cells and its two nuclei may, in rare cases, be separated by a wall to form two uninucleate neck canal cells.

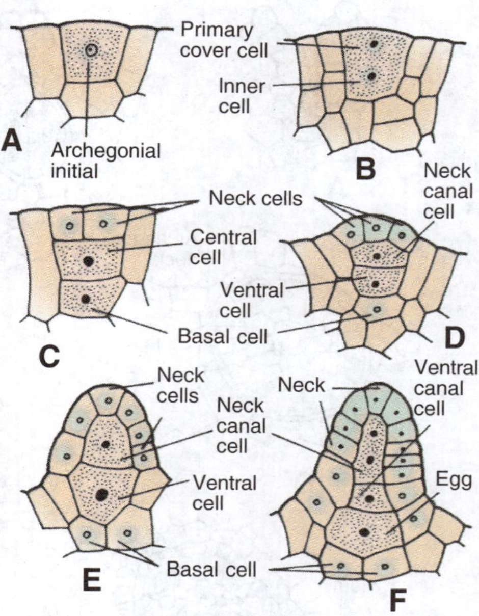

Fig. 11.11. (A–F). Development of archegonium in *O. claytoniana* (After Campbell)

EMBRYOGENY

As a result of fertilisation a diploid zygote is established. It is surrounded by a dense cytoplasm, which secretes a wall, thus establishing a diploid cell called the **zygote** or the **oospore.** It is lying in the venter and is protected by the surrounding prothallial tissue.

The zygote divides (Cross, 1931) by a wall parallel to the long axis of the archegonium (vertical wall) into two almost equal cells (Fig. 11.12, B). The two cells thus formed divide each by a wall at right angles to the first wall to form four cells (Fig. 11.12, C). This is the quadrant stage. The epibasal quadrants give rise to the leaf and the stem and the two hypobasal ones give rise to the foot and the root. The anterior epibasal quadrant forms the leaf and the posterior develops into stem. Similarly, the anterior hypobasal quadrant gives rise to the root and the posterior to the foot. The quadrants divide transversely to form an octant stage. Subsequent divisions are not very regular. Cross (1931) stated that in ***O. cinnamomea*** the stem, leaf and the root develop from that half of the octant which is nearer to the archegonial neck. The foot develops from the other half. The foot is well developed and may perform haustorial functions. The root apical initial appears endogenously and by its activity gives rise to the primary root. The development of the embryo is slow and it retains its globular form for a sufficiently long time.

O. cinnamomea

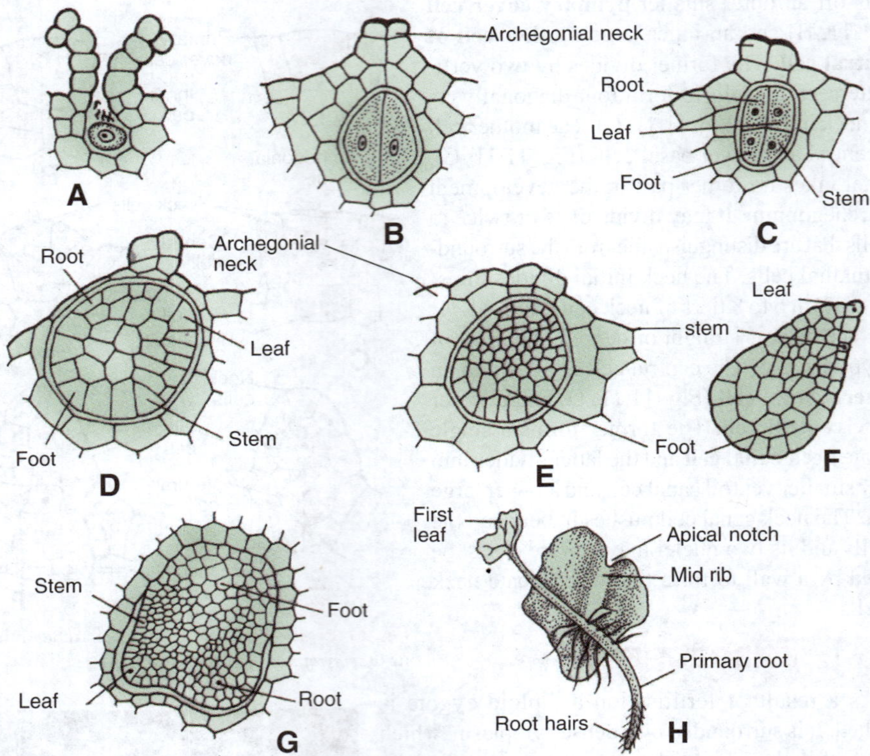

Fig. 11.12. (A–H). Stages in the development of Embryo in *Osmunda.* A, C, D, F, H, *O. claytoniana*; B, E, G, *O. cinnamomea.* (A, C, D, F, H, After Campbell; B, E, G, After Cross).

The leaf initial makes its appearance as three-sided apical cell in the anterior epibasal octant. The young leaf soon apperars as a conical protuberance (Fig. 11.12, F) on one side. The first leaf does not come out of the surrounding prothallial tissue until quite late. The tetrahedral root apical cell grows into a massive primary root. The stem apex appears on one side of the leaf apex in the posterior epibasal half and grows into the underground stem after the primary leaf and root had established themselves. The primary leaf has broad lamina with irregular margin. The venation is furcate. It possesses stomata on both the sides. The vascular bundle of the petiole is of collateral type and that of the root is diarch. The vasculature of the young rhizome is protostelic and becomes dictyoxylic at a later stage (after producing 3 or 4 leaves).

APOGAMY AND APOSPORY

Brown (1920) and Sarbadhikari (1939) reported apogamy and apospory in ***O. regalis*** and ***O. javanica***. Sarbadhikari (1939) reported aposporous production of gametophytes from the sporophytic tissue (see Chapter I). He found that in ***O. javanica*** the aposporous gametophytes produced antheridia but no archegonia. The sporophytes develop apogamously as vegetative outgrowths from the tissue of the prothallus.

Economic importance

Osmunda species are of economic importance as roots and stems of ***Osmunda*** provide fibre which is used for growing and cultivating epiphytic plants especially orchids. In addition a few species are cultivated as ornamentals for their beautiful foliage.

Taxonomic Position

Division	:	Filicophyta
Class	:	Protoleptosporangiopsida
Order	:	Osmundales
Family	:	Osmundaceae
Genus	:	***Osmunda***
Species	:	***regalis***

QUESTIONS

ESSAY TYPE

1. Osmundaceae is regarded as a connecting link between Eusporangiate and Leptosporangiate ferns. Discuss. Are the modern Osmundaceae living fossils? **(Kanpur, M.Sc., 2003)**
2. Decribe the family Osmundaceae. **(Rohilkhand, 2000)**
3. Discuss the orgin of pith with special reference to Osmundaceae.
4. Osmundaceae is intermediate between Leptosporangiate and Eusporangiate. Comment.
5. Give a brief account of the sporophyte of ***Osmunda***.
6. Describe the life history of ***Osmunda***.
7. Give an account of the anatomy of the stem and root of ***Osmunda***.
8. Give a brief account of asexual reproduction in ***Osmunda***.
9. Give a brief account of sexual reproduction in ***Osmunda***.

SHORT ANSWER TYPE

10. Write notes on :
    (*a*) ***Osmunda***
    (*b*) Antheridium in ***Osmunda***.
    (*c*) Development of Sporangium in ***Osmunda***.
    (*d*) Intermediate position of Osmundaceae.
    (*e*) Affinities between Ophioglossales and Osmundales.
    (*f*) Sporangium in ***Osmunda***.
    (*g*) Prothalius in ***Osmunda***.
11. Distingiush between
    (*a*) Stem of ***Gleichenia*** and ***Osmunda***.
    (*b*) Fertile fronds of ***Botrychium*** and ***Osmunda***.
12. Draw a labelled diagram of the ***Osmunda*** leaf showing fertile and sterile pinnae.
13. Give the points of resemblance between ***Osmunda*** and leptosporangiate ferns.

OBJECTIVE TYPE

14. Select the correct answer :
    (*i*) Which of the following is the monotypic genus of the family Osmundaceae ?
    (*a*) ***Osmunda***
    (*b*) ***Leptopteris***
    (*c*) ***Todea***
    (*d*) ***Thamnopteris***
    (*ii*) ***Osmundales is*** a link between
    (*a*) Leptosporangiate and Eusporangiate
    (*b*) ***Selaginella*** and ***Lycopodium***
    (*c*) ***Eligulatae*** and ***Ligulatae***
    (*d*) Pteridophytes and Gymnospsms
    (*iii*) Dimorphic leaves are present in
    (*a*) ***Osmunda regalis***
    (*b*) ***O. cinnamomea***
    (*c*) ***O. javanica***
    (*d*) ***O. claytoniana***.
    (*iv*) Fertile pinnae are bladeless, slender and bear many sporangial clusters in
    (*a*) ***Osmunda***
    (*b*) ***Ophioglossum***
    (*c*) ***Botrychium***
    (*d*) ***Pteris***
    (*v*) Development of Sporangium in ***Osmunda*** resembles that of
    (*a*) ***Ophioglossum***
    (*b*) ***Lycopodium***
    (*c*) ***Selaginella***
    (*d*) None of the above.

Anogramma leptophylla

CHAPTER 12

Leptosporangiopsida

- Introduction
- Classification
- Characteristics of the members of family **Schizaeceae** (***Schizaea, Anemia, Mohria, Lygodium, Actinostachys***)

Introduction

This class includes about 232 genera and 8,680 species. They grow luxuriantly in the tropical rain forests, but are also met with in the temperate regions of the world. They love to grow under moist, cool and shaded places. Some of them also grow under extremes of habitat conditions. A number of them grow under xerophytic conditions e.g., ***Drynaria, Hypodematium crenatum, Dryopteris chrysocoma, Adiantum incisum, Woodsia elongata*** etc. A few forms like ***Marsilea, Azolla, Salvinia*** and ***Pilularia*** are hydrophytes. Majority of them are mesophytes. They also form a dominant epiphytic vegetation in the forests. Majority of them are perennials and only a few are annuals. They range in size from extremely small forms like ***Anogramma leptophylla*** (4 to 5 cm. in height) to huge tree-like forms of the ***Cyatheacea.*** Their ability to live under varied environmental conditions and the property of the leaf of perform a dual function of photosynthesis and reproduction are considered to be responsible for their widespread occurrence. A detailed account of the general features is given in Chapter 8 and it is needless to dwell upon them again in this chapter.

Drynaria *Anogramma leptophylla* *Dryopteris chrysocoma*

Woodsia elongata *Hypodematium crenatum*

Regarding their apical growth, the stem apices of all leptosporangiate ferns grow with the help of a single triangular apical cell with three cutting faces e.g. ***Dryopteris, Polypodium*** except in some ferns where the apical cell has two cutting faces e.g. ***Pteridium aquilinum.*** The apical cell may be differentiated on an apical dome e.g. ***Asplenium, Polypodium, Dryopteris*** or it may be situated on a small protuberance at the base of an apical depression or concavity e.g. ***Pteridium aquilinum.*** In this species the bilateral symmetry of the apical cell is manifested in the bilateral anatomy of the rhizome. The leaves of all leptosporangiate ferns have apical cells with two cutting faces.

Classification

The ferns included under this class have been variously classified by the taxonomists. Engler and Diels (1936) treated it as a sub-class and divided it into three orders : (*i*) **Filicales**; (*ii*) **Marsileales**; and (*iii*) **Salviniales**. Reimers (1954) and **Smith** (1955) also followed the same system. These three orders have been recognised on account of the nature of spores (whether homosporous or heterosporous), types of sporocarps, endosporic or exosporic gametophytes, and their habitats. The Marsileales and Salviniales are heterosporous and aquatic. Christensen (1938) divided this class into two orders (*i*) **Filicales** and (*ii*) **Salviniales**. He included the **Marsileales** in a separate family ***Marsileaceae*** and included it under the order **Filicales.** Marsileales deserve the rank of a separate order as they possess a number of distinct features, which take them away from the homosporous form. Pichi-Sermolli (1958) included all the ferns (including Eusporangiopsida, Protoleptosporangiopsida and Leptosporangiopsida) under the class **Filicopsida** and divided it into seven sub-classes. The leptosporangiate ferns were included by him under the sub-class Filicidae, which was divided into 15 orders and 32 families. These include the orders (*i*) **Gleicheniales** fam. ***Glicheniaceae;*** (*ii*) **Schizaeales** (***Schizaeaceae***); (*iii*) **Hymenophyllales** (***Hymenophyllaceae***); (*iv*) **Matoniales** (***Maroniaceae***); (*v*) **Platyzomales** (***Platyzomaceae***); (*vi*) **Polypodiales** (***Dipteridaceae, Cheiropleuriaceae, Grammitidaceae, Polypodiaceae***); (*vii*) **Cyatheales** (***Cyatheaceae, Lophosoriaceae***); (*viii*) **Dicksoniales** (***Dicksoniaceae, Dennstaedtiaceae, Lindsaeaceae***); (*ix*) **Pteridales** (***Pteridaceae, Negripteridaceae, Sinopteri-daceae, Cryptogrammaceae, Gymnogrammaceae, Actinopteridaceae, Adiantaceae, Parkeriaceae, Vittariaceae***); (*x*) **Davalliales** (***Davalliaceae, Oleandraceae***); (*xi*) **Aspidiales** (***Aspleniaceae, Thelypteridaceae, Athyriaceae, Aspidiaceae, Lomariopsidaceae***); (*xii*) **Blechnales** (***Blechnaceae***); (*xiii*) **Loxsomales** (***Loxomaceae***); (*xiv*) **Plagiogyriales** (***Plagiogyriaceae***); **and** (*xv*) **Hymenophyllopsidales** (***Hymenophyllopsidaceae***). Pichi-Sermolli included the order Marsileales (with the family ***Marsileaceae***) under the sub-class ***Marsileidae*** and the order ***Salviniales*** (with the family ***Azollaceae*** and ***Salviniaceae***) under the sub-class **Salvini** idae. This system is more elaborate and seems to be more natural. He has not indicated the genera included in his families. Ching (1940) splitted the family ***Polypodiaceae*** into 33 smaller families. Copeland (1947) visualised 19 families included under order **Filicales** of class **Filicinae**. Reimers (1954) included seven families under the order Filicales. These are ***Gleicheniaceae, Schizeaceae, Matoniaceae, Hymenophyllaceae, Dicksoniaceae; Cyatheaceae*** and ***Polypodiaceae.*** He included only family ***Marsileaceae*** under ***Marsileales*** and families ***Azollaceae*** and ***Salviniaceae*** under order ***Salviniales.*** Holttum (1949) also recognised the order **Filicales, Marsileales** (***Marsileaceae***) and **Salviniales** (***Azollaceae*** and ***Salviniaceae***). He recognised fourteen families in the order **Filicales**. These include ***Schizaeceae, Gleicheniaceae, Hymenophyllaceae, Dicksoniaceae, Matoniaceae, Dipteridaceae, Cyatheaceae, Dennstaedtiaceae*** (with 11 sub-families), ***Adiantaceae, Polypodiaceae, Thelypteridaceae, Grammitidaceae, Protocyatheaceae,*** and ***Plageogyriaceae.***

Cronquist ***et al*** (1966) recognised the class **Polypodiatae** and divided it into eight sub-classes (see chapters 1 and 8). He included the leptosporangiate ferns in three sub-classes named

(*i*) **Polypodiidae**, (*ii*) **Marseleidae** and (*iii*) **Salvinidae**. These has been further divided into orders and families. The sub-class Polypodiidae has been divided into fourteen orders : (*i*) **Schizaeales** (from ***Schizaeaceae***); (*ii*) **Pteridales** (families ***Pteridaceae, Negripteridaceae, Smopteridaceae, Crypotgrammataceae, Parkeriaceae, Gymnogrammaceae, Adiantaceae, Vittariaceae*** and ***Actinopteridaceae***); (*iii*) **Dicksoniales** (***Dicksoniaceae, Dennstidiaceae, Lindsaeaceae***);

(*iv*) **Davalliales** (***Davalliaceae, Olendaceae***); (*v*) **Hymenophyllales** (***Hymenophyllaceae***); (*vi*) **Gleicheniales** (***Platyzomaceae, Stromatopteridaceae, Gleicheniaceae***); (*vii*) **Cyatheales** (***Cyantheaceae, Lophorsoriaceae***); (*viii*) **Loxsomales** (***Loxomaceae***); (*ix*) **Aspidiales** (***Thelypteridaceae, Aspleniaceae, Athyriaceae, Aspidiaceae, Lomariopsidaceae***); (*x*) **Blechnales** (***Blechnaceae***); (*xi*) **Matoniales** (***Matoniaceae***); (*xii*) **Polypodiales** (***Dipteraceae, Cheiropleuriaceae, Polypodiaceae, Grammitidaceae***); (*xiii*) **Plagiogyriales** (***Plagiogyriaceae***); (*xiv*) **Hymenophyllopsidales** (***Hymenophyllopsidaceae***). The sub-class **Marsilidae** includes the order **Marsileales** with one family ***Marsileaceae.*** The sub-class **Salvinidae** includes the order **Salviniales** with two **families**, ***Salviniaceae*** and ***Azollaceae.***

Biehorst (1971) included all the Leptosporangiate ferns under class **Filicopsida** and divided it into four orders : (*i*) **Noegerathials** (family ***Noegerathiaceae***), (*ii*) **Filicales** (families ***Psilotaceae, Stromatopteridaceae, Gleicheniaceae, Schizaeaceae, Anemiaceae, Lygodiaceae, Osmundaceae, Hymenophyllaceae, Matoniaceae, Platyzomaceae, Dipteridaceae, Cheriopleuraceae, Polypodiaceae, Vittariaceae, Cyatheaceae, Pteridaceae, Parkeriaceae, Davalliaceae, Aspleniaceae, Aspidiaceae, Blechnaceae, Hymanophyllopsidaceae, Loxsomaceae, Plagiogyriaceae***); (*iii*) **Marsileales** (***Marsileaceae***); ***and*** (*iv*) **Salviniales** (***Salviniacea*** **and** ***Azollaceae***).

Reimer's system of classification for class leptosporangiopsida has been followed in this text because of its simplicity. This system is balanced as far as the size and number of groups is connected. It is much easier for the students to follow it.

Family : SCHIZAECEAE

There is no general agreement regarding the limits of this family, which is considered to be the most primitive family of the order filicales. According to Pichi-Sermolli (1958) the family includes five living genera that had a common origin. These genera include ***Schizaea*** (28 species), ***Anemia*** (98 species), ***Mohria*** (three African ***species***), ***Lygodium*** (39 species) and ***Actinostachys*** (13 species). ***Actinostachys*** was previously regarded as sub-genus of ***Schizaea*** but exclusive researches of Bierhorst on the gametophyte and embryo of ***Actinostachys*** have brought out certain difference that support its elevation to a separate genus. Bierhorst (1971) includes only two living genera (***Schizaea*** and ***Actinostachys***) in the family ***Schizaeaceae.*** He excludes all, except Coenozoic spore species, fossil genera from the family. Reed (1948) recognised four families (***Schnizaeaceae, Mohriaceae, Lygodiaceae*** and ***Anemiaceae***); some recognised only ***Schizaeceae, Anemiaceae*** and ***Anemiaceae***); and ***Lygodiaceae;*** whereas others consider ***Schizaeaceae*** and ***Anemiaceae*** as valid families. In this text only one family with five living genera is recognised. These genera are closely related to each other in their anatomy and gross morphology, but differ in the position of sporangia and form of gametophytes. Genus ***Schizaea*** is briefy described along with other genera of the family.

Fig. 12.1. (A–B). *Schizaea pennula.* A. a complete plant showing habit; B. a fertile pinna showing 4 rows of sporangia. Also note the hair arising from the receptacular area.

Distribution

The genus ***Schizaea*** is mostly tropical in distribution with the exception of ***S. pusilla*** which is found in North temperate regions (Sphagnum bogs in North America). ***S. pennula*** (Fig. 12.1) is known from India. The genus ***Actinostachys*** is found in abundance in New Caledonia with the exception of one species which is restricted to southern Florida. The genus ***Mohria*** with its three species is known only from Africa. Majority of the species of ***Anemia*** are found in tropical America with only one species known from India, the rest are pan-tropical in distribution.

Habit and Habitat

The members of this family are herbaceous perennials that perennate by means of underground rhizomes. They are essentially terrestrial plants that frequent xerophytic habitats. A few of them love to grow under moist and shaded conditions (species of ***Schizaea*** and ***Actinostachys***).

Schizaea sp.

Anemia

Mohria

Actinostachys

Lygodium

Schizae a pusilla

External Characters

The plant body is distinguished into roots, stem and leaves. The stem is an underground rhizome that may be creeping (***Lygodium***) or oblique upright (***Schizaea, Anemia***) or erect and ascending (***Actinostachys***). The rhizome in all the genera except ***Anemia*** are dichotomously branched. In ***Anemia*** the branching is lateral. The rhizomes bear simple and unbranched hair in ***Schizaea, Actinostachys, Lygodium*** and ***Anemia.*** In ***Mohria*** the rhizome is covered within glandular scales. The rhizomes are anchored to the substratum by means of adventitious root that may be sparingly branched and have distinct root caps covering the root tips.

The leaves are circinately coiled in young condition and are usually spirally arranged except in ***Lygodium*** where they are arranged in the alternate rows on the upper surfaces of the rhizome. The leaves vary in length from two centimetres (***Actinostachys sp.***) to thirty metres or more in ***Lygodium.*** They are characteristic in having a power of unlimited growth (a feature in which it resembles the stem). The petiole and the rachis are flexible and can twine around the neighbouring shrubs

Fig. 12.2. ***Schizaea elegans.*** A complete plant showing habit.

and trees. The rachis is further branched into primary and secondary branches. The former bear a tuft to apical hair. The latter bear the leaflets or the pinnules that are arranged in a pinnate manner. The pinnules may be entire serrate or deeply incised. In some species the secondary branches are further dichotomously branched and each branch bears a tuft of terminal deeply incised or palmatisect leaflets. The leaves are bipinnate in ***Anemia*** and ***Mohria.*** The leaves are dichotomously branched in ***Actinostachys.*** In ***Schizaea pusilla*** the leaves are simple. In ***S. bifida*** (Fig. 12.3, A) the leaves are forked once or twice, whereas in ***S. dichotoma*** (Fig. 12.3, B) they are flat and repeatedly dichotomously branched and may be up to 50 cm. tall and 25 cm. broad. In ***S. elegans*** (Fig. 12.2) the leaves are simple, fan-shaped and deeply incised. The venation is open dichotomous with free vein endings, except in ***Lygodium*** and ***Anemia*** where the vein ends are joined by marginal veins. Some species of ***Anemia*** show a reticulate pattern of venation.

Schizaea bifida *Schizaea dichotoma*

All the pinnules of a leaf are potential sporophylls in ***Mohria*** and ***Schizaea elegans*** (Fig. 12.3). In other genera there is segregation between fertile and sterile regions of the leaf. In ***Anemia colimensis*** the lowermost pair of pinnae is fertile and the sporophylls have well developed lamina. In ***A. aspera*** and ***A. perrieriana*** the lowermost pair of fertile pinnae are held erect. In ***A. rutifolia*** two pairs of lower pinnae are fertile and in ***A. millefolia*** the leaves are ***dimorphic***. In ***Schizaea dichotoma*** (Fig. 12.3, B), ***S. pusilla*** and ***S. rupestris*** the plants bear separate sterile and fertile leaves. In ***S. dichotoma*** the fertile leaves (Fig. 12.3, B) have dichotomously branched leaf lamina and each branch is terminated by sporangia bearing pinnae that are characteristically arranged in two parallel rows. In ***S. melanesica, S. intermedia, S. pennula*** (Fig. 12.1) and ***S. laevigata*** the sporangia bearing pinnae are arranged in penicellate tufts with sporangia arranged

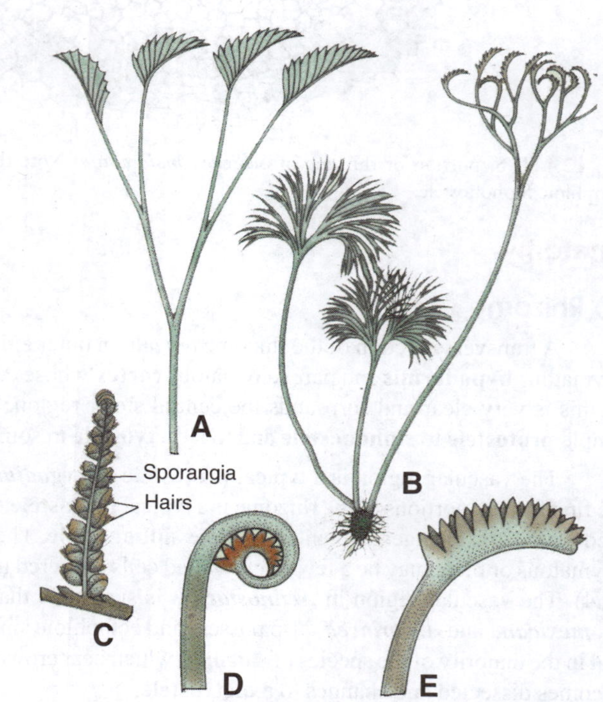

Fig. 12.3. (A—E). *Schizaea. A Schizaea bifida*; *B.S. dichotoma* showing habit; C. Part of a fertile pinna enlarged to show the marginal position of the sporangia; D—E. cirinately coiled leaves of *S. rupesteris.* (A. after Diels, B. after Lotsy; C. after Bauer; D, E, after Bower).

in four apparent rows. In ***S. pennula*** the sterile leaves have long petioles with sickle-shaped pinnae arranged in terminal penicellate tufts (fig. 12.1, A).

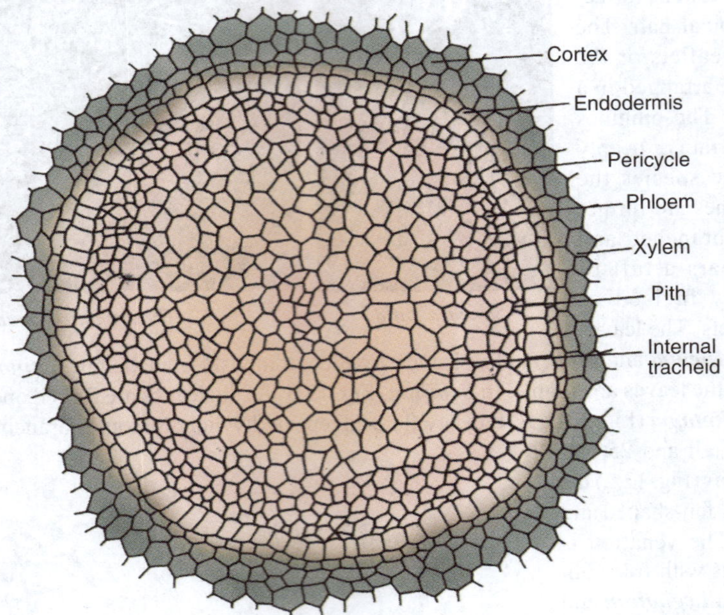

Fig. 12.4. T. S. portion of rhizome of *Schizaea molacceana.* Note the tracheids in the centre of pith. It is an ectophloic Siphonostele.

Anatomy

(i) Rhizome

A transverse section of the rhizome reveals an outer **epidermal** layer followed by the scleren-chymatous **hypodermis** and parenchymatous **cortex** whose cells are full of starch grains. The endo-dermis is very clear and surrounds the central stelar region. The stelar organisation varies from a simple **protostele** to **siphonostele** and to a **dictyostele** in some genera.

The vascular region is a typical **protostele** in ***Lygodium. In Schizaea*** the vascular region in the first formed portions of the rhizome is a typical **protostele.** In the later developed parts it becomes medullated and changes into an **ectophloic siphonostele.** The pith may be homogenous and paren-chymatous or there may be a few thick-walled cells scattered in it. These are the tracheidal cells (Fig. 12.4). The vascular region in ***Actinostachys*** is similar to that ***Schizaea.*** Some species of ***Anemia*** (***A. mexicana*** and ***A. coriacea***) also possess and ectophloic siphonostele. In all the species of ***Mohria*** and in the majority of the species of ***Anemia*** (which bear crowded leaves) the ectophloic siphonostele becomes dissected and changes to a **dictyostele**.

(ii) Root

The anatomy of the root (Fig. 12.5) in all the genera has a **diarch** stele surrounded by the pericycle, distinct endodermis and a cortex distinguished into inner sclerenchymatous cortex and an outer parenchymatous cortex. The root epidermis, before producing root hair, has shorter cells or the **trichoblasts** and longer cells. The root grows by means of a single apical cell. The transverse section of root of ***S. pusilla*** presents beautiful and a more or less regular pattern (Fig. 12.5) of tissue arrange-ment. The xylem consists of 4—6 cells (tracheids) and is **diarch**. The phloem bundles that alternate with xylem are composed of 4—6 sieve cells. The **pericycle** and **endodermis** consists of 6 cells each. The endodermal cells are large and barrel-shaped and possess distinct **casparian strips** along the inner tangential and radial walls. Next to endodermis is the **inner cortex** that consists of 6 cells with thick walls. The outer cortex is made up of one or two layers of cells each having seven or eight cells that thick walled. The epidermis is made up eight or more thin-walled cells (Fig. 12.5) and gives out root hair. It is usually a permanent layer in ***Schizaea.***

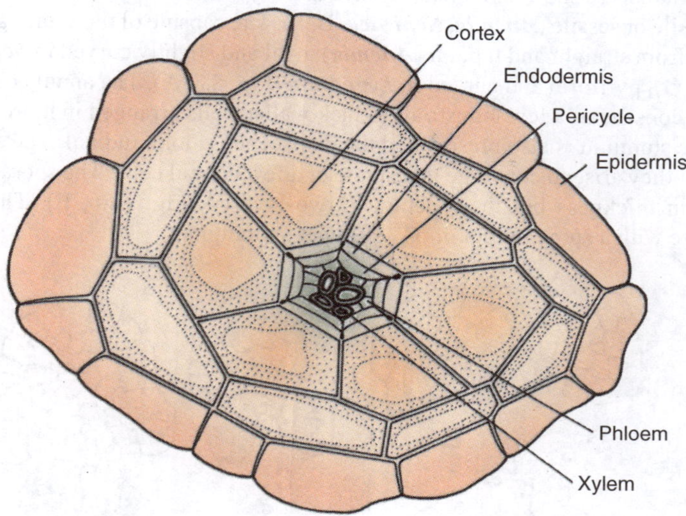

Fig. 12.5. T. S. root of *Schizaea pusilla.*

(iii) Leaflet

The leaflet or the pinnule has the usual upper and lower epidermal layers with stomata restricted to the lower epidermis. The mesophyll is undifferentiated and consists of loosely arranged chlorophyllose cells. The vascular bundle is concentric. The xylem forms a central core surrounded by phloem which in turn is surrounded by pericycle and endodermis.

Reproduction

The sporophyte reproduces asexually by means of spores that are produced within sporangia. The sporangia are borne on the fertile leaflets or pinnules called the **sporophylls.** The sporangia arise singly and do not form sori. They arise on either side of the vein in ***Schizaea, Actinostachys*** and ***Anemia*** and form two rows. (Fig. 12.3, C). In ***S. pennula*** (Fig. 12.1, B) they appear to form four rows. The sporangia arise on the dorsal side of the sporophyll either on the margin or a little close to the margin. The position of a sporangium is characteristic in ***Lygodium*** and ***Mohria.*** In both these genera the sporangium arise singly close to the end of each vein. Regarding their protective devices the margins of the sporophylls (fertile pinnule) turn or curve upwards to protect the sporangia in ***Schizaea*** and ***Mohria.*** There is no true indusium. In ***Anemia*** the

S. pennula

sporangia are not completely covered by the curved margins of sporophylls. They are partly protected. In ***Lygodium*** a true indusium arises as an outgrowth from the sporophyll tissue and covers the single sporangium. The term **monangial sorus** is used to express this condition.

Structure of Sporangium

The sporangia of ***Schizaeceae*** present certain interesting features in the form and structure. The sporangial stalks are short and thick (***Mohria*** and ***Lygodium***, Fig. 8.8, A, E) or the sporangia may be sub-sessile or sessile (***Anemia, Schizaea,*** B, F). The capsule of the sporangium is large and varies in shape from straight and tubular (***Anemia***), oval and slightly curved in ***Schizaea*** (Fig. 8.8, B; Fig. 12.6, F, G), pyriform and curved in ***Lygodium*** (Fig. 8.8. A). The annulus in all these cases is terminal and consists of a few industrial or thick walled cells arranged in more than one row of cells. A definite stomium is also present and the dehiscence is longitudinal. The annulus cells are so arranged that they arise to converge to a single or a few central cells. The sporangial wall is one layer of cells in thickness but the tapetum is two-layered (Fig. 12.6, E). The sporangia are leptosporangiate with a spore output of 64—128 per sporangium.

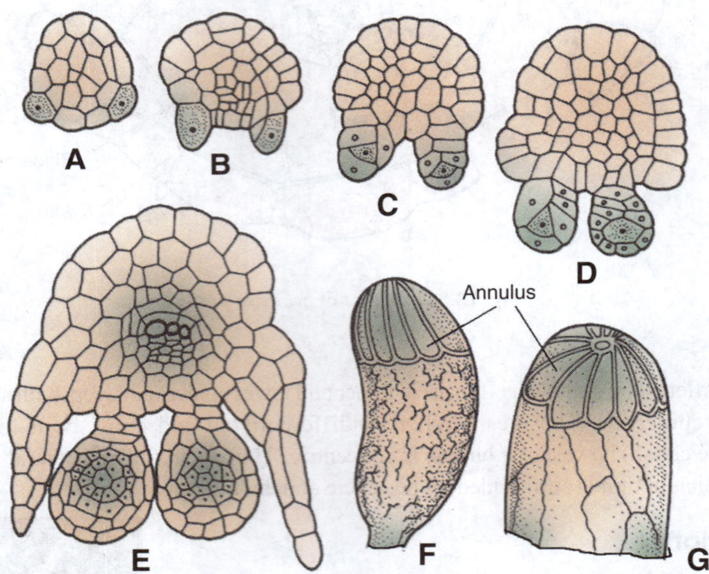

Fig. 12.6. A—E. Stages in the development of sporangium in *Schizaea rupestris*; F, G. Mature sporangia with terminal annulus. (After Bower).

The spores are usually triangular, tetrahedral and trilete (Fig. 8.10). In ***Anemia nigerica*** the triangular spores have mucronate and rounded corners. The **sexine** is striate. In ***Lygodium smithianum*** the proximal surface of the triangular spores is granulose whereas the distal one is tuberculate. The tubercles have granulose surface (Fig. 8.10).

Prothallus or the Gametophyte (Fig. 12.7 to 12.9)

There is a great variation in the development and structure and form of the prothallus among the ***Schizaeaceae.*** In ***Anemia, Lygodium*** and ***Mohria*** the spore divides into two equal cells of which the distal ones remains inactive throughout. The proximal cell divides by wall perpendicular to the first to cut off a small lateral **rhizoidal initial** and a large **prothallial initial** (Fig. 8.11). The prothallial initial divides by a series of walls parallel to the first wall and grows into a germ filament that elongates in a direction parallel to the polar axis of the spore (facing the proximal pole). The primary rhizoid elongates parallel to the equatorial axis of the spore. In ***Actinostachys*** and some species of ***Schizaea*** the spore divides without any polarity in cell divisions and direction of growth. It results in a plate of cells in which a meristematic cell is differentiated in one of the peripheral marginal cells. Further growth of the prothallus is in the direction of the meristematic cell. In ***Lygodium, Mohria*** and ***Anemia*** the gametophyte is cordate, flat and lop-sided. Further growth is asymmetrical and a notch, with apical meristem situated in it, is formed. There is a distinct cushion behind the distinct notch. The prothalli are bisexual but rarely unisexual. The antheridia develop from superficial cells towards the lower end of the cushion or along its lateral sides. The archegonia develop on the cushion.

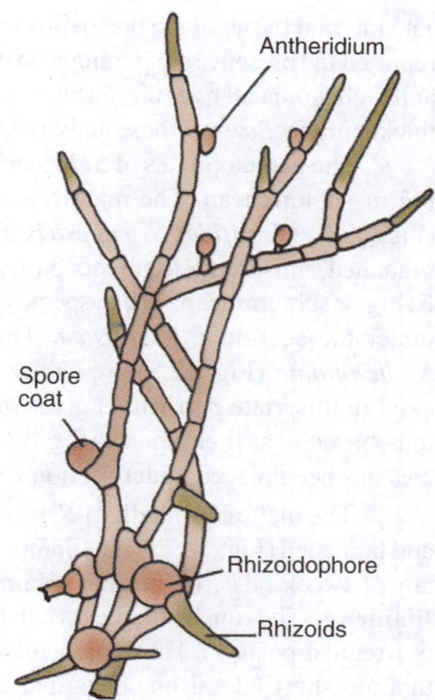

Fig. 12.7. Filamentous prothallus of ***Schizaea pusilla.***

The gametophyte of ***Actinostachys*** has been studied in detail by Bierhorst (1965, 1966, 1968) for six species. The mature prothalli are tuberous and subterranean. The prothalli are several cell layers in thickness and are profusely covered with multicellular rhizoids that arise from epidermal cells. The prothallus grows by means of a many celled apical meristem. In old gametophyte some surface buds develop and grow into short lateral branches or the so-called primary axes which on separation from the parent prothallus may develop into new prothalli. Specialised buds also arise from the tips of rhizoids and even for archegonial necks and develop vegetatively into new prothalli. The mature prothallus shows distinct foldings that give it a ridged and grooved appearance on surface. This is due to differential thickening.

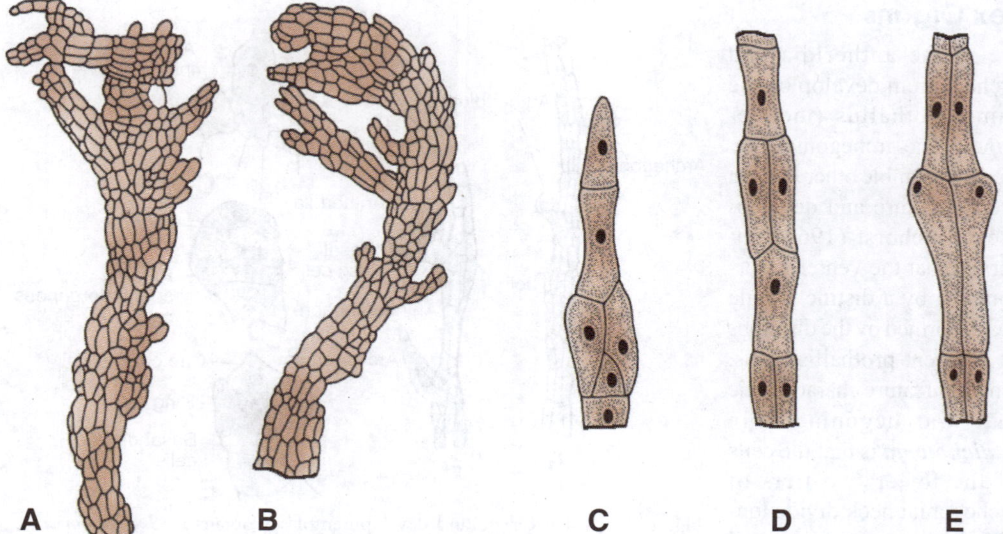

Fig. 12.8. (A–E). Gametophytes of *Schizaea dichotoma.* A–B. Portions of mature gametophytes; C, D, E. Portions of gametophytes a little behind the growing points (After Bierhorst).

The internal tissue of the prothallus is parenchymatous and is mycorrhizal. Bierhorst (1971) has also reported the presence of sporangia on the tips of rhizoids. The prothalli are bisexual, and bear antheridia and archegonia all over the surface. The gametophytes posses, besides apical meristem, a peripaheral thickening meristem, whose activity leads to the characteristic tuberous shape of the mature prothallus.

The gametophytes of ***Schizaea*** may be surface living or subterranean. The mature prothalli of ***S. pusilla*** (Fig. 12.7), ***S. bifida, S. rupestris*** and ***S. fistulosa*** are branched, uniseriate, filaments partly green and surface living or subterranean. These species have been included under the section ***Eu-schizaea.*** The gametophytes of ***S. dichotoma*** (Fig. 12.8) are subterranean and have an axial multiseriate part which gives off branches that are uni-serrate near their apices (Fig. 12.8, A, B). This species has been placed under section ***Lophidium.***

Schizaea fistulosa

The mature prothalli of ***S. pusilla*** are filamentous and branched (Fig. 12.7). The filaments are uniseriate and are of two kinds : (*i*) **runner filaments** and (*ii*) **bush filaments**. The runner filaments can be distinguished by their longer cells, greater length, horizontal or irregular position. These filaments bear two kinds of short branches. These are the **rhizidiophores** that are short lateral branches that are lateral in position and bear unicellular **rhizoids** and have swollen cells that are filled with an endophytic fungus. The other types are called the **antheridiophores** (Fig. 12.7) and are short or few celled lateral branches which bear antheridia. The **archegonia** are borne on **bush filaments** that are short, erect branches (Fig. 12.7). The prothalli are monoecious.

Bierhorst gave a detailed description of the structure of mature gametophyte of ***S. dichotoma*** (Fig. 12.8). The prothalli are underground and have frequent shaded localities. The main axis is multiseriate with uniseriate tips. It is branched, the branches are also multiseriate and become uniseriate at the apices. The growth is apical. The prothallus has a characteristic jointed appearance.

Sex Organs

The antheridia and archegonium develop on the same prothallus (monoecious). The archegonia (Fig. 12.10) resemble other ferns in their structure and development. Biehorst (1967) observed that the venter is surrounded by a distinct sterile jacket formed by the divisions of adjacent prothallial cells. Another feature characteristic of archegonia in ***S. dichotoma*** is that the cells of the lower two tiers of archegonial neck divide longitudinally into inner small and outer large cells. The neck at this level consists of eight cells (4 large and

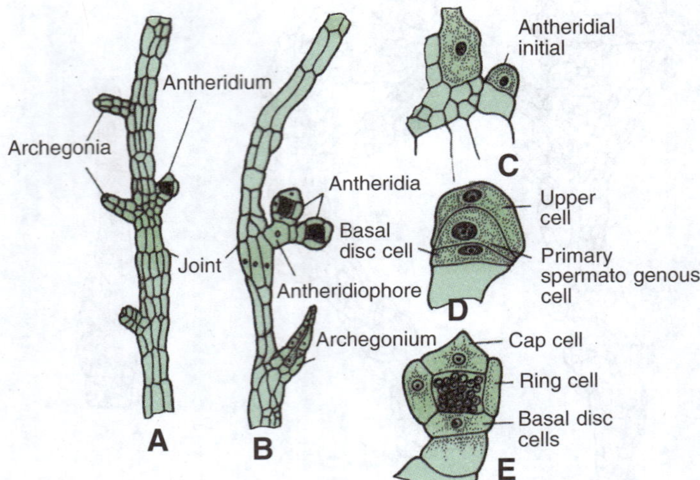

Fig. 12.9. (A–E). Origin and developmentof sex organs in *S. dichotoma* A. Portion of prothallus bearing archegonia and an antheridium; B. Portion of prothallus bearing two antheridia and an archegonium; C. Formation of an antheridial initial; D. Young antheridium with a single spermatogenous cell, a terminal cell and a basal cell ; E. A, Mature antheridium (All after Bierhorst).

4 small), a feature also found in some bryophytes where the archegonial neck is two-layered in the lower portion.

Lygodium japonicum

The antheridia are of emergent type in the family and may be borne on the ventral surface or laterally in the thalloid prothalli or on special branches called the antheridiophores. The antheridial jacket consists of **a basal cell, a ring cell** and one or two cap cells. The number of antherozoids varies from 25 to 156.

Embryology

The development of embryo has not been studied in detail. Biehorst (1971) studied it in ***Actinostachys*** whereas Vladesco (1935) studied it in ***Lygodium japonicum*** and ***Anemia phyllitidis.*** In the former it is similar to the other leptosporangiate ferns. In ***Anemia*** the epibasal cell divides by a median longitudinal wall and the hypobasal by a transverse wall. Further details were not observed, but are believed to be similar to other ferns.

DISCUSSION

The filamentous nature of the prothallus in some species of ***Schizaea*** has led to a considerable discussion regarding their evolutionary status. The uniseriate and branched gametophyte of ***S. pusilla*** and ***S. bifida*** resemble some branched algae and were regarded as primitive because the filamentous habit is unaffected by environmental changes. Such a line of thought was given up due to following reasons :

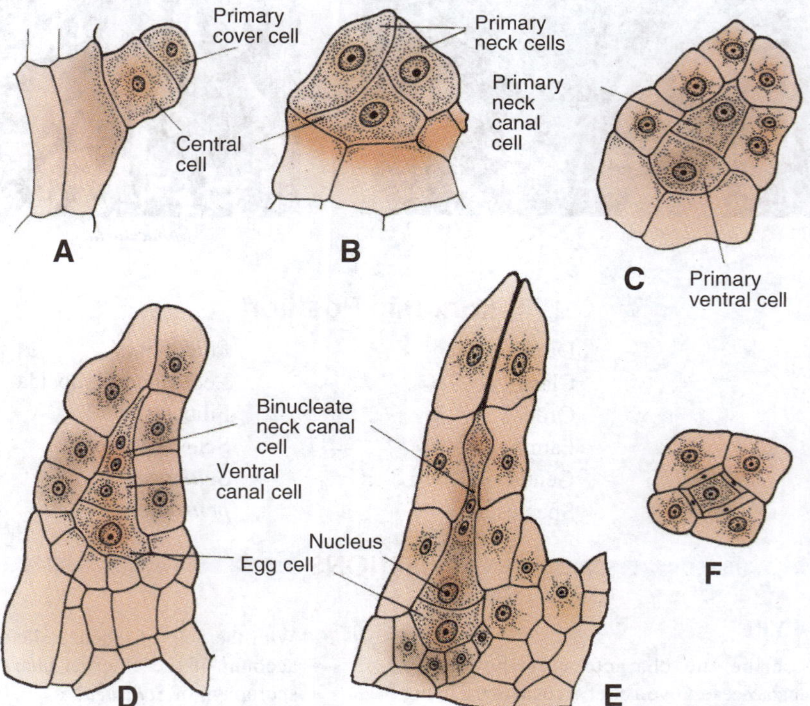

Fig. 12.10. (A—E). Stages in the development of archegonium in *S. dichotoma* (After Bierchorst).

1. Other members of Schizaeaceae *e.g.* ***Lygodium***, ***Mohria***, ***Anemia*** and ***Actinostachys*** possess cordate prothalli with a distinct mid-rib.
2. The massive prothalli of the liverworts do not show a filamentous tendency. Similarly massive, and subterranean prothalli of primitive vascular plants like ***Lycopodium*** and ***Ophioglossum*** never exhibit a filamentous trend of growth.
3. The primitive vascular plants like ***Psilotum*** and ***Tmesipteris*** have fleshy subterranean gametophytes that never exhibit filamentous tendencies.

Bierhost (1967) supported this line of thought and compared filamentous prothalli of some species of ***Schizaea*** to an algal order ***Sphacelariales***. The origin of branches and their position is similar in ***S. dichotoma*** and ***Sphacelaria***. Bierhorst recognised three kinds of gametophytes in the genus ***Schizaea***. These are :

1. The uniseriate filamentous types that may be subterranean or epiterranean (*e.g.* ***S. bifida***, ***S. rupestris***, ***S. fistulosa***)
2. The multiseriate and branched underground prothallus of ***S. dichotoma***.
3. The radially symmetrical, unbranched or rarely branched or irregular and tuberous prothalli of ***Actinostachys.*** They are subterranean.

Bierhorst (1967) has regarded the fleshy subterranean and cordate prothalli to be primitive.

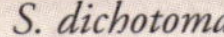

S. dichotoma

Sphacelaria

Taxonomic Position

Division	:	Filicophyta
Class	:	Leptosporangiopsida
Order	:	Filicales
Family	:	Schizacaceae
Genus	:	***Schizaea***
Species	:	***pennula***

QUESTIONS

ESSAY TYPE

1. Describe the characters of the family Schizacaceae. Give a brief account of the living genera of the family.
2. With the help of labelled sketches, give an account of the external characters of the sporophyll of ***Schizaea***.
3. Give a brief account of reproduction in Schizaeaceae.

4. Describe the gametophyte of ***Schizaea*** with labelled diagrams.
5. Briefly describe the life-cycle of ***Schizaea***.

SHORT ANSWER TYPE

6. Write short notes on :
   (*a*) Fertile pinna in ***Schizaea***
   (*b*) Spores
   (*c*) Prothallus
   (*d*) Development of Sex organs.
   (*e*) Taxonomic position of ***Schizaea***.
7. Draw labelled diagrams of the following :
   (*a*) Complete plant of ***Schizaea***.
   (*b*) Mature sporangia in ***Schizaea***
   (*c*) Spores in ***Schizaea***
   (*d*) Filamentous prothallus.
8. Describe the development of antheridium / archegonium in ***Schizaeaceae***.
9. Give the names of the living genera of ***Schizaeaceae***.
10. Distinguish between :
    Schizaea and ***Lygodium***

OBJECTIVE TYPE

11. Select the correct answer :
    (*i*) Schizaeaceae is represented by—genera.
    (*a*) Two
    (*b*) Four
    (*c*) Three
    (*d*) Five.
    (*ii*) In which member of the family true indusium is present ?
    (*a*) ***Schizaea***
    (*b*) ***Lygodium***
    (*c*) ***Mohria***
    (*d*) ***Actinostachys.***
    (*iii*) Dichotomous branching is the characteristic feature of
    (*a*) ***Schizaea bifida***
    (*b*) ***S. rupestris***
    (*c*) ***S. elegans***
    (*d*) ***S. dichotoma.***
    (*iv*) Stele in ***Schizaeaceae*** is
    (*a*) Protostele
    (*b*) Ectophloic Siphonostele
    (*c*) Endophloic Siphonostele
    (*d*) Solenostele.
    (*v*) The prothallus is filamentous entous in
    (*a*) ***Schizaea***
    (*b*) ***Lygodium***
    (*c*) ***Mohria***
    (*d*) ***Actinostachys.***

CHAPTER 13

Sticherus flabellatus

Family Gleicheniaceae Genus *Gleichenia*

- Introduction
- Description of ***Gleichenia*** Smith

Introduction

The number of genera included in this family varies with taxonomists. Christensen (1938) recognised five genera (***Gleichenia***, ***Stromatopteris***, ***Dicranopteris***, ***Sticherus*** and ***Platyzoma***). Ching (1940) and Naka (1950) recognise seven or eight genera. Copeland (1947) included six genera (***Hicriopteris*** in addition to above five). Holtum (1957) excluded ***Platyzoma*** from this family and recognised only ***Gleichenia***, ***Dicranopteris*** and ***Stromatopteris***. Some authors like Diels (1900) included only ***Gleichenia*** and ***Stromatopteris***. Bierhorst

(1971) included ***Stromatopteris*** in a separate family ***Stromatopteridaceae.*** Abbot (1954) described a fossil genus called ***Oligocarpia***. The family was represented in the past in the Mesozoic era and is a pimitive family of the order Filicales.

The genus ***Gleichenia*** is discussed in detail in this text.

Gleichenia *Dicranopteris* *Platyzoma* *Sticherus flabellatus* *Stromatopteris* *G. longissima* *Sticherus* *Gleichenia linearis* *Hicriopteris*

GENUS–GLEICHENIA Smith

According to Smith and Bierhorst (1971) the genus included about 130 species. Christensen (1938) recognised about 10 species that are found in the southern hemisphere. Some of the species of this genus *e.g* ***Gleichenia linearis***, ***G. pectinata*** etc., have been transferred to the genus ***Dicranopteris***. In this text they are regarded as species of ***Gleichenia***. The genus is represented in India by two ***G. linearis*** (= ***D. linearis***) and ***G. glauca***. The former is a widely distributed species and is very common in the tropical and subtropical regions. In India it is found in Southern India, Western Ghats, Eastern Himalayas and Kumaon Hills. ***G glauca*** is found only in the Eastern Himalayas.

They usually ocur in exposed localities and form dense and bushy growths. All the species are terrestrial and xerophytic.

SPOROPHYTE

The sporophyte is distinguished into a long, dichotomously branched, prostrate and subterranean rhizome that bears leaves on the upper surface and adventitious roots on the lower or under-surface. The leaves of this genus are characteristic and arise at longer intervals on the rhizome. They appear to arise in a single row but closer examination reveals them to be borne in three rows. They range in length from a few feet to more than 50 metres. In ***G. linearis*** (Fig. 13.1 B) and many other species the leaves exhibit repeated forking. They appear to be dichotomously branched, but actually it is a pseudodichotomy because the growth of rachis is arrested (Fig. 13.1) and the two pinnae or leaflets below continue to grow. In ***G. longissima*** the first pair of leaflets below the arrested tip of the rachis grows to a limited extent and then stops growing. So the leaf in this species, bears only a pair of leaflets or pinnae (Fig. 13.1, A). In ***G. linearis***, ***G. pectinata***, ***G. dichotoma*** and some other species the pair of first leaflets terminates in a dormant apex and the growth is continued by another pair of leaflets that arise in air below the arrested tip. The growth is continued in this manner and in some species the leaves attain a size of more than fifty metres. The

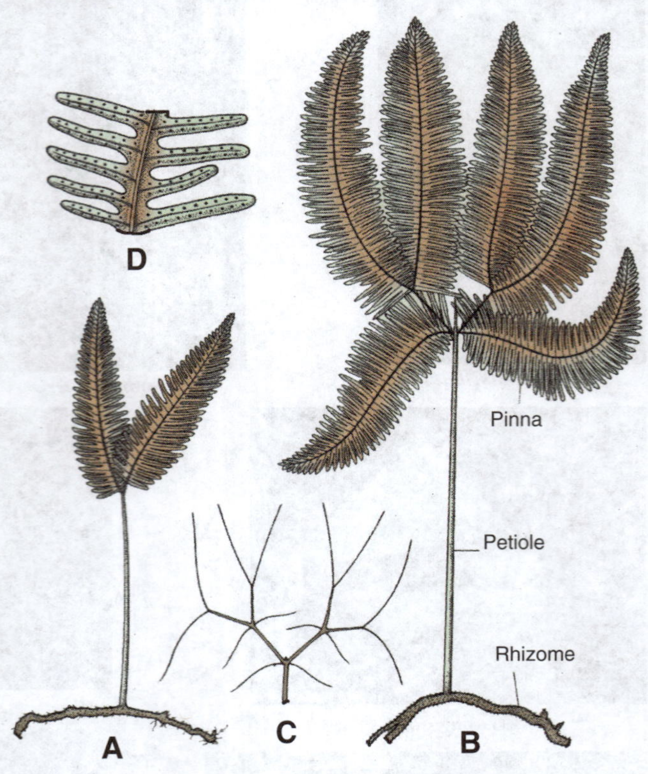

Fig. 13.1. *Gleichenia* A. *G. longissima* with a rhizome bearing a leaf with two pinnae that have a limited growth. There is no further branching in this case; B. *G. linearis*, a plant bearing a leaf with indefinite growth. Note the branching; C. Stereodiagram showing pattern of branching of leaf in *G. linearis*; D. Portion of pinna showing sori.

arrested apical buds are protected by hair or scales or by stipule-like out-growths. In ***G. linearis*** (Fig. 13.1, B) the branches of the rachis are subtended by **accessory** or basal branches that bear pinnules. Sometimes the branches are of unequal size. The smaller branch bears a pair of pinnae at its tip and does not grow further. The large branch continues to grow in the same fashion. This type of branching gives a zigzag appearance to the leaf. The whole leaf which simulates an aerial branch is called the **frond**.

The terminal divisions or branches of the frond bear lobed pinnae or leaflets. The incisions of the lobes almost reach the midrib so that they appear as pinnules. Each pinnule is traversed by a vein that gives off lateral veinlets which may fork dichotomously or may remain simple. The venation is open.

The adventitious roots are branched and arise from the under-surface of the rhizome and may be irregularly arranged or may be disposed in three regular rows.

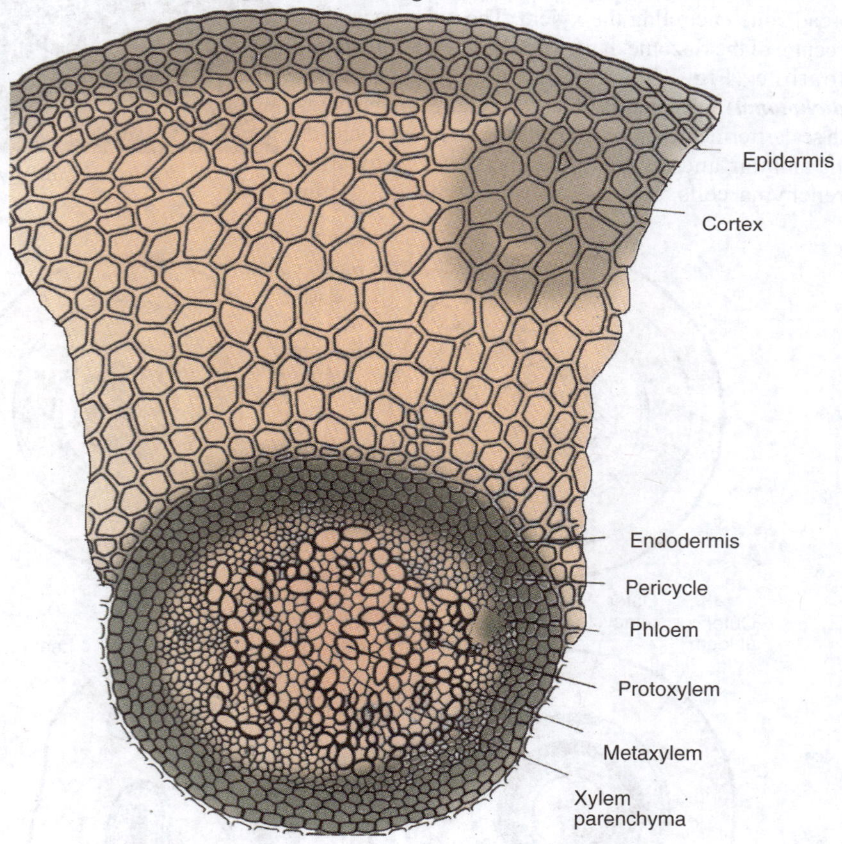

Fig. 13.2. T. S. portion of stem of *Gleichenia linearis.*

Anatomy

(a) Rhizome

A cross section through the rhizome (Fig. 13.2) reveals the following structure from outside within :

Epidermis. It is made up of a single layer of colourless and thin-walled cells. In some species the cells of the epidermis are sclerenchymatous and brown in colour. It is not perforated by any stomata.

Cortex. It is composed of many layers of cells and forms a broad zone of tissue sandwiched between epidermis and the endodermis. In ***G. linearis*** (Fig. 13.2) the whole of the cortex is sclerenchymatous. The cells are brown in colour. In ***G. dichotoma*** there are a few layers of thin-walled cells of the cortex external to the endodermis (Fig. 13.3 B). The rest of the cortex is sclerenchymatous.

Stele or the Vascular Region. The stelar organisation in majority of the species is protostelic except in ***G. pectinata*** where it is amphiophloic siphonostele in the internodal region and a solenostele in the nodal region (Fig. 13.3, C.D.). In ***G. linearis***, ***G. dichotoma*** and a number of other species the stele is a mixed protostele (Fig. 13.3. A, B). It is surrounded on the outside by a single layered endodermis whose cells possess distinct casparian bands. Next to endodermis is a few layered pericycle. It is composed of thin walled cells and completely surrounds the phloem (Fig. 13.2). The phloem consists of large and thin-walled sieve tubes and phloem parenchyma. It forms a broad zone encircling the xylem. The xylem occupies the centre of the rhizome. It may be circular (***G. flabellata***) in outline (Fig. 13.3. A) or lobed (***G. linearis***, ***G dichotoma***) and consists of larger metaxylem tracheids with scalariform thickenings, smaller protoxylem tracheids with annular and spiral thickenings and thinwalled parenchyma cells. A cross section reveals as if the

Gleichenia dichotoma

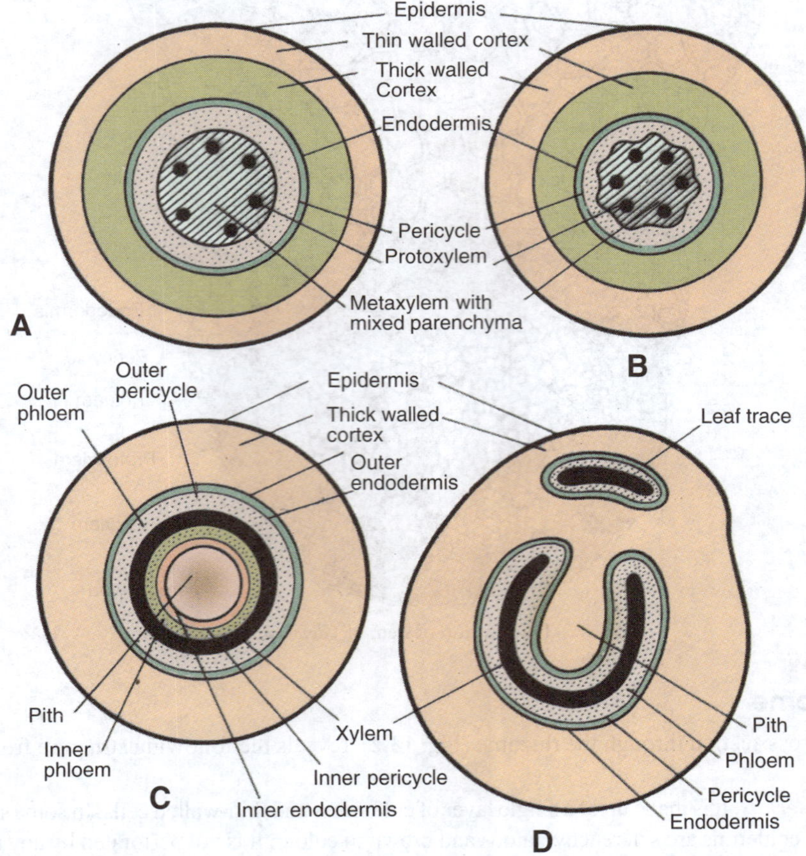

Fig. 13.3. Diagrammatic cross sections through the rhizomes of some species of ***Gleichenia.*** A. ***G. flabellata***; B. ***G. dichotoma***; C. T. S. through internode of ***G. pectinata*** showing amphiophloic siphonostele; D. T.S. through node of ***G. pectinata*** showing a solenostele.

metaxylem and protoxylem elements are embedded in the matrix of thin-walled cells. Such a protostele is often designated **mixed protostele.** The protoxylem occurs in the form of small patches situated a little distance behind the periphery of the xylem. The arrangement of the xylem is, therefore, mesarch.

In ***G. pectinata*** a cross section through the internode reveals an amphloic siphonostele. In the centre there is a distinct sclerenchymatous pith surrounded by inner endodermis and inner pericycle (Fig. 13.3. C). Next to the inner pericycle is the inner phloem, which completely encircles the inner periphery of the xylem. The xylem is mesarch. The xylem tracheids are mixed with parenchyma cells. The outer phloem, outer pericycle and outer endodermis encircle the xylem towards its outer circumference. A section through nodal region reveals a solenostelic stele because the continuity of the amphiophloic siphonostele is broken by leaf gap left by the departing leaf trace. (Fig. 13.3. D)

(*b*) Petiole

The leaf trace that enters the petiole is C-shaped (Fig. 13.4, A.B). In most of the species

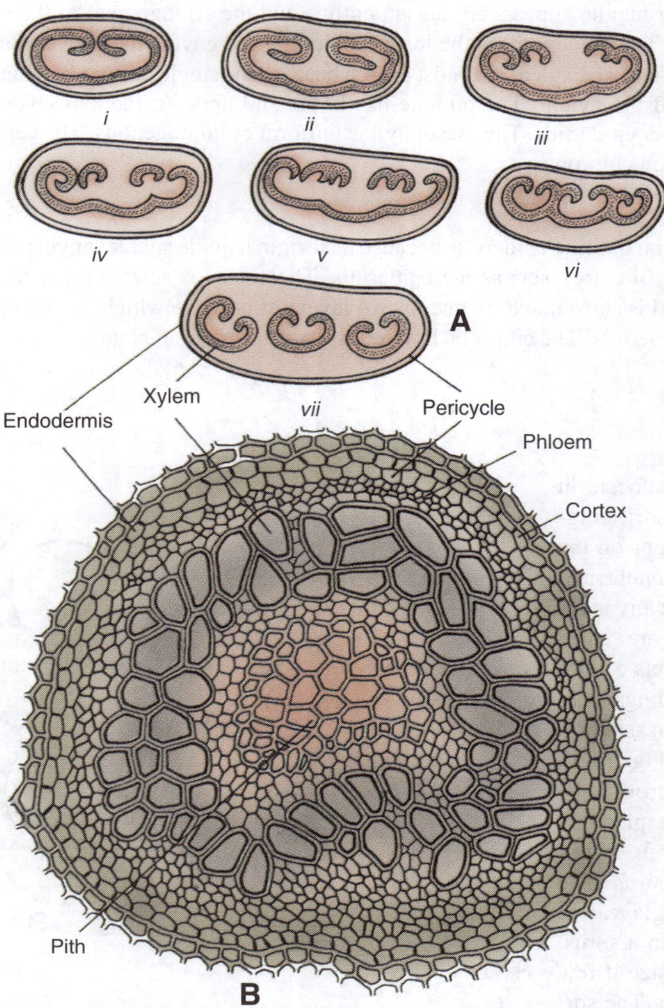

Fig. 13.4. (A—B). A. A series of cross sections through rachis of *G. linearis* showing extramarginal type of origin of pinna traces; B.T.S. part of petiole of *G. dicarpa* cut near the base. Note the contraction of the xylem which results in the meeting together of the two margins (A, after Bower; B, after Dr. Kidston)

(*G. linearis*, *G. pectinata*, *G. dichotoma*) it remains undivided throughout the length of the petiole and divides into three branches just below the first pair of leaflets or pinnae. In ***G. linearis*** the petiolar bundle consists of curved C-shaped xylem with ends curved inwards (Fig. 13.4, A) and surrounding a central parenchymatous region or the pith. The protoxylem occurs in the form of small patches towards the concave side of the xylem arch. The number of protoxylem masses may be three (one median and two lateral) in smaller bundles and more than three in larger bundles. The xylem is surrounded towards its convex side by phloem which encircles it completely and then invaginates inwards around the bent edges and continues upwards towards apex of the arch. The pericycle surrounds the phloem and is made up of three or four layers of thin-walled cells. The endodermis completely encircles the bundle and does not invaginate the xylem arch. Invagination of the endodermis has been reported in some species of ***Gleichenia***. In ***G. cryptocarpa*** the leaf trace bundle divides into three, immediately on entering the petiole. In this case the petiole has three vascular bundles.

The structure of the vascular bundle remains unaltered in the rachis and its branches. In the pinnule the vascular bundle appears circular in outline and the arrangement of the vascular tissues becomes collateral. The phloem faces the lower epidermis and covers the convex lower face of the semilunar xylem strand. The xylem is almost exarch because the single protoxylem mass is present on the upper surface of the xylem. The pinnule has lower and upper epidermal layers with stomata restricted to the lower epidermis. The mesophyll is uniform or homogenous. The cells enclose large air spaces and contain chloroplasts.

(*c*) Root

The root has no definite epidermis because the original epidermal layer gets sloughed off and the outermost layer of cortex acts as the epidermis. The cortex is sclerenchymatous. The stele is diarch to tetrach and is surrounded by one or two layers of pericyle which in turn is encircled by a single layer of endodermis. The end cells have conspicuous casparian bands.

Reproduction

(*a*) Sori

The sporangia occur in groups called the sori (ring-sorus). The sori occur on the ventral side of the pinnules and are not protected by any indusium (naked). They are seated on the lateral veinlets a little behind the vein endings or in the middle of the vein and form two distinct rows (Fig. 13.5). The sporangia arise on a circular receptacle or placenta and vary in number from four to fifteen in each sorus. The size of the sporangia varies with their number in a sorus. They are larger in size if few and smaller if more. The sori with few sporangia may have them arranged in a uniseriate

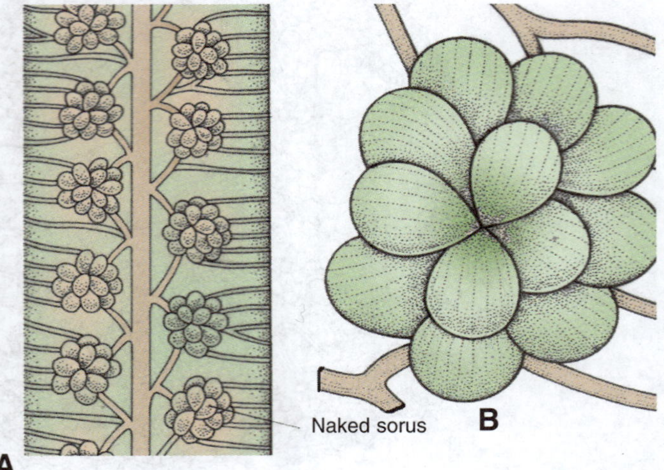

Fig. 13.5. (A–B). Sori of **Gleichenia**. **A.** A pinnule of *G. pectinata* bearing sori on veins; **B.** A sorus of *G. pectinata* enlarged to show the arrangement of sporangia (After Bower)

manner without any sporangia in the centre *i.e.* the centre of the sorus is vacant. In some species the sorus has 4 to 8 sporangia arranged in the centre (Fig. 13.6). There may be overlapping of the sporangia in the centre. A fertile pinnule may have one to many sori on its ventral surface. In ***G. polypodioides*** the fertile pinnules contract and their margins bend downwards so as to partially cover the sori. Species possessing epidermal hair or scales offer protection to the sori. The sporangia constituting a sorus are compactly arranged and give a superficial appearance of a synangium, but actually they are not organically fused and are easily separable.

G. polypodioides

(*b*) Sporangium

Structure

The sporangium is approximately pear-shaped structure that may be sessile or subsessile (Fig. 13.7). The capsule has a single layer of **wall cells or jacket cells.** Some of the jacket cells develop thick walls and form a distinct annulus whose position is oblique transverse. Next to the jacket layer are two layers of the tapetal cells. The outer layer of tapetum is composed of tubular and flattened cells that are not so conspicuous. The inner layer of tapetum is composed of large and elongated cells with distinct nuclei and cytoplasmic contents (Fig. 13.7, F). The cells wedge in between the central mass of spore mother cells. Later the tapetal cells disorganise to form a plasmodial fluid that nourishes the young spores formed as a result of tetrad division of the mother cells. The number of spores per sporangium varies from 128-1, 024.

The sporangium dehisces by means of a vertical slit caused by the contraction of the annulus cells. The line of dehiscence in sporangia arranged along the margin of the sorus and with no sporangia in the centre faces inwards *i.e.* towards the centre. The position of this line of dehiscence is subject to variation in case there are central sporangia in sorus.

Development

Bower (1899) studied the development of the sporangium in ***Gleichenia.*** Before the commencement of the sporangial development there is formation of an elevated and a circular receptacular surface on the placenta on the ventral surface of the sporophyll. Sporangia arise from the superficial cells of the receptacle and develop simultaneously. In ***G. dichotoma*** the marginal cells of the receptacle enlarge (Fig. 13.7, A, B) and behave as **sporangial initials.** The development is leptosporangiate. The sporangial initial cuts off segments along its three sides. The segments cut off earlier contribute to the stalk of the sporangium whereas

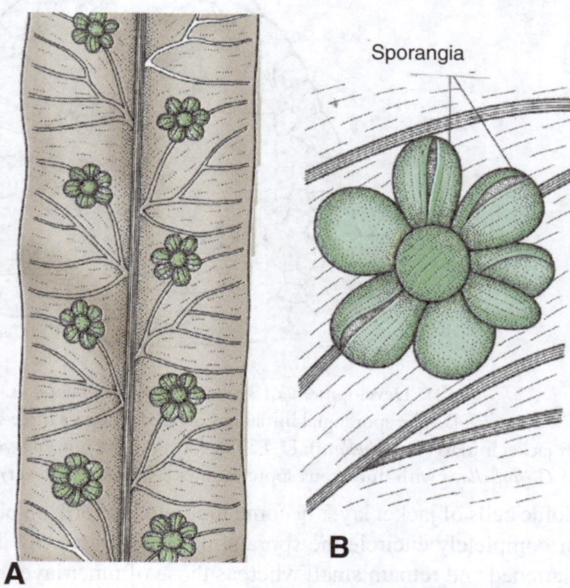

Fig. 13.6. (A–B). Sori *Gleichenia linearis.* A. A pinnule bearing sori on veins; B. A sorus enlarged to show the arrangement of sporangia (After Bower).

the three segments cut off last of all form the three **primary jacket cells.** The fourth jacket cell is formed by the periclinal division of the initial (Fig. 13.7, C) This division differentiates a central **archesporial cell** and the surrounding **primary jacket cells.** The archesporial cell divides by means of periclinal and anticlinal division to form a two-layered tapetum enclosing a **primary sporogenous cell** (Fig. 13, 7, D, E) which by further divisions gives rise to 32-256 spore mother cells. The jacket layer divides by means of anticlinal walls to form a complete jacket around the tapetum and the spore mother cells.

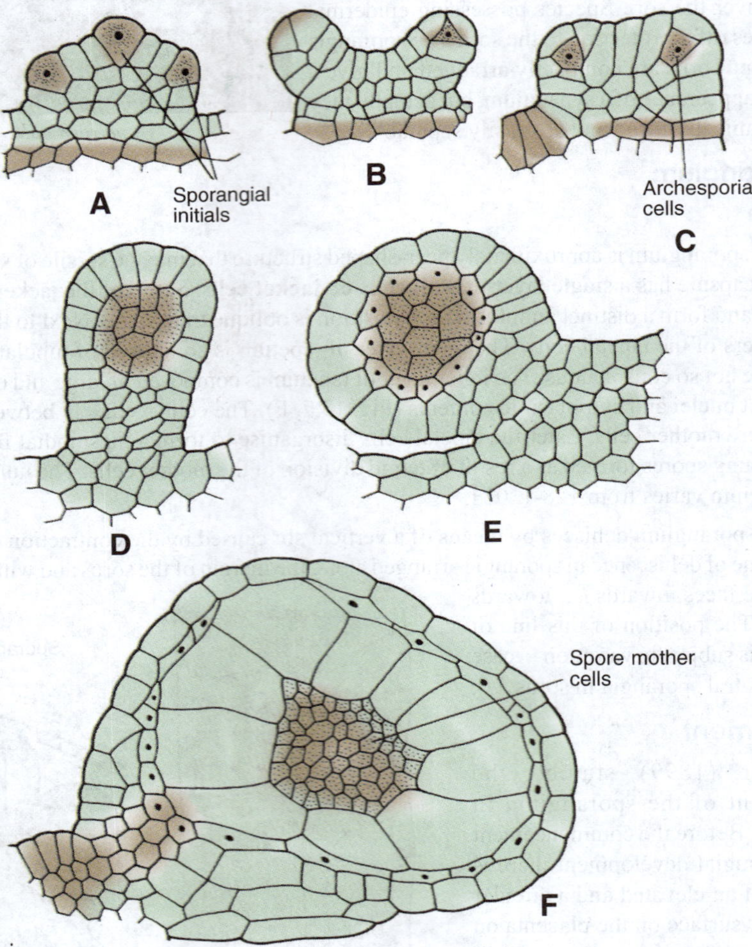

Fig. 13.7. (A–F). Development of sporangia in *Gleichenia*. A. A young sorus bearing three sporangial initials (*G. linearis*); B. The sporangial initial has divided along its three sides (*G linearis*); C. Archesporial cell surrounded by jacket initials (*G. flabellata*); D, E. The formation of tapetal layers (two) in *G. circinata* ;. F. A mature sporangium of *G. flabellata* with numerous spore mother cells. (After Bower)

Some cells of jacket layer become indurated to form an oblique transverse annulus that may partially or completely encircle the sporangium (see chapter 8). The cells of the outer tapetal layer become flattened and remain small whereas those of inner layer enlarge and wedge between the developing spore mother cells. Their nuclei may divide and the cells of this tapetal layer may become multinucleate. Later their cell walls disorganise and the contents form a nourishing fluid called the **plasmodial fluid.** Finally the spore mother cells undergo meiosis and form haploid spores.

GAMETOPHYTE

The spore is the pioneer structure of the gametophyte generation. It germinates to give rise to the prothallus which is monoecious.

G. linearis

Spore

Two types of spores are formed in ***Gleichenia.*** These are: (*i*) the **trilete type** or the **tetrahedral type,** and (*ii*) the **monolete type** or the **bilateral type**. The former is formed in ***G. glauca***, ***G. circinata***, ***G. linearis***, ***G. pectinata*** and others. The monolete type is formed in ***G. bifidus***, ***G. vestitus***, and ***G. volubilis***. The spore wall is two layered. The outer exine or the exosporium is smooth. The intine is thin. The spores are smaller in size and have scanty cytoplasm and a single nucleus. There are no chloroplasts. Food is stored in the form of oil drops.

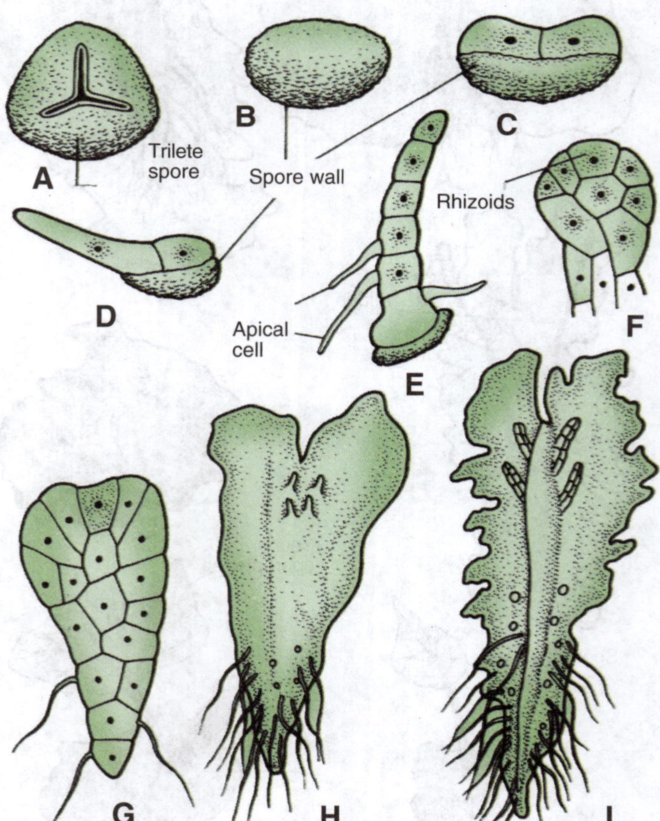

Fig. 13.8. (A–I). Gametophyte generation of *Gleichenia.* A. Trilete or tetrahedral type of pore of *G. linearis*; B–C. Monolete or bilateral type of spores of *G. bifidus*; D–G. Early stages of spore germination in *Gleichenia*; H–I. Mature prothalli of *Gleichenia bifidus.* (After Stokey)

Germination of the Spore

The germination of spores was studied by Rauwendoff in 1879. Later Stokey (1950) studied it in some species of ***Gleichenia*** (***G. bifidus***, ***G. volubilis***, ***G. glauca***). The spores on sowing absorb water and swell up, oil drops become smaller in size but increase in number. The chloroplasts start appearing. Starch grains also appear. The outer spore coat or the exine ruptures after 15-21 days of sowing. As a result the contents enclosed by the intine project out and later divide by a transverse wall into a lower rhizoidal cell and an upper prothallial cell (13.8, C). The rhizoidal cell grows into the first rhizoid (Fig. 13.8, D). The upper or the prothallial cell may divide by transverse walls into a filament of 5 or 6 cells (Fig. 13.8, E). The cells contain sufficient chloroplasts. The terminal cell of the filament divides by oblique walls to cut off a two-sided apical cell whose activity leads to the formation of a narrow plate of cells (Fig. 13.8, F). Further development is quite slow and results in the development of an elongated cordate prothallus, with a distinct midrib (Fig. 13.8, I)

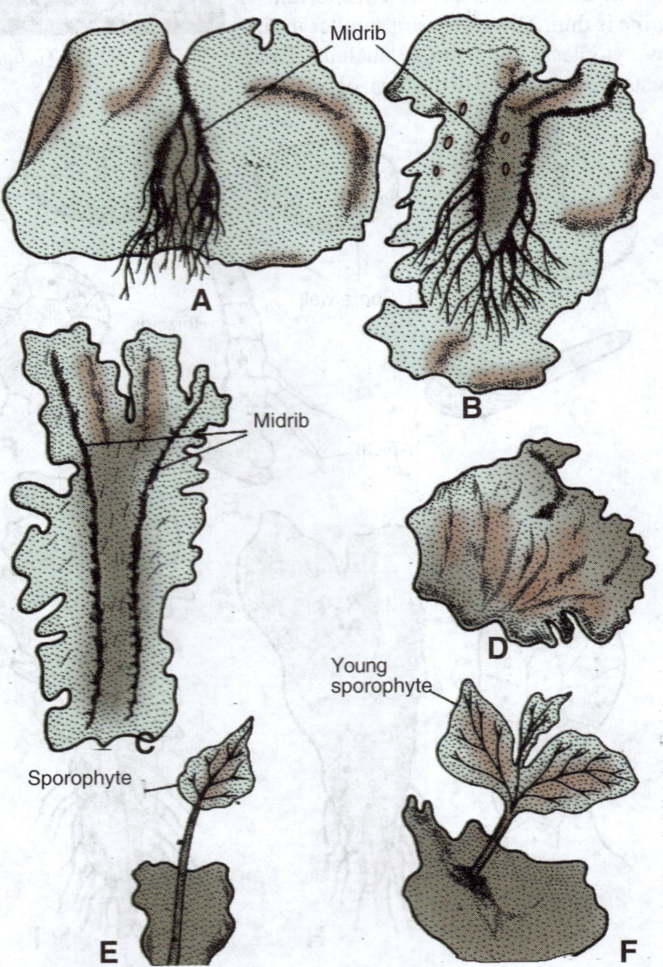

Fig. 13.9. (A—F). **Prothalli of *Gleichenia*;** A. *G. polypodioides*; B. G. *pectinata*; C. G. *linearis*; D. Prothallus of *G. linearis* bearing young sporophyte; E. A mature cotyledon of sporophyte of *G. linearis*; F. Prothallus of *G. pectinata* bearing a young sporophyte. (After Campbell)

Mature Prothallus

Fully developed prothalli show a great variation in form and shape (Fig. 13.9 A–C). They are dorsiventral, epiterranean, green, and cordate structures with fimbriated or convoluted margin. There is a distinct midrib which appears in the form of a narrow groove on the dorsal side but projects conspicuously on the ventral side. Numerous rhizoids arise from the ventral side. Older gametophytes of some species (***G. glauca*** and ***G. linearis***) may branch, each branch having its own apical cell. The archegonia develop on the midrib. The prothallus in ***G. polypodioides*** (Fig. 13.9, A) is short and broad. It resembles that of ***Adiantum*** and ***Dryopteris***. It has a thick cushion behind the apical notch. Archegonia develop in this cushion on the ventral surface. The mid-rib is not prominent in this species. Presence of an endophytic fungus has been reported in older gametophytes. Formation of adventitious branches has been reported in some species. These are a means of vegetative propagation. Two-celled hair that are glandular in nature have been reported to arise in the archegonial region of prothalli of some species.

Sex organs or Gametangia

Both the antheridia and archegonia develop on the same prothallus (monoecious). The sex organs develop on the ventral side but some antheridia may develop on the dorsal surface. The prohalli are **protandrous.**

Antheridia

Stokey (1950) reported that the antheridia appear after the prothallus is two or three months old. They develop on both the dorsal and the ventral surfaces and are usually formed on the midrib, but may also develop on the wings and rarely on the margins. The antheridia vary greatly in size and are usually spherical or globular in shape. Each antheridium has a single-layered wall of thin-walled cells. There is distinct, triangular opercular cell. The number of cells composing its wall varies with the size of the antheridium (Fig. 13.10). They are smallest in ***G. polypodioides*** (Fig. 13.10, F) and largest in ***G laevigata*** (Fig. 13.10, A, B). The mature antheridium of the latter species measures 100μ across and produces several hundred multiflagellate spermatozoids. The antheridia may be sessile or subsessile.

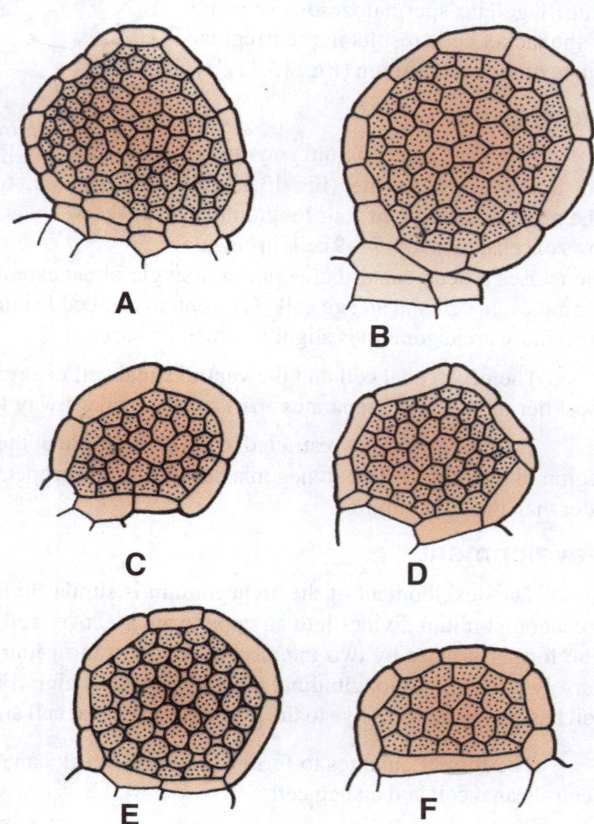

Fig. 13.10. (A–F). Size and number of wall cells in the antheridia of the genus Gleichenia. A—B. *G. laevigata* ; *C. G. pectinata* ; D—E. *G. linearis* ; F. *G. polypodioides.* (After Campbell)

The antheridia absorb water, their contents swell up and throw off the opercular cell thus making an exit for the spermatozoids to ooze out.

Development

The antheridium develops from a single superficial cell (Fig. 13.11, A) that increases in size and divides by an oblique transverse wall to cut off an antheridial initial (Fig. 13.11B). The antheridial initial divides by three oblique walls to cut off **three jacket initials** and a central **primary androgenial cell** (Fig. 13.11, C—E). The jacket initials undergo anticlinal divisions to form 6—12 jacket cells enclosing a central, triangular opercular cell. The **primary androgonial cell** divides several times to form several hundred **spermatocytes** or **androcytes** (Fig. 13.11, F, G). The protoplasts of these androcytes metamorphose into multiflagellate spermatozoids. Twisting of the jacket cells results in the irregular shape of the antheridium (Fig. 13.11, H).

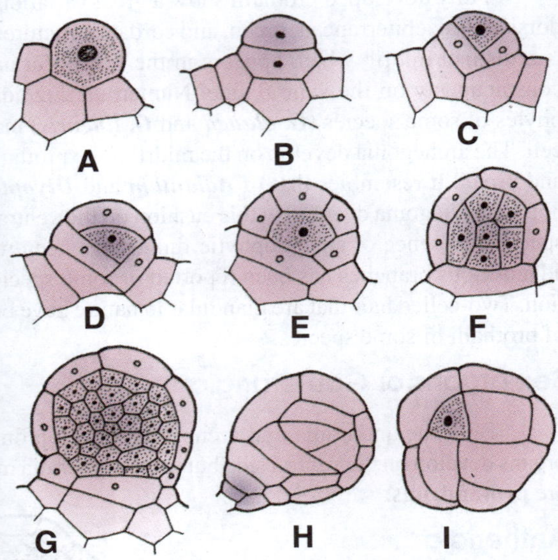

Fig. 13.11. (A—1). Stages in the development of antheridium in *Gleichenia pectinata.* A—F. Stages in development; G. Mature antheridium in section; H. Surface view of mature antheridium; I. Apex of antheridium showing triangular apical cell. (After Campbell)

Archegonia

A mature archegonium consists of a distinct neck and a venter (Fig. 13.12,G). The neck is made up of four longitudinal rows of cells, each row 7—9 cells in height. It encloses a neck canal that contains a single binucleate neck canal cell. The venter has a single ventral canal cell and an egg cell. The venter is embedded in the tissue of the prothallus. The neck in the mature archegonium is slightly bent at its base.

The neck canal cell and the ventral canal cell disorganise prior to fertilisaton and the uppermost tier of neck cells separates apart so as to make a way for the spermatozoids to enter.

The archegonia are restricted to the ventral side of the prothallus and are found near the apical region of the midrib. The archegonia develop in an acropetalous succession and appear three weeks later than the antheridium.

Development

The development of the archegonium is similar to that of other leptosporangiate ferns. The archegonial initial divides into an upper primary cover cell and a lower central cell (Fig. 13.12, B). The former divides by two intersecting walls to form four primary neck cells which divide transversely to form four longitudinal rows of neck cells high (Fig. 13.12, C—G). Meanwhile the central cell has divided to give rise to the primary neck canal cell and the primary ventral cell (Fig. 13.12, E)

The former enlarges to form a binucleate neck canal cell. The latter divides to form a small ventral canal cell and an egg cell.

EMBRYOGENY

The fertilized egg or the oospore lies embedded in the venter. It is the pioneer structure of the sporophytic generation. The first division of the zygote is parallel to the long axis of the archegonium forming an **epibasal cell** and a **hypobasal cell** (Campbell, 1940). The second division is transverse thus forming a **quadrant stage.** The epibasal half is nearer to the apex of the prothallus and gives rise to the stem and the cotyledon. The hypobasal half forms the foot and the root. The foot is quite conspicuous and serves as a haustorium. The primary root grows downwards into the soil and the cotyledon pushes its way upwards through the apical notch and grows straight (Fig. 13.9, F). The primary root does not grow through the foot but remains independent of it. The stem apex grows by means of a single apical cell that is tetrahedral in shape. Details of embryo development are not available in ***Gleichenia.*** The embryogeny appears to be similar to other leptosporangiate ferns.

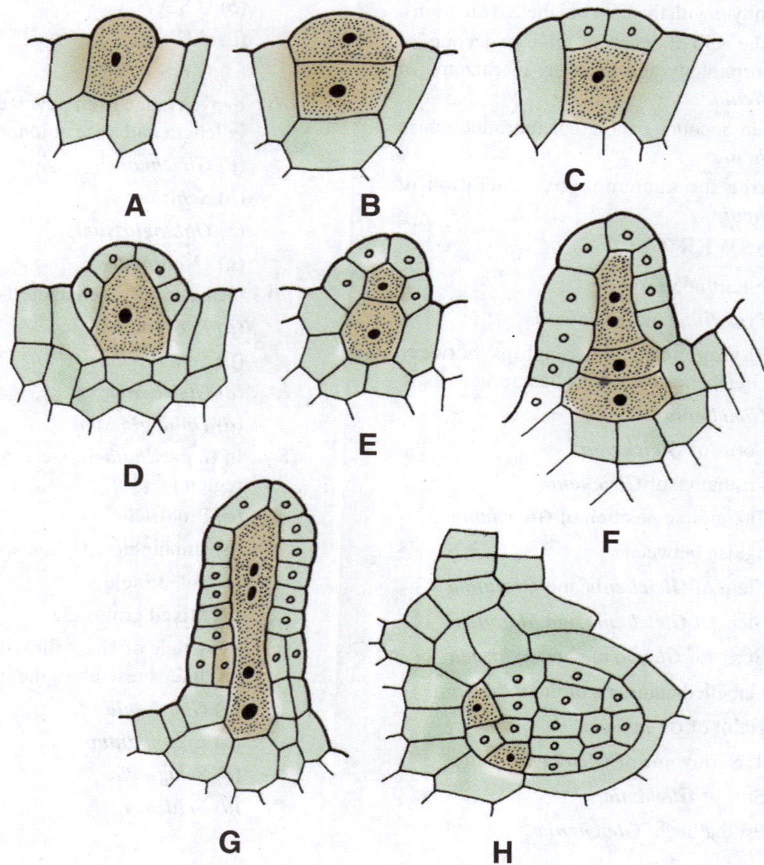

Fig. 13.12 (A—G). Stages in the development of archegonium in *G. pectinata.* H. Archegonium with young embryo. (After Campbell)

Taxonomic Position

Division	:	Filicophyta
Class	:	Leptosporangiopsida
Order	:	Filicales
Family	:	Gleicheniaceae
Genus	:	**Gleichenia**
Species	:	**linearis**

QUESTIONS

ESSAY TYPE

1. Give a brief account of morphology and anatomy of ***Gleichenia.***
2. Describe the characteristic feature of Gleicheniaceae. Give an account of the sporophyte with the help of labelled diagrams.
3. With the help of diagrams, give an account of the morphology and anatomy of rhizome of ***Gleichenia.***
4. Give an account of asexual reproduction in ***Gleichenia***.
5. Describe the gametophytic generation of ***Gleichenia***.

SHORT ANSWER TYPE

6. Write short notes on:
   (*a*) ***Lygodium*** and ***Gleichenia***
   (*b*) Phylogenetic relationships between Gleicheniaceae and Adiantaceae
   (*c*) ***Gleichenia***
   (*d*) Sorus of ***Gleichenia***
   (*e*) Prothallus of ***Gleichenia***
   (*f*) Taxonomic position of ***Gleichenia***
7. Distinguish between
   (*a*) Stem of ***Gleichenia*** and ***Osmunda***
   (*b*) Stem of ***Gleichenia*** and ***Marsilea***
   (*c*) Sorus of ***Gleichenia*** and ***Schizaea***
8. Draw labelled diagrams of the following:
   (*a*) Habit of ***Gleichenia***
   (*b*) T. S. rhizome of ***Gleichenia***
   (*c*) Sori of ***Glichenia***
   (*d*) Prothallus of ***Gleichenia***

OBJECTIVE TYPE

9. Select the correct answer:
   (*i*) ***Gleichenia*** is represented by two species in
   (*a*) India
   (*b*) U.S.A
   (*c*) U.K.
   (*d*) U.S.S.R.
   (*ii*) Leaves range from few feet to 50 metres in length and arise at longer intervals in
   (*a*) ***Gleichenia***
   (*b*) ***Schizaea***
   (*c*) ***Ophioglossum***
   (*d*) ***Osmunda***
   (*iii*) "Ring sorus" is characteristic of
   (*a*) ***Lycopodium***
   (*b*) ***Selaginella***
   (*c*) ***Gleichenia***
   (*d*) ***Ophioglossum***
   (*iv*) In ***G. pectinata*** the stele in the internodal region is
   (*a*) Protostele
   (*b*) amphiphloic siphonostele
   (*c*) Solenostele
   (*d*) Mixed protostele
   (*v*) In which of the following genera the prothallus resembles the liverworts?
   (*a*) ***Gleichenia***
   (*b*) ***Lycopodium***
   (*c*) ***Selaginella***
   (*d*) ***Schizaea.***

Dryopteris Spikes

CHAPTER 14

Family Polypodiaceae Subfamily Dryopteridioideae

- Introduction
- Description of ***Dryopteris*** Adanson

Introduction

The classification of this family is undecided. A number of systems of classification have been proposed but difference of opinion are there. None of them has been agreed upon by all the taxonomists. In this text Christensen's (1938) system has been followed. He has subdivided this family into fifteen sub-families and has included about 170 genera and 7,000 species. He has, however, agreed that these sub-families can be raised to the status of families. These sub-families can be grouped together because of three characters : (*i*) Sporangia have long stalks ; (*ii*) the capsules are biconvex ; and (*iii*) the annulus is vertical and incomplete. Representatives of sub-families ***Dryopteridioideae*** (***Dryopteis***), ***Pteridoideae*** (***Pteridium*** and ***Pteris***), ***Polypodioideae*** (***Polypodium***) and ***Gymnogrameoideae*** (***Adiantum***) are discussed in this text.

Dryopteris

Sub-Family DRYOPTERIDIOIDEAE Christensen

This sub-family includes 25 genera (Christensen, 1938) and 1,500 species. Detailed account of the life history of ***Dryopteris*** the type genus is given in this chapter.

GENUS–DRYOPTERIS Adanson

Distribution

The genus includes 250 species that are distributed in the North temperate, tropical and subtropical regions of the world. Nayar and Kaur (1974) listed 39 species from India. S.K. Basu (1978) enumerated 41 species of ***Dryopteris*** from India. They grow well in Western, North-Western and Eastern Himalayas. ***Dryopteris ramosa*** and ***D. blanfordii*** are very common in Kashmir. ***D. chrysocoma*** is very common in India and frequents exposed places and hill tops between 1,540—2,700 metres above sea level. ***D. hirteps*** and ***D. pulvinulifera*** are other common species met with in the Eastern Himalayas.

SPOROPHYTE

External Morphology

The plant represents the **sporophyte** generation and is distinguished into **root, rhizome** and the **leaves** (Fig. 14.1). Leaves or the fronds are the only above ground portions of the plant. The other organs are subterranean.

Rhizome

It is underground and may be creeping and dorsiventral (***D. rigida***) or short, stumpy, obliquely placed and semi-erect (***D. chrysocoma),*** (Fig. 14.1). It is sparingly branched or unbranched and is densely covered with brown, soft, broad and often fimbriated scales. The older parts of the rhizome are beset with bases of old and fallen leaves. These make the surface of rhizome uneven. It grows by means of an apical bud that is protected by scales and young circinately coiled leaves.

Leaf

The leaves come above ground and attain a height of 15—50 or more centimetres. A characteristic feature of the leaves is that they are coiled in a **circinate manner** (like a watch spring) when young. The young leaves are covered with ramenta and as they grow old they unfurl into a pinnately compound leaf. In older leaves the scales are found only on the petioles. They fall away form other parts of the leaf. The older and pinnately compound leaves are clearly distinguished into a **petiole** and the **lamina**. The lamina may be **unipinnate** (***D. chrysocoma***) or **bipinnate** (***D. rigida***). Sometimes the **pinnae** are deeply incised into **pinnules** (Fig. 14.1) and the leaf appears bipinnate, but it is actually unipinnate. The pinnae or the **leaflets** are smaller at the base and go on becoming longer towards the middle and smaller towards the apex of the leaf. This gives the **leaf** a spindle-shaped appearance. The margins of the pinnules are dentate or crenate. The petiole continues into the lamina as **rachis** and bears pinnae. Each pinnule is traversed by a **mid-rib** which bears **furcate** lateral veins. Every pinna can bear **sporangia** (Fig. 14.1.) except in ***D. cochleata*** where there is segregation into sterile and fertile pinnae.

Fig. 14.1. (A–B). *Dryopteris chrysocoma* A. Sporophyte showing habit; B. A pinnae as seen from ventral side. Note the kidney-shaped sori.

Root

The primary root is short-lived and is replaced by adventitious roots that rise endogenously from the rhizome. They are branched and possess a root cap. They are thin, black and wiry structures that anchor the rhizome to the substratum and also absorb water and other nutrients from the soil.

Anatomy

Rhizome

A transverse section of rhizome (Fig. 14.2, A–B) is irregular in outline and reveals the following structure :—

Epidermis

It is single layer of rectangular cells that become sclerenchymatous in the older rhizome. The cutile is thick.

Hypodermis

It is made up of a few layers of scleranchymatous cells that give mechanical strength to the rhizome.

Ground Tissue.

It is wholly parenchymatous. The cells are polygonal in outline and enclose small intercellular spaces. They store food material in the form of starch and protien. Here and there some tanniferous cells are also present.

Stele

It is a typical **dictyostele** composed of a varying number of oval or circular **meristeles** arranged in an interrupted ring in the ground tissue (Fig. 14.2 A, B). The stele, if extracted as a whole from the ground tissue, appears as a cylindrical meshwork of vascular strands (Fig. 8.4). The stem stele gives off leaf traces to the leaves and the leaf trace bundles from the base of the petiole give off vascular strands to the roots.

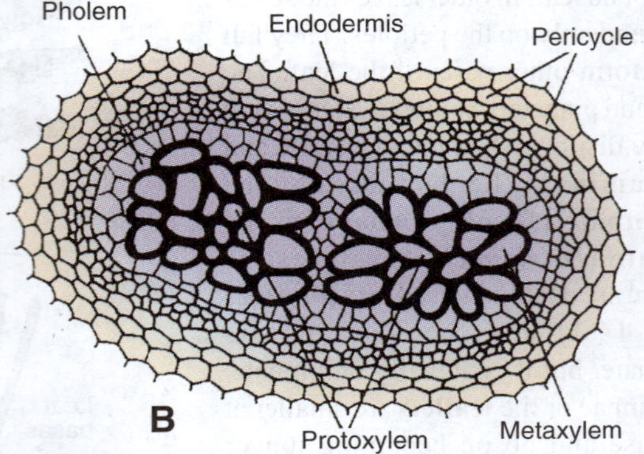

Fig. 14.2. A—B *Dryopteris*. A. T.S. rhizome showing dictyostele; B. A meristele showing detailed internal structure.

Meristele

An individual **meristele** is bordered by a single layer of **endodermis** followed by a single layer of **pericycle**. Both these layers are made up of parenchymatous cells. The cells of endodermis have distinct **casparian strips** and store food in the form of starch grains. Next to the pericycle is **phloem** tissue which completely surrounds the central mass of **xylem**.

The phloem can be distinguished into **protophloem** and **metaphloem**. The former occurs as a single layer of **sieve cells** and parenchyma next to the pericycle, whereas the latter forms two or three layers of sieve cells and phloem parenchyma. **Companion** cells are absent.

The xylem is also distinguishable into **protoxylem** and **metaxylem**. The former has smaller **tracheids** that are spiral or reticulate. It is surrounded by metaxylem tracheids that are larger in diameter and are **pitted**. The pits are transversely elongated giving a **scalariform** appearance. The xylem is thus **mesarch**. The vessels are absent. The meta and proto-xylem are separated by a layer or two of parenchymatous cells. The phloem and the xylem are also separated by a layer or two of parenchyma. The tracheids are spindle shaped i.e., with tapering ends. Secondary growth is absent.

Root

A transverse section of the roots (Fig. 14.3) is roughly circular in outline and reveals the following structure :—

Epidermis

The original epidermal layer is sloughed off during downward growth of the root through soil. It is replaced by the outermost layer of the cortex. This is usually termed as **exodermis** or **rhizodermis** or **piliferous** layer. It is made up of thin-walled cells that bear unicellular **root hair** in the **root hair zone**.

Cortex

It is made up of several layers of cells and is distinguishable into two types : (*i*) **parenchymatous cortex** which is next to the epidermis and consists of 5—7 layers of thin-walled, polygonal cells that enclose small intercellular spaces. The cells store food and also help in translocation of water absorbed by root hair to the stelar region. (*ii*) **Sclerenchyamtous cortex** that forms the central mechanical cylinder external to the vascular region. These cells possess pitted thickenings and are polygonal in shape. They are devoid of cytoplasmic contents.

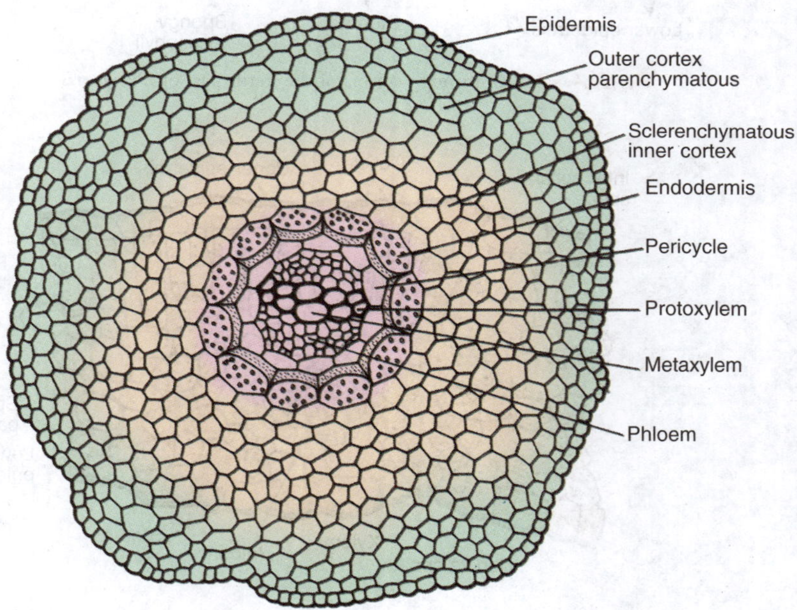

Fig. 14.3. T. S. root of *Dryopteris chrysocoma.*

Endodermis

It is distinct and is made up of barrel-shaped cells that are filled with starch and possess **casparian strips** on the radial and inner horizontal walls.

Pericycle

It is made up of a single layer of thin-walled cells.

Stele

It is **diarch** and consists of a central plate of xylem with **metaxylem** in the centre and **protoxylem** at either end (**exarch**). Phloem forms two conspicuous groups alternating with the xylem. **Pith** is absent. Secondary growth is absent. The lateral roots originate from definite cells of the endodermis that are opposite protoxylem groups.

Leaf

The internal structure of the leaf (Fig. 14.4–14.6) will be considered separately for the **pinnule** and the **petiole**.

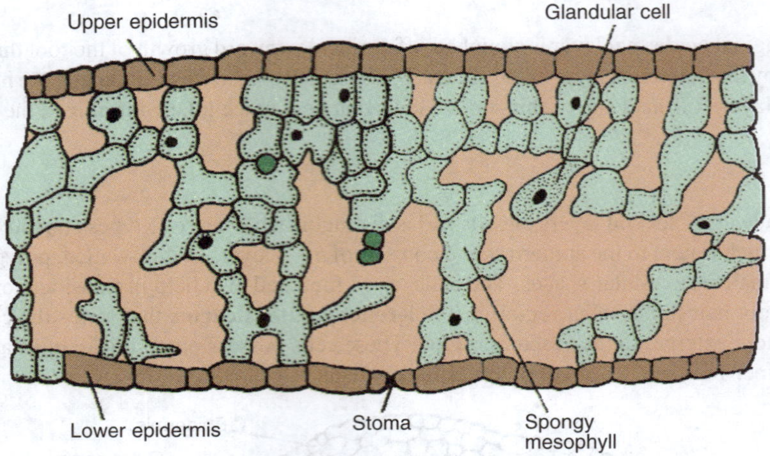

Fig. 14.4. (a) V. S. through lamina of the sterile part of *Dryopteris.*

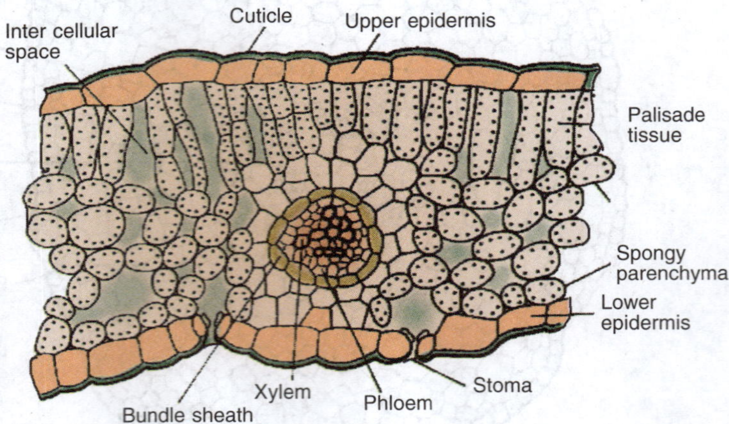

Fig. 14.4 (b). V. S. leaf lamina showing detailed internal structure of the sterile pinnule of *Dryopteris*

Pinnule

A vertical section of the sterile pinnule (Fig. 14.4 a, b) reveals the two **epidermal layers** (dorsal and ventral) whose cells are parenchymatous and contain chloroplasts. The stomata may be restricted to the lower epidermis or may be present on both the layers. There is no distinction of the **mesophyll** into a palisade and spongy parenchyma. The cells of the mesophyll are parenchymatous, lobed, and chlorophyllose. They enclose small or large intercellular spaces. Distinct **glandular outgrowths** arise from some mesophyll cells. The glands store water.

The vascular bundle of the mid-rib may be **concentric** or **bicollateral**. The bundles of smaller veins are only collateral with phloem present towards the lower side of the xylem.

Petiole

Transverse section of the petiole is almost semi-circular in outline due to the presence of a groove on one side (Fig. 14.5). It reveals an outer **epidermis** which bears **ramenta** or the scales. Next to it is 3—7 layered **hypodermis** which is wholly sclerenchymatous and forms a peripheral mechanical cylinder that resists the bending strain. The rest is the **ground tissue** which is parenchymatous. It is composed of polygonal cells that store food and also enclose small intercellular spaces. The stelar organisation is **dictyostelic** and consists of varying number of **meristeles** that are almost circular in outline (Fig. 14.6) and have a structure exactly similar to those of rhizome meristeles.

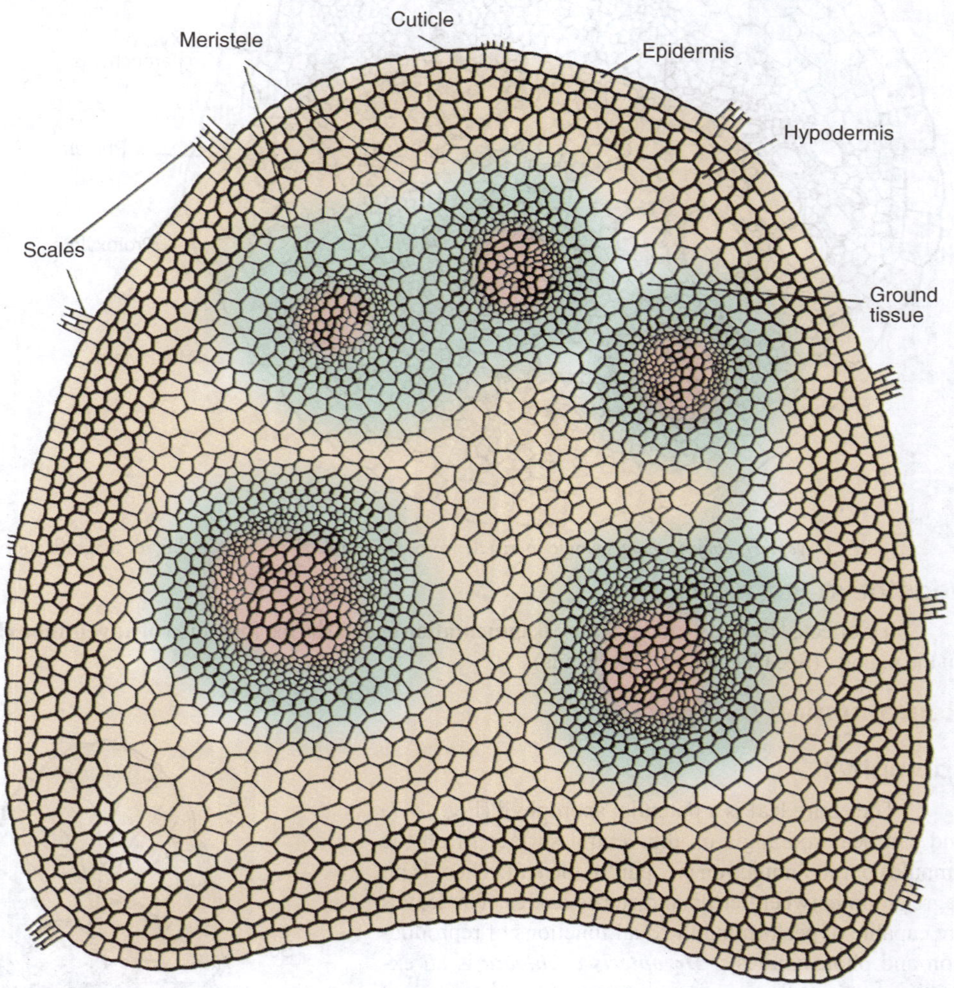

Fig. 14.5. T. S. through basal region of petiole of *Dryopteris chrysocoma* showing detailed internal structure.

REPRODUCTION

Reproduction is effected by vegetative means as well as by the formation of **spores**, which are produced within sac-like structures called the **sporangia**. The sporangia are grouped into definite structures called the **sori (singular sorus)**. The sori are borne on the ventral surface of the pinnules which are called the **sporophylls**.

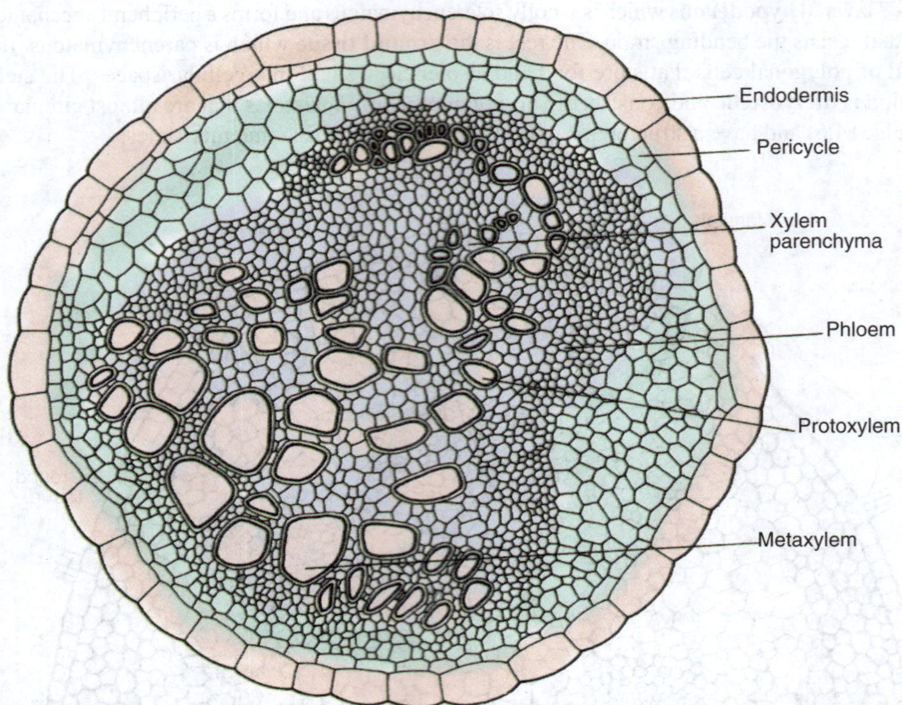

Fig. 14.6. ***Dryopteris chrysocoma.*** A single meristele from the petiole as seen under higher magnification.

Vegetative Propagation

It is effected by two methods : (*i*) by **fragmentation** and (*ii*) by the formation of **adventitious buds** on the petioles (See Chapter VIII for details).

Reproduction by Spores

Sporophyll

The entire leaf is a potential sporophyll (Fig. 14.7) and can bear sporangia on the ventral surface of all the pinnules or on pinnules on any part of the leaf. There is no segregation between fertile and sterile leaves. The leaves are capable of performing the dual functions of reproduction and photosynthesis. ***Dryopteris cochleata*** is an exception because there is segregation between sterile and fertile parts of a leaf.

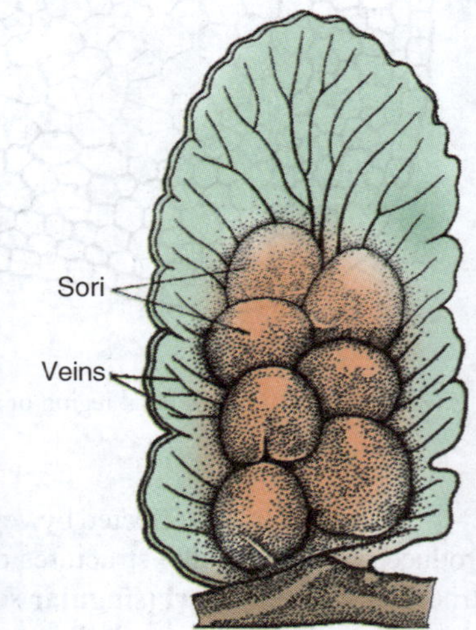

Fig. 14.7. A pinnule of *Dryopteris* as seen from ventral side. Note the position of sori.

Sorus

A group of sporangia is called a sorus. It is covered by an outgrowth called the **indusium.** In ***Dryopteris*** there is true indusium because it is a special outgrowth from the tissue called the **placenta** or the **receptacle** (Fig. 14.9). The sorus arises singly upon the **anadromic** branch of a forked lateral vein (Fig. 14.8). This vein arises from lateral vein and passes through the receptacle and continues its course forward beyond the sorus. A mature sorus is, therefore, borne on the vein and not at its end. The receptacle or the placenta is a group of cells that arise by the active division of the cells on the ventral side of the pinnule, at a place where the sorus has to develop. The sori are greyish when young but when mature they change in colour to brown or dark brown.

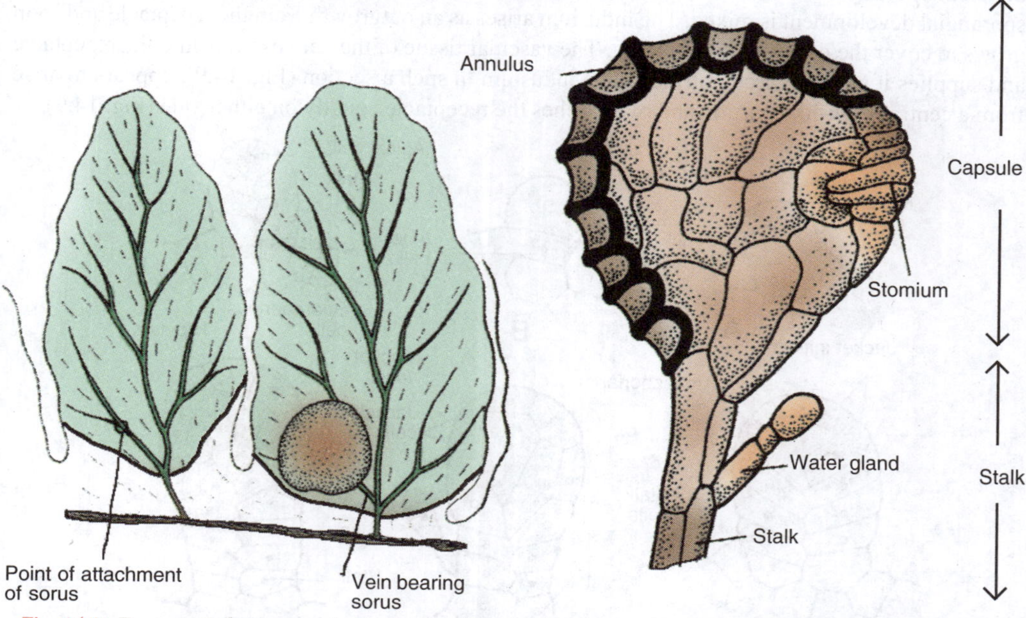

Fig. 14.8. ***Dryopteris chrysocoma.*** **A.** Fertile pinnules showing the position of sorus on a vein; **B.** A sporangium.

Internal Structure of the Sporophyll

A vertical section of the sporophyll passing through the vein (Fig. 14.9). bearing the sorus, shows the following structure :

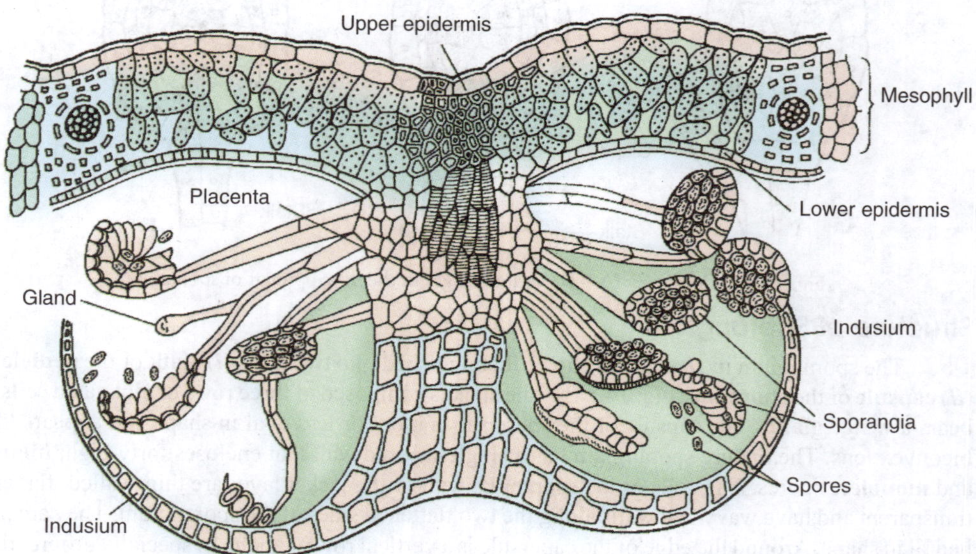

Fig. 14.9. V. S. fertile pinnule of *Dryopteris chrysocoma.*

The pinnule has the usual structure as described before (Fig. 14.4). In addition it shows a swollen and placenta-like **receptacle** on the ventral side, in the region of the anadromic branch of the forked lateral vein. The receptacle appears as a distinct mound of thin-walled cells that are capable of division. Sporangial initials arise from this tissue and develop into mature sporangia. The sequence of sporangial development is mixed. The indusium arises as an outgrowth from the receptacle and soon grows to cover the developing sporangia. The vascular tissue of the vein extends into the receptacle and supplies it with required nutrition. The indusium in such a section (Fig. 14.9), appears to arise from a central stalk-like portion and over-arches the receptacle equally on either side (Fig. 14.9).

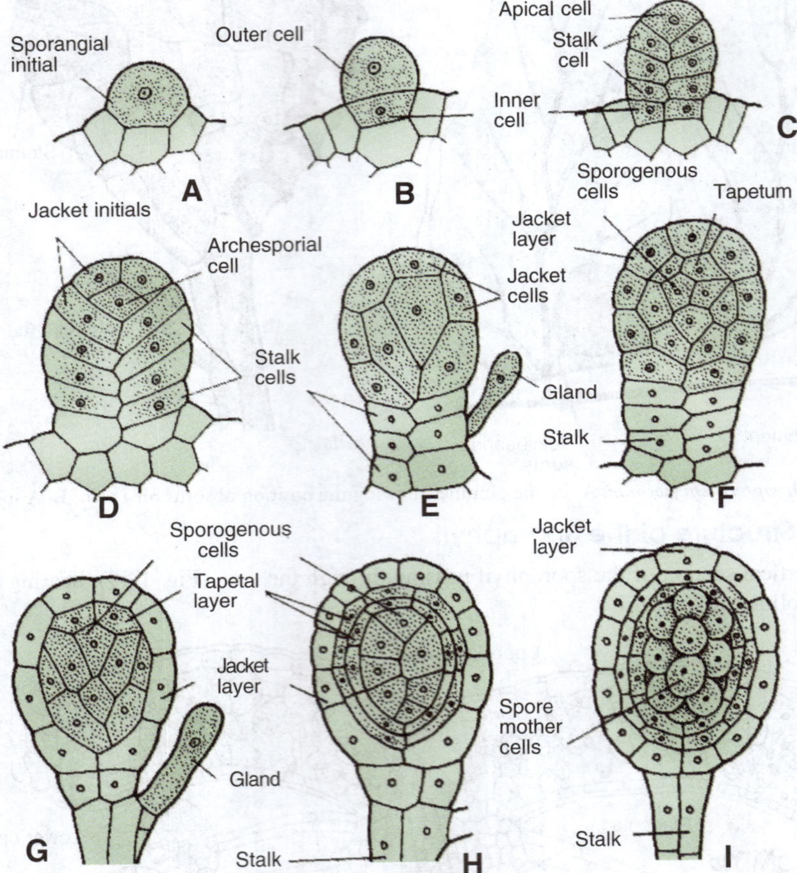

Fig. 14.10. *Dryopteris* (A—I). Various stages in the development of sporangium.

Structure of Sporangium

The sporangium in ***Dryopteris*** can be distinguished into two parts (*i*) **stalk** or the **pedicle** and (*ii*) **capsule** or the **spore sac** (Fig. 14.8 B). The stalk is composed of three rows of elongated cells and bears a **water gland**. The capsule or the spore sac is more or less oval in shape and appears like a biconvex lens. The mature sporangium has a single-layered wall that encloses forty-eight **bilateral** and **monolete** spores. The cells of the sporangial wall or the jacket layer are thin-walled, flattened, transparent and have wavy cell walls along the two flattened sides of the sporangium. They are polyhedral in shape. Around the edge of the capcrsule is a vertical row of about 16 specially differentiated cells with specially thickened walls. It is the **annulus**. It extends over about two-thirds of the circumference of the capsule connecting the sides and forms an incomplete ring of large cells. The radial and inner tangential walls of the cells are thickened. The outer walls are thin. The remaining about one third of the edge of the sporangium between the beginning of the stalk and the end of the annulus is composed of a small group of long, flat and thin-walled cells. It is called the **stomium**. Two cells in the stomium region are narrow and radially elongated. These are called the **lip cells.** The three thin-walled cells above the lip cells are called **epistomium**, whereas the two thin-walled cells below the lip cells are called the **hypostomium**. The annulus and the stomium help in the dehiscence of the sporangium and dispersal of spores. The stomium cells permit easy cleavage.

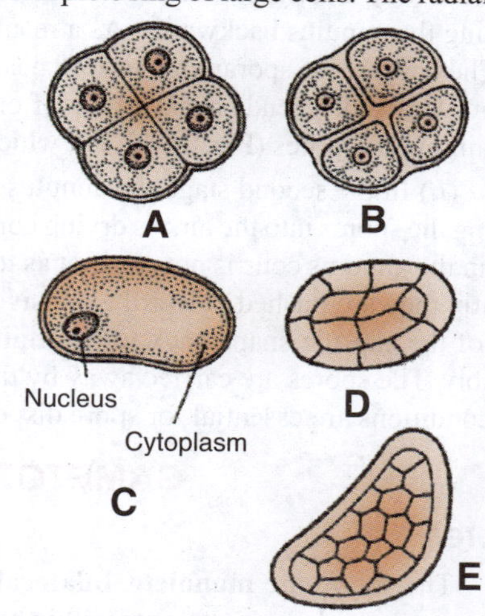

Fig. 14.11. (A—E). Spore tetrads and spores of *Dryopteris chrysocoma.* A—B. Tetrads of spores; C. Spore in section; D—E. Spore as seen in surface view.

Development of the Sporangium

It is **leptosporangiate** and has been discussed in detail in Chapter VIII (Fig. 8.9). It is exactly similar in ***Dryopteris*** with the exception that in ***Dryopteris*** there are 12 spore mother cells which undergo reduction division to form 48 spores. In ***Dryopteris*** one of the stalk cells gives rise to a water gland (Fig. 14.10).

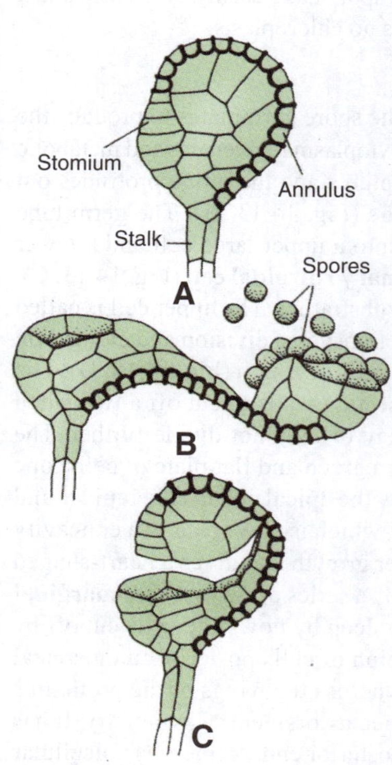

Fig. 14.12. (A—C). Stages in the dehiscence of the sporangium.

Sporogenesis

The process of differentiation of haploid **spores** from the diploid spore mother cells is called **sporongenesis**. The spore mother cells round off and separate from one another. The tapetal cells start disorganising. Each spore mother cell has a diploid nucleus. During **sporogenesis** the diploid nucleus undergoes **meiosis** to form four haploid nuclei in each mother cell. As a result of simultaneous laying down of the walls between the nuclei, **tetrads** of haploid cells are formed (Fig. 14.11). These tetrads of haploid cells are called the spore tetrads and are **tetrahedral** in form. Later the spores secrete their own walls and separate from each other. The parent spore mother cell wall degenerates and spores separate from each other. The spore mother cells are the last cells of sporophyte generation and the haploid spores are the first or pioneer structures of the gametophyte generation.

Dehiscence of Sporangium (Fig. 14.12). The sporangia normally open in the dry weather. The annulus and stomium are directly concerned in the process. The former functions as a mechanical spring and the latter as a weak spot permitting easy cleavage. The dispersal of spores takes place in two stages:

(*i*) In the first stage the sporangium breaks open at the stomium. The opening is brought about by the annulus. The mature sporangium dries out. It loses water all over its surface including the annulus cells which at first are full of water lose it only through their thin outer walls. The volume of water in the annulus cells thus decreases. Consequently their thin outer walls are drawn inwards. The combined suction force set up over the entire annulus causes tension bending the annulus backwards. As a result the stomium ruptures between the lip cells (Fig. 14.12, A). The body of the sporangium is torn nearly half across the middle. The annulus slowly straightens out and gradually bends backward as if on a hinge ; carrying with it the upper half of the capsule containing the spores (Fig. 14.12, B), which are exposed.

(*ii*) In the second stage the annulus suddenly snaps back to its original position like a spring hurling the spores into the air. As drying continues a stage is reached when the cohesive force of water within the annulus cells is not so great as to hold the outer thin walls in a sucked-in position. Consequently they are pushed outward explosively. As a result the annulus along with the ruptured upper part of the capsule snaps back to its original position throwing the spores with a jerk, into the air forcibly. The spores are carried away by the wind. They remain viable for a pretty long period. The dry conditions are essential for spore dispersal.

GAMETOPHYTE GENERATION

Spores

The spores are **monolete**, **bilateral** (Fig. 8.10), minute, dusty and dark coloured structures. The spore protoplasm is enclosed within two spore coats called (*i*) the outer **exine** or **exosporium** and (*ii*) inner **intine** or **endosporium**. The exine is covered by another coat called the **perine** or **perispore**. The exine is composed of two layers, the outer **ectine** and inner **endine.** The ectine is made up of radially elongated rods called the **columellae**. The upper free ends of columellae are fused to form a layer called **the tegillum**. The tegillum is smooth in ***D. penthri*** (Fig. 8.10) and spinulose in ***D. faeterra*** (Fig. 8.10). It is reticulate in ***D. chryscomma*** (Fig. 14.11, E). The spore has a scanty cytoplasm and a single nucleus (Fig. 14.11, C). The cytoplasm stores food but has no chloroplasts.

Germination of the Spore

Under suitable conditions of moisture and temperature the spore germinates to produce the prothallus (Fig. 14. 13). It absorbs water and swells up, the latent cytoplasm rejuvenates and metabolic activities start. The nucleus becomes prominent. The exine ruptures and the intine protrudes out along with enclosed contents in the form of a small **germ tube** (Fig. 14.13, B). The germ tube elongates, its cytoplasm develops chloroplasts and it divides into a upper large cell and a lower small cell. The lower cell loses chloroplasts and acts as the **primary rhizoidal cell** (Fig. 14.13, C). It gives out the first rhizoid which establishes contact with the substratum. The upper cell is called the **prothallial cell** and contains many chloroplasts. By further transverse division and elongation it develops into a green, uniseriate filament 3—5 or even more cells in length (Fig. 14.13, D). The terminal cell of the filament divides by two oblique and intersecting walls to cut off a two-sided apical cell (Fig. 14.13, E). The lower cells of the filament may or may not divide further. The segments cut off by the apical cell divide and redivide to form a green and flat plate of cells, one cell in thickness (Fig. 14.13, F). The young segments cut off by the apical cell divide rapidly and cells cut off towards their outer face, overgrow the apical cell, which is now lying in a concavity or a depression. The depression becomes deeper during further growth and gives a heart-shaped appearance to the prothallus. The single apical cell is replaced by a series of two or more **marginal meristematic cells**. The apical notch has become sufficiently deep by now. The cells cut off by the marginal meristem lead to the formation of a central cushion of cells posterior to the apical notch. The cushion is two or more cells or more cells thick, whereas the wings of the prothallus remain one cell in thickness. The prothallus has by now assumed a dorsiventral symmetry. It has distinct anterior side (towards the apical notch) and a narrow posterior end. Numerous unicellular rhizoids arise from the ventral side of the posterior end. They grow into the soil and besides anchoring the prothallus, also absorb water and other nutrients from the soil.

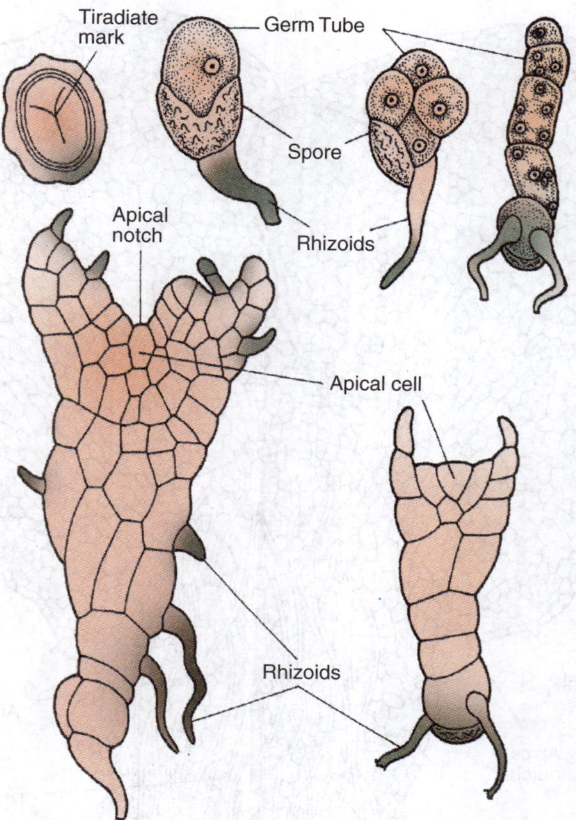

Fig. 14.13. (A—F). *Dryopteris filix-mas.* Stages in the germination of spore and formation of young prothallus.

Under unfavourable circumstances and crowded conditions the spore germinates to give rise to a branched and filamentous prothallus. Such prothalli are starved and bear only male sex organs (antheridia). If transferred to favourable conditions normal cordate form is regained and the prothallus becomes bisexual.

Mature Prothallus

The adult fern **prothallium** or **prothallus** is a small, green, flat, surface-loving, thallus-like object (Fig. 14.14). It is, somewhat, heart-shaped in outline. It varies in size from 5 or 6 to 13 mm in diameter. It is thin in texture. Structually the notch is massive and cushion-like. It is several layers of cells thick. All the prothallial cells are thin-walled, elongated and polygonal or hexagonal in shape. The cytoplasm forms a thin, lining layer within the cell wall and encloses a central vacuole. The single nucleus is embedded in the cytoplasm which contains many small, discoid chloroplasts. Being green the prothallus is autotrophic in its mode of nutrition. There are no intercellular spaces between the cells. The growing point is located in the notch. Many delicate, brown, hair-like thin-walled unicellular **rhizoids** arise from the lower surface of the cushion especially from its hinder end. The rhizoids are unbranched. They penetrate the soil and thus serve to attach the prothallus to the substratum. Besides they absorb moisture and mineral nutrients from the soil for the growth of the prothallus. The prothallus grows only in moist, shady habitats. It is the self-supporting haploid plant of the gametophyte generation and is concerned with **sexual reproduction**. It does not live for a long time and cannot withstand drought. Its survival is dependent upon moisture. In fact, it is less adapted for life on land as compared with the leafy fern plant (sporophyte).

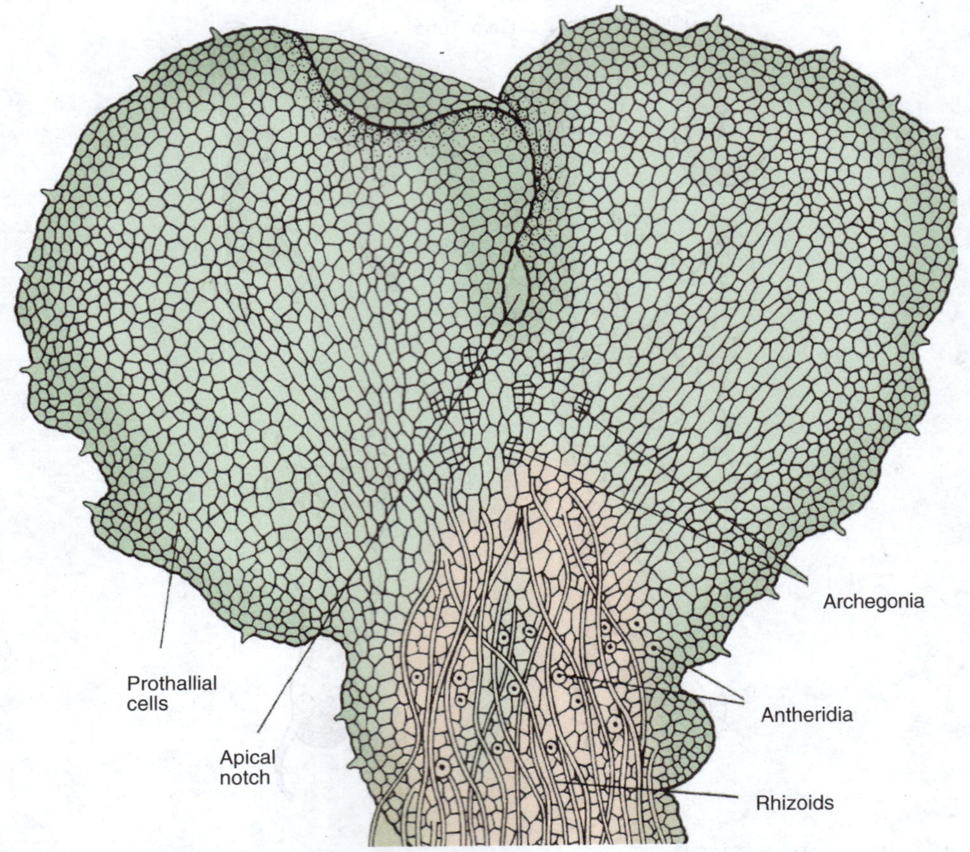

Fig. 14.14. A mature prothallus of *Dryopteris* as seen from ventral side. (After Kny)

Sex Organs

The prothallus is monoecious and bears sex organs on the ventral side. In rare cases the prothalli are dioecious, but such prothalli may bear archegonia at a later stage. The **antheridia** (male sex organs) arise among the rhizoids towards the posterior side of the prothallus and are emergent. The **archegonia** develop in the central cushion behind the apical notch. The necks of archegonia are bent towards the antheridia and protrude above the surface of the prothallus. The ventral position of the sex organs exposes them directly to moist air and water, which are necessary for their development and dehiscence. The monoecious prothalli are always **protandrous**. This device is helpful in bringing about cross fertilization.

Antheridium

The antheridia are small, sessile and globular structures. Each antheridium consists of a wall and is made up of three tabular cells (Fig. 14.15, C). These three cells are called : (*i*) **first ring cell**, (*ii*) **second ring cell**, and (*iii*) **an opercular or cap cell**. The cap cell forms a lid, which gets lifted during dehiscence and allows the spermatozoids to escape. The second ring cell or the **circular cell** forms the middle portion of the wall and extends completely around the antheridium. The first ring cell or the **funnel cell**, extends around the basal part of the antheridium. Besides these three cells, there is the fourth cell or the **basal cell** which can be regarded as the single celled stalk of the antheridium. The contents of the wall cells of a mature antheridium include a little cytoplasm, nucleus, mitochondria, few chloroplasts and vacuoles full of phenol contents. The antheridial wall encloses about thirty-two **spermatazoids** (Fig. 14.15, E).

Structure of Spermatozoid

The spermatozoid of ***Dryopteris*** is a large, multiflagellate and a coiled structure which is mostly nuclear in orgin and has a prominent posterior **vesicle** (Fig. 14.15, F). The liberated spermatozoid has 2 ½ to 3 spirals and consists of a **motor apparatus** and a nucleus. The motor apparatus is composed of a number of **basal granules** or **centrioles** and a mitochondrial part. The flagella arise from this part. The nuclear portion is surrounded by cytoplasmic sheath. The coiled or nuclear part moves along an irregularly spiral path with the help of flagella. The vesicle consists of a homogenous ground substance with fat globules, 4—6 plastids and a number of smaller granules embedded in it. Before entering the archegonial neck the spermatozoids discards its vesicle.

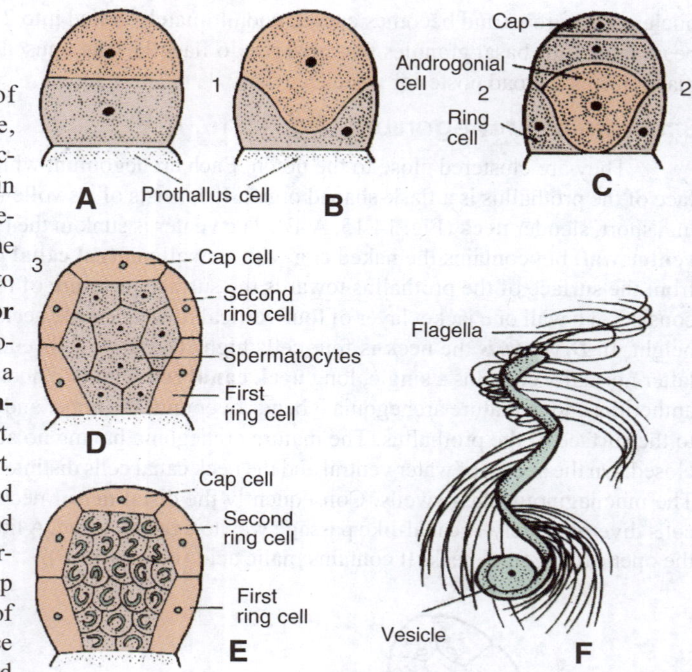

Fig.14.15. (A—E). Stages in the development of antheridium in *Dryopteris;* F. A spermatozoid.

Dehiscence of Antheridium

Dehiscence is effected in the presence of water which is absorbed by the mucilaginous antheridial wall and consequently by the spermatid walls. This results in swelling of spermatids and antheridial wall. The cells become turgid and tension develops due to hydrostatic pressure. This tension it relieved by the cap cell either being tited upwards or detached. The spermatozoids enclosed a thin membrane extrude out ***en masse***. This extrusion is further assisted by the inward swelling and shortening of the wall cells. The membrane soon dissolves and the spermatozoids become free to swim.

Development of Antheridium

It develops from a superficial cell of the prothallus which bulges out and divides by a transverse wall to form a basal cell and an **antheridial initial**. The antheridial initial divides by a transverse wall (Fig. 14.15, A). Later this wall undergoes curvatures. In ***Dryopteris*** it curves so much that it almost touches the basal cell (Fig. 14.15, B, C) and looks like a funnel. This wall results in the formation of an upper **dome cell or central cell** and a lower **first ring cell**. The central cell divides again by a curved wall (Fig. 14.15, C, 2—2) into an outer **jacket cell** and central **primary androgonial cell** (Fig. 14.15, C). The **jacket cell** divides again by a periclinal wall (Fig. 14.15, C, 3—3) thus forming an upper **cap** or **cover cell** and **second ring cell**. The wall 3—3 curves and touches the wall 2—2 (Fig. 14.15, D). Thus constituted the wall of the antheridium has three cells which enclose a central primary andogonial cell that divides and redivides to form 16 **sperm mother cells**. (Fig. 14.15, D), which divide again into 32 **spermatids**, whose protoplasts metamorphose into multiflagellate spermatozoids (Fig. 14.15, E). The sperm mother cells loose their chloroplasts and have dense cytoplasmic contents and distinct nuclei. Later two **polar granules** appear on either side of the nucleus. The mother cells now divide into spermatids so that each has a single polar granule. The granule later enlarges into a band-shaped motor apparatus and becomes attached to one side of the nucleus. The nucleus also grows and becomes curved and ultimately coiled into 2—3 coils. The motor apparatus bears numerous basal granules that give rise to flagella. The unused food, some plastids and cytoplasm forms a broad posterior vesicles.

Structure of Archegonium (Fig. 14.16, A-B)

They are clustered close to the notch. Each archegonium which projects from the lower surface of the prothallus is a flask-shaped organ. It consists of a swollen base, the **venter** and a projecting, short, slender **neck** (Fig. 14.15, A-B). The venter is sunk in the tissue of the prothallus. It has no **venter wall** but contains the naked **egg** with a small **ventral canal cell** above it. The neck projects from the surface of the prothallus towards the surface moisture of the soil and is slightly curved. It consists of a wall or a jacket layer of four vertical rows of sterile neck cells. Each row is 3—7 cells in height. In ***Dryopteris*** the neck is four cells high. The neck cells enclose a central **neck canal**. The latter normally contains a single, long **neck canal cell** with two nuclei. The neck points towards the antheridia and in mature archegonia it bends over until the upper and lies almost parallel to and close to the surface of the prothallus. The mature archegonia having no access to fluid water will remain closed. On the access of water ventral and the neck canal cells distintegrate and become mucilaginuos. The mucilaginous mass swells. Consequently the distal tier of neck cells is ruptured (B) and neck cells diverge widely. A canal-like passage way to egg is formed. A little of the mucilage exudes from the open archegonial neck. It contains malic acid (Fig. 14.17 F).

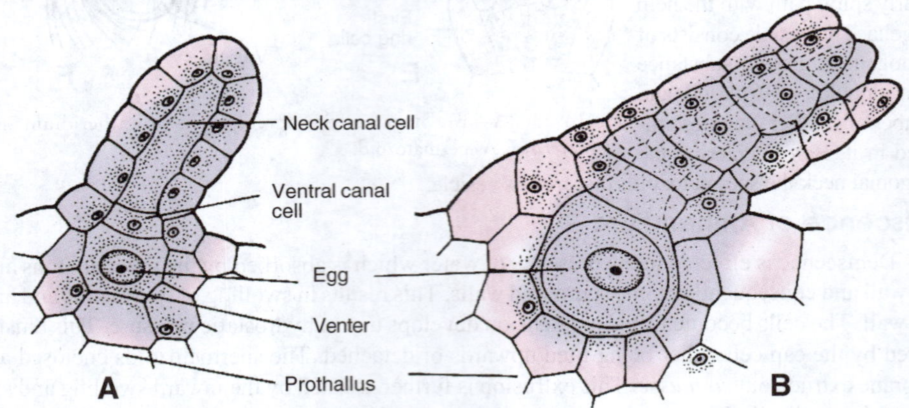

Fig. 14.16. (A—B). *Dryopteris* Fully developed archegonium; B. Mature and dehisced archegonium.

Development of Archegonium

The archegonium develops from a single superficial cell the *archegonial initial* (Fig. 14.17A), which divides by a transverse wall into an upper **primary cover cell** and a lower **mother cell or the central cell** (Fig. 14.17 B). The lower cell divides again by a transverse wall forming three cells, the upper primary cover cell, middle **central cell** and a lower **basal cell.** (Fig. 14.17, C). The cover cell divide by two interesting walls into four diagonally arranged **primary neck cells** (Fig. 14.17 D). Meanwhile the central cell is divided by a transverse wall into a upper **primary neck canal cell** and a lower **primary ventral cell** (Fig. 14.17, E). The primary neck cells divide transversely to form a neck of 3—5 or 7 cells in height. The neck canal cell grows in between the neck cells and forms a single neck canal cell whose nucleus divides into two (Fig. 14.17, E) so that it becomes binucleate. The primary ventral cell divides transversely into an upper smaller **ventral canal cell** and a lower larger **egg cell** (Fig. 14.17, E). The neck becomes curved due to the unequal growth of its cells.

Fertilization

It occurs in the presence of fluid water between the lower surface of the prothallus and the soil. Fertilisation in ferns presents no difficulties. Both kinds of sex organs are in contact with the moist earth and open on to the lower surface of the prothallus. If young prothalli are growing in the company of older ones, the thin film of water surrounding them will contain numerous sperms. Some of these are attracted by the malic acid diffusing into the water from the mucilage excluded by the open necks of archegonia of the older prothalli. Thus attraced they swim towards the centre of diffusion. Once in the neighbourhood of the open archegonial neck some of them make their way down the canal to the egg in the **venter** (Fig. 14.17 G). One of these enters the egg to accomplish fertilisation. The male and the female nuclei fuse. The fertilised egg secretes a wall around it to become a **zygote** (Fig. 14.17 H). There are a number of archegonia on each prothallus but only one is fertilized. Why? It may be that the prothallus secretes a chemical which repels the sperms after the first zygote is formed. This prevents more egg from being fertilized. There are, however, reports where more than one archegonia are fertilised in one prothallus (Becquerel, 1931; Vladesco, 1935-37). Such cases are no doubt rare. After fertilisation the growth of the prothallus stops. The cells of the archegonial venter proliferate and form a cap like **calyptra** around the fertilised egg.

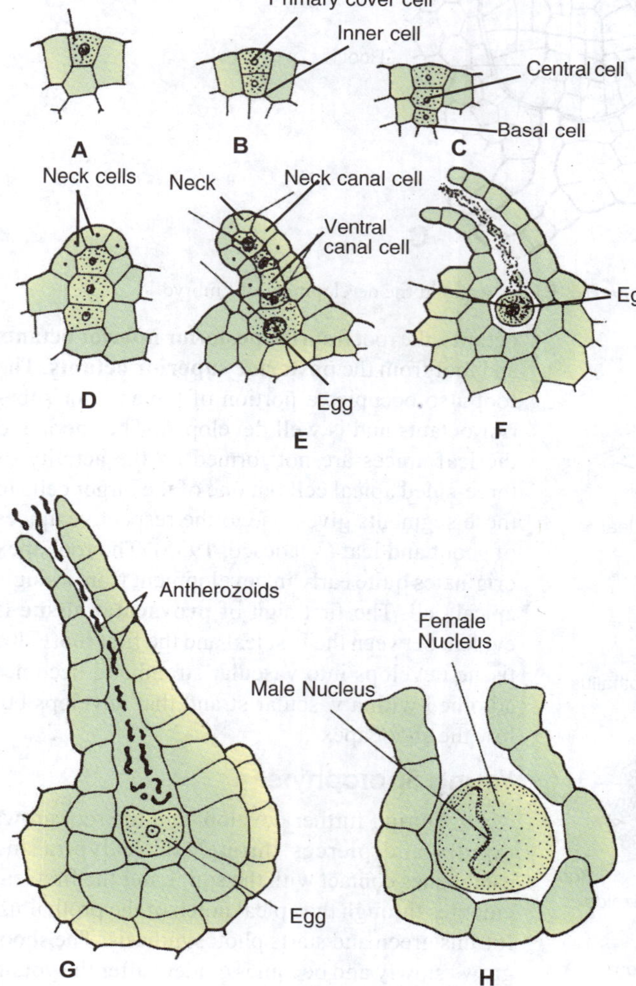

Fig. 14.17. (A—H). **Dryopteris.** A.-F. Various stages in the development of an archegonium. *G -H.* Fertilisation.

Development of Embryo

The zygote divides by a wall parallel to the long axis of the archegonium (Fig. 14.18, B) forming two unequal cells. The smaller cell **which** is towards the apex of the prothallus is the **epibasal cell** and the larger is the **hypobasal cell**. The second wall is laid in a transverse plane and forms a **quadrant stage**. The third wall is also transverse and forms an **octant stage**. Further divisions result in the formation of a 16-celled and 32 celled embryo. At the 32-celled stage (Fig. 14.18, C) the differentiation of various organs of the young sporophyte may become evident. The shoot arises from the **anterior superior octants**, the first leaf from the **anterior inferior**

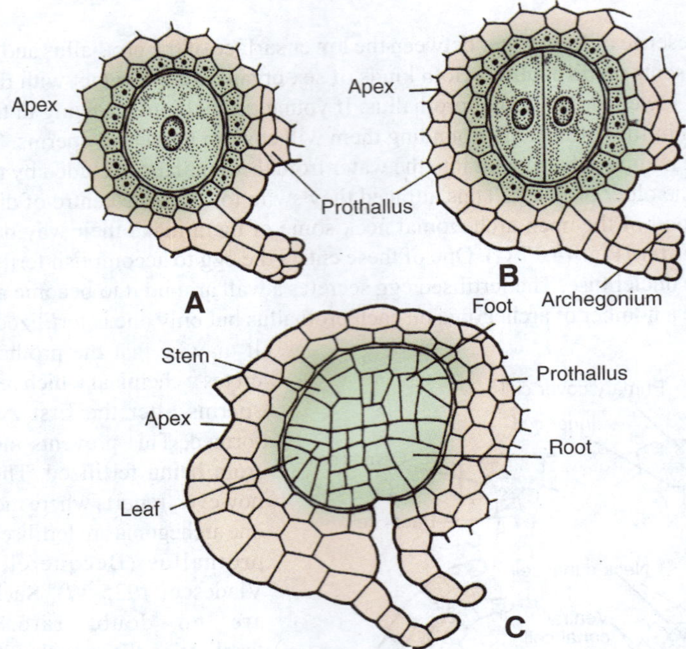

Fig. 14.18. (A—C). *Dryopteris.* Earlier stages in the development of Embryo.

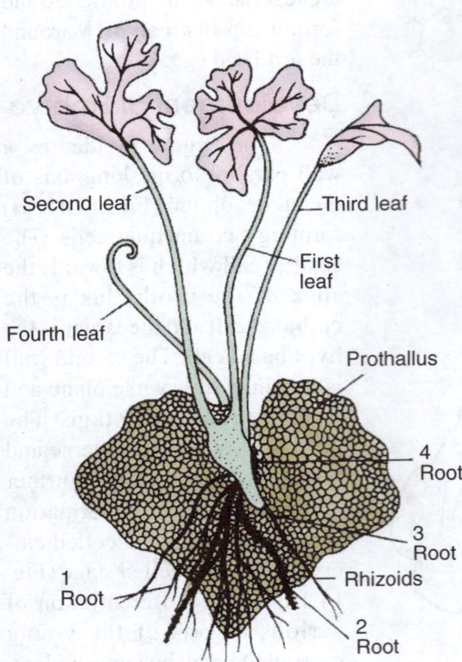

Fig. 14.19. Prothallus of **Dryopteris** bearing a young sporophyte.

octants, the root from the **posterior inferior octants** and foot from the **posterior superior octants**. The foot also occupies a portion of the anterior superior octants and is well developed. The shoot and the leaf apices are not formed by the activity of three-sided apical cell but one of the larger cells in these segments gives rise to the respective apices of shoot and leaf (Vladesco, 1935). The root apex originates quite early in development from a single apical cell. The first sign of **provacular tissue** is evident between the first leaf and the first root. This tissue develops into vascular strand and becomes adjoined with a vascular strand that develops below the shoot apex.

Young Sporophyte

During further development the root grows rapidly and pierces through the calyptra and establishes contact with the soil. Later the first leaf emerges through the apical notch of the prothallus. It truns green and starts photosynthesis. The shoot grows slowly and becomes evident after the young sporophyte has produced a few young and juvenile leaves and primary roots (Fig. 14.19). The shoot grows underground and bears adventitious roots and young fronds.

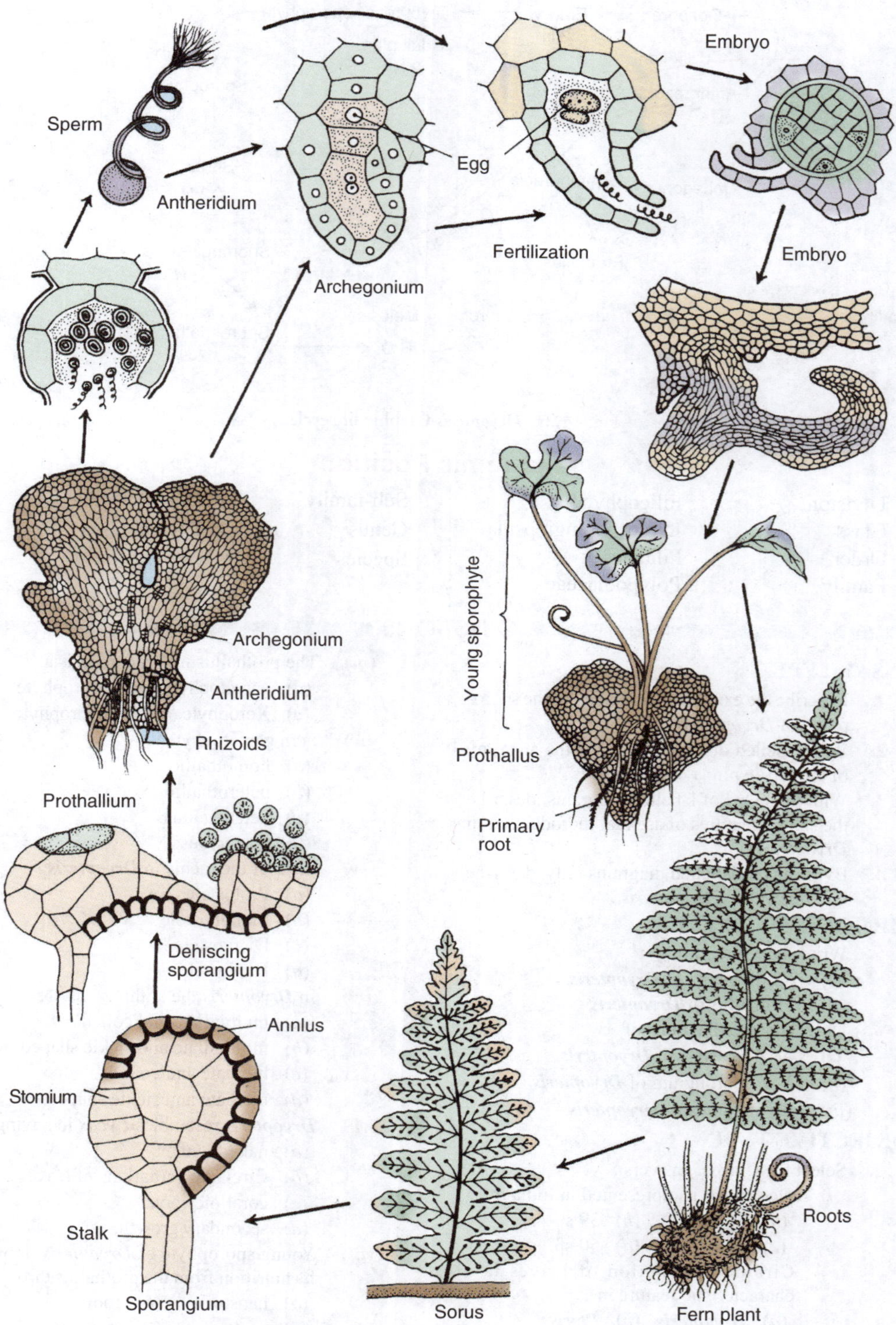

Fig. 14.20. A diagrammatic representation of life cycle of *Dryopteris*

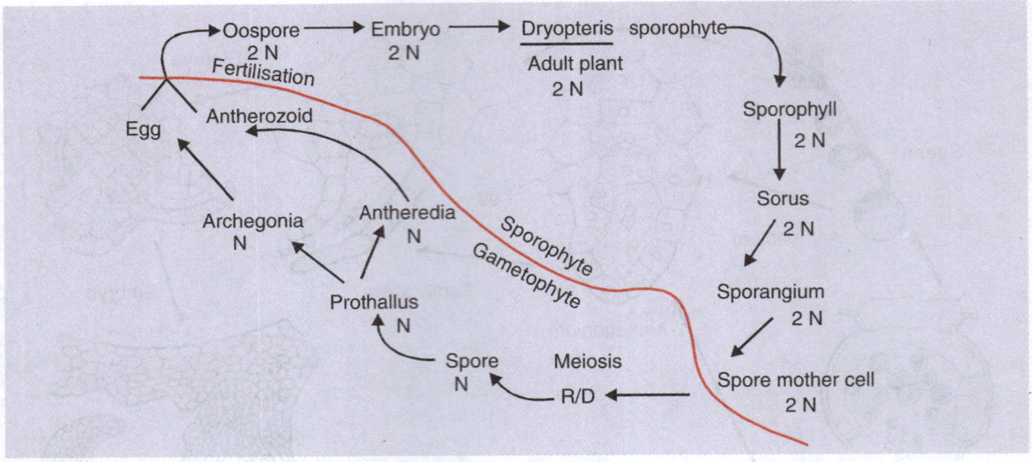

Fig. 14.21. *Dryopteris*-Graphic life cycle.

Taxonomic Position

Division	:	Filicophyta	Sub-family	:	Dryopteridioideae
Class	:	Leptosporangiopsida	Genus	:	***Dryopteris***
Order	:	Filicales	Species	:	***chrysocoma***
Family	:	Polypodiaceae			

QUESTIONS

ESSAY TYPE

1. Describe the external merphology of the sporophyte of ***Dryopteris.***
2. With labelled diagrams describe the anatomy of the fertile pinna of ***Dryopteris.***
3. With the helps of labelled diagrams, describe the salient features of asexual reproduction in ***Dryopteris***
4. By means of labelled diagrams only, describe the life cycle of ***Dryopteris.***

SHORT ANSWER TYPE

5. Write short notes are
   (*i*) Fertile pinna of ***Dryopteris.***
   (*ii*) Prothallus of ***Dryopteris***
6. Draw labelled diagrams of :
   (*i*) A Dictyostele of ***Dryopteris.***
   (*ii*) T.S. Sporongium of ***Dryopteris.***
   (*iii*) Fertile pinna of ***Dryopteris.***

OBJECTIVE TYPE

7. Select the Correct answers :

(*i*) *Dryopteris* is represented in India by
(*a*) 30 sp. (*b*) 39 sp
(*c*) 41 sp (*d*) 50 sp.

(*ii*) Circinate vernation of leaves is a characteristic feature in
(*a*) ***Dryopteris*** (*b*) ***Pteris***
(*c*) ***Pteridium*** (*d*) all of the above.

(*iii*) The prothallus in ***Dryopteris*** is a
(*a*) gametophyte (*b*) Sporophyte
(*c*) Xerophyte (*d*) Hydrophyte.

(*iv*) Fern gametophyte is
(*a*) homothallic
(*b*) heterothallic
(*c*) heteroecious
(*d*) autoecious.

(*v*) Stele of the rachis in ***Dryopteris*** *is*
(*a*) Protostele
(*b*) Siphonostele
(*c*) Dictyostele
(*d*) Haplostele.

(*vi*) In ***Dryopteris*** the anthrozoids are
(*a*) multiciliate and coiled
(*b*) multiciliate and sickle shaped
(*c*) biciliate and coiled
(*d*) biciliate and sickle shaped.

(*vii*) ***Dryopteris*** resembles ***Cycas*** in having
(*a*) naked ovule
(*b*) Circinate vernations of leaves
(*c*) coralloid roots
(*d*) secondary growth.

(*viii*) Young sporophyte of ***Dryopteris*** draws its nutrition from the prothallus through
(*a*) haustoria (*b*) foot
(*c*) roots (*d*) rhizoids.

Pteridium aquilinum

CHAPTER 15

Family Polypodiaceae Subfamily Pteridoideae

- Introduction
- Description of genus ***Pteridium aquilinum*** (L. Kuhn)
- Description of genus ***Pteris*** Linn.

Introduction

The absence of inducium is the main character istic feature of this sub family. It includes 12 genera and about 278 species. ***Polypodium, Platycerium, Pteridium*** and ***Pteris*** are some of the important ferns belonging to this group. The life histories of ***Pteridium*** and ***Pteris*** are discussed in this chapter.

GENUS–PTERIDIUM AQUILINUM (L) Kuhn

It is widely distributed and a cosmopolitan monotypic genus that has several varieties and sub-species. It is widely distributed in India along entire Himalayan tract that grows well at altitudes between 1,000 and 3,000 metres. The species is found commonly on forest floors, edges of forests, exposed mountain slopes and on open grass lands. It is a very hardy plant and once established it does not permit other ferns to grow in the area and becomes an obnoxious field pest. It reaches its climax in the moist tropical regions where it forms evergreen and permanent shrubby growth. It occurs almost all over the world except the arctic zone and the temperate South America.

SPOROPHYTE

Morphology

Rhizome

The rhizome is subterranean and grows 6 to 15 centimetres below the surface of the soil (Fig. 15.1 A). The rhizome is long, creeping and profusely branched. Watt (1940) described the branching of the rhizome in *Pterdium aquilinum.* He distinguished three types of branches that arise from the sympodial rhizome. These are :—

1. ***Long Shoots.*** They arise from the parent axis and grow in the same direction thus penetrating deep in the soil. They have longer internodes and do not bear leaves.

2. ***Intermediate Shoots.*** After their origin from the parent axis they grow obliquely upwards for some distance and then run horizontally. They do not bear leaves.

3. ***Short Shoots.*** They also grow obliquely upwards and then run horizontally a little below (2—10cm) the soil surface. They bear leaves and have short internodes.

Branches of varying lengths that fall within the range of these three types have also been reported. The profusely branched rhizome presents a complicated system of branching and affords an efficient anchorage to the species. It is profusely covered with multicellular hair. The scales are absent.

The young plant has an unbranched creeping rhizome with 4—10 alternating leaves. It grows under the soil and soon bifurcates dichotomously into two equal dichotomies. Each dischotomous branch bears a leaf near its point of origin from the main axis. Both these branches grow deep into the soil and bear leaves that are arranged in a spiral manner. The axis continues to branch

Fig. 15.1 (A—B). A. Portion of a sporophyte of *Pteridium aquilinum* showing habit; B. A. pinna showing position of sori.

dischotomously, but these dichotomies are unequal. The shorter branch bears a leaf. It appears short and stumpy so that at maturity it looks like a bud at the base of a leaf. Watt (1940) regarded these branches as short shoots. The longer dichotomous branch was perhaps regarded by him as a long shoot.

Leaves

The young plants bear leaves on the main axis but as the plants grow older the leaves seem to be restricted to the short and thick branches. The older portions of rhizome are beset with bases of old leaves and hair. They arise on the upper side of the rhizome and are borne in an alternate manner. They arise at an appreciable distance from each other and are circinately coited when young. A mature leaf may vary in length from 2—12 feet. It is tripinnately compound and has a distinct petiole that is as long as the pinnately divided lamina. The portion of the petiole that extends into the lamina is called the rachis. The lower branches of the rachis are longer and gradually decrease in leng thtowards the apex so as to give a deltoid or a conical appearance to the laminar portion (Fig. 15.1.A).

The pinnules are traversed by a large and prominent mid-rib from which arise lateral veins. The veins run obliquely upwards and are furcate *i.e.*, they branch dichotomously (Fig. 15.1, B). The pinnules are firm and rough to touch (sub-coriaceous). The petioles are covered with unbranched and multicellular hair.

Roots

They arise at irregular intervals all along the ventral or lower surface of the rhizome and are sparingly branched. They are adventitious and endogenous in origin.

Anatomy

Rhizome

Gottlieb (1959) studied the anatomy of the rhizome in ***Pteridium aquilinum*** right from the seedling stage and recognised six well marked stages in the ontogeny of the stele (Fig. 15.2.). These are :—

Pteridium aquilinum

(i) Protostelic stage. A week-old stem that is hardly 6—8mm. long and bears one or two young leaves, has a single protostelic vasculature. The xylem forms a solid central core.

(ii) Siphonostelic stage. A five-to-six-week old stem is almost upright and bears 2—5 leaves. It has siphonostelic vascular cylinder with only outer endodermis, pericycle and phloem (ectophloic siphonostele). This condition persists till the stem is about 2½ to 3 months old.

(iii) Solenostelic stage. As the young stem takes its course down into the soil and establishes itself as a rhizome the siphonostele is broken into two separate and unequal strands. The rhizome is about 5—10mm. in length and is more than 2½ to 3 months old.

(iv) Dictyostelic stage. After the third stage when the rhizome has achieved a length of about 1.5—2.0mm the two larger or lower strands break up into a number of smaller strands or meristeles. They are arranged in a ring and some of them have a direct connection with the leaf traces and are formed due to the appearance of leaf gaps whereas others have no connection with the leaf or branch traces and appear due to 'stelar perforation'. At this stage the rhizome starts branching and bears a number of short shoots with leaves.

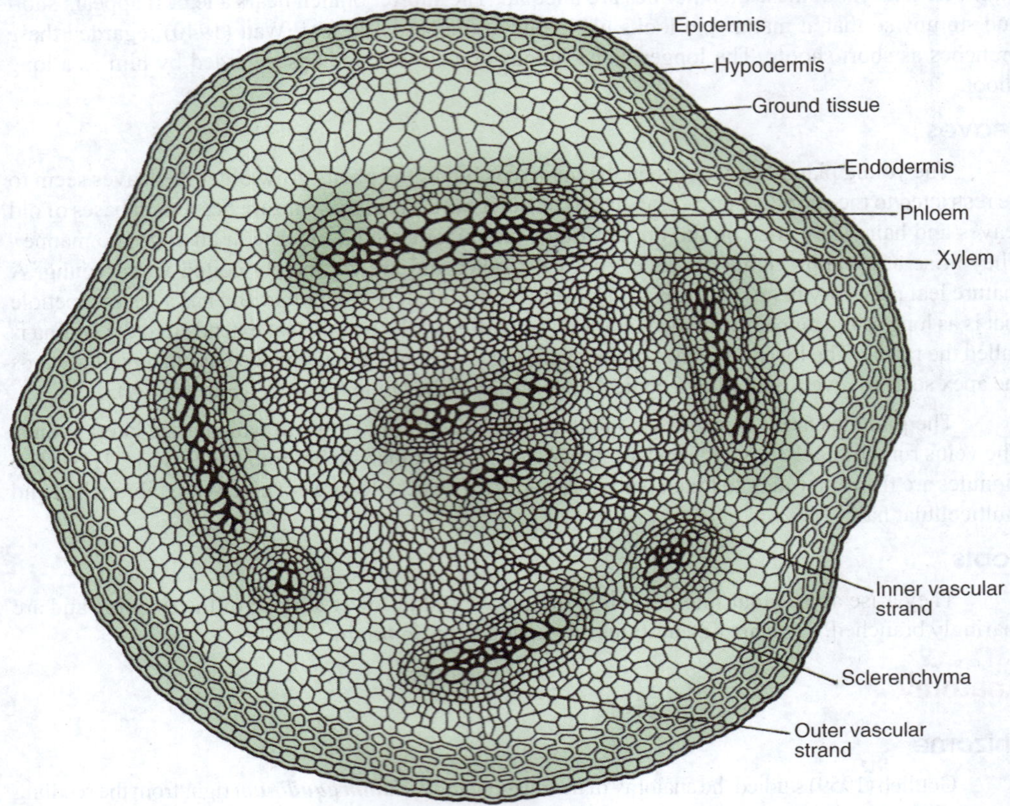

Fig. 15.2. A transverse section through mature rhizome of ***Pteridium aquilinum*** showing detailed internal structure.

(v) First Medullary Bundle Stage. This stage when rhizome is about 6 cm. or more long and has achieved the age of 4 months. A medullary bundle originates from the inner side (towards pith) of one of the lower peripheral bundles. This bundle separates from the peripheral bundle and forms the first medullary bundle.

(vi) Second Medullary Bundle Stage. This stage starts when the rhizome is 6—12 months old and is up to 20 cms long. A second medullary bundle arises from one of the lower peripheral bundles and separates from it as a second medullary bundle. This is the adult condition of the rhizome. This stage persists in older rhizome and there is no further alteration in the stelar organsation. At this stage the vasculature appears like the **dicyclic *dictyostele*** (Fig. 15.2). The peripheral ring of vascular bundles or the meristeles is separated from the medullary meristeles by an incomplete patch of sclerenchyma. The meristeles of the outer ring are small and larger in number whereas there are only two or rarely more meristeles in the inner ring.

Each meristele (Fig. 15.3). has an outer ring of the endodermis. The cells have casparian bands on their radial walls. Next to this is the pericycle. It may be composed of one or two layers of thin walled cells surrounding the phloem. The phloem has an outer ring of smaller sieve elements and larger parenchyma cells that constitute the **protophloem** and an inner ring of larger sieve elements or the **metaphloem.** The phloem has no companion cells and phloem fibres, and almost completely surrounds the mesarch xylem. The phloem and xylem elements resemble those of ***Dryopteris*** in their structure and arrangement.

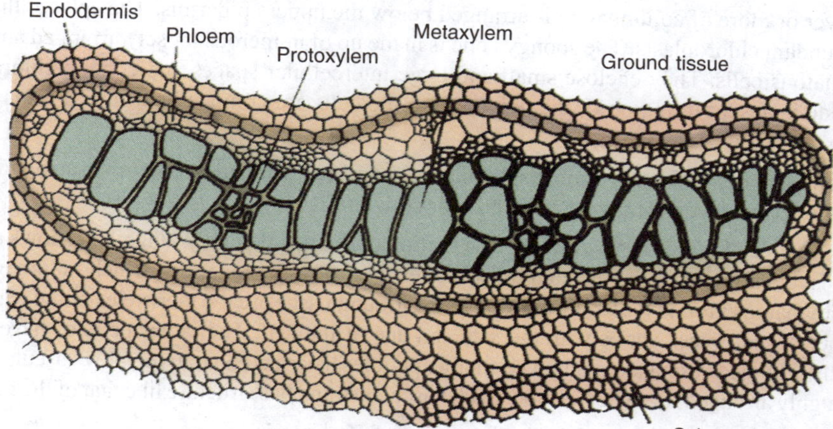

Fig. 15.3. ***Pteridium aquilinum.*** A meristele in detail.

The entire stelar region is embedded in the parenchymatous ground tissue which is surrounded by the sclerenchymatous hypodermis, which forms the peripheral strengthening tissue (Fig. 15.2). It is encircled by the outermost layer of thin-walled or thick-walled cells called the epidermis. The outer walls of the epidermis are cutinised.

The structure of the root resembles that of ***Dryopteris*** in all its details. It is diarch (Fig. 15.4.C).

Leaf

(*i*) Pinnule. A T.S. through the pinnule reveals that the internal structure is similar to that of a bifacial leaf. There are the usual two layers of epidermis with stomata restricted to the lower epidermis (Fig. 15.4, B). In between the two epidermal layers is the mesophyll tissue, which is clearly distinguishable into palisade parenchyma and the spongy parenchyma. The former

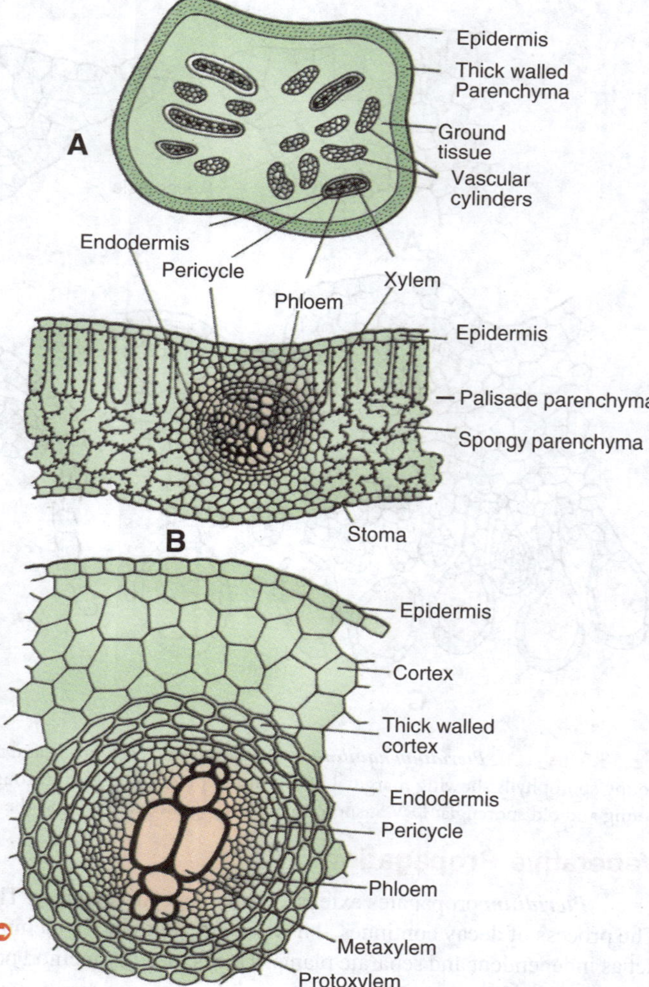

Fig. 15.4. (A–C). ***Pteridium aquilinum.*** **A.** T.S through petiole (diagrammatic) showing internal structure; **B.** V. S. Pinnule ; **C.** T. S. root.

forms a layer or more of columnar cells arranged below the upper epidermis. The cells of this layer contain abundant chloroplasts. The spongy zone is made up of numerous loosely arranged and lobed parenchymatous cells. They enclose small and large intercellular spaces and contain chloroplasts. The palisade is absent in region of the veins where the cells are compactly arranged so as to give strength to this region.

The vascular bundles are almost spherical in shape and each one of them is surrounded by a distinct endodermal layer. It may be concentric or collateral.

(*ii*) Petiole (Fig. 15.4, A). A transverse section of the petiole of a young plant shows an outer epidermal layer encircling a few layered thick sclerenchymatous hypodermis, which in turn encircles the parenchymatous ground tissue. Embedded in the ground tissue are the meristeles or the leaf traces. The number of leaf traces varies with the size and age of the leaf. The larger petioles have several haphazardly arranged vascular strands. The smaller petioles have a comparatively few vascular strands that are roughly arranged like a horse-shoe. Each vascular strand has structure like that of the meristele.

The pattern of apical growth is similar to that of ***Dryopteris.***

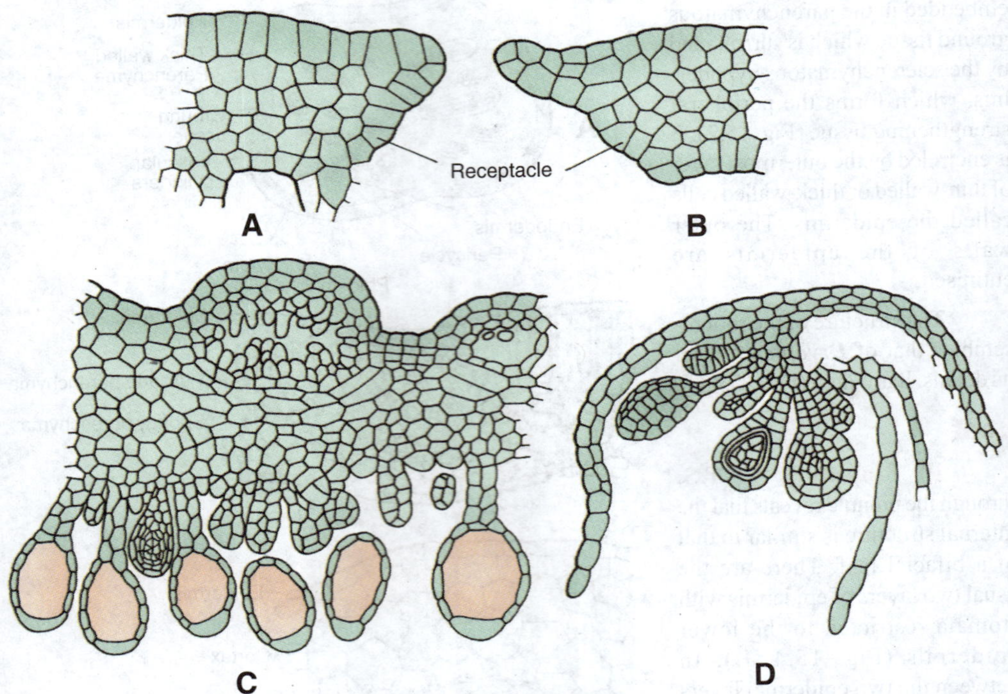

Fig. 15.5. (A—D). *Pteridium aquilinum*. Development of receptacle and the sorus. A, B. Vertical sections through young sporophylls showing marginal origin of receptacle; C. AV. S. through the continuous sorus showing mixed young and old sporangia; D.V.S. sporophyll showing mixed sorus and the flaps (After Bower)

Vegetative Propagation

Pteridium propagates extensively by vegetative means. The older parts of the rhizome decay. The process of decay continues and when it reaches the branching region, the branches separate and act as independent and separate plants. This is the only method prevalent in ***Pteridium*** and leads to a gregarious habit.

Sporophylls and Sori

Every leaf in ***Pteridium*** is a potential sporophyll as there is no segregation of reproductive and vegetative leaves. The ultimate segments of the lamina or the pinnules bear sporangia on their margins. The sporangia develop on marginal receptacles (Fig. 15.1, B) and are grouped in a continuous or a confluent and a linear type of sorus. It is also called a **coenosorus.** As a matter of fact numerous smaller sori occur so close to each other that they lose their identity and appears as one long sorus disposed along the two lateral margins of the fertile pinnules. A vertical section of the sorus (Fig. 15.5, A—D) reveals that the receptacle is traversed by a vascular strand that runs underneath the sporangia and connects the free ends of the veins. The sorus is protected by two flaps that grow a little to cover the sporangia from the lower end. The upper indusial flap (Fig. 15.5) is formed by the reflexed margin of the pinnule and is well developed. The lower indusial flap is true indusium where the upper flap is a false indusium because it is simply a revolute margin of the pinnule and not a specially developed structure. The lower or the true indusium is not well developed and is a thin sheet of tissue made up of a single layer of thin-walled cells.

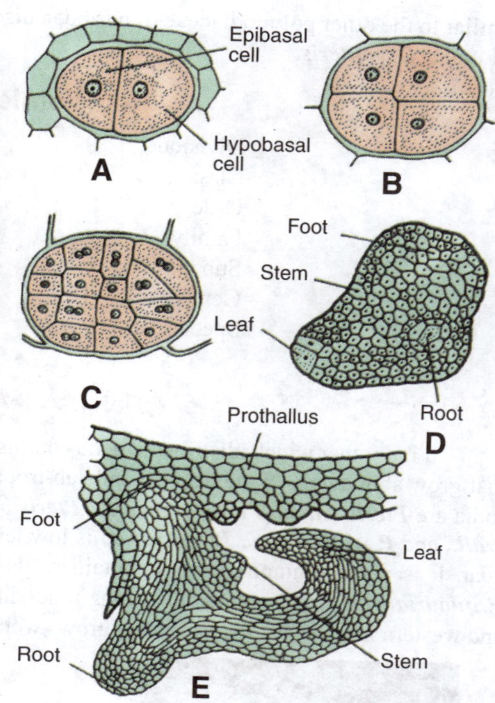

Fig. 15.6. (A-E). *Pteridium aquilinum.* Various stages in the development of embryo.

The receptacle originates from a row of wedge-shaped initials that occupy the margin of the pinnules (Fig. 15.5, A, B). These initials divide and redivide to form a continuous receptacle which bears sporangia. The adaxial or the upper indusium appears as a superfical outgrowth from the marginal segments of the leaf a little above the receptacular initials. It soon overgrows and envelops the developing receptacle. The lower indusium arises late and is comparatively less developed. The sequence of sporangial development appears to be basipetalous in a young sorus but it soon changes to a mixed type. In some varieties the order of development remains basipetalous (***P. aquilinum* var, *caudatum***), and the sorus is gradate.

P. aquilinum var. *caudatum*

Sporangium

Structure, development and mode of dehiscence of the sporangium are similar to that of ***Dryopteris.*** The number of spores produced per sporangium is also 48. Manton (1950) reported the haploid chromosome number as $n = 52$.

Gametophyte

The spores are tetrahedral and lack **perispore.** Their mode of germination and the structure of prothallus is similar to that of ***Dryopteris.*** The position, structure and development of sex organs is also similar to ***Dryopteris.***

The development of the zygote and the details of segmentation pattern leading to the formation of a mature embryo and then to the young sporophyte (Fig. 15.6, A—E) are also similar to the other polypodiaceae. It has been discussed in detail in Chapter 8 and also in Chapter 14 under ***Dryopteris.***

Taxonomic Position

Division	:	Filicophyta
Class	:	Leptosporangiopsida
Order	:	Filicales
Family	:	Polypodiaceae
Sub family	:	Pteridioideae
Genus	:	***Pteridium***
Species	:	***aquilinum***

GENUS– PTERIS Linn.

It is another widely distributed genus of this sub-family. It is represented by about 250 species that grow abundantly in the tropical and sub-tropical regions of the world. The common species in India are ***Pteris vittata, Pteris cretica L., Pteris biaurita L., P. quandriaurita Retz., P. stenophylla Wall.***, and ***P. wallichiana. Pteris vittata*** is low level fern and brings out new leaves throughout the year. It is very common along mountain walls and grows up to 1,200 metres above sea level. ***P. quadriaurita*** grows abundantly along road-sides and the valleys throughout the North-Western and western Himalayas. ***Pteris cretica*** grows well from 1,200—2,400 metres above sea level.

P. vittata

P. cretica

P. stenophylla

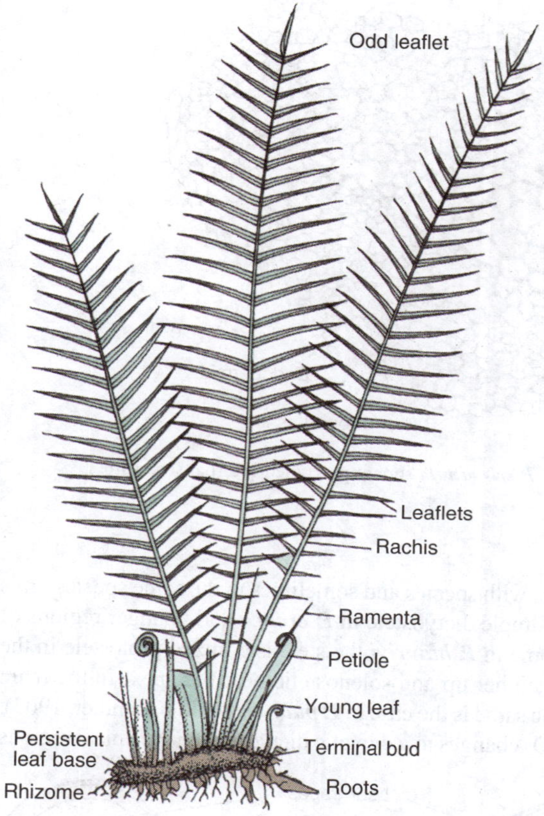

Fig. 15.7. A complete plant of *Pteris vittata*.

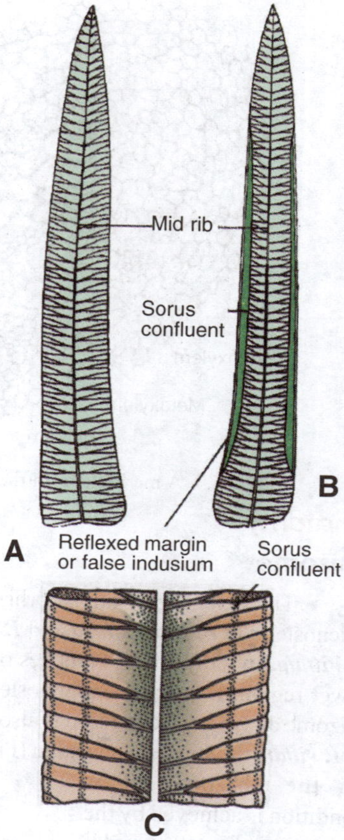

Fig. 15.8. ***Pteris vittata*** A. a sterile pinna; B. A fertile pinna showing sorus; C. Portion of B enlarged.

The plants have creeping rhizomes in ***P. biaurita***, ***P. vittata*** (Fig. 15.7) and ***P. grandiflora***. In ***P. cretica*** the rhizome is branched short, stumpy and semi-erect. The rhizome is covered with scales. Hair are absent. Roots arise from the lower surface of the rhizome or they may arise all over the surface as in ***P. biaurita.***

P. biaurita

The leaves are pinnately compound (Fig. 15.7) rarely digitate. The petioles are covered with scale. Venation is of open furcate type (Fig. 15.8). The leaves in ***P. vittata*** are once pinnate (Fig. 15.7). The pinnae are smaller near the base, larger near the middle and again go on decreasing in length towards tha apex so as to give a spindle-like appearance to the frond. Leaf apex is occupied by an odd leaflet or pinna. Venation in this case is open furcate. Every pinna is traversed by a central midrib which gives off lateral veins that bifurcate near the tip. The pinnae are sessile and are broad at the base and gradually taper towards the apex (Fig. 15.7). The leaves are bipinnate in ***P. biaurita.*** The leaflets or pinnules are coriaceous or rough to touch.

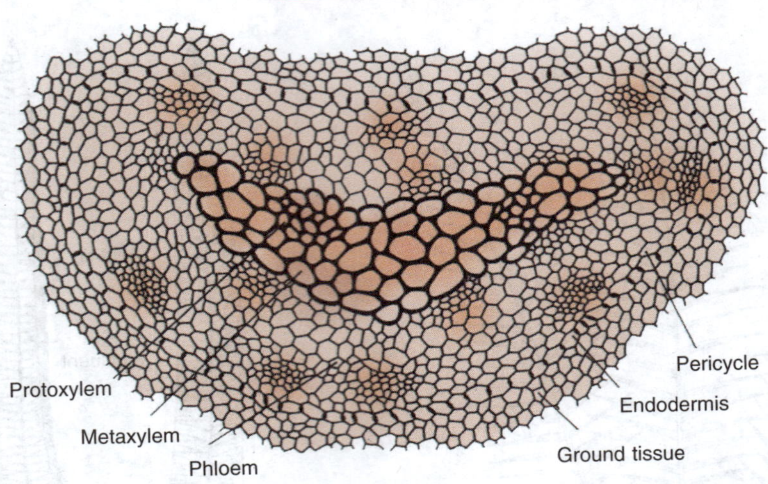

Fig. 15.9. A meristele from the rhizome of ***Pteris vittata*** showing detailed internal structure.

Anatomy

Rhizome

The stelar organisation of rhizome varies with species and sometimes in the same species . It is solenostelic in ***P. grandiflora*** and ***P. vittata***, a simple dictyostele in ***P. cretica*** and younger regions of ***P. Vittata.*** In the younger branches of the rhizome in ***P. biaurita*** the stele is a mixed protostele in the lower region. It becomes siphonostelic a little higher up and solenostelic near the apex. In the main rhizome dictyostelic condition is also found. The same is the case in ***P. palmata*** willd (Chandler, 1905). In ***P. vittata*** the stelar organisation (Fig. 15.10, D) changes to a dicyclic dictyostele in the apical regions of the rhizome. This condition is achieved by the separation of small, circular or oval strands or root traces from the outer surface of the parent meristeles that in turn were derived by the stelar perforation of the solenostelic arc. These changes lead to the formation of irregularly arranged root traces around 5—5 large meristeles (Fig. 15.10, D). There are no medullary strands. Structure of meristele of rhizome is shown in Fig. 15.9.

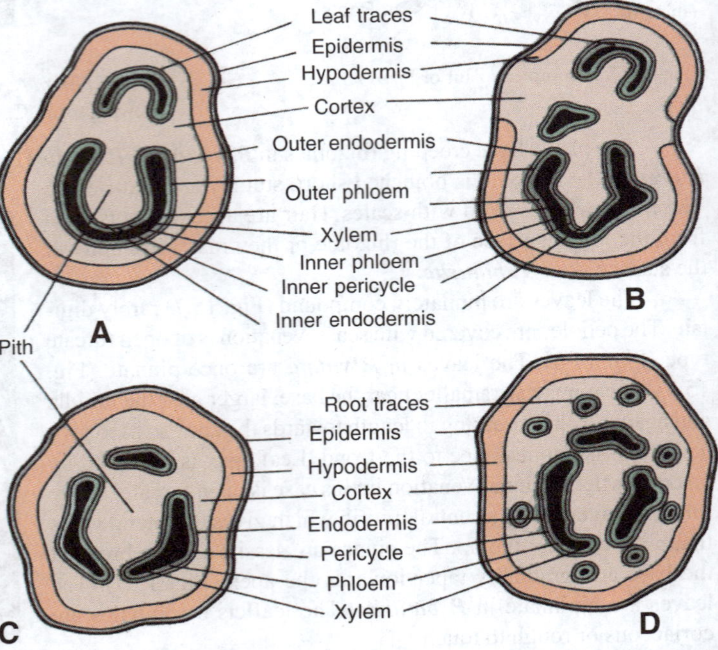

Fig. 15.10. (A–l). ***Pteris vittata.*** Cross sections of rhizome at various levels showing changes in stelar organisation.

Pinnule : The pinnule has upper and lower epidermal layers (Fig. 15.11). In ***P. cretica*** the upper epidermis has larger cells with less sinuous walls. The stomata are restricted to the lower epidermis which has smaller cells and the more sinuous walls. The mesophyll may or may not be differentiated into palisade paenchyma and spongy parenchyma. The midrib region has single concentric type of vascular strand (*P. vittata*) with distinct endodermis. The bundle sheath extensions are prominent and occur as groups of thick-walled cells below the upper and above the lower epidermis. Palisade and spongy tissue is absent around the mid-rib.

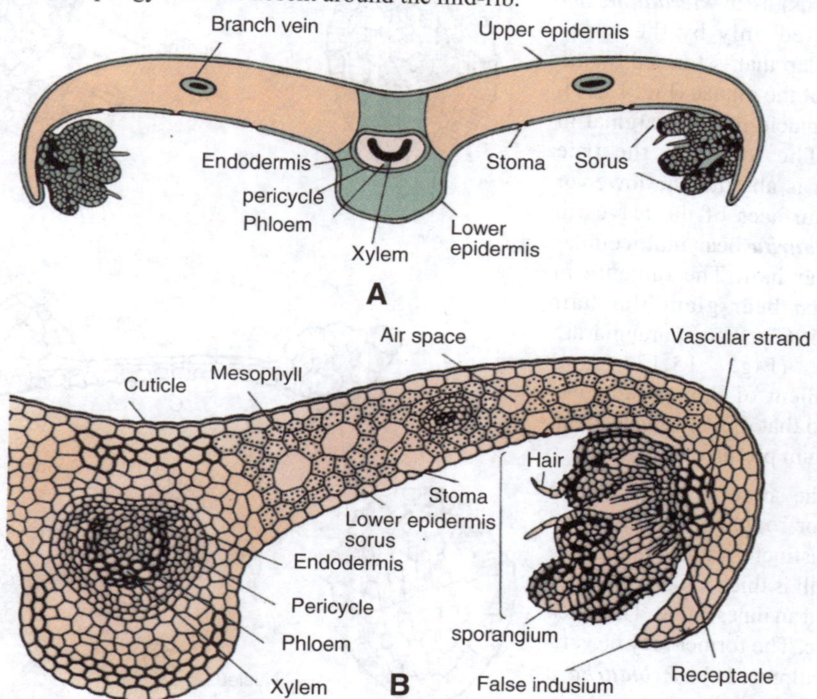

Fig. 15.11. (A–B). *Pteris vittata.* A. Outline figure of T. S. fertile leaflet showing internal structure, false indusium and sporangia borne on receptacle; B. portion of A in detail.

Petiole : The petiole (Fig. 15.12) is traversed by a single C-shaped (*P. vittata*) or U-shaped or V-shaped leaf trace. In *P. biaurita* the U-shaped leaf trace enters the petiole and becomes V-shaped higher up. The xylem has two adaxial hooks. In the rachis the petiole trace gives off strands into its pinnae if the leaf is unipinnate or it divides to give rise to secondary and tertiary pinna traces in bipinnate leaves (*P. biaurita*). The rachis traces are marginal in origin and are usually flat U-shaped or shallow arc-like. In *P. cretica* the C-shaped leaf trace divides into two strands which unite into a single V-shaped strand higher up in the petiole. The hooked

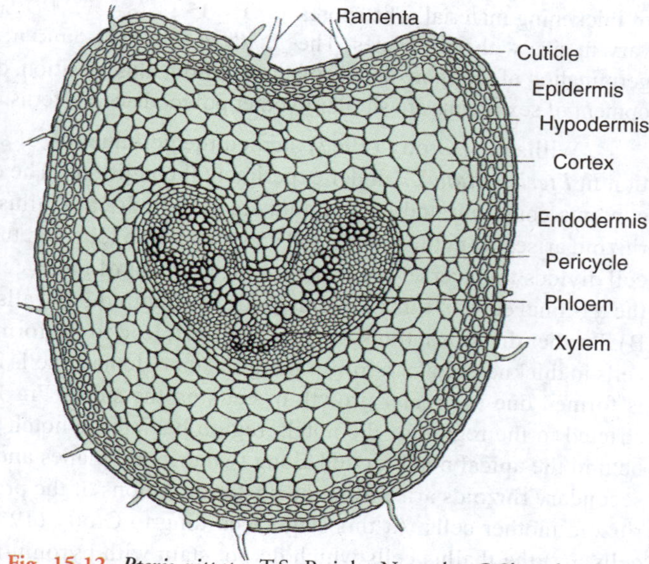

Fig. 15.12. *Pteris vittata.* T.S. Petiole. Note the *C-Shaped vascular* bundle and the protoxylem hooks.

xylem of the leaf trace, the marginal orgin of the pinna and rachis traces and the solenostelic tendency in the stem stele are the primitive features of the genus. The root is diarch.

Sorus and Spore: The sorus is of the continuous type (coenosorus) as in ***Pteridium***; but is protected only by the upper indusial flap that is formed by the margins of the pinnae (Fig. 15.11). The receptacle is intramarginal in origin. The lower or the true indusium is absent. The lower or ventral surfaces of the leaves in ***Pteris biaurita*** bear multicellular and linear hair. The ramenta in ***P. cretica*** bear glandular hair (Singh, 1963). The sporangia are umixed (Fig. 15.13) and development of the sporangia is similar to that of ***Dryopteris***. Each sporangium produces 48 spores.

The spores (Fig. 15.13) are bluntly or roughly triangular and have a distinct triradiate mark. The spore wall is thick and has an outer exine and an inner intine. They lack perispore. The former may be variously sculptured. In ***P. biaurita*** it bears many irregular bands of thickening (Singh, 1963). In ***P. cretica*** the exine has variously bent bands of thickening material. The spores vary in size with the species. The germination of the spore and formation of prothallus, position, distribution and structure and development of sex organs are similar to other polypodiaceous ferns.

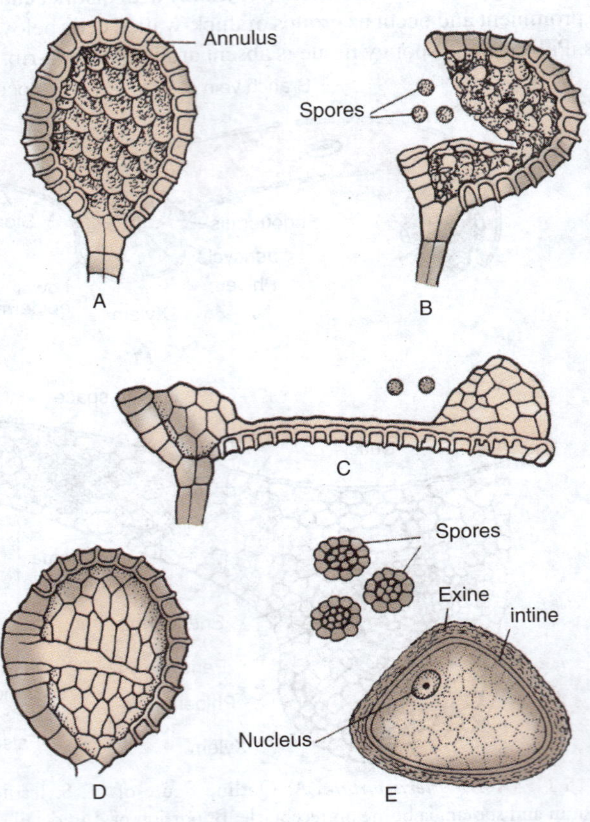

Fig. 15.13. (A—E). *A-D.* Structure and dehiscence of sporangium in *Pteris vittata* ; E. Structure of spore.

William, J. Crotty (1967) studied the germination of spores and rhizoid cell differentiation in ***Pteris vittata*** L., under experimental conditions. The exine ruptures and the uninucleate contents come out and grow into a small cylindrical thallus cell (Fig. 15.14A). A unicellular rhizoid arises from the thallus cell and grows into long and tubular, primary rhizoid. The thallus cell divides transversely to form a short filament of green cells (Fig. 15.14, B). One or more of the terminal cells of this filament divide by longitudinal walls to form a plate of cells (Fig 15.14, B). Further divisions, transverse and vertical, lead to the formation of a bigger plate one layer of cells in thickness. Later a row of marginal meristematic cells is distinguished and an apical notch is formed due to greater growth of cells on the sides (Fig 15.14, C) Divisions are mainly restricted to the region of the notch, region behind the notch and in the lateral wings. The cells behind the apical notch divide along two or more planes and form a thick central cushion. The secondary rhizoids arise from the posterior regions of the prothallus. Mature thallus cells act as rhizoid mother cells. At this stage according to Crotty (1967) the prothallus has two types of cells : (*i*) the thallus cells which do not stain with pyronin-methyl green and remain living an hour after the staining process, (*ii*) the rhizoidal cells which readily stain with pyronin-methyl green. The stain kills the cell cytoplasm. The rhizoidal cells differentiate and grow into secondary rhizoids.

The structure and development of sex organs are similar to ***Drypoteris.*** The development of embryo (Fig. 15.15) is also similar to that of ***Dryopteris*** in general.

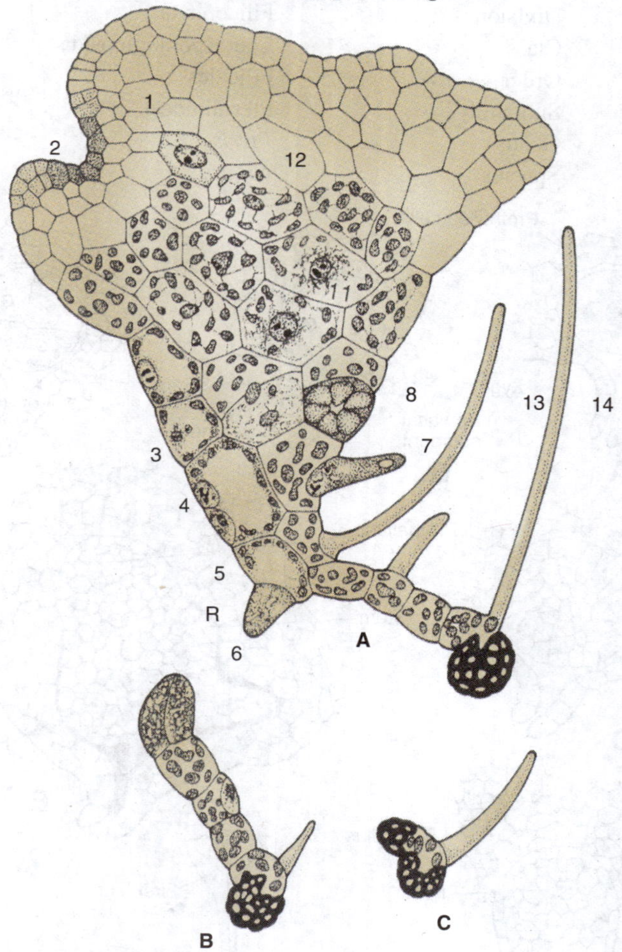

Fig. 15.14. (A—C). *Pteris vittata.* Stages in the germination of spore and formation of rhizoid cell (After Crotty William, 1967).

Economic Importance

Both ***Pteridium*** and *Pteris* are of little importance. These are noxious weeds and are toxic to live stock. Generally ***Pteridium aquilinum*** grows on land intended for cultivation and grazing. However, the plants have been in use since

10th century (*a*) as a source of potash for soap and glass making; (*b*) fronds are used as fuel and thatching material for cottages; (*c*) as medicine as antihelmintics ; (*d*) in brewing ; (*e*) as a dye preparation of leather and (*f*) for paper and board manufacture.

Taxonomic Position

Division	:	Filicophyta
Class	:	Leptosporangiopsida
Order	:	Filicales
Subfamily	:	Pteridioideae
Genus	:	*Pteris*
Species	:	*vittata.*

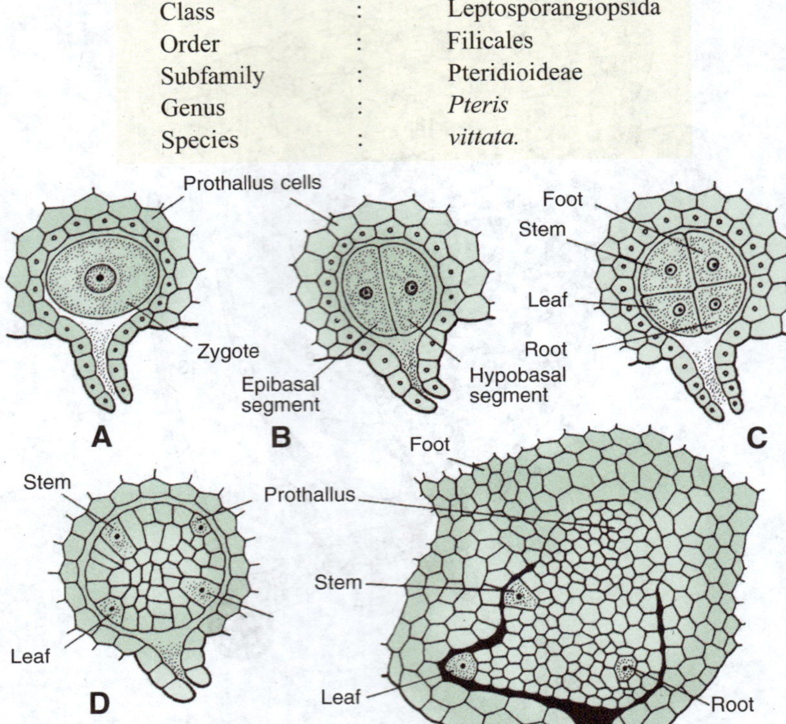

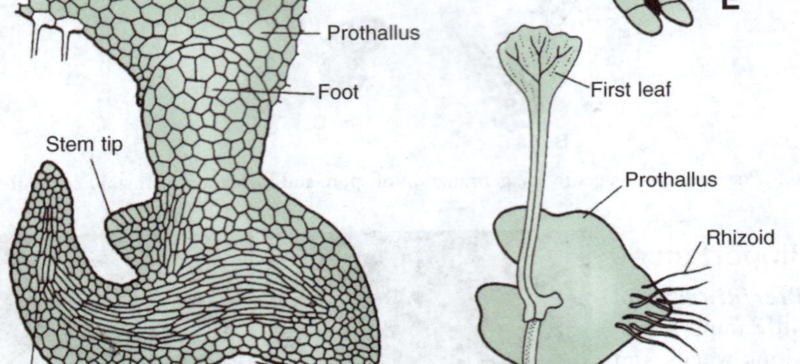

Fig. 15.15. (A–G). ***Pteris vittata.*** Various stages in the development of embryo.

QUESTIONS

ESSAY TYPE

1. Describe the significant features of the life history of a fern studied by you :
2. By means of labelled diagrams only, illustrate the life history of a fern.
3. Describe the development of reproductive structures in ***Dryopteris/Pteris.***
4. Describe the structure of Archegonia of ***Lycopodium, Selaginella, Equisetum, Marsilea*** and ***Pteridium***. Discuss their evolution.
5. Explain the alternation of generation with the help of an example of a pteridophyte.
6. Describe the gametophyte of ***Pteris.***
7. Draw graphic life cycles of ***Pteris*** and ***Selaginella.*** List the major differences between the saprophyte and gametophyte generations of both the genra.
8. With the help of suitable diagrams, Compare the gametophytes of ***Psilotum*** and ***Pteris.***
9. With the help of suitable diagrams, describe the structure of mature gametophyte of ***Pteris*** and explain the process of fertilization in it.
10. Explain the development of embryo in ***Pteris.***
11. Give a brief account of the structure and development of sex organs in ***Dryopteris*** or ***Pteris*** or ***Pteridium.***
12. Describe the morphology of sporophyte in ***Pteridium*** *or* ***Pteris.***
13. Describe the morphology of sporophyte in ***Pteris.***
14. With the help of labelled diagrams, give an account of the anatomy of the sporophyte of ***Pteris.***
15. Compare the gametophytes of ***Dryopteris, Pteridium*** and ***Pteris.***

SHORT ANSWER TYPE

16. Write short notes are
    (*i*) Prothallus of ***Pteridium***
    (Allahabad, 2000)
    (*ii*) Spore dispersal in ***Pteridium***
    (*iii*) Sporangium of ***Pteridium***
    (*iv*) Mechanism of sporangial dehiscence and spore dispersal ***Dryopteris*** in ***Pteridium***
    (*v*) Spores of pteridophytes studied by you.
    (*vi*) Fronds in ***Pteridium***
    (*vii*) Ramenta
    (*viii*) Anatomy of ***Pteris*** petiole
    (*ix*) Sporongium of ***Pteris***
    (*x*) Anatomy of the sporophyte in ***Pteris***
    (*xi*) Leaf of ***Pteris***
    (*xii*) Germination of spores in ***Pteris***
    (*xiii*) Function of Annulus
    (*xiv*) True and False indusia of ***Pteridium.***
    (Allahabad, 2001)
17. Draw labelled diagrams of :
    (*i*) T.S. Sporophyte of ***Pteridium***
    (*ii*) T.S. Petiole (rachis) of ***Pteridium***
    (Allahabad, 2002)
    (*iii*) Sporangium of ***Pteridium***
    (*iv*) T.S. ***Pteridium*** Rhizome
    (*v*) Sporophyte of ***Pteridium aquilinum***
    (*vi*) A meristele of ***Pteridium***
    (*vii*) T.S. Fertile leaflet in ***Pteris.***
    (*viii*) T.S. of sporophyll of ***Pteridium***
    (Allahabad, 2003)
18. Why water is necessary for completion of life cycle of pteridophytes? **(Allahabad, 1995)**
19. How many spores are present in the sporangium of ***Pteridium*** and why?
20. How will you distinguish between
    (*a*) T.S. Petioles of ***Marsilea*** and ***Pteridium***
    (*b*) True and False indusia of ***Pteridium.***
    (*c*) Rhizome of ***Dryopteris*** and ***Pteris***
    (*d*) Fertile pinna of ***Pteris*** and ***Pteridium***
21. What is the advantage to have sex organs on the under surface of prothallus of ferns?
    (Allahabad, 2000)
22. Briefly discuss the dictyostele with relevant examples.
23. Explain the structure and function of prothallus in ***Dryopteris/Pteris/Pteridium.***
24. Describe the position and structure of sporangis in ***Pteris/Dryopteris/Pteridium.***
25. Describe the structure of mature sporangium in ***Pteris*** and the process of spore dispersal in it.
26. Describe the morphology of leaves in ***Pteris***
27. What is the relative position of sex organs in *Pteris* gametophyte?
28. How many spores are present in the sporongium of ***Dryopteris***.
29. How many spores are present in sporangium of ***Pteris/Pteridium.***
30. In which pteridophyte medullated stele is present?
31. Write notes on
    (*a*) V.S. of fertile leaf of ***Pteris***
    (*b*) T.S. rachis of ***Pteris***.
    (Nagpur Univ., 2005)

OBJECTIVE TYPE

32. Fill in the blanks:—

(*i*) Young leaves of ***Pteridium*** show —— vernation.

(*ii*) Fern prothallus is a means of ——— reproduction.

(*iii*) The leaves of ***Pteris*** when young are covered by———.

(*iv*) Meristeles are present in the ——— of ferns.

(*v*) ***Pteris*** belongs to the family ——— order ——— class ———.

(*vi*) Confluent sorus is found in ——— .

(*vii*) The number of spores in a sporangium of ***Pteridium*** is ——— .

37. Select the Correct answers :

(*i*) The sorus in ***Pteris*** is
- (*a*) discontinuous and circular
- (*b*) continuous and linear
- (*c*) discontinuous and vermiform
- (*d*) discontinuous and reniform.

(*ii*) Ferns differ from bryophytes in having
- (*a*) motile sperm
- (*b*) alternation of generations
- (*c*) archegonia
- (*d*) Vascular tissue.

(*iii*) Mesearch xylem is characteristic of
- (*a*) ***Lycopodium***
- (*b*) ***Selaginella***
- (*c*) ***Equisetum***
- (*d*) ***Dryopteris.***

(*iv*) Coenosorus is present in
- (*a*) ***Lycopodium***
- (*b*) ***Selaginella***
- (*c*) ***Equisetum***
- (*d*) ***Pteris.***

(*v*) In which of the following a sporangium has 48 spores in it?
- (*a*) ***Lycopodium***
- (*b*) ***Selaginella***
- (*c*) ***Equisetum***
- (*d*) ***Pteris.***

(*vi*) Spores of the fern germinate to form
- (*a*) thick welled zygospore
- (*b*) as embryo
- (*c*) prothallus
- (*d*) new plant.

Polypodium vulgare

CHAPTER 16

Family Polypodiaceae Subfamily Polypodioideae

- General characteristics of Polypodideae
- Characteristics of genus ***Polypodium*** L.
- General Characteristic of Gymnogrammeoideae
- Characteristics of genus ***Adiantum*** L.

GENERAL CHARACTERISTICS OF POLYPODIOIDEAE

It is the largest sub-family of polypodiaceae. It includes about 1,200 species (Christensen, 1938) that are distributed in the tropical and sub-tropical regions. ***Polypodium vulgare*** occurs as a common epiphyte in the British Flora. The common species that occur in India are ***Lepisorus nudus*** Hook, ***L. excavatus*** (Bory) Ching, ***L. kashyapii*** (Mehra), ***L. clathratus*** (Clarke) Ching; ***Polypodium microrhizoma***, ***P. amoenum*** wall, ***P. subamoenum***

Clarke, ***P lineare*** Thunbg : ***Phymatodes oxyloba*** (Wall) Ching, ***P. stracheyi*** Ching, ***P. ebenipes*** (Hook) Ching, ***P. malcadon*** (Hook) Ching; ***Drynaria mollis*** Bedd.; ***Loxogramme involuta*** Wall. and ***Microsorium membranaceum*** (Den.) Ching.

The sporophytes have creeping or semi-erect and branched or sparingly branched subterranean rhizome. The leaves may be simple or pinnate compound. Venation may be free or reticulate. The sori arise on the ventral surface of the sporophylls and are as a rule naked (exindusiate) and globose or oblong in shape. They are usually borne terminally (***Polypodium***) or subterminally on the free veins. In ***Phymatodes*** the sori occur on anastomosing veins (Pleosori) or on a plexus of veins (dictyosori). In ***Hymanopteris***, ***Lepisorus*** and ***Laxogramme*** the sori are fused into coenosori or confluent sori. The coenosori may be arranged parallel to the mid-rib (drymoglossoid), ***e.g.***, ***Drymotaenium*** Makino or parallel to main veins (Gymnogrammeoid), ***e.g.***, ***Loxogramme.*** In ***Cheiropleuria*** the sporangia are scattered all over the lower surface of the sporophylls. Such ferns are called **acrostichoid.** In ***Platycerium devaux*** there are two kinds of leaves : (*i*) humus collecting or **nest-leaves** (*ii*) pendent sporophylls, that are once or many times forked and bear sporangia on their ventral surfaces. The sporangia occur in patches. The fronds are covered with stellate hair. In ***Drynaria*** also there are two kinds of leaves : (*i*) brown, sessile, slightly lobed and humus collecting scale leaves and (*ii*) petiolate, green pinnatified or pinnate compound fertile leaves. The rhizome is usually dictyostelic and the roots are diarch. Spores lack a perispore and may be bilateral or tetrahedral. The development and structure of prothallus and sex organs is similar to ***Dryopteris***. The embryo also follows the same course of development.

P. amoenum

Microsorium membranaceum

Cheiropleuria

Fig. 16.1. *Polypodium sp.* A. Portion of a plant showing habit, B, A pinna bearing sori.

Genus ***Polypodium*** is briefly discussed as the type genus of thes sub-family.

Genus–POLYPODIUM L.

It is widely distributed and a common genus that occurs along the tropical and subtropical regions of the world. A few species are, however, found in the northern temperate regions (***P. vulgare***). It includes about 50 species. A few species have been included under the genus ***Phymatodes.*** In India it occurs commonly as an epiphyte on tree trunks and branches. The common Himalayan epiphytic species are ***P. microrhizoma***, ***P. amoenum***, ***P. subamoenum, P steewartii***, ***P. lachnopus***, and ***P lineare.*** They grow on tree trunks or branches of ***Quercus***, ***Cedrus***, ***Abies***, ***Rhododendron*** and rarely ***Pinus wallichiana.***

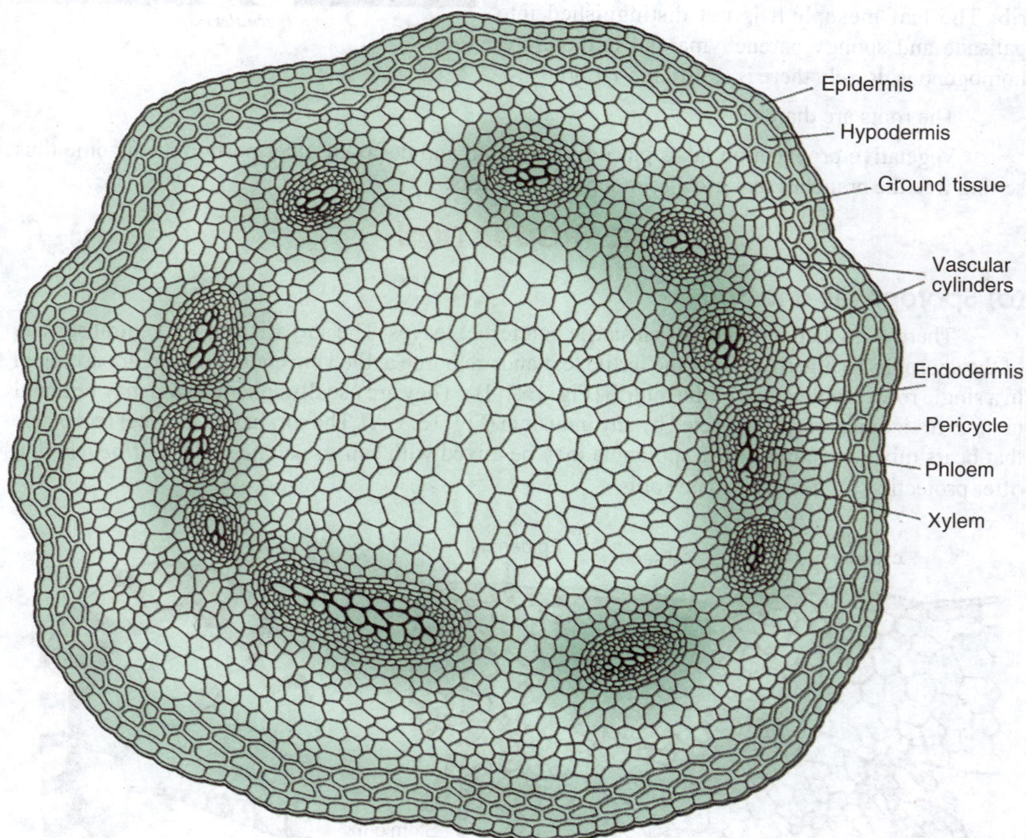

Fig. 16.2. ***Polypodium.*** T.S rhizome showing detailed internal structure.

Morphology

The rhizome is generally creeping (Fig. 16.1) and branched; sometimes fleshy, semierect and glaucous. The rhizome is covered with peltate, lanceolate and brownish scales or ramenta. The leaves may be simple and spathulate (***P. lineare***) or simple or deeply incisedand pinnate. Venation may be free and furcate or reticulate and anastomosing. The older portions of the rhizome are covered with bases of old leaves. The younger leaves are circinately coiled. The leaf lamina is covered with multi-cellular hair that arise from the epidermis. The adventitious roots are in large numbers and are branched. The roots orginate endogenously and come out on the lower side of the rhizome.

Anatomy

P. scouleri

The anatomy of the rhizome (Fig. 16.2), root and the sterile lamina is similar to that of ***Dryopteris.*** The stomata in the leaves are restricted to the lower epidermis and are irregularly distributed. In ***P. lineare*** Thunbg, ***P. scouleri*** Hook. et Grew, and ***P. vulgare*** Rehfus., the stomata are globular and are present even on the veins. The leaf epidermis is covered with club-shaped and multicellular hair in ***P. lineare*** (Singh, 1963). Non-club shaped hair occur around the mid-rib. The leaf mesophyll is not distinguished into palisade and spongy parenchyma and is therefore, homogenous. Rarely there is a, differentiation.

The roots are diarch.

Vegetative propagation takes place by the death and decay of older portion of rhizome thus setting free the branches that grow as independent plants.

REPRODUCTION

(a) Sporophylls and Sori

There is no distinction between sterile and fertile leaves. The sori arise on the ventral surface of the sporophyll (Fig.16.1, B) or the fertile leaf and are borne at the vein ends. The sori are arranged in a single row on either side of the midrib (Fig. 16.1, B). They are usually spherical and may be small or large in size and are not protected by any indusium (Fig. 16.3, A). The sori have a distinct receptacle that bears mixed sporangia. The sporangia may be mixed with hair or peltate scales (***P.lineare***) that offer protection to the exindusiate sorus.

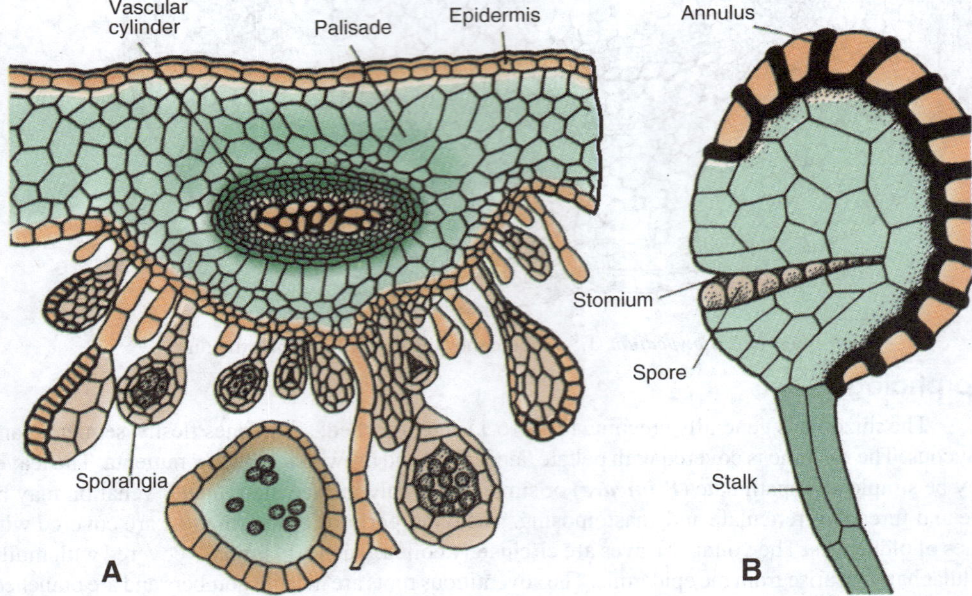

Fig. 16.3. (A—B). *Polypodium sp.* A. V.S. through sorus. There is no indusium; B. A sporangium.

(b) Sporangia

The structure (Fig. 16.3, B) and development (Fig. 16.3, A and 8.9) of the sporangium is leptosporangiate type and is exctly similar to that of ***Dryopteris.*** Each sporangium produces 48 dark brown, big, elliptic (***P. lineare, P vulgare, P. pteropus***) with flattened one side and thin-walled spores. The size of the spore is ususlly 0.057 × .041μ in ***P. lineare*** (Singh, 1963).

P. pteropus

Gametophyte

The development and structure of prothallus and sex organs is exactly similar to that of ***Dryopteris.***

The development of embryo is also identical to other polypodiaceous ferns.

Taxonomic Position

Division	:	Filicophyta
Class	:	Leptosporangiopsida
Order	:	Filicales
Family	:	Polypodiaceae
Sub family	:	Polypodioideae
Genus	:	***Polypodium***
Species	:	***vulgare***

Sub-Family Gymnogrammeoideae

The sub-family includes abpout 27 genera that are of varied habits and habitats and possess uncertain interrelationship. Some of them come quite close to the pteridoideae. The sporophytes are usually terrestrial except for ***Ceratopteris*** Brongiart which is aquatic. The plants vary in their height from a few centimetres (***Anogramma leptophylla***) to a few feet. The rhizome may be upright or creeping and is sparingly branched. It is covered with scales or ramenta.

Ceratopteris

Anogramma leptophylla

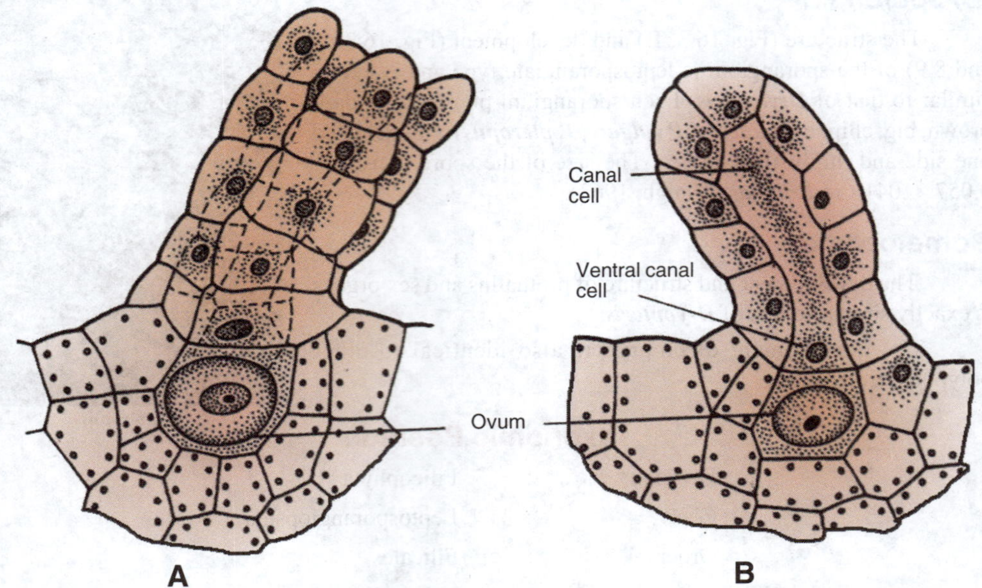

Fig. 16.4. (A—B). ***Polypodium vulagare.*** **A.** Closed and mature archegonium; **B.** Archegonium with open neck (After Strasburger)

The leaves are arranged spirally or alternate and may be simple or compound. In ***Craspedodictyum*** Copeland and ***Bommeria*** Fournier, the leaves are palmately compound. In ***Syngramma*** J. Smith, the leaves are simple, lanceolate or ovate with the free ends of the furcate veins united by a continuous, intramarginal veins. In ***Adiantum***, ***Coniogramma Pityrogramma*** and ***Cheilanthes*** the leaves are decompound. The sori are found on the ventral surface of the fertile leaves and are superficial, without true indusia, but protected by false indusia (***Adiantum***, ***Pellaea***). The sori are either seated on the vein endings (***Adiantum***), sometimes confluent into an uninterrupted (***Pellaea***, ***Cheilanthes***) or interrupted sori that are without a connecting commissure. In ***Onychium*** the connecting commissure is present. Such a commissure is a single intramarginal vein that connects free vein endings. This vein passes through receptacle of the sorus. ***Ceratopteris*** is an interesting member of this family (some include it in a separate family called ***Parkeriaceae***) in which the sporangia occur singly at the vein ends and are spherical and subsessile. The fronds are dimorphic. In ***Aspleniopsis***, ***Syngramma***, ***Coniogramma*** the sporangia are mixed with paraphyses. Paraphyses are absent in other genera, ***e.g.***, ***Gymnopteris.*** Linear and oblong sori that occur along the veins are also known in some ***e.g.***, ***Pterozonium*** Fee, ***Coniogramme*** Fee. Sporangia have a mixed arrangement on the receptacle and are shortly stalked or sessile. The spores are tetrahedral and globose. They lack perispore. The gametophytes and sex organs are similar in development and structure to other polypodiaceae.

The stem anatomy shows variations and may be siphonostelic, solenostelic or even dictyostelic. In ***Adiantum*** all the three types of steles are found at various levels on the same stem.

Some of them are epiphytic, a large number grow under moist and shaded localities and a few are xerophytic and grow on exposed place, ***e.g.***, ***Cheilanthes albomarginata. Ceratopteris*** is aquatic.

Christensen (1938) divided the sub-family in 5 tribes. A brief information about ***Adiantum*** L., which belongs to the tribe Adianteae is given here.

Genus–ADIANTUM L.

Adiantum commonly known as ***maiden hair fern*** is a widely distributed genus with about 200 species that grow abundantly in the tropical and subtropical regions of the world and love moist and shaded localities. The common species met with in India are ***A. capillus-veneris*** L. (Fig. 16.5, A) ***A. pedatum*** L., ***A. incisum*** Forsk., ***A caudatum*** Linn., ***A. venustum*** Don., ***A lunulatum*** Burnhand., ***A. edgeworthii*** Hook. The Indian species occur all along the Himalayan range and form a conspicuous part of the fern flora. Nayar (1961) studied the morphology of 24 Indian species of ***Adiantum.***

A. capillus-veneris

Bommeria

Cheilanthes

Coniogramma

Gymnopteris

Pellaea

Onychium

Pterozonium

The Sporophyte : MORPHOLOGY

Rhizome

The rhizome in the genus ***Adiantum*** is covered with paleae and may be erect, ***e.g.***, ***A. caudatum***, ***A. edgeworthii*** and ***A. philippense***, or semi-erect ***e.g.***, ***A. pedatum***, or creeping as in ***A. capillus-veneris*** (Fig. 16.5, A), ***A. thalictroides***, ***A. pectinatum***, ***A. peruvianum*** etc. The rhizome is hard and brown in the erect and semi-erect species and light brown and soft in the creeping forms. The paleae covering the rhizome vary in their shape or outline. They may be lanceolate (***A. pedatum*** and

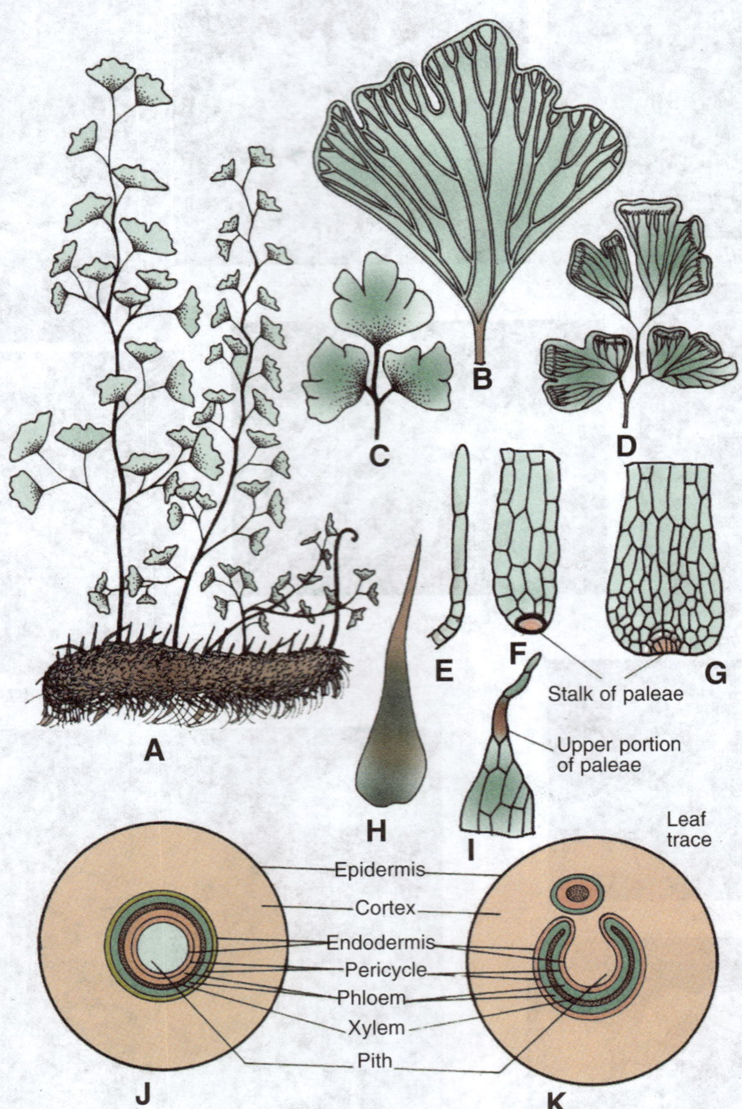

Fig. 16.5. (A—K). *Adiantum* A. A complete plant of *Adiantum capillus-veneris*; B.A pinnule showing venation; C. Sterile pinnules; D. Fertile pinnules as seen from ventral side; E-I. Stages in the development of palea (After Nayar and Kaur); J. T.S. rhizome of *A. rubellum* showing a siphonostele; K. Solenostele in *A. pedatum*.

A. philippense) to ovate lanceolate (***A. venustum***, ***A. caudatum***). The paleae are attached to the rhizome by a short and broad stalk (Fig. 16.5 E-I). The base of the paleae may be variously auricled on either side of the stalk. The apex of the paleae is usually long and drawn out (Fig. 16.5, H). Its margin may be smooth (***A. capillus-veneris***, ***A venustum***) or dentate (***A. pectinatum***). The paleae develop as unicellular outgrowths which undergo transverse division to form a 5-10 cells high uniseriate hair (Fig. 16.5, E). The cells of this hair are longer towards the apex. The cells at the base of this hair divide longitudinally thus making it flat and multiseriate. Soon it assumes a lanceolate appearance. The basal cell develops into a short stalk (Fig. 16.5, G).

Leaf

Adiantum has characteristic leaves that are, as usual, circinately coiled in the bud condition. The leaf has a dark brown to black and shining petiole which is traversed by a median groove devoid of the marginal ridges, on the adaxial side. The petiole is covered with paleae that may extend into the rachis and the leaflets (pinnae) as in ***A. caudatum.*** In some species the paleae are restricted to the base of the petiole only. Nayar (1961) has reported the presence of multicellular nonglandular hair all over the petiole and the rachis in some species. In some species such hairs are restricted to the adaxial groove. The base of the petiole is continuous with the cortex of the rhizome (Ching, 1957; Nayar, 1961). The petiole and the rachis are hard in all the 24 species found in India.

The leaves may be one pinnate, ***e.g.***, ***A. caudatum***, ***A. edgeworthii*** and ***A. philippense***; two to more pinnate in ***A. capillus-veneris*** (Fig.16.5, A), ***A. thalictroides***, ***A. pectinatum***, ***A. pedatum*** and others. In ***A. caudatum*** the unipinnate leaves bear pinnae in alternate manner on either side of the strong rachis. The pinnae are stalked. The rachis may terminate in a pinna or may be elongated bearing a vegetative bud at the tip. Under favourable conditions this vegetative bud may develop into a daughter plant (when the leaf tip touches the ground) which in turn may, repeat the process thus leading to a well known 'walking habit'. In the bi-or tripinnate species the leaflets are called the pinnules. In ***A. pedatum*** the petiole forks at the tip forming a pair of diverging rachises. Each of these two rachises further branches on the side facing each other. These branches bear terminal and lateral pinnules. In ***A. capillus veneris*** the rachis divides pinnately and the ultimate branches bear pinnules in an alternate manner. Each branch terminates in a pinnule. The terminal pinnule usually differs in shape and size from the lateral pinnules. The venation is free and dichotomous in all species (Fig. 16.5, B). The veins spread in a fan-like manner in the lamina.

Root

The roots are adventitious and arise in clusters from the underside of the rhizome, in the creeping forms (Fig. 16.5, A) and from the base of the rhizome in the errect forms. They are black and wiry structures and are branched.

ANATOMY

A. Rhizome

In a transverse section the rhizome presents an irregular or wavy outline (Fig. 16.7). It shows the following tissue systems:—

Epidermis. It forms the outermost layer of thin-walled or slightly thick-walled cells. The cells are generally smaller in size and possess brownish walls. Externally the epidermal layer is lined with a thick cuticle.

Cortex or Ground Tissue. Next to the epidermis lies the extensive ground tissue that may be parenchymatous or partly parenchymatous and partly sclerenchymatous. In species (***A. peruvianum***, ***A. trapeziforme***, ***A pectinatum***) scattered group of sclerenchymatous cells occur in the ground tissue.

In *A. pedatum* and *A. caudatum* there is a distinct sclerenchymatous hypodermis below the epidermis. Next to it is the parenchymatous ground tissue whose cells may be small or large and enclose small intercellular spaces. In some species the entire ground tissue is composed of small parenchymatous cells with brown walls and there is no well defined sclerenchymma.

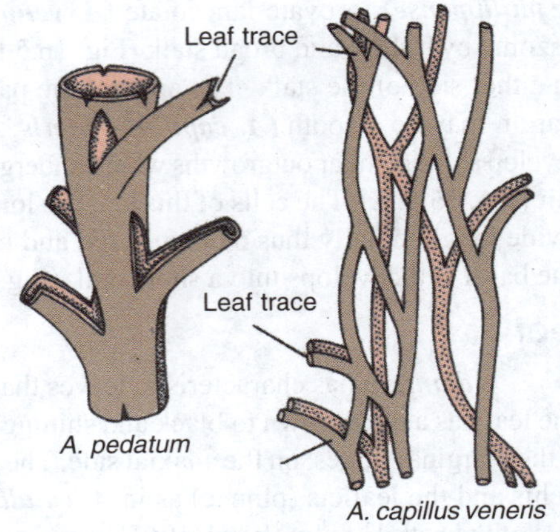

Fig. 16.6. Stereodiagrams of the Stelar organization in *A. pedatum* and *A. capillus-veneris.* (After Nayar)

Stele

The stelar organisation in ***Adiantum*** varies in the species of ***Adiantum.*** Nayar (1961) studied the anatomy of 24 species of ***Adiantum*** and gave a vivid description of the variations in the stelar system of the rhizome.

In ***Adiantum rubellum*** the stelar cylinder is **amphiphloic siphonostelic** (Fig. 16.5, J) at certain regions ***i.e.***, there is a complete and uninterrupted stelar cylinder with outer endodermis enclosing outer pericycle, outer phloem, xylem, inner phloem, inner pericycle and inner endodermis. In the centre there is a pith. At places where the **leaf trace** departs a **leaf gap** is left behind and the intact siphonostelic cylinder is broken at one point, thus changing to a **solenostele** (Fig 16.5, K). In ***A. pedatum***, ***A. flabellulatum***, ***A. hispidulum***, ***A. venustum***, ***A. nobile***, ***A. pectinatum*** and many others the stelar cylinder is solenostelic.

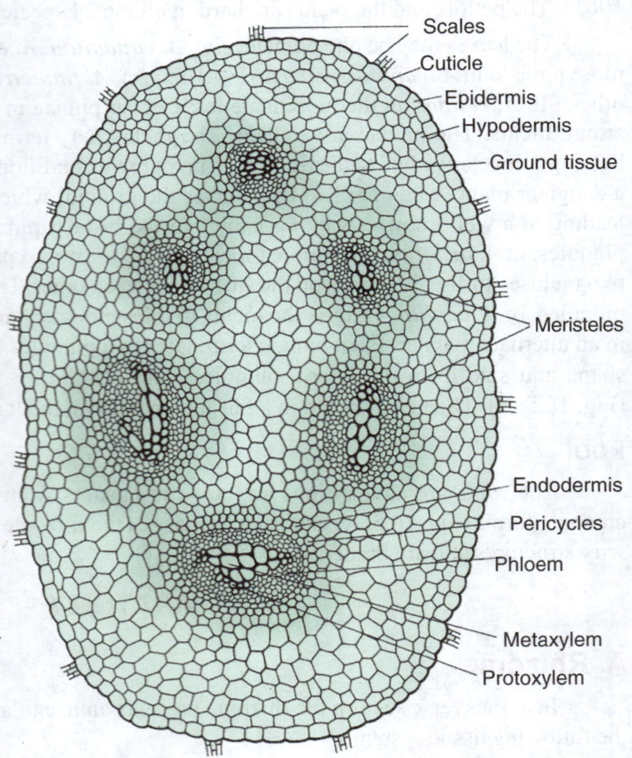

Fig. 16.7. *Adiantum capillus-veneris* L. T.S. rhizome showing a dictyostele with six meristeles.

In ***A. capillus-veneris*** (Fig.16.7), ***A. thalictroides***, ***A. caudatum***, ***A edgeworthii*** and ***A. philippense*** stelar cylinder is **dictyostelic.** It is dissected by many closely placed and overlapping leaf gaps arranged in one or more close spirals. In ***A. capillus-veneris*** the stelar cylinder in the young plants is **solenostelic** because the leaf gaps or lacunae are small. In the older plants the gaps expand excessively and cut the stelar cylinder into a reticulum with long meshes (Nayar, 1962).

The root traces arise indiscriminately from the outer surface of the stelar cylinder and have no connection with the leaves or branches. They are diarch and acquire a sclerenchymatous sheath in the outer cortex of rhizome.

B. Leaf

Petiole

A transverse section of the petiole is almost circular with distinct adaxial groove (Fig. 16.8). Next to the single layered epidermis is one to may layered hypodermis (Fig. 16.6), next to which lies the ground tissue. Number of vascular strands entering the petiole depends upon the number of traces given off by the stele to each leaf (Nayar, 1962). In A. ***caudatum***, ***A. edgeworthii*** and ***A. phillipense*** only a single gutter-shaped vascular strand enters the petiole. In ***A. capillus-veneris*** and many other species two band-shaped vascular strands enter the petiole (Fig. 16.8, A). During their course upwards they unite and form one bundle (Fig. 16.8, B). The xylem bands of the vascular bundles are often slightly curved with the concavity facing outwards and is thickest in the middle, becoming gradually thin towards either margin where the protoxylem is loacted (exarch). In ***A. bausei*** a small patch of included parenchyma is present in the xylem band. The ends of the xylem band may become hooked in some species (A. ***peruvianum, A. tenerum*** and ***A. polyphyllum***).

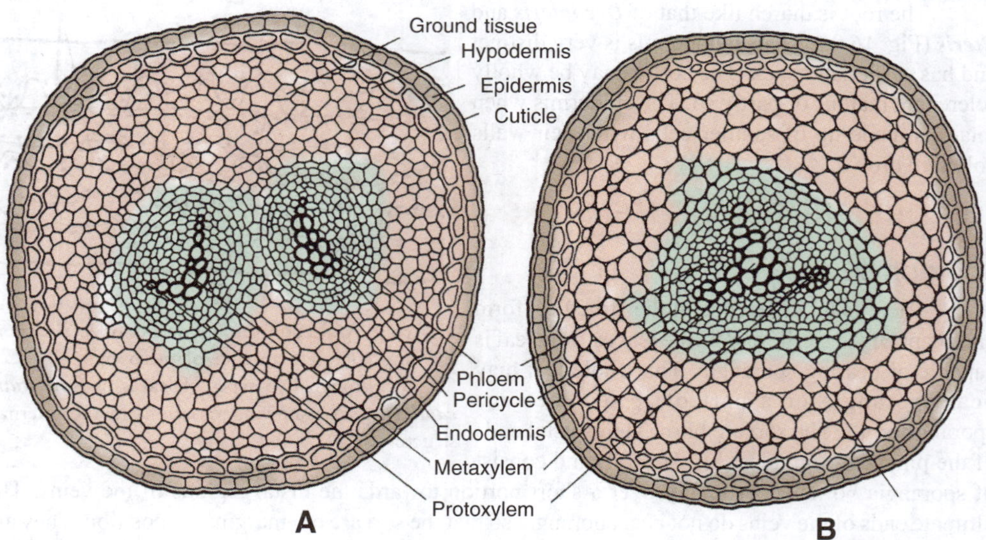

Fig. 16.8. (A—B). *Adiantum capillus-veneris L.* T.S. through the base of the petiole showing two meristeles; B. T.S. through the petiole a little above the base showing a single meristele which is three angled. The two meristeles have fused together to form one trace.

Lamina

The lamina exhibits an internal structure usual for the mesophytic leaves. It has the two epidermal layers bounding an undifferentiated mesophyll whose cells are armed. The epidermal layers are covered by a thin cuticle. In some species, ***e.g., A. capillus-veneris***, ***A. thalictroides*** and ***A. pedatum*** there is only one or two layered mesophyll (Fig. 16.9A, B, C). In ***A. pedatum*** the mesophyll is entirely lacking in some regions of the lamina and at such places the two epidermal layers are contiguous. Both the upper and the lower epidermis are usually chlorophyllous. The epidermal cells over the veins are nonchlorophyllous, highly elongated and narrow with tapering ends. The walls of these cells are hyaline and pronouncedly thickened. In ***A. macrophyllum*** groups of thick-walled cells occur scattered throughout the upper epidermis. The stomata are irregularly distributed thorughout the surface of the leaf.

In *A. flabellulatum* the stomata are aggregated to areas near the veins. Both uniseriate and multiseriate hair have been recorded on the epidermis. In *A. caudatum* and *A. hispidulum* the epidermis bears small paleae or ramenta. In species with thin lamina the veins are not surrounded by any bundle sheath (*A. pedatum*). In species with thick lamina the veins are surrounded by a sheath of sclerenchyma which extends towards either epidermis to merge with the specialised epidermal cells over the veins (Nayar, 1962), *e.g., A. macrophyllum.* The phloem in the vascular bundle of the vein lies towards the lower epidermis and xylem towards the upper epidermis.

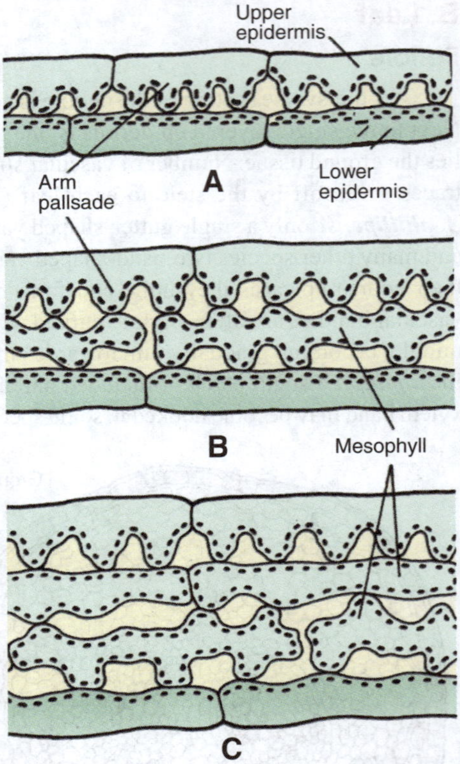

Root

The root is diarch like that of *Dryopteris* and *Pteris* (Fig. 16.9.a). The endodermis is very distinct and has casparian bands. The cortex may be wholly sclerenchymatous or partly so. The epidermis when intact is made up of smaller cells with their walls coloured brown.

Fig. 16.9. A. Section through leaflet of *A. pedatum* showing complete absence of mesophyll; (B—C). Section through two types of leaves of *Adiantum capillus-veneris* showing variations in the internal structure.

SPORE PRODUCING MEMBERS

Sporophylls

There is no difference in the structure, form and shape of the sterile and fertile leaves. Any leaf is capable of bearing sporangia. The whole leaf may bear sporangia or a part of it may become sporangiferous. The sori are borne at the distal ends of the pinnae or pinnules (Fig 16.5, D) and consist of sporangia borne superficially over a short portion towards the distal regions of the veins. The ultimate ends of the veins do not bear sporangia so that the sori are not marginal in position. They are submarginal. The complete sporangia bearing region of the leaflet or the pinnule folds downwards and usually loses chlorophyll during the course of its development. It forms the **false indusium.** The sporangia usually develop on the infolded portion of the pinnule but may in some cases spread slightly on the surface covered by the folding (Nayar, 1962). The sporangia may develop at the distal ends of all the veins of a pinnule (*A. philippense*) and form a continuous sorous or the sporangia develop only at the distal ends of some veins in which case small separate sori are formed. In the sori of *A. tenerum*, *A. bausei* and *A. rubellum* the sporangia are interspersed with small paraphyses (Nayar, 1962). The paraphyses are small uniseriate hair-like structures with a one or two-celled stalks bearing a globular cell with dense hyaline contents.

Sporangia (Fig. 16.10)

The sporangium has two well defined regions : (*i*) the **stalk** and (*ii*) the **head** or the **capsule** (Fig. 16.10 A, B) The stalk is made up of three rows of cells, each row about 4 cells long (Nayar, 1962). The sporangial stalk is primarily made up of two rows of cells, the third row is formed secondarily during sporangial development from one of the basal cells of the sporangial head. The head or the capsule is shaped like a biconvex lens (Fig. 16.10). It has a single layered wall enclosing the spores. The wall has a distinct vertical **annulus** of 12-24 cells long. The annulus is separated from the stalk by one or two ordinary thin-walled cells. The **stomium** is also not continuous with the annulus and the stalk. From the annulus it is separated by two (***A. capillus-veneris***) to six (***A. thalictroides***) wall cells, whereas from the stalk it is separated by two or three wall cells. The rest of the sporangial wall is composed of a few large cells that have a regular outline. The sporangial stalk lacks water gland.

The development of the sporangium is exactly similar to that of ***Dryopteris.***

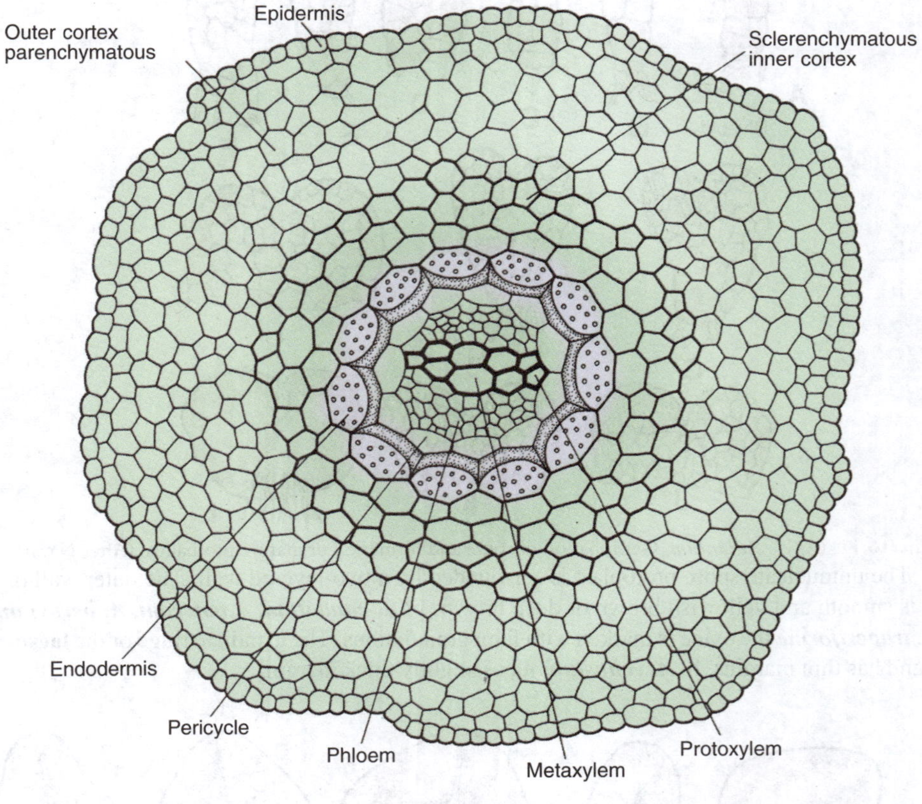

Fig. 16.9a. **Adiantum** T.S. root.

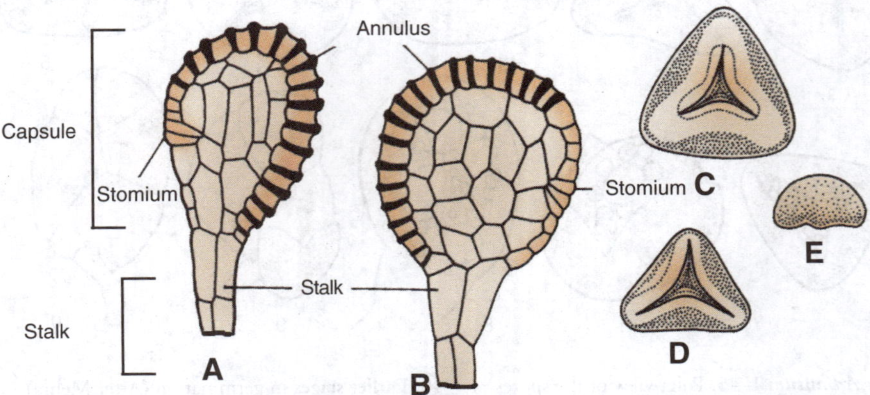

Fig. 16.10. Sporangia (A–B) and spores (C–E) of *Adiantum.* (After Nayar)

GAMETOPHYTE

Spores

The spores are the pioneer structure of the gametophyte generation. The spores in ***Adiantum*** are tetrahedral (Fig. 16.11, 1-5) and have a triangular and with concave sides.

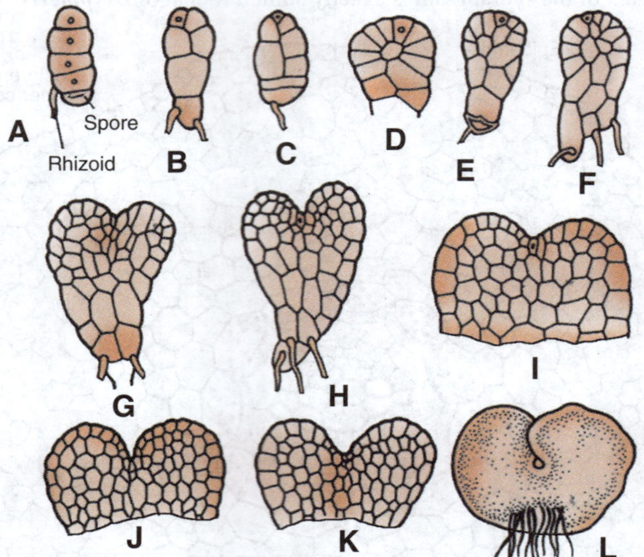

Fig. 16.11. (A–L). *Adiantum.* Germination of spore and formation of mature prothallus. (After Nayar)

The uninucleate spore protoplast is surrounded by a two-layered wall. The outer wall or the **exine** is smooth and yellowish brown or deep brown. In ***A. caudatum***, ***A.tenerum***, ***A. peruvianum*** and ***A. trapeziforme*** the exine is marked with faint granulations. The triradiate ridge or the **laesura** is short and has thin margins. In ***A. venustum*** it has slightly thick margin.

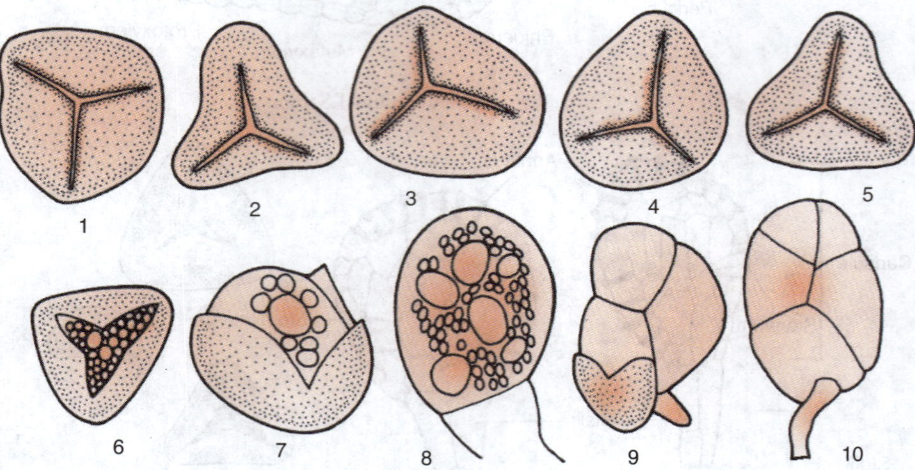

Fig. 16.11. (*a*) *Adiantum* 1—5. Polar view of the spores ; 6—10. Earlier stages in germination (After Mehra)

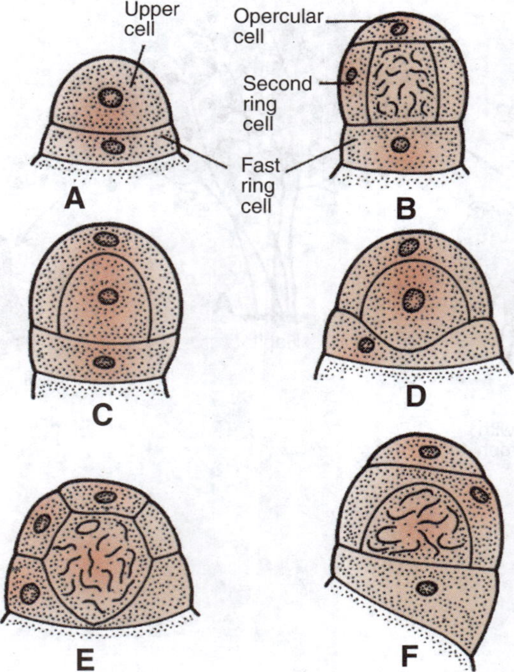

Fig. 16.12. (A–F). Stages in the development of antheridium in ***Adiantum lunulanum.***

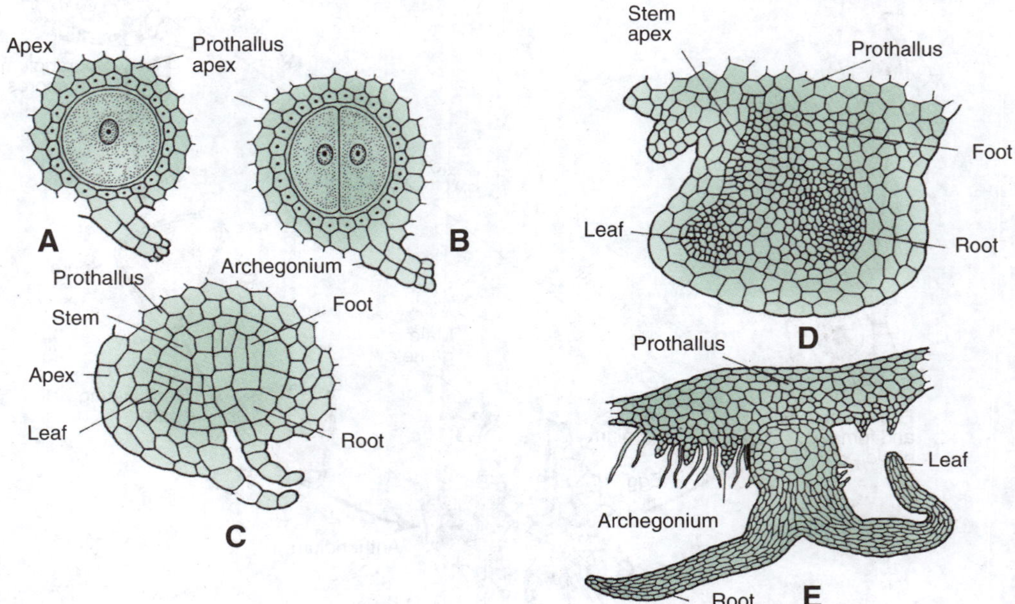

Fig. 16.13. (A–E). Various stages in the development of embryo in ***Adiantum concinnum*** (A–C) and ***A. capillus-veneris*** (D–E).

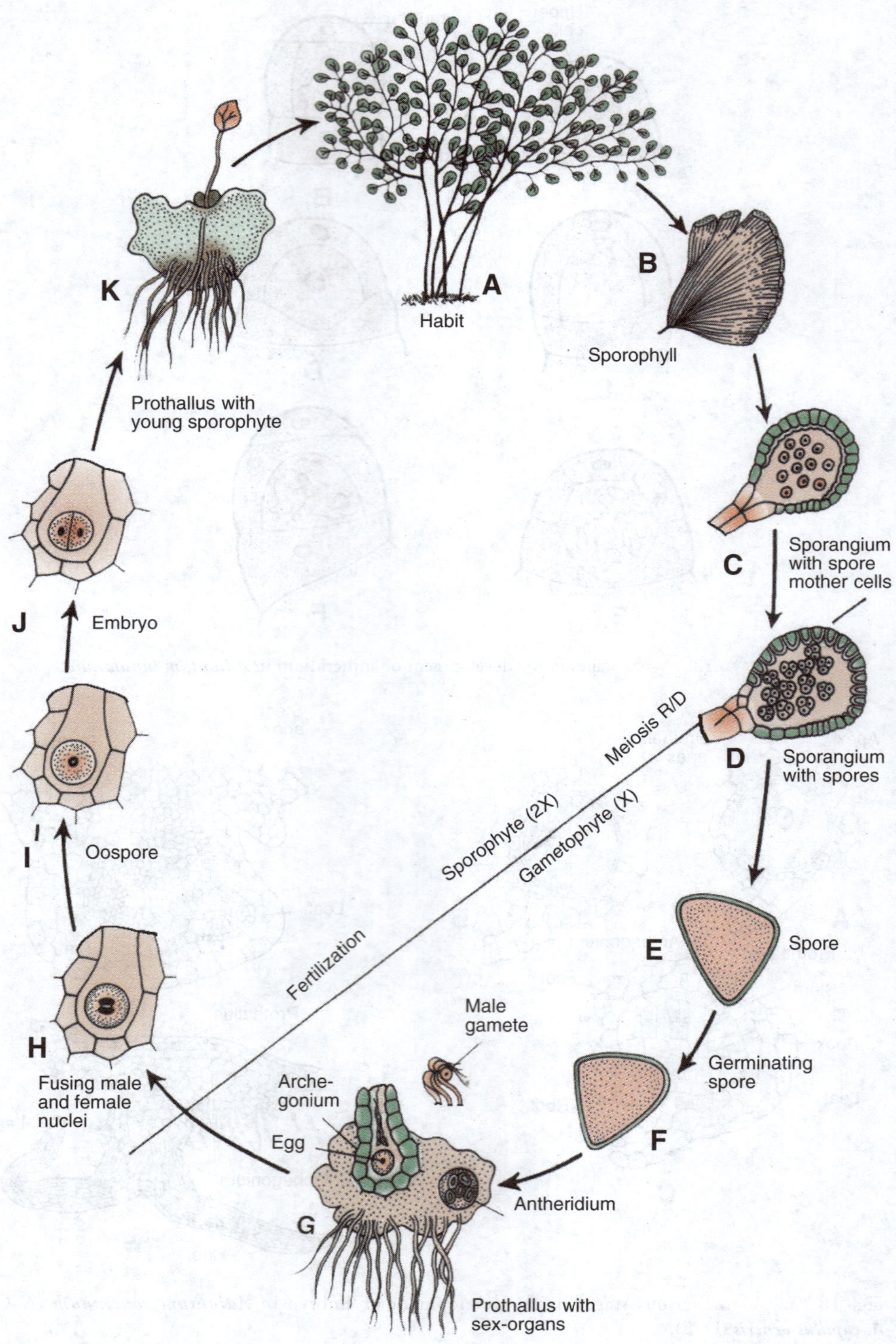

Fig. 16.14. Life-cycle of *Adiantum*

Germination of the spore

The spores germinate on falling on a suitable substratum. The exine ruptures at the triradiate ridge and the intine protrudes out in the form of a germ tube. It encloses the protoplasts. A lateral rhizoid arises from the germ tube. The germ tube becomes a filament of 4-6 cells long (Fig. 16.11 *a*, A,B). The cells develop chloroplasts and become green. The cells are barrel-shaped. The terminal cell of the filament divides by two oblique walls to cut off a three-sided meristematic (apical) cell. The lower cells of the germ filament may divide longitudinally, either all of them (***A. bausei***, ***A. tenerum*** and ***A. vogeyeria***), or only a few upper ones lying below the apical cell. The apical meristematic cell divides along the three faces and leads to the formation of a spathulate plate of cells (Fig. 16.11 *a*, E, F). During further divisions the apical cell becomes narrow or almost spindle-shaped (Fig. 16.11 *a*, E, F). It persists till the prothallus assumes its characteristic cordate form (Fig. 16.11*a*, G, H). Later it is replaced by an apical meristem of two cells lying in a deep apical notch (Fig. 16.11*a*, J, K.). The cells behind the apical notch divide in horizontal plane to form a thick cushion. The wings of the prothallus are only one layer of cells in thickness and are usually parenchymatous. In some species (Nayar, 1962) collenchymatous cells have been noticed at the corners.

The structure and development of an antheridium (Fig. 16.12) and archegonium is similar to those of ***Dryopteris.*** The development of embryo also follows the same course as of ***Dryopteris.*** There are minor differences as illustrated in Fig. 16.13, A-E.

Normally the prothallus bears only one sporophyte but two or more sporophytes may develop under cultural conditions (Nayar, 1962). Apogamy has been reported in ***A. philippense*** (Mehra, 1949).

Taxonomic Position

Division	:	Filicophyta
Class	:	Leptosporangiopsida
Order	:	Filicales
Family	:	Polypodiaceae
Sub family	:	Gymnogrammeoideae
Tribe	:	Adiantae
Genus	:	***Adiantum***
Species	:	***rubellum***.

QUESTIONS

ESSAY TYPE

1. Describe the life cycle of ***Adiantum*** in short.
2. Differentiate between the internal structure of ***Nephrolepis*** (***Pteris***) and ***Adiantum.***
3. With the help of labelled diagrams, describe the structure of gametophyte of any fern studied by you.
4. Give a comparative account of the prothallus of ***Equisetum*** and ***Adiantum.***
5. Compare the gametophyte of ***Lycopodium*** and ***Adiantum.*** Give the diagrams wherever necessary.
6. Explain the life cycle of ***Adiantum*** by labelled diagrams only. Classify ***Adiantum***.
7. Write detailed note on Phylogenetic relationship between Adiantum and Gleichemaceae.
8. Write on essay on evolution of sorus in ferns
9. Give on illustrated account of the morphology of the sporophyte in ***Polypodium.***
10. By labelled diagrams only, describe the life-cycle of ***Polypodium.***
11. With the help of diagrams, describe the gametophyte of ***Polypodium.***
12. What is alternation of generations? Discuss it with the help of the life-cycle of ***Polypodium.***

13. Give an account of internal structure of the rhizomes of ***Polypodium*** and ***Adiantum***.

14. Explain the structure and function of ***Adiantum.***

15. Compare the prothallus and sex organs of ***Equisetum*** and ***Adiantum***

16. Compare the asexual spores of ***Selaginella*** with that of ***Adiantum***

17. What is stele? Name and describe the structure of stele present in rhizome of ***Adiantum*** and ***Equisetum.***

18. Compare the Spore producing organs of ***Selaginella*** with that of ***Adiantum.***

SHORT ANSWER TYPE

19. Briefly describe the ***sporangium*** of ***Adiantum*** with a labelled diagram.

20. Describe in brief the maiden hair fern. **(Lucknow, 2003)**

21. Name the genus called maiden hair fern and why? **(Himachal Pradesh, 2000)**

22. Draw labelled diagram of the following :
    (*i*) T. S. petiole of ***Adiantum*** **(Lucknow, 2001, 2003, 2005)**
    (*ii*) Fertile pinna of ***Nephrolepis*** **(Lucknow, 2003)**
    (*iii*) T. S. Rhizome of ***Adiantum*** **(Himachal Pradesh, 2000)**
    (*iv*) T. S. ***Adiantum*** root **(Garhwal, 2000)**
    (*v*) T. S. ***Adiantum*** leaf through spores,

23. Write short notes on
    (*a*) Mature prothallus of ***Adiantum.*** **(Himachal Pradesh, 2000)**
    (*b*) Indusium
    (*c*) Sporangium of ***Adiantum***
    (*d*) Sorus of ***Adiantum***
    (*e*) Venation in Ferns.

24. Filamentous Gametophytes are reduced rather than primitive. Why?

25. Write the classification of ***Adiantum.***

26. Describe the structure of antheridium in a fern you have studied. **(Kanpur, 2001)**

27. Draw labelled sketches of the spore bearing organs of ***Selaginella***, ***Equietum*** and ***Adiantum.***

28. Compare the spore bearing organs of ***Lycopodium, Selaginella and Adiantum.***

29. Distinguish between
    (*i*) Sori of ***Adiantum*** and ***Pteridium.***
    (*ii*) Sori of ***Polypodium*** and ***Adiantum***
    (*iii*) Root of ***Pteris*** and ***Adiantum***

30. Describe the structure of antheridium in ***Polypodium.***

OBJECTIVE TYPE

31. Fill in the blanks :
    (*i*) Ramenta is found in—of—.
    (*ii*) Spores of pteridophytes are—.
    (*iii*) False indusium is found in—.
    (*iv*) ***Adiantum*** belongs to subfamily—and family—.
    (*v*) ***Polypodium*** belongs to subfamily—and family—.
    (*vi*) ***Polypodium*** is an—.
    (*vii*) ***Adiantum*** is also known as—.
    (*viii*) In ***Polypodium*** the sori are borne—.
    (*ix*) Sori are borne on ventral side of the leaf in—
    (*x*) In ***Adiantum***, the cortex in the root is wholly or partly—.

32. Select the correct answer :
    (*i*) Most of the species of ***Polypodium*** are
    (*a*) epiphytes
    (*b*) halophytes
    (*c*) mesophytes
    (*d*) xerophytes.

    (*ii*) Leaves of ***Polypodium*** are
    (*a*) simple
    (*b*) compound
    (*c*) simple and spathulate or deeply incised and pinnate
    (*d*) absent

    (*iii*) Sori are not protected by indusium in
    (*a*) ***Lycopodium***
    (*b*) ***Dryopteris***
    (*c*) ***Pteris***
    (*d*) ***Polypodium.***

    (*iv*) Maiden hair fern is the name given to
    (*a*) ***Pteris***
    (*b*) ***Adiantum***
    (*c*) ***Polypodium***
    (*d*) ***Dryopteris.***

(*v*) Palae covers the rhizome in
(*a*) ***Dryopteris***
(*b*) ***Polypodium***
(*c*) ***Pteris***
(*d*) ***Adiantum.***

(*vi*) Amphiphloic siphonostele is present in the rhizome of
(*a*) ***Adiantum rubellum***
(*b*) ***A. pedatum***
(*c*) ***A. nobile***
(*d*) ***A. pectinatum.***

(*vii*) Stelar organisation varies in the rhizome of the species of
(*a*) ***Dryopteris***
(*b*) ***Pteris***
(*c*) ***Polypodium***
(*d*) ***Adiantum.***

(*viii*) Mesophyll is completely absent in the leaves of
(*a*) ***A. nobile***
(*b*) ***A. pectinatum***
(*c*) ***A. pedatum***
(*d*) ***A. caudatum.***

(*ix*) In ***Adiantum*** the position of the sori is
(*a*) laternal
(*b*) terminal
(*c*) axial
(*d*) basal.

(*x*) The cortex in the root is wholy or partly Sclerenchymatous in.
(*a*) ***Pteris***
(*b*) ***Polypodium***
(*c*) ***Adiantum***
(*d*) ***Pteridium.***

Hymenophyllum tunbridgense

CHAPTER 17

Hymenophyllaceae

- Characteristics and a brief account of family **Hymenophyllaceae**
- Characteristics and a brief account of family **Matoniaceae**
- Characteristics and a brief account of family **Cyatheaceae**
- Characteristics and a brief account of family **Dicksoniaceae**

Characteristics and a Brief account of Family–HYMENOPHYLLACEAE

They are also known as the 'Filmy ferns'. The family includes about 650 species that were previously included under two genera, namely, ***Hymenophyllum*** (with 300 species) and ***Trichomanes*** (with 350 species). Copeland (1938) recognised 33 genera. In 1947, Copeland described a new genus ***Rosenstokia*** thus raising the number to 34. ***Hymenophyllum, Trichomanes, Didymoglossum, Crepidomanes*** and ***Cardiomanes*** are the common genera. Holttum (1949) recognised only two genera ***Hymenophyllum*** and ***Trichomanes.*** Smith (1955) recognised three genera (***Hymenophyllum, Trichomanes*** and ***Cardiomanes***). Those species in which the sorus is protected by a bilabiate indusium are put under the genus ***Hymenophyllum,*** those with a cup-like indusium are put under genus ***Trichomanes. Cardiomanes*** includes only one species (***C. reniforme***), that has characteristic kidney-shaped leaf lamina. A fossil genus ***Hymenophyllites*** has also been described from the upper Carboniferous. The number of genera can be increased if we take into consideration a number of other factors. Copeland (1938) has dwelt upon these characters in detail. Two common representatives of the family in India are ***Hymenophyllum exsertum*** (Fig. 17.1) and ***Didymogllossum insigne*** (= ***Trichomanes bipunctatum***) (Fig. 17.2).

Fig 17.1. *Hymenophyllum exsertum.* **A.** A portion of a plant showing habit; **B.** Part of a leaf showing marginal sori.

Hymenophyllum *Trichomanes* *Crepidomanes*

Cardiomanes *C. reniforme*

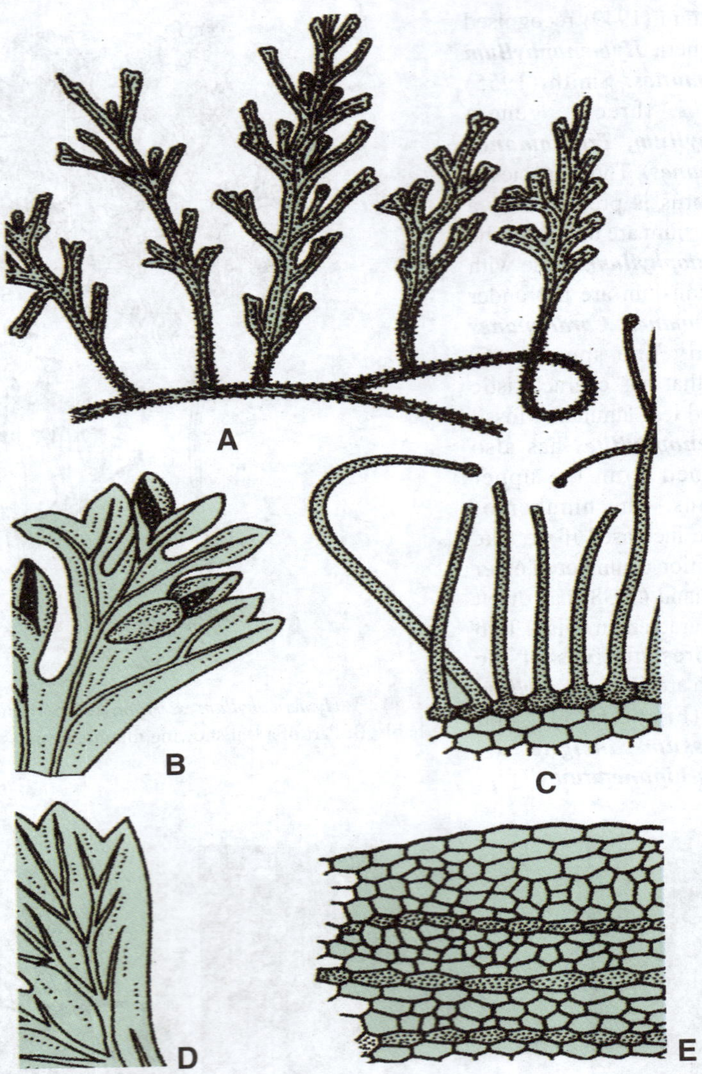

Fig. 17.2 (A—E). *Didymoglossum insigine.* A. Part of a plant showing habit; B. Part of a leaf showing marginal sori; C. Section of lamina showing long and unicelluar setae; D. Portion of lamina showing setae arising along the midrib; E. Section of the lamina showing false veinlets. (All after Bir)

The Hymenophyllaceae love to grow under moist and shaded localities and are found in abundance in the tropical rain forests. They grow as epiphytes or on dripping rocks (***Didymoglossum insigne***). A few species are terrestrial and able to tolerate dry conditions. Holloway (1923) described some species of ***Hymenophyllum*** that shrink during the drier months of the year and again recover during the rainy season.

The **morphological features** of the sporophyte and the gametophyte can be listed as under :

1. The rhizome is generally dorsiventral, creeping, narrow, long, subterranean and branched. It is covered with brown scales or setae (Fig. 17.2, A). The leaves are usually borne in two rows on the upper side of the rhizome. The roots are adventitious and may be arranged in two rows or are borne irregularly on the inner side of the rhizome. In some species the roots are absent. The roots

have no root caps . The fronds show repeated branching. The finer branches or the pinnules have a single vein. The venation is dichotomous and open type. The lamina has decurrent base e.g. ***Didymoglossum insigne*** (Fig. 17.2). A number of long, unicellular and dark brown setae arise from the epidermal surface of the lamina and along the midrib and veins (Fig. 17.2, C). The members of this family possess delicate leaves and most of them have one cell thick lamina. ***Didymoglossum*** and ***Crepidomanes*** possess false veinlets in the leaves (Fig. 17.2, D-E). These are series of elongated cells filled with brownish contents. Such false veins are absent in ***Trichomanes*** and ***Hymenophyllum. Cardiomanes reniforme*** has simple and kidney-shaped leaves. ***Hymenophyllum cruentum*** has also simple leaves with serrate margins. The simple leaves in ***Trichomanes goebelianum*** are 3 to 4 milimetres in length. This is the smallest filmy fern.

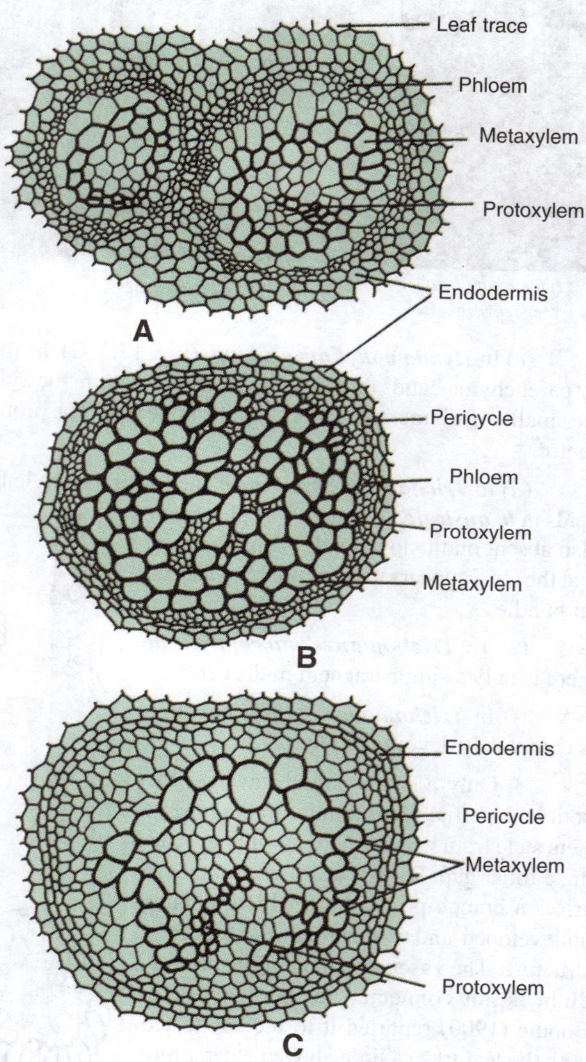

Fig. 17.3. Anatomy of rhizome in Hymenophyllaceae ; A. T.S. portion of rhizome of *Hymenophyllum dilatatum*; B. T.S. portion of rhizome of *Trichomanes scandens*; C. T.S. portion of rhizome of *H. scabium.* (After Boodle)

2. The **anatomy of the rhizome** reveals a single-layered epidermis, which is followed by a narrow sclerenchymatous, or partly sclerenchymatous and partly parenchymatous, or wholly parenchymatous cortex. The stele is a **protostele**. The leaf traces do not leave any foliar gap. The protostele of the stem shows considerable variations in structure. The stele is surrounded by a continuous layer of endodermis followed by one or more layers of pericycle. The phloem also completely surrounds the xylem. The following variations have been noted in various species :—

(*a*) In ***Hymenophyllum dilatatum*** and ***H. demissum*** the metaxylem forms a ring around a central parenchyma in which lie embedded few protoxylem tracheids (Fig. 17.3, A) so that the xylem is mesarch.

(*b*) In ***Trichomanes scandens*** (Fig. 17.3, B) the xylem forms a central core with a little parenchyma in the centre. The protoxylem forms peripheral groups (exarch). Phloem completely encircles the xylem.

Hymenophyllum cruentum *H. australe* *H. demissum*

(*c*) In ***Hymenophyllum scabium*** (Fig. 17.3, C) the metaxylem ring is interrupted or separated by parenchyma bands into two plates. In ***H. australe*** the xylem elements of the lower metaxylem arc are smaller in diameter than those of the upper arc. The protoxylem is represented by a tracheid in the centre.

(*d*) In ***Trichomanes muscoides*** the lower metaxylem arc, which is composed of smaller tracheids in ***h. australe,*** is absent. The phloem is also absent on the lower side with the result that the stele appears to be a collateral vascular bundle.

(*e*) In ***Trichomanes microphyllum*** there is only a single tracheid in the centre.

(*f*) In ***Trichomanes motleyi*** the stele is completely devoid of xylem.

3. Only a single strand traverses the petiole. Its structure is similar to the type of stem stele from which it departs without bearing a foliar gap. From the axil of a leaf there arises a branch primordium which remains undeveloped and is regarded as a vestigial structure. The vascular strand of the primordium is not connected to the stem stele. Boodle (1900) reported it to the connected with the leaf trace a little above its departure from the stem stele.

4. Majority of the species have a lamina or leaf blade composed of a single layer of cells. The region of the midrib is no doubt several cells thick and possesses a reduced vascular strand that is composed of a few xylem and phloem elements. In some cases the vascular tissue is absent and only **pseudoveins** are present. These are made up of elongated and brown cells. Such leaves are designated as **filmy type** leaves. In ***Hymenophyllum dilatatum*** the leaf lamina is made up of several cell layers. The cells have no intercellular spaces between them.

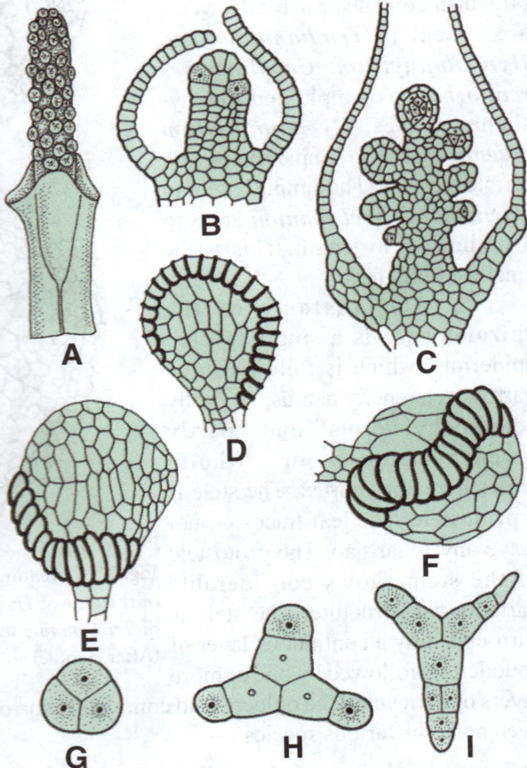

Fig. 17.4 (A—I). *Hymenophyllaceae.* A. Surface view of the sorus of *Trichomanes tenerum.* Note the elongated placenta bearing sporangia; B. L.S. sorus of *Trichomanes* sp. bearing receptacle with two sporangial initials; C. L.S. mature sorus of *Trichomanes* sp. showing basipetalous development of sporangia; D—F. Sporangia of *Hymenophyllum dilatatum*; G—I. Earlier stages of spore germination in *Hymenophyllum acanthoides.* (A, after Goebel; D—F, after Bower, G—I after Stokey).

5. The roots may be monarch or diarch.

6. The **sporangia** are grouped in distinct marginal sori. They develop on a marginal receptacle or placenta. The sori receptacle is usually elongated and is seated on a vein ending. The vein continues into the elongated receptacle. The receptacle in ***Trichomanes*** grows by means of an intercalary basal merisem and grows into a elongated bristle like process that extends beyond the leaf margin and the indusium (Fig. 17.4, A). Such a prominent growth of the receptacle is lacking in ***Hymenophyllum.*** In ***Didymoglossum insigne*** the receptacle forms a short bristle like process (Fig. 17.2, B). In ***Trichomanes*** the sporangia originate in a basipetalous succession (Fig. 17.4, B, C). This is due to the presence of a basal meristem. When mature these sporangia overlap each other. Such species have, therefore, gradate sori. Those species which lack a basal meristem or in which such a meristem is not very active, the sporangia develop simultaneously, e.g., ***Hymenophyllum.*** The sorus is surrounded by a well developed indusium. The indusium is cup-shaped in ***Trichomanes*** and ***Cardiomanes*** and bilabiate in ***Hymenophyllum, Didymoglossum*** (Fig. 17.5, B) and ***Crepidomanes.*** The bilabiate indusium consists of two flaps that cover the receptacle from both adaxial and abaxial faces. It develops along with receptacle and is one cell layer thick. The indusium covers and protects the sorus until the sporangia dehisce and liberate the spores.

7. The **development of the sporangium** is similar to that of ***Gleichenia.*** A mature sporangium has single-layered wall, two-layered tapetum and 32—512 spores. The annulus is obliquely vertical in position (Fig. 17.4, D—F, 17.5) and brings about the dehiscence of the sporangium. The line of dehiscence is lateral and the process of opening is very slow. It is followed by a sudden closing of the flaps. Such a method of dehiscence is also met with in advanced leptosporangiate ferns. In some species of ***Trichomanes*** (***T. pinnatum***) only 32 spores are produced. The sporangium has a short, few celled stalk, which is formed by the first few segments of the sporangial initial. The largest number of spores (512) are produced in ***H. tunbridgene,*** then follow ***Cardiomanes reniform*** (256), ***Hymenophyllum dilatatum*** (128), ***Trichomanes crispum*** (64), ***T. rigidum*** (32—64), ***T. spicatum*** (48) and ***T. pinnatum*** (32—48).

H. tunbridgense *Hymenophyllum dilatatum*

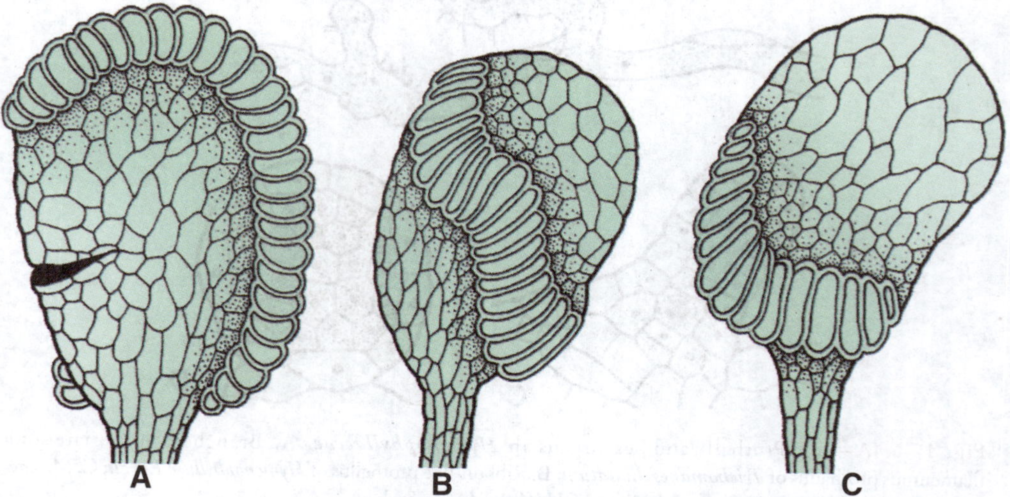

Fig. 17.5 (A—C). Sporangia of *Hymenophyllum* (After Bower)

8. Stokey (1940) and Holloway (1944) recognised two kinds of (Fig. 17.6, A, B) gametophytes in the Hymenophllaceae. These are :

(*a*) Irregularly branched and ribbon-like **prothalli** (Fig. 17.6, B) of ***Hymenophllum*** and ***Cardiomanes.*** These are one cell thick. Sometimes the margins of these strap-shaped prothalli may be more than one cell thick because they bear sex organs. Both types of sex organs are borne on the same prothallus. Such gametophytes grow by means of a distinct apical cell, but they never acquire a heart-shaped or cordate form. In some cases the margins of the prothallus bear sterigma like outgrowths that terminate in small multicellular gemmae. The gemmae act as means of vegetative propagation. The rhizoids are unicellular and arise from the margins. The sex organs are borne on lateral or near their margins and on the ventral side. Antheridia may also appear on the dorsal side (***Hymenophyllum***). The archegonia are borne together at localised places along the margins of the prothallus.

(*b*) In ***Trichomanes*** the prothalli are profusely branched and have uniseriate filaments (Fig. 17.6, A) that look like branched filaments of green algae. They are mycorrhizal and bear unicellular rhizoids. The antheridia arise directly on any lateral branch. The archegonia are produced on small lateral branches called the archegoniophores.

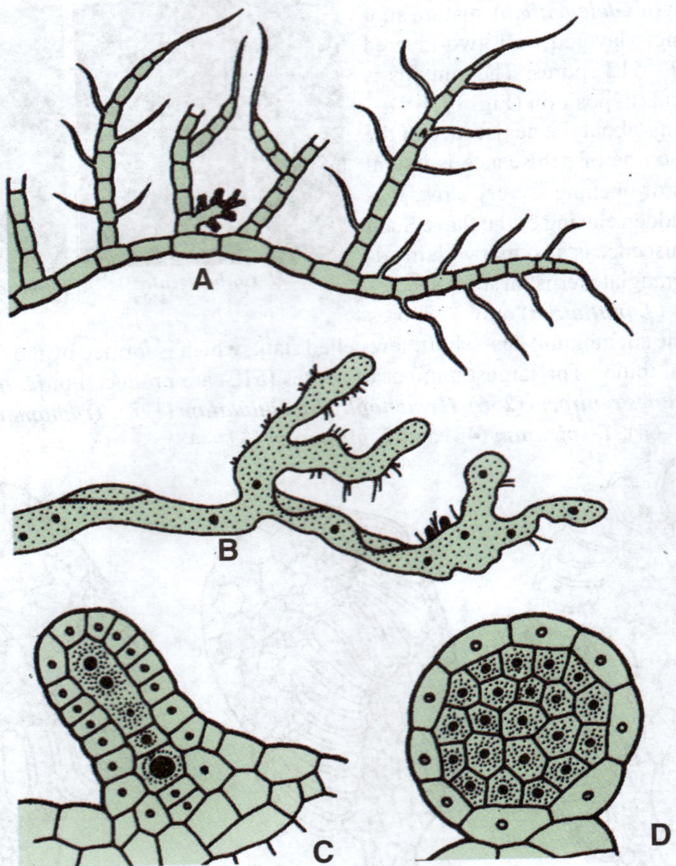

Fig. 17.6 (A—D). Prothalli and sex organs in *Hymenophyllaceae*. A. Branched, uniseriate and filamentous prothallus of *Trichomanes dilatatum*; B. Ribbon-like prothallus of *Hymenophyllum Kurzii*; C. Mature archegonium of *H. Kurzii*; D. Antheridium of *H. Kurzii*.(After Stokey)

9. The **development of the antheridia and archegonia** is similar to that of ***Gleichenia*** and ***Schizaea.*** The antheridia have a single layered jacket composed of many cells and with a distinct opercular cell. They produce about one hundred spermatozoids. The archegonia have 4 longitudinal rows of neck cells, each row 6-9 cells high. There is a binucleate neck canal cell.

10. **Development of embryo** has been studied in detail in ***Cardiomanes reniforme*** (Holloway, 1994). It differs in some respects from those of other leptosporangiate ferns. The first division of the zygote is at right angles to the long axis of the archegonium (Fig. 17.7, A) and not parallel as is the case in other leptosporangiate ferns. The cell towards the archegonial neck is the **epibasal cell**, the lower cell is the **hypobasal cell**. The two cells divide by walls at right angles to the first wall and form cells (Fig. 17.7, B). This is the quadrant stage. Further divisions are not regular and lead to the formation of a globular mass (Fig. 17.7, C) without any clear distinction of apical cells of the primary organs. Further divisions are not regular and lead to the formation of a globular mass (Fig. 17.7, C) without any clear distinction of apical cells of the primary organs. Further divisions lead to the appearance of initials of the first leaf and the primary root (Fig. 17.7, D) which bear no relation to the epibasal or the hypobasal region of the embryo. The shoot apex appears late of the base of the leaf. There is a well developed foot. Holloway (1944) stated that in some embryo it appeared as if the root, the leaf and the shoot are developed from the epibasal half, but he was not definite about this relation. The root originated endogenously. The first procambial strands appear between the root and the leaf. The shoot grows slowly and its procambial strands develop late and join with that of the root and the leaf.

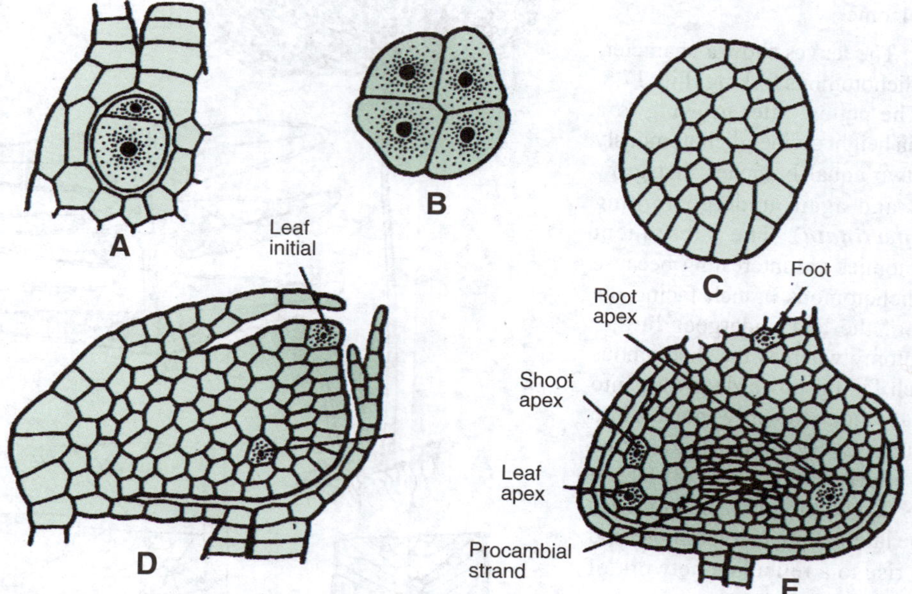

Fig. 17.7 (A—E). Stages in development of Embryo in *Cardiomanes reniforme.* (After Holloway)

Taxonomic Position

Division	:	Filicophyta
Class	:	Leptosporangiopsida
Order	:	Filicales
Family	:	Hymenophyllaceae
Genus	:	***Hymenophyllum***
Species	:	***exsertum***

Characteristics and a brief account of Family–MATONIACEAE

Matonia

The family is represented by two genera ***Matonia*** (Fig. 17.8) and ***Phanerosorus,*** each of which includes two species. ***Matonia*** is commonly found in Malaya, New Guinea and Borneo. ***Phanerosorus*** occurs only in Sarawak and New Guinea. The family has a restricted distribution now, but it was quite common and abundant in the Triassic. It became widespread and cosmopolitan during the Jurassic and Cretaceous. The fossil genus ***Matonidium*** (***M. wiesneri***) has been placed in this family because of its characteristic fronds. ***Laccopteris*** (***L. woodwardii***, ***L. polypodioides***) has also been referred to this family.

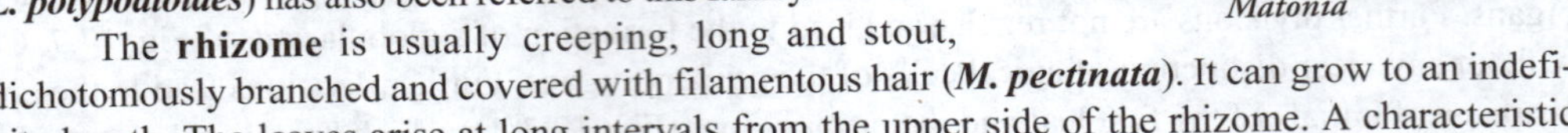

The **rhizome** is usually creeping, long and stout, dichotomously branched and covered with filamentous hair (***M. pectinata***). It can grow to an indefinite length. The leaves arise at long intervals from the upper side of the rhizome. A characteristic about ***Matonia pectinata*** is the occurrence of young and old leaves without any reference to stem tip i.e. they occur intermingled. The leaves may grow as high as 2 metres. The roots arise from the lower surface of the rhizome.

The **leaves** show a characteristic dichotomous forking (Fig. 17.8, A). The petiole after growing to a certain height divides dichotomously into two equal branches that grow apart and again undergo forking (***M. pectinata***). The subsequent dichotomies are interesting because the dichotomous branch facing the base of the leaf undergoes further dichotomy whereas the dichtomous branch facing upwards grows into pinna which is deeply incised into pinnules. Every pinnule is traversed by a midrib which gives off lateral veins that again fork. The branches of the lateral veins an astomose and give rise to a reticulate network of veins (Fig. 17.8, C, D).

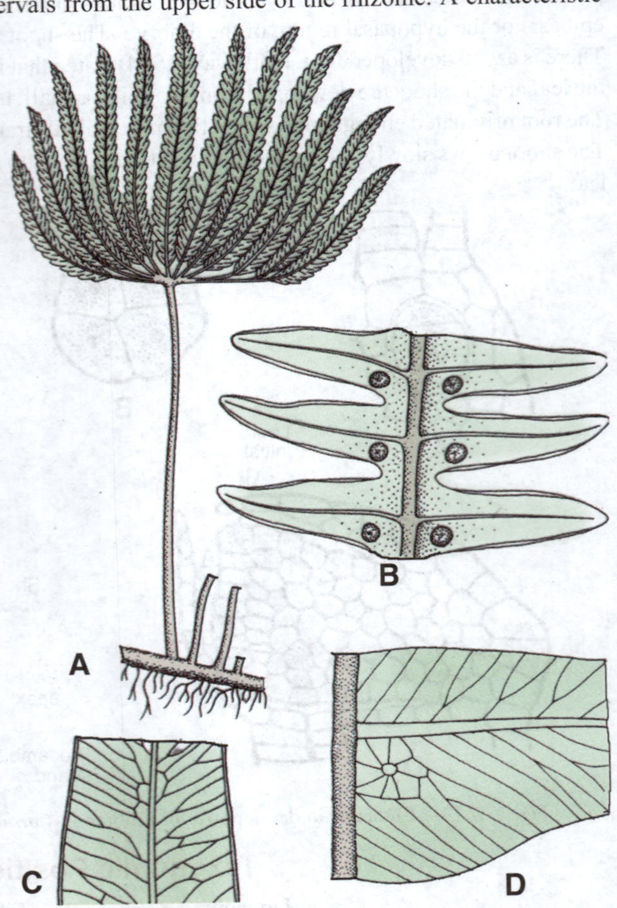

Fig. 17.8. (A—D). *Matonia pectinata*. A. Part of a sporophyte (greatly reduced); B. A fertile pinna bearing sori (enlarged); C—D. Sterile pinnae showing venation (C, after Seward, D, after Diels)

The **young rhizome has a protostelic stele.** As it grows older the protostele changes into an amphiophloic siphonostele at the internode and a solenostele at the node, where a leaf trace departs leaving a leaf gap. Further up this central siphonostele is encircled by another amphiophloic siphonostelic cylinder so that the stele becomes polycyclic. There is a parenchymatous ring between the two concentric steles. Further ontogenetic changes lead to the development of third siphonostelic ring around the second, thus making it a tricyclic stele. The innermost cycle of the stele may be protostelic or siphonostelic. At the nodes the outer two stele become solenostelic due to the departure of leaf traces that leave leaf gaps. The three concentric rings of stele appear as if embedded in a parenchymatous ground tissue.

M. pectinata

The outer and the middle siphonostelic cylinders remain free from each other in the internodes of the rhizome but are connected with each other at the nodes. They are connected by means of a leaf trace. The leaf trace gives out from both its edges a downwardly projecting tissue called as "Compensating tongue" (Tansley and Lulham, 1904). The inner siphonostele has only outer endodermal layer. The outer two rings have both outer and inner endodermal layers followed by outer and inner pericycle and phloem rings (Fig. 17.9). The xylem is mesarch and is mixed with parenchyma cells.

The leaf trace is C-shaped and runs undivided throughout the length of the petiole. Its structure is like that of a meristele. It gives off pinna trace whose origin is extramarginal. The roots are triarch.

The **position and distribution** of the sori is similar to that of ***Gleichenia.*** The sori occur on either side of the midrib of the pinnule and develop on distinct receptacles. The number of sori may be few or many. It depends on the size of the leaves and leaflets. A sorus contains 6—9 sporangia in a ring around the receptacle thus giving it a spherical shape (Fig. 17.8, B). Every sorus is covered by a thick and hemispherical indusium (Fig. 17.10., E) which withers when sporangia are mature. The sori are seated on the lateral veins. The indusium develops from the centre of the receptacle and is coriaceous or leathery in texture. It forms an umbrella-like covering over the sporangia (Fig. 17.10, E). The sporangia develop in a leptosporangiate manner and possess short stalk. Each sporangium produces 64 spores. The indusium is shed after the maturaton of the sporangia, thus exposing them. The sporangia dehisce by means of an obliquely vertical strip of annulus. It causes a transverse dehiscence slit around the sporangial jacket.

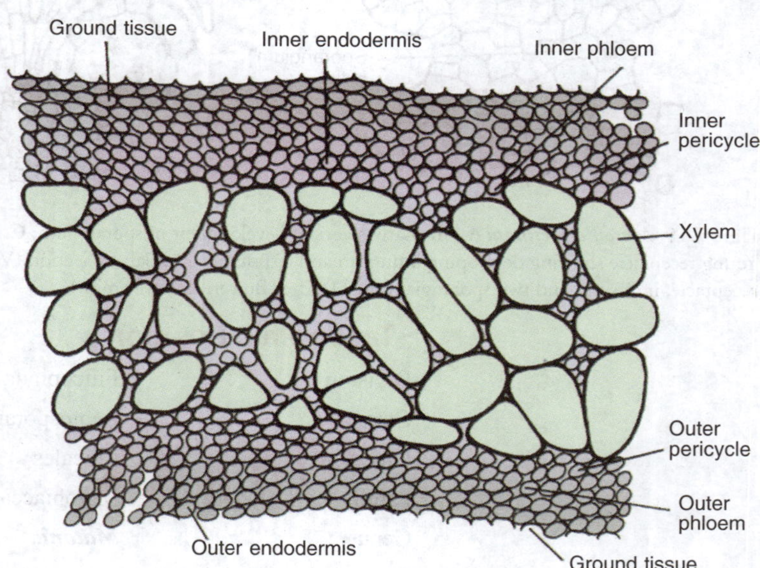

Fig. 17.9. T.S. portion of outermost ring of siphonostele of *Matonia pectinata* showing details of vasculature and other tissues around it.

Stokey and Atkinson (1952) studied the gametophytes and sex organs of ***Matonia.*** Their structure and development is similar to that ***Gleichenia.*** They also observed some stages in the development of embryo. The first division wall of the zygote is parallel to the long axis of archegonium. These stages indicate that the root, foot, cotyledon and stem quadrants are differentiated quite early during the development of the embryo, most probably at the octant stage.

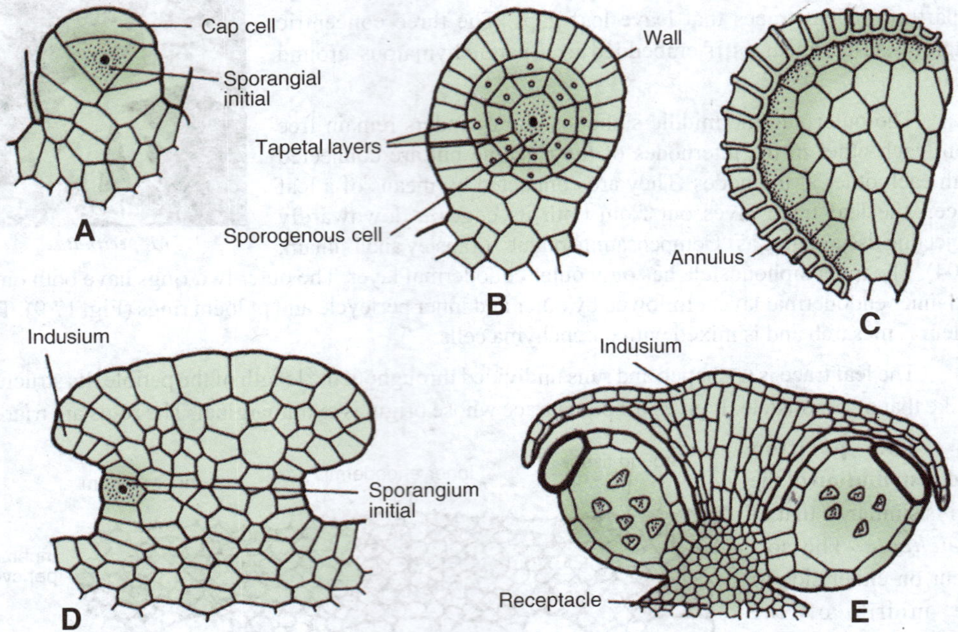

Fig. 17.10. *Matonia pectinata* A—B. Early stages of development of sporangium; C. A mature sporangium; D. Young receptacle showing developing indusium and a sporangial initial; E. Section (V.S) through a sorus showing receptacle, indusium and two sporangia. (A—D, After Bower, E, after Smith).

Taxonomic Position

Division	:	Filicophyta
Class	:	Leptosporangiopsida
Order	:	Filicales
Family	:	Matoniaceae
Genus	:	***Matonia***
Species	:	***pectinata***

Characteristics and a brief account of Family–CYATHEACEAE

This family includes most of the tree ferns. Previously all the tree ferns were included in this family but later genera like ***Diksonia*** and ***Cibotium*** were excluded from this family and placed in a separate family (Dicksoniaceae) because of their marginal sori. The Cytheaceae include tree ferns with superficial sori. Bower (1923) included only three genera in this family (***Chythea, Alsophila,*** and ***Hemetelia***). He created a new family ***Protocyatheaceae*** to include ***Lophosoria*** and ***Metaxya,*** both of which are monotypic and have simple, superficial and naked sort. The cyatheaceae includes about 700 species (Cyathea = 300 ; ***Alsophila*** = 300 ; ***Hemetelia*** = 100). Christensen (1938) in-

Alsophila Cythaea Cythaea medullaris Lophosoria C. spinulosa Lophosoria quadripinnata Metaxya

cluded ***Lophosoria*** and ***Metaxya*** in the family Cyatheaceae and thus recognised 5 genera. The members of this family achieve their best growth in tropical rain forests. They are localised in distribution. In India ***Cyathea spinulosa*** is found in the Eastern Himalayas.

The protocyatheaceae include smaller ferns as compared to the Cyatheaceae ***Lophosoria quadripinnata*** has a short, upright, rhizome which is about 3 feet high and 3—4 inches in diameter. The leaves are borne spirally and 6—10 feet long and multipinnate with open venation. It is found in the Western tropics. ***Metaxya rostrata,*** the second monotypic genus of this family, is restricted to South America and consists of creeping rhizomes bearing large (up to one metre long) leaves that are unipinnate. The sori are naked. The sporangia are large and shortly stalked. The annulus is vertical and complete. There is a distinct lateral stomium.

The **sporophytes** of the various species of Cyatheaceae vary in height. They reach a maximum height of 60—80 feet. The stem is aerial, erect, radial unbranched or bears lateral branches. In some species the stem is short, conical and of low stature. The branching stems of ***Alsophila aculeata*** and ***Cyathea mexicana*** give the plant a shrubby habit. The short and stumpy stems of some Cyathaeceae may bifurcate near the apex (***Cythaea medullaris***) and the two branches are of equal size. The continuity of the pith into the two equal limbs indicates that the branching is of dichotomous type. The **stem** is densely covered with hair and scales. In ***C. medullaris*** and ***C. spinulosa*** the bulk of thickness of the stem is composed of adventitious roots that mass around the trunk in thick covering (Fig. 17.11). They usually mass around the bases or a little distance up the stem. The bases of old leaves also add to the diameter of the stem. Characteristic hexagonal scars of fallen leaves can be seen in the upper portions of stem (Fig. 17.11). The region of the stem above the leaf scars is clothed with decaying leaf bases of old leaves. The stem shows no secondary growth.

The **leaves** are large, in some cases (***Cyathea medullaris***) up to 4 metres long. They may be rarely simple (***C. sinuata***) but commonly repeatedly pinnate. Leaves are usually 3 to 4 times pinnate but bipinnate (***Alsophila taentis***) and unipinnate (***A. phegopteroides***) conditions are also found. The surface of the petiole as well as that of the stem is covered by chaffy scales. Venation is usually of open fucate type. Young leaves are circinately coiled.

Fig. 17.11. *Cyathea spinulosa* showing tree trunk with adventitious roots and a crown of leaves. Only young leaves are shown. Note the scars of fallen leaves.

The **internal structure of the stem** is highly complicated and in some species (***Alsophila excelsa***) it exhibits recapitulation of the stelar theory. They mature stems of all the genera have a polycyclic dictyostele. The outermost ring of meristeles (that make up the dictyostele) is made up of varying number of broad strands. Each strand is enclosed within plates of sclerechyma. The leaf traces arise from the lower margins of the foliar gaps and pass obliquely outwards through the cortex. In addition to these leaf traces there are also present accessory vascular strands in the pith. These are called the accessory medullary strands. Each medullary strand has structure like that of the meristele and is surrounded by sclerenchyma patches. These strands anastomose irregularly and may have blind endings or may pass into the petiole along with the leaf traces. The medullary bundles are present in almost all the ***Cyatheaceae.*** In ***C. imrayana*** in addition to the accessory medullary strands there are also present accessory cortical strands. They may also anastomose freely and may end blindly or get connected with the leaf trace bundles. They follow an irregular course. Their origin is quite independent of the general vascular system. The occurrence of these accessory medullary and cortical strands gives the stem a polycylic dictyostelic condition. The accessory cortical strands are not present in all the species. They are absent in ***Cyathea medullaris*** and ***Hemiterlia setosa.***

In *Alsophila excelsa* the **anatomy of the stem** was studied from the sporeling stage (Gwynne-Vaughan, 1903). The stem has a protostelic cylinder at its base. The leaf traces from the first few nodes do not leave any foliar gaps . At the eighth node, the stele becomes a solenhostele, pith appears and the leaf traces leave leaf gaps (Fig. 17.12, C). The first few levels at the summit receive only one leaf trace. Later the traces divide into 2 or three strands. The tenth leaf receives 4 strands. The accessory medullary bundles in the stem appear at the level of about the tenth leaf ***i.e***, when the stele has changed into a dictyostele. Ogura (1927) reported the same sequences of stelar ontogeny in a species of ***Cyathea*** (Fig. 17.12, A, B). The stems have usually a narrow cortex.

The **sporangia** are grouped in distinct sori that are arranged in a single series on either side of the midrib of the pinnule. Only one sorus is borne on a lateral vein. Its position on the vein varies with species. It may be nearer to the midrib as in ***Cyathea serra, C. medullaris*** (Fig. 17.13 A), ***C. sinuata*** and ***C. spinulosa.***

In ***Hemitelia horrida*** they lie at the vein endings and nearer to the margin of the pinnule. They are never exactly marginal in position. The sori in ***Alsophila*** are naked (Fig. 17.13, D). A sorus bears numerous sporangia that develop in a basipetalous succession and are mixed with hair (Fig. 17.13, E). The receptacle is conical and raised above laminar surface. The sori in this genus resemble with ***Gleichenia*** with respect to their position, orientation of sporangia and presence of hair. In ***Hemitelia*** the receptacle is elongated and has indistinct scale-like indusium at its base. In ***Cyathea*** the receptacle (Fig. 17.13, E) is globose or elongated and is sufficiently raised above surface. The indusium is well developed and arise as cup-like structure around the base of the receptacle and envelops the entire sorus when it is young. At maturity the indusium is torn irregularly at its summit and withers away leaving its persistent base (Fig. 17.13, E). In ***C. medullaris*** the indusium develops very slow and young sporangia are protected by hair. In ***C. dealbata*** and ***C. spinulosa*** the indusium develops rapidly and soon covers the receptacle bearing young sporangia (Fig. 17.13, B—C). There are no hair on the young receptacle of these species (Fig. 17.13 B). The young receptacle in this case has a distinct cell at the apex (Fig.

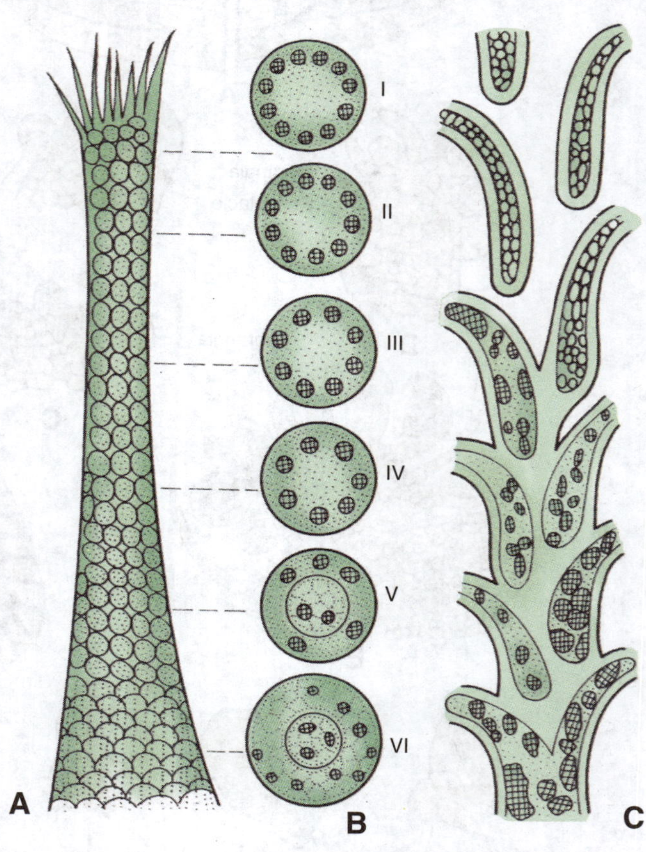

Fig. 17.12 (A—C). ***Anatomy of Cyatheaceae.*** **A.** Stem of a species of *Cyathea*; **B.** Transverse section of the stem shown in A, at various levels; **C.** Stereodiagram of vasculature of stem of *Alsophila excelsa* as seen in median longitudinal section. (A. B. after Ogura, and C after Gwynne Vaughan)

17.13, B). Bower (1899) does not regard it as an apical cell. The sporangia develop from distinct initial cells that appear in a basipetalous succession (Fig. 17.13, B, C) on the receptacle. The sporangial development follows the usual leptosporangiate course. A young sporangium has single-layered wall (Fig. 17.13, E), two-layered tapetum and from four (***C. dealbata***) to sixteen (***C. medullaris***) spore mother cells. The mature sporangia are smaller in size and have a distinct stalk and a head. The former is made up of 4 vertical rows of cells. The head or the capsule has obliquely vertical **annulus** (Fig. 17.13, F) that incompletely encircles the sporangial capsule. There is a distinct **stomium** that marks the point at which the transverse dehiscence slit will appear.

C. dealbata

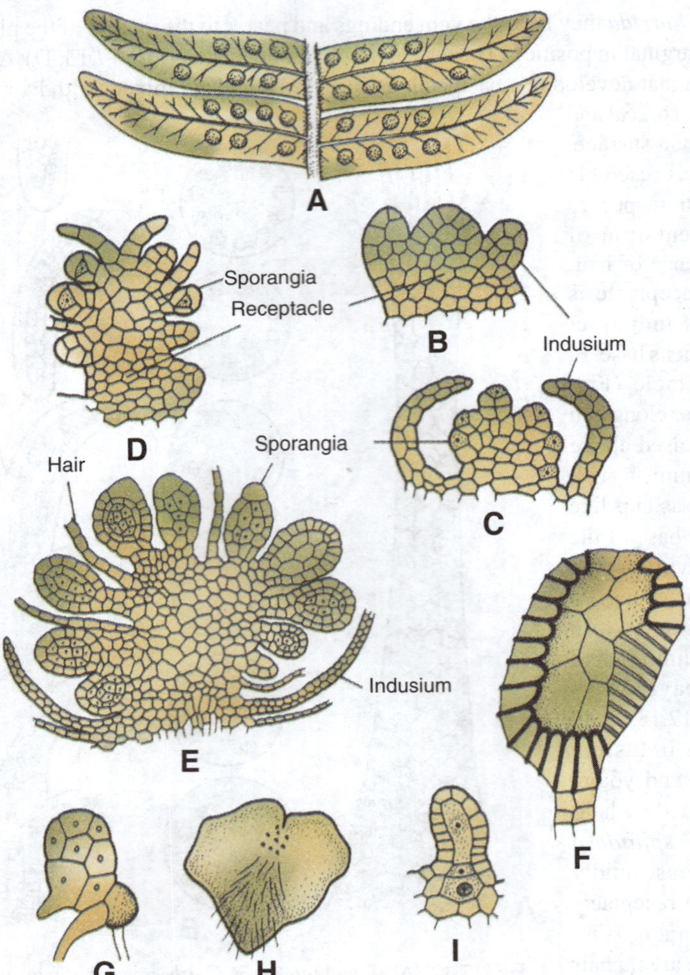

Fig.17.13. (A—I). *Cyatheaceae* A. Sporophylls of *C. medullaris*; B. A young sorus of *C. spinulosa*; C. A young sorus of *C. spinulosa* with well developed indusium and developing sporangia; D. V.S. young sorus of *Alsophila atrovirens*; E. V.S. mature sorus of *C. spinulosa* showing indusium, hair and sporangia; F. Sporangium of *C. spinulosa;* G. A young prothallus of *C. muricata*; H. Mature prothallus of *C. tussacii*; I. Mature archegonium of *C. medullaris.* A. after Smith ; G—I. after Stockey)

Germination of spore and development of prothallus has been studied by Stokey (1930). The normal course of development shows no features of special interest. The spore divides into the lower primary rhizoidal cell and the upper prothallial cell (Fig. 17.14, A). The former gives rise to the primary rhizoid and the latter grows into a short filament of green cells (Fig. 17.14, B). The cells divide longitudinally to form a flat plate of green cells (Fig. 17.14, C). The apical cell is distinguished in this plate-like gametophyte and adds to its cells (Fig. 17.14, C, D). Later it is replaced by a group of meristematic cells. The mature prothallus is cordate (Fig. 17.14, E) and has a cushion of several cell layers thick behind the apical notch. Stokey (1930, 1951) reported the presence of scale-like hair that arise from the ventral surface of the prothallus (Fig. 17.14, E). Such hari are not known in any other family of the leptosporangiate ferns.

The **sex organs** are borne on the vental side. Some times antheridia may occur on the dorsal side. They usually develop on the posterior end. The archegonia develop on the ventral side on the cushion just behind the apical notch. The development of the antheridium is similar to other leptosporangiate ferns, but cyatheaceous antheridia possess two **opercular** or **cap cells** (Fig. 17.15, H) and a single **stalk** cell. The antheridial initial divides by an oblique transverse wall into a lower smaller stalk cell (Fig. 17.15, B) and an upper larger **antheridial cell.** The former divides no further. The latter divides by a transverse wall into a lower first **ring cell** and an **upper cell** (Fig. 17.15, C). This transverse wall then bulges downwards forming a curved or a deeply concave wall (Fig. 17.15, D). The wall does not touch the stalk cell. Later the upper cell divides by slightly curved wall, which touches the second wall (Fig. 17.15, E, 3—3). The antheridium has now an **upper dome cell, a first ring cell,** a **stalk cell** and a **primary androgonial cell.** The **dome cell** divides by a transverse wall (Fig. 17.15, F), which later bulges downwards and touches the third cell (Fig. 17.15, F, G).

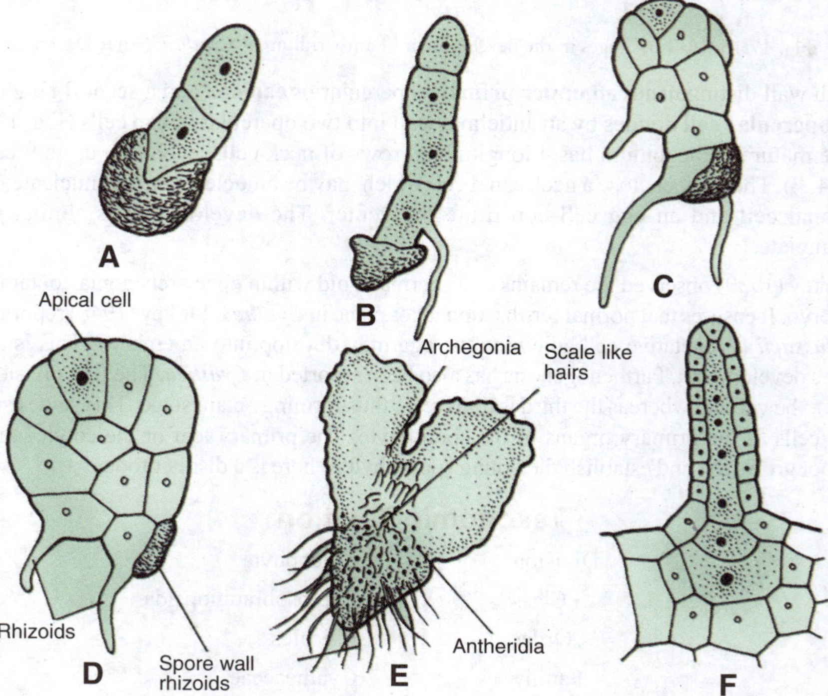

Fig. 17.14. A—E. Stages in the germination of spore and formation of prothallus in *Cyathea*; F. A mature archegonium.

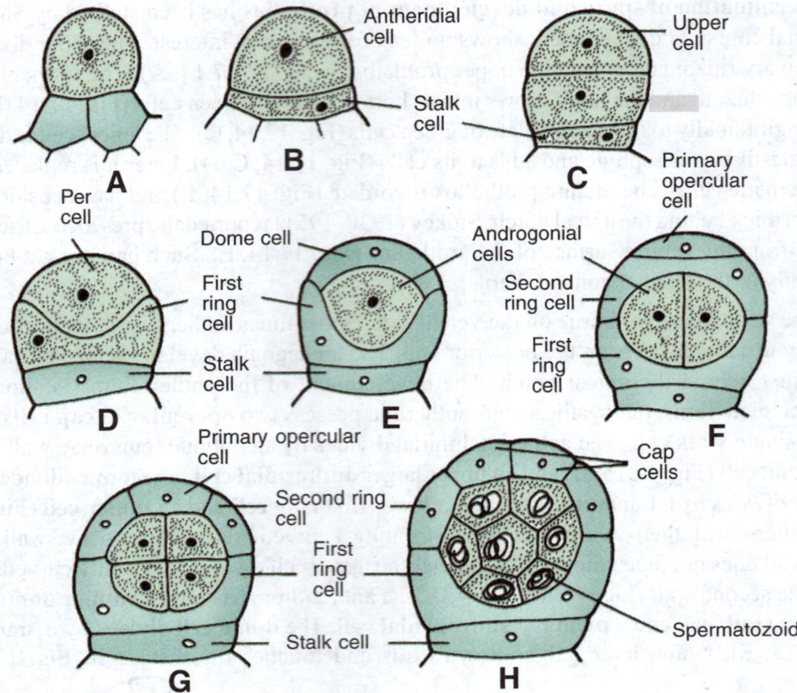

Fig. 17.15. (A–H). Stages in the development of antheridium of *Cyathea.* (After Davie)

This fourth wall distinguishes an **upper primary** opercular or **cap cell** and a **second ring cell.** The **primary opercular cell** divides by an anticlinal wall into two opercular or cap cells (Fig. 17.15, H).

The mature archegonium has 4 longitudinal rows of neck cells, each row up to 9 cells high (Fig. 17.14, F). The neck enclose a neck canal cell which may be binucleate or four nucleate. A single ventral canal cell and an egg cell constitute the venter. The development is similar to other leptosporangiate.

Stokey (1930) observed the remains of a spermatozoid within open archegonia containing two-celled embryo. It ensures that normal fertilisation takes place in ***Cyathea.*** Stokey (1948) reported that in ***Cyathea tussacii a*** vegetative cell adjoining the egg may develop into an embryo. This is a case of apogamone development. Parthenogenesis has also been reported in ***Cyathea.*** The first division is also visualised to be vertical whereas the third is transverse thus forming octant stage. This stage establishes the apical cells of the primary organs of the sporophyte. The primary leaf or the cotyledon and the primary root grow first and establish the young sporophyte. There is a distinct foot.

Taxonomic Position

Division	:	Filicophyta
Class	:	Leptosporangiopsida
Order	:	Filicales
Family	:	Cyatheaceae
Genus	:	***Cyathea***
Species	:	***spinulosa***

Characteristics and a brief account of Family–DICKSONIACEAE

The family includes terrestrial ferns, most of them being tree-like in habit. The stem is upright, radial, sometimes lofty; rarely prostrate, massive and tuberous (***Cibotium***). The stem is covered with numerous unbranched and filamentous hair. Scales are absent. In ***Cibotium*** the hair may attain a length of 3-5 centimetres. The leaves form a crown at the summit of the stem and are very long. Their bases are hidden among the hair. The leaves are pinnately divided and have coriaceous or leathery texture, sometimes, dimorphic. The stem stele is **dictyostele.** The sori are marginal in position and are present on the vein endings. They are protected by a cup-shaped or a bilabiate indusium (Fig. 17.16, D). The sporangia have an obliquely vertical annulus and dehisce transversely. The spores are tetrahedral, perispore is absent, and are produced from 32-64 per sporangium. The germination of the spores is similar to that of Cyatheaceae. The mature gametophyte (Fig. 17.18, B) is cordate (***Cibotium schiedii***) but differs from cyatheaceae in lacking in hair. The prothallus bears numerous, unicellular rhizoids on the ventral surface. Davie (1951) studied the development of antheridia in ***Dicksonia*** and stated it to be similar to that of ***Cyathea.*** The archegonia have long and curved necks. The necks are composed of 4 longitudinal rows of cells, each row 7-10 cells high. They are curved (Fig. 17.18, E) towards the posterior side of the prothallus. Their structure development is similar to that of Cyathea. The position of the sex organs is also similar to the Cyatheaceae. The embryos develop from the fetilised egg. Apogamy is also known (Stokey, 1930).

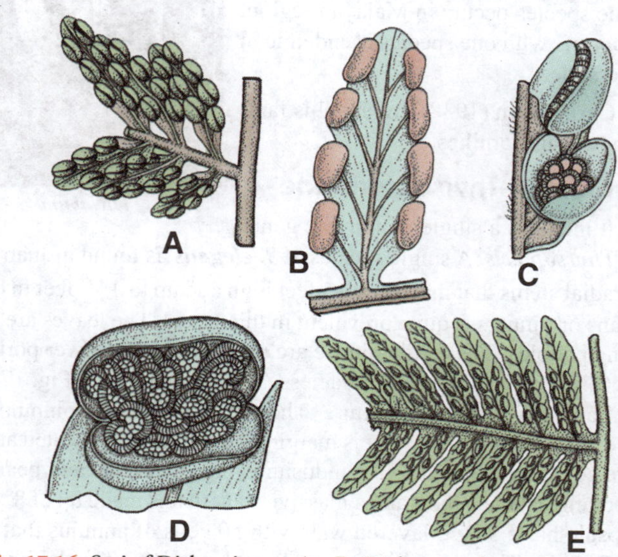

Fig. 17.16. Sori of Dicksoniaceae. A—B. *Dicksonia arborescens.* A. Sori borne on the lower part of the pinna of first order; B. Sori borne on the margins of the pinna of the second order; C. Lower portion of a deeply divided pinna showing arrangement of sori in *Cibotium barometz*; D. Part of pinna of *Cibotium barometz* showing two sori with indusia. (After Hooker)

Cibotium

Cibotium schiedii

The members of this family grow well in the tropical regions of the world. ***Dicksonia*** with about 17 species is restricted to the Southern hemisphere. ***Cibotium*** with about 13 species is found in Central America, Hawaiian island and East Africa. ***Cystodium*** with one species occurs in Malayan region. ***Thrysopteris*** with one species is endemic in Juan Ferandez.

Christensen (1938) divided this family into two sub-families.

A. Subfamily–Thyrsopterideae

It includes a single monotypic genus called ***Thyrsopteris.*** A single species of ***T. elegans*** is found in Juan Fernadez. The sporophytes have erect, radial stems that may be 3—5 feet high and up to 1½ feet in diameter. Vegetative propagation by means or runners is quite prevalent in this genus. The leaves are 3-4 times pinnate and have thick and long petioles. The fertile pinnae are restricted to the lower portion of the leaf and have reduced blades (Fig. 17.17, A) or even bladeless. The sterile pinnae (Fig. 17.17, A) are restricted to the tip of the leaf and are leathery in texture. The fertile region has pinnules with reduced lamina and each pinnule terminates in sorus that is marginal in origin and is seated at vein ending (Fig. 17.17, A). The sori are covered with a cup-like indusium (Fig. 17.17, B) and bear basipetally arranged sporangia. The sporangia have a short and a massive stalk that is made up of 8 vertical rows of cells. The head or the capsule has a single-layered wall with 50 cells of annulus that form a 'curiously twisted hoop' (Fig. 17.17, F). There is a definite and lateral stomium. The dehiscence of the sporangium is lateral. It produces 48-64 spores. The rhizome is siphonostelic. A number of accessory medullary strands appear in the pith, thus making it polycyclic.

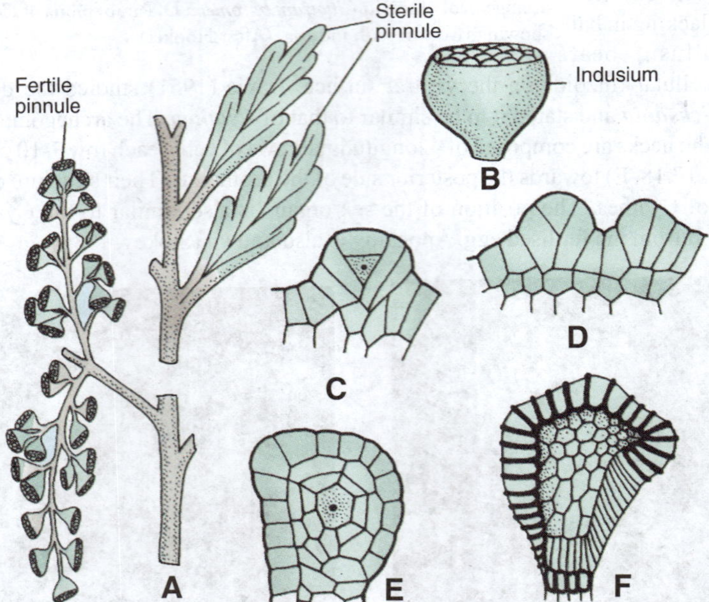

Fig. 17.17 (A—F). *Thyrspoteris elegans.* A. Portion of a leaf of *Thyrsopteris elegans* bearing fertile and sterile pinnules; B. A sorus with cup-shaped indusium; C—E. Stages in the development of sporangium; F. A mature sporangium. (After Bower)

B. Subfamily–Dicksonioideae

Diels (1898) includes only two genera (***Cibotium*** Kaulfuss and ***Dicksonia*** L' Heritier) in this sub-family. Christensen (1938) included ***Culcita*** Presl. and ***Cystodium*** J.Sm. also in this family. He agreed no doubt with Copeland in excluding ***Culcita*** (***Balantium*** auct, non Kaulfuss) from this family and associating it with ***Dennstaedtia.*** it has nine species.

Dicksonia which includes about 17 species (some species have been transferred to ***Dennstaedtia*** that have lofty, erect , radial stems which bear dimorphic leaves. Numerous adventitious roots arise from the stem and form a thick felt around the basal regions. The leaves are pinnate compound. The sori are strictly marginal (Fig. 17.16, A) is position and are protected by a bilaniate indusium (Fig. 17.16, C). The outer value of the indusium is formed by a slight alternation of the leaf tooth.

Cibotium includes about 13 species most of which have tree—like habit with aerial stems reaching 1-5 metres in height. Some have creeping massive rhizome e.g. ***C. barometz*** (L) Sm. The leaves are arranged in a spiral fashion. The stem is ***richly*** clothed with dense, unbranched and up to 5 centimetres long hair. There are no scales. The steler organisation is an amphiphloic siphonostele, amphiphloic solenostele or even dictyostelic (Fig. 17.18, F). There are no medullary and cortical strands. The cortex may be wholly sclenrenchymatous or wholly parenchymatous or partly sclerenchymatous and partly parenchymatous. The sori are marginal in position but may shift slightly inwards at maturity. They are gradate and two sporangia develop basipetalously. The indusium is bilabiate (Fig. 17.16, D). One of the tips develops from the abaxial side of the leaf where as the other forms the adaxial side. The development of the sporangia is leptosporangiate with 64 spores.

Hypolepis *D. punctilobula.* *C. barometz* *Dennstaedtia* *Saccoloma*

The sporangial stalks are long and annulus is obliquely vertical with a lateral stomium. The sporangia dehisce by means of transverse slit. The prothallus and the sex organs have almost the same structure and follow the same course of development as in ***Cyathea.*** Embryo may develop apogamously or may develop from a fertilised egg.

Bower (1923) also included a third sub-family ***Dennstaedineae*** in this family. He included the following genera in this sub-family : ***Dennstaedtia*** Bernhardi; ***Microlepia*** Prest; ***Leptolepia*** Mattenius; ***Saccoloma*** Kaulfuss; ***Hypopteris*** Bernhardi. The sub-family is characterised by creeping rhizomes bearing repeatedly pinnate leaves with strictly marginal sori. The stems are covered with hair. Scales are present only in ***Saccoloma.*** The indusium is bilabiate with upper flap larger than the lower. The vasculature in ***Dennstaedia*** and ***Microlepia*** is siphonostelic. In ***D. dissecta*** there are two concentric siphonosteles. The leaf traces leave foliar gaps so that the vasculature at the node is solenostelic (perforated siphonostele).

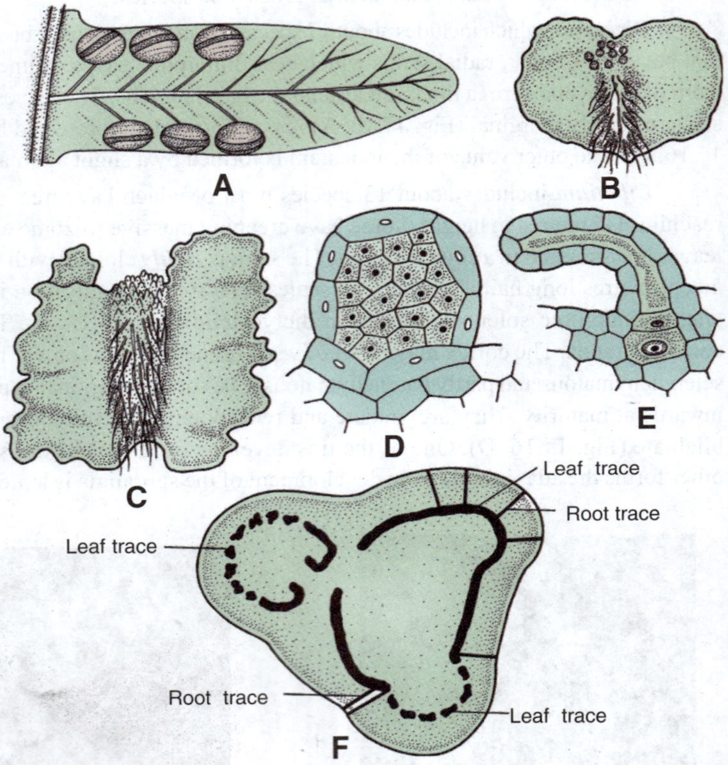

Fig. 17.18. (A–F). *Dicksoniaceae*. A. Pinnule of ***Cibotium chamissoi*** bearing marignal sori; B. A. mature prothallus of *C. schideii*; C. same as B; D—E. Antheridium and Archegonium of ***C. menziesii*** and ***C. schiedii,*** respectively; F.T.S (diagrammatic) through stem of C. ***barometz*** showing three leaf traces and two root traces. (A. After Smith; B. After Stokey, F. After Ogura)

The leaf trace is flat and ribbon-like bundle that remains undivided. It leaves a leaf gap which is narrow and closes very soon, *e.g.*, ***D. punctilobula.*** In ***D. adiantoides*** localised thickening appears on the inside of the margin of the leaf gap and gives off a projecting ridge into the vascular cylinder of the stem. This vascular ridge is continuous from one leaf case to the next and so on. In ***D. rubiginosa,*** the vascular ridge is separate and may divide into a number of separate strands that traverse the vascular cylinder. These strands maintain a connection with the leaf gap. In ***Saccoloma elegans*** the stele is polycyclic dictyostele with three rings.

The marginal sori have sporangia arranged in a basipetalous manner (***D. punctilobula, D. apiifolia***) but in ***D. rubiginosa*** the sporangia are not arranged in any regular sequence and the sorus appears to be mixed. In ***Microlepia*** the sporangia are also basipetalous in arrangement but variations are not ruled out. The annulus shows slight variations. In some species of ***Dennstaedtia*** it is vertical and is not interrupted by the stalk because it is diverted to one side. In other species of the same genus it is interrupted by the sporangial stalk. The dehiscence is transverse. The number of spores varies between 32—64.

In *Saccoloma* the stem is massive, upright and radial bearing large unipinnate compound leaves. Venation is parallel and may be forked. The sori appear to be intramarginal. The indusium is bilabiate with unequal tips. Bower regards the sori to be truly marginal in origin and gradate.

Hypolepis has mixed sori that are naked. Bower considers it proper to put it under the genus *Polypodium.*

Taxonomic Position

Division	:	Filicophyta	Division	:	Filicophyta
Class	:	Leptosporangiopida	Class	:	Leptosporangiopsida
Order	:	Filicales	Order	:	Filicales
Family	:	Dicksoniaceae	Family	:	Dicksoniaceae
Sub-family	:	Thrysopteridae	Sub-family	:	Dicksoniodeae
Genus	:	***Thrysopteris***	Genus	:	***Dicksonia***
Species	:	***elegans***	Species	:	***apiifolia***

QUESTIONS

ESSAY TYPE

1. Describe different types of sori found in Leptosporangiate ferns and discuss the evolution of these sori.

2. Describe the morphological features of the sporophyte and gametophyte of Hymenophyllaceae.

3. Describe the morphology and Anatomy of the sporophyte of ***Matonia.***

4. Give an account of the structure and development of sporangium of ***Matonia.***

5. Describe ***Cyatheaceae*** in brief.

6. Briefly describe the structure and development of antheridium in ***Cyathea.***

7. Compare the prothallus of ***Hymenophyllum ,*** ***Matonia*** and ***Cyathea***

8. With the help of diagrams, describe the sori of Dicksoniaceae.

9. Compare the fertile pinnae of ***Thrysopteis*** and ***Dicksonian.***

10. Cytheaceae hold an intermediate position in Filicales. Discuss.

SHORT ANSWERS TYPE

11. Classify the following genera :

(*a*) ***Hymenophyllum***
(*b*) ***Matonia***
(*c*) ***Cyathea***
(*d*) ***Thrysopteris***
(*e*) ***Dicksonia***

12. Name the genus known as filmy fern.

13. Name the plant known as Tree fern.

14. In which fern ribbon like prothallus is found ?

15. Draw labelled diagrams of sori of Dicksoniaceae

16. Write short notes on :

(*i*) ***Matonia***
(*ii*) ***Hymenophyllum***
(*iii*) Characteristics of sori in Adiantaceae and cyatheaceae
(*iv*) Variations in positions of Annulus.
(*v*) Characteristics of sori of Cyatheaceae
(*vi*) Prothallus in Cyatheaceae and Hymenophyllaceae

17. Describe main types of Sori found in filicales **(Kanpur M.Sc., 2003)**

OBJECTIVE TYPE

18. Select the correct answer :

(i) Members of which family are known as filmy ferns ?
(*a*) Matoneaceae
(*b*) Cyatheaceae
(*c*) Dicknoneaceae
(*d*) Hymenophyllaceae

(ii) The sori in ***Hymenophyllum*** are borne on
(*a*) margins of the sporophylls
(*b*) dorsal surface of sporophylls
(*c*) ventral surface of the sporophylls
(*d*) the rachis of the sporophylls.

(iii) Dorsoventral creeping narrow subterranean and branched rhizomes are characteristic of
(*a*) ***Pteris***
(*b*) ***Adiantum***

(c) ***Matonia***
(d) ***Hymenophyllum.***

(iv) Stele in the rhizome of ***Hymenophyllum*** is
(a) dictyostele
(b) plectostele
(c) protostele
(d) amphiphloic siphonostele.

(v) Ferns commonly known as tree ferns belong to family
(a) Hymenophyllaceae
(b) Cyatheaceae
(c) Matoniaceae
(d) Polypodeaceae.

(vi) Dichotomous forking of leaves is characteristic of
(a) ***Hymenophyllum***
(b) ***Cyathea***
(c) ***Matonia***
(d) ***Pteris.***

(vii) Gradate sori are characteristic of the plants of the family
(a) Hymenophyllaceae
(b) Cyatheaceae
(c) Matoniaceae
(d) Dicksoniaceae.

(viii) ***Cyathea*** is commonly known as
(a) Horsetail
(b) Water fern
(c) Filmy fern
(d) Tree fern.

Pilularia globulifera

CHAPTER 18

Pterophyta–Leptosporangiopsida –Marsileales

- Characteristic features of genus ***Marsilea L.***
- Characteristic features of genus ***Regnellidium*** Linn
- Characteristic features of genus ***Piluaria* sp.**
- Discussion

The order Marsileales of the class Leptosporangiopsida includes a small group of aquatic or semi-aquatic ferns. They are all included in a single family ***Marsileaceae.***

The family ***Marsileaceae*** is represented by 3 living genera. They are ***Pilularia, Regnellidium*** and ***Marsilea.*** All these are aquatic or semi-aquatic in habit. ***Pilularia*** has 6 species. ***Regnellidium*** has one. ***Regnellidium*** is characteristic in being the only seedless vascular plant that produces latex (Labourian, 1952). ***Marsilea*** is represented by 53 living and 10 fossil species. It is usually considered as a type.

GENUS–MARSILEA L.

Distribution

Marsilea is world-wide in distribution. However, it is rich in species in the warmer parts of the world such as tropical Africa and Australia. It is well represented in the northern hemisphere. About 9 living species have been recorded from India. Of these ***Marsilea minuta*** is the commonest. It is practically found all over India. In Punjab it is very common after the rains. ***M. brachypus, M.quadrifolia, M. rajasthanensis*** and ***M. aegyptiaca*** are the other important Indian species. Four species have been reported from Rajasthan.

Habitat

Some species such as ***M. minuta*** and ***M. quadrifolia*** are hydrophytic. They grow submerged or partially out of water. The sporocarps are produced under water. Some species are amphibious. They grow after the rains in temporary shallow ponds and puddles. They fruit only under dry, terrestrial conditions when water dries up e.g. ***M. vestita*** and ***M. aegyptiaca.*** A few species of ***Marsilea*** are **xerophytic**. They grow on dry land (***M. condensata, M. rajasthanensis***). All the species are rooted either in the mud or soil.

SPOROPHYTE

External Morphology

The ***Marsilea*** plant is a sporophyte. Apparently it looks like a four-leaved clover plant. It is differentiated into stem, roots and leaves (Fig. 18.1). The younger parts in some species are covered with dense hairs. The hairs are multicellular and unbranched.

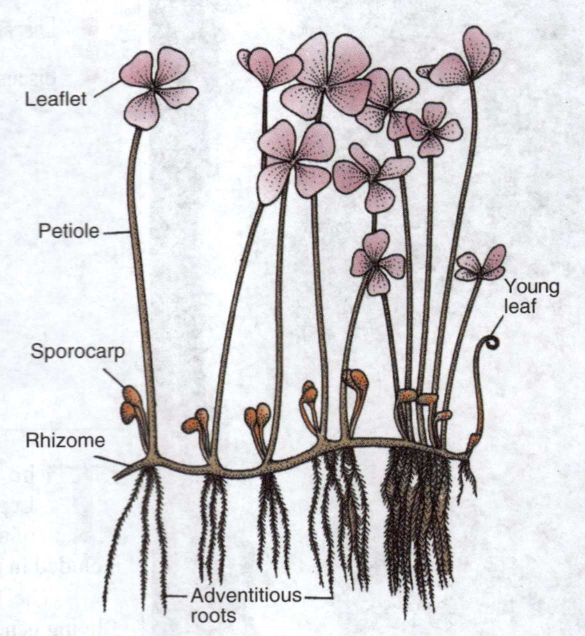

Fig. 18.1. Portion of sporophyte of *Marsilea minuta* showing habit.

Stem

The stem is long and slender (Fig. 18.1). It creeps either on the surface like a **stolon** or slightly below the surface of the soil like **rhizome**. It has an indefinite power of growth and is branched. The branches arise at the bases of leaves and are extra-axillary in position arising in the lateral or oblique position. They run in all directions and may get rooted at the nodes. In this way single plant may cover an extensive area about 20 metres in diameter or even more. The stem is divisible into distinct nodes and internodes. The internodes are long in the hydrophytic plants and short in the sub-terrestrial or xerophytic species.

Marsilea minuta

M. vestita

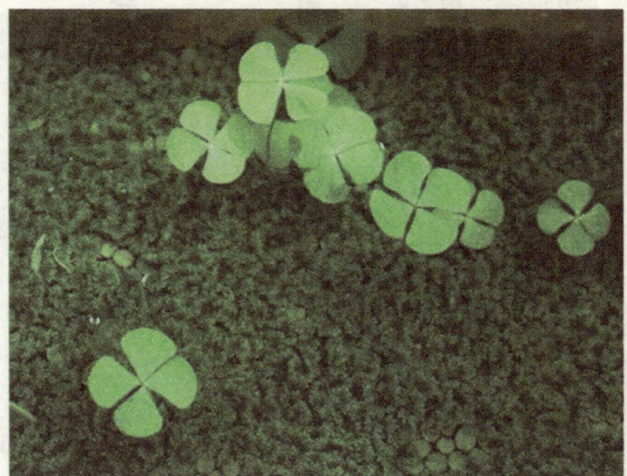
M.quadrifolia

Pilularia

Regnellidium

Marsilea

Roots

The primary root is short-lived. All others are adventitious and arise gradually at the nodes (Fig. 18.1) on the underside of the stem. Occasionally they may arise from the internodal region. They are thin and may be branched or unbranched. Their number at each node may be one or more. The branching is monopodial. The branch roots are arranged in the two rows and are developed in an acropetalous succession.

Leaves

The leaves arise from the upper side of the stem alternatively and are arranged in two rows (Fig. 18.1). They are petiolate and compound. The petioles of the submerged plants are long, thin and flexible with the lamina floating on the surface of water. The leaves of plants growing on mud or land have upright short petioles with the lamina held in a spreading position. The compound lamina is divided into four leaflets of the same size. They spring from the tip of the petiole so that the leaf apparently looks **quadrifoliate**. Actually it is **quadrijugate** as the four leaflets are arranged in a pinnate manner. Two of them form the distal pair and the other two form the proximal pair (Fig. 18.2, F). The leaflets of the proximal pair are really alternate. Occasionally the number of leaflets may be 5 or 6 or even 8 instead of the usual number 4 . In outline the leaflets are obovate or obcuneate to wedge-shaped. The margin is variable. In many species it is entire. In ***M. minuta*** and other hydrophytic species the outer margin of the leaflet varies from entire to crenate or toothed, whereas in the terrestrial species (***M. aegyptiaca*** and ***M. rajasthanensis***) it varies from crenate to slightly deeply lobed (Gupta, 1962). The venation is reticulate and multicostate. It is a closed reticulum. The veins of the leaflet divide repeatedly in a dichotomous manner. The branch veins are linked by cross connections and are also united near the margin by loop veins (Fig. 18.2). The leaflets exhibit sleeping movements. At night the pinnae become folded upwards. The young leaves have **circinate ptyxis** in the bud.

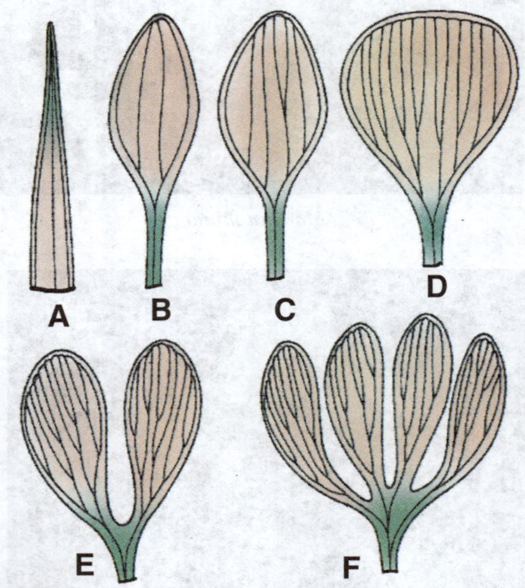

Fig. 18.2. A—F *Marsilea* Ontogeny of the pinnae. The venation is dichotomous. The vein endings form loops so that ultimate vein ends are fused. (After Braun)

Sporocarps

When water begins to dry up, the leafy plant bears special , usually bean-shaped asexual reproductive bodies called the **sporocarps**. The sporocarps are stalked and bisporangiate. The stalk may be long or short and is called the **peduncle** or **pedicel**. The shape, size and contents of the sporocarp and the attachment of its pedicel to the petiole of the leaf vary cosiderably. For details on these points refer to asexual reproduction.

Tubers

The stem towards the end of the growing season in some species (***M. hirsuta, M. minuta***) bears resting propagation bodies called the tubers. The tubers function as perennating structures sprouting on the advent of the season favourable for growth.

ANATOMY

(a) Stem (Fig. 18.3)

A transverse section of the mature stem or rhizome is circular in outline and reveals the following structures :—

1. Epidermis

It forms the outermost and continuous single layer of cells. The epidermal cells are thick-walled and compactly arranged.

2. Cortex

It is differentiated into three regions, the **outer cortex**, the **middle cortex**, and the **inner cortex**. The outer cortex is parenchymatous and may be one to several cells in thickness. The parenchyma cells are compactly arranged. Here and there among them occur a few tannin cells. The outer cortex of compactly arranged parenchyma cell serves to maintain the cylindrical form of the stem. The **middle cortex** which lies internal to the outer cortex consists mainly of air cavities. The aerenchyma is an air storage tissue consisting of a single layer of air chambers arranged in a ring. The chambers are separated from one another by single layered partitions consisting of thin-walled parenchymatous cells. The **inner cortex** is a solid tissue of several cells in thickness. The outer cell layers of the inner cortex are thick-walled or sclerotic constituting the sclerenchyma. The inner layers of cells are compactly arranged and parenchymatous. Some of these cells may contain tannin, others are filled with starch. In the xerophytic species (***M. aegyptiaca***) the air chambers in the middle cortex become obliterated.

3. Stele

The vascular cylinder is medullated. The **xylem** is in the form of a ring. There is phloem on both sides of the ring-shaped xylem. Such a stele is called **amphiphloic siphonostele**. Both the inner and the outer phloem are covered by pericycle which is single layered. The pericycle in turn is flanked by endodermis. The siphonostele is thus limited externally and internally by endodermis . The disposition of the various tissues from outward to inwards is **outer endodermis, outer pericycle, outer phloem, xylem, inner phloem, inner pericycle** and **inner endodermis**. The **medula** or **pith**, which is delimited by inner endodermis, occupies the centre. It is parenchymatous in the plant growing in water. It becomes more or less sclerotic in plants growing in mud or damp soil (Fig. 18.3). In the

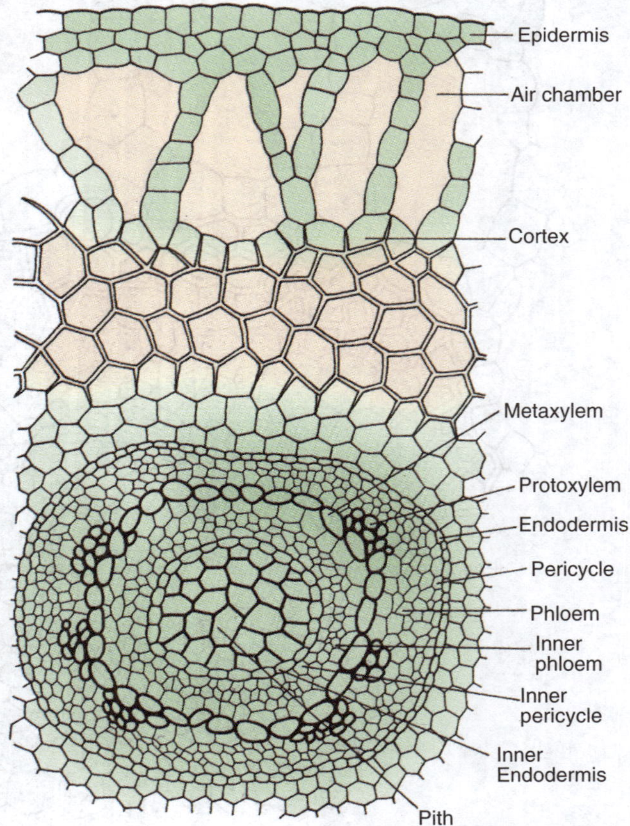

Fig. 18.3. T. S. through older portion of the rhizome of ***M. minuta*** showing sclerenchymatous pith.

xerophytic species it is sclerenchymatous. ***M. vestita*** has exarch xylem with well defined protoxylem elements. ***M. aegyptiaca*** has mesarch protoxylem (Gupta and Bhardwaja). In ***M. quadrifolia*** and others protoxylem elements are not conspicuous. In a section passing through or near the node, the vascular cylinder consists of two parts, an arc-like or C-shaped stele and a small curved vascular strand. The latter represents a **leaf trace**. The inner and outer layers of endodermis, pericycle and phloem become continuous at the ends of the arc-like stele. This is an **amphiphloic solenostele**. Vessels have also been reported in some species of ***Marsilea.***

(*b*) Petiole (Fig. 18.4)

There is the single layered epidermis made up of rectangular cells. Beneath it are a few layers of thin-walled cells that constitute the **outer cortex.** This is followed by the aerenchyamtous **middle cortex** which consists of a ring of air chambers. The air chambers are separated by single layered partitions of thin-walled parenchyma cells. The air spaces tend to disappear in drier habitats. The inner cortex is a solid, compact tissue several cells deep. The outer cell layers are sclerenchymatous (Fig. 18.4). The

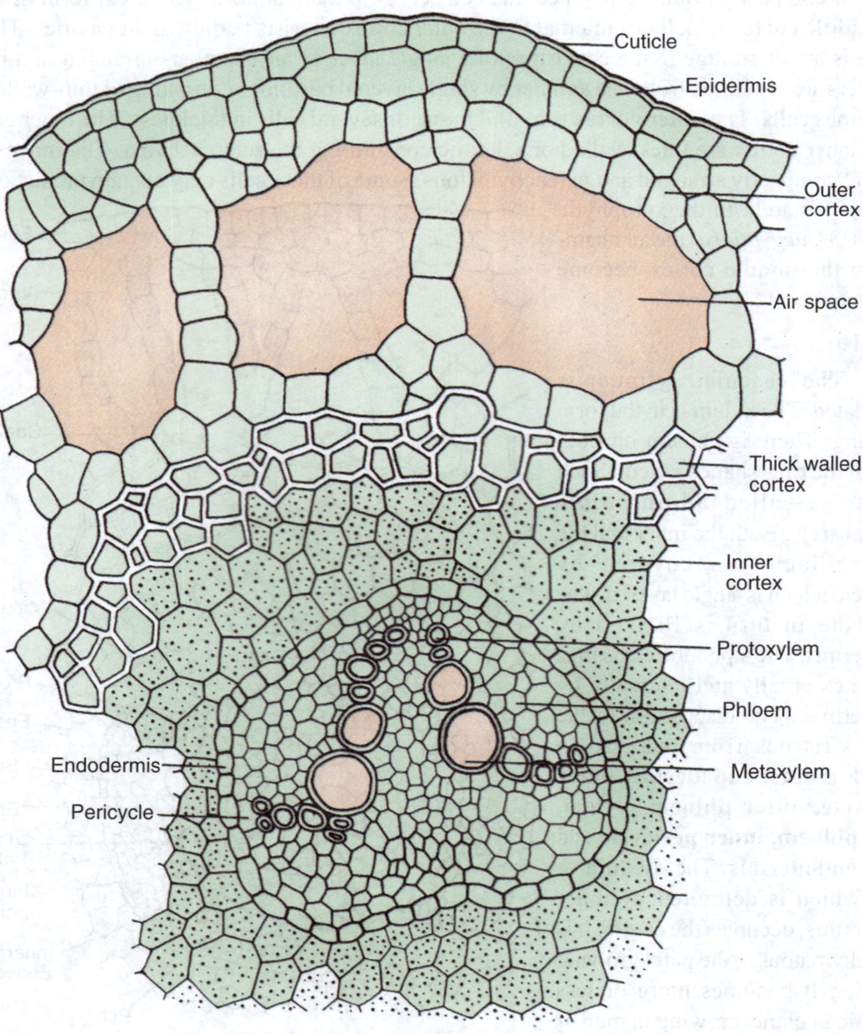

Fig. 18.4. T. S. petiole of *M. minuta.*

inner layers of cells are parenchymatous and are filled with starch. Here and there tannin filled cells also occur. The inner cortex is delimited by a single layered **endodermis**. Within the endodermis is the stele which is somewhat triangular in outline. It lies in the centre and has a single **vascular bundle**. The xylem part of the bundle is shaped like the letter 'V' with opening towards the axis. The arm has sometimes more large tracheids in the middle with a few smaller ones at either end. Between the two arms of the xylem is the xylem parenchyma. Surrounding the xylem is the **phloem** followed by the **pericycle.** The pericycle is bounded by the endodermis.

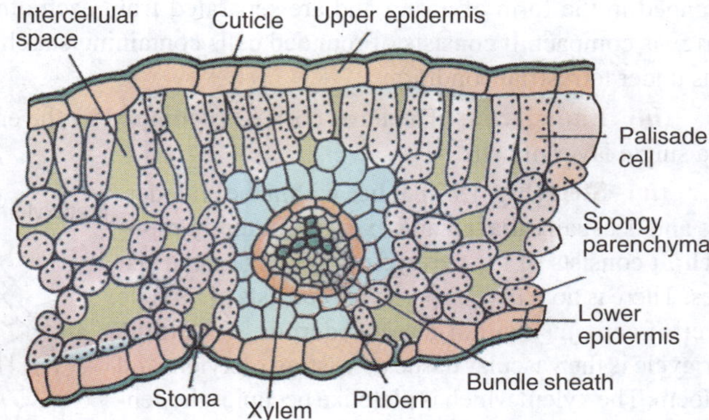

Fig. 18.5. V.S. pinna or leaflet of *M. minuta.*

(c) Leaflet. (Fig. 18.5)

A transverse section of the leaflet reveals the following structure : —

(*i*) **Epidermis.** It consists of upper and lower epidermal layers each made up of a single layer of parenchymatous cells. Their continuity is interrupted by slightly sunken stomata. In the hydrophytic species with floating leaves the stomata are restricted to the upper epidermis only. As the water dries up and the leaflets become aerial the stomata develop in the lower epidermis as well. According to Gupta the stomata are fewer but bigger and the epidermal cells have more wavy walls in plants growing under drier habitats than those found in water.

(*ii*) **Mesophyll.** It lies between the upper and lower epidermis and is differentiated into **palisade** and **spongy parenchyma**. The former lies beneath the upper **epidermis** and consists of columnar cells rich in chloroplasts. The spongy parenchyma which faces the lower epidermis consists of rounded cells. Just beneath the lower epidermis are the large air chambers. They are separated by septa. Submerged leaves, however, show no distinction into the palisade and spongy tissue.

(*iii*) **Vascular bundle.** Embedded in the mesophyll tissue are vascular bundles. They are concentric in nature. Each bundle has a central core of xylem surrounded by phloem. Externally bundle is bounded by the single layered endodermis. It is distinct.

(*d*) **Root.** (Fig. 18.6). In a transverse section it consists of the piliferous layer, cortex and stele.

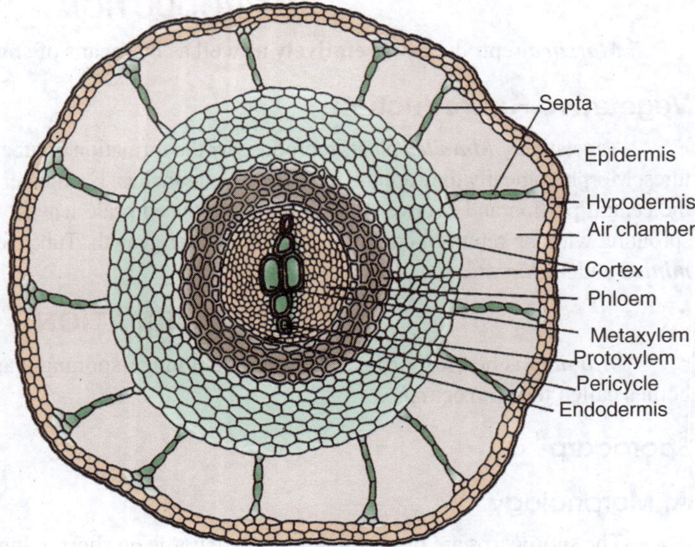

Fig. 18.6. T. S. root of *M. minuta.*

***(i)* Piliferous layer.** It is the surface layer consisting of compactly arranged biconvex cells. The outer walls of the cells are thickened.

***(ii)* Cortex.** It lies within the piliferous layer and is several layers thick. It shows distinction into **outer cortex** and **inner cortex**. The outer cortex is **aerenchymatous**. The large air chambers are arranged in the form of a ring and are separated from each other by longitudinal septa. The inner cortex is compact. It consists of rounded cells containing starch. They often become sclerenchymatous under terrestrial conditions.

***(iii)* Endodermis.** The inner cortex is delimited by the endodermis. It is distinct and consists of a single layer of cells.

(iv)* Stele** (Fig. 18.7). It lies within the endodermis and is usually **diarch** and **exarch** but rarely **monarch**. It consists of the pericycle and the vascular tissues. There is no pith. The pericycle consists of a single layer of cells just within the endodermis. Within the pericycle is the vascular tissue consisting of xylem and phloem. The xylem which is platelike occupies the centre of the stele. In the middle of the elongated xylem plate are two large and two small metaxylem elements. The protoxylem is **exarch**. It consists of two small masses of still smaller protoxylem cells one at each end and touching the pericycle. On either side of the xylem plate is the band-like phloem. There are thus two xylem bundles and two phloem. They are situated on alternate radii. Recently White reported the occurrence of vessels in the roots of certain species of ***Marsilea.

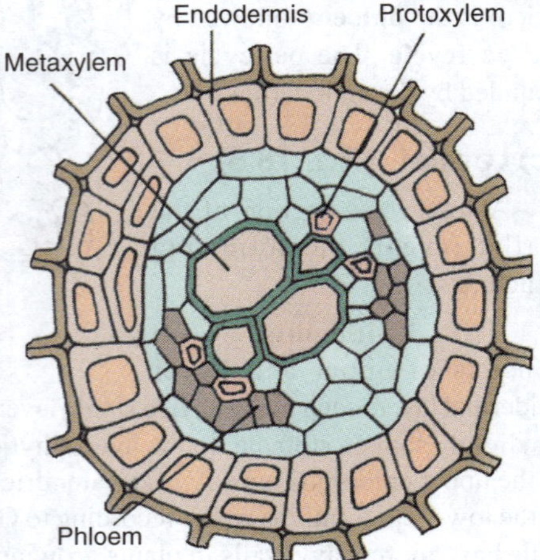

Fig. 18.7. Portion of stele of root of *Marsilea minuta* showing diarch xylem.

REPRODUCTION

Marsilea reproduces **vegetatively** as well as by means of **spores.**

Vegetative Reproduction

Vegetatively ***Marsilea*** plants reproduce by the formation of tubers. They are small, bud-like structures. Morphologically they are the modified side branches. Each tuber contains reserve food material in the central portion and is covered with minute reduced scale leaves. It serves as a perennating organ, sprouting with the return of conditions favourable for growth. Tuber formation has been reported in ***M. minuta*** and ***M. hirsuta.***

SPORE FORMATION

Marsilea is heterosporous. The micro and mega-sporangia are borne in special bean-shaped bodies called the **sporocarps**.

Sporocarp

A. Morphology

The sporocarps are bisporangiate and are borne on short or long stalks called the **peduncles** or **pedicels**. They appear on the plant when it has reached a certain stage of maturity. In some species they appear when the water in the ponds begins to evaporate rapidly and its level falls considerably, but are always produced under water. In others they are produced when water completely dries up and the plants become terrestrial. The pedicel or peduncle of the sporocarp is attached to the petiole. The petiole and the grouping of the sporocarps varies considerably. Gupta (1961) puts the various species of ***Marsilea*** under three well defined categories according to the mode of attachment of pedicel with the petiole :

(i) Pedicels directly inserted on the petiole in a linear sequence on the same side (Fig. 18.8, A). Examples are ***M. polycarpa*** and ***M. subangulata.*** Both are foreign species. No Indian species belong to this category. In ***M. coromandelica*** (Fig. 18.8, D) and ***M. vestita*** the peduncle of the solitary sporocarp is directly attached to the base of the petiole.

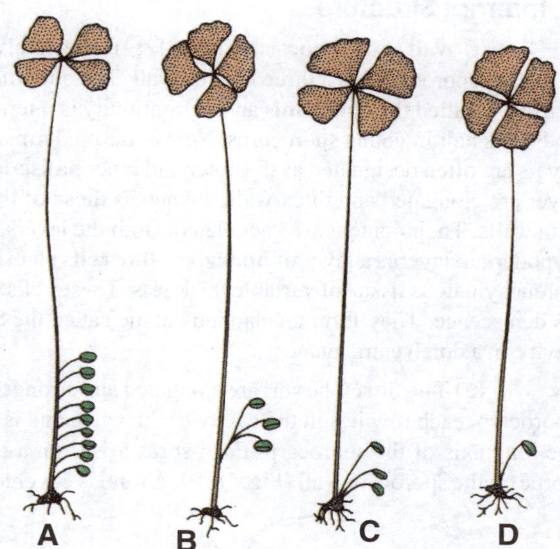

Fig. 18.8. Mode of attachment of the sporocarps to the petiole; A. *Marsilea polycarpa*; B. *M. quadrifolia*; C. *M. minuta*; D. *M coromandelica.*

(ii) Pedicels united with one another and then jointly inserted on the petiole. The common Indian examples is ***M. quadrifolia*** (Fig. 18.8, B)

(iii) Pedicels free or slightly connate and attached to the base of the petiole. The common Indian species, ***M. minuta,*** is best example of this type of attachment (Fig. 18.8, C)

Many variations from the above-mentioned three types have been recorded in nature. Conditions intermediate between the first and second types were recorded in ***M. quadrifolia*** (Gupta, 1961). When young, the sporocarps are soft, green and covered with hairs. At maturity they become brown to dark brown, hard and nut-like. The hairs persist in the terrestrial species but disappear at maturity in the aquatic species. In shape the sporocarps are oval or beanshaped in ***M. minuta*** and other hydrophytic species, but squarish or rectangular in ***M. aegyptiaca, M. rajasthanensis*** or other xerophytic species (Gupta, 1961). Just beyond the point of attachment of the sporocarp to the stalk (raphe) are often present one or two teeth-like protuberances, at the back in the median plane. They are known as the **horns** or **teeth**. In ***M. minuta*** both the teeth are prominent. ***M. quadrifolia*** has the lower tooth conspicuous and the upper blunt. In ***M. aegyptiaca*** the upper tooth is blunt and prominent . The lower is absent. In ***M. uncinata*** the teeth are spine-like projections, with the upper one more conspicuous than the lower. There are some species in which teeth on the sporocarp are lacking.

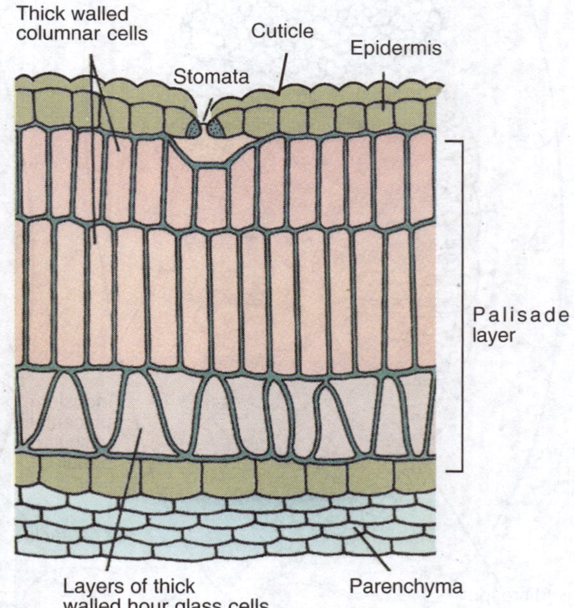

Fig. 18.9. T. S. through the wall of sporocarp of *Marsilea minuta.*

B. Internal Structure

(*i*) Wall of the Sporocarp (Fig. 18.9). The bivalved sporocarp has a thick and resistant wall. A mature sporocarp has a three layered wall. The outermost layer is made up of broad and columnar cells. It is called the **epidermis** and its continuity is interrupted by stomata. It is covered with cuticle and bears hair in young sporocarps. Next to the epidermis is the two layered **hypodermis**. These two layers are often designated as the outer and inner **palisade layers,** respectively. The cells of the outer layer are elongated and thick walled whereas those of the inner layer are more elongated and have thin walls. Their content are vacuolated. Both the layers have chloroplasts in their cells. Next to the hypodermal layer is a layer of **hourglass-like cells** whose cells are also thick. Inner to this layers is parenchymatous tissue of variable thickness. These cells gelatinise in a mature sporocarp and help in its dehiscence. They form a gelatinous tissue called the **Sporophore** that runs around the sporocarp cavity, in a dorsiventral plane.

(*ii*) The Sori. The sori are elongated and arranged in two alternating rows in the cavity of the sporocarp, each row lies in the cavity of the valve and is situated dorsiventrally and transversally to the long axis of the sporocarp. Each sorus arises on a ridge-like placenta or the receptacle that is borne on the sporocarp wall (Fig. 18.10, A) and is covered by membranous indusium made up of two

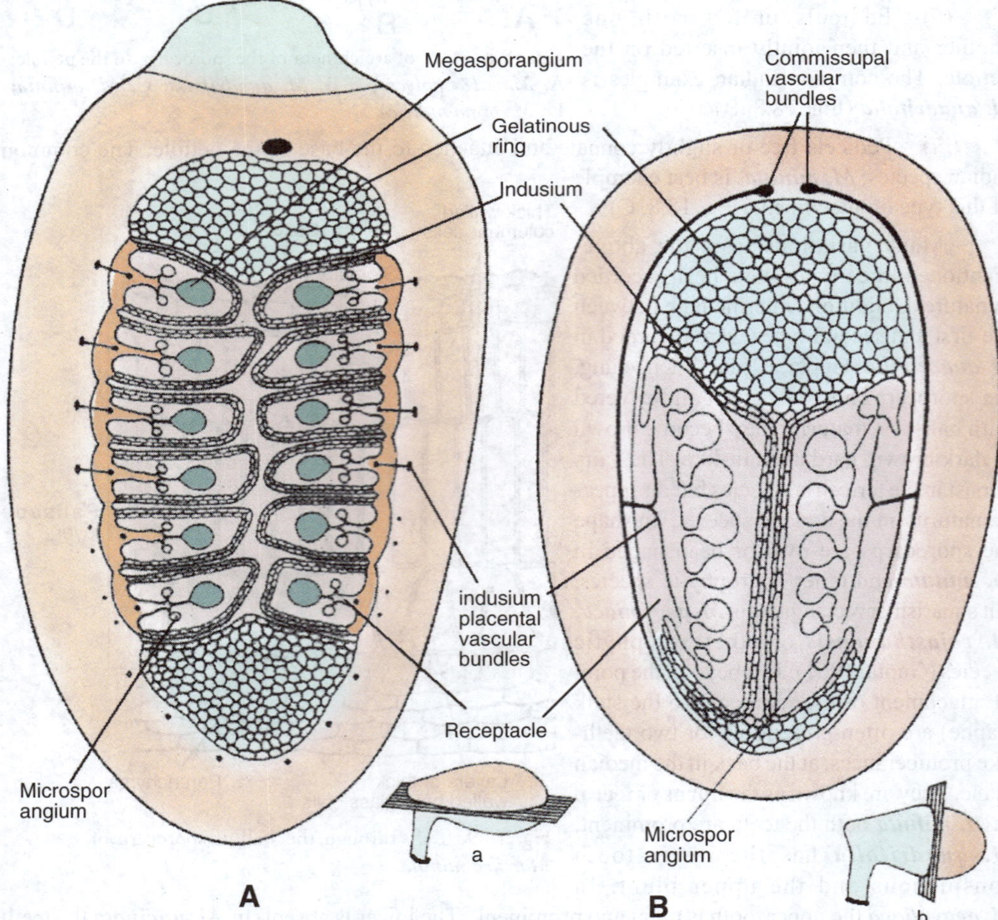

Fig. 18.10. (A-B). Internal structure of sporocarp. A. Horizontal section through sporocarp; B. Vertical transverse section through sporocarp (*a* and *b* indicate planes of sections).

layers of cells. The sori overlap each other and the indusia of adjacent sori are partially fused. The number of sori in a sporocarp varies from two (***M. aegyptiaca***) to twenty (***M. quadrifolia, M. vestita***). There are 11-12 sori in ***M. minuta.*** Each sorus bears both **mega** and **microsporangia.** The former are short-stalked and are arranged in a row at the tip of the receptacle, whereas the latter are long stalked and arise on the sides. Their number in a sorus varies with species. In ***M. minuta*** a sorus has 4-8 megasporangia and 8-13 microsporangia. In ***M. aegyptiaca*** there are 5-16 **mega** and 9-19 **microsporangia.** Megasporangia are sometimes absent in sori of ***M. minuta, M. rajasthanensis*** and ***M. vestita.***

Mehra and Loyal (1960) reported three biotypes of ***M. minuta:*** (*i*) A biotype with normal micro and megasporangia. The megasporangia have one megaspore ; (***ii***) Sporocarps with abnormal microsporangia that have different shapes and sizes of microspores and normal megasporangia with one megaspore; (***iii***) a biotype with abnormal mega- and microsporangia. The former have many sterile megaspores, the latter have also many microspores of different shapes and sizes. They recorded $\boldsymbol{n} = 20$ in ***M. minuta.***

In a horizontal section the sori are seen to be arranged in two alternating rows. (Fig. 18.10, A). This section reveals the exact number of sori in the sporocarp. The megasporangia are seen at the apex of the receptacle and develop first whereas microsporangia are on the sides and develop later. The sorus is thus gradate and sporangia develop in a basipetalous manner. The placental vascular bundles and commissural bundle are cut in a transverse plane. The gelatinous ring appears in the form of two big masses at either end. The pedicel is also cut transversely.

A vertical transverse section reveals only two sori (Fig. 18.10, B). The sori reveal many megasporangia and only two or three microsporangia at the sides. The attachement of the indusium to the outer side of the placental ridge is clear. The stalk is not cut in this section. The commissural bundles, the placental bundles and placental branches are very clear. The sporophore appears as two distinct masses on the dorsal and ventral sides.

In a vertical longitudinal section (Fig. 18.11) or sagittal section the pedicel is also cut longitu-

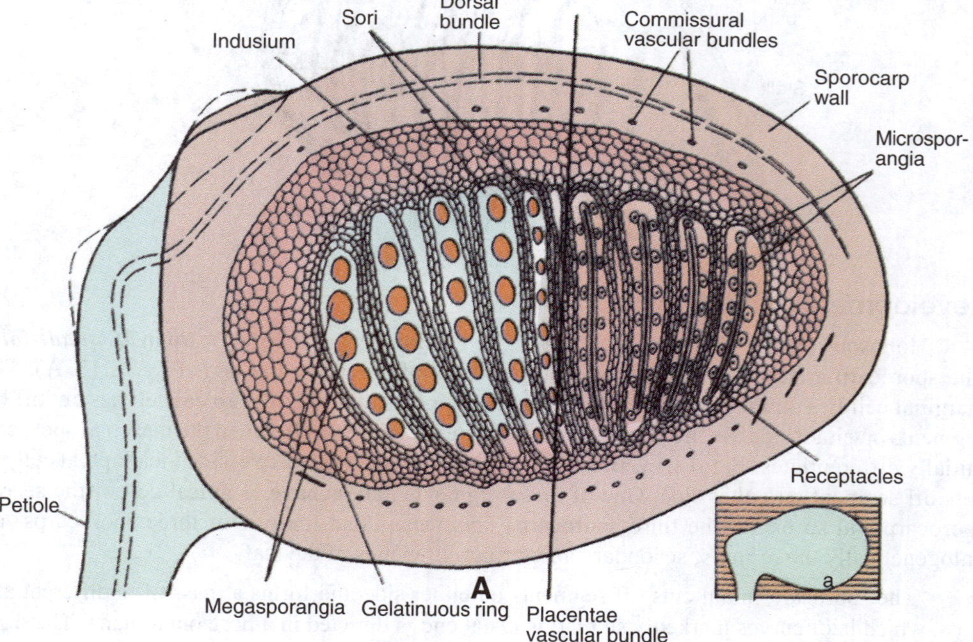

Fig. 18.11. A vertical longitudinal section through the sporocarp.

dinally and the gelatinous ring is seen to surround the sori, which are seen to contain either mega-or microsporangia depending upon whether the section passes through the former or the latter. The dorsal vascular bundle is seen in continuation with vascular strand of the pedicel. The commissurals are cut transversely and can be seen around the gelatinous ring. A section passing through the microsporangia shows the placental bundles also.

Vascular Supply of Sporocarp (Fig. 18.12)

The vascular bundle of the pedicel of the sporocarp runs upwards and without undergoing any change reaches the point of attachment of the body of sporocarp and the pedicel. Now it takes a bend near the lower tooth of the sporocarp (Fig. 18.12) and continues forwards along the upper or the dorsal side of the sporocarp. In this region it is called the dorsal bundle (Fig. 18.12). The dorsal bundle gives off **lateral bundles** (commissural bundles) all along its length. The number of lateral bundles corresponds to the number of sori. These lateral bundles arise in pairs and supply the sori situated in either valve of the sporocarp. After giving off the last pair of lateral bundles the dorsal bundle bifurcates into two branches that supply the right and left valves of the sporocarp (***M. minuta,*** Puri and Garg, 1953). The lateral bundles during their course downwards towards the ventral face of the sporocarp bifurcate into two branches. This bifurcation is effected somewhere near the middle of their course downwards. The adjacent branches of the bifurcated laterals anastomose near the ventral edge of the sporocarp and form a closed network (Fig. 18.12). A small **placental bundle** arises from each lateral bundle near its point of bifurcation. It enters the receptacle and bifurcates into two placental bundles that traverse the entire receptacle along the dorsal and ventral sides. The placental bundles are not given off from the first and the last pairs of lateral bundles, because there are no receptacular ridges in this area.

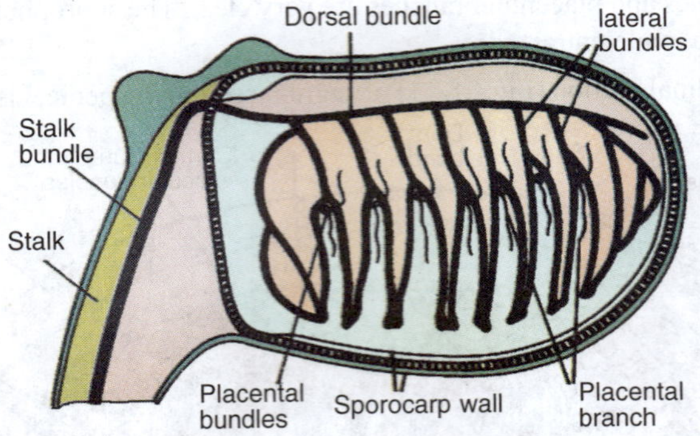

Fig. 18.12. Showing vascular supply of the sporocarp.

Development of the Sporocarp

Johnson (1898) gave a vivid account of the development of the sporocarp in ***M. quadrifolia.*** The sporocarp arises from one of the marginal cells of a very young leaf (Fig. 18.13, A). This marginal cell lies on the inner face of the leaf and appears after the leaf apical cell has cut off 6-8 segments on either side. So the young leaf is just a mound of few cells when the marginal sporocarp initially differentiates (Fig. 18.13, B). This cell behaves just like the two-sided leaf apical cell and cuts off segment on either side. One of these segments may behave as apical cell of the second sporocarp and so on for the third sporocarp. This means that the two or three sporocarps are ontogenetically the primary, secondary and tertiary branches of the leaf.

The sporocarp initial cuts off segments on either side and forms a mass of undifferentiated cells, which later curves markedly so that its distal end is directed in a horizontal plane. The distal portion grows further and assumes an elongated shape. It grows into a sporocarp. The segments cut off by the apical cell divide further and distinguish the outer epidermal layer followed by a two layered hypodermis. The rest of the cells are thin-walled and enclose small air spaces. This region corresponds to the leaf segment. Two rows of **soral mother cells** appear on the ventral side of the young sporocarp. The soral mother cells are distinguished from the marginal cells of the very young sporocarp (Fig. 18.13, C). Later, involutions appear on the surface of the sporocarp (Fig. 18.13, C-E) facing downwards (ventral) and form two alternating rows of canals. These are the soral canals. Each soral canal is lined on its inner face by a tissue called the indusium. This tissue later develops into a two layered indusium. The soral mother cells divide to form the receptacular surfaces that are also arranged in two alternating rows. The receptacle of each row faces the indusium (Fig. 18.13, E). The soral canals open externally by small pores, but during further growth these pores enclosed by the greater growth of the shoulders of the sporocarp, lateral to the sori. Each soral receptacle develops sporangial initials in a basipetalous manner a little before the closure of the soral canals (Fig. 18.13, F). The first sporangial initials appear at the tip of the receptacle and develop into megasporangia.

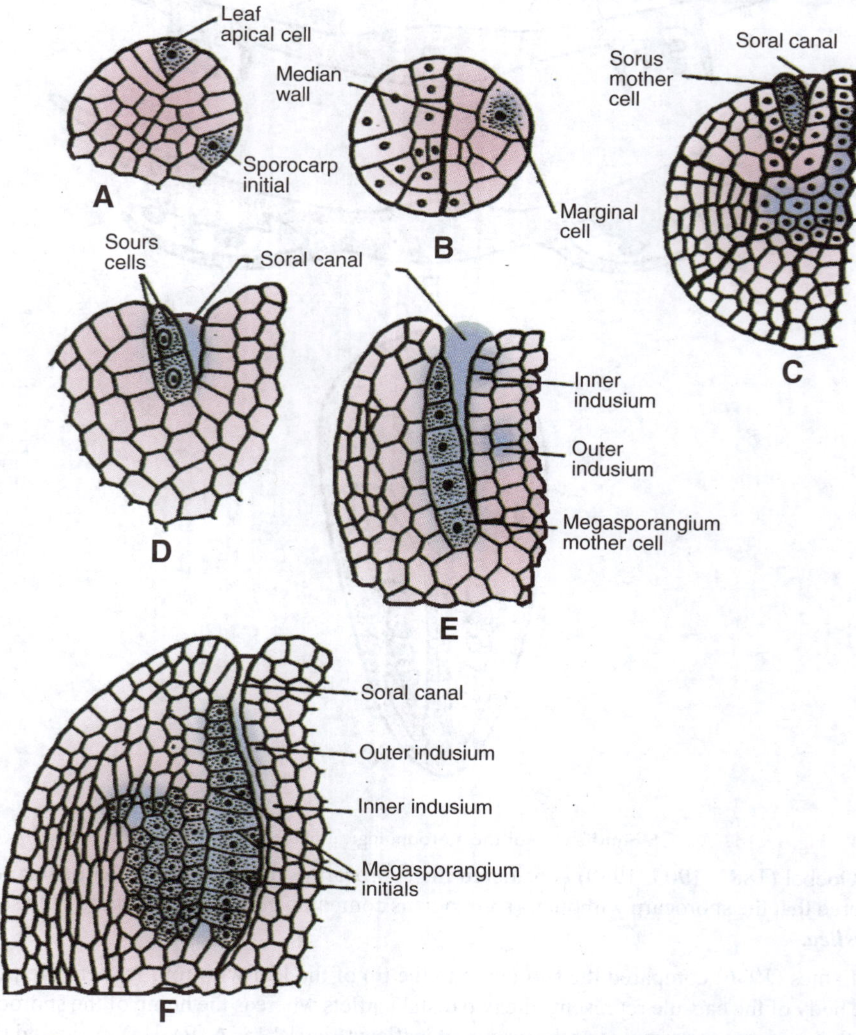

Fig. 18.13. Stages in the development of sporocarp in *M. quadrifolia.* (After Johnson)

The other initials appear along the sides and grow into microsporangia. The megasporangia thus develop first and are older than the microsporangia.

Morphological Nature of the Sporocarp

Two main interpretations have been put forth to explain the morphology of the sporocarp of ***Marsilea.*** These are :

(*a*) The Laminar or Leaf Segment Hypothesis. This hypothesis believes that the sporocarp is a modified lateral, fertile segment of the leaf or frond. It has been supported by majority of the workers whose views are given below :—

Russow (1872) and Busgen (1890) regarded the sporocarp of ***Marsilea*** to be composed to two leaflets whose ventral surface face each other.

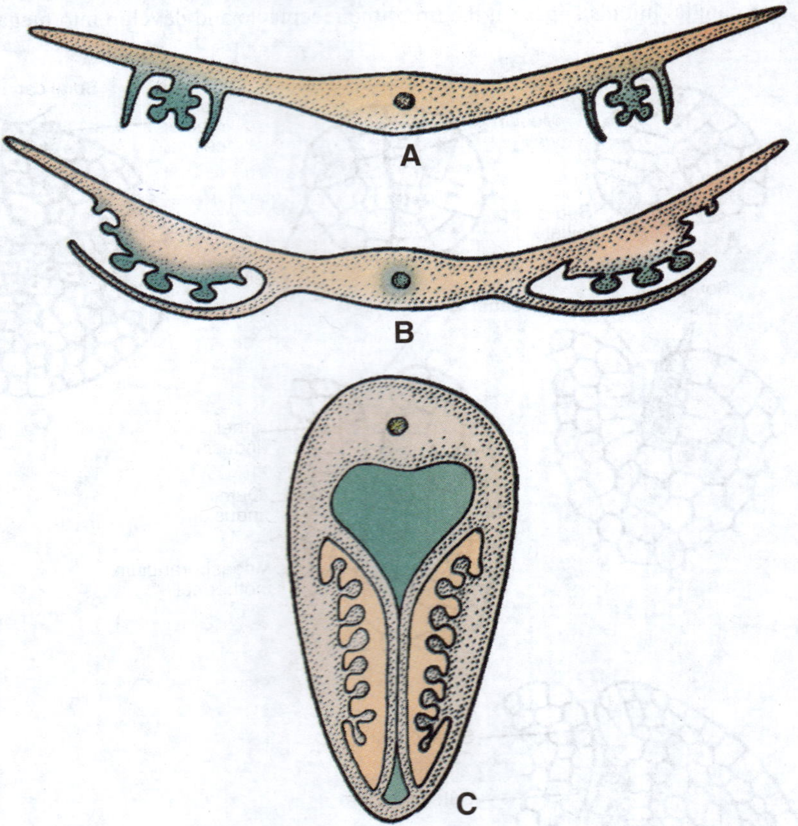

Fig. 18.14. (A—C). Smith's view of the morophological nature of sporocarp. (After Smith)

Goebel (1882, 1905, 1930) considered the sporocarp to be a fertile leaflet or the pinna. He considered that the sporocarp without its sporangia is comparable to a single leaflet of the sterile leaf of ***Marsilea.***

Eames (1936) compared the sporocarp to the tip of the leaf with four leaflets. He postulated that the body of the capsule represents the two distal leaflets whereas the hump of the sporocarp with two teeth represents the remains of the proximal leaflets (Fig. 18.15, A, B). He considered the hump as reduced proximal pair of leaflets. His conclusions are based upon the similarity of course of vascular bundles in the sporocarp and the four leaftlets of a sterile leaf. In ***M. quadrifolia*** the main vascular bundle sends two vestigial branches into the lower tooth. This suggests that the hump represents the remains of the proximal pair of leaflets.

Smith (1938, 1955) compares the sporocarp of ***Marsilea*** to an enfolded leaflet (Fig. 18.14, C) or a pinna that is traversed by a single midrib (Fig. 18.14, A—C) that gives off a number of lateral veins on either side. He linked such an enfolded pinna to the enfolded leaflet of a leptosporangiate homosporous fern (Cyatheaceae) bearing gradate sori on its abaxial side (Fig. 18.14, B—C). The gradate sori in ***Marsilea*** and presence of distinct indusia support Smith's view point that the sporocarp of ***Marsilea*** is comparable to the enfolded pinna of a Cyatheaceous fern (Fig. 18.14, A—C).

Takhtajan (1953) stated, "the sporocarp of the Marsileales may be considered a more specialised fertile segment of a leaf of the type common to Schizeaceae."

Puri and Garg (1953) have discussed in detail the literature regarding the morphological nature of the sporocarp (Fig. 18.15, C). While commenting upon its morphological nature they state, "The sporocarp is equivalent to a single leaflet which consists of as many pinnules as the number of commissural bundles (lateral bundles), the latter forming the midrib of the former". They further observed that in ***M. minuta*** the sporocarp trace diverges from one end of the V-shaped vascular bundle of the petiole. Such a vascular supply is characteristic of the leaflets and hence the laminar nature of the sporocarp is fully supported.

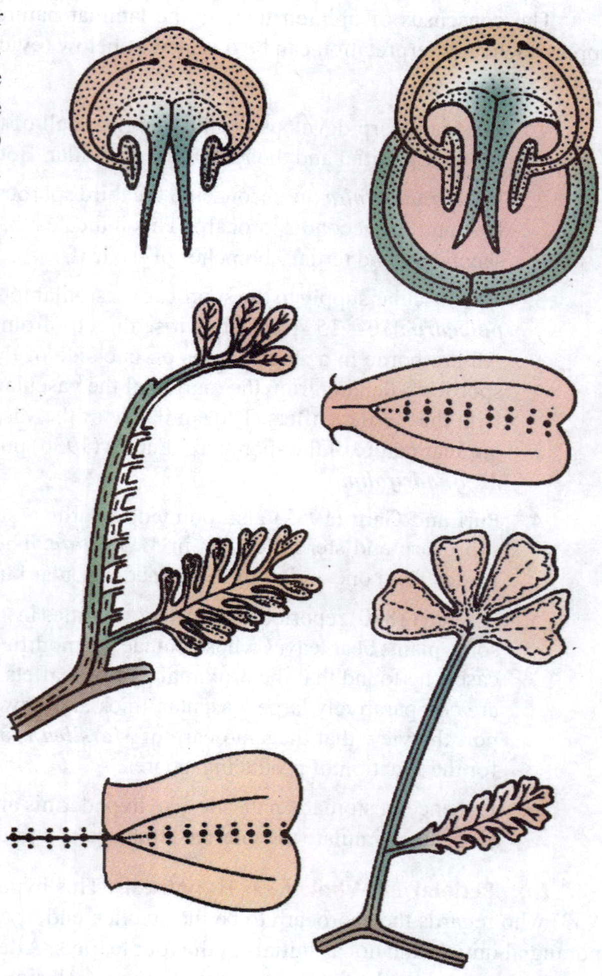

Fig. 18.15. A—F. Morphological nature of sporocarp. A—B. Eames view of morphological nature; C. Representing Puri and Garg's view; D—F. Representing Gupta's viewpoint.

Gupta (1962) also supported the laminar nature (Fig. 18.15, D—F) of the sporocarp. He regards the sporocarp to be a leaflet with as many marginal lobes as the lateral bundles. He does not regard the sporocarp to be a compound structure formed by the concrescence of the pinnules. In his opinion the pinna or the leaflet was only lobed and not divided into pinnules as is regarded by Puri and Garg. In his own words, "to the author it appears that the sporocarp is a leaflet which consisted of as many lobes as the number of commissural bundles. The sori are situated marginally in the depressions of these lobes (Fig. 18.15, E) rather than on their sides. Consequently they get their vascular supply from the midrib bundles, that is, at the point of bifurcation of the commissurals themselves."

The consensus of opinion favours the laminar nature of the sporocarp and the points that support such an interpretation can be outlined as below (evidences in favour of laminar or leaf segment theory) :—

1. The sporocarp develops from a marginal cell of the leaf. The early segmentation of the sporocarp initial and the leaf initial is similar. Both have two-sided apical cell.
2. In ***M. quadrifolia*** the second and the third sporocarps arise from the marginal cells of the first and the second sporocarp. This indicates that the three sporocarps are the primary, secondary and tertiary branches of the leaf.
3. The vascular supply to the sporocarp is similar to that of the leaflets of a sterile leaf. In ***M. polycarpa*** 10—15 sporocarps arise directly from the petiole in an acropetalous manner. All are borne in a single series on one side of the petiole. The vascular supply to each sporocarp departs from the margin of the vascular bundle of the petiole. Same is the case with the sterile leaflets. The similarity of the vascular supply is clearly indicative of the laminar nature of the sporocarp. Eames (1936) pointed out such a similarity in the case of ***M. quadrifolia.***
4. Puri and Garg (1953) also pointed towards a similar nature of vascular supply to the sporocarp and sterile leaflets in ***M. minuta.*** The vascular supply in both the cases diverges from one end of the V-shaped vascular bundle of the petiole.
5. Busgen (1890) reported certain abnormalities in ***Marsilea.*** He observed that in ***M. hirsuta*** some plants bear leaves whose pinnae get modified into sporocarp-like structures. In some cases, he found that the proximal pair of leaflets becomes very small and the distal ones are comparatively large, vascular, thick and brown. Such teratologies in nature also support the view that the sporocarp of ***Marsilea*** is a modified lamina or leaflet, specialised for the function of producing spores.
6. Presence of stomata, palisade-like hypodermis and the inner parenchymatous tissue with air spaces simulate the internal structure of a pinna.

(*b*) Petiolar or Whole Leaf Hypothesis. This hypothesis was put forth by Johnson (1898, 1933) who regards the sporocarp to be the swollen end of a sterile leaf whose marginal cells act as sporangial initials and not as initials of the four leaflets. After a detailed study of the development of the sporocarp and the leaf, Johnson came to the conclusion that sporocarp develops from an initial with two cutting faces. The behaviour of the initial is exactly similar to that of leaf initial. The leaflets or the four pinnae of the sterile leaf develop not from a single initial but from many five-sided marginal cells. The ontogeny of the pinnae is different from that of the sporocarp. This led Johnson to believe that sporocarp is comparable to the whole leaf and not to a single pinna or a leaflet . The second point which led Johnson to put forth his hypothesis was that the primary sporocarp may give rise to secondary sporocarp or even a tertiary sporocarp which throws light on its being a twice or thrice compound foliar structure. The third point that favours his hypothesis is given his own words, "The wall of the capsule, including the vascular bundle system are developed entirely or practically so, from the four sections in each segment dorsal to the marginal cell. Hence the two valves into which the capsule splits at bursting cannot be homologised with the divisions of the lamina, since these are developed from the numerous sections formed on both sides by the continued activity of the marginal cell. For this reason also any seeming similarity in the branching of the vascular bundle systems of the two organs can have no meaning in the direction of homology". After discussing his viewpoint Johnson concluded, "according to our present knowledge we may consider the capsule as the swollen end of a petiole in which the marginal cells are devoted to the formation of the sporangia instead of a lamina."

SPORANGIA

Marsilea is heterosporous. The **megasporangia** or **macrosporangia** produce **megaspores** (one in each megasporangium) and the **microsporangia** produce **microsprores** (32—64 per microsporangium). The sporangia develop in a basipetalous manner on the receptacle. The megasporangia develop first at the top of the receptacle and are, therefore, older (Fig. 18.16). The microsporangia develop later along the sides of the receptale (Fig. 18.16). Both the types of sporangia remain shut within the sporocarp till it dehisces.

A. Megasporangia

***(a)* Structure.** The megasporangium is an almost spherical structure with a short stalk. It is larger than the microsporangium and positioned at the tip of the receptacle (Fig. 18.16, A). It has a single layered jacket of thin-walled cells. There is no annulus. Next to the wall are two layers of tapetal cells (Fig. 18.16, A) that surround the megaspore mother cells. At maturity the tapetal cells disorganise to form the plasmodial fluid (Fig. 18.16, B, C) and all but one megaspore mother cells disorganise. The surviving or the functional mother cell undergoes meiosis to form 4 haploid cells out of which 3 degenerate and one grows big to form the megaspore. Kolhatkar (1938), Eames (1906) and Smith (1938, 1955) observed that all the mother cells divide meiotically to produce up to 32 or even 64 spores. Out of these all, except one, degenerate. The surviving one functions as the megaspore (Fig. 18.16, C).

In abnormal sporocarps more than one megaspores may be formed. At maturity the normal megasporangium has one-layered wall and contains one large megaspore that almost fills its cavity.

The megaspore escapes by the distinegration of the megasporangial wall.

***(b)* Development** (Fig. 18.17, A—E and 18.16. A—C). The megasporangia develop from superficial cells at the tip of the receptacle. A superficial cell acts as a megasporangial initial (Fig. 18.17 A) and divides by transverse wall into an outer and inner cell. The outer cell develops

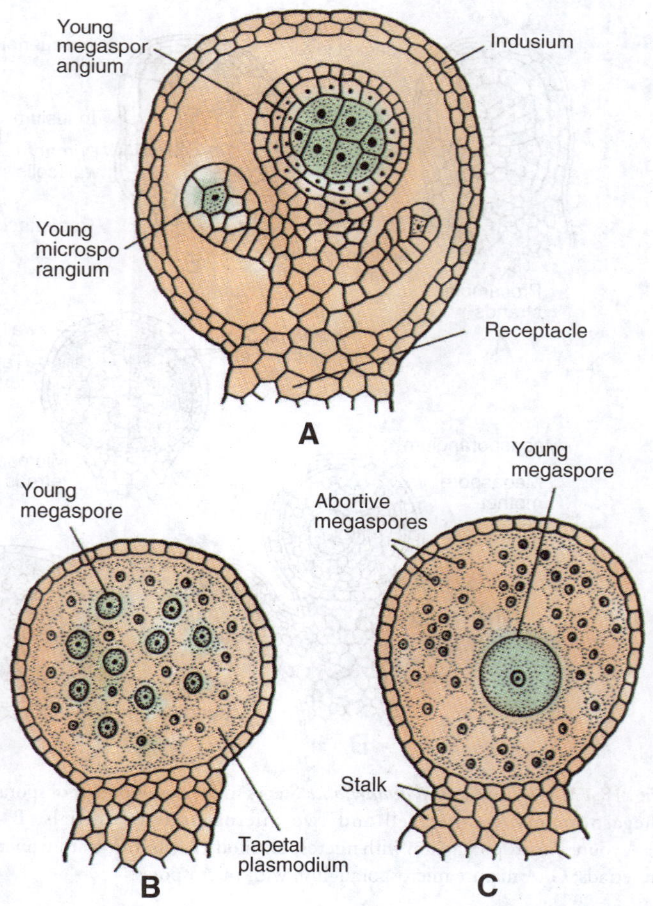

Fig. 18.16. (A—C). *Marsilea quadrifolia. A.* Sorus of *M. quadrifolia* showing indusium, receptacle, megasporangium and two young microsporangia (horizontal section); B—C. Megasporangia.

into an entire sporangium. It divides by three successive diagonal divisions to distinguish a tetrahedral apical cell with three cutting faces. The apical cell cuts off two segments along its three cutting faces. These segments constitute the stalk of the sporangium (Fig. 18.17, A). The apical cell now divides by an arched periclinal wall towards its outer face (Fig. 18.17, A). This puts an end to the apical growth of the sporangium. This last periclinal division distinguishes an outer smaller **primary jacket** cell and an inner tetrahedral **archesporial** cell (Fig. 18.17, A). The Jacket cell divides by anticlinal walls to form a single layered wall. The archesporial cell divides by periclinal wall to cut off **primary tapetal cells** (Fig. 18.17, B) which by further periclinal and anticlinal divisions gives rise to a two-layered **tapetum**. The central primary sporogenous cell divides to form a mass of 8—16 spore mother cells. These divide by meiosis to form tetrads of spores. In all 32 to 64 spores are produced. Out of these 31 or 63 degenerate and only one functions (Fig. 18.16) as the **megaspore**. It grows bigger in size at the expense of others. According to some authors all but one megaspore mother cell degenerate. Boterberg (1956) in some species of ***Marsilea*** and Machlis and Rawtisher Kunkel (1967) in ***7M. vestita*** report the formation of 16 mother cells which form tetrads. Out of these 15 tetrads abort and three megaspores of the sixteenth tetrad also abort leaving behind only one functional megaspore. It grows in size and fills the megasporangium.

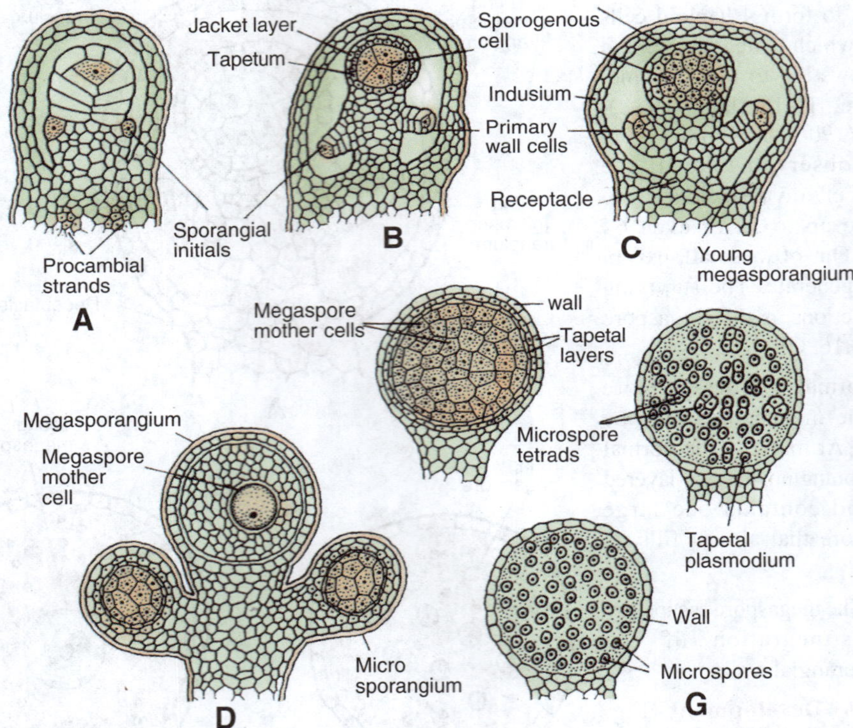

Fig. 18.17. A—G. *Marsilea quadrifolia.* Stages in the development of sporangium. A. T.S. young with one terminal megasporangial mother cell and two microsporangial initials; B—D. Latter stages of development; E. A young microsporangium with microspore mother cells; F. A mature microsporangium with microspores arranged in tetrads; G. A mature microsporangium with microspores.

Microsporangia

The microsporangia develop along the sides of the receptacle and are smaller in size; have longer stalks and oval capsules. The structure is similar to that of the megasporangium except that it has 32—64 microspores. The number of microspores varies from 40—64 in ***M. minuta*** (Gupta, 1962), from 12 to 32 in ***M. rajasthanensis*** (Gupta 1962) and 10—32 in ***M. codensata.***

The development of the microsporangia is similar to that of megasporangia except that the microsporangia have larger number of microspores.

During sporangial development certain changes take place in the wall of the sporocarp. The cells of the hypodermal layer divide by periclinal divisions into two layers. The cells of the outer layer divide anticlinally and indergo considerable elongation. The cells of the inner layer also elongate. The cell walls of both these layers and the epidermis become considerablly thick and hard thus forming strong outer covering . The parenchymatous cells around the receptacles and the receptacular tissue may undergo gelatinisation.

Dehiscence of the Sporocarp (Fig. 18.18)

As the sporangia mature the parenchymatous tissue of the receptacle and the parenchyma tissue next to sporocarp wall become gelatinous. Under natural conditions the sporocarp may open after two or three years of its maturation. Its outer thick and hard rind is resistant to drying and injury and stands in the way of its early dehiscence. The opening is probably due to partial decay of the hard and stony wall of the sporocarp. The spores of ***Marsilea*** remain viable for 20-30 or even 50 years and therefore a long delay in the opening of the sporocarp is not detrimental to the genus. In laboratory a sporocarp can be opened by giving a slight cut at one end. The sporocarp imbibes water through this cut end. The gelatinous ring or the sporophore also absorbs water and swells. After about 15-30 minutes the sporophore protrudes out through the split or cut portion of the sporocarp. The sporophore or the gelationus ring expands and pushes out of the split sporocarp. It also pulls along with it the sori that are attached to it. The sporophore becomes free from the sporocarp wall except along its dorsal side; One of its ends, usually on the vertral side, breaks and the sporophore starts straightening. It comes out of the sporocarp in a looped condition and ultimately straightens in the form of a long

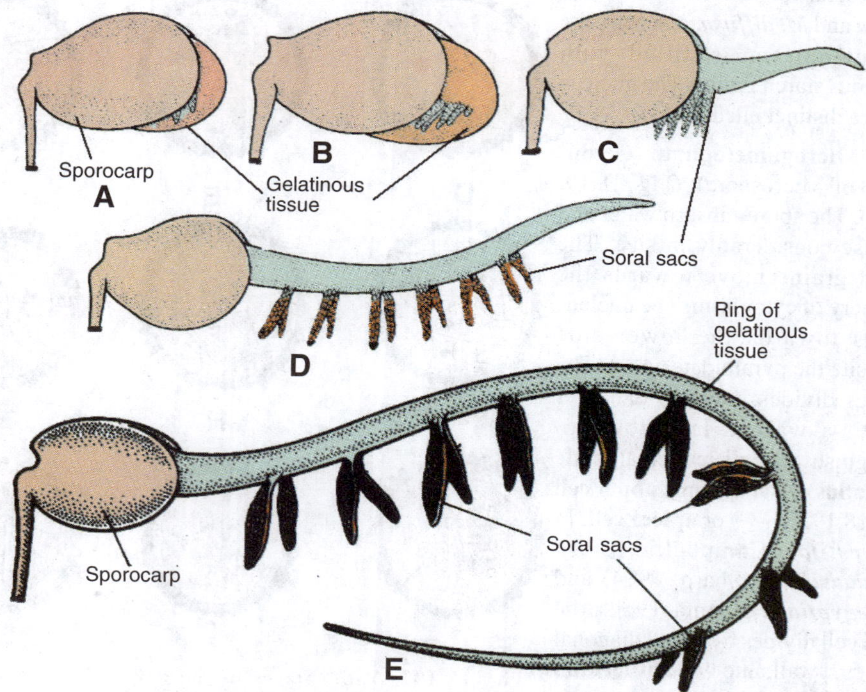

Fig. 18.18. (A—E). Various stages in the dehiscence of sporocarp.

gelatinous cylinder bearing a number of sori (that varies with species) in two alternating rows (Fig. 18.18, A—C). The sori appear as sac-like structures and the contained sporangia can be clearly seen. This whole process of liberation of contents, under artificial conditions takes 1-2 hours or even more.

The indusia or the walls of the soral sacs gelatinise and liberate the sporangia. Later the sporangial walls also gelatinise and the spores are liberated in the surrounding water. This whole process usually requires five to six hours after the extrusion of the sporophore. The spores usually do not germinate within the sporangia and do so only after being liberated.

GAMETOPHYTE GENERATION

The microspores and the megaspores germinate separately to produce male (microgametophyte) and the female (megagametophyte) gametophytes. The spores are, therefore, the pioneer structures of the gametophyte or sexual generation.

Microspores

The microspore is almost yellow in colour and more or less globose in shape, with a faint triradiate ridge towards the pyramidate apex (Fig. 18.19A). The microspores vary greatly in diameter. Usually they are 60 μ in ***M. minuta,*** 60 to 75 μ in ***M. quadrifolia*** and about 45 μ in M. ***aegyptiaca*** (Gupta, 1962). The uninucleate cytoplasm is surrounded by the inner cellulose cell wall called the **intine** or **endosporium** which in turn is surrounded by the outer **exine** or **exosporium**. The exine is transparent and may be variously sculptured. Erdtman (1960) studied in detail the wall structure of the microspore in ***M. hirsuta*** and ***M. diffusa,*** respectively. The cytoplasm is filled with numerous starch grains. The nucleus shows a distinct nucleolus.

Microgametophyte (Germination of Microspore), (Fig. 18.19, A—L). The spores absorb water and increase considerably in size. The starch grains move towards the periphery of cytoplasm. The nucleus moves towards the lower side (opposite the pyramidate apex). The nucleus divides into two daughter nuclei, a wall is laid down to distinguish a small lenticular **male prothallus cell** and a large **upper cell** (Fig. 18.19, B—C) or **apical cell**. In ***M. vestita*** (Campbell, 1892), ***M. quadrifolia*** (Sharp, 1914) and ***M. aegyptiaca*** (Gupta 1962) the apical cell divides by a wall diagonal to the first wall, into two **antheridial initials** (Fig. 18.19, D, E). In these

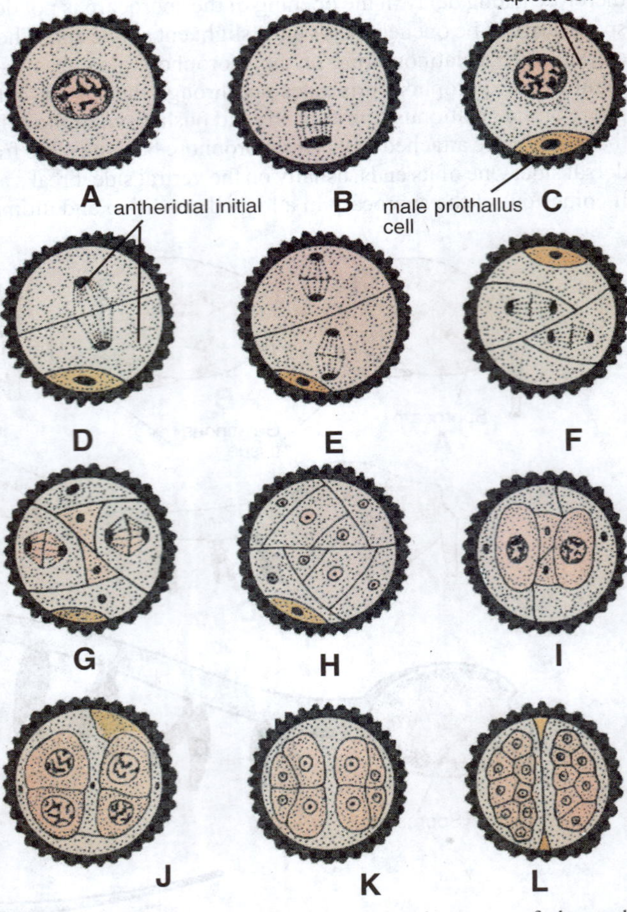

Fig. 18.19. (A–L). Stages in the development of the male gametophyte in *M. quadrifolia* (After Sharp)

species there is only one prothallial cell. Belajeff (1898) and Demalsy-Feller (1956) reported that in ***M. elata*** and ***M. diffusa*** the upper cell divides to cut off another prothallial cell. Kolhatkar (1937) also described the formation of a second prothallial cell in his ***Marsilea*** from Poona (***M. poonensis***). There seems to be a variation among the various species of ***Marsilea*** regarding the number of prothallial cells. Whatever be the case the larger cell or the apical cell after cutting off one or two prothallial cells divides by a diagonal wall into two antheridial initials.

Each antheridial initial develops into an antheridium containing spermatozoids. Their further develop-ment is similar. Each one of them divides by a periclinal wall (Fig. 18.19, F) to form an outer wall or **jacket cell** and a wedge-shaped sister cell. The latter divides by another periclinal wall laid across its angle into a second smaller jacket cell and a large outer cell (Fig. 18.19, G). The large outer cell divides again by a periclinal wall to cut off an outer or peripheral **third jacket cell** and a central cell or the **primary androgonial cell** (Fig. 18.19, H). So each antheridium has three jacket cells and a primary androgonial cell at this stage. During further development the primary androgonial cell in each antheridium divides simultaneously into 16 **androcytes** (Fig. 18.19, I—L) whose protoplasts metamorphose into multiflagellate spermatozoids.

All these changes take place within the microspore wall so that the microgametophytes are **endosporic.** During the formation of the androcytes the prothallial cells and the jacket cells of the antheridia disintegrate. The two masses of the androcyted lie free within the spore wall (Fig. 18.19. L).

Spermatozoids

The exosporium of microspore ruptures at the triradiate ridge. The two androcyte masses are visible through the endosporium which also ruptures and liberates the contents. Usually the 16 spermatids are liberated ***en masse*** but sometimes the mass breaks and we usually see subunits of 4 spermatids (Rice and Laetsch, 1967). Each spermatid cell becomes active and performs slow revolutions which later increase in frequency. The membrane of the spermatids dissolve and releases the spermatozoids. This release is effected by the separation of the spermatozoid from lateral vesicle which remains within spermatid for about a minute and then disintegrates.

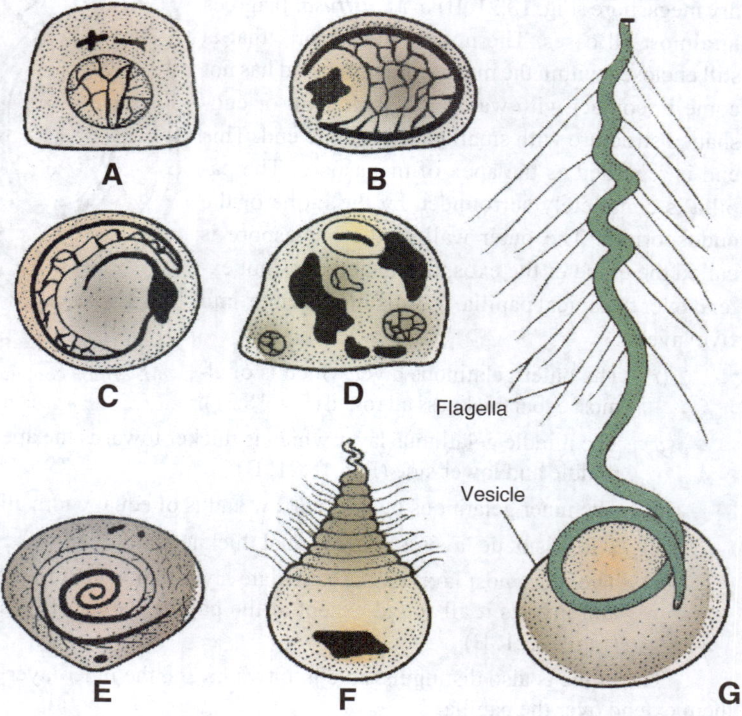

Fig. 18.20. A—E. Stages in the formation of a spermatozoid; F—G. Mature spermatozoids (After Sharp)

The liberated spermatozoid (Fig. 18.20, F,

G) has two main parts : (*i*) the posterior cytoplasmic vesicle and (*ii*) an anterior nucleus and flagella bearing coil containing 9—11 gyres. Mizukami and Gall (1966) recorded the number of flagella to be more than 100. The vesicle contains cytoplasm with starch grains and mitochondria scatered in it. The anterior coiled portion of the spermatozoids consists of : (*i*) a continuous band of mitochondrial nature, (*ii*) an elongated nucleus and (*iii*) a series of microtubules that separate the centrioles or the basal granules from the nuclear and mitochondrial portion. The mitochondrial and the nuclear portions are helically coiled. The nuclear helix is behind the mitochondrial helix and is slightly towards its inner side. The flagella are inserted in two rows along the mitochondrial and nuclear portions of the helix. They arise from paired basal granules (centrioles) that are also arranged in two rows. The microtubules are present between the flagella insertions and helix (mitochondrial as well as nuclear). Some flagella arise in the mitochondrial part of the helix and some from the nuclear part. The flagella are absent in the anterior most region of the coil.

After the sperm reaches the destination, its motility is lost and it uncoils (Fig. 18.20, G). After some time the vesicle is also discarded.

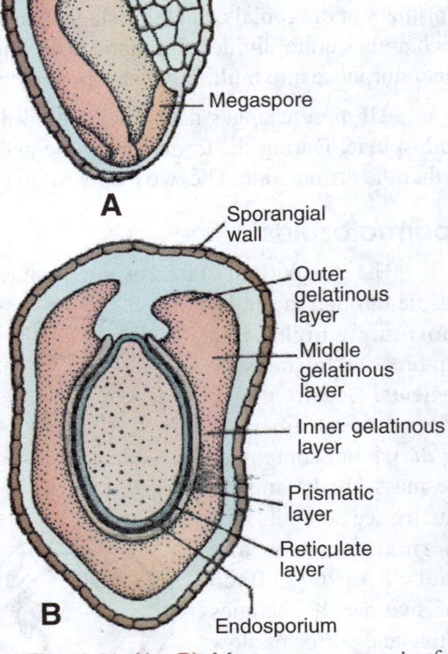

Fig. 18.21. (A—B). Mature megasporangia of *M. diffusa* containing fully mature megaspore. (After Boterberg).

Megaspore

Boterberg (1956) described the structure of the dry megaspore (Fig. 18.21, B) in ***M. diffusa.*** It agrees in almost all cases. The mature megaspore, that is still enclosed within the megasporangium and has not come in contact with water, is a large oval or egg-shaped structure with small papilla at one end. This end is regarded as the apex of megaspore. The papilla is completely surrounded by the intine or the endosporium. The outer wall of the megaspore is called the exine or the exosporium and does not extend over the apical papilla. It is distinguishable into five layers :

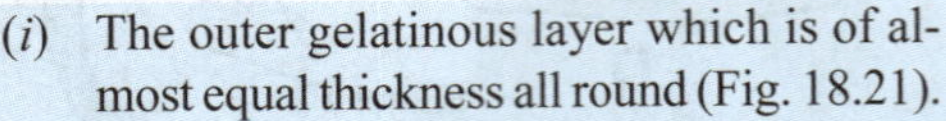

(*i*) The outer gelatinous layer which is of almost equal thickness all round (Fig. 18.21).

(*ii*) The middle gelatinous layer which is thicker towards the apex and narrower towards the middle and lower side (Fig. 18.21, B).

(*iii*) The inner gelatinous layer is narrow and is of equal width all round (Fig. 18.21, B).

(*iv*) The prismatic layer is also of equal thickness all round.

(*v*) The innermost layer or the reticulate layer that is next to the intine or the endosporium and invests it all round except at the papilla. None of these layers covers the papilla (Fig. 18.21, B).

The intine is also distinguishable into two layers, the inner layer and the outer layer. Both of them extend over the papilla.

The nucleus is located in the apical papilla (Fig. 18.22A) and is surrounded by finely granular cytoplasm. The basal region of the spore is filled with coarse granules of starch, albuminous. substances and oil globules that are embedded in the cytoplasm of this part (Fig. 18.22).

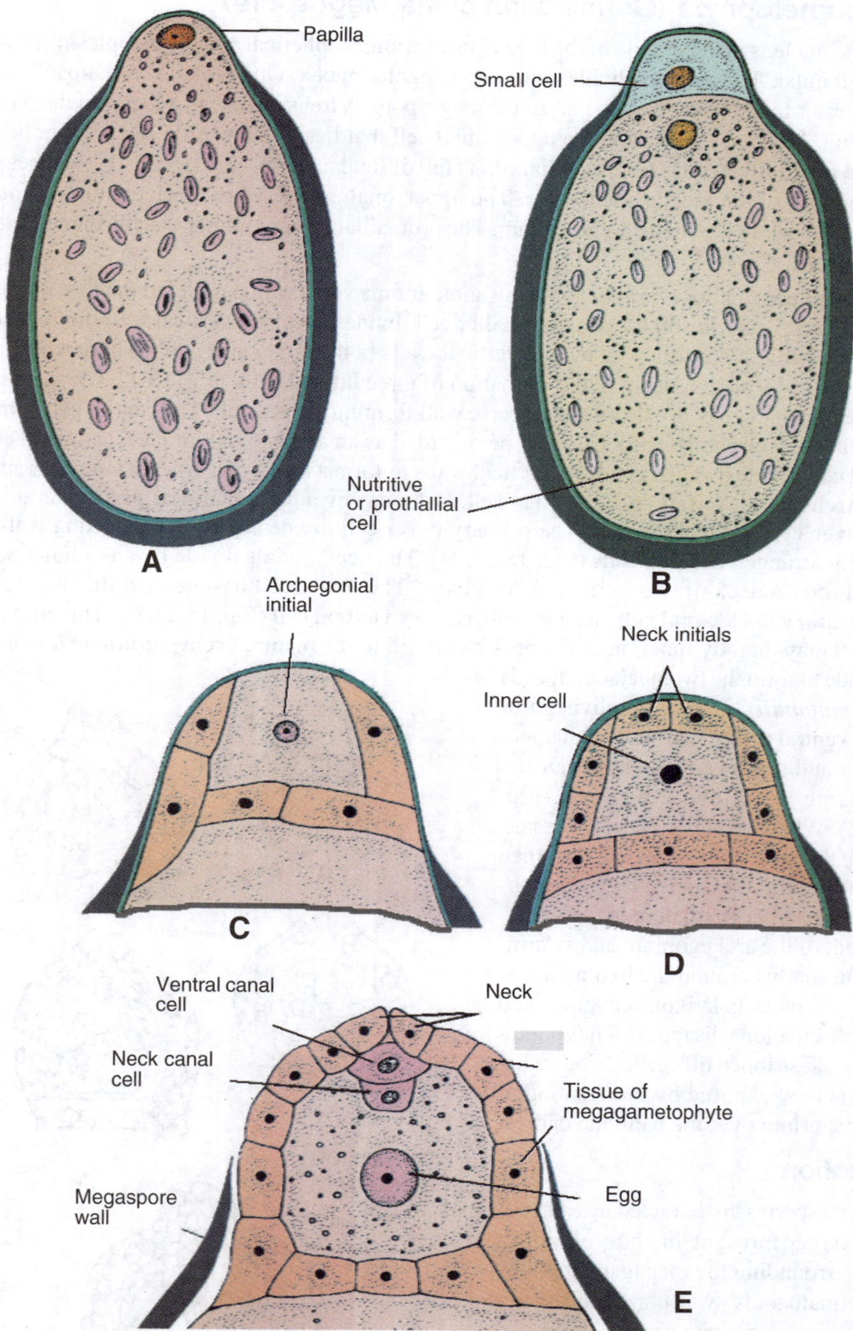

Fig. 18.22. Germination of megaspore of *Marsilea vestita.* A. Undivided megaspore. Note that the nucleus lies in the papilla; B. Two-called gametophyte. The first wall separates the papillate part from the lower larger prothallial cell; C. Archegonial initial surrounded by cells of the gametophyte; D. A developing archegonium; E. A mature archegonium. (After Campbell)

Megagametophyte (Germination of the Megaspore)

The nucleus in the papilla of the megaspore becomes spherical and the cytoplasm surrounding it more granular. The nucleus divides into two daughter nuclei. One remains in the papilla and the other migrates to the lower bigger part of the megaspore. A transverse wall appears at the base of the papilla (Fig. 18.22, B) separating the upper small cell that lies in the papilla and a lower larger cell called the **prothallial cell**. The prothallial cell is full of food material (Fig. 18.22, B) and occupies the greater portion of the megagametophyte. The upper smaller cell, whose exact nature is not understood, develops into the single archegonium. The prothallial cell does not divide further and serves as a nutrient cell.

The upper smaller cell may divide by another transverse wall parallel to the first one and thus separate a **basal cell**. In some cases the smaller cell divides first by a vertical wall thus cutting off a first **lateral cell**. This is followed by two vertical walls in the larger sister cell. These walls intersect the first vertical wall and result in the formation of three **lateral cells** (Fig. 18.22. C), surrounding a **central cell,** which may divide by a transverse wall forming a basal cell. The central cell surrounded by three lateral cells and a basal cell may be regarded as an archegonial initial. The lateral cells and the basal cell divide by horizontal and vertical walls to form a tissue one layer in thickness surrounding the **archegonial initial** or the **central cell**. It divides by a periclinal wall to form an upper **primary cover cell** and a lower cell. The primary cover cell divides by two intersecting walls into 4 quadrately arranged **neck initials** (Fig. 18.22, D). The neck initials divide by an oblique periclinal wall and form a neck of two tiers of 4 cells each. The lower cell has meanwhile divided into an upper **primary neck canal cell** and a lower **primary ventral cell** (Fig. 18.22, E). The primary **neck canal cell** may directly function as the neck canal cell of the mature **archegonium** (***M. vestita***) or it may divide to form the two **neck canal cells** (***M. drummondii***). Transverse division of primary ventral cell divides it into an upper lenticular and small **ventral canal cell** and lower, large and globose **egg cell** (Fig. 18.22, E). When all these changes are taking place the megaspore wall had absorbed ufficient water and assumed the characteristic structure described before. After the development of the archegonium and swarming of the sperms around the hydrated megaspore, the outer gelatinous envelope and the middle envelope disappear. The sporangial wall had stripped off earlier. The archegonium is now exposed by the bursting of the endosporium over the papillate end.

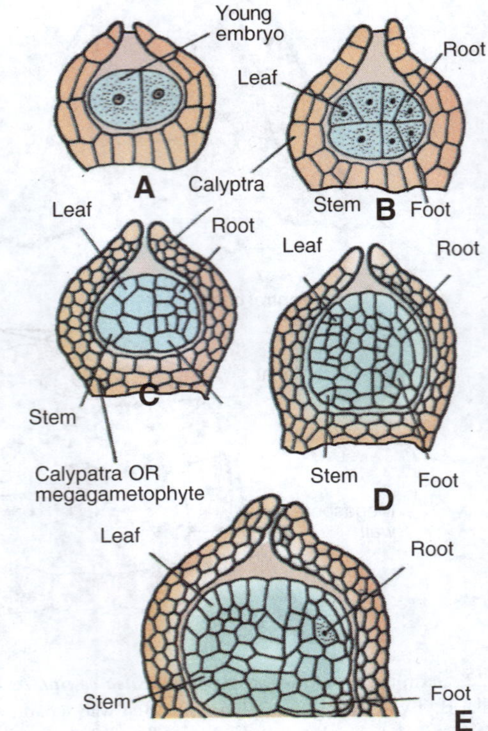

Fig.18.23. (A—E). Stages in the development of embryo in *M. vestita.* (After Campbell).

Fertilization

The sperms are attracted by a chemical substance present in the gelatinous sphere surrounding the megagametophyte. The spermatozoids swarm around the hydrated megaspore containing the megagametophyte. The penetrate the gelatinous envelop, move slowly through it and on reaching the more fluid inner part of the swollen exine, they swim actively and reach the open archegonial necks. Only one spermatozoids enters the open neck and with its anterior end foremost penetrates the egg cytoplasm. According to Atkinson (1913) the posterior portion of the sperm touches the nuclear membrane of the egg nucleus and later it enlarges and enters the membrane thus effecting **syngamy.** Most of the sperms remain embedded in the mucilage around the egg. They straighten their coils, show senesecence and die.

After, or even during fertilisation the gelatinous layers of exine form a homo-geneous matrix with numerous sperms embedded in it.

Parthenogenetic development of the embryo has been observed by Shaw (1897) and Nathansohn (1900) in ***M. macra*** and by Strasburger (1907) in ***M. drummondii.*** Gupta (1962) states, "it has been noticed in our laboratory by Senn in ***M. rajasthanensis, M. aegyptiaca*** and ***M. minuta***". In such cases the unfertilised egg develops into an embryo. Strasburger (1907) reported that the megaspores of ***M. drummondii*** were diploid and they produced diploid megagametophytes that bear archegonia with diploid eggs. These diploid eggs develop into diploid embryo with $2n = 32$.

EMBRYOGENY

After fertilisation the zygote in ***M. vestita*** takes 24 hours to develop into a young sporophyte bearing the first leaf. The first wall of the zygote appears within an hour. It is parallel to the long axis of the archegonium (Fig. 18.23, A) and divides the zygote into two daughter cells of unequal size. The **hypobasal cell** is smaller than the **epibasal cell.** In majority of the leptosporangiopsida the epibasal cell is smaller than the hypobasal cell.

The second wall is at right angles to the long axis of the archegonium and results in the formation of 4 cells. This is the quadrant stage. The quadrants bear a definite relation to the primary organs of the embryo. The shoot and the leaf develop from the epibasal half (Fig. 18.23, B) whereas the foot and the root develop from the hypobasal quadrants. The octant walls are not laid down in a uniform manner. The epibasal quadrants divide by a median wall into four epibasal cells. The cells nearer the

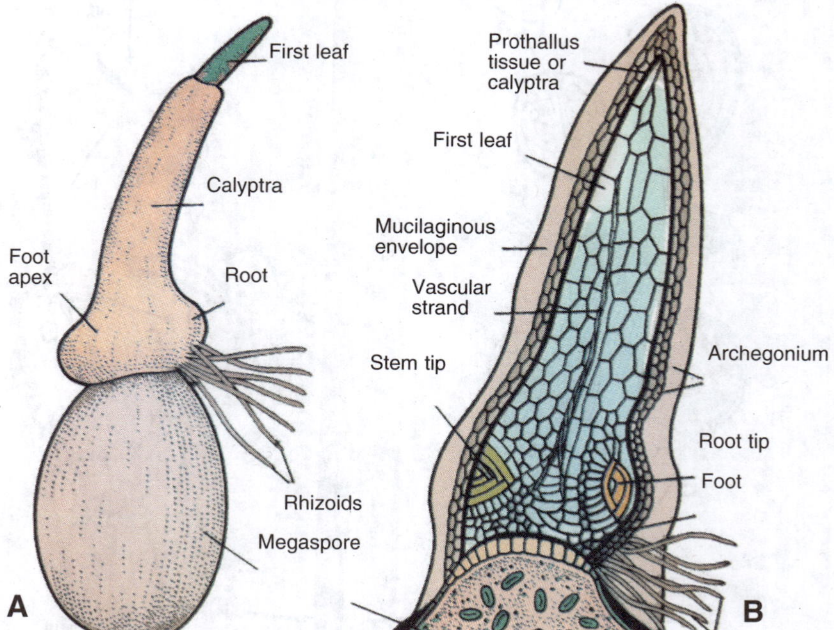

Fig. 18.24. (A—B). *Marsilea.* **A.** Mature female gametophyte bearing young embryo that has pierced through the calyptra; **B.** Median longitudinal section through megagametophyte and development of embryo surrounded by sheath or calyptra. (After Sachs).

archegonial neck give rise to the **first leaf** or the **cotyledon** whereas the posterior ones or those away from the archegonial neck give rise to the shoot (Fig. 18.23, B, C). The former are also called the **anterior octants** and the latter as **posterior octants.** The octants walls dividing the hypobasal quadrants are not median in plane but diverge strongly from the median line. The anterior hypobasal octants give rise to the root and the posterior to the foot (Fig. 18.23, C, D).

The primary octants giving rise to the leaf divide in a regular fashion and soon become quite distinct (Fig. 18.23, E). Further divisions lead to the formation of a long and a cylindrical structure that tapers gradually towards the apex (Fig. 18.23, B). When all these changes are taking place the tissue surrounding the archegonium divides to form a two or three-celled **thick calyptra** (Fig. 18.23, A—E). The thickness of the calyptra at the base of the archegonium remains one layer of cells only (Fig. 18.23, A—E). The calyptra grows along with the embryo so that latter is enclosed within it even at a stage when all the primary organs have taken shape. The first leaf or the cotyledon develops no lamina and remains cylindrical. The stem or shoot apical cell develops from one of the cells the posterior epibasal octant (Fig. 18.23, D; 18.24, B). Adjoining cell may develop into a second leaf.

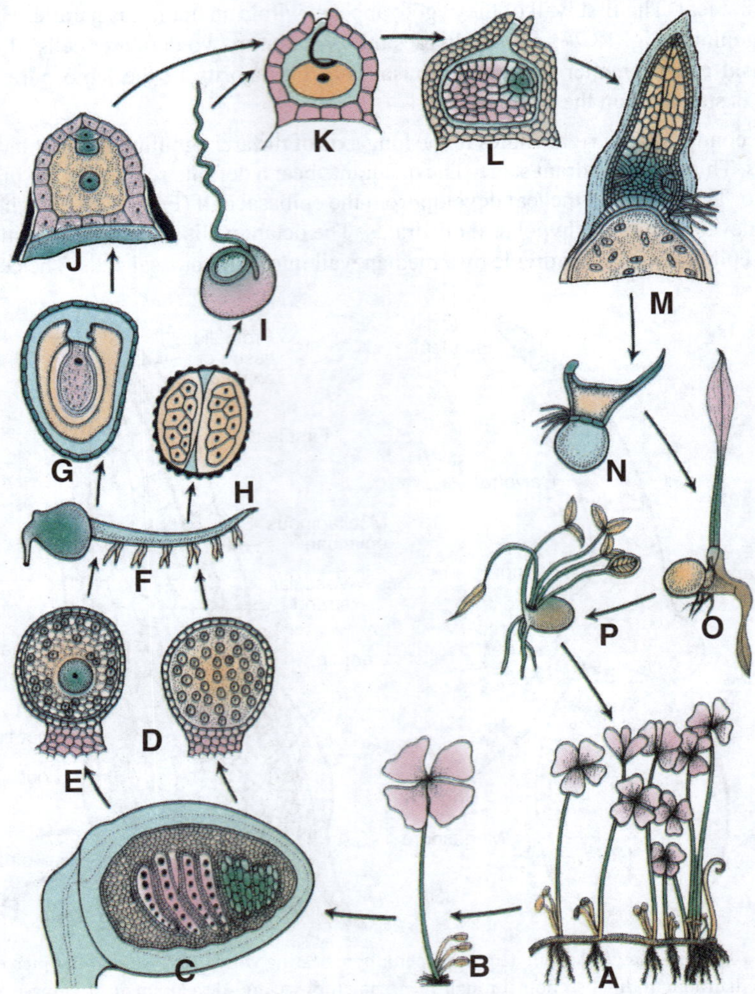

Fig. 18.25A. *Marsilea.* A diagrammatic representation of life cycle.

The first root develops from an apical cell distinguished in the anterior hypobasal quadrant (Fig. 18.24, B). By the activity of this apical cell, the root soon grows vertically downwards and penetrates the soil. By this time the cells of the calyptra become green and numerous rhizoids may develop from the basal layer of the calyptra (Fig. 18.24, A).

The foot develops from the posterior hypobasal octants. The cells divide regularly and then irregularly to form a small root (Fig. 18.23, E). The basal region of the young sporophyte is in direct contact with the food stored in the prothallial cell and performs the nutrient absorbing or haustorial function.

The first leaf soon pierces through the calyptra and appears as an awl-shaped or cylindrical structure without any lamina (Fig. 18.25, N). It is not circinately coiled. The first few leaves are of this type. The leaves that appear later are spathulate (Fig. 18.25, O) and then two lobed (Fig. 18.25, P), four lobed and ultimately leaves with distinct four leaflets or pinnate appear. These are cicinately coiled in young condition.

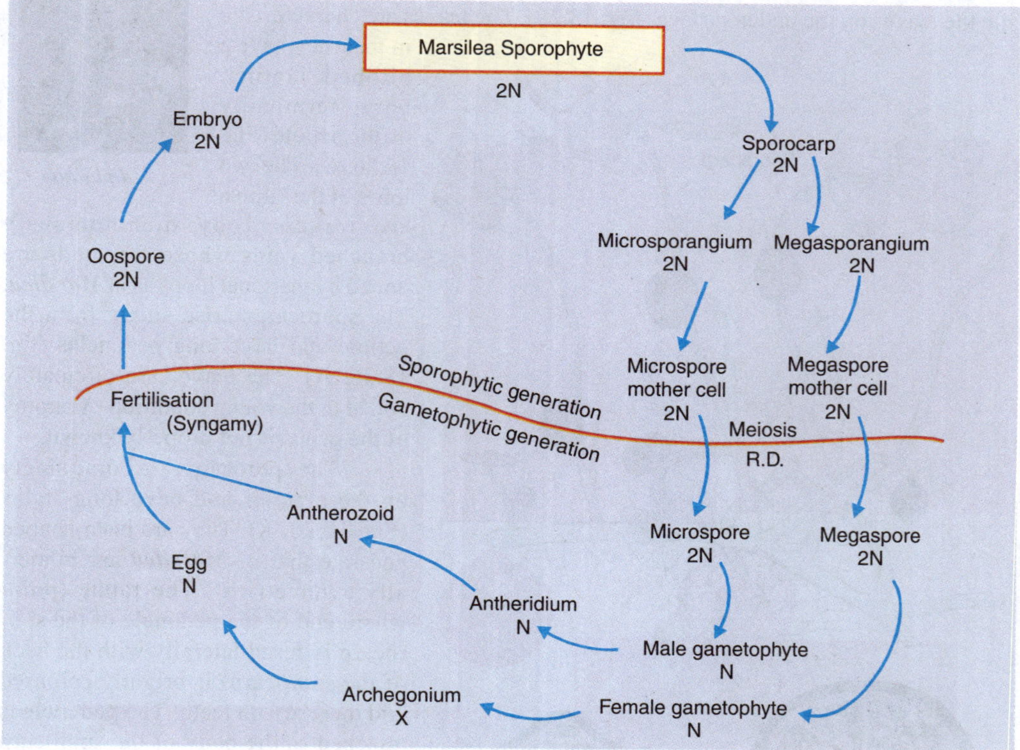

Fig. 18.25B. Graphic life-cycle *of Marsilea.*

Economic Importance

The sporocarps of some species of ***Marsilea*** are edible because of their nutritious value. Bailey (1892) and Bancroft (1893) reported that sporocarps of ***M. drummondii*** yield a starch that can be made into a paste and later cooked into cakes that were eaten by the aborigines in Australia. The sporocarps of ***M. sylvatrix*** have a great nutritious value and were a means of subsistence for the Burke expedition into the interiors of Australia.

Taxonomic Position

Division	:	Pterophyta
Class	:	Leptosporangiopsida
Order	:	Marsileales
Family	:	Marsileaceae
Genus	:	***Marsilea***
Species	:	***quadrifolia***

GENUS–REGNELLIDIUM Lind.

R. diphyllum

This genus is represented by a single species ***R. diphyllum*** Lind. that occurs in Brazil. It has a creeping and profusely branched rhizome bearing two alternating rows of leaves on its upper surface and cluster of roots opposite the leaves on the under surface (Fig. 18.26). The leaves are characteristic in that they have a bilobed lamina, borne terminally on the petiole (Fig. 18.26, A). The two lobes of the lamina are traversed by dichotomously branched veins whose free ends are united by marginal loops as in ***Marsilea***. The sporocarps arise singly from the petiole and have long peduncles (fig. 18.26, A). The leaves are circinately coiled in the young condition. Anatomy of the genus is not properly known.

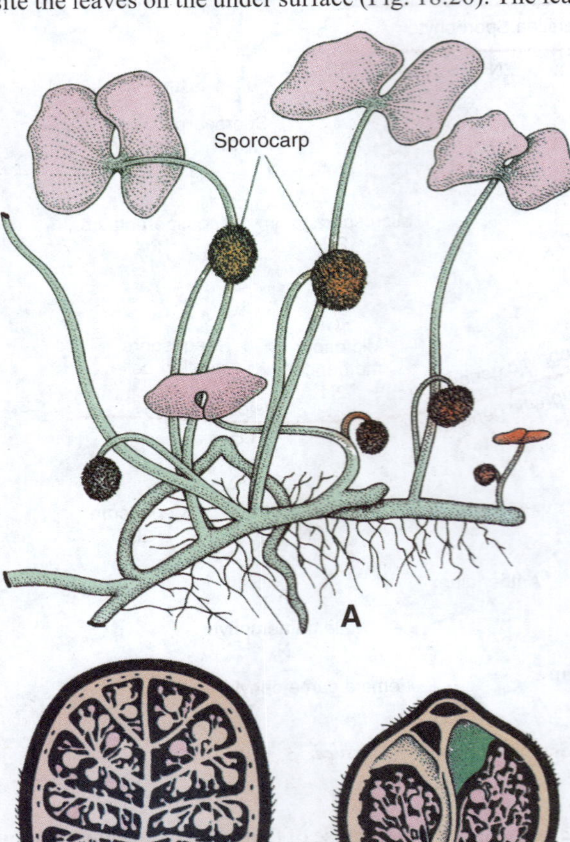

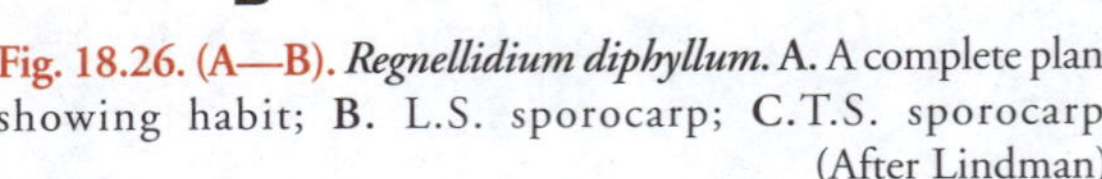

Fig. 18.26. (A—B). *Regnellidium diphyllum.* A. A complete plant showing habit; B. L.S. sporocarp; C.T.S. sporocarp. (After Lindman)

The sporocarps are borne singly on the petioles and have long stalks (Fig. 18.26, A). They are bean-shaped and like that of ***Marsilea*** are bilaterally symmetrical. The raphe (point where end of the peduncle of the sporocarp is fused laterally with the back of the sporocarp) is brightly coloured and there are no teeth. The peduncle is attached to the body of the sporocarp near the middle of the longer side. A longitudinal section of the sporocarp (Fig. 18.26, B) reveals clearly the number and arrangement of the sori, which is similar to that of ***Marsilea.*** The sporocarp has two valves and the sori are not surrounded by any gelatinous ring or the sporophore. Each sorus has a stalked receptacle surrounded by a distinct indusium. The indusium is attached to the lateral sides of the receptacle and not to the sporocarp wall. In this respect ***Regnellidium*** resembles most of the leptosporangiate ferns. The sporangia are attached to the radial walls of the soral sac (Fig. 18.26, B, C). Nothing is known about the details of the development of sporangia.

Dehiscence of Sporocarp

The sporocarp can be made to dehisce in the laboratory by giving a small cut at one end. They absorb water, swell up, the cut portion enlarges. Usually a distinct split first appears in the ventral portion and is followed by two or more lateral splits running parallel to the veins. The sporangia embedded in a mucilaginous mass emerge in occasional spurts through the splits. No definited gelatinous ring is observed. There is neither an indication of distinct sori coming out. This is just like ***Pilularia*** and unlike ***Marsilea.***

The microsporangia containing about 54 to 58 **microspores** (Fig. 18.27, D,E), often emerge entire and usually dehisce readily, in the presence of water. A longitudinal slit appears in the sporangial wall and spores come out in a gelatinous vesicle (Fig. 18.27, A). The vesicle absorbs water, expands several times, becomes thin in consistency and after about four hours the microspores disperse in water.

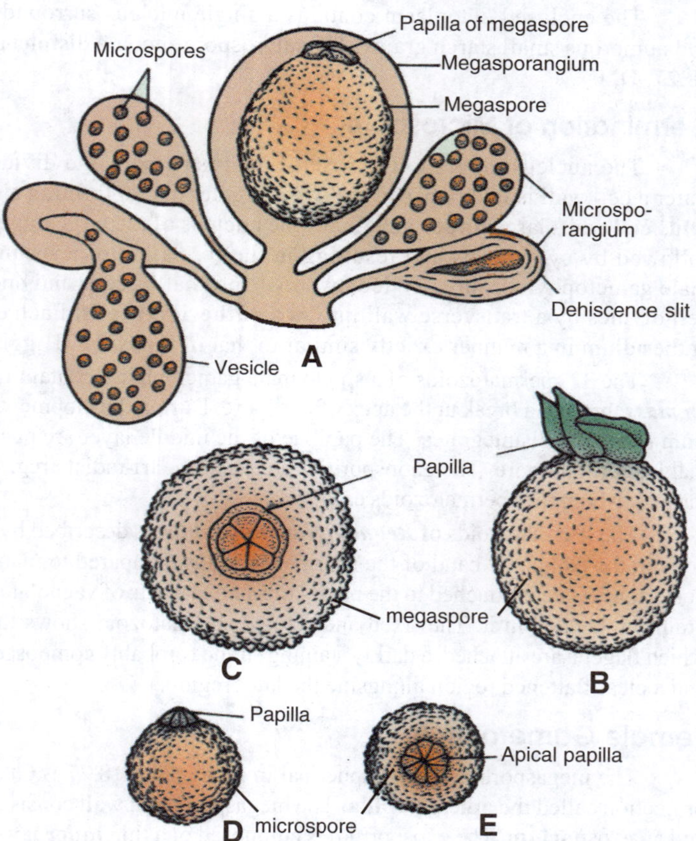

Fig. 18.27. (A–E). ***Regnellidium*** A. Dehisced sporocarp showing a megasporangium with one megaspore and microsporangia with microspores; B—C. Two megaspores; D—E. Two microspores.

The megasporangia contain a single megaspore (Fig. 18.27, A) surrounded by a gelatinous layer that absorbs water and swells greatly. The megasporangial wall absorbs water and bursts, thus liberating the megaspore. The liberated megaspore (Fig. 18.27, B—C) is surrounded by a gelatinous layer. After germination the megaspores rise to the surface of water and get freed from the gelatinous envelope.

Usually a sporocarp contains 160 megasporangia and about 1,920 microsporangia. The time required for germination of the spores varies considerably and depends upon the amount of water available. Higginbotham (1941) observed that the sperms were released after 22 hours after placing the cut sporocarp in water. The archegonia were fully formed on the female prothallus. This indicates that there is a possibility of the later formed archegonia.

Male Gametophyte

The microspores are almost spherical and unicellular surrounded by a thick wall. Higginbotham described the microspore wall as composed of an inner **endsporium or entine** and outer **exosporium** or **exine**. The exine is composed of three layers :

1. An inner thin layer surrounding the endosporium.
2. The middle "**prismatic**" or **vesiculate** layer that bears numerous papillate outgrowths.
3. The outer layer.

The enclosed cytoplasm contains a single nucleus surrounded by vacuolated cytoplasm and numerous small starch grains. The microspore shows a distinct and papillate apical end (Fig. 18.27, D, E).

Germination of Microspore

The nucleus moves to the base of the microspore and divides into two daughter nuclei. Later a cell wall is laid between the two daughter nuclei forming a small lenticular **male prothallus cell** and a large **upper cell.** Later the nucleus of the large upper cell divides again into two, followed by cytokinesis and resulting in the formation of a **second prothallial cell.** Now the male gametophyte is three-celled, two male prothallus cells and one large upper cell. The upper cell divides by a transverse wall into two **antheridial cells.** Each one of them develops into an antheridium in a manner exactly similar to that of ***Marsilea*** (Higginbotham, 1941).

The 32 spermatozoids of a single male gametophyte (containing two antheridia) are released ***en mass*** through a break at the apex of the spore. During development the outer layer of the exosporium gradually disintegrates. The papillae of the middle layer are pushed outwards at the apex thus making a round pore. The endosporium breaks at the tri-radiate region to complete the perforation through which the spermatozoids escape.

The spermatozoids of ***Regnellidium diphyllum*** as described by Higginbotham, have only 2-5 coils in the elongated band of the spermatozoid (as compared to many in ***Marsilea***). The spherical posterior vesicle is attached to the basal coils and consists of vacuolate cytoplasm with several starch grains embedded in it. The fixed and stained spermatozoid shows the several flagella, a strand to which flagella are attached, a darkly staining strand (probably composed of blepharoplast and nucleus) and a clear flattened region alongside the later region.

Female Gametophyte

The megaspore is almost spherical in shape (Fig. 18.27 B,C) and has 5 to 6 angled, conical projection called the **apical papilla.** The megasporangial wall consists of an inner thin endosporium and an **exosporium.** The exosporium is composed of a thin **inner layer,** a thick middle or **prismatic layer** and an outer **papillose layer.** The exosporium according to Chrysler and Johnson (1939) and Higginbotham (1941) extends over the apical papilla (in ***Marsilea*** it is only the endosporium that extends over the papilla). The megaspore is tightly packed with a number of starch grains and other smaller granules of fat and protein. The nucleus lies in or a little below the **apical papilla,** where the cytoplasm is clear and devoid of starch grains.

The megaspore germination is exactly similar (see Higginbotham, 1941) to that of ***Marsilea*** (Fig. 18.28, A—J).

Fertilisation

The developing female gametophyte is surrounded by a uniform and homogenous layer of mucilage. Spermatozoids swim towards the female gametophyte and swarm around the mucilage layer and then penetrate it and swim about in the mucilages. There is no funnel as in ***Marsilea.*** The sperm, sometimes many of them, enter the open neck of the archegonium. One is able to swim to the egg and effect fertilisation.

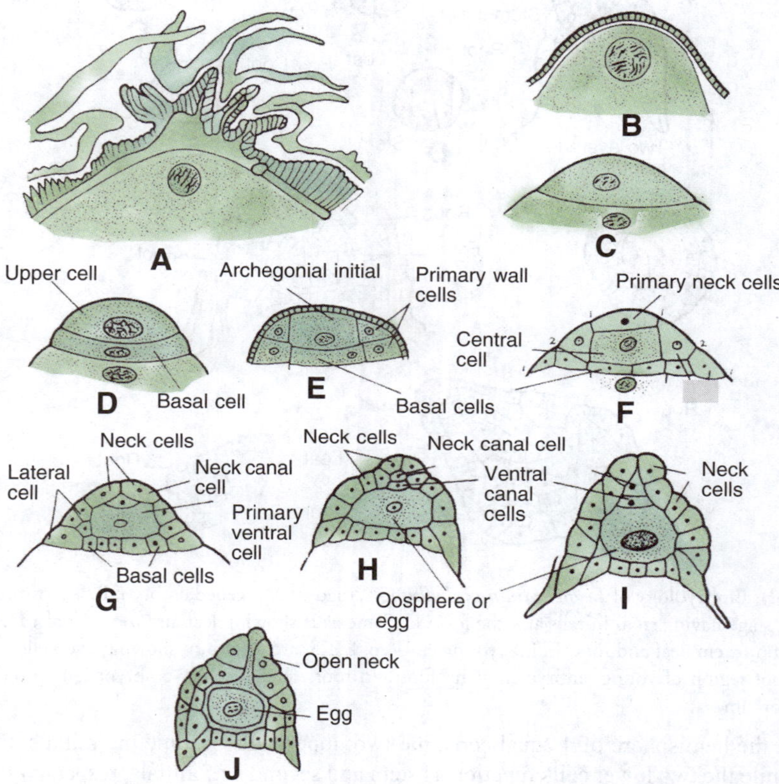

Fig. 18.28. (A–J). ***Regnellidium.*** Development of female gametophyte and archegonia. A. L.S. apex of megaspore before germination; B. Apical papilla of megaspore with nucleus; C. First division of megaspore showing upper smaller cell; D. Showing second division; E. Formation of lateral cells cutting off a central archegonial initial; F. Formation of primary neck cell and a central cell; G. Formation of neck cells and neck canal cell; H. Formation of ventral canal cell and egg; I. Mature archegonium; J. Archegonium with open neck. (All after Higginbotham)

Embryology

The peripheral or lateral cells surrounding the feritlised egg become two-layered and form a well defined **calyptra.** The following steps lead to the formation of the mature embryo (Fig. 18.29, 18.30) :

1. The fertilised egg divides first by a vertical wall (wall parallel to the longitudinal axis of archegonium) into two unequal cells (Fig. 18.29, A).
2. The second wall, as observed in three embryo by Higginbotham (1941), is also vertical, forming two equal cells in one hemisphere and two unequal cell in the other (In ***Marsilea*** & ***Pilularia*** the second wall is transverse).
3. The third wall is laid in a transverse plane and an eight-celled embryo is formed in which four cells of one hemisphere are equal in size whereas in the other hemisphere there are two large and two small cells (Fig. 18.29).

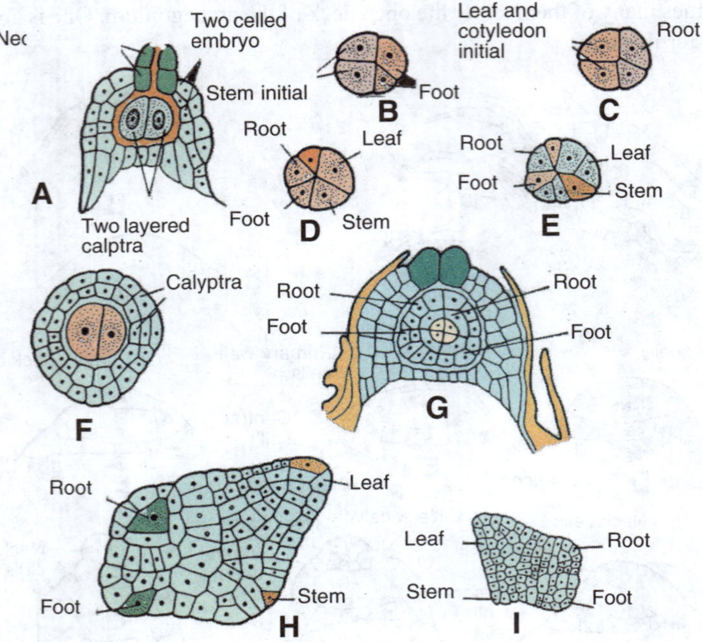

Fig. 18.29. (A–I). **Embryology of *Regnellidium diphyllum*;** **A.** diad or two-celled embryo; **B.** T.S. through embryo showing octant stage having stem initials and the foot; **C.** Same as B showing leaf and root initials; **D.** L.S. octant stage showing root, stem, leaf and root; **E.** L.S. young embryo; **F.** T.S. archegonium showing two-celled embryo; **G.** T.S. through root region of young embryo showing foot and root; **H.** L.S. older embryo; **I.** L.S. older embryo. (After Higginbotham)

4. In the hemisphere of 4 equal cells, the two upper cells become the initials of first leaf, while the two lower cells function as stem and second leaf initials, respectively.
5. The upper larger cell, in the unequally divided hemisphere, functions as the root initial whereas the two lower cells develop into the foot of the embryo (Fig. 18.29).
6. The initials of both the second leaf and the stem are three sided in the beginning. The later formed leaves grow by means of two-sided apical cells.
7. The first leaf initial grows rapidly to form a conical structure (Fig. 18.29). It is surrounded by an elongating calyptra (Fig. 18.30). The first root grows by means of tetrahedral apical cell and forms a small conical protuberance on one side (Fig. 18.30, A). The stem develops slowly, by a three-sided apical cell. A number of rhizoids grow out from the calyptra.
8. The vascular strands develop in the young root and leaf than in the stem tip. They ultimately become joined.
9. The first leaf elongates and breaks the calyptra, and is followed by the root. The first leaf is flattened and almost linear or elliptical. The second leaf is almost spherical. The third, fourth and fifth leaves are two-lobed. The second and third roots appear opposite the second and third leaves. (Fig. 18.30 E)

Loyal (1962) obtained abnormal young sporophytes from the material he obtained from Almora. He observed abnormal sporelings which exhibited "duplication of the primary organs." He categorised them into four types (*i*) Sporelings with two cotyledonary leaves and one primary root; (*ii*) Sporelings with two cotyledonary leaves and two separate primary roots; (*iii*) Sporelings with two shoot primordia; and (*iv*) sporeligs with branched first leaf. Embryology of ***Regnellidium*** was also studied by Higginbotham (1941), Bloom (1955) and Voth (1956).

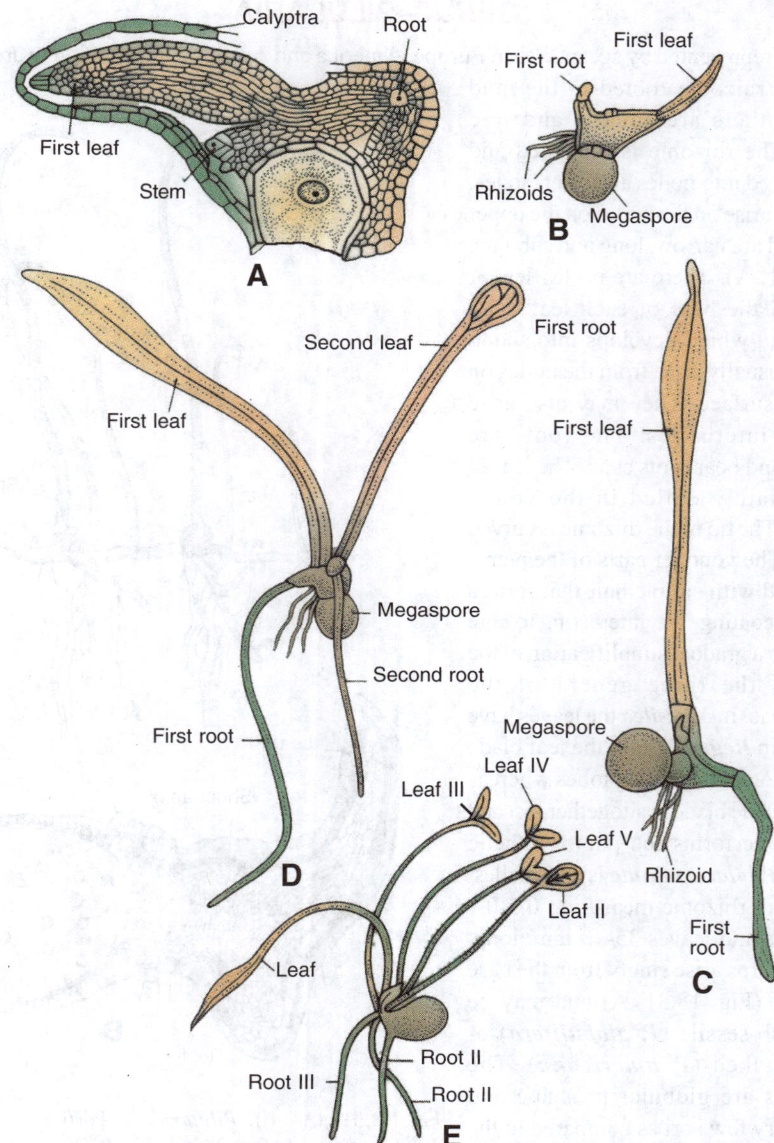

Fig. 18.30. (A–E). Later stages of embryology in *Regnellidium.* A. L.S. older embryo still within the calyptra; B. The embryo breaking through calyptra; C. A weak old plant; D. A young sporophyte with two leaves and two roots; E. Young sporophyte with 5 leaves (about 4 weeks old), (After Higginbotham, 1941).

Loyal (1962) counted 2n =38 in ***Regnellidium.***

Taxonomic Position

Division	:	Pterophyta
Class	:	Leptosporangiopsida
Order	:	Marsileales
Family	:	Marsileaceae
Genus	:	***Regnellidium***
Species	:	***diphyllum***

GENUS–PILULARIA

It is represented by six species in Europe, America and Australia. Some of them are amphibious with their rhizome rooted in the mud whereas others are aquatic and free floating. The rhizome is branched and distinguished into nodes and inter-nodes. The leaves arise on the nodes on the upper surface and are narrow, long and subulate (Fig. 18.31, A). There are no leaflets or pinnae. At the base of each leaf there arises a bud which develops into shoot. The roots usually arise from the nodes on the under surface. They may also arise from the internodes. The roots are branched and bear roots caps. The leaves are circinately coiled in the young condition. The tip of the rhizome is curved upwards. The younger parts of the plants are covered with simple hair that form a protection coating. It is interesting to note that there is a gradual simplification of the leaves in the three genera of the Marsileaceae. In ***Marsilea*** the leaves have 4 leaflets, in ***Regnellidium*** the leaf blade is deeply notched into two lobes whereas in ***Pilularia*** the blade is altogether lost and the rachis performs the photosynthetic functions. ***Pilularia minuta*** is the smallest species with rhizome measuring 0.5-0.7 mm. across and leaves 35-40 mm. long. The sporocarps arise singly from the base of the leaf (Fig. 18.31, A) and may be sessile, sub-sessile (***P. globulifera***) or shortly stalked (***P. americana***). The sporocarps are globular in shape and possess very few sori as compared to the other two genera.

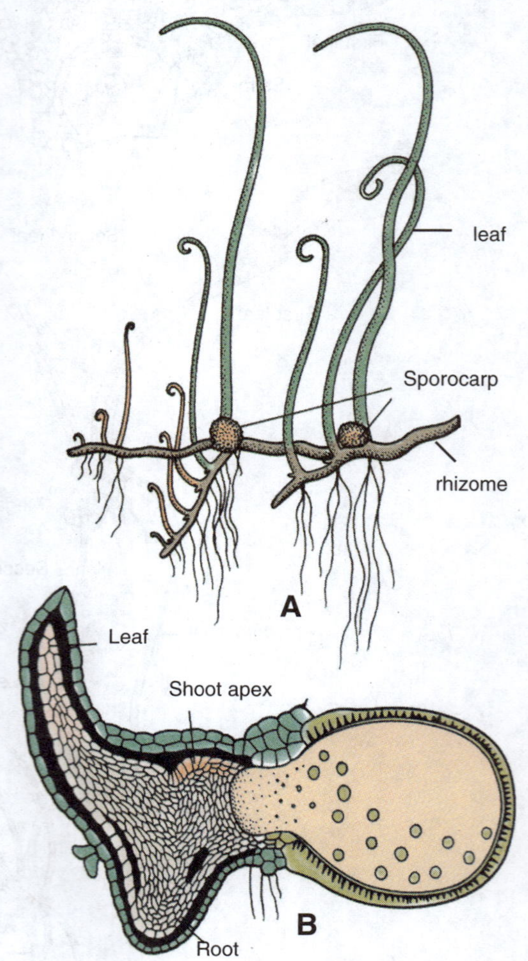

Fig. 18.31. (A—B). *Pilularia.* **A.** Portion of sporophyte of *P. globulifera* (After Meunier); **B.** Young sporophyte of *P. globulifera* (After Campbell).

The rhizome (Fig. 18.32) has the same internal organisation as that of ***Marsilea.*** The stele is slightly reduced. The inner endodermis is absent or poorly developed in some species.

The root (Fig. 18.33) is diarch. The cortex is lacunate. The lacunate or the air spaces possess special spirally coiled cells that offer mechanical strength.

The sporocarp is spherical (Fig. 18.31) and has a simple internal structure (Fig. 18.34). A longitudinal section reveals four loculi or sori. The wall of the sporocarp (Fig. 18.34, A) consists of an external epidermis bearing hair and interrupted by stomata. Next to it are two layers of

P. americana

P. globulifera

hypodermal cells. The cells are thick-walled. The outer layer of hypodermis is composed of smaller, thick-walled cells, whereas the inner layer is made up of elongated and palisade-like thick-walled cells. Next to hypodermis are a few layers of thin-walled cells that later on gelatinise and form a mucilage which on absorbing water swells up and helps in dehiscence of the sporocarp and liberation of contents.

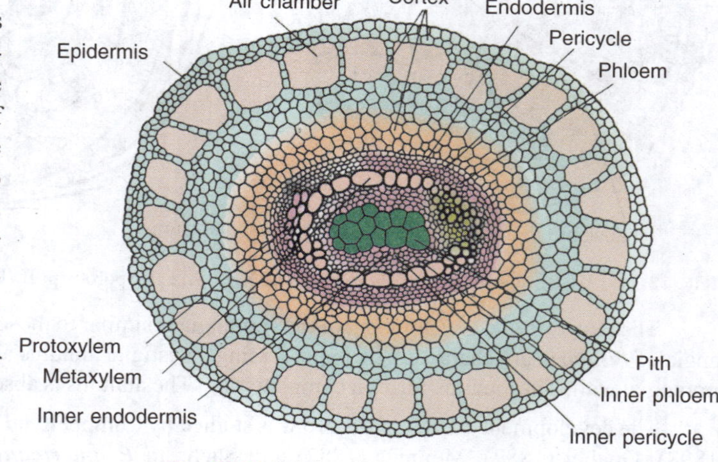

Fig. 18.32. ***Pilularia*** T.S. through Rhizome.

There are four sori (Fig. 18.34, A) that contain numerous sporangia developed on four peripheral receptacles. Each sorus has a many layered indusium (Fig. 18.34, A). The megasporangia contain only one large megaspore each, whereas the microsporangia contain numerous microspores per sporangium (Fig. 18.34, A, B). In a transverse section, which appears to be longitudinal, only two sori are visible. They extend throughout the diameter of the sporocarp. The microsporangia arise in distal groups on the receptacle (Fig. 18.34, A, B) whereas the megasporangia arise near the base. There is no gelatinous ring or the sporophore in the sporocarp of ***Pilularia.***

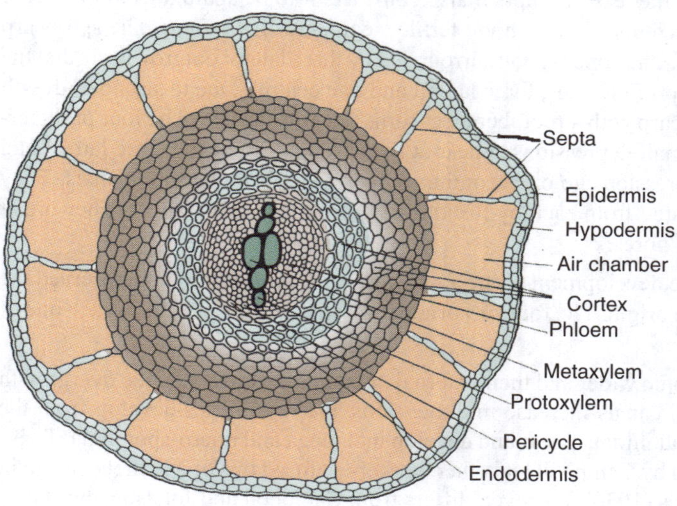

Fig. 18.33. ***Pilularia.*** T.S. through root.

The soft or parenchymatous tissue of the sporocarp (inner layers of wall, receptacles and indusia) disorganise and become mucliginous. The mucilage absorbs water, swells up and exerts pressure on the sporocarp wall. The wall breaks into four quarters by means of longitudinal slits. The sporangia containing spores are liberated along with the mucilage.

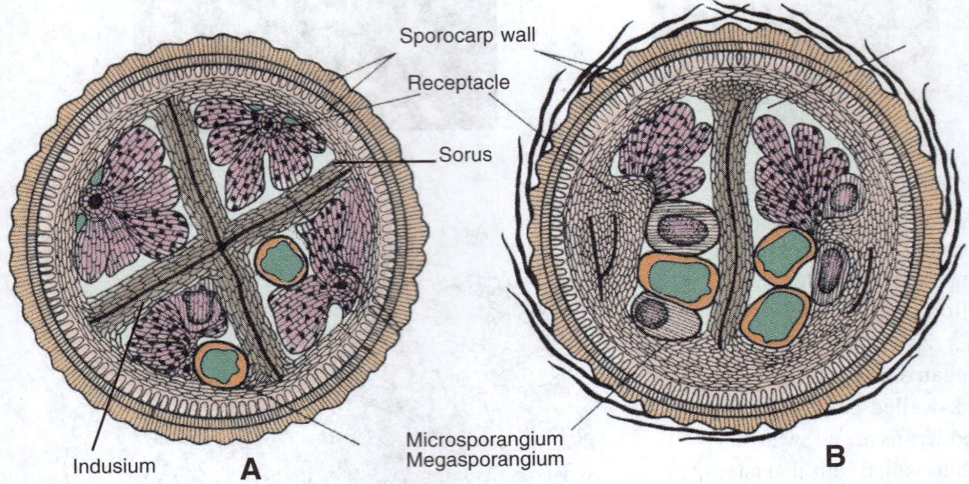

Fig. 18.34. (A—B). Sporocarps of *Pilularia globulifera*. A. L.S. sporocarp; B. T.S. sporocarp (After Luerssen)

The structure and development of the sporangia is similar to those of ***Marsilea.*** The megasporangia of ***Pilularia*** are characteristic in possessing a vestigial annulus along the apical region of the sporangial wall. The annulus forms a complete ring. The annulus is absent in the other two genera.

The development of the sporocarp was studied by Campbell in ***P. americana*** and by Johnson (1898), Goebel (1882), Meunier (1887) and Sachs in ***P. americana***. The sporocarp initial in ***P. americana*** arises towards the inner face of the very young leaf. A marginal cell of the leaf acts as an initial cell of the sporocarp. It functions as an apical cell and divides actively so that the sporocarp appears as distinct projection on the leaf. Campbell also considers it to be analogous to the fertile spike in ***Ophioglossum,***and also compares it with the fertile segments of ***Aneimia.*** The sporocarp assumes the shape of a blunt cone. The cone is soon surrounded by three lateral outgrowth so that the sporocarp tip consists of distinct prominences, three lateral and one central. Due to stronger growth on the outside the tip of the sporocarp with 4 protuberances turns towards the leaf. The four protuberances are free at the top. Later small depressions appear at the base of each of the three lateral and central protuberances. These depressions go on becoming deeper and form the soral canals. They become four deep cavities separated from each by fused margins of the protuberances. They are at first open at their apices by small pores.

Johnson (1898) studied the development of the sporocarp in ***M. globulifera*** and observed the four protuberances to be lateral in origin. The four sori originate as distinct apical cells in each one of these lateral lobes.

Later the soral canals become wider and their external openings are closed by the overgrowth of the adjacent tissues. The soral canals increase in dimensions, the receptacles develop from the peripheral cells. Soon sprorangial initial appear and develop into mega and micro-sporangia. These four protuberances were regarded by Campbell and other workers as fused tips of 4 leaflets that go in the formation of sporocarp. Eames (1936), however, differs from Campbell and Johnson and states, "the prominences are not, however, related to the fundamental nature of the sporocarps but to the presence of soral pits." He further adds, "that these projections are related to the sori and not to the leaflets entering into the make up of capsule is evident in the fact that in ***P. minuta,*** where there are two sori, there are two prominences (not four, as should be the case if teeth represent valves). Further, the prominences are not at the apex but on the ventral margin."

The germination of spores is similar to ***Marsilea.*** In ***Pilularia*** the germinating microspores cut off a small prothallial cell and an upper cell. The prothallial cell divides to form a small **basal cell,** which is also present in some species of ***Marsilea,*** but not in all. The rest of the stages are similar except that in ***Pilularia*** the central cell is cut off by a dome-shaped wall which is not the case in ***Marsilea.*** A distinct cap cell is also distinguishable in ***Pilularia***. The two groups of androcytes are not clearly recoginisable and appear as a single mass. Campbell (1918) pointed out a close similarly between the antheridium of ***Pilularia*** and the polypodiaceous ferns. The spermatozoids consist of a globular posterior vacuole and a short coiled anterior part the bears many flagella. The flagella bearing nuclear and mitochondrial portions is very much reduced as compared to ***Marsilea.***

The megaspore in ***Pilularia*** does not posses a distinct apical papilla. The apical portion is only narrower than the basal part, which is filled with food material and forms the nutrient supplying part of the gametophyte. The nucleus is present in the upper narrow part. The development of the megagametophyte is almost similar in ***Marsilea*** and ***Pilularia.*** The few difference are (i) the central cell of archegonium is surrounded only by two lateral cells and not three as in Marsilea, (ii) the neck cells of the archegonium are much longer than those in ***Marsilea,*** (iii) the nucleus of the egg cell is larger than that of ***Marsilea,*** and (iv) the tissue surrounding the archegonium increases suddenly in size whereas in ***Marsilea*** this increase is gradual and slow.

The development of embryo (fig. 18.31, B) presents no difference, in ***Pilularia*** and ***Marsilea.***

Taxonomic Position

Division	:	Pterophyta
Class	:	Leptosporangiopsida
Order	:	Marsileales
Family	:	Marsileaceae
Genus	:	***Pilularia***
Species	:	***globulifera***

DISCUSSION

Two views are prevalent regarding the **evolutionary status of the family *Marsileaceae.*** These are :

1. According to one view ***Marsilea*** is regarded as the primitive most member of the family. ***Regnellidium*** is considered to be intermediate and ***Pilularia*** is stamped as the advanced and recent derivative. Such a conclusion is based purely on morphological grounds. The supporters of this view (Eames, 1936; Chrysler and Johnson, 1939 ; Higginbotham, 1941) took into account the leaf form, number of leaflets, number of sori and number of sporangia as criteria for determining the evolutionary status of the three genera. They regarded that ***Marsilea*** with larger number of sporocarps per leaf, larger number of sori and sporangia per sporocarp and larger number (four) of leaflets per leaf, is the primitive or ancestral form. ***Regnellidium*** comes next in this respect and ***Pilularia*** with no leaflets and with only 1-4 sori and fewer sporangia is the advanced member.

2. Copeland (1947) regarded ***Pilularia*** as an ancestral form of the family and considered ***Regnellidium*** and ***Marsilea*** to be derived genera. His conclusions are based on the fact that in some species of ***Pilularia*** (***P. minuta***) the number of sporangia per sorus is limited. In ***P. minuta*** a sorus has one mega- and two micro-sporangia. The sporangial development is basipetalous

i.e., the sorus is gradate. In ***Marsilea*** Eames (1936) stated that there is tendency for the sori to become mixed.

Fossil record also lends some support to Copeland's view. Sporocarps resembling those of ***Regnellidium*** were reported from the tertiary rocks of Mohgaon Kalan and Sausar in India by Sahni (1943). These extinct sporocarps possessed thinner walls as compared to the many-layered walls in the present-day sporocarps of the genus. Complexity in wall structure may be regarded as specialisation and not reduction (Loyal, 1962). ***Marsilea*** is represented by Deccan Intertrappean is suggestive of its relict nature (Loyal, 1962). ***Marsilea*** is represented by eight fossil species from the cretaceous and teritiary rocks (Reeds, 1954). ***Pilularia novaezealandica*** was described by Reeds (1954) from the Jurassic rocks of New Zealand which are decidedly older than the cretaceous. This meagre fossil record throws some light on the primitive position of ***Pilularia.***

Loyal (1962) has also summed up cytological data as supporting Copeland's view. ***Pilularia carolineana*** has $n = 10$ which is regarded as a primitive feature. ***Regnellidium*** has $n = 19$ ($2n = 38$) and ***Marsilea*** has $n = 20$ ($2n = 40$). ***Regnellidium*** is regarded as an aneuploid derivative of present-day ***Marsilea*** (Loyal, 1962). These two genera also come closer to each other in the vascular organisation of their sporocarps and in the development of the primary organs of the young sporophyte. The greater number of species of ***Marsilea*** and their cosmopolitan distribution coupled with the wide range of genetic variability in the genus (Mehra and Loyal, 1959) are other point of advancement besides the complex organization of its sporocarps and sori.

In case we focus our attention towards embryology of the genus ***Marsilea,*** we find that the first few leaves of the young sporophyte are awl-shaped and cylindrical like those of ***Pilularia.*** The quadrifoliate condition is established only after a few awl-shaped and bladeless leaves are produced. In ***Pilularia*** this condition persists throughout life whereas in ***Marsilea*** its occurrence in the young sporophyte may be regarded as recapitulation of an ancestral character. In ***M. aegyptiaca*** the number of sori varies between two and six (Gupta, 1962) but this number is not characteristic of the genus and increases to 12 (normal sporocarps of ***M. minuta***) or even 15-16; in abnormal sporocarps of ***M. minuta.*** This tendency towards decrease in the number of sori in ***M. aegyptiaca*** can be regarded as reversion to the ancestral character. In ***Pilularia*** the sporocarps have 2-4 sori.

The facts stated above clearly speak of ***Pilularia*** being the relict form whereas ***Regnellidium*** and ***Marsilea*** are derived. One may also agree with Copeland (1947) when he said, "that the surviving genera do not represent a series but are ends or fragments of a related series."

QUESTIONS

ESSAY TYPE

1. Discuss the morphological nature of sporocarp of ***Marsilea.*** **(U.P. College, 2000)**
2. Explain the life-cycle of ***Marsilea*** with the help of labelled diagrams only.
3. With labelled sketches only, illustrate the structure of Sporocarp of ***Marsilea***
4. Describe the structure of Sporocarp of ***Marsilea*** and a note on mode of dispersal of its spores. **(Meerut 2000)**
5. Describe the spore producing organs of ***Marsilea*** and discuss briefly its morphological nature.
6. Describe the anatomical features of rhizome and petiole of ***Marsilea.***
7. Describe the life history of ***Marsilea.***
8. With the help of labelled diagrams, describe the structure of sporocarp of ***Marsilea.***
9. With the help of labelled diagrams only show the structure of sporocarp of ***Marsilea*** and give a short account of the mode of its spore dispersal.
10. Describe the structure and morphological nature of the sporocarp of ***Marsilea.*** **(Poorvanchal, 2003)**
11. Describe the development of micro- and megaspores in sporocarp of ***Marsilea.***
12. Describe the structure of ***Marsilea*** sporocarp in different planes of sections.
13. With the help of labelled diagrams describe the external and internal structure of sporocarp of ***Marsilea.***

14. Draw neat and labelled diagrams to illustrate the structure of sporocarp of ***Marsilea*** in all three planes i.e. H.L.S, V.L.S and T.S. **(Allahabad, 2001)**

15. How sporophyte development takes place in ***Marsilea*** ? Describe.

16. (*a*) Compare the gametophyte of ***Marsilea*** and ***Selaginella***.

(*b*) Define Stele. Describe the types and structure of stele in rhizomes of ***Marsilea*** and ***Adiantum.***

17. Explain sporophytic generation of ***Marsilea.***

18. Describe the process of reproduction in ***Marsilea.***

19. Bring out the structure of mature sporocarp of ***Marsilea*** with the help of suitable diagrams and briefly discuss the foliar theory of its origin.

20. Draw a labelled diagram of H.L.S. ***Marsilea*** sporocarp and explain the lamina theory of its origin as suggested by Puri and Garg.

21. Describe the aerial stem of ***Equisetum*** and the rhizome of ***Marsilea*** as seen in T.S. Mention their similarities and differences. Draw labelled diagrams.

22. Describe the general features of ***Marsilea.*** In what respect do they differ from ***Equisetum*** ? Discuss. **(Lucknow, 2003)**

23. Describe with example the range of variations in the sporocarp structure of water ferns with special reference to the evolutionary tendencies noted there in.

24. Describe the general characteristics of ***Regnellidium*** and ***Pilularia.*** Compare the sporocarps of these plants with that of ***Marsilea.***

25. Describe the life cycle of ***Regnellidium*** with the help of labelled diagrams.

26. Give a brief account of the sporophyte of ***Pilularia.*** How does it differ from ***Marsilea***?

27. Describe the gametophytes of ***Regnellidium*** and ***Pilularia.***

28. With the help of labelled diagram only compare the structure of spore bearing organ of ***Nephrolepsis, Marsilea*** and ***Lycopodium*** **(Lucknow, 2004)**

SHORT ANSWER TYPE

29. Describe briefly the following:

(*a*) (*i*) Aquatic and terrestrial characters of ***Marsilea.***

(*ii*) Sperms of ***Selaginella***, ***Equisetums*** and ***Marsilea.***

(*iii*) Morphological nature of sporocarp of ***Marsilea*** according to Puri and Garg. **(Garhwal, 2000)**

(*b*) Morphological nature of sporocarp of ***Marsilea*** **(Kanpur, 2000; Lucknow, 2000; Awadh, 2003)**

(*c*) What are the characteristic features of plants of ***Marsilea*** growing in water and on land surface?

(*d*) Structure of mega-and micro-gametophyte of ***Marsilea.***

(*e*) ***Marsilea*** rhizorme. **(Lucknow, 2003)**

30. Differentiate between:

(*a*) Roots of ***Marsilea*** and ***Equisetum***

(*b*) Antherozoids of ***Equisetum*** and ***Marsilea***

(*c*) T. S. rhizomes of ***Equisetum*** and ***Marsilea***

(*d*) T. S. of petiole of ***Marsilea*** and ***Pteridium***

(*e*) Ecto-and amphiphloic siphonostele

(*f*) Rhizome of ***Marsilea*** and stem of ***Selaginella.***

(*g*) Rhizome of ***Marsilea*** and Rhizome of ***Regnellidium***

(*h*) Plants of ***Regnellidium*** and ***Marsilea***

(*i*) Plants of ***Pilularia*** and ***Marsilea.***

31. Draw labelled diagrams of the following:

(*i*) Internal structure of ***Marsilea*** sporocarp (***a***) T. S. (***b***) L. S. **(Kerala, 2001)**

(*ii*) T. S. petiole of ***Marsilea*** **(Lucknow, 2001, Poorvanchal, 2000, 2001, 2003)**

(*iii*) T. S. Stem of ***Marsilea*** **(Poorvanchal, 2002)**

(*iv*) T. S. ***Marsilea*** rhizome **(Kanpur, 2000; U.P. College, 2000; Lucknow, 2001, 2004; Awadh, 2000, Allahabad, 2000; Lucknow, 2002, Allahabad, 2002)**

(*v*) Dehisced sporocarp of ***Marsilea***

(*vi*) Internal Structure of Rhizome of ***Marsilea***

(*vii*) Venation and Vernation in ***Marsilea*** leaf

(*viii*) T. S. Root of ***Marsilea***

(*ix*) Leaflet of ***Marsilea***

(*x*) V. T. S. and V. L. S. sporocarp of ***Marsilea*** **(Lucknow, 2000, 2003)** (*xi*) H. L. S. sporocarp of ***Marsilea*** **(Kanpur, 2003; Poorvanchal, 2003)**

(*xii*) Stele in ***Marsilea***

(*xiii*) Spermatozoid of ***Marsilea***

(*xiv*) Female gametophyte of ***Marsilea*** **(Kanpur, 2002)**

(*xv*) Development of embryo in ***Marsilea*** **(Garhwal, 2000)**

31. Write short notes on the following :

(*i*) Internal structure of ***Marsilea*** rhizome

(*ii*) Puri and Garg (1953)

(*iii*) Sporocarp of ***Marsilea***

(*iv*) Structure of leaf of ***Marsilea***

(*v*) Dehiscence of sporocarp of ***Marsilea***

(*vi*) Young embryo of ***Marsilea***

(*vii*) Sporophore

(*viii*) Male gametophyte of ***Marsilea***

(*ix*) Internal structure of ***Marsilea*** sporocarp

(*x*) Female gametophyte of ***Marsilea*** **(Gorakhpur, 2003)**

(*xi*) Rhizome of ***Marsilea***

(*xii*) Amphiphloic Siphono Stele

(*xiii*) External morphology of ***Marsilea*** plant

(*xiv*) Fertilisation in ***Marsilea*** **(Lucknow, 2001; Allahabad, 2002)**

(*xv*) Structure and dehiscence of sporocarp of ***Marsilea***

(*xvi*) Petiole of ***Marsilea***

(*xvii*) Structure of gametophytes of ***Marsilea***

(*xviii*) Structure of sporocarp in ***Regnellidium.***

32. Describe the structure of the megagametophyte of ***Marsilea.*** **(Kanpur, 2001)**

33. Describe the development of embryo in ***Marsilea.***

34. Describe the development of female gametophyte in ***Marsilea***.

35. Draw the sporocarp of ***Marsilea*** as seen in L. S.

36. Discuss the morphology of Leaflet in ***Marsilea.***

37. Give the distribution of ***Marsilea.***

38. Compare steles of ***Selaginella*** stem and ***Marsilea*** rhizome.

39. Describe the sporophyte morphology.

40. Give the V. L. S. of ***Marsilea*** Sporocarp. Discuss morphological nature of sporocarp. **(Himachal Pradesh, 2000)**

41. Give the botanical name of water fern.

42. (*a*) The megaspore of ***Marsilea*** is enclosed by a hard covering. Yet the structure is not called a seed. Why? Compare the leaves of ***Marsilea***, ***Regnellidium*** and ***Pilularia.***

(*b*) Compare the strobilus of ***Equisetum*** and sporocarp of Marsilea **(Kumaon, 2004)**

43. Write Critical notes on :

(*a*) vascular supply in sporocarp of ***Marsilea***

(*b*) Sporocarp of ***Marsilea*** **(Kanpur M.Sc., 2000; Poorvanchal, 2002)**

(*c*) Sporocarp of ***Pilularia.***

(*d*) Sporocarp of ***Regnellidium***.

(*e*) Structure of gametophyte of ***Marsilea*** **(Poorvanchal, 2002)**

44. Name a hydrophytic ***pteridophyte***

45. Name a ***pteridophyte*** which has amphiphloic siphonostele.

46. Name the type of stele found in ***Marsilea*** petiole. **(Poorvanchal, 2002)**

47. Give the classification of ***Marsilea***

48. Name four heterosporous forms of aquatic ***pteridophytes***.

49. In which plant amphiphloic stele is found and where?

50. Name the stele found in ***Marsilea*** rhizome. **(Poorvanchal, 2001)**

51. Give the name of reproductive body attached on ***Marsilea*** plant.

52. What type of stele occurs in ***Marsilea*** rhizome? **(Kanpur, 2000)**

53. Name a ***pteridophyte*** growing in water. **(Kanpur, 2000)**

54. Mention two characters common to ***Cycas*** and ***Marsilea***. **(Poorvanchal, 2000)**

55. What is the role of sorophore in ***Marsilea*** Sporocarp?

56. Mention two differences between female gametophyte of ***Selaginella*** and ***Marsilea.***

57. Write down the structure of ***Marsilea*** antherozoid.

58. What comes out after the dehiscence of sporocarp of ***Marsilea***?

59. Which pteridophyte is called as 'water fern'?

60. Explain in two lines:

(*a*) The most accepted view about morphological nature of ***Marsilea*** sporocarp.

(*b*) Significance of gelatinous ring in ***Marsilea*** sporocarp.

61. How does sporocarp of ***Marsilea*** differ from a seed?

62. Give two important points of male gametophyte of ***Marsilea***.

63. Where are plants of ***Pilularia*** found?

64. Name the species of ***Regnellidium*** found in Brazil?

65. Where are the sporocarps of ***Regnellidium*** borne?

66. How do sporocarps of ***Regnellidium*** and ***Pilularia*** dehisce?

67. What is the chromosome number of the ***Regnellidium***?

68. Describe the leaf of ***Pilularia***.

69. In what type of habitat ***Marsilea*** plants bearing sporocarps grow? **(Allahabad, 2003)**

70. Differentiate between sporoearp of ***Marsilea*** and ***Azolla***. **(Kanpur, M.Sc., 2003)**

71. Asexual reproduction in ***Marsilea*** **(Garhwal Univ., 2005)**

72. Draw the H.L.S. and T.S of ***Marsilea*** sporocarp and explain them. **(Kuvempu Univ., 2005)**

73. Describe the internal sturcture of rhizome of ***Marsilea***. **(Mysore Univ., 2004)**

74. Explain morphological nature of ***Marsilea*** sporocarp. **(Nagpur Univ., 2005)**

75. Write short note on sporocarp of ***Marsilea***. **(Garhwal Univ., 2005)**

OBJECTIVE TYPE

76. Select the correct answer :

(i) The development of sporangia in ***Marsilea*** is
- (*a*) Eusporangiate type
- (*b*) Leptosporangiate type
- (*c*) Polysporangiate type
- (*d*) Archesporangiate type.

(ii) Who has described the morphological nature of ***Marsilea*** sporocarp?
- (*a*) Prof. S.R. Kashyap
- (*b*) Prof. V. Puri
- (*c*) Prof. B. Sahni
- (*d*) Prof. J.C. Bose.

(iii) Sporocarp of ***Marsilea*** is
- (*a*) Unisporangiate
- (*b*) Bisporangiate
- (*c*) Quadisporangiate
- (*d*) Multisporangiate.

(iv) Amphiphloic siphonostele is found in the rhizome of
- (*a*) ***Equisetum***
- (*b*) ***Marsilea***
- (*c*) ***Selaginella***
- (*d*) None of the above.

(v) In ***Marsilea*** archegonium the number of neck canal cells is usually
(*a*) 1 (*b*) 2
(*c*) 3 (*d*) 4.

(vi) Type of development of embryo in ***Marsilea*** is
- (*a*) Endoscopic
- (*b*) Exoscopic
- (*c*) Lateral
- (*d*) Amphiphloic.

(vii) While the number of leaflets in ***Marsilea*** is usually four the number of leaflets in ***Regnellidium*** is
(*a*) 1 (*b*) 2
(*c*) 3 (*d*) 4.

(viii) The leafblade is altogether lost in
- (*a*) ***Marsilea***
- (*b*) ***Regnellidium***
- (*c*) ***Pilularia***
- (*d*) ***Adiantum.***

(ix) Which of the following does not occur in India?
- (*a*) ***Azolla***
- (*b*) ***Marsilea***
- (*c*) ***Regnellidium***
- (*d*) ***Pilularia***

(x) Which of the following species of ***Marsilea*** grows on dryland?
- (*a*) ***M. minuta***
- (*b*) ***M. quadrifolia***
- (*c*) ***M. aegyptiaca***
- (*d*) ***M. rajasthanensis.***

(xi) Usually the number of leaflets in ***Marsilea*** is
- (*a*) 1
- (*b*) 2
- (*c*) 3
- (*d*) 4.

(xii) ***Marsilea*** rhizome has
- (*a*) No Cortex
- (*b*) Outer Cortex only
- (*c*) Outer and inner Cortex
- (*d*) Outer, middle and inner Cortex.

(xiii) The middle cortex in ***Marsilea*** is
- (*a*) Parenchymatous
- (*b*) Chlorenchymatous
- (*c*) Aerenchymatous
- (*d*) Sclerenchymatous.

(*xiv*) Laminar or Leaf segment Hypothesis of the nature of sporocarp was first proposed by

(*a*) Russow
(*b*) Goebel
(*c*) Eames
(*d*) Smith.

(*xv*) Johnson (1898) proposed that sporocarp of ***Marsilea*** is a

(*a*) Modified stem
(*b*) Modified leaf
(*c*) Modified petiole
(*d*) Modified root.

(*xvi*) The number of microspores produced per microsporangium in ***Marsilea*** is

(*a*) 128
(*b*) 56
(*c*) 32 – 64
(*d*) 1 – 4.

(*xvii*) What comes out of ***Marsilea*** sporocarp after dehiscence?

(*a*) Microspores
(*b*) Megaspores
(*c*) Both microspores and megaspores.
(*d*) Ring of gelatinous tissue containing soral sacs.

(*xviii*) In which of the following fern genera the spermatozoid has posterior cytoplasmic vertical and anterior nucleus and flagella?

(*a*) ***Pteris*** (*b*) ***Nephrolepis***
(*c*) ***Marsilea*** (*d*) ***Adiantum.***

(*xix*) In ***Regnellidium*** the number of microspores per micro sporangium is .

(*a*) 32 (*b*) 54 to 58
(*c*) 64 (*d*) 128.

77. Fill in the blanks:

(*i*) Bisporangiate sporocarp is found in ———.

(*ii*) Laminar hypothesis of the sporocarp of ***Marsilea*** was supported by ——— .

(*iii*) Stele in ***Marsilea*** rhizome is ———.

(*iv*) The type of stele found in rhizome of ***Marsilea*** is ———.

(*v*) Common name of ***Marsilea*** plant is ———.

(*vi*) The leaflet is altogether lost in ——— .

(*vii*) The sporocarp of ***Marsilea*** contains both ——— and ——— .

(*viii*) ***Pilularia*** does not occur in ——— .

(*ix*) ***Regnellidium*** is ——— and is found in ——— .

(*x*) Bilobed ***lamina*** is present in ——— .

Giant *Salvinia*

CHAPTER 19

Pterophyta–Leptosporangiopsida—Salviniales

- Distinguishing features
- Genus *Salvinia* Guettard
- Genus *Azolla* Lamark

Distinguishing Features

The order Salviniales of the class leptosporangiopsida is a small assemblage of aquatic and free floating ferns. The order is characterised by the following features :

1. The order includes two genera (*Azolla* and *Salvinia*) that are free floating in water. *Azolla* has unbranched and pendent roots. Roots are absent in *Salvinia.*
2. The sporangia are borne within sporocarps whose morphology is quite different from that of *Marsilea* because a sporocarp is

represented by a single sorus. The indusium of the sorus is the sporocarp wall.

3. The sporangia have no annulus and develop in a basipetalous manner.
4. The microsporangia contain numerous microspores and the megasporangia have only one megaspore each. The microspores in ***Azolla*** are embedded in **massulae** that are studded with anchor-shaped bodies called the **glochidia.** The massulae are formed from the tapetal plasmodium.
5. The gametophytes are much reduced.
6. The stem is horizontal and much branched with a reduced stele.
7. The leaves are not circinately coiled in the bud condition.

Christensen (1938), Reimers (1954), Pichi-Sermolli (1953) and Bonnet (1957) divided the order into two families each represented by a single genus. ***Salvinia*** Guettard with its 12 species belongs to the family ***Salviniceae*** and ***Azolla*** Lamarck with 6 species is put under the family ***Azollaceae***. Copeland (1947), Eames (1936) and Bower (1928) recognised only one family ***Salviniaceae*** and placed both the genera under it. ***Azollopsis tomentosa*** and ***A. coccoides*** are two extinct species of this genus that have been found in Upper Cretaceous of Montana, U.S.A. Twenty-five species of ***Azolla*** are also known in fossil forms from various horizons in U.S.A., Britain and India. ***Azolla intertrapa*** was collected from the tertiary Chert bed in some localities in India by Sahni and Rao. ***Salvinia*** is known in the form of compressions of leaves found in the Middle Tertiary of some regions, but not in Australia.

Family Salviniaceae is represented by genus ***Salvinia***, which is discussed here. The family Azollaceae represented by the genus ***Azolla*** is discussed after the genus ***Salvinia***.

S. auriculata

GENUS–SALVINIA Guettard

It is represented by about twelve species most of which are found in the African countries. ***Salvinia natans***, ***S. molesta*** and ***S. auriculata*** occur in India. ***Salvinia natans*** is found in abundance in the Dal lake at Kashmir (Srinagar). It is an annual species, others are perennial. ***S. molesta*** is found in Kerala State.

SPOROPHYTE GENERATION

External Characters

The sporophyte (Fig. 19.1, A) has a herbaceous branched, horizontal and floating rhizome. The rhizome is beset with whorls of leaves that cover it from view. The leaves arise in clusters of three (Fig. 19.1, A) and the point from where they arise may be called the node. Two leaves of the cluster are above the surface of water and the third one is submerged. The floating and submerged leaves differ in morphology, the former being green, hemispherical, or even oblong in shape. Their upper surface is papillose and covered with stiff hair that save it from being wetted. The lower surface is brownish and also covered with hair of the same colour. These leaves may assume a boat-like shape in some species. The submerged leaves are long and filiform (Fig. 19.1, B). They are segmented into numerous thin and hair-like segments that are densely clothed with hairs. They appear like roots and most probably act as balancers. The hair covering their surface are multicellular and may absorb water and other nutrients. The tips of the hair in some species have spongy tips which indicate their function as water absorbers. A bud arises at the stem node between the floating and the submerged leaves. The floating leaves are arranged in two alternate rows, the third row is of the submerged leaves. The three leaves of a node alternate with the three leaves of the next node and so on, so that there are actually six rows of leaves. The roots are absent. Lateral branches from the stem arise from the node.

Loyal and Grewal (1967) discussed the morphology of ***S. auriculata*** and pointed out an interesting feature that the symmetry of the stem is bilateral in the internodal region and radial in the nodal region. They are of the opinion that the bilateral symmetry of the internodal region is derivable from the radial symmetry. The stem is beset with numerous multicellular hair with "sharply pointed terminal cells." They regard the submerged leaf as submerged organ. Bonnet (1955) regarded it as an axis bearing root axes. In ***S. auriculata*** the submerged organ arises from between the main stem and the lateral branch and is disposed towards the ventral side. The floating leaves in this species are inserted slightly in an alternate manner and not opposite (Loyal and Grewal, 1967). One of the two leaves is smaller in size so that they are anisophyllous. The leaves are conduplicate in young condition. Venation is closed.

Vegetative propagation is effected by fragmentation. The fragile rhizomes and their lateral branches easily break and the separated parts of the sporophyte develop into new individuals. Like this the plants spread rapidly under favourable conditions. In ***S. auriculata*** the vegetative propagation is the chief source of reproduction as its spores are not viable.

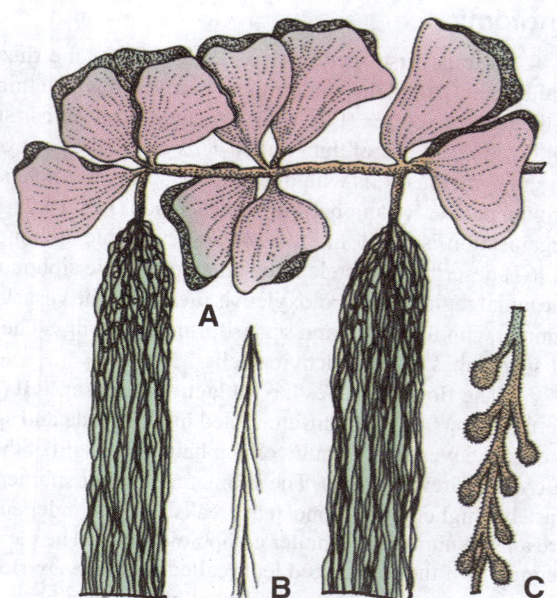

Fig. 19.1. (A–C). ***Salvinia auriculata*** Aubl. A. Portion of a plant showing habit; B. Submerged leaf; C. Submerged leaf with sporocarps.

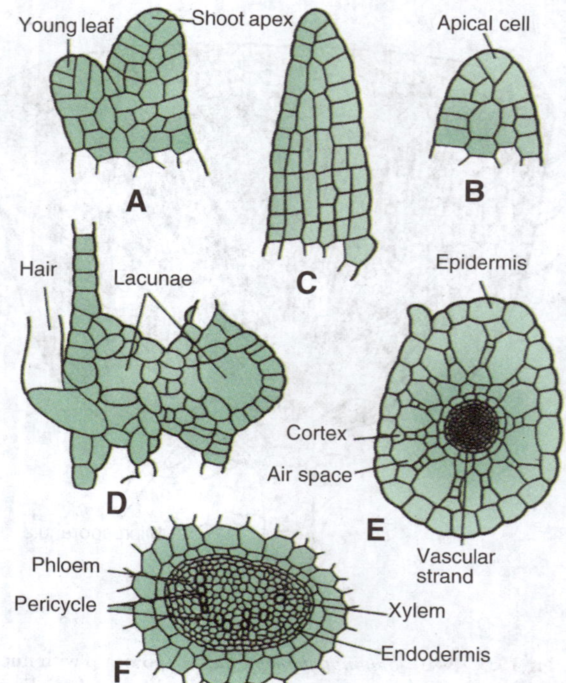

Fig. 19.2. (A–F). ***Salvinia natans.*** A. Stem apex as seen in horizontal section; B. Leaf apex of a young leaf; C. L.S. of a leaf segment of submerged leaf; D. Section of aerial leaf showing air spaces; E. T.S. of stem; F. T. S. of stelar region showing details.

Anatomy

A transverse section (Fig. 19.2, E,F.) of the rhizome reveals a distinct outer layer of thin-walled cells called the epidermis. It is covered with a thin cuticle and there are no stomata. Next to it is the lacunate cortex (Fig. 19.2, E). The central stele is surrounded by a distinct layer of endodermis followed by a layer of thin-walled cells called the pericycle (Fig. 19.2, F). The centre is occupied by parenchymatous cells with one or more scattered tracheidal cells. The xylem is represented by a few broken patches (Campbell) of the tracheids (Fig. 19.2., F). Phloem surrounds the xylem. The stelar organisation is much like that of an ectophloic siphonostele. In ***S. auriculata***, Loyal and Grewal (1967) described the stele to be an amphiphloic siphonostele in the internode and a siphonostele in the nodal region. The protoxylem is mesarch. The vessels are absent and the tracheids possess, spiral, annular, annular-spiral and scalariform thickenings. The phloem lacks companion cells and consists of sieve tubes and parenchyma cells.

The floating leaves have a **lacunate mesophyll** (Fig. 19.2, D) sandwiched between two epidermal layers. It is not differentiated into palisade and spongy parenchyma. Both the epidermal surfaces are covered with multicellular hair. The origin of the leaf trace, in ***S. auriculata*** is extramarginal (Loyal & Grewal, 1967). The submerged leaf or submerged axis also reveals a siphonostele near its base but higher the siphonostele breaks into two independent steles and as we proceed upwards there are a large number of smaller ectophloic steles. The parent stele near the base is amphiphloic. In the segments of the submerged leaf (called root axes by Bonnet) the xylem is eccentric and consists of one or two tracheids. Two patches of phloem can be made out on either side of the xylem. A single layered pericycle and endodermis are also distinguishable.

The stem grows by means of a three-sided apical cell (Fig. 19.2, A) that cuts off segments from all its sides. Each segment divides periclinally into a dorsal and a ventral cell.

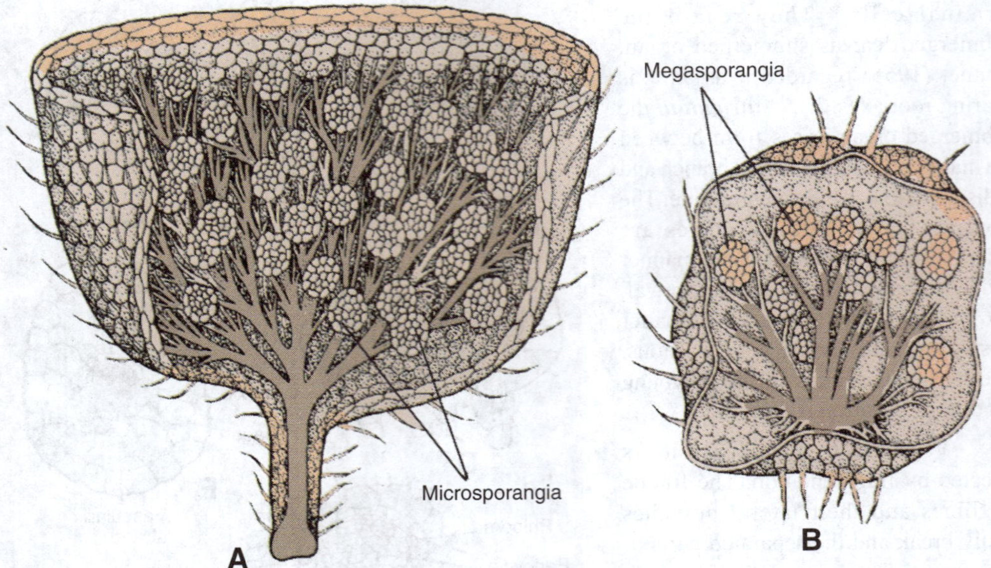

Fig.19.3. (A–B). *Salvinia oblongifolia.* A. Sporocarp with microsporangia ; B. Sporocarp with megasporangia. (After Martins from Eames, 1936).

Reproduction

Sporocarp

S. oblongifolia

The sporocarps may be borne in clusters or in rows on the segment of submerged leaves (Fig. 19.C). Their number, in ***S. natans*** varies from 4—25 or even more. A sporocarp is borne at the tip of a segment. In a group, therefore, the sporocarps are sympodially arranged. The first formed sporocarps are megasporangiate (Fig. 19.3, B) and contain up to 15 or more megasporangia. The later formed sporocarps are usually microsporangiate and are comparatively larger in size and number. They contain many microsporangia that are borne on branched receptacle (Fig. 19.3, A). The sporocarps may be spherical, globose, or flattened in shape. The surface is usually smooth and covered with hair in young specimens. The flattened sporocarps of some species have ridged sides. The megasporangiate sporocarps in ***S. natans*** have unbranched receptacles (Fig. 19.4, A) whereas in ***S. oblongifolia*** the receptacle is branched (Fig. 19.3, B). The wall of the sporocarp is comparable to the indusium of the sori in ***Marsilea***. It is two-layered thick (Fig. 19.5, E,F). At the tips the wall layers increase in thickness and are composed of elongated and closely packed cells (Fig. 19.5, E,F). Lower down the two layers enclose small and large lacunae or air chambers. The lacunae near the base of the sporocarp are traversed with filamentous trabeculae that arise from the receptacular surface (Fig. 19.5, E,F).

From the base of the sporocarp arises a thick column like receptacle that is unbranched in the megasporangiate sporocarps of ***S. natans*** and branched in the similar sporocarps of ***S. oblongifolia***. The receptacle in the microsporangiate sporocarps of all species is branched (Fig. 19.3, A, 19.4, A). The receptacle is traversed by the vascular strand of the leaf segment on which it is borne terminally (Fig. 19.5, E, F).

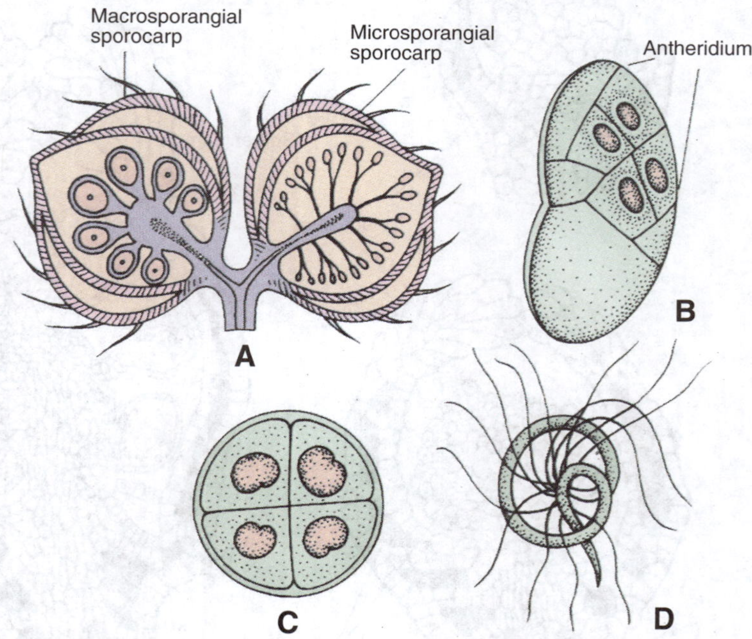

Fig. 19.4. *Salvinia natans*. (A–D). **A.** Sporocarps with receptacles and sporangia. The receptacle is unbranched in megasporangiate sporocarp; **B.** Microgametophyte with antheridium; **C.** Microgametophyte as seen from above; **D.** A spermatozoid. (A, after Luerssen; B—D, after Campbell).

The sporocarp develops from a single apical cell (Fig. 19.5, A) that cuts off segments along its two sides. It soon forms a sorus primordium (Fig. 19.5, B). The primordium is soon surrounded by an outgrowth on all sides. This is the **indusium** (Fig. 19.5, B.C). which later covers the entire sorus. The superficial cells of sorus act as sporangial initials. The cells derived from the soral primordium grow to form a receptacle with sporangia arising from its distal end as well as the sides (Fig. 19.5, C—F). In the megasporangiate sporocarps only up to 25 sporangia develop whereas in the microsporangiate sporocarps the receptacle becomes branched and many sporangia develop (Fig. 19.3, A) in it. The development of the sporangia is strictly leptosporangiate.

In ***S. auriculata*** (Loyal and Grewal, 1967) the wall of the sporocarp is three-layered. The innermost layer is perforated by stomata and has cells with sinuate cell walls. Presence of small 3—5 celled hair also been recorded on this layer (Loyal and Grewal, 1967). The stomata are restricted to the basal region. In ***S. natans*** they are scattered all over the surface. Air chambers are present between the inner and middle layers. Number of trabeculae arise from the inner layer and connect it with the middle layer. The outermost layers are hexagonal and bear many multicellular hair. There are no air spaces between the outer two layers.

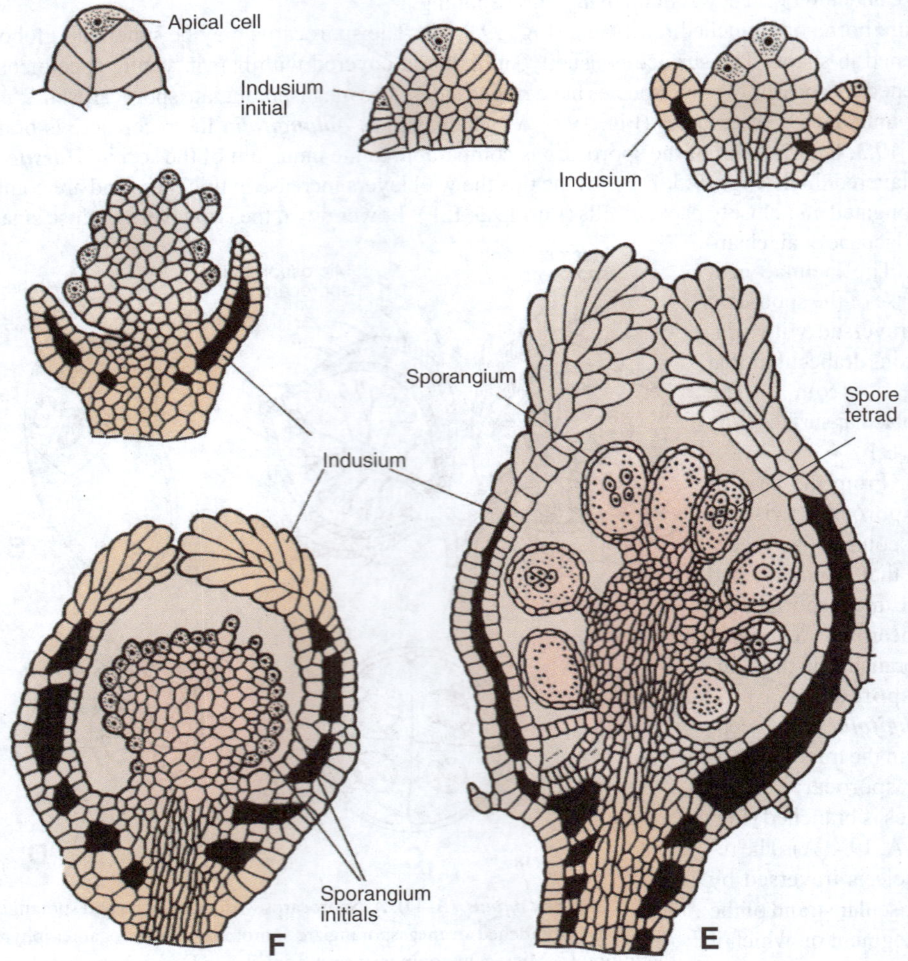

Fig. 19.5. (A—F). *Salvinia natans.* Stages in the development of sporocarp and sporangia. (After Zawidski)

Sporangia

The megasporangium has a short **stalk** and an ovoid or globose head of the **capsule** (Fig. 19.6, A). The capsule has a single-layered wall made up of thin-walled cells. Next to the wall is a single layered tapetum, which encloses eight megaspore mother cells that divide meiotically to form 32 megaspores. All the megaspores except one degenerate. The surviving or the functional megaspore enlarges in size and stores sufficient food. The cytoplasm of the aborted megaspores and the degenerated tapetal cells gathers around the functional megaspore and hardens to form a thick and vaculate **perispore** (Fig. 19.6, A, B). The perispore is thicker towards the apex of the spore and forms a complex pollen-chamber-like structure of a gymnosperms seed (Fig. 19.6, B). The perispore at the base of this chambered region becomes slightly raised and forms a small protuberance or a mound with three lateral flaps (Fig. 19.6, B). This mound also encloses a three-sided cavity with three raised processes or ridges. The three flaps of the central mound lie exactly above the triradiate mark of the megaspore and separate during spore germination. The wall of the megaspore has the usual two layers, the exine which is surrounded by the perispore and the intine which is thin and encloses the uninucleate spore cytoplasm. The nucleus lies surrounded by granular cytoplasm at the tip of the megaspore. The rest of the megaspore is full of more fluid cytoplasm that is filled with coarse granules of stored food. The megasporangia have no annulus.

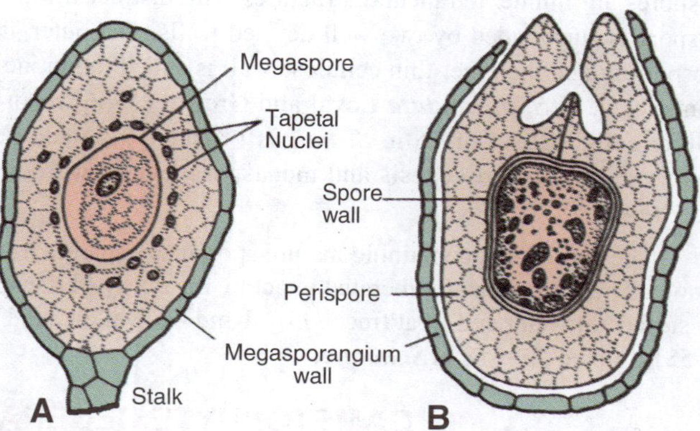

Fig. 19.6.(A–B). ***Salvinia natans.*** **A.** Megasporangium with the surviving megaspore surrounded by plasmodial mass; **B.** Mature megaspore with perispore. (A, after Kundt ; B. after Arnoldi)

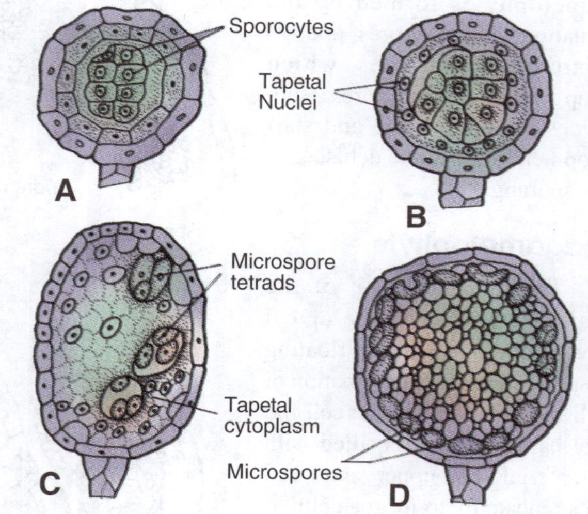

Fig. 19.7. ***Salvinia natans.*** **A—D.** Stages in the development of microsporangium and formation of microspores. (After Yasui)

The structure of the ***microsporangium*** (Fig. 19, D) is similar to that of the megasporangium except from the differences pointed out here. The developing microsporangium contains 16 microspore mother cells (Fig. 19.7, A.B). As a result of sporogenesis 64 microspores are produced (Fig. 19.7, D). None of them degenerates and forms a plasmodial fluid. The microspores are irregularly scattered throughout the sporangial cavity and are embedded in the tapetal cytoplasm that forms the plasmodial fluid. Later the microspores migrate to the periphery of the tapetal plasmodium. The plasmodial fluid hardens along with the maturation of the spores and forms an alveolar mass with microspores embedded along its periphery (Fig. 19.7, D). This hardened alveolar mass is termed as **massula.** The microspores are minue, tetrahedral structures with distinct triradiate marks at their apices. Each microspore is surrounded by two well defined walls. The outer, thick wall is called the exine or the exosporium and the inner thin cellulose wall is called the intine or endosporium. The cytoplasm is uninucleate. In ***S. auriculata*** Loyal and Grewal (1966) recorded the presence of 8 microspore mother cells instead of 16 in ***S. natans*** (Yasui, 1911) and ***S. auricullata*** (Deshpande, 1943). Abnormal microsporogenesis and megasporogenesis were reported in ***S. auriculata*** by Loyal and Grewal.

The sporocarps containing mature sporangia separate from the leaf segments and sink to the bottom. The sporangia are liberated intact by the decay of the sporocarp wall and again come up to the surface of water and float freely. Loyal and Grewal counted $2n = 45$ in ***S. auriculata*** and Loyal (1958) counted $2n = 18$ in ***S. natans.***

GAMETOPHYTE GENERATION

Heterospory leads to the development of two kinds of gametophytes : (1) the megametophytes formed by the germination of megaspores, and (2) the microgametophytes which develop from the microspores. Both the types are endosporic and start development prior to the dehiscence of the sporangia.

Megagametophyte

The first division of the nucleus, which lies in the apical portion of the horizontally floating megaspores, leads to the formation of a small and a lenticular upper cell and a large basal cell that is filled with food material. The upper small cell divides repeatedly to form a cellular tissue of the gametophyte. It divides first by an anticlinal division to form two unequal cells (Fig. 19.8A)

Since the megaspore is swimming in a horizontal direction, this division appears to form an upper and a lower cell. The upper cell is larger. Further divisions in these cells result in the formation of a thick and lobed apical cushion that breaks through the spore wall and the perispore (Fig. 19.9, A). The three lobes of the perispore separate and diverge. The archegonia appear on this apical cushion. The first

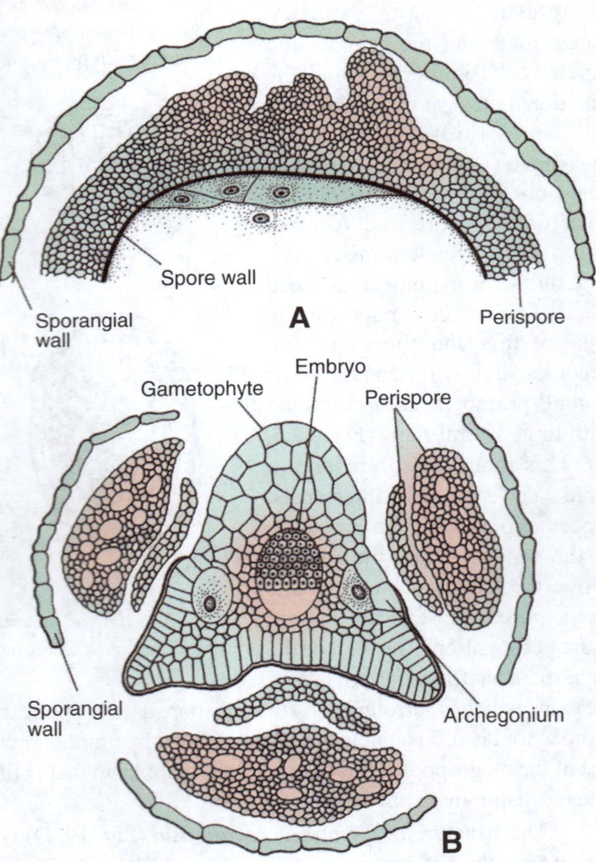

Fig. 19.8. (A–B). *Salvinia natans*. A. Median L.S. through apex of megaspore showing portion of megasporangial wall perispore and early divisions in the apex of the megaspore. B. Transverse section passing through the apical region of the germinated megaspore showing the sporangial wall, six lobes of the perispore and the gametophyte, which appears triangular in outline and has two young archegonia embedded in it. Young embryo is also seen. (After lasser)

archegonium develops in the large segment and later, two more develop on either side of the first (Fig. 19.9, B). The longer axis of the archegonia, in a floating megagametophyte, is parallel to the surface of the water. The apical cushion at this stage has protruded out of the megaspore wall and appears triangular in a cross section (Fig. 19.8, B). It has a small, upper protuberance or a mound and a marginal meristem around the lower side. The meristem develops into two wings that extend background over the sides of the megaspore wall and serve as balancing organs for the floating megagametophyte.

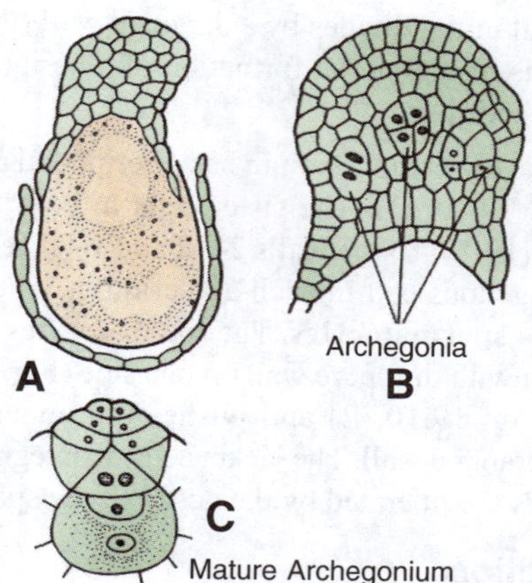

Fig. 19.9. (A–C). *Salvinia natans.* Female gametophyte. A. Showing apical cushion and the lower multinucleate portion; B. Upper portion with three archegonia; C. A. mature archegonium.

Meanwhile the nucleus of the lower larger cell below the apical cushion has undergone free nuclear division. There is no wall formation and the nuclei lie scattered in the cytoplasm (Fig. 19.9, A) of this cell. This multinucleate cell acts as reservoir of food.

The tissue of the apical cushion just around the archegonia divides in such a manner so as to orientate the archegonia through an angle of 90°. The necks of the archegonia now point upwards. The cushion may develop further and enlarge in size and grow bigger than the megaspore. Its cells turn green and if fertilisation fails to take place, more archegonia develop along the sides of the first three.

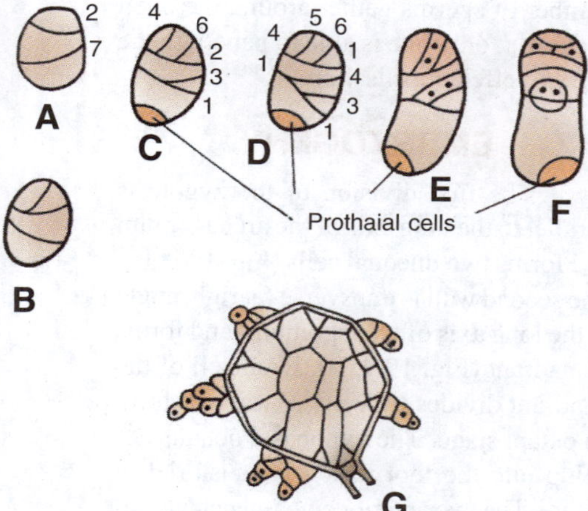

Fig. 19.10. (A—G). *Salvinia natans* A—F. Stages in the development of male gametophyte; G. Dehiscence of sporangium and escaping male gametophyte.

A mature archegonium (Fig. 19.9, C) has a neck made up of three tiers of 4 cells each or of 4 longitudinal rows, each row three cells in height. The neck encloses a neck canal occupied by a single binucleate neck canal cell. The venter is large and contains an upper smaller ventral canal cell and a lower large egg cell with a prominent female nucleus (Fig. 19.9, C). The development of the archegonium is similar to that of ***marsilea***.

The portion of the female gametophyte below the apical cushion is covered by the spore wall and remains of sporangial wall (Fig. 19.9, A).

Microgametophyte

The microspores complete their germination within the microsporangium. They remain embedded in the massulae during their course of development. The contents of the microspores divide into three cells by two transverse (Fig. 19.10, A, 1-1, 2-2) divisions. The lowest cell of the row divides by another transverse or oblique transverse wall (Fig. 19.10, B) to form a small lenticular **male prothallial cell.** Further divisions take place only in the two upper cells (Fig. 19.10, C—E), which act as antheridial initials ; the lowest cell and the prothallial cell remain undivided. Each antheridial initial divides by a diagonal wall (Fig. 19.10, C, 3—3) that intersects the walls 1—1 and 2—2. This results in the formation of a peripheral smaller jacket cell and a large **sister cell** in the upper antheridium.

The lower antheridium has a peripheral cell parallel to the wall 2—2 and large triangular sister cell (Fig. 19.10, D). The sister cells in both the antheridia divide by walls 5 and 6 which again intersect (19.10, E) the walls 2—2 and 1—1 respectively, thus forming two sterile jacket cells and a spermatogenous cell in each antheridium. The **spermatogenous cell** of each antheridium forms a group of 4 **spermatocytes.** The spermatocytes are not surrounded by jacket cells on all sides, but are in contact with the spore wall on one side (Fig. 19.10, E). The undivided basal cell elongates considerably (Fig. 19.10, E) and pushes the upper two antheridial cells out of the massula and the microsporangial wall. The jacket cells disintegrate and the spermatids containing multiflagellate spermatozoids are liberated by the decay of the spore wall.

Fertilisation

The neck canal cell and the ventral cell disorganise and the neck cells spread apart to create a passage for the spermatozoids to enter and effect fertilisation. Large number of sperms gather around the open necks, but only one is able to penetrate the egg and effect fertilisation.

EMBRYOGENY

The first division of the zygote is parallel to the long axis of the archegonium and forms two unequal cells (Fig. 19.9, B). The second wall is transverse (at right angle to the long axis of archegonium) and forms a quadrant (Fig. 19.9, B). Each cell of the quadrant divides by vertical walls to form an octant stage. The hypobasal octants develop into the foot which is haustorial in nature. The two anterior epibasal octants develop into the first leaf. Out of the two posterior epibasal octants one gives rise to the stem apex, whereas the other is functionless. There is no root in ***Salvinia***. The upper and the lower octants divide by periclinal walls to form a central plate of eight cells. The cells of this plate divide repeatedly to form a column-like structure in the centre. It connects the foot with the leaf and the stem apex (Fig. 19.11. A). Growth of this central column pushes the leaf and the stems out of the gametophytic tissue (Fig. 19.11, B). Some workers interpret that there is an embryonic root that does not grow further and mixes with the foot. They call such a root as **vestigial root.** Others

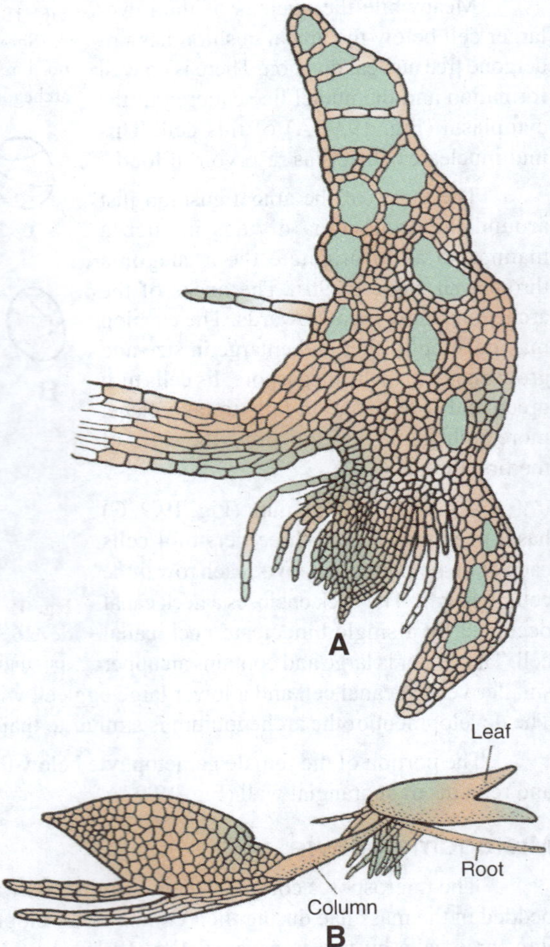

Fig. 19.11. (A–B). *Salvinia natans.* A. Young Sporophyte with an elongated column, two leaves and stem tip; B. Side view of the young Sporophyte attached to the gametophyte.

interpret that the column and the foot both represent a single enlarged foot of the embryo. The leaf segment grows into a cordate or a sagittate leaf raised on the column. The stem apex is present below the leaf and faces the gametophyte (Fig. 19.11, A). The column is traversed by a central vascular strand that extends from the foot to the leaf base, where it bifurcates and sends one branch to the leaf and the other to the stem apex. The stem apex divides actively and forms the rhizome which bears two floating leaves. Later the rhizome bears clusters of three leaves, two out of which are floating and the third is submerged and l filamentous. It is unbranched and thread-like at this stage.

Taxonomic Position

Division	:	Pterophyta
Class	:	Leptosporangiopsida
Order	:	Salviniales
Family	:	Salviniaceae
Genus	:	***Salvinia***
Species	:	***natans***

GENUS–AZOLLA Lamarck

Azolla is represented by thirty-one species out of which six are extant species. ***Azolla pinnata*** is found in India, Britain and North America. Taxonomically the species of ***Azolla*** are placed in four sections of the genus. Only two out of these are well known (***Rhizosperma*** and ***Azolla***). These sections are based on the characters of massulae and megaspore apparatus as follows :—

1. Section ***Azolla*** : Massulae have hooked or anchor-shaped glochidia. Megaspore apparatus has three floats in the swimming apparatus.

   It is represented by many species, *i.e.*, ***A. filiculoides***, ***A. microphylla***, ***A mexicana***, etc.
2. Section ***Rhizosperma*** : Massulae either with straight, simple or branched glochidia or without them. Megaspore apparatus has 9 floats, *e.g.*, ***A. prisca***, ***A. turgica***, ***A. nona***, etc.
3. Section ***Antiqua***. Glochidia not known. Megaspore has 9 floats of equal size. Represented by one fossil species.
4. Section ***Filifera.*** Glochidia hair-like, coiled or knobbed. Megaspores not known. Represented by one fossil species only.

In India, the genus is represented by ***Azolla pinnata*** which is widely distributed and occurs free floating on the surface of water in road side ponds, pools, reservoirs and lakes. It prefers undisturbed and standing waters. The plants resemble small and delicate moss gametophyte and cover the

A. filiculoides

A. microphylla

A. mexicana

Azolla pinnata

entire surface of water. Young plants have green or bluish green leaves which turn red at maturity so that they can be recognised from a distance.

SPOROPHYTE GENERATION

External Characters

The sporophytes (Fig. 19.12, A) are small in size and are distinguished into stem, leaves and roots. The stem is horizontal and profusely branched. It is thickly covered with leaves the overlap each other and are arranged in an alternate manner. The roots arise from the underside of that stem and are unbranched. The leaves are small in size and are divided into two lobes that are usually of equal size. The dorsal lobe is aerial and green. It is thick and sterile. The ventral lobe is thin, almost colourless and submerged in water. The bilobed leaves arise from the dorsal surface of the stem and are arranged in two alternative rows. The dorsal or the upper lobe has its dorsal side facing towards the sky. It encloses large mucilage filled cavities that harbour filaments of ***Anabaena azollae*** Strasb., that lives as long as the leaf containing it. Oes (1913) observed that there is a symbiotic relationship between the alga and the plant. The alga fixes free nitrogen of the air for the plant and later gives it shelter and food. The same species of ***Anabaena*** occurs in the leaves of ***Azolla*** all over the world, which indicates that dispersal by means of vegetative parts plays a greater role than the spores. The younger portion of the stem is upwardly curved. The leaves are not circinately coiled in the bud condition. The submerged lobe of the leaf is believed to absorb water. It is mostly one layer of cells in thickness except the basal part which is thicker and encloses a row of air chambers.

Vegetative propagation is effected by means of fragmentation of the stem and separation of the lateral branches. This rapid method of multiplication leads to enormous increase in the number of plants. ***Azolla filiculoides*** is annual. The closely packed leaves enclose small spaces between them. They are filled with air and are one of the means to keep up the buoyancy of the plant. Small papillae can also be seen on the upper surfaces of the aerial lobes.

Anatomy

The stem of ***Azolla*** like that of ***Salvinia*** grows by means of a three-sided apical cell that cuts off segments from all the three sides. A transverse section of the mature stem, reveals a distinct epidermis followed by a parenchymatous cortex that is 4 to 8 cells in thickness. Unlike ***Salvinia*** there are no lacunae or air spaces in the cortex. The endodermis and pericycle are derived from a common mother cell layer and are not very distinct. The pericycle is also one layered. The vascular cylinder

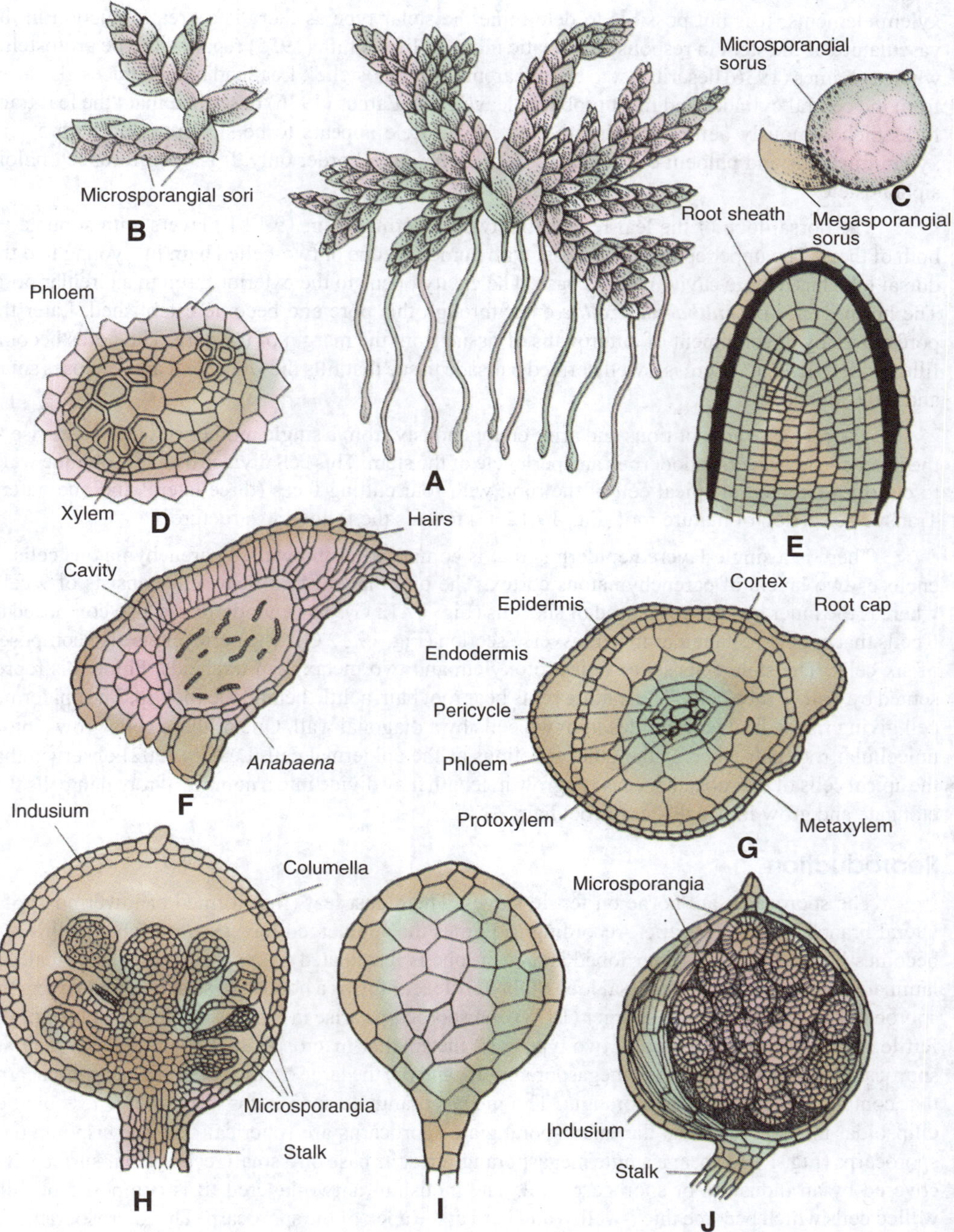

Fig. 19.12. (A—J). ***Azolla pinnata.*** A. Complete plant showing habit; B. Portion of A-bearing microsporocarp; C. Mega-and Microsporangial sori; D. T.S. through stem stele; E. L.S. root apex; F. L.S. dorsal lobe of the leaf; G. T.S. root; H. Microsporangiate sporocarp in section; I. A. microsporangium; J. Sporocarp with microsporangia.

(Fig. 19.12 D) consists of a few xylem tracheids and phloem elements that are twice the number of xylem elements. It is not possible to determine the stelar type as there is a great reduction in the vasculature of the stem in response to aquatic mode of life. Smith (1955) regards it to be protostelic, whereas Eames (1936) describes it to be 'apparently siphonostelic'. Leaf and branch traces also arise from the central cylinder and most probably leave gaps. Eames (1936) described that "the leaf trace forks dichotomously before entering the leaf." The stele appears to be siphonostelic with 5—10 xylem tracheids and phloem encircling it towards the outer border only. It is, therefore, ectophloic siphonostele.

The dorsal lobe of the leaf reveals the two epidermal (Fig. 19.12 F) layers with stomata on both of them. The upper epidermis is beset with numerous one or two-celled hair. In a young leaf the dorsal lobe has a large cavity near its base. The cavity opens to the exterior through a circular pore. The hormogones of ***Anabaena azollae*** enter through this pore and become established. Later the pore closes by development of outgrowths of tissue from the margin of the pore. The pores become filled with mucilage. There is a well defined palisade tissue that fills the whole leaf and encloses small and large air spaces.

The roots are adventitious and arise endogenously from a single mother cell that gives rise to the inner cortical layer, endodermis and pericycle of the stem. This cell divides by a few oblique walls to cut off a tetrahedral apical cell of the root, with four cutting faces (three lateral and one outer). Transverse section of mature root (Fig. 19.12, G) reveals the following structure :

There is a single-layered epidermis that is composed of thinwalled parenchymatous cells. It encloses two layers of parenchymatous cortex. The outer layer of cortical cells consists of 9 cells whereas the inner layer is composed of six cells (Fig. 19.12 G). The endodermis is also composed of 6 cells that appear rectangular in a transverse section (Fig. 19.12 G). The pericycle is also composed of six cells. The stele consists of four protoxylem and two metaxylem tracheids. Phloem is represented by 4 sieve tube cells. The mature roots bear root hair, a little behind the root tip. The epidermal cell giving rise to root hair divides into two cells by a diagonal wall. One of these cells grows into a unicellular root hair whereas the other functions as the epidermal cell. Leavit (1902) observed that the apical cells of the roots that cease to grow in length may divide into a number of root hair cells that elongate and grow into a cluster of root hair.

Reproduction

The sporocarps are borne on fertile leaves. The oldest leaf (first formed or lowermost) of a lateral branch is always fertile. According to Eames the submerged lobe (ventral lobe) of this leaf becomes bilobed or nearly four lobed and each lobe is terminated by a sporocrap. A marginal and laminate outgrowth from the dorsal lobe of the fertile leaf forms a hood-like structure that covers the sporocarps. According to Strasburger (1873), the sporocarps arise in the axil of the dorsal lobe of the fertile leaf. The sporocarps are of two types : (*i*) the smaller macrosporangiate or megasporangiate sporocarps that contain only one megasporangium, and (*ii*) the large or microsporangiate sporoacarps that contain numerous microsporangia. The megasporangiate sporocarps are more or less oval or ellipsoidal in shape whereas the microsporangiate sporocarps are spherical. The megasporangiate sporocarps (Fig. 19.13) bear a single megasporangium at its base on a small receptacular surface. It is covered by an indusium or sporocarp wall. The indusium is two-layered. It is composed of thin-walled cells which become thick-walled and hard at the apex of the sporocarp. The microsporangiate sporocarps (Fig. 19.12) bear numerous microsporangia that have long unbranched stalks arising from a central cushion-like or even elongated **receptacle** or **columella** (Fig. 19.12, A).

The development (Fig. 19.13, A—F) of the sporocarp follows a very simple course. The young megasporangiate sporocarps become recognisable when a single megasporangial initial has cut off a few segments that form its stalk (Pfeiffer, 1907). A circular protuberance composed of row of cells appears at the base of the sporangial stalk. This is the primordium of indusium or the sporocarp wall. This ring of cells soon overgrows the developing sporangium and surrounds it on all sides, except the apex (Fig. 19.13, B). The cells of the indusium divide periclinally and make it two-layered thick. A

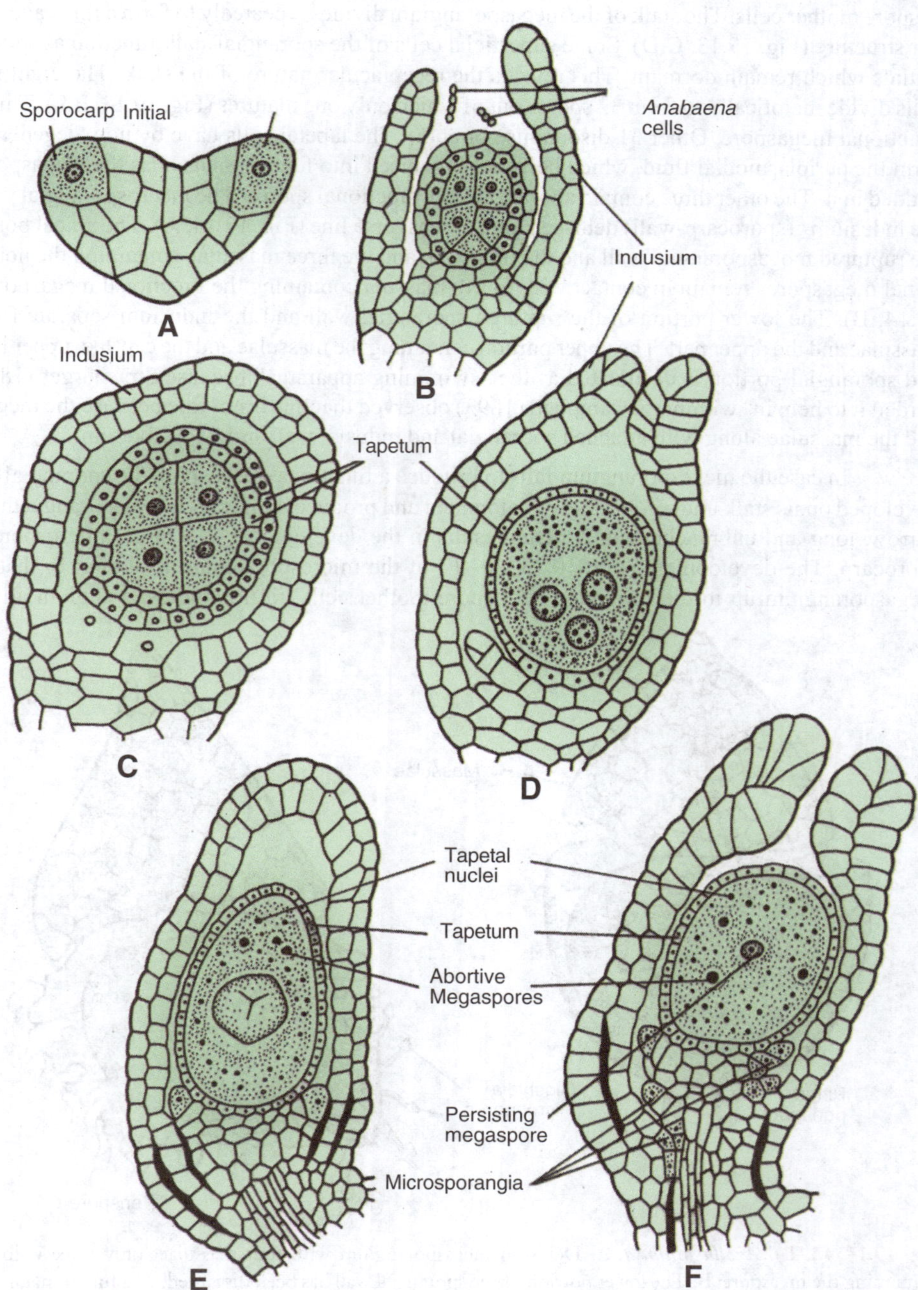

Fig. 19.13 (A—F). *Azolla.* Stages in the development of sporocarp and megasporangium; A—E. A. *pinnata*; F. *A.caroliniana* (F. after Pfeiffer)

few segments or hormogonia of ***Anabaena azollae*** are also present within the indusium (fig. 19.13, B). The free margins of the indusium, at the top, grow towards each other and close the opening (Fig. 19.13, C,D.). The megasporangial initial develops in a leptosporangiate manner into a mature megasporangium which has a single layered wall, a single layered **tapetum** that encloses eight megaspore mother cells. The stalk of the megasporangium divides repeatedly to form a thick and columnar structure (Fig. 19.13, C,D). Some superficial cells of the sporangial stalk function as sporangial initials which remain dormant. This reveals the receptacular nature of the stalk. The eight mother cells divide meiotically to form 32 spores out of which only one matures (Fig. 19.13, E,F). This is the functional megaspore. Other 31 disorganise or abort. The tapetal cells have by now degenerated to form the periplasmodial fluid, which becomes organised into four massulae has the megaspore embedded in it. The other three contain aborted or non-functional spores. The macrosporangial wall and the indusium (sporocarp-wall) dehisce along a transverse line (Fig. 19.14, A). The apical portion of the ruptured megasporangial wall and the indusium and the three massulae containing the non-functional megaspores remain in contact with fourth massula containing the functional megaspore (Fig. 19.14, B). The lower portion of the ruptured sporangial wall and the indusium separate from the massulae and the upper part. The upper portion containing the massulae and the cap-like upper indusial and sporangial portion is designated as the "swimming apparatus" because Strasburger (1873) regarded it to help in swimming. Campbell (1893) observed that this is not true because the megaspore and the massulae along with attached sporangial and indusial wall sink to the bottom.

In case the megasporangium fails to produce a functional megaspore the sporangial initials developed on its stalk undergo further development and produce numerous microsporangia that have narrow, long and unbranched stalks. This results in the development of a larger microsporangiate sporocarp. The development (Fig. 19.15, A—F) of the microsporangium is similar to that of the megasporangium up to the formation of 8 spore mother cells. In the microsporangium all the 32

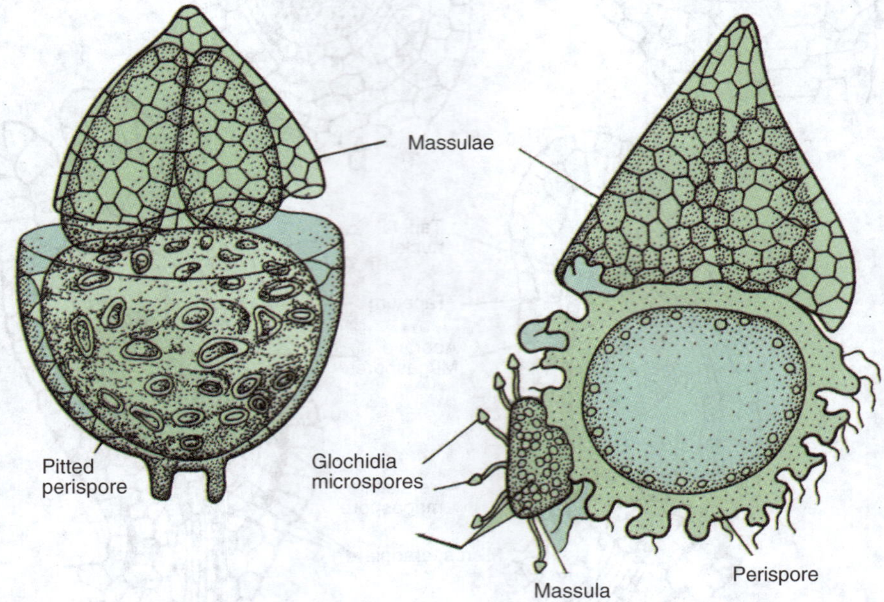

Fig. 19.14. (A–B). ***Azolla pinnata.*** **A.** Dehisced megasporangium with four massulae (only three visible), One containing the mgaspore; **B.** The lower portion of the sporangial wall has been discarded. The functional megaspore with its massula is attached to the other massulae. Also note the massula with microspore and glochidia attached to it. (After Bernard)

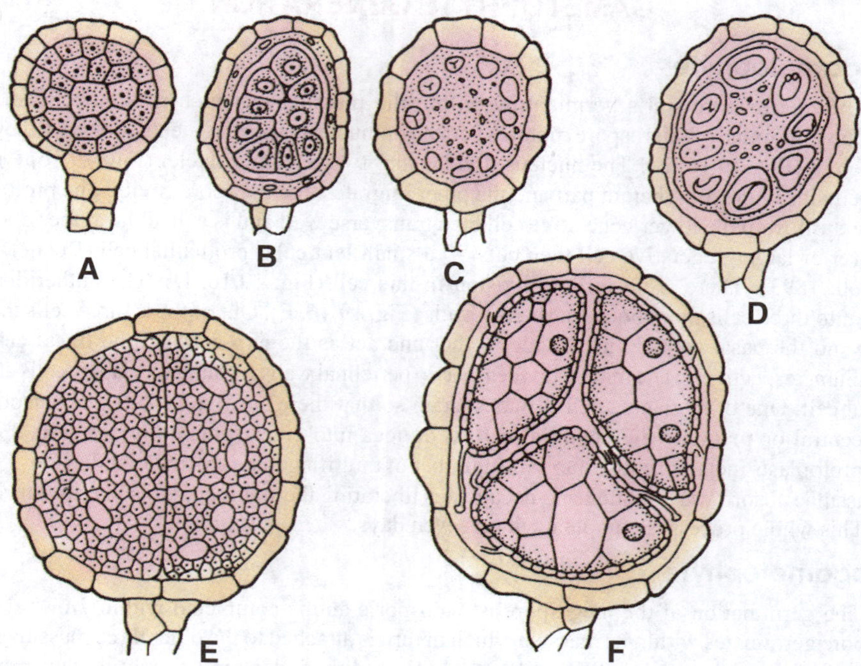

Fig. 19.15. (A–F). *Azolla pinnata.* Stages in the development of microporangium. (After Hannig)

microspores remain functional (Fig. 19.15). The tapetum breaks down to form the plasmodial fluid. The microspores move to the periphery of the plasmodial fluid (Fig. 19.15, D). The plasmodium now becomes organised into 4 or more massulae, each containing more than one microspores embedded in it (Fig. 19.15, F). The cytoplasm of the massula is alveolar. In ***Azolla filiculoides*** and ***A. caroliniana*** numerous hooked processes arise from the surface of each massula. These are called the **glochidia** (Fig. 19.16 A). Glochidia are absent in some other species. As soon as the microspores mature the sporangial walls dehisce and the massulae containing microspores lie free in the cavity of the sporocarp. Later the indusium (sporocarp wall) also ruptures and the massulae float away under water. The massulae studded with glochidia attach themselves with macrosporal massulae (Fig. 19.14, B). This attachment is effected by means of glochidia. The microsporocarps sink to the bottom before dehiscence. The massulae are liberated at the bottom and swim under water.

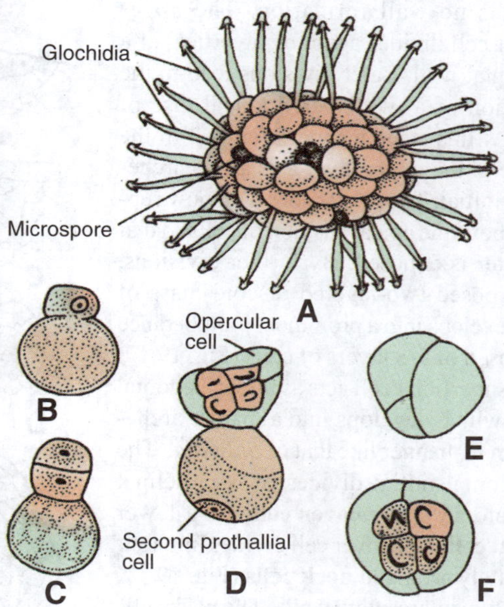

Fig. 19.16. (A—F). *Azolla pinnata.* A. Massula containing microspores and glochidia; B—F. Stages in the development of male gametophyte. (After Smith).

GAMETOPHYTE GENERATION

Microgametophyte

The micro-spore starts germinating within the massula in which it is embedded. Before germination the exine of microspore ruptures and the uninucleate spore contents surrounded by initine protrude out (Fig. 19.16, B). The nucleus divides into two daughter nuclei (Fig. 19.16, C) one of which remains in the protuberant part and the other migrates into the spore cavity. The protuberance is then cut off from the lower vegetative cell by a transverse wall and is called the antheridial initial. The lower or larger vegetative cell then cuts off a small lenticular prothallial cell (Belajeff, 1898 ; Campbell, 1893). There is no further division in this cell (Fig. 19.16, D). The antheridial initial divides into three cells by oblique transverse walls (Fig. 19.16, E). Out of these three cells the distal, smaller and the basal cells do not divide further and act as the **cap cell** and the **basal cell** of the antheridium, respectively. The middle divides by two periclinal walls forming two jacket cells enclosing a central cell. One of the jacket cells divides again so that there are five jacket cells surrounding a single central or **primary androgonial cell.** It divides into eight **spermatocytes** (Fig. 19.16, D) whose protoplasts metamorphose into same number of multiflagellate spermatozoids (Fig. 19.16, F). The massulae absorb water, soften and decay, thus liberating the spermatozoids into the surrounding water. This whole process is complete within seven days.

Megagametophyte

The germination of the megaspore is endosporic and is completed within 7 or 8 days. The megaspore germinates within its massula which in turn is attached to the other three massulae and the upper half of the sporangial wall (Fig. 19.14, A, B) and the indusium (swimming apparatus). The nucleus of the megaspore enlarges in size and divides into two. A cell wall is laid down between the two daughter nuclei, cutting off a small, lenticular apical cell and large basal cell which is filled with starch grains. The basal cell nucleus undergoes free nuclear division thus making it multinucleate. There is no wall formation. The upper smaller cell divides again by a vertical or a horizontal wall. Later divisions lead to the formation of a two-layered apical disc of cells. At this stage a superficial cell in the middle of the upper layer acts as an archegonial initial. The megapsore wall now ruptures above the apical cushion of cells so that the latter is exposed. By further divisions, this exposed two-layered discoid mass of cells develops into a prominent protuberance composed of 5-8 layers of cells (Fig. 19.17, A). A superficial cell acts as the archegonial initial which develops into a mature archegonium in manner like that of *Salvinia.* The archegonial initial divides by a periclinal wall into an upper **cover cell** and a lower **central cell.** The cover cell divides into four quadrately arranged **neck cells** (Fig. 19.17, B), by two intersecting walls. The neck cells divide by transverse walls to form three or four tiers of 4 cells each (Fig. 19.17, B, D). These constitute an archegonial neck consisting of 4 longitudinal rows of cells, each

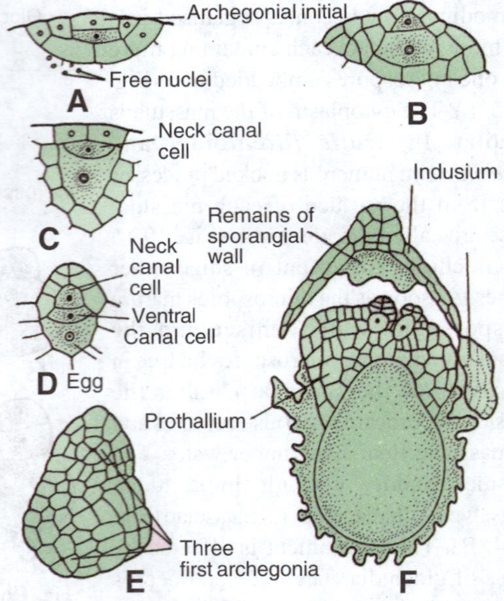

Fig. 19.17 (A—F). Development of archegonium and the female gametophyte in *A. pinnata.* (After Campbell)

row 4 cells high. The central cell divides into an upper **primary neck cell** and a lower **primary ventral cell** (Fig. 19.18, B). The former grows between neck tiers and forms a single, uninucleate neck canal cell or divides to form two neck canal cells. The primary ventral cell divides into (Fig. 19.17, D) an upper small, ventral canal cell and a lower large egg cell (Campbell, 1893). Smith (1955) reported that the primary ventral cell may function directly as an egg cell. The mature archegonium opens in the usual manner forming an open neck for the spermatozoids to swim down to the egg.

In case the first archegonium is fertilised, no more archegonia develop. In case of failure of fertilization more archegonia can develop.

The growing apical cushion (with its archegonia) protrudes out of the massula enclosing it and pushes aside the other three massulae which, however, remain, attached to it (Fig. 19.17, F). The female gametophyte of ***Azolla*** does not become green.

EMBRYOGENY

The zygote or the oospore increases in size and divides by a transverse or an oblique longitudinal wall (Fig. 19.18, A 1-1) into two nearly equal cells. The second wall (Fig. 19.18, B) is either vertical or transverse, depending upon the plane of the first division and forms a four-celled embryo (quadrant stage). The four primary organs of the mature embryo or young sporophyte can be traced back to each cell of the quadrant stage. The two epibasal cells (nearer to archegonial neck) give rise to the first leaf and the stem, whereas the two hypobasal cells give rise to foot and the first root (Fig. 19.18, C). Except the foot the other three organs grow by means of an apical cell. The quadrants divide further to form an octant stage. The apical cells of the primary organs differentiate from their respective quadrants at a thirty-two celled stage (Fig. 19.18, C). The first leaf and the stem develop almost at the same rate. The root develops slow. The foot develops into a small cylindrical organ (Fig. 19.18, F). The first leaf overarches the stem and soon grows into a visible segment on one side (Fig. 19.18, D). The growing embryo throws off the so-called "swimmig apparatus" and the embryo rises to the surface of water along with the megagametophyte after the development of air cavities in the first leaf. The procambial strands develop into the first vascular strand that bifurcates below the stem apex giving a branch to the stem and the first leaf. It is continuous below into the first root (Fig. 19.18, F).

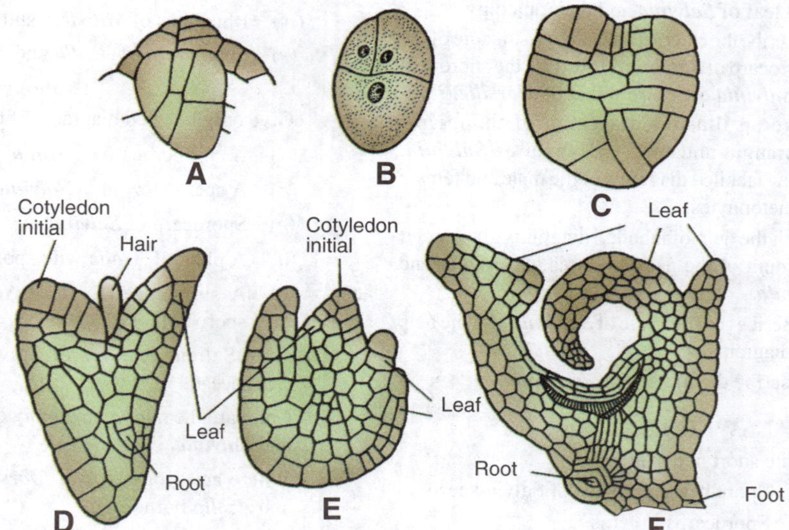

Fig. 19.18. (A–F). Stages in the development of embryo in *A. pinnata* (After Campbell)

Chromosome numbers

In *Azolla pinnata* Loyal (1973) reported $2n = 44$ from the root tips. In ***Salvinia natans*** he reported $2n = 18$ from the root tips (Loyal, 1973).

Economic Importance

R.N. Singh (1977) has reported the use of ***Azolla*** as a precious green compost for rice cultivation. An increase of 30 to 38% in grain yield has been reported from fields incorporated with ***Azolla.***

Taxonomic Position

Division	:	Pteriophyta
Class	:	Leptosporangiopsida
Order	:	Salviniales
Family	:	Azollaceae
Genus	:	***Azolla***
Species	:	***pinnata***

QUESTIONS

ESSAY TYPE

1. Give a brief account of Salviniaceae. How does the sporocarp of ***Salvinia*** differ from that of ***Azolla***?
2. Give a brief account of Salviniaceae and discuss the affinities of the group.
3. Give an illustrated account of the sporocarp of ***Azolla.***
4. Describe salient features of ***Salvinia*** along with the external morphology of its sporophyte.
5. Give an illustrated account of the stem, root and leaf of ***Salvinia*** and their anatomy.
6. Describe the external and internal structure of the sporocarp of ***Salvinia***. How does the sporocarp of ***Salvinia*** differ from that of ***Marsiliea***?
7. Give an illustrated account of the microsporangia and megasporangia of ***Salvinia.*** Draw labelled diagrams of the male and female gametophytes.
8. With the help of labelled diagrams give a brief account of the gametophytes of ***Salvinia*** and ***Azolla.***
9. Describe the life cycle of ***Salvinia*** with the help of diagrams.
10. Describe the life history of ***Azolla.***

SHORT ANSWER TYPE

11. Write short notes on :
    (*i*) Spore bearing organs of Salviniaceae
    (*ii*) Sporocarps in ferns
    (*iii*) Sporocarp of ***Azolla***
    (*iv*) Male gametophyte of ***Azolla***
    (*v*) Female gametophyt of ***Azolla.***
    (*vi*) Sporocarp in ***Salvinia***
    (Rohilkhand, 2000, 2003)
    (*vii*) Gametophyte or ***Azolla.***
    (*viii*) Give identifying characters of ***Azolla.***
    (Kanpur M.Sc., 2002)
12. How would you distinguish between :
    (*i*) Sporocarps of ***Azolla*** and ***Salvinia.***
    (*ii*) ***Marsilea*** and ***Salvinia***
    (*iii*) Rhizomes of ***Salvinia*** and ***Azolla***
    (*iv*) Rhizomes of ***Marsilea*** and ***Azolla***
    (*v*) Sporocarps of ***Azolla*** and ***Marsilea.***
    (Kanpur M.Sc., 2003)
13. Give only labelled diagrams of the following:
    (*a*) L. S. Sporocarp of ***Azolla***
    (*b*) A spermatozoid of ***Salvinia***
    (*c*) Sporocarp of ***Salvinia***
    (*d*) A plant of ***Azolla*** with sporocarp
    (*e*) A submerged leaf of ***Salvinia*** with sporocarp.
14. Give Salient features of Salviniaceae and Azollaceae.
15. Compare the microgametophytes of ***Marsilea*** and ***Salvinia.***
16. Where are sporocarps of ***Marsilea, Salvinia*** and ***Azolla*** borne?
17. Describe briefly the embryology in ***Salvinia.***

18. Give an account of megasporangium of ***Salvinia***. How a megaspore develops into female gametophyte? **(Kanpur M.Sc., 2005)**

19. Describe biological significance of ***Azolla***. **(Kanpur M.Sc., 2003)**

OBJECTIVE TYPE

20. Select the correct answer :

(*i*) Which of the following is an aquatic pteridophyte?

(*a*) ***Azolla***
(*b*) ***Ophioglossum***
(*c*) ***Equisetum***
(*d*) ***Lycopodium.***

(*ii*) Which one of the following is a free floating fern?

(*a*) ***Marsilea***
(*b*) ***Adiantum***
(*c*) ***Azolla***
(*d*) ***Equisetum.***

(*iii*) Which of the following statements is correct?

(*a*) The first formed Sporocarps in ***Salvinia*** are microsporangiate.
(*b*) The first formed sporocarps in ***Salvinia*** are megasporangiate.
(*c*) Both microsporangiate and macrosporangiate sporocarps in ***Salvinia*** are produced simultaneously.
(*d*) In ***Salvinia*** sporocarps are absent.

(*iv*) In ***Salvinia***

(*a*) all the leaves of a cluster are submerged.
(*b*) all the leaves of a cluster are tree floating
(*c*) two of the leaves in a cluster are above the cluster and the third are submerged.
(*d*) the leaves are produced independently.

(*v*) The stele in rhizome of ***Salvinia*** is

(*a*) Endophloic siphonostele
(*b*) ectophloic siphonostele
(*c*) amphiphloic siphonostele
(*d*) Solenostele.

(*vi*) Which of the following encloses filaments of ***Anabaena***?

(*a*) ***Marsilea***
(*b*) ***Salvinia***
(*c*) ***Azolla***
(*d*) ***Lycopodium.***

(*vii*) Vegetative propagation in ***Azolla*** takes place by

(*a*) fragmentation of the stem and separation of the branches
(*b*) budding
(*c*) perennating leaves
(*d*) Corms.

Calamites – A fossil sphenophyte

CHAPTER 20

General Discussion

- Fossil Lycopods
- Fossil Sphenophytes
- Biotechnology–Application in Pteridophytes
- Economic importance of Pteridophytes
- Diversity and distribution of Pteridophytes in India

FOSSIL LYCOPODS

It has been mentioned earlier in the Chapter-4 that division Lycophyta is represented both by living as well as extinct genera and has a very long evolutionary history. Members of Division Lycopyta have been observed to extend from the Palaeozoic era to the present day. All the modern living lycopods are basically herbaceous plants, however, most of the late palaeozoic lycopods, particularly their carboniferous relatives were large trees like *Lepidodendron* and *Sigillaria*. These forest tree-like (Arborescent) lycopods were of the height of upto 40 meters and even more in some cases. Their contribution in the present day economy specially in the form of coal cannot be ignored. It is estimated that more than 3/4th of the North-American and European Westphalian coals may consist of biomass derived from arborescent lycopods. These trees were heterosporous and in some forms megaspores with surrounding megasporangia and megasporophylls looked almost seed like. One of the important features of palaeozoic lycopods, is in the nature of secondary growth which was mainly concentrated to cortex region only. This is unlike in other trees where secondary growth is generally concentrated in the xylem region. Apart from heterosporous arborescent lycopods, fossils of several homosporous, herbaceous lycopods are also reported such as *Protolepidodendron* and *Baragwanathia*. Both herbaceous and arborescent fossil lycopods are described here in brief.

(i) Herbaceous, homosporous fossil lycopods :

Protolepidodendrales. Protolepidodendrales are the group of fossil herbaceous homosporous plants having subterranean, aerial dichotomously branched axes covered with ligulate or eligulate microphylls. The group had an apparently worldwide distribution during Devonian and early Carboniferous periods. They were probably ancestral to later lycophytes like Lepidodendrales and Selaginellales. *Protolepidodendron* represents the type genus of this group. *Baragwanathia longifolia* from silurian of Australia along with *Yarravia* is the oldest fossil which can be assigned under Protolepidodendrales. The plant body of *Baragwanathia longifolia* (Fig. 20.1) is slightly larger than modern day lycopods. The stem is covered with slender leaves. Sporangia are reniform, found in association with ordinary leaves present in certain area of stem representing cone of the plant. Anatomically stem stele is star shaped, Actinostele, similar to *Asteroxylon.*

Baragwanathia longifolia–Impressions on rocks

Fig. 20.1 Reconstruction of *Baragwanathia longifolia.*

Plants of *Protolepidodendron.* (Fig. 20.2) were of about 30 meters height and 1 cm in diameter. Plant body consists of both prostrate as well as aerial axes. Prostrate branches are dichotomously branched giving rise to aerial axes. Like other lycopods whole plant body is covered with small leaves which are bifurcated at their apices, a unique feature of *Protolepidodendron* leaves. Some of the leaves have radially elongated sporangia on their upper surface. Triangular protostele is found in the stem which is uncommon among modern lycopods.

Fig. 20.2. Reconstruction of *Protolepidodendron* tree

(ii) Arborescent, heterosporous fossil lycopods :

Lepidodendron is the name assigned to ancient arborescent lycopod also known as the "*Scale Tree*". (Fig. 20.3) It was one of the most common carboniferous tree (360 to 286 million years ago) lycopods, capable of growing upto more than 130 ft. with supporting trunk measuring upto 6ft or even more in diameter. The genus was originally described for stem surfaces showing the characteristic spirally arranged diamond shaped, vertically positioned structures known as "leaf cushions". Stem anatomy showed the proportion of xylem wood is relatively small because secondary growth is mainly confined to cortical zone only. During their life cycle the trees shed parts of their outer bark and basal part of the trunk and large stems typically show various stages of decortication.

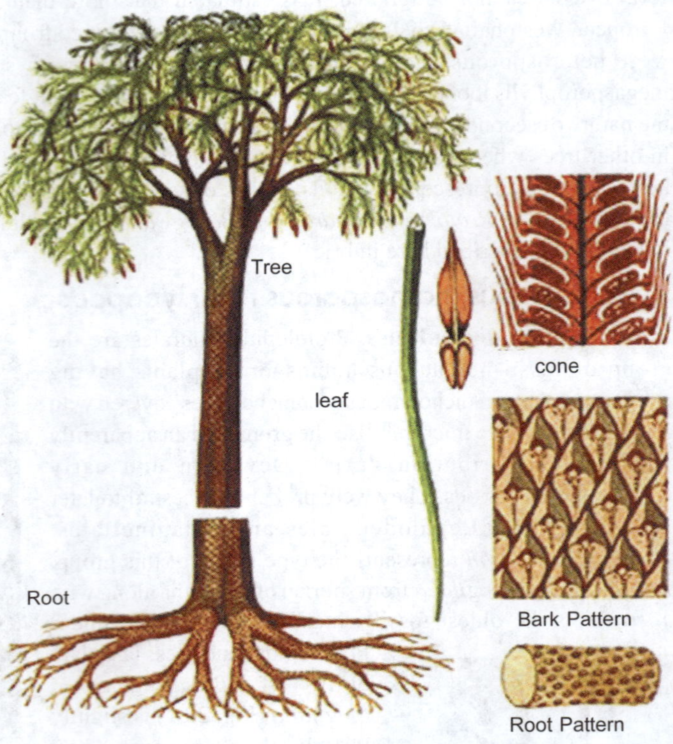

Fig. 20.3. *Reconstruction* of *Lepidodendron tree*

Rooting pattern of the plant indicates presence of humid environment. Root fossils of the lepidodendrons are known as ***Stigmaria.*** In this way the genus has been described under several generic names on the basis of branching patterns of aerial stems. It can be lateral branching or dichotomous branching. Large branch scars of lateral branches have been described as *Ulodendron*. *Lepidophyllum* under the name, long narrow and stiff leaves have been described. In this leaves of the ultimate branches being smaller and needle like. The plants were heterosporous so their strobili were described as *Lepidostrobus* and *Lepidocarpon*. *Lepidostrobus* could gain a length of upto 1m. The isolated sporophylls are assigned under the name *Lepidostrobophyllum*. The tree became almost completely extinct by the end of Westphalian.

Lepidodendron–A fossil lycopod

In brief *Lepidodendron* is a reconstructed single genus assigned with following generic names

Trunk & Branches	:	*Lepidodendron, Lepidophloios*
Branch scars	:	*Ulodendron*
Leaves	:	*Lepidophyllum*
Roots	:	*Stigmaria*
Strobili	:	*Lepidostrobus, Lepidocarpon*

Lepidophloios differs from *Lepidodendron* in having diamond-shaped leaf cushions that are wider than high but also spirally arranged. This tree had the same type of rooting and foliage system as that of *Lepidodendron.*

Fossilized branch scar of ***Lepidodendron*** tree

The trunk of *Lepidodendron* was basically unbranched for most of its length except at the high ups where it is dichotomously branched. The branches were covered with microphylls which upon falling, leave characteristic scars on stems. Large branch scars of lateral branches have been described as *Ulodendron.* Presence of leaf cushions (elevated leaf bases) is also an important characteristic of *Lepidodendron* stems. The Lepidodendrids were ligulate. Immediately above the leaf scars, ligule scars can also be seen within a pit. Anatomy of stem indicates that major girth of the stem is due to the presence of well developed secondary cortex and periderm (Fig. 20.4). Stele is quite reduced. Protoxylem were exarch, secondary xylem present in relatively small amount. Only primary phloem are seen, secondary phloem has not been found (Eggert & Kanenoto, 1977). Pith represents the centre core of the stem. Periderm is quite well developed-a marked feature of lycopods. Periderm cells are normally radially aligned, quite longer than broad with sharply tapered end.

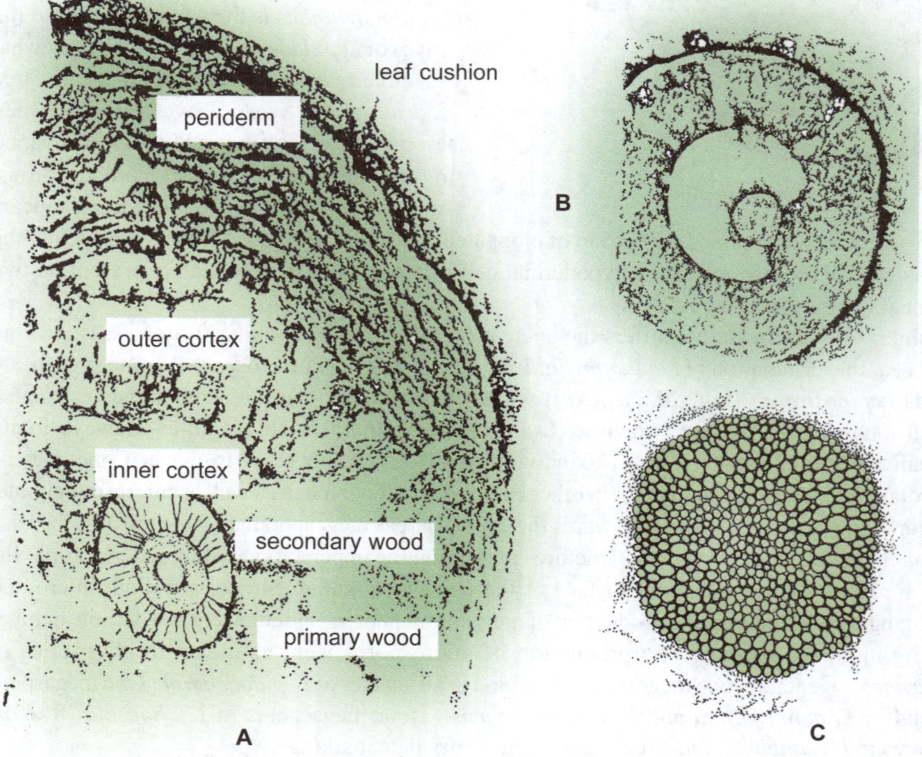

Fig. 20.4. ***Stem anatomy*** of *Lepidodendron*
(*A*) ***L. scleroticum***, branch (about 4 min diameter) in cross section (*B, C*) ***L. vasculare***, stem and stele
(Figure after : (A) Andrews ; (B,C) Bierhorst)

Stigmaria is the generic name given to the rooting system of arborescent lycopods i.e., *Lepidodendron, Lepidophloios* and *Sigillaria*. Though the name *Stigmaria* has been assigned to fossil root casts (Fig. 20.5.) of various pieces of the Pennsylvanian period regardless of the form genera and species. *S. ficoides* is a single monotypic genus. The woody, a few times bifurcating axes have been observed to bear hollow tube-like appendages which had the function of roots. Such rooting systems are known as rhizomorphic. Very few modern lycopods except *Isoetes* have rhizomorphic rooting system. *Stigmaria* rootlet beds are very common and often occur underneath coal seams. These rootlets are of about 1cm in diameter. Root hairs were not found on *Stigmaria*. Anatomically, *Stigmaria* is very close to roots of *Isoetes*. A flattened vascular tube is frequently visible on the end of *Stigmaria*, located just off center of the axis of root and running longitudinally down its length, which provided fluids and perhaps nutrients to main tree in life (Fig. 20.6.)

Fig. 20.5. *Stigmaria ficoides* –Root casts.

Fig. 20.6. *Stigmaria*–Root T.S.

Lepidophylloides is the name assigned to the leaves of *Lepidodendron* tree which somewhat resemble needles of the modern spruce tree *Cedrus deodar*. Size of leaves varied between few millimeters to a meter. Size of leaves varied in different species of fossil lepidodendrons. Anatomically a single vascular strand appears to be present which comprised of a group of elongated tracheids surrounded by phloem. Below the epidermis well developed extensive hypodermal cells are present. The stomata are present on two longitudinal furrows on underside only.

Similar to other modern conifers the ancient *Lepidodendron* reproduced by cones that were attached near the termination of *Lepidophylloides*. Since plants were heterosporous some strobili are known as *Lepidostrobus* (Fig. 20.7 *a, b, c*)) whose length could reach to about 1 meter. Isolated Lepidostrobus sporophylls are known as *Lepidostrobophyllum* which are almost always found individually indicating they were attached individually to the tree rather than in cluster. Cones may be microsporangiate, megasporangiate or in other case it can be both mega as well as microsporangiate. Microsporangia were usually placed towards the apex whereas megasporangia were towards the base of the cone. Sporangia are elongated structures present on the upper surface of ligulate sporophylls. Ligules are present at the adaxial surface at the point of bending. Structurally both micro and megasporangia are similar but the production of megaspores per sporangium were in a definite number depending upon the species. The production of microspores were abundant in all species of *Lepidodendron.* Generally one megaspore was found in all species of *Lepidodendron.* One megaspore was found in *L. monosperma* and *L. braidewoodensis*, four megaspores in *L.foliaceous,* 8 to 16 megaspores in *L.veltheimanum.* All the spores have triradiate markings.

Lepidocarpon (Fig. 20.7(*d*)) is a megasporophyll holding a megasporangium with only single functional megaspore. The megasporophyll encloses the megasporongium completely thus giving

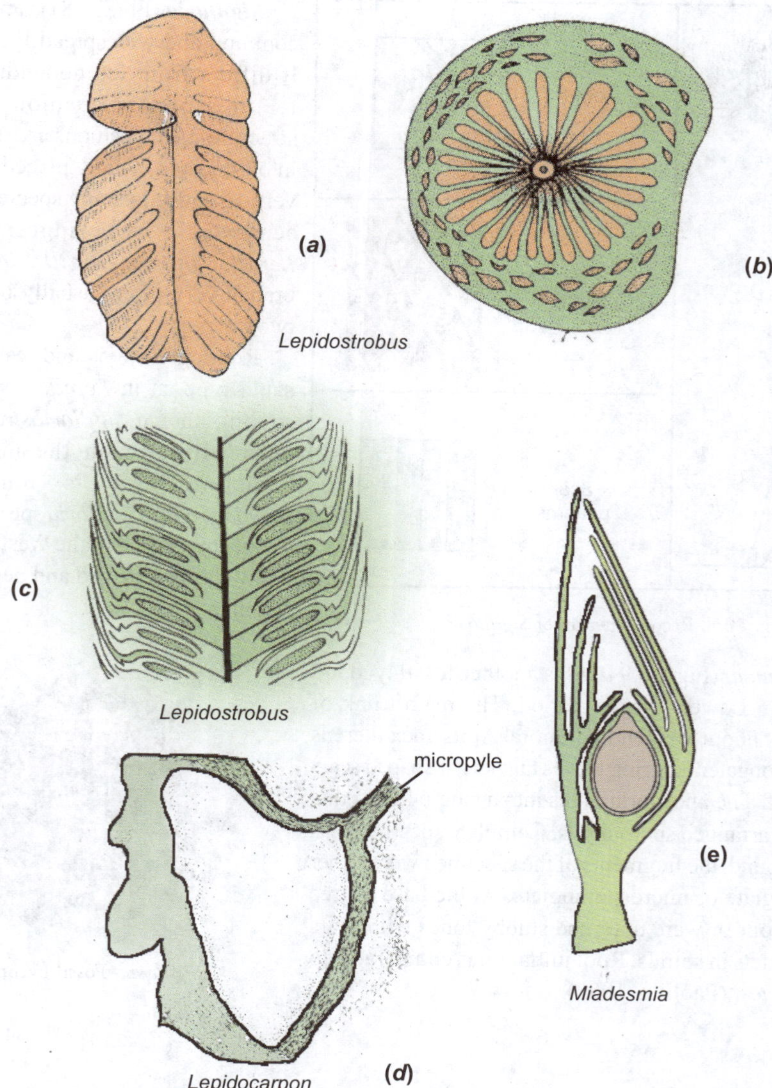

Fig. 20.7. Fructifications of fossil lycopods (*a*) ***Lepidostrobus*** in L.S. (*b*) ***Lepidostrobus*** in T.S. (*c*) ***Lepidostrobus***, diagrammatic section (*d*) ***Lepidocarpon*** (e) ***Miadesmia***

(Figures after : (*a, d*)Bierhorst; (*b*) Ziller ; (*c*) Sporne ; (*e*) Benson)

optimal protection. The whole unit (megasporophyll + megasporangium) is shed from the cone. *Lepidocarpon* thus already pretty much shows a seed like morphology. The sporangium in *Lepidocarpon* is an elongated sac like structure that differs from *Lepidostrobus* which is broad at the base tapering towards the apex. Presence of embryo has been reported by Philip (1979) in *Lepidocarpon*.

Miadesmia (Fig. 20.7(*e*)) is another female fructification of *Lepidodendron* known from lower carboniferous period. Structurally *Miadesmia* is similar to *Lepidocarpon* except that the megasporangium is covered with many sporophyll flaps. However, it can't be said with certainty that *Miadesmia* is a lepidodendrid fructification.

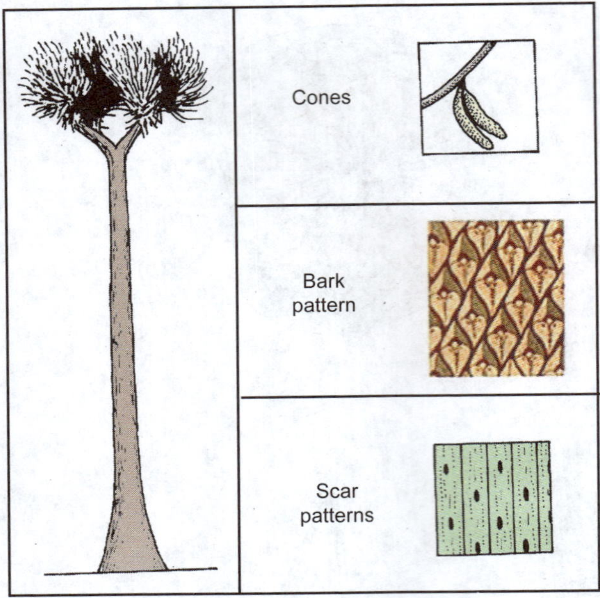

Fig. 20.8. Reconstruction of *Sigillaria.*

Sigillaria (Fig. 20.8) is another very common and widespread lycopod tree. It differs from lepidodendron in the features of leaf cushion and cone structure. Leaf cushions are rhomboidal in outline and usually placed in distinct vertical rows; in some species rows are separated by ridges. Unlike *Lepidodendron, Sigillaria,* did not branch very often, usually only one or two times.

Rooting systems and leaves are the same type as in *Lepidodendron*. The strobili, known *Sigillariastrobus,* were borne directly on the main stem. Structurally it is much alike *Lepidostrobus. Sigillaria* species became extinct by the end of the Westphalian but a few forms survived and persisted into Permian.

Pleuromeia (Fig. 20.9 *a, b*) is another fossil lycopod reported from Lower Triassic period. The main trunk of the plant is of about 1m height or more. At its apex there is a crown of elongated tapering leaves that left lipshaped scars on falling off. The apex terminates into a single large cone with spirally arranged sporophylls. Complete specimen has never been found, the fragments of the specimen were either megasporangiate or microsporangiate. At the base of tree trunk were four or were upturned stubby lobes with scars of many rootlets in spirals. Root initiation is comparable to those in *Isoetes*. (Paolillo, 1982).

Sigillaria- Fossil Lycopod

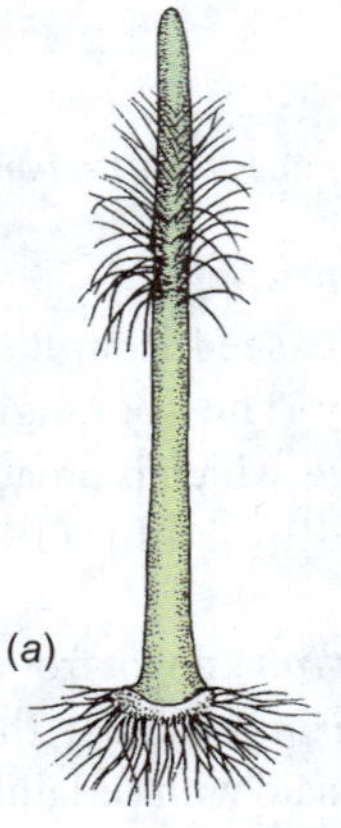

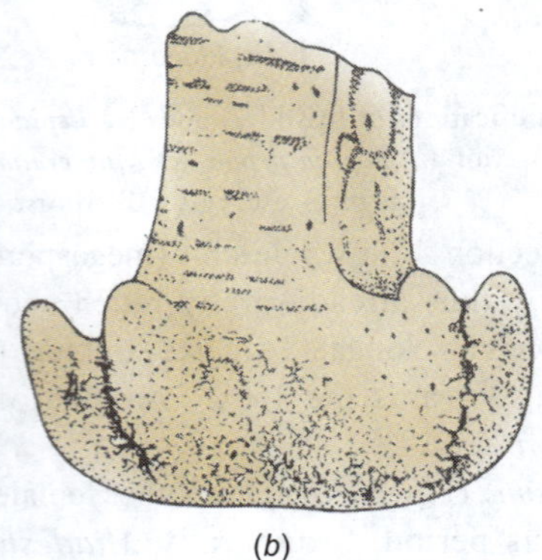

Fig. 20.9. Reconstruction of *Pleuromeia*
(*a*) ***Plant*** reconstructed (*b*) lower part of trunk.
(Figure after : (a) Hirmer (b) Bierhorst)

Classification

Division : Lycophyta
Class : Ligulopsida
Order : Lepidodendrales
Family : Lepidodendraceae
Genus : *Lepidodendron, Stigmaria, Lepidocarpon, Sigillaria.*

FOSSIL SPHENOPHYTES

Sphenophytes or Articulates are a group of spore plants with articulate axes and leaves standing in whorls. Fructification typically consists of strobili. There are two major groups among sphenophytes, the Calamitales and Sphenophyllales.

Calamitales

The order Calamitales was at its peak of development during the upper carboniferous along with lepidodendrales and were found in the coal measures and swamp forests. The order become extinct at the end of Permian.

Fig. 20.10. (*a*) Reconstruction of *Calamites.*

Fig. 20.10. (*b*) *Calamites* trunk exhibiting vertical robbing.

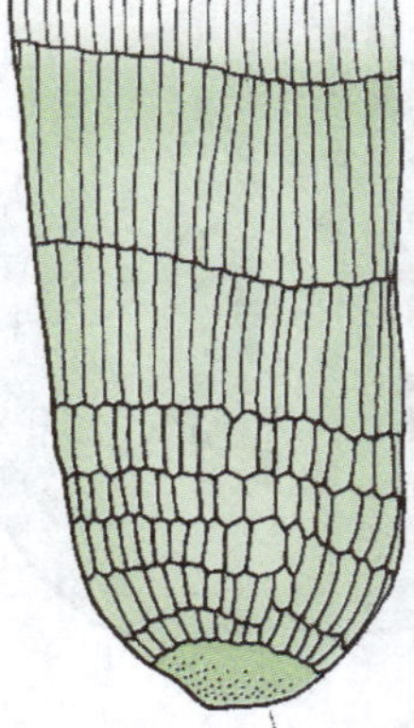

Fig. 20.10. (*c*) *Calamites suckowi*, a pith cast (after Jogman and Kukuk)

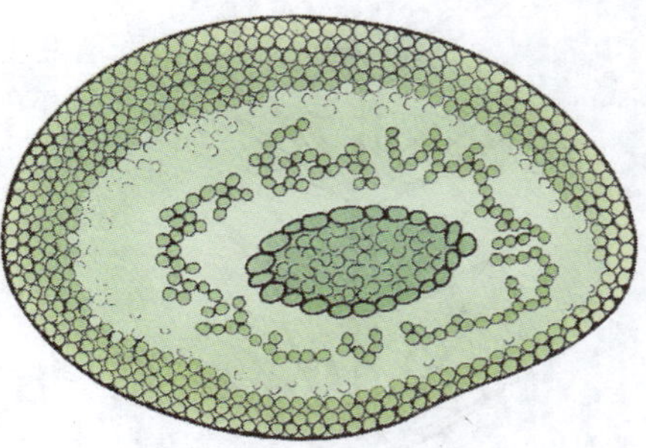

Fig. 20.10. (*d*) Transverse section of twig of *Calamites* (after Andrews)

Calamites is a genus of extinct arborescent horsetails to which the modem horsetails (*Equisetum*) are closely related. Unlike their modern herbaceous relatives, these plants were medium sized trees, growing upto a height of around 20 m. The name *Calamites* was originally given to the pith casts of the hollow stems, but is now also used for the whole plant. Following are the form genera under which the part of the plant is described.

Stem	:	*Calamites*
Leaves	:	*Annularia* and *Asterophyllites*
Root	:	*Asteromyelon*
Fructification	:	*Calamostachys*.

The plant body of *Calamites* consisted of subterranean axes with adventitious roots arising at the holes (fig. 20.10 *a*). From the horizontal axes aerial shoot arose. The trunks of *Calamites* had a distinctive, segmented, bamboo-like appearance and vertical ribbing (fig. 20.10 *b*). The branches, leaves and strobili were all borne in whorls. The leaves were needles shaped with upto 25 per whorl.

The trunk of the tree produced secondary xylem during secondary growth. The vascular cambium was unifacial, producing secondary xylem towards the stem centre, but not the secondary phloem. The trunk and stems were hollow wooden tubes, like their modern relatives. When these trunks broke or buckled, these were filled with sediments forming pith casts of stem (Fig. 20.10 (*c*)). A transection of twig of *Calamites* may be studied in figure 20.10 (*d*).

Fig. 20.11. ***Annularia***, reconstructed. (after Abbott)

Annularia is a genus of calamitalean foliage. These leaves were found at the node in the form of rosette at one plane (Fig. 20.11). *Annularia* leaves were ovate to spatulate in shape, similarly the shapes of rosette could be circular to oval in outline.

Asterophyllites (Fig. 20.12) is another types of foliage of calamitalean plants. The narrow needle shaped leaves were positioned in rosette at the node but all leaves were directed towards the apex of the shoot. The leaf whorls were laterally compressed.

Asteromyelon is the name given to calamitalean roots. Probably these roots arise both from underground and aerial axis but without node and internodes. Anatomically a well preserved pith with exarch primary xylem can be seen. The wedges of xylem are less prominent as compared to xylem wedges of stem. Prominent cortex was also present in the section. (Fig. 20.13)

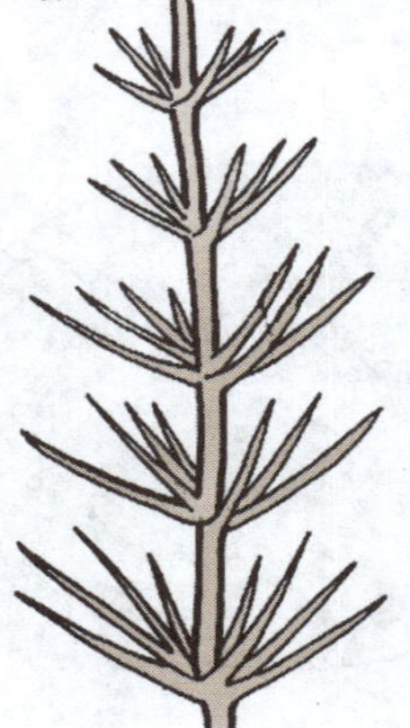

Fig. 20.12. ***Asterophyllites,*** reconstructed (after Abbott)

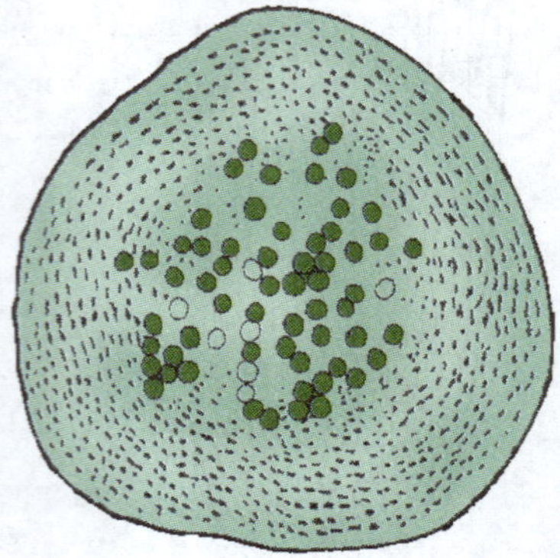

Fig. 20.13. ***Asteromyelon*** T.S. root (after Andrews)

Calamostachys is the generic name given to one of the ructification type of calamitaleans. The strobilus axes are articulated like all other axial parts of the plant and bear bracts at the nodes. The sporophyll holding sporangia are inserted in between on the internodes. The exact mode of attachment of sporophylls is often difficult to recognise in compressed materials (Fig. 20.14 (*a*)). *Calamostachys binneyana* (Fig. 20.14 *b, c*) is a homosporous plant of about 7.5 cm diameter and 3.5 cm long and had six sporangiophores with 12 bracts per whorl. *C.casheana* is a heterosporous plant, similar in general organization to *C. binneyana.*

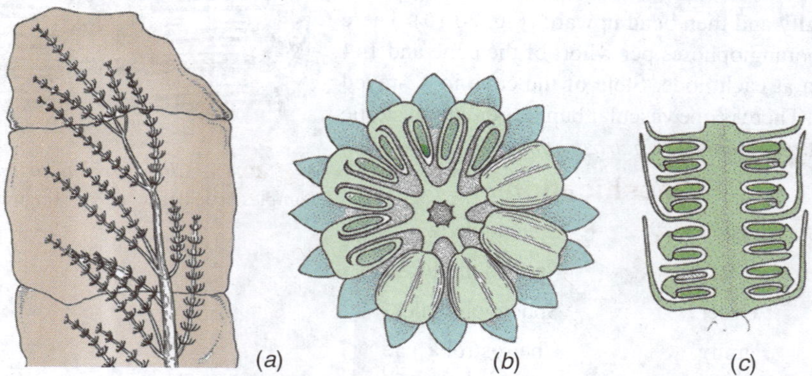

Fig. 20.14. Arborescent Sphenophytes : Fructifications (*a*) ***Calamostachys ludwigi***, a profile of fossil showing a branch bearing numerous cones (*b*) ***C. binneyana***, transection (diagrammatic) (*c*) ***C. binneyana,*** transection (in longisection)

(Figures after : (*a*) Weiss ; (*b*) Zimmermann ; (*c*) Andrews)

Calamites reproduce by means of spores, which were produced in small sacs organized into underground cones. They probably also reproduced by means of underground rhizomes. This is the only group of trees of their period known to have clonal habit.

Classification

Division	:	Sphenophyta
Class	:	Calamopsida
Order	:	Calamitales
Family	:	Calamitaceae
Genus	:	*Calamites*

Hyenia (Hyenia) *Protohyenia, Hyenia* and *Calamophyton* belong to mid Devonian period fossils. These are considered as primitive ferns by some palaeobotanists, though their taxonomic position is uncertain. *Hyenia* had horizontal rhizome from which aerial axes arose.

The aerial part of the plant bore whorls of fertile and sterile appendages. Sterile appendages, function as leaves of the plant. There were two lateral parts of fertile appendages in *Hyenia*. Each segment of lateral appendage bore sporangia.

As far taxonomic position is concerned *Hyenia* is considered to be related to Cladoxylales (Schweitzer , 1972), because of lack of whorled arrangement of appendages, type of branching and lack of knowledge of stem structure.

SPHENOPHYLLALES

Cheirostrobus (Fig. 20.15)

One of the most complex fructification among Pteridophytes is of *Cheirostrobus pettycurrensis* (Scot, 1897) from carboniferous deposits at Pettychur of Scotland. Length of the cast is about 10 cm with diameter of about 4 cm. At each whorl of the cone there are 12 basic units each of which within a few millimeter from cone axis divided into upper fertile and lower sterile segments. The fertile and sterile segment is again divided into 3 sub units. The fertile unit is peltate at its distal part and bears four elongated sporangia of about 1cm. The sterile segment units extend out horizontally and then bend upward (Fig. 20.15). There are 12 sporangiophores per whorl of the cone and 144 sporangia at each node. Stele of the cone in 12-rayed structure. There is one vascular bundle per unit of sterile and fertile segments of *Cheirostrobus pettycurrensis*.

Fig. 20.15. ***Cheirostrobus pettycurrensis***, a complex articulate cone. (Figure after : Scot)

Classification

Division	:	Sphenophyta
Class	:	Sphenophyllopsida
Order	:	Sphenophyllales
Family	:	Cheirostrobaceae
Genus	:	*Cheirostrobus*

BIOTECHNOLOGY–APPLICATION IN PTERIDOPHYTES

Pteridophytes are represented by about 305 genera and 13,000 species in the world (Anonymous, 1999; H.P. Singh, 2003). Though they are dominated by spermatophytes in modern day flora, still they attract ones attention due to their beautiful foliage. Therefore plants of this group are mainly used for horticultural purposes only. Though they are thought to provide a number of medicines, fibres, building materials, food etc. however, with time the knowledge about usefulness of Pteridophytes have been recognized and resulted in its thorough exploitation. This has caused many ferns to become endangered and need special attention for its conservation. It is here multiplication of economically important ferns/pteridophytes have been started through spore and tissue culture using biotechnological means.

Gottlich Haberlandt (1902), a German botanist, was the first scientist to give an idea about cell and tissue culture. Cellular totipotency was proved by Muir & co-workers (1958), when he raised a complete tobacco plant from a single, isolated cell. They also devised Nurse technique for tissue culture. Konav & Nataraja (1965) proved in Buttercup, that any part of callus can give rise to embryoids. Philip R. White is one of the pioneers in the tissue culture technique. But most of the history of tissue culture is related to important plants only (Gautheret, 1983). Application of Biotechnology using tissue culture in the group of economically important pteridophytes is a recent phenomenon (Hennen & Sheehan, 1978, Hickok et. al., 1987, Byrene & Caponetti, 1992, Amoroso & Amoroso, 1998, 2003, Manickam et al, 2003). Tissue culture techniques are being used to multiply and conserve various pteridophyte species specially ferns.

Tissue Culture Technique

Tissue culture Technique requires four basic components. The first and foremost is availability of proper and healthy material to be cultured. This culturing process requires specific nutrient medium; aseptic conditions for keeping cultured material and the last proper aeration for it.

Selection of suitable and balanced nutrient medium (Table-1) is very important for the success of any tissue culture technique. The medium can be solid, semisolid or in liquid form. Solid form of nutrient medium is prepared by use of 1% Agar-Agar or gelatine. The basal medium contains 2-4% sugar, vitamins, minerals, auxin or any growth regulators. Coconut milk & yeast extract can also be added to basal medium in certain cases. Since nutrient medium is very good for growth of bacteria and fungi, it is very necessary to be sterilized. All the apparatuses and the media are sterilized completely by autoclaving them at 120°C for 15-20 min. Some of the important nutrient medium and their composition is given in Table 1. Inoculation chamber and inoculation needle (material) has to be dust free. They have to be sterilized by different methods, particularly by using ultraviolet radiations. Material being used for culture has to be surface sterilized. For proper aeration, culture materials are plugged by sterilized cotton and sterilized air is supplied.

TABLE 1

Composition of Murashige & Skoog's (MS) medium and B5 medium (White's medium)

Salts	MS		B_5	
A. Macronutrient	mg/1	mM	mg/1	mM
NH_4NO_3	1650	20.6	–	–
KNO_3	1900	18.8	2500	25
$CaCl_22H_2O$	440	3.0	150	1.0
$MgSO_47H_2O$	370	1.5	250	1.0
KH_2PO_4	170	1.25	–	–
$(NH_4)_2SO_4$	–	–	134	1.0
$NaH_2PO_4H_2O$	–	–	150	1.1
B. Micronutrient	mg/l	μM	mg/1	μM
KI	0.83	5.0	0.75	4.5
H_3BO_3	6.2	100	3.0	50
$MnSO_44H_2O$	22.3	100	–	–
$MnSO_4H_2O$	–	–	10	60
$ZnSO_47H_2O$	8.6	30	2.0	7.0
$Na_2MoO_42H_2O$	0.25	1.0	0.25	1.0
$CuSO_45H_2O$	0.025	0.1	0.025	0.1
$CoCl_26H_2O$	0.025	0.1	0.025	0.1
Na_2.EDTA	37.3	100	37.3	100
$FeSO_47H_2O$	27.8	100	27.8	100
Sucrose (g.)	30,000		20,000	
pH	5.7–6.0		5.5–5.8	
C. Vitamins				
Inositol	100		100	
Nicotinic acid	0.5		1.0	
Pyridoxin HCl	0.5		1.0	
Thiamine HCL	0.1		10.0	
D. Amino acid				
Glycine	2.0			
E. Plant Growth Regulators				
NAA/IAA/2,4,-D	0.01-10.0		0.01-10.0	
Kinetin	0.04–10.0		0.04–10.0	

Now small portion of material or plant tissue (Shoot tip, root tip, embryo, single cell, spore) are inoculated in the test tube (Figure 20.16, 20.17) containing the nutrient medium. The test tubes are

Fig. 20.16. Culture jars topped with aluminium foils and paper held in place with rubber bands.

Fig. 20.17. Culture tubes topped with cotton plug.

then placed in inoculation room at 20-25°C temperature. By 4-6 weeks callus is formed. The callus is further differentiated into root and shoot if auxin and cytokinin have been supplied in proper ratio. Skoog & Miller proved that auxin brings about root formation and cytokinin induces shoot formation.

Fern and Tissue Culture

Ferns have been used extensively in the ornamental plant industry for landscaping and indoor purposes. The beauty of their gently slopping leaflets makes an ideal accent in any setting. Other uses of ferns are for human food, handicraft raw material, biofertilizer, medicine, etc. (Zamora et al., 1986). It is in this context that propagation of ferns *in vivo* and *in vitro* for ornamental plant industry and for plant research is being done by several workers using spores or vegetative parts. Micropropagation of *Pteris vittata* spores and *Nephrolepis biserrata* runners using different media formulations have been done by Aimee Julianne Anico-Parr (2000). Tissue culture technique has been applied mostly in ornamental plant industry and in case of pteridophytes specially ferns as this

Fig. 20.18. *Pteris vilttata* prothallus (PT) in agar medium.

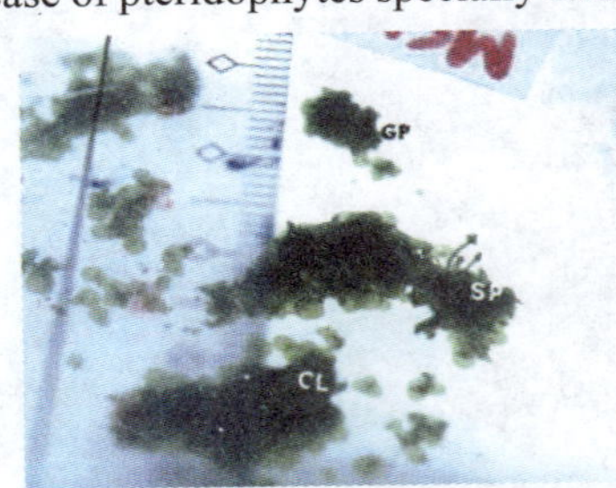

Fig. 20.19. *Pteris vittata* gametophyte clumping (GP), callus (CL) and frond (F).

Fig. 20.20. *Pteris vittata* rooted sporophyte [Root (R), Callus (CL) and frond (F)].

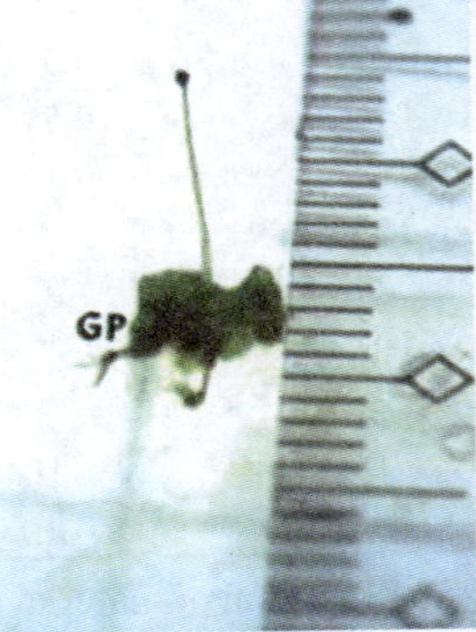

Fig. 20.21. *Pteris vitta* sporophyte still attached to gametophyte.

technique generates more plants from a single mother plant in a shorter period of time (Kyte and Kleyn, 1996)

Fig. 20.22. ***Pteris vitata*** Sporophyte with root (black) clumps (BC) before root formation.

It has been observed that germination of *Pteris vittata* spores and elongation of resulting sporophytes are best done with ½ strength MS medium containing, 125 mg/l NaH_2PO_4, 5mg/l thiamine-HCl, 500 mg/l myoinositol, 20 g/l sucrose, 1mg/l BAP, 1 mg l NAA solidified with 8 g/l bacto-agar. The growth of *Pteris vittata* spores in ½ strength MS with 30 g/l sucrose solidified with 8 mg/l bacto-agar indicated that this fern is not very sensitive to presence or absence of growth hormones. Spontaneous appearance of callus in culture was observed. Sporophytes appeared from both callus and gametophyte cultures. Rooting of sporophytes is best at 1/2 strength MS with 10 mg/l sucrose solidified by 8 g/l agar. (Figure 20.18, 20.19, 20.20, 20.21, 20.22)

Nephrolepis biserrata runner tips embedded into medium with ½ strength mg and 30 mg/l sucrose solidified with 8 g/l bacto-agar produced buds and shorts. Rooted plantlet from buds using their medium was observed within 63 days. Callus formation was prolific in ½ strength MS, 125 mg/l NH_2PO_4, 5mg/l thiamine HCl, 500 mg/l myoinositol, 20g/l sucrose, 2mg/l BAP/mg/l NAA. Rooted plantlets from callus observed within 94 days in same medium with growth regulators adjusted to 1mg/l BAP and 0.1 mg/l NAA (Figure 20.23). Runner tips were observed to be best explant in most of the cases (Beck and Caponetti, 1983) and half strength MS. Salt is best medium for fern culture. Similar experiments have been done with spores of *Dicksonia antarctica* but were found to be difficult to germinate young cut fronds of *Angiopteris* are also being used in tissue culture industry (Figure 20.24).

Fig. 20.23. ***Nephrolepis biserrata*** runner tip showing callus (c) all over the runner (R) and shoot (S).

In one of the early studies spores of *Osmunda cinnamomea* L. were planted on Knudson's medium in course of investigation of an orchid embyo culture. It was observed that they germinated in a few days rapidly forming thin heart shaped green prothalli. This produced green calluses on the upper surfaces after 2 months. The callus grew slowly or died on this medium. When transferred to the same medium supplemented by B-vitamin in appropriate concenterations, the calluses grew rapidly and continued to do so. (Morel & Wetmore, 1951). Chromosomal studies show the prothallial callus to be haploid.

Fig. 20.24. ***Angiopteris sp.*** callus (C) formation from part of young cut frond.

Similar experiments were made to conserve and propagate one of the endangered economically important ornamental ferns *Platycerium grande* (fee) C. Pred. The giant staghorn fern of Philippines. Suitable culture media were developed for better spore germination, gametophyte and sporophyte development of *P.grande* for mass propagation of the fern (Amoroso & Amoroso, 2003). Spores were collected from fertile fronds of *P.grande,* then they were separated from sporangia, sterilized in 10% sodium hydrochloride solution before culturing in a culture medium (Zamora *et. al.,* 1992) containing agar dissolved in 500ml sterile distilled water supplemented with 0.25 g ammonium nitrate, 0.1g potassium phosphate, 0.05g magnesium sulphate, 0.05 g calcium chloride and 0.1g ferric chloride of pH between 5.5-5.8. Spores were found to germinate 19 days after inoculation in culture, 75% of the spores had germinated and developed into gametophyte. Sporophyte (plantlets) developed from prothalli seven months after inoculation. Similarly several studies are underway on tissue culture of economically important pteridophytes especially those which are endangered.

ECONOMIC IMPORTANCE OF PTERIDOPHYTES

It is generally thought that the pteridophytes are quite useless members of the plant kingdom, or at the best they can be considered to be of limited horicultural importance. It is because the deleterious effects of fern growth are well discussed, but their useful effects are largely ignored. Various members of Pteridophytes specially ferns are found to provide food, fibres, medicines etc. almost associated with every aspect of human life. In India, various papers have been published on economically useful pteridophytes specially in the field of medicines (Kapur & Sarin, 1978, Puri & Arora, 1961, Single & Viswanathan, 1996, Kirtikar & Basu, 1985). Gurung (1988) described useful pteridophytes of Nepal. Powell (1976) provides a detailed summary of the ethnobotany of New Guinea and lists many references to pteridophytes. Recently Singh (2000) highlighted economic viability of pteridophytes in India. In the following paragraphs economic importance of pteridophytes in various fields have been discussed..

Pteridophytes as food

Ferns are found to be good source of starch, greens and additives. Starch has been obtained from pulpy apical part of the caudex of certain species of *Cyathea, Angiopteris* and *Marattia.* Sporocarps of *Marsilea drummondi* (Australian swamp fern) are grounded into paste to prepare bread. These are considered to be starvation food used by tribals in various parts of the world. Fern greens like fresh uncurled frond of *Diplazium esculentum, Pneumatopteris sorgerensis* are sold in low land markets in New Guinea. *Diplazium* is eaten in Philippines and in Malaysia. Other freshly edible ferns are *Ampelopteris prolifera, Angiopteris evceta, Ceratopteris, Pteridium esculentum, Helminthostachys* etc. Rhizome of *Blechnum orientale* and pith of *Cyathea khasyana, C. nilgirensis, C. spinulosa* and many other species are cooked and eaten by tribals in many parts of the country. A salt has been prepared from the residual ash of *Asplenium atrobryum* (Croft & Leach, 1984) that contain higher percentage of Potassium.

Pteridophytes as fibre, handicraft & construction Material

Several species of ferns produce soft, sultry hairs on its surface that are being used to stuff pillows in many parts of the world. These species include *Cystodium sorbifolium, Ceiba pentandra* etc. Stem of *Cyathea magna, C.angienis* are used as picket fences in the highlands of New Guinea. Stems and rachises of ferns are often reinforced with strong sclerotic vascular strands or the outer layer itself is extremely tough and durable. Therefore several tree ferns are good fencing material. Even baskets are being made from *Lygodium* stems. *Equisetum debile* is used to clean cooking and dining utensils. The stem of this plant contains crystals of silica and fine abrasive action of the crystals make it useful cleaning agent.

Pteridophytes as Horticultural Plants and ritual items

Pteridophytes specially ferns occupy a special place almost in all green houses and gardens. Leafless green dichotomously branched "whisk-fern" *Psilotum* is one of the common plants every one likes to have in their surroundings. In urban gardens, ornamental ferns are common. Tree ferns *Cyathea contaminans, C. felina* and *C. magna* are encouraged to grow in gardens in many parts of world. Ferns like *Angioperis* and *Marattia* are of high aesthetic value. They find their place in garden because of their beautiful appearance and eloquence.

Ferns are often used as personal decoration materials , either casually or for ceremonial occasions. Wagner & Grether (1948) showed that *Selaginella* and *Dicranopteris* are two such species. *Lycopodium* is used as head dress ornamentation on ceremonial ocassions.

Almost every garden has plants like several species of *Nephrolepis* and *Pityrogramma calomelanos* (the silver backed fern), although these are present as reasonably attractive weeds rather than actively encouraged. "Maiden hair fern" – *Adiantum cuneatum, A. tenerum. A. trapeziform* and other cultivar varieties often supplement the garden beauty.

Various species of *Selaginella* have been used by green houses as border plants. *S. lepidophylla* is sold as dried plant on roadsides local markets and the plant rejuvenates once gets wet with water. Another pteidophyte i.e., *Lycopodium* has the status of Ground ring as far as horticultural aspects are concerned. Both the pendulous epiphytic Lycopods as well as creepers, the ground lycopods (Club moss) has aesthetic values. *Lycopodium* is being used on various ceremonial occasions.

Pteridophytes as weed

Salvinia . The waterfern is one of the trouble some and dangerous weeds that propagates very easily in any water reservoir. It can block free flow of water. This plant has long been known as a pest in tropical and sub-tropical countries (Anon, 1979, Kleinchmidt, 1973, Wild, 1961). Its eradication by physical means is quite difficult or impossible. Introduction of save predators (Hentry & Perichard, 1982) as its biological control agent would help to reduce its growth and multiplication in water bodies. Terrestrial fernweeds, like *Pteridium sp., Sphaerostephanos, Christella, Nephrolepis etc*. are also some of the troublesome weeds because of their habit of quick spreading by long creeping rhizomes. The braken fern *Pteridium aquilinum* is capable of rapid colonization of bare grounds by establishment of numerous sporelings, especially after fire (Gliessman, 1978) and this is a particular problem in tropical regions where the burning of grasslands is a common practice. The plant is toxic to livestock although the animals seem to do poorly in areas heavily infested with braken fern.

Pteridophyte as Biofertilizer

Azolla-a heterosporous water fern which contains endophytic cyanobacterium *Anabaena azollae* in its leaf eavity, used as biofertilizer in rice fields. Out of six known species of *Azolla* (*A.niloticas, A. mexicanas A. microphylla, A. rubra, A. pinnata* & A *caroliniana*). *A. pinnata* is common in our country. A rice farmer is sure of at least two good successive crops whenever *Azolla* plants in full bloom occupy the rice fields. It is referred as goddess of fertility. Its maximum cultivation is being done in nurseries (microplots 20m^2) which were harvested and dried to be used as green manuring. Mostly in rice fields it can be used either prior to rice plantation or after transplantation of rice followed by water drainage and incorporation of *Azolla* (Singh, 1980). Azolla is also a heavy metal resistant plant and shows good tolerance against heavy metals viz. As, Hg, Pb, Cu, Cd, Cr etc.

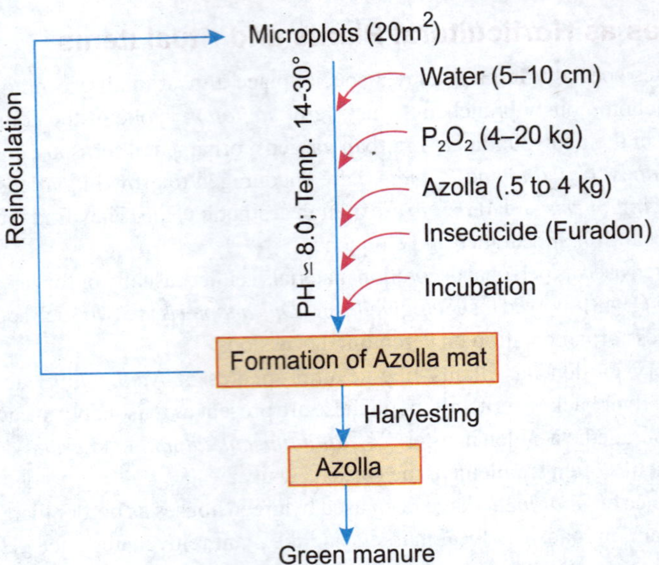

Fig. 20.25. Mass cultivation of *Azolla.*

Pteridophyte as Medicines

While going through the literature available on medicinal use of plants many names of pteridophytes come across which are being used in medical purposes. It has been known since time immemorial that *Dryopteris filix-mas*-a fern has been used for the treatment of tapeworm. Almost all parts of the plant be it root, stem or frond of pteidophytes are being used as medicine. Some important diseases and their treatment with pteridophytes are being discussed here.

Stomach pains

Holdsworth and Giheno (1955) record that a species of *Lycopodium* is chewed in the central high land of new Guinea to induce vomiting after food poisoning or acute stomach ache. *Lycopodium clavatum* as well as *Lygodium longifolium* are used for treatments of stomach ache and diarrhoea.

Fever, headaches & colds

Pteridium aquilinum is used to treat tooth ache and mouth infection (Powell, 1976 b). Leaves of *Cyclosorus* is used for treatment of nasal infection. *Selaginella flabellata* is used to control fever, headaches and menstruation (Blackwood, 1935).

Boils, Ulcers, Wounds

Futscher (1959) reports *Gleichenia linearis* (*Dicranopteris linearis*) is being bound externally onto wounds. Leaves of *Pteris ensiformis* and *Aspidium latifolium* and roots of *Dryopteris milneana* are being applied to boils, ulcers and arrow wounds. Hot fronds of *Polystichum* sp. are applied to groin swellings by tribals. Fronds of fleshy terrestrial fern *Angiopteris evecta* was bound on to fractured limbs in same countries (Scheifenhovel, 1978).

Menstruation, childbirth & contraception

Pteris ensiformis is used to control menstruation whereas *Pteris tripartita* is used internally during child birth in Bougainville. *Lygodium dichotomum* is used as a contraceptive, the roots and stem being taken internally.

DIVERSITY AND DISTRIBUTION OF INDIAN PTERIDOPHYTES

There are 305 genera and 13000 species of Pteridophytes all over the world (Anonymous, 1999). India with a highly variable climate has a rich flora of Pteridophytes. Fern and its allies flourish well in tropical, sub-tropical, and temperate forests. As far as diversity and distribution of Pteridophytic flora are concerned, there are about 1100-1200 species of ferns and its allies that are known to occur in our country (Chandra, 2000 ; Dixit, 1984; MOEF, 1999). The total number of species present in various phytogeographical zones of India are maximum in Eastern Himalayas with 607 species followed by 560 in N.E India, 443 species in N.W Himalayas, 349 in Western Ghats with minimum number in Central India having only 112 species reported (Manickam, 2003). There are 235 endemic Pteridophytic flora in India, out of which 201 are fern species and 34 species are its allies. These findings indicate that Eastern Himalayas, and the Eastern and Western Ghats of South India are the two important zones which harbour maximum number of Pteridophytes. Western Ghats of South India provide better space and climate for pteridophytic flora of the region. It is because like Himalayan region it receives greater precipitation as compared to Eastern Ghats. The rugged terrain and abundant rainfall support a diverse and unique flora in the region. The shade and moisture loving pteridophytes thrive between 700 m and 2000 m altitude. Altitudinal gradiant are directly related to precipitation. Epiphytes prefer high altitude forests and only robust, species like *Drynaria quelsifolia* are found at lower levels. About one third of ferns in South India are either epiphytic or epilithic in nature, the rest are terrestial species. About two third of *Lycopodium sp.* and half of the *Selaginella sps.* are epiphytic in this region. Most of the species of *Equisetum* , *Isoetes* and *ferns* prefer to grow in aquatic habitats or in marshes (Theuerkauf, 1994). Similarly Himalayan range also supports almost all the pteridophytic species growing in India. Sepcies of *Psilotum, Lycopodium, Selaginella, Equisetum* and many fern species like *Pteridium, Pteris, Adiantum* etc. all of them are found at various altitudes in the Himalayan belt.

Lycopodium which is represented by 33 sps. in India spread mostly in South Indian hills and Eastern Himalayas. Twelve species of *Lycopodium* have been reported from Darjeeling and Sikkim (Mehra and Bir, 1964) region, where as nine species are known to occur in Kodaikanal (South India). *L. clavatum* and *L. cernuum* are some of the common creeping species; *L. phelgmaria, L. squarrosum, L. linifetium* and *L. funiform* are the epiphytic species which mostly occur in tropical and sub-tropical forest of India. Simiarly71 species of *Selaginella* are found in India and majority of them inhabit in tropical and sub-tropical forests and some in the plains of India. Species like *S. biformis, S. adunca, S. monospora, S. caulescens, S. nepalensis* are known to grow in Eastern Himalayas. *S. bryopteris, S. adunca, S. subdiaphana* and *S. chrisocaulos* are found in the upper Gangetic plains. Many species of *Selaginella* are found in central India (*S. bryopteris, S. radicata, S. rupestris*). *S. involvens, S. radicata* and *S. catarctrum* are common south Indian species. Among *Equisetum, E. arvense* is cosmopolitan species and can assume many growth forms. In India, *E. debile, E. diffusum, E. arvense, E. robustum, E. palustre* and *E. ramosissimum* are common species mostly reported from northern India. *E. diffusum* and *E. ramosissimum* are widely distributed throughout Himalayas from Shimla to Darjeeling between 1500-2100 m altitude. *E. debile* is found in the plains of U.P., Punjab, Bengal, M.P. as well as in South India, *E. arvense* and *E. palustre* are restricted to eastern Himalayas only.

Fern flora of India is represented by 305 genera and about 10,000 species (Holttum, 1949). They are spread right from the equator to the arctic regions. The Himalayan range supports its luxuriant growth, where more than 1000 species have been reported (Dixit, 1984). *Ophioglossum*-the “Adder’s tongue” fern is a composolitan one. *O. vulgatum* has been found to grow in clay-sandy soil. The species can be seen frequently along the river banks, streams, water reserviors in North India. It occurs from Chamba to Darjeeling, ascending upto 9000 ft. *O. aitchisoni* can be seen during rainy seasons only. *O. nudicaule* grows in Bangalore, Belgaon, Palghat, Palai and Kumbnad. *O. pendulum* is a common Indian epiphytic species of the genus. Six species of *Botrychium* has been reported from

India found at various altitude of Himalayas. There are *B. virginianum, B. lannginosum, B. lunaria, B. multifidum, B. daucifolium* and *B. ternnatum.* Five species of *Osmunda* are known to inhabit in north and north-western Himalayas. These are *O. regalis, O. claytoniana, O. javanica, O. japonica,* and *O. cinnamomea. O. japonica* is a cultivated species in South India *O. claytoniana* is a very common high altitude species very common on the exposed places of Kedarnath at an altitude of 10,000 ft. Two species of *Gleichenia* (*G. linearis* and *G. glauca*) are found in our country. *G. linearis* is very common in tropical and sub tropical belt of India. *G. glauca* is restricted to Eastern Himalaya only. Species of *Dryopteris, Pteridium, Pteris* and *Adiantum* are well represented in India. They are cosmopolitian genera, many of them are cultivated and grown for ornamental purposes. Most of them prefer Himalayan region for its luxuriant growth. 41 species of *Dryopteris* are found in India all from Himalayan ranges (Basu, 1978). *Pteridium* occupies entire Himalayan belt between 1000-3000 m altitude, *P. aquilinum* is monotypic and cosmopolitan.

Pteris is represented by about 250 sps. that grow abundantly in the tropical and sub-tropical belt of world. *P. vittata* is a low level fern which can be seen along the mountain walls and grows upto 1200 m altitude. *P. cretica* can be seen above 1200 m altitude. *P. quadriaurita* can be seen along road-sides and valley throughout the North-western and W. Himalayas. *Polypodium* and *Adiantum* occupy the tropical and sub-tropical belt for their luxuriant growth. Most of the species of *Polypodium* in India are epiphytic in nature found in the Himalayas. Some common ones are *P. microrhizoma, P. amoenum, P. subamoenum , P. steewartii* etc. which grow on the trunks and branches of *Quercus, Cedrus, Abies* and *Rhododendron* etc. *Adiantum,* the maiden hair fern, grows well in Himalayan region. It is very conspicious part of fern flora of Himalayas. Some common Indian species are *A. capillus-veneris, A. pedatum, A. incisum, A. caredatum* etc. These species prefer shady and moist habitats. They are important part of many small gardens maintained as ornamental plants. Filmy ferns are represented by two common Indian species i.e. *Hymenophyllum exsertum* and *Didymoglossum.* These species love moist and shaded places and grow abundant in tropical rains forests. Filmy ferns of South India have been studied by Madusoodanan *et. al.* (2003).

Marsilea, Salvinia and *Azolla* are the common species found aquatic or semi aquatic in and around the water bodies. *Marsilea* is represented by 9 species whereas *Salvinia* by 12 species. *A. pinnata* is a widely distributed aquatic pteridophyte, can be seen as free floating plant in road side ponds, water reservoirs, lakes etc. *M. minuta* and *M. quadrifoltia* are common hydrophytic species and very common all over India. *M. condensata* has been reported from Rajasthan and other parts of the country. *Salvinia natans* is found in abundance in Dal Lake at Kashmir. *S. molesta* is reported from Kerala, Both *Salvina* and *Azolla* are aquatic free floating ferns.

QUESTIONS

ESSAY TYPE

1. Describe the characteristic features of Lepidodendrales. **(Kanpur M.Sc., 2002)**
2. Give identifying characters of Lepidodendron. **(Kanpur M.Sc., 2002)**
3. Write down an essay on economic importance of Pteridophytes.
4. Write an essay on diversity of pteridophytes in India.
5. What are biotechnological applications in pteridophytes. Discuss.
6. *Lepidocarpon* and *Miadesmia* are the fructifications of the true seed plants. Critically comment on the statement **(Allahabad M.Sc., 2000)**
7. Mention the names, classification and distinguishing featuers of diverse pteridophytes collected from Allahabad region and suggest the remedies for their conservation. **(Allahabad M.Sc., 2000)**
8. Give a detailed account of Calamitales. **(Rohilkhand M.Sc., 2003)**
9. Stigmarian appendages are roots. Discuss.
10. Give an account of ***Lepidodendron***. **(Kuvempu Univ., 2005)**
11. Describe anatomy of ***Lepidodendron*** stem. **(Mysore Univ., 2004)**

SHORT ANSWER TYPE

10. Compare the armour of leaf cushion of ***Lepidodendron*** with that of ***Bothrodendron.*** **(Kanpur M.Sc., 2005)**

11. Mentioning the respective age differentiate between ***Miadesmia*** and ***Lepidocarpon.*** **(Kanpur M.Sc., 2005)**
12. Draw a labelled sketch of L.S. of ***Cheirostrobus.*** **(Kanpur M.Sc., 2005)**
13. Compare ***Hyenia*** with ***Calamites.*** **(Kanpur M.Sc., 2004)**
14. Compare ***Miadesmia*** with megasporophyll of ***Selaginella.*** **(Kanpur M.Sc., 2004)**
15. Draw a labelled diagram of ***Lepidocarpon***. **(Kanpur M.Sc., 2004)**
16. Write medicinal importance of Pteridophytes. **(Kanpur M.Sc., 2003)**
17. Write names of fructifications of ***Calamitales.*** **(Kanpur M.Sc., 2003)**
18. Differentiate between ***Sigilaria*** and ***Pleuremia***. **(Kanpur M.Sc., 2002)**
18. Write notes on the following
    (*a*) Seed habit in Lepidodendrales. **(Rohilkhand M.Sc., 2000, 2003)**
    (*b*) ***Cheirostrobus***
    (*c*) ***Calamites***
    (*d*) ***Lepidodendron***
    (*e*) Calamitian cones
    (*f*) Fructifications of Calamitales **(Rohilkhand M.Sc., 2000)**
    (*g*) ***Sigillaria***
    (*h*) Fossil Lycopods **(Rohilkhand M.Sc., 2000)**

OBJECTIVE TYPE

20. Select the correct answer :

(*i*) Which of the following was a tree?
- (*a*) ***Baragwanathia***
- (*b*) ***Lepidodenidron***
- (*c*) ***Calamites***
- (*d*) ***Annularia***

(*ii*) ***Lepidocarpon*** is a fossil of
- (*a*) Stem
- (*b*) Root
- (*c*) Microspores
- (*d*) Ovule

(*iii*) The root fossil of Lycopods is described under the generic nature.
- (*a*) ***Lepidocarpon***
- (*b*) ***Stigmaria***
- (*c*) ***Lepidodendron***
- (*d*) ***Miadesmia***

(*iv*) Fossils of tree Lycopods have been assigned to
- (*a*) Cretaceous
- (*b*) Carboniferous
- (*c*) Permian
- (*d*) All of the above.

(*v*) ***Cheirostrobus*** has been obtained as fossil from
- (*a*) India
- (*b*) Pakistan
- (*c*) South Africa
- (*d*) Scotland

(*vi*) Tissue culture has been generally used for
- (*a*) Lycopods
- (*b*) Ferns
- (*c*) Sphenophytes
- (*d*) None of the above

(*vii*) A salt has been prepared from the ash of
- (*a*) ***Pteris vittata***
- (*b*) ***Marsilea drummondi***
- (*c*) ***Lycopodium cernuum***
- (*d*) ***Asplenium atsobryum***

(*viii*) Which of the following is used for preparation of bread?
- (*a*) ***Asplenium atrobryum***
- (*b*) ***Lycopodium cernuum***
- (*c*) ***Marsilea drummondi***
- (*d*) ***Pteris vittata***

(*ix*) ***Lygodium dichotomum*** is used for
- (*a*) Stomach disorders
- (*b*) Boils and wounds
- (*c*) Food
- (*d*) Contraception

Bibliography

Abraham, A, (1958), *Mem. Ind. Bot.Soc.* **1** : 24-29

—and C.A. Ninan (1962). *Jour. Indian Bot. Soc.* **41**(3) : 339—421

—and—(1965). *Caryologia*, **13**(3) : 537-539.

Agashe, S.N. (1968). *Amer. Fern. Jour.*, **58** : 74-77 (Phloem in Equisetum).

Aghion, D. (1962). *C.R. Acad. Sci.* **255** : 993-5

Albaum, H.G. (1938), *Amer. Jour. Bot.* **25** : 124-133

Allen, R.G. (1911). *Trans. Wisconsin Acad. Sci.* **17** : 1-56

Allison, H.E. (1951). *Nature* (*London*). **168** : 301

—(1952). *Ibid.* **169** : 79

—(1952). *Ibid.* **169** : 1019

—(1952). *Ann. Bot.* (*Lond.*) *N.S.*, **16** : 165

—(1953). *Ibid.* **17** :37

—(1953). **17** : 447

—(1954). *J. Expt. Bot.* **3** : 406

—(1954). *Nature* (*Lond.*). **173** : 1032

—(1954). *Ann. Bot.* (*Lond.*) *N.S.* **18** : 499

—(1995). *Ibid.* **19** : 247

—(1956). *J. Expt. Bot.* **7** : 4

—(1956). *Ibid.* **7** : 1

—(1959). *Nature* (*Lond.*) **184** : 1575

—and A. Saweykowska (1960). Ibid. **186** : 813

Alston, A.H.G. (1956). *Taxon* **5** : 23-25 (Polypodiaceae)

Ananier. A.R. 91957. *Bot. Zh.* **42** : 691-702

Aderkas, Von. P. (1986). *Ann. Bot.* **57** : 505

Andrews, H.N. (1960). *Cold Spr. Harb-Symp. quant Biol.* **24** : 217-234

Andrews, H.N. (1961). *Studies in Palaeobotany. John Wiley N.Y.*

—(1964). *Sciences*, **142** : 925-31

—(1960). *Paleobotanist.* **7** : 85-89

—& Kasper Andrews (1972). *Amer. Jour. Bot.* **59** : 658-659 (abstracts)

Atkinson, L.R. (1972). *Ibid.* **59** : 675 (Abstracts)

—(1960). *Phytomorophology*, **10** : 26-36

—(1960). *Ibid.* **10** : 351-367

—(1962). *Ibid.* **12** : 264-88

—(1965). *Amer. Fern. Journal*, **55** : 81-83

—(1973). *Bot. Jour. Linn. Soc.* 67 (1) : 73-90

—A.G. Stokey (1964). *Phytomorphology*. **14** : 51-70

Andus, L.J. (1959).*Jour. Linn. Soc.* (*Bot*). **56** : 177-187

Axelrod, D.I. (1959), *Evolution*. **13** : 264-275

Baishya, A.K. & Rao, R.R. (1982). Ferns and Fern allies of Meghalaya State, Jodhpur India pp. 1-162

Ball, E. (1960). *Phytomorph*. **10** : 377-396

—(1963). *Nature* (*London*). **197** : 103

Banks, H.P. (1968). "*The early history of Land Plants*". Yale Univ. Symp. on Evolution and Environments. New Haven Conn.

Banks, H.P. (1975). *Taxon*. **24** : 401-413

—and Leclerg. S. & Hueber, F.M. (1975). *Palaentographia Americana* (8) **48** :77-127

Barber, H.N. (1957). *Proc. Lonn. Soc. N.S.W.*, **82** : 201-209

Barklay, B.D. (1931). *Bot. Gaz.* **91** : 452-461

Barringtor, D.S. (1990). *Ann. Mo. Bot. Gard.* **77** : 297

Barghoorn, E.S. (1957). *Mem. Goel. Soc. America*. **2**(67) : 75-86

Bartoo, D.R. (1929). *Bot. Gaz*, **81** : 322-281

Baxter, R.W. (1972). *Amer. Journ. Bot.*, **59** : 659 (Abstracts)

—and G.A. Leisman (1967). *Ibid* **54** : 748-754

Beck, C.B. (1960). *Brittonia*, **12** :351-368

—(1966). *Boli. Rev*. **45** : 379-400

—(1966). *Taxon*, **15** : 337-339

Bell, P.R. (1959), *Journ. Linn. Soc.* (*Bot.*) **56** : 188-203

—(1960). *New Phytol*. **59** : 53-59.

Bell, P.R., (1959). *Journ, Linn. Soc.* (*Bot.*) **56** : 188-203

—(1960), *New Phytol*. **59** : 53-59.

Bell, P.R. (1970). *Sci. Progress Oxford*, **58** : 27-45

—(1975a). *J. Cell Sci.*, **17** : 141-153

—(1975b). *Endeavour* **34** : 19-22

—(1976). *Jour. Linn. Soc.* **73** : 47-78

—and Richards. R.M. (1958). *Nature* (*London*). **182** : 1748-49

Beer, R. (1913). *Ann. Bot.* **27** : 643-659

Bergad, R.D. (1972). *Amer. Jour. Bot.* **59** : 659 (Abstracts)

Bhamdie, S. (1975). *Jour. Indian Bot. Soc.* **36** : 491-501 (Isoetes)

—,S. (1962a). *Proc. Indian Acad. Sci.* **56** : 56-76 (Isoetes)

—,S. (1962b). *Agra Univ. Res. Jour.* **11** : 233-236 (Isoetes)

—,and Puri, U. (1963). *Mem. Indian Bot. Soc.* **4** : 55-63 (Shoot apex in Lycopadiales)

Bhamble, S. (1963a). *Proc. Nal. Inst. Sci. India*, **29** : 169-190 (Isoetes)

—,S. (1963b). *Proc. Indian Acad. Sci.* **58** : 153-164 (Isoetes)

—,S. (1965a). *Ibid.* **61** : 242-252 (Lycopodium)

—,S. (1965b). *Labdev Journ. Sci. & Tech.* **3** : 143-144 (Isoetes)

—,S. & Sharma, R. (1965) *Ibid*. **3** : 71 (Isoetes).

—,S. (1966). *Curr. Sci*. **35** : 46-48 (Selaginella)

—,S. (1970a). *Journ. Indian Bot. Soc*. **50** : 56-62 (Isoetes)

—,S. (1979). "*Recent Researches in Pteridophytes Plant Science*". New Delhi (ed. by Bir. S.S.) : 34-39.

—,S. & Madan, P. (1980). *Proc. Indian Acad, Sci*. **89** : 29-35 (Botrychium)

—,S. &—,P (1982) *Fern Gaz*. **12** : 215-223

—,S. &Puri V.(1985) *Trends in Plant Research* (ed. C.M. Govil and V. Kumar), pp. 55-81. Bishen Singh & M. Pal, Dehradun.

—,S. & Malik, A.K. (1988). *Res. J. Pl. Environ*. **4** : 17-25

—,S. & Gupta, S. (1989) S.S. Bir Commemoration Volume, pp. 643-648.

—,S. (1990). Secondary Growth in Pteridophytes in *: Growth Patterns in Vascular Plants* (ed. M. Iqbal). Dioscorides Press Inc. Oregon, U.S.A.

Bhardwaja, Y. (1935). *Curr. Soc*. **3** : 300

Bir, S.S. (1959). *Journ. Indian Bot. Soc*. **38** : 528-539

—(1963). *Bull. Bot. Surv. India*, **5** : 151-161

—(1966-67). *Palynological Bull*. **II and III** : 3-11

Bir, S.S. (1976a). Advances in Pollen-spore Research **2** : 92-119

—(1976b.). Taxonomy of Indian Pteridophytes. *In Recent Advance in Botany* (Edt. P. Kachroo), pp. 70-115.

—(1977). *Bull. Bot. Surv. India* **19** : 323-329

—(1983). Cytogenetics of Pteridophyta. In *Genetical Research in India*. New Delhi pp. 49-60.

—(1987). *Indian Fern Journal*. **4** : 104-150

—and S.M. Vasudeva (1973). *Proc. 60th Indian Sci. Cong. Abstracts*. 402 Bierhorst. D.W. (1953) *Amer. Jour. Bot*. **40** : 649-658

—(1954). *Virginia J. Sci*. **5** : 72-78

—(1954). *Amer. Jour. Bot*. **41** : 274-281

—(1954). *Ibid*. **41**: 732-739

—(1956). *Phtomorph*. **6** : 176-184

—(1958). *Amer. Jour. Bot*. **45** : 534-537

—(1959). *Amer. Jour. Bot*. **46** : 170-179

—(1960). *Phytomorphology*, **10** : 249-305

—(l965). *Bull. Torrey Bot. Club*, **92** : 475-488

—(1966). *Amer Jour. Bot*. **53** : 1 23-133

—(1967). *Ibid*. **54** : 538-549

—(1972). *Ibid*. **59** : 675 (Abstracts)

—(1973). *Bot. Journ, Linn. Soc*. **67** : 45-58

—(1977). *Aner. Journ. Bot*. **64** : 125-152

Bohra, D.R. & Sharma, B.D. (1978). *Ameghiniana* **15** : 321-326.

Bohra, D.R. & Sharma, B.D. (1979). *Ann. Bot*. **44** : 749-756.

H.C., (1957) Morphology of Plants, New York

Bose, M.N. & Roy, S.K. (1963). *Palaeobotanist*. **12** : 226-228

Bonnet, A.L.M. (1957). *Rev. Cytl. Biol. Veg*. **28** : 1-88 (Azolla).

Braithwaite, A.F. (1964). *New Phytol.* **63** : 293-305

Britton, D.M. (1991). *Can. J. Bot.* **69** : 634

Britton, D.M. & Brunton, D.F. (193). *Can. J. Bot.* **71**, 1076

Breekon (1972). *Amer. Journ. Bot.* **59** : 675 (Abstracts, Spore Morphology of Lycopods)

Brook, K.E. and S.S. Tepfer (1972.) *Ibid.*, **59** : 675 (Abstracts, Sex expression in *Selaginella*)

Brown, D.F.M. (1964). *Nova Hedurigia*, **16** : 1-154

Brownlie, B. (1961). *Trans. Roy. Soc. N.Z. Bot.*, 1 1-4

Bruchman, H. (1909, 1913). *Flora*, **99** : 12-15 ;104: 180-224 ; 105 : 337-346

Camp, W.H. and M.M., Hubbard (1963). *Amer, Jour. Bot.* **50** :235-243

Campbell, D.H. (1891). *Ann. Bot. Lond.* **5** : 231-258

—(1921). *Amer.Jour. Bot.* **83** : 303

—(1918). *Mosses and Ferns*. New York, 3 Auft.

—(1932). *Amer. Nat.*, **66** : 481

—(1940). *The evolution of land plants*. California

Camponetti. J.D. (1972). *Amer. Jour. Bot.*, **59** : 675

Chaloner, W.G. (1970). *Biol. Rev. Camb. Phil. Soc.*, 45 : 357-377

Chandra, P. (1963). *Curr. Sci.* **32** : 34-35

Chandra, P. (1979). *J. Bombay Nat. Hist. Soc.* **74** : 640-650

—,(1981). *Kalikasan Phillip. J. Boil.* 10 (2-3) : 177-189

—,(1982). *Nova. Hedwigia*, **36** : 241-247

Chatterji, J. (1964). *Phytomorphology*, **14** : 451-457

—and H.V. Mohan Ram (1969). *Bot. Notis.* **121** : 471-490

Chatterji, J. and B.K. Nayar (1970). *Bot. Soc. Linn. Jour*, **63** : 265-276

Chopra, H.A. and D.S.Loyal (1973). *Proc. Ind. Sci. Cong 60th Diamond Jubilee Session*, 404 (Abstracts)

Cheadle, V.I. (1956). *Amer. Journ. Bot.* **43** : 719-731

Ching, R.C. (1940). *Sunyatesenia*, **5(4)** : 201-268

Ching, R.C. (1954). *Acta. Phytotax. Sin.* **3** : 93-98

—(1963). *Ibid.* **8** : 289-335

—(1964). *Ibid.* **9** : 41-85

—(1965). *Ibid.* **10** : 1-5

—(1966). *Ibid.* **11** : 11-16

—(1974). *Ibid.* **13** : 96-98

—(1978a). *Ibid.* **16** (3) : 1-19

—(1978b). *Ibid.* **16** (4) : 16-37

—(1984). *Indian Fern Journal* **1** : 47-50

—, Lin **Y.X**. & Wu, **S.K**. (1983). *Acta. Bot. Yunnan.* **5** : 1-23

Christensen, C. (1939). *Filicae Chap. XX in Verdorn's Mannual of Pteridology*. The Hague, 522-550

Chruch, A.H. (1919). *Thalassiophyta and sub-aerital trasnmigration*. Oxford.

Clutter, M.E. and I.M. Sussex (1965). *Bot. Gaz.*, **126** : 72-78

Copeland, E.B. (1938). Genera Hymenophyllacearum. *Phillip Journ. Sci.* **67** : 1-108.

—(1947). Genera Fillcum, Chronica Botanic Company, Waltham, Mass.

Crene, F.W. (1953). *Amer. Fern. Journ.* **43** : 159-180

Crotty, W.J. (1967). *Amer. Journ. Bot.* **14** : 105-117

Cusick, F. (1954). *Ann. Bot. Lond.* **18** : 171-181. (Morphogenesis in *Selaginella wildenovii*)

Cutter E. G. (1954). *Nature*, **171** : 440-441

—(1955). *Ann. Bot. Lond. N.S.* **19** : 485-499

—(1956). *Ibid.* **20** : 143-65

—(1957). *Ibid.*, **21** : 343-72

—and Wardlaw (1963). *Nature*, **199** : 985-987

D'Amato, F., and S. Avanzi, Caryologia, **18**(3) : 383-394

Davie, J.H. (1951). *Amer. Jour. Bot.* **38** : 621-628

De Maggio, A.E. (1963). *Jorn. Linn. Soc. Bot.* **18** : 361-76

—(1964). *Proc. Nat. Acad. Sci.* U.S.A. **52** : 852-859

—and Wetmore, R.H. (1961). *Amer. Jour. Bot.* **48** : 551-65

De-Sloover, J.L. (1961). *Cellule*, **62** : 103-116

—(1963). *Ibid.* **63** : 31-57

—(1964). *Ibid*, **64** : 147-200

Devdas. S.C. and Beck, C.B. (1912). *Amer Jour. Bot.* **501** : 557-567

Devi, S., and B.K. Nayar (1969). *Jour. Sen. Vol.* pp. 209-222 edited by H. Santapau et al. *Bot. Soc. Bengal*, Calcutta

Dhir, K.K. (1980). *Bibliotheca Pterido.* **1** : 1-58

—.& Sood Ajay (1981). *Ibid.* **2** : 1-99

—.&Saiki, Y. (1984). *Nova Hedwigia.* **39** : 169-175.

Dickson, F.G. (1946). *Ohio Jour. Sci.*, **46** : 73-107

Dickson, F.G. (1946). *Ohio Jour. Sci.* **46** : 109-141

Dixit, R.O. (1984). *A census of the Indian Pteriodophytes*. Flora of India. Series IV. B.S.I. Howrah

Dixit, R.D. & Vohra, J.N. (1984). A dictionary of Indian Pteridophytes. B.S.I. Howrah

Duckett. J.G. (1970). *New Phytology*, **69** : 333-346 (Spore size in *Equisetum*)

Duncan, R.E. (1941). *Amer. Fern. Journ.* **28** : 921-931

—(1943). *Bot. Gaz.* **105** : 202-211

Eggert, D.A. (1961). *Palaenotographia*, **B-108**, : 43-92

Eicher, D.L. (1968). *Geologic Time Seale*, Prentice Hall, Englewood Cliffs, N.J.

Emberger, L. (1949). *Ann. Sci. Nat. Bot. Ser.* 11., **10** : 131-144

Engler, A. & Diels. L. (1936). Syllabus der Pflanzenfamilien. Autt. II Gebgindr Borntraeger, Berlin

Erdtman, G. (1957). *Pollen and Spore Morphology and Plant Taxonomy*. Almqvist and Wiksells, upsala, Sweden

Farlow, W.G. (1874). *Quart, Jour. Micr. Soc.* **14** : 206-72

—(1874). *Bot. Zeit.* **32** : 108-183

—(1889). *Ann. Bot.* **2** : 382-386

Farmer, J.B. and Digby, L. (1907). *Ann. Bot.* **21** : 161-199

Farrar, D.R. (1967). *Science.* **155** : 1266-1267

—and Wanger, W.H. (1968). *Bot. Gaz.* **129** : 210-219

Ford, S.O. (1904). *Ann. Bot. London.* **18** : 589-605

Fraser—Jenkins. C.R. (1984). *Bull. Bot. Mus. nat. Hist.* (*Bot.*) **12** : 36-76

—, (1986). *Ibid.* **14** : 183-218

— & Khullar, S.P. (1985). *Indian Fern J.* **2** : 1-16

Ganders, F.R. (1972). *Bot. Jour. Linn. Soc.* **65** : in Press

Goebel, K. (1930). *Organographie der Pflanzen, 3rd ed. Pt. 2 Bryophyta-Pteridophyten*. Fischer Jena.

Good, C.W. and T.N. Taylor (1972). *Amer. Jour. Bot.* **59**(6) : 617-626

Gopal, B. (1968). *Tropical Ecology*. **9** : 153-170

—(1969a). *Ibid.* **10** : 278-291

—(1969b). *Amer. Fern, J.* **59** : 150-152

Goswami, H.K. (1978). *Acta Societatis Botanicorum Poloniac* **47** : 307-315

Graham, L.E. (1996). *J. Plant. Res.* **109** : 241

Gupta, K.A. (1962). *Marsilca. Monogr*, No. 2, C.S.I.R. New Dehli.

Gullvog, B.M. (1968). *Grana Polynol.*, **8** : 24 (fine structure of spores)

Gullvog. B.M. (1969). *Phytomorphology*, **19** : 82-92 (Spores of Equisetum)

Gurung, V.L. (1985). *Fern Gaz.* **13** : 25-32

Hall, J.W. and N.P. Swanson (1968). *Amer. Journ. Bot.* **55** : 1055-1061. (Azolla)

Hartel, K. (1936). *Beitr. Biol. Pfl.*, **25** : 125

Hartman, M.E. (1931). *Bot. Gaz*, **91** : 252-276

Harvery Gibbson, R.J. (1934, 1896, 1897, 1902). *Ann. Bot. Lond.* **8** : 133-206 ; **10** : 77-88 ; **11** : 123-155 ; **16** : 449-466

Hauke, R.L. (1963). *Beth, Nova Hadwigia.* **8** : 1-123

—(1957). *Bull. Torrey Bot. Club.* **84** : 178-181

—(1968). *Ibid.* **95** : 341-345

Hewkins, L.A. (1907). *Ohio. Nat.* **7** : 122-128

Hawitson, W. 91962). *Ann. Missouri Bot. Gdn.* **49** : 57-93

Hill J.B. (1914). *Bot. Gaz.* **58** : 61-85

Hill. L.V. (1967). *Conad, Journ. Bot.* **45** : 1179-1191

Holloway, J.E. (1918). *Trans. New Zelanad Inst.*, 50 : 1-44

—(1935). *Ann. Bot.* **49** : 513-519

—(1939). *Ibid.* (n.s.) **3** : 313-336

—(1944). *Trans. Royal Soc. N.Z.* **74** : 196-206

Holttum, R.E. (1947). *J. Lin. Soc. Bot.* **53** : 123-158

—(1954). *A. revised flora of Malaya*, Vol **II**. Ferns of Malaya. Government Printing Officer, Singapore

—and U. Sen (1961). *Phytomorphology*, **11** : 406-420

—(1949). *Biol. Rev.* **24** : 267-298

—(1961). *Advance. Sci. London.* **18** : 234-242

Holttum, R.E. (1965). *Kew. Bull.* **19** : 463-487

—(1967). *Phytomorphology*. **17** : 36-41

—(1969). *Bluinca* **17** : 5-32

—(1973a). *Bot. Journ. Linn. Soc.* **67** (1) : 1-10

—(1973b). *Ibid.* **67** (1) : 173-189

—(1982). Deeication. In : *Flora Malesiana*, Series II. Pteridophtya. Vol I (Editors C.G.G.J. Van Steenis & R.E. Holttum) Martinus Nijhoffi, DR. W. Junk Publishers. The Hague (Botson) London pp. 7-20.

—(1984). *Indian Fern. J.* **1** : 31-39

Hopping, C.A. (1956). *Proc. Roy.Soc. Edinb.* **66B** : 10-28

Horner, H.T. Jr. and and H.A. Arnott (1961). *Bot. Gaz.*, **124** : 371-383

Hoskins, J.H., and S. Oswa (1956). *Amer. Jour. Bot.* **43** : 36-46

Hotta, Y. and S. Oswa (1958). *Exp. Cell Res.* **15** : 85-94

—and T. Sakakai (1959). *Deva. Biol.*, **1** : 65-78

Hueber, F.M. (1964). *Men. Torry Bot. Club.* **21** : 5-9

—and H.P. Banks (1967). *Taxon*, **16** : 81-85

James, C, French (1972). *Amer. Jour. Bot.* **59** : 224-227

Jayasckora, R.D.E. and Bell P.R. (1959). *Planta*, **94** : 1-14

Jennings, J.R. and D.A. Eggert (1972). *Amer. Jour. Bot.* **59** : 676 (Abstracts)

Johnson, D.S. (1898). *Bot. Gaz.* **26** : 1-24

Jones, C.E. (1905). *Trans. Linnn. Soc. Lond.*, **7** : 15-36

Jyotsna, C. and H.Y., Mohan Ram (1968). *Bot. Notisr*, **121** : 371-400

Kachroo, P, and B.K. Nayar (1952). *Journ. Univ. Gauhati*, **3** : 163-171

—and (1953), *Phytomorphology*, **3** : 240-248

Kachroo, P. (1955). *Jour. Ind. Bot. Soc.*, **34** : 11-20

—(1955). *Bull. Bot. Soc. Bengal.* **9** : 44-59

—(1955). *Sci. and Cult.* **20** : 402-403

—(1956). *Ibid.* **22** : 103-105

—(1963). *Jour. Ind Bot. Soc.* **42** : 109-194

Karrfolt, E.E., and D.A. Eggert (1972). *Amer. Jour. Bot.*, **59** : 679 (Abstracts)

Kashyap, S.S. (1914). *Ann. Bot.* **28** : 163-181

Kato, Y. (1965). *Bot. Mag.* **78** : 187-193

—(1965). *Cytologia*, **30** : 67-74

Kato, M. (1990). *Bot. J. Linn. Soc.* **102** : 303

Kaufman, P.B. et al. (1972), *Amer Journ. Bot.*, 646 (Abstrats)

Kaur, S. (1961). *Sci. and Cult.*, **27** : 347-350

Kaur, S. and Raza, F. (1983). Bibliography of Indian Pteridology.

Kaur S. (1974). *Bot. Jour. Linn. Bot. Soc.* **68** : 153-162

—(1978). *Geophytology*. **9** : 91-95.

—(1979). In Progress in Plant Research Vol. 1. Lucknow.

—(1984). *New Botanist* **11** : 1-76

Kawasaki, T. (1956, 1957, 1959). Studies on sexual generation of ferns 1-3. *Journ. Jap. Bot.* **31** : 139-143 ; 333-338 ; **32** : 177-184 ; 332-336 ; **34** : 86-90

Kawasaki, T. (1970), Reproductive Biology of Pteridophyta. IV. *Ibid.*, **63** : 153-169

Khare, P.B. & Kaur, S. (1987) *Indian Fern. J.* **4** : 70-73

Khare, P.B. & Kaur, S. (1980) New Botanist, **7** : (in press)

Khullar, S.P. (1984). *Indian Fern J.* **1** : 89-95

—,S.P. (1981). In contemporary trends in Plant Sciences, New Delhi pp. 302-307

—, & Gupta, S.C. (1980). *Nora Hedwigia*, **32** : 415-443.

—, Sharma, S.S. & Singh. Paramjit (1983). *Nora Hedvigia*, **38** : 661-667

Kidston. R., and D.T. Gwyne-Vaughan (1907, 1908, 1909, 1910, 1914). *Trans. Royal Soc. Bot. Ebnb*. **45** : 759 ; **46** ; 213-232 ; **46** : 651-667 : **57** : 455-477 ; **50** : 469-480

Klew Kowski, E.J. Jr. (1969). *Bot. J. Linn. Soc*. **62** : 347-359

—(1969). *Ibid.*, **62** : 360-377

—(1970). *Ibid.*, **63** : 153-169

—(1970). *Ibid.*, **63** : 170-176

—(1971). *Bioscience*, **21** : 317-322

—(1971). *Jour. Hered.*, **62** : 367-370

—(1972). *Evolution*, **26** : 66-73

—(1973). *Amer. Jour. Bot*. **60** : 146-154

Knoblock, I.W. (1969). *Amer. Journ. Bot.*, **56** : 646-653

Knox, E.M. (1950). *Trans. Bot. Soc. Ebinb. Proc*. **35** : 211-357

Kondo, T. (1962). *Bull. Schizovoka Univ.*, **13** : 239-267.

Kranz, H.D. & Huss Var (1996). *Plant Syst. Evil*. **202** : 1

Kraussel. R., and H. Wayland (1935). *Palentographica*. **80** : 1-46,

—and—(1936). *Ibid.*, **81** : 114-126

Kulp, J.L. (1961). *Science*, **133** : 1105-1114.

Labouriau, L.G. (1952). *Rev. Brazil Boil.*, **12** : 181-183

Lam, H.J. (1955). *Acta Bota, Neerlandica*, **4**(3)

—(1957). *Blumea*, **8** : 528-531

Lang, W.H. (1909). *New Phytol.*, **8** : 1-12

—(1925). *Tsans, Roy. Soc. Ebinb*. **54** : 253-279

—(1927). *Ibid.*, **55** : 443-455.

—(1937). *Philo Trans*, **B**, **227** : 425-491

—and Cookson, I.C. (1930). *Ibid.* B. **219** : 133-162

—and—(1935). *Ibid.*, **224** : 241-249

Lawson, A.A. (1971). *Trans. Roy. Soc. Ebinb*. **51** : 785-794 ; **53** : 93-113

Leclerecq. S. (1954). *Svensk Bot. Tidskrift*, **49** : 301-315

—(1956). *Evolution* **10**(2) : 109-114.

Lloyd, R.M. and E.J. Klewoki (1970). *Biotropica*, **2** : 129-137

—(1972). *Amer. Jour. Bot*. **59** : 676 (Abstracts).

—(1972). *Ann. Missouri Bot. Garden*. **61** : 318-331

Love, A & Love, D. (1958). Nucleus **1** (1) : 1-10

Lovis, J.D. (1973). *Bot. Jour. Linn. Soc*. **67** (1) : 211-227

Loyal, D.S. (1958). *Curr. Sci.*, **27** : 3357-3358.

—(1961). *Panjub. Univ. Res. Bull.*, **13** : 25-30

—and R.K. Grewal (1966). *Cytologia*, **31** : 330-338.

—& Harvinder, K. Chopra (1976). *Phytomorphology*. **26** : 355-358

—&—(1977). *Curr. Sci*. **46** : 89-91

—& Ratra, R. (1979). *Jour. Indian Bot. Soc*. **58** : 233-240

Lyon, A.C. (1964). *Nature London*, 1082-1083.

Macchlis,L., and Rawtischer-Kunkel, E. (1967). *Amer. Jour. Bot.*, **54** : 689-694

—and—(1963). Cytol., **15** : 97-138

Mahabale, T.S. (1938). *Jour. Bomb. Univ.*, **6** : 62-75

—(1938). *Cour. Sci*, **7** : 62-63

—(1945). *Jour. Univ. Bomb.*, A **14** : 42-49

—(1950). *Jour. Ind. Bot. Soc.*, **29** : 31-33

—(1950). *Nature*, 165 : 410-411

—(1954). *Palaeobotanist*, **3** : 33-37

—(1963). *Jour. Ind. Bot. Soc. Mem.*, **4** : 51-54

Mamay, S.H. (1950). *Ann. Mo. Bot. Gdn.*, **37** : 409-476

Manton, I (1950). *Problems of Cytology and Evolution in the Pteridophyta.* Cambridge University Press.

—(1953). *Symp. Soc. Exp. Boil.*, **7** : 175-185

—(1954). Cytological notes on one hundred sepcies of royalayan ferns. In Holttum 1954, pp. 623-627 (appendices).

—(1958). *Jour. Linn. Soc. Lond.*, **56** : 74-92

—(1959). Cytological information on the ferns of West Tropical Africa. In Alston 1959, pp. 71-85

—(1964). *Ann. Bot. Lond.*, **6** : 283-292

Marschal, C.C. (1925). *Bot. Gaz.*, **79** : 85-94

Martens, P. (1951). Cellule, **54** : 103-32

—(1966). *Trends in Plant Morphogenesis.* Longmans, 155-169.

Mehra, P.N. (1939). *Ferns of Mussorie.* The Univ. of Pb. Lahore.

—(1944). *Proc. Nat. Acad. Sci* **14** : 189

—(1961). Presidential address, 48th *Ind. Sci.Cong.*, *Roorkee*, 1-24.

—(1968). *Conquest of Land and evolutioary Patterns in early Land Plants.* Sir Albert Charles Mem. Lecture. Birbal Sahni Inst. of Palaeobot. Lucknow.

Mehra, P.N., and S.S. Bir. (1957). *Curr. Sci.*, **26** : 151-152

—and—(1958). *Proc. Nat. Inst.Sci., India*, **24B** : 47-53

—and—(1960). *Amer. Fern. Journal.*

—and K.R. Khanna (1959). *Journ. Genetics.*, **56** (**2**) : 1-14

—and Loyal, D.S. (1956), *Curr. Sci.*, **25** : 363-364.

—and—(1956). *Ann. Bot. n.s.*, **20** : 545-552

—and—(1958). *Nature.* **181** : 577

—and—(1959). *Res.Bull.Panjab Univ.* (N.S.). **10** : 357-374.

—and—(1965). *Caryologia*, **18** (**3**) : 461-498.

—and S.C. Verma (1957). *Curr. Sci.*, **26** : 55-56

—and—(1957). *Nature*, **180** : 715-717

—and—(1960). *Caryologia.* **13** : Fase. 2 (in Carsodistampa)

—and H.K. Palta (1971). *Phytomorphology*, **21** (**4**) : 361-375

Merker, H., (1958, 1959). *Bot Notiser*, **111** : 608-618 ; 112 : 441-462

—(1961). *Ibid.* **114** : 88-102.

Mester, M.R. (1972). *Amer. Jour. Bot.*, **59** : 679-680 (Abstracts).

Momose, T (1964). *Journ. Jap.*, **39** : 225-235

—(1964). *Ibid.*, **39** : 305-311
—(1955). *Ibid.*, **40** : 12-18 ; 48-54 ; 303-306.
—(1965). *Ibid.*, **40** : 140-147 ; 277-281
Morel, G., and Westmore (1951). *Amer. Journ. Bot.*, **38** : 141-143.
Namboodri, K.K., and C.B. Beck (1968). *Amer.Jour.Bot.*, **55** : 447-457 ;
—and—(1968). *Ibid.*, **55** : 458-463 ; 464-472
Naf, U. (1956). Growth, **20** : 91-105.
—(1959). *Nature*, **184** : 798-800.
—(1965). *Plant. Physiol.*, **40** : 888-890
Nayar, B.K. (1954). *Phytomorphology* ; **4** : 411-426
—(1956). *J.Ind. Bot. Soc.*, **35** : 333-343
—(1958). *Bul. Nat. Bot. Gdn.* No. **29**.
—(1960). *Journ. Indian Bot. Soc.*, **39** : 259-277
—(1960). *Curr. Sci.*, **29** : 380-2.
—(1961). *Bul. Nat. Bot. Gdn.* (India). No. **52** : 1-40.
—(1962). *Journ. Linn. Soc. Lond.*, **58** : 185-199
—(1964). *Ibid.*, **58** : 449-460.
—(1962). *Bot. Gaz.*, **123** : 223-232.
—(1962). *Bull. Nat. Gdn. Bot, (India)*, **67** : 1-13 ; 68 : 1-35 ; 75 : 1-14.
Nayar, B.K. and S., Kaur (1969). *Phytomorphology*, **19** : 179-188
Nayar, B.K. and S., Kaur (1971). *Bot. Rev.*, **37** : 295-396.
—and Bajpai, N. (1964). *Journ. Linn. Soc.* (Bot.), **59** : 63-76
—and—(1970). *Amer. Fern Jour.*, **60** : 107-118
—and P. Chandra (1963). *Jour. Ind. Bot. Soc.*, **42** : 392-400
—and—(1966). *Palyn. Bull.* **2** : 92-95
—and S. Chandra (1965). *Bull. Nat. Bot. Gdn.*, **117** : 1-98
— and S. Devi (1963). Pollen et spores **5** : 354-372.
— and — (1964, 65, 67, 68). Grana Palynol., **5** : 80-120 ; **5** : 222-246 ; **5** : 342-395 ; **6** : 121-127; **7** : 568 -600 ; **8** : 158-203.
Nishida M. (1955). Phytomorphology. 5 : 449-456.
— (1965). Journ Jap. bot. 40: 16-171.
— and T.Sakuma (1961). Ibid., 36 : 142-152.
Norstog, K (1965). Amer. Jour. Bot., **56 :** 993-999.
Ogura, Y. (1927.Bot. Mog. Tokyo, **41**: 172-180
Oliver, J.C (1972). *Amr. Fern Journal*, **62** : 16-20.
Pal, S. (1959). *Curr. Sci.,* **28 (11) : 455.**
Panigrahi, G and Manton. I (1958). *Journ. Linn.Soc. Lord.*, **55** : 729-743.
Pant, D.D (1962). *Proc. Summer Sch. Darjeeling.* 276-301
Pant,D.D. and Bharati Mehra (1964). *Proc. Natt. Inst. Sci. India,* **30** : 92-98
—and P.F Kidwai (1968). **Ann. Bot. N.s.**, **32** : 601-608
—and P.K. Khare (1974). *Proc. Royal Soc.Lond.*, **186B** : 102-121
—and D.D.Mishra (1973). *Proc. Ind. Sci. Cng.* **60** : 407 (Abstracts)

—and,D.D. andMishra. L. (1976). *Palaentographica* **155B** : 129-139

Pant, D.D. and Mishra. L(1977). *Ibid.* (in PressP).

Pan, D.D., Nautyal; D.D. & Mishra, D.R. (1984). *Phyta Monograph.* **1** : 1-111

Payne, W.W. (1972). *Amer Journ. bot.,* **59** : 676 (Abstracts)

Peterson, R.L., A.E. Star and D.f. Fairbrothers (1972). Ibid., **59** : 676 (Abstracts).

Pichi-Sermolli, R.E.G. (1958). *Uppsala Univ.* **Arsskr., 6** : 70-90

—(1959). *vistas in botany*, W.B Turrill, London.

Pichi-Sermolli, R.E.G. (1965). *index Filicum 1934-1960. Regnum Veg.* **37** (i-xiv) : 1-370

—(1970). Webbia **26** : 129-194

—(1972). Ibid. **26** : 491-536

—(1973).*Bot, Jour. Linn. Soc.* **67** Suppl. 1 : 11-40

—(1977). *Webbia.* **31** (2) : 313-512

Portanen, J.N., and C.R. Portanen (1963). *Canad. Jour. Bot.,* **41** : 1657-1661

Puri, V. (1955). *Recent advances in plant morphology*, Agra univ. extension lectures, Agra.

Pfeiffer, N.E. (1922). *Bull Mo. Bot. Gdn.* **9** : 79-232

Ramanujam, C.G.K.: and W.N. Stewart (1969). *Palaeontographica* Bd. **127** : Abr. B, 159-16 (Apidocarpon constip)

Rashid, A. ((1970). Phytomorphology **20** ; 255

—(1972). Phy Physiol. **66** : 277

—(1978). Pfl. Physiol. **86** : 203.

Rice, H.V. and Laetsch, W. M. (1967). *Amer. Jour. Bot.,* **54** : 856-866

Rogers, L.M. (1927). *La Cellulae*, 37 ; 315-353

Sahni, B. (1923). *jour. Ina. Bot. Soc.* **8** : 185-191

—(1925). *Phil. Trans.,* **B 218** : 143-176

—(1928). *Ibid.,* **B 221** : 1-37

Sarbadhikari, P.C. (1924), *Ann. Bot. Lord.,* **38** : 1-26

—(1927). Ibid., **41** : 1-35

Satterthwait, D.A. and Schopf, J.W. (1972). *Amer. Jour. Bot.,* **59** : 373-376.

Saxena, G. (1992). *Proc. Nat. Acad. Sci. Indian (Biol).* **62** : 499

Schedlbauer, M.D. (1972). *ibid.,* **59** : 676-677.

Schraudoif, H. (1967). *Planta,* **76** : 37-46.

Sharma, B.D., & Bohra, D.R. (1976, 1977, 1985). *Phytomorphology.* **26, 27, 35** : 411-414, 141-195, 167-172

Sharma, B.D., Suthar, O.P. & Bohra, D.R. (1986). *Indian Fern* **J. 3** : 82-86

Shattuck, C.H. (1910). *Bot. Gaz.,* **49** : 19-40

Sharp, L.W. (1914). *Bot. Gaz.,* **58** : 419-431

Singh, P. K. (1977). Proccedings Indian Science Congress (64th Session of Agricultural Science, 977P. p. 177-118)

Sinha, T.C.N.(1963). Journ. ind. Bot soc., XLII : 475-544

Sinha, B. M. B. and Verma, A.K. (1971). *Proc. Indi. Sci. Cong.,* 431

—and—(1977). *Caryologia.* **30** : 265-271

Slade, Brenda, f. (1971). *New phytol.,* **70** : 879-884.

Smith, R.W. (1900). *Bot. Gaz.*, **29** : 228-58 ; 333-46

Sledge, W.A. (1982). *bot. Jour. Linn. Soc.* **84** : 1-30

Sossountozov, I (1954). *Plant. Physion.*, **7** : **1-15.**

—(1957). Phyton (Buenos Aires). **8** : 79-96

Spessard, E.A. (1928). *Bot. Gaz.*, **85** : 323-333

Sporne, K.R. (1949). *New phyton.* **48** ; 259-276

—(1956). boil. Rev., **31** : 1-29

—(1958). *Proc. Linn. Soc. Lond.* **69** : 75-84

Steevs, T.A. and I.M.Sussex (1952). *Nature*, 170-672

—and Wetmore, R.H. (1952). *Phytomorphy*, **3** : 339-354.

Sterl, W.N. (1939). *Bot. Rev.*, **5** : 43-453

—(1933). *Bot, Gaz*, **95** : 164-167

—(1951). *Bot. Rev.*, **17** : 90-104

Stewart, R.R. (1945), *Bull. Torrey. Bot* .CI **72** : 399-426

—(1972). Floraof Pakistan, Karachipp 1-21.

—1984). *Indian Fern J.* **1** : 41-45

Stocky, A.G. (1951). *phytomorphy.*, **1** : 39-58.

—and L.R. Atkinson (1952, a,b,c). *Ibid.*, **2** : 1-9 ; 9-15 ; 105-113, 135-150.

—and—(1954,a,b,c). *Ibid.*, **4** : 165-172 ; 192-201 ; 310-315.

—and—(1954,a,b,c). *Ibid.*, **6** : 19-40 ; 239-249 ; 249-261.

—and—(1957). *Ibid.*, **7** : 275-292.

—and—(1958). *Ibid.*, **8** : 391-403

Stokey, A.G (1918). *Bot. Gaz.* **45** : 97-102

—(1930). *Ibid.*, **90** : 1-45

—(1940). *Ibid.*, **101** : 759-790.

Stokey, **A.G.** (1942). *Ibid.*, **103** ; 559-569

—(1945). *Ibid.*, **106** : 402-411.

—(1948), *Ibid.*,**109** : 363-380.

Suranga, K.R. (1966). Indian Fossil Pteridophytes. C.S.I.R. New Delhi

Suthar, O.P. & Sharan , B.D. (1986). *Indian Fern J.* 3 : 46-52

Sussex. I.M.(1966). the origin and development of heterospory in vascular plants. *Trends in plant Morphogenesis.* 140-152.

Takashashi, c. (1962). *Cytologia.* **17** : 79-96

Thurston, E.L (1969). *Amer. Ferojour.*, **59** : 68-79.

Treub, M (1884). *Ann. Jour. Bot. Butienz.*, **4** : 107-138.

—(1890). *Ibid.*, **8** : 1-37

Tryon, A.F. (1964). *Amer. Jour, Bot.* **91** : 939-942.

Vashistha, B.R. (1927). *Jour. Ind. Soc.*, **6** : 8-30

Verma, S.C. (1956). *Curr. Sci.*, **25** : 289-399.

—(1957). *Cytologia.*, **22** : 393-403.

—(1958). *Acta Bot, Nederland.*, **7** : 629-634.

—(1960). *Caryologia.*, **31** : 274-284.

—(1965). *Cytologia.*, **30** : 173-174

—and S.P. Khuller (1966). *Phytomorphology*, **16** : 302-314.

Voeller, B.R. (1964). *Collognes Centre Natl. Rech. Sci., Gif- sar-Yvette*, NO. **123** : 665-684

Wanger, W.H.Jr. (1952). *Univ. Publ. Bot.*, **6** : 1-212

—(1954). *Evolution,* **8** : 103-118

—(1955). *Rhodora.*, 57 : 219-240

—(1966). *Plant Biology Today* pp. 164-184. Belmont Calif.

—(1969). *Systematic Biology.* **1692** : 67-90

—(1973). *Bot. Jour. Linn. Soc.* **67** (1) : 245-256

—(1977). *Brittoma.* **29** : 54-63

—(1979). *Taxon.* **28** : 87-95

—, F.S. Wagner and L.D. Gomez (1972). *Amer. Jour .Bot.*, **59** : 677

Walker, A. (1955). *Watsonia.*, **3** : 193-200

Walker, T.G. (1958). *Evolution.* **12** : 82-92

Walton, J (1949). *Trans. Roy. Soc. Edinb.*, **61** : 729-736

Ward, A. (1914). *Flora.*, **106** : 237-263.

Ward, M. (1954). *phytomorph.,* **4** : 1-17.

Webster, T.R. and Steeves, T.A. (1964). *Canad. Jour. Bot.*, **42** : 1665-1676

Whillier, D.P. (1966). *Amer. Journ,. Bot.* **51** : 730-736.

—(1972). *Ibid.*, **59** : 677

White, R.A. (1970). *Amer. Jour. Bot.* **57** : 458

Whilliams,S. (1933). *Trans. Roy. Soc. Edinb.*, **57** : 711-737

Walf, P.G. (1997). *Am J. Bot.* **84** : 14-29

Yabe, Y & K. Yasui (1914). *Boi. Mag. Tokyo*, **37** : 233-245.

Yadav, B.L. & Kaur, R (1985). *Indian Fern J.* **2** : 93-97

Yasyu, K. (1911). *Ann. Bot.*, **25** : 469-483

Zamora, P. (1958). *Comparative Anatomy of the protoxylem elements of several species of selaginella.* Thesis cornell Univ.

Zimmermann. W. (1930). *Die phylogenie der Planzen.* Gustav, Fischer, Jena.

Zimmermann. W. (1952). *The Palaeobotanist*, **1** : 456-470.

Blackwood, B. (1935). *Both side of the Buka passage.* Oxford, Clarendon Press.

Futscher, O. 1959. Microbibliotheca Anthropos. 30

Henty, E.E. & Pritchard, G.H. 1982. *Bot. Bull.* 7, Div. of Botany, Lae 1-195

Gliessman, S.R. (1978.) *Amer. Fern J.* 68 : 41-44.

Singh, H.B. (2003) *Pteridology in the New Millennium* (S. chandra & Srivastava M. eds.). Kluwer Acad. Publishers. 420-446.

Gurang, V.L. (1988). Useful Pteridophytes of Nepal Himalayas. Ad. Plant Sci. 1 : 67-76.

Powell, J.M. (1976 a) Sci. in New Guinea, 4 : 173-201.

Powel J. M (1976 b). Part III. Ethnobotany In Paijmans, K. ed : New Guinea Vegetation 106-183.

Kleinschmidt, H.E. (1973). Advisory leaflet. 1181 *Div. of pl. industry*, Queensland.

Wild, H. (1961). *Kirkia* 2 : 1-66.

Holdsworth, D.K. & Giheno, J. (1975). Sci. in New Guinea. 3 : 191-198.

Anonymous, (1978). *Advisory leaflet* 1427. 1-7, *Div. of pl. industry, Queensland.*

Singh., S., Dixit, R.D. & Sahu, T.R. (2005). *Indian Journal of Traditional Knowledge* 4. (4) : 392-395.

Morel, G. & Wetmore, R.H. 1951. *American Journal of Botany*, 38 (2) : 141-143.

Aimee Julianne Anico Parr. (2000). Micropropagtion of ferns, Institute of Biol, College of Sci. University of Phillipines.

Kapur, S.K. & Sarin Y.K. (1978). Indian Drugs. 15 : 87-90

Kirtikar, K.R. & Basu, B.D. (1935) *Indian Medicinal Plants.* 4 vols. Ed. Lalit Mohan Basu, Allahabad.

Puri, G.S. & Arora, R.K. (1961). *Indian Foresters.* 87 : 179-183.

Singh, H.P. & Viswanathan, M.V. (1996). *J. Econ. Taxon Bot.* 12 : 24-35.

Amoroso, CB & Amoroso, V.B. (2003). *Pteridology in the new Millennium.* S. Chandra & Srivastava, M. (eds.) Kluwer Acad. Publishers.

Byrene, T.E. & Caponetti, J.D. (1992). *Amer. Fern* 82 : 1-11.

Amoroso, CB & Amoroso, UB. (1998). *Central Mindanao Univ. Graduate J.* 44-50.

Hickok, L.G, Warne, T.R. & Slocum, M.K. 1987. *Amer. J. Bot.* 74 : 1304-1316.

Hennen, G.R. and Sheehan, T.J. 1978. *Hortscience*, 13 : 245.

Zamora, P.M. & Co. L. 1986. *Asia Life Sci.* 2 : 88-97.

Zamora, P.M., Chaimongkol, S. Amoroso, CB & Marzan, M. 1993. *Philipp J.Sci.* 122. 223-253.

Manickam, V.S., Vallinayagam, S. & Johnson, M. 2003. *Pteridology in the New Millennium* (eds : S. Chandra & Srivastava M. J. Kluwer Acad. Publishers.

Gautheret, R.J. (1983). *Bot. Mag. Tokyo* 96 : 393-401.

White, P.R. (1963). *The Cultivation of Animal & Plant cells,* Ronald Press, New York.

Beck, M.J. & Caponetti, J.D. 1983. *Amer. J. Bol.* 70 (1) ; 1-7.

Schweitzer, H.J. (1972) *Palaentographica*, 137-154.

Chandra, S. & Srivastava, M. (eds) (2003). *Pteridology in New Millennium.* NBRI Golden Jublee Volume. Kluwer Acad. Publisher. 520 pp.

Chandra, S. (2000). *The Ferns of India, International Distributors,* Dehradun.

Dixit, R.D. 1984. A Census of Indian Pteridophytes *Bot. Surv. of India* Publ. DOEn, Govt. of India Howrah, 1-177.

MOEF. 1999- National Policy and Management Action Strategy on Biodiversity, 74 pp.

Manicum, V.S. (2003) online article.

Chaudhary, N.P. (1973). The Pteridophytic flora of Upper Gangetic Plain. Navyug Traders. N. Delhi.

Kumari, A. and Srivastava, R.B. (2003), Pteridophytic Flora of Northern Part of Bihar (India). Adjacent to Indo-Nepal Border pp. 164-175 (*Pteridology in the New Millennium*). Kluwer Acad Publ.

Theuerkauf, W.D. (1994) BGC News, 2(3) 54-55 pp.

Singh, Samar and Panigrahi, G.. (2005), Fern & Fern Allies of Arunachal Pradesh.

Hameed, C.A., Rajesh, K.P. & Madhusoodanan, P.V. (2003) *Filmy ferns of South India,* Penta Book Publ. & Distributors. Calicut.

Panigrahi, G. (1995). *Pl Sci. Res.* 17 : 43-56.

Panigrahi, G. (1998). Biodiveristy Conservation : Problem and Prospects (Patra H.K. eds) Bhubneswar.

Panigrahi, G. (2000). *Pl. Sci. Res.* 20 : 1-45

Nayar, B.K. & Geevarghese (1993) *Fern Flora of Malabar.* Indus Publ. Co. N. Delhi.

Reference Books

Arber, A. (1950). *The Nature philosophy of Plants form.* Cambridge.

Arnold, C.A. (1947). *An Introduction to Paleobotany.* New York and London

Bierhorst, D. W. (1971). *Morphology of Vascular Plants. McMilan Company,* New York.

Bold, H.C. (1957). *Morphology Plants,* New York.

Bower, F.O. (1923-28). *The ferns*, Vol. 1-3. Cambridge.

—(1935). *The Primitive land plants.* New York.

—(1908). *The origin of land flora*, Mc. Milan, London

Chandra S and Srivastava 19 (eds.) 2003. *Pteriodology in New Millinium,* NBRI Golden Jublee Volume, Kluwer Acad. Publisher. 520 pp.

Chandras. 2000. *The ferns of India*. International Distributor, Dehradun.

Copeland, E.B. (1974) *Genera Filicum.* Chronica Botanica. U.S.A

Develoryas. T. (1962). *Morphology and evolution of Fossil Plants*, New York.

Dittmer, H.J. (1964) *Phylogeny and form in the plant kingdom.* Toronto, N.Y. and and London

Doyle, W.T.. (1970). *The Biology of higher Cryptogams*, London.

—(1964). *Non Vascular Plant. Form and Function.* Belmont. California.

Eames, A.J.(1936). *Morphology vascular plants (lower groups).* New York and London.

Foster, A.S. and Gifford. E.M. (1959). *Comparative Morphology of Vascular Plants*. San Francisco.

Gifford E.M. Jr. & Foster, A.S. (1986). *Morphology and Evolution of Vacular Plants*, Freenian, N.Y.

Hauke, R.L (1963). *A taxonomic Monograph of the genus Equisetum sub-germs Hipochast, Beih Nova Hedwigia,* 8 : 1-123.

Haput, A.W. (1953). *Plant Morphology.* N.Y. Toronto, London. Manton, I (1950). *Problem of Cytology and evolution in Pteridophyta.* Cambridge university Press.

Love , A Love, D & Pichi sermolli, R.E.G. (1977). *Cytotaxonomica Atlas of the Pteridophytes*. J Gamer Vaduz.

Lovis, J.D. (1977). Evolutionary Patterns and Processes in ferns. In : *Advances in Botanical research.* Academic Press, London **4** : 229-415.

Mehra, P.N.& Asha Gupta (1986). *Gametophytes of Himalayan Ferns*. pp. 1-206

Rashid, A. (1999). *An Introduction to Pteridophyta* : Diversity Development. Differentiation.Vikas Publishing House Pvt. Ltd. New Delhi. 426 p.

Parihas, N.S. (1965). An Introduction to Embryophyta Vol. II Pteridophyta - Central Book Depo. Allahabad.

Reed, C.F. (1952), *Index Isotales. Bot-Soc. Brat*. **27** : 1-72

— (1965-1966), Index Selaginellarum. *Mem. Soc.Brati* **18** : 1-287.

— (1966). Index Prelotales, *Bot. Soc. Brot*. **40** : 71-96.

— (1971, 1972). Index to *Equisetophyta* (Parts 1 and 2). Recentes. Reed Herbarium Baltimore, Maryland 128 pages & Fossils Reed Herbarium Ballimore, Maryland 402 pages.

Reimers, H.(1954). Pteridiphyta. In *Engler's Syllabus der Pflazenfamilien*. 12th Edition **1** : 269-311. Berlin.

Smith, G.M. (1955). *Cryptogamic Botany*, Vol. 2. New York and London.

Sporne, K.R. (1966). *The Morphology of Pteridophytes*. London.

Tryon, R.M & Tryon H.N. (1982). Ferns & Allied Plants Springer-Verlag, New York.

Stewart, W.N. & Rothwell, G.W. (1993) *Plaeobotany & Evolution of Plants*. Cambridge Univ Press. Cambridge.

Wardlaw, C.W. (1955). *Embryogenesis in plants*, London.

—(1968). *Morphogenesis in plants. A contemporary study*. London.

Zimmermann, W. (1959). Phylogenie der Pfanzen. Jens Stuttgarat.

INDEX

M

N

O

T

V

W

Y

Z

S